Introduction to Robust Estimation and Hypothesis Testing

Introduction to Robust Estimation and Hypothesis Testing

Fifth Edition

Rand R. Wilcox

Professor of Psychology
University of Southern California
Los Angeles, California, United States

ACADEMIC PRESS
An imprint of Elsevier

Academic Press is an imprint of Elsevier
125 London Wall, London EC2Y 5AS, United Kingdom
525 B Street, Suite 1650, San Diego, CA 92101, United States
50 Hampshire Street, 5th Floor, Cambridge, MA 02139, United States
The Boulevard, Langford Lane, Kidlington, Oxford OX5 1GB, United Kingdom

Notices

Knowledge and best practice in this field are constantly changing. As new research and experience broaden our understanding, changes in research methods, professional practices, or medical treatment may become necessary.

Practitioners and researchers must always rely on their own experience and knowledge in evaluating and using any information, methods, compounds, or experiments described herein. In using such information or methods they should be mindful of their own safety and the safety of others, including parties for whom they have a professional responsibility.

To the fullest extent of the law, neither the Publisher nor the authors, contributors, or editors, assume any liability for any injury and/or damage to persons or property as a matter of products liability, negligence or otherwise, or from any use or operation of any methods, products, instructions, or ideas contained in the material herein.

Library of Congress Cataloging-in-Publication Data
A catalog record for this book is available from the Library of Congress

British Library Cataloguing-in-Publication Data
A catalogue record for this book is available from the British Library

ISBN: 978-0-12-820098-8

For information on all Academic Press publications
visit our website at https://www.elsevier.com/books-and-journals

Publisher: Katey Birtcher
Editorial Project Manager: Sara Valentino
Production Project Manager: Beula Christopher
Designer: Bridget Hoette

Typeset by VTeX

Printed in the United States of America

Last digit is the print number: 9   8   7   6   5   4   3   2   1

Working together to grow libraries in developing countries

www.elsevier.com • www.bookaid.org

à mon seul œuf

Contents

Contents

Preface

There are many new and improved methods in this 5th edition. The R package written for this book provides a crude indication of how much has been added. When the 4th edition was published, it contained a little over 1200 R functions. With this 5th edition, there are now over 1700 R functions.

All of the chapters have been updated. For the first three chapters the changes deal with software issues and some technical issues. Major changes begin in Chapter 4. One of the major additions has to do with measuring effect size. Coverage of this topic has been expanded considerably. For example, Chapter 5 now describes six robust measures of effect size when comparing two independent groups. Confidence intervals for all six are returned by the R function ES.summary.CI. Each provides a different perspective on how the groups compare. When there is little or no difference between the groups, all six tend to give very similar results. But when the groups differ substantially there is a possibility that measures of effect size can differ substantially. In effect, multiple measures of effect size can help provide a deeper and more nuanced understanding of data, as illustrated in Section 5.3.5. Easy-to-use R functions have been added for measuring effect size when dealing with multiple groups.

There have been advances even at the most basic level. Consider, for example, the goal of computing a confidence interval for the probability of success when dealing with a binomial distribution. There are now clearer guidelines on when it is advantageous to use the Schilling–Doi method versus the Agresti–Coull method. When comparing two independent binomial distributions, more recent results point to using a method derived by Kulinskaya et al. (2010). A related issue is computing a confidence interval for the probability of success given the value of a covariate. It is known that when this is done via a logistic regression model, a slight departure from this model can yield inaccurate results. A method for dealing with this concern is covered in Chapter 11.

New functions related to regression have been added that include robust methods for dealing with multicollinearity and robust analogs of the lasso method. The coverage of classification (machine learning) methods has been expanded. Included is a function making it a simple matter to compare the false positive and false negative rates of a collection of techniques.

There are new results on comparing measures of association as well as measures of variation. New techniques related to ANOVA-type methods are covered in Chapters 7 and 8 and new ANCOVA techniques are covered in Chapter 12. A variety of new plots have been added as well.

As was the case in previous editions, this book focuses on the practical aspects of modern, robust statistical methods. The increased accuracy and power of modern methods, versus conventional approaches to ANOVA and regression, is remarkable. Through a combination of theoretical developments, improved and more flexible statistical methods, and the power of the computer, it is now possible to address problems with standard methods that seemed insurmountable only a few years ago.

Consider classic, routinely taught and used methods for comparing groups based on means. A basic issue is how well these methods perform when dealing with heteroscedasticity and non-normal distributions. Based on a vast body of literature published over the last 60 years, the situation can be briefly summarized as follows. When comparing groups that have identical distributions, classic methods work well in terms of controlling the Type I error probability. When distributions differ, they might continue to perform well, but under general conditions they can yield inaccurate confidence intervals and have relatively poor power. Even a small departure from a normal distribution can destroy power and yield measures of effect size, suggesting a small effect when in fact for the bulk of the population there is a large effect. The result is that important differences between groups are often missed, and the magnitude of the difference is poorly characterized. Put another way, groups probably differ when null hypotheses are rejected with standard methods, but in many situations, standard methods are the least likely to find a difference, and they offer a poor summary of how groups differ and the magnitude of the difference. A related concern is that the *population* mean and variance are not robust, roughly meaning that an arbitrarily small change in a distribution can have an inordinate impact on their values. In particular, under arbitrarily small shifts from normality, their values can be substantially altered and potentially misleading. For example, a small shift toward a heavy-tailed distribution, roughly meaning a distribution that tends to generate outliers, can inflate the variance, resulting in low power. Thus, even with arbitrarily large sample sizes, the sample mean and variance might provide an unsatisfactory summary of the data.

When dealing with regression, the situation is even worse. That is, there are even more ways in which analyses, based on conventional assumptions, can be misleading. The very foundation of standard regression methods, namely estimation via the least squares principle, leads to practical problems, as do violations of other standard assumptions. For example, if the error term in the standard linear model has a normal distribution, but is heteroscedastic, the least squares estimator can be highly inefficient, and the conventional confidence interval for the regression parameters can be extremely inaccurate. Another concern is assuming that a linear model adequately models the association under study. It is well established that this is not always the case. New advances aimed at dealing with this issue are described.

A seemingly natural way of justifying classic methods is to test assumptions, for example, test the assumption that data are sampled from a normal distribution. But all indications are that this strategy is unsatisfactory. Roughly, methods for testing assumptions can have inadequate power to detect situations where it is beneficial to use a robust method instead.

In 1960, it was unclear how to formally develop solutions to the many problems that have been identified. It was the theory of robustness developed by P. Huber and F. Hampel that paved the road for finding practical solutions. Today, there are many asymptotically correct ways of substantially improving on standard ANOVA and regression methods. That is, they converge to the correct answer as the sample sizes get large, but simulation studies have shown that when sample sizes are small, not all methods should be used. Moreover, for many methods, it remains unclear how large the sample sizes must be before reasonably accurate results are obtained. One of the goals in this book is to identify those methods that perform well in simulation studies, as well as those that do not.

Although the goal is to focus on the applied aspects of robust methods, it is important to discuss the foundations of modern methods, which is done in Chapters 2 and 3, and to some extent in Chapter 4. One general point is that modern methods have a solid mathematical foundation. Another goal is to impart the general flavor and aims of robust methods. This is important because misconceptions are rampant. For example, some individuals take the term robust to mean that a method controls the Type I error probability. This is just one component of modern robust methods.

A practical concern is applying the methods described in this book. To deal with this problem, easy-to-use R functions are supplied, many of which are not readily available when using other software packages. The package of R functions written for this book can be obtained as indicated in Section 1.8 of Chapter 1. With one command, all of the functions described in this book become a part of your version of R. Illustrations, using these functions, are included.

The book assumes that the reader has had an introductory statistics course. That is, all that is required is some knowledge about the basics of ANOVA, hypothesis testing, and regression. The foundations of robust methods, described in Chapter 2, are written at a relatively non-technical level, but the exposition is much more technical than the rest of the book, and it might be too technical for some readers. It is recommended that Chapter 2 be read or at least skimmed, but those who are willing to accept certain results can skip to Chapter 3. One of the main points in Chapter 2 is that the robust measures of location and scale that are used are not arbitrary, but were chosen to satisfy specific criteria. Moreover, these criteria eliminate from consideration the population mean, variance, and Pearson's correlation.

From an applied point of view, Chapters 4–12, which include methods for addressing common problems in ANOVA and regression, form the heart of the book. Technical details are kept to a minimum. The goal is to provide a simple description of the best methods available, based on theoretical and simulation studies, and to provide advice on which methods to use. Usually, no single method dominates all others, one reason being that there are multiple criteria for judging a particular technique. Accordingly, the relative merits of the various methods are discussed. Although no single method dominates, standard methods are typically the least satisfactory, and many alternative methods can be eliminated.

Finally, the author would like to thank James Algina, Gui Bao, James Jaccard, H. Keselman, Boaz Levy, J. Magadia, and Allan Campopiano for catching typos in earlier editions as well as bugs in earlier versions of the R software. A special thanks goes to Patrick Mair and Felix Schönbrodt, whose efforts helped provide easy access to the R functions described in this book.

Ancillary links:

Students can access accompanying information on the companion site located here:

> https://www.elsevier.com/books-and-journals/book-companion/9780128200988

Registered instructors can request to access electronic review copies and teaching resources here:

> https://educate.elsevier.com/9780128200988

## CHAPTER 1

Introduction

Introductory statistics courses describe methods for computing confidence intervals and testing hypotheses about means and regression parameters based on the assumption that observations are randomly sampled from normal distributions. When comparing independent groups, standard methods also assume that groups have a common variance, even when the means are unequal, and a similar homogeneity of variance assumption is made when testing hypotheses about regression parameters. Currently, these methods form the backbone of most applied research. There is, however, a serious practical problem: Many journal articles have illustrated that these standard methods can be highly unsatisfactory. Often, the result is a poor understanding of how groups differ and the magnitude of the difference. Power can be relatively low compared to recently developed methods, least squares regression and Pearson's correlation can yield a highly misleading summary of how two or more random variables are related, the probability coverage of standard methods for computing confidence intervals can differ substantially from the nominal value, and the usual sample variance can give a distorted view of the amount of dispersion among a population of participants. Even the population mean, if it could be determined exactly, can give a distorted view of what the typical participant is like.

Although the problems just described are well known in the statistics literature, many textbooks written for non-statisticians still claim that standard techniques are completely satisfactory. Consequently, it is important to review the problems that can arise and why these problems were missed for so many years. As will become evident, several pieces of misinformation have become part of statistical folklore, resulting in a false sense of security when using standard statistical techniques. The remainder of this chapter focuses on some basic issues related to assuming normality, as well as some basic issues associated with the classic analysis of variance (ANOVA) test and least squares regression. But it is stressed that there are additional issues that will be discussed in subsequent chapters. One appeal of modern robust methods is that under general conditions they can have substantially higher power compared to more traditional techniques. But from a broader perspective, the main message is that a collection of new and improved methods is now available that provides a more accurate and more nuanced understanding of data.

1.1 Problems With Assuming Normality

To begin, distributions are never normal. For some this seems obvious, hardly worth mentioning, but an aphorism given by Cramér (1946) and attributed to the mathematician Poincaré remains relevant: "Everyone believes in the [normal] law of errors, the experimenters because they think it is a mathematical theorem, the mathematicians because they think it is an experimental fact." Granted, the normal distribution is the most important distribution in all of statistics. But in terms of approximating the distribution of any continuous distribution, it can fail to the point that practical problems arise, as will become evident at numerous points in this book. To believe in the normal distribution implies that only two numbers are required to tell us everything about the probabilities associated with a random variable: the population mean μ and population variance σ^2. Moreover, assuming normality implies that distributions must be symmetric.

Of course, non-normality is not, by itself, a disaster. Perhaps a normal distribution provides a good approximation of most distributions that arise in practice, and there is the central limit theorem, which tells us that under random sampling, as the sample size gets large, the limiting distribution of the sample mean is normal. Unfortunately, even when a normal distribution provides a good approximation to the actual distribution being studied (as measured by the Kolmogorov distance function described later), practical problems arise. Also, empirical investigations indicate that departures from normality that have practical importance are rather common in applied work (e.g., Hill and Dixon, 1982; Micceri, 1989; Wilcox, 1990a). Even over a century ago, Karl Pearson and other researchers were concerned about the assumption that observations follow a normal distribution (e.g., Hand, 1998, p. 649). In particular, distributions can be highly skewed, they can have heavy tails (tails that are thicker than a normal distribution), and random samples often have outliers (unusually large or small values among a sample of observations). Skewed distributions turn out to be a much more serious concern than once thought. Details are summarized in Chapters 4 and 5.

The immediate goal is to indicate some practical concerns associated with outliers and heavy-tailed distributions. Outliers and heavy-tailed distributions are a serious practical problem because they inflate the standard error of the sample mean, so power can be relatively low when comparing groups. Fisher (1922), for example, was aware that the sample mean could be inefficient under slight departures from normality. Even two centuries ago, Laplace was aware that the sample median can be more efficient than the sample mean (Hand, 1998). The earliest empirical evidence that heavy-tailed distributions occur naturally stems from Bessel (1818). Data analyzed by Newcomb (1886) again indicated that heavy-tailed distributions are common. Modern robust methods provide an effective way of dealing with these problems.

A classic way of illustrating the effects of slight departures from normality is with the *contaminated* or *mixed normal* distribution (Tukey, 1960). Let X be a standard normal random variable having distribution $\Phi(x) = P(X \le x)$. Then, for any constant $K > 0$, $\Phi(x/K)$ is a normal distribution with standard deviation K. Let ϵ be any constant, $0 \le \epsilon \le 1$. The *mixed normal* distribution is

$$H(x) = (1 - \epsilon)\Phi(x) + \epsilon\Phi(x/K), \tag{1.1}$$

which has mean 0 and variance $1 - \epsilon + \epsilon K^2$. (Stigler, 1973, finds that the use of the mixed normal dates back at least to Newcomb, 1882, p. 382.) In other words, the mixed normal arises by sampling from a standard normal distribution with probability $1 - \epsilon$; otherwise sampling is from a normal distribution with mean 0 and standard deviation K.

To provide a more concrete example, consider the population of all adults, and suppose that 10% of all adults are at least 70 years old. Of course, individuals at least 70 years old might have a different distribution from the rest of the population. For instance, individuals under 70 might have a standard normal distribution, but individuals at least 70 years old might have a normal distribution with mean 0 and standard deviation 10. Then the entire population of adults has a mixed (or contaminated) normal distribution, with $\epsilon = 0.1$ and $K = 10$. In symbols, the resulting distribution is

$$H(x) = 0.9\Phi(x) + 0.1\Phi(x/10), \tag{1.2}$$

which has mean 0 and variance 10.9. Moreover, Eq. (1.2) is not a normal distribution, verification of which is left as an exercise.

To illustrate problems that arise under slight departures from normality, we first examine Eq. (1.2) more closely. Fig. 1.1 shows the standard normal and the mixed normal probability density function corresponding to Eq. (1.2). Notice that the tails of the mixed normal are above the tails of the normal, so the contaminated normal is said to have heavy tails. It might seem that the normal distribution provides a good approximation of the contaminated normal, but there is an important difference. The standard normal has variance 1, but the mixed normal has variance 10.9. The reason for the seemingly large difference between the variances is that σ^2 is very sensitive to the tails of a distribution. In essence, a small proportion of the population of participants can have an inordinately large effect on its value. Put another way, even when the variance is known, if sampling is from the mixed normal, the length of the standard confidence interval for the population mean, μ, will be over three times longer than it would be when sampling from the standard normal distribution instead. What is important from a practical point of view is that there are location estimators other than the sample mean that have standard errors that are substantially less affected by heavy-tailed distributions. By "measure of location" is meant some measure intended to represent the typical participant or object, the two best-known examples being the mean and the median. (A more formal definition is given in Chapter 2.) Some of these measures have relatively short confidence intervals

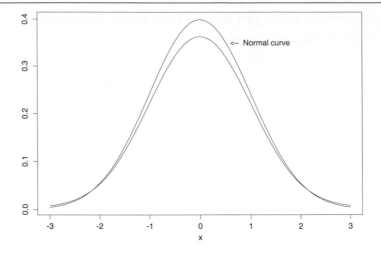

Figure 1.1: Normal and contaminated normal distributions.

when distributions have a heavy tail, yet the length of the confidence interval remains reasonably short when sampling from a normal distribution instead. Put another way, there are methods for testing hypotheses that have good power under normality, but that continue to have good power when distributions are non-normal, in contrast to methods based on means. For example, when sampling from the mixed normal given by Eq. (1.2), both Welch's method and Student's method for comparing the means of two independent groups have power approximately 0.278 when testing at the 0.05 level, with equal sample sizes of 25 and when the difference between the means is 1. In contrast, several other methods, described in Chapter 5, have power exceeding 0.7.

In an attempt to salvage the sample mean, it might be argued that in some sense the mixed normal represents an extreme departure from normality. But based on various measures of the difference between two distributions, they are very similar, as suggested by Fig. 1.1. For example, the *Kolmogorov distance* between any two distributions, F and G, is the maximum value of

$$\Delta(x) = |F(x) - G(x)|,$$

the maximum being taken over all possible values of x. (If the maximum does not exist, the supremum or least upper bound is used.) If distributions are identical, the Kolmogorov distance is 0, and its maximum possible value is 1, as is evident. Now consider the Kolmogorov distance between the mixed normal distribution, $H(x)$, given by Eq. (1.2), and the standard normal distribution, $\Phi(x)$. It can be seen that $\Delta(x)$ does not exceed 0.04 for any x. That is, based on a Kolmogorov distance function, the two distributions are similar. Several alternative methods are often used to measure the difference between distributions. (Some of these are discussed by Huber and Ronchetti, 2009.) The choice among these measures is of interest when dealing with theoretical issues, but these issues go beyond the scope of this book. Suffice it to say that the difference between the normal and mixed normal is again small. Gleason (1993) discusses the difference between the normal and mixed normal from a different perspective and also concludes that the difference is small.

Even if it could be concluded that the mixed normal represents a large departure from normality, concerns over the sample mean would persist, for reasons already given. In particular, there are measures of location having standard errors similar in magnitude to the standard error of the sample mean when sampling from normal distributions, but that have relatively small standard errors when sampling from a heavy-tailed distribution instead. Moreover, experience with actual data indicates that the sample mean does indeed have a relatively large standard error in some situations. In terms of testing hypotheses, there are methods for comparing measures of location that continue to have high power in situations where there are outliers or sampling is from a heavy-tailed distribution. Other problems that plague inferential methods based on means are also reduced when using these alternative measures of location. For example, the more skewed a distribution happens to be, the more difficult it is to get an accurate confidence interval for the mean, and problems arise when testing hypotheses. Theoretical and simulation studies indicate that problems are reduced substantially when using certain measures of location discussed in this book.

When testing hypotheses, a tempting method for reducing the effects of outliers or sampling from a heavy-tailed distribution is to check for outliers, and if any are found, throw them out and apply standard techniques to the data that remain. This strategy cannot be recommended, however, because it yields incorrect estimates of the standard errors, for reasons given in Chapter 3. Indeed, the estimate of the standard error based on this approach can differ substantially from an estimate that is technically correct, as will be illustrated at the end of Section 3.3.2. (Also see Bakker and Wicherts, 2014.) Generally, the method for down-weighting or eliminating extreme values must be taken into account when estimating a standard error.

Yet another problem needs to be considered. If distributions are skewed enough, doubts begin to rise about whether the population mean is a satisfactory reflection of the typical participant under study. Fig. 1.2 shows a graph of the probability density function corresponding to a mixture of two chi-squared distributions. The first has four degrees of freedom, and the second is again chi-squared with four degrees of freedom, only the observations are multiplied by 10. This is similar to the mixed normal already described, only chi-squared distributions are used instead. Observations are sampled from the first distribution with probability 0.9, otherwise sampling is from the second. As indicated in Fig. 1.2, the population mean is 7.6, a value that is relatively far into the right tail. In contrast, the population median is 3.75, and this would seem to be a better representation of the typical participant under study.

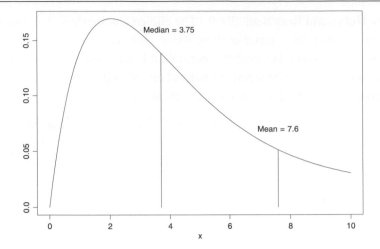

Figure 1.2: Mixed chi-squared distribution.

A seemingly natural strategy is to test the assumption that a distribution is normal, and if not statistically significant, assume normality. But numerous papers indicate that this strategy in particular, and the general strategy of testing assumptions, is unsatisfactory. A related strategy is to test for normality, and if non-normality is detected, test hypotheses based on some alternative parametric family of distributions, such as the family of lognormal or exponential distributions. In particular, one might choose a family of distributions based on which family of distributions offers the best fit to the observed data. But in the context of a general linear model, extant results do not support this approach (e.g., Keselman et al., 2016). Perhaps there are situations where this approach performs reasonably well, but this remains to be determined.

1.2 Transformations

Transforming data has practical value in a variety of situations. Emerson and Stoto (1983) provide a fairly elementary discussion of the various reasons one might transform data and how it can be done. The only important point here is that simple transformations can fail to deal effectively with outliers and heavy-tailed distributions. For example, the popular strategy of taking logarithms of all the observations does not necessarily reduce problems due to outliers, and the same is true when using Box–Cox transformations instead (e.g., Rasmussen, 1989; Doksum and Wong, 1983). Other concerns were expressed by Thompson and Amman (1990). Better strategies are described in subsequent chapters.

Skewness can be a source of concern when using methods based on means, as will be illustrated in subsequent chapters. Transforming data is often suggested as a way of dealing with skewness. More precisely, the goal is to transform the data so that the resulting distribution is approximately symmetric about some central value. There are situations where this strategy is reasonably successful. But even after transforming data, a distribution can remain severely skewed. In practical terms, this approach can be highly unsatisfactory, and assuming that it performs well, can result in erroneous and misleading conclusions. When comparing two independent groups with say a Student's t test, the assumption is that the same transformation applied to group 1 is satisfactory when transforming the data associated with group 2. A seemingly better way to proceed is to use a method that deals well with skewed distributions even when data are not transformed and when the distributions being compared differ in the amount of skewness.

Perhaps it should be noted that when using simple transformations on skewed data, if inferences are based on the mean of the transformed data, then attempts at making inferences about the mean of the original data, μ, have been abandoned. That is, if the mean of the transformed data is computed and we transform back to the original data, in general, we do not get an estimate of μ.

1.3 The Influence Curve

This section gives one more indication of why robust methods are of interest by introducing the influence curve as described by Mosteller and Tukey (1977). It bears a close resemblance to the *influence function*, which plays an important role in subsequent chapters, but the influence curve is easier to understand. In general, the *influence curve* indicates how any statistic is affected by an additional observation having value x. In particular, it graphs the value of a statistic versus x.

As an illustration, let $\bar{X}$ be the sample mean corresponding to the random sample $X_1, \ldots, X_n$. Suppose we add an additional value, x, to the n values already available, so now there are $n + 1$ observations. Of course this additional value will in general affect the sample mean, which is now $(x + \sum X_i)/(n + 1)$. It is evident that as x gets large, the sample mean of all $n + 1$ observations increases. The influence curve plots x versus

$$\frac{1}{n+1}\left(x + \sum X_i\right), \tag{1.3}$$

the idea being to illustrate how a single value can influence the value of the sample mean. Note that for the sample mean, the graph is a straight line with slope $1/(n + 1)$, the point being that the curve increases without bound. Of course, as n gets large, the slope decreases, but in practice, there might be two or more unusual values that dominate the value of $\bar{X}$.

Now consider the usual sample median, M. Let $X_{(1)} \leq \cdots \leq X_{(n)}$ be the observations written in ascending order. If n is odd, let $m = (n + 1)/2$, in which case, $M = X_{(m)}$, the mth largest order statistic. If n is even, let $m = n/2$, in which case, $M = (X_{(m)} + X_{(m+1)})/2$. To be more concrete, consider the values

2 4 6 7 8 10 14 19 21 28.

Then $n = 10$ and $M = (8 + 10)/2 = 9$. Suppose an additional value, x, is added, so that now $n = 11$. If $x > 10$, then $M = 10$, regardless of how large x might be. If $x < 8$, $M = 8$, regardless of how small x might be. As x increases from 8 to 10, M increases from 8 to 10 as well. (For more results on the finite sample properties of the median, see Park et al., 2020, as well as Rousselet and Wilcox, 2020.)

The main point is that in contrast to the sample mean, the median has a bounded influence curve. In general, if the goal is to minimize the influence of a relatively small number of observations on a measure of location, attention might be restricted to those measures having a bounded influence curve. A concern with the median, however, is that its standard error is large relative to the standard error of the mean when sampling from a normal distribution, so there is interest in searching for other measures of location having a bounded influence curve, but that have reasonably small standard errors when distributions are normal.

Also notice that the sample variance, s^2, has an unbounded influence curve, so a single unusual value can inflate s^2. This is of practical concern because the standard error of $\bar{X}$ is estimated with $s/\sqrt{n}$. Consequently, conventional methods for comparing means can have low power and relatively long confidence intervals due to a single unusual value. This problem does indeed arise in practice, as illustrated in subsequent chapters. For now, the only point is that it is desirable to search for measures of location for which the estimated standard error has a bounded influence curve. Such measures are available that have other desirable properties as well.

1.4 The Central Limit Theorem

When working with means or least squares regression, certainly the best-known method for dealing with non-normality is to appeal to the central limit theorem. Put simply, under random sampling, if the sample size is sufficiently large, the distribution of the sample mean is approximately normal under fairly weak assumptions. A practical concern is the description "sufficiently large." Just how large must n be to justify the assumption that $\bar{X}$ has a normal distribution? Early studies suggested that $n = 40$ is more than sufficient, and there was a time when even $n = 25$ seemed to suffice. These claims were not based on wild speculations, but more recent studies have found that these early investigations overlooked two crucial aspects of the problem.

The first is that early studies looking into how quickly the sampling distribution of $\bar{X}$ approaches a normal distribution focused on very light-tailed distributions, where the expected proportion of outliers is relatively low. In particular, a popular way of illustrating the central limit theorem was to consider the distribution of $\bar{X}$ when sampling from a uniform or exponential distribution. These distributions look nothing like a normal curve, the distribution of $\bar{X}$ based on $n = 40$ is approximately normal, so a natural speculation is that this will continue to be the case when sampling from other non-normal distributions. But more recently it has become clear that as we move toward more heavy-tailed distributions, a larger sample size is required.

The second aspect being overlooked is that when making inferences based on Student's t, the distribution of T can be influenced more by non-normality than the distribution of $\bar{X}$. In particular, even if the distribution of $\bar{X}$ is approximately normal based on a sample of n observations, the actual distribution of T can differ substantially from a Student's t distribution with $n - 1$ degrees of freedom. *Even when sampling from a relatively light-tailed distribution*, practical problems arise when using Student's t, as will be illustrated in Section 4.1. When sampling from heavy-tailed distributions, even $n = 300$ might not suffice when computing a 0.95 confidence interval via Student's t.

1.5 Is the ANOVA F Robust?

Practical problems with comparing means have already been described, but some additional comments are in order. For many years, conventional wisdom held that standard ANOVA methods are robust, and this point of view continues to dominate applied research. In what sense is this view correct? What many early studies found was that if groups have *identical* distributions, Student's t test and, more generally, the ANOVA F test are robust to non-normality in the sense that the actual probability of a Type I error would be close to the nominal level. Tan (1982) reviews the relevant literature. Many took this to mean that the F test is robust when groups differ. In terms of power, some studies seemed to confirm this by focusing on standardized differences among the means. To be more precise, consider two independent groups with means μ_1 and μ_2 and variances σ_1^2 and σ_2^2. Many studies have investigated the power of Student's t test by examining power as a function of

$$\delta = \frac{\mu_1 - \mu_2}{\sigma},$$

where $\sigma = \sigma_1 = \sigma_2$ is the assumed common standard deviation. What these studies failed to take into account is that small shifts away from normality, toward a heavy-tailed distribution, lower δ, and this can mask power problems associated with Student's t test. The important

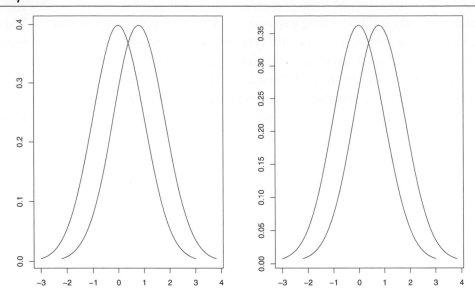

Figure 1.3: Small changes in the tails of distributions can substantially lower power when using means. In the left panel, Student's t has power approximately equal to 0.94. But in the right panel, power is 0.25.

point is that for a given difference between the means, $\mu_1 - \mu_2$, modern methods can have substantially more power.

To underscore concerns about power when using Student's t, consider the two normal distributions in the left panel of Fig. 1.3. The difference between the means is 0.8, and both distributions have variance 1. With a random sample of size 40 from both groups, and when testing at the 0.05 level, Student's t has power approximately equal to 0.94. Now look at the right panel. The difference between the means is again 0.8, but now power is 0.25, despite the obvious similarity to the left panel. The reason is that the distributions are mixed normals, each having variance 10.9.

More recently, it has been illustrated that standard confidence intervals for the difference between means can be unsatisfactory, and that the F test has undesirable power properties. One concern is that there are situations, where, as the difference between the means increases, power goes down, although eventually it goes up. That is, the F test can be *biased*. For example, Wilcox (1996a) describes a situation involving lognormal distributions, where the probability of rejecting is 0.18, when testing at the $\alpha = 0.05$ level, even though the means are equal. When the first mean is increased by 0.4 standard deviations, power *drops* to 0.096, but increasing the mean by 1 standard deviation, power increases to 0.306. Cressie and Whitford (1986) show that for unequal sample sizes, and when distributions differ in skewness,

Student's t test is not even asymptotically correct. More specifically, the variance of the test statistic does not converge to one as is typically assumed, and there is the additional problem that the null distribution is skewed. The situation improves by switching to heteroscedastic methods, but problems remain (e.g., Algina et al., 1994). Yet another concern has to do with situations where distributions differ in skewness. This can result in inaccurate confidence when using methods based on means, as illustrated in Section 5.2. The modern methods described in this book address these problems.

1.6 Regression

Outliers, as well as skewed or heavy-tailed distributions, also affect the ordinary least squares regression estimator. In some ways, the practical problems that arise are even more serious than those associated with the ANOVA F test.

Consider two random variables, X and Y, and suppose

$$Y = \beta_1 X + \beta_0 + \lambda(X)\epsilon,$$

where ϵ is a random variable having variance σ^2, X and ϵ are independent, and $\lambda(X)$ is any function of X. If ϵ is normal and $\lambda(X) \equiv 1$, standard methods can be used to compute confidence intervals for β_1 and β_0. However, even when ϵ is normal, but $\lambda(X)$ varies with X, probability coverage can be poor, and problems get worse under non-normality. There is the additional problem that under non-normality, the usual least squares estimate of the parameters can have relatively low efficiency, and this can result in relatively low power. In fact, low efficiency occurs even under normality when λ varies with X. There is also the concern that a single unusual Y value, or an usual X value, can greatly distort the least squares estimate of the slope and intercept. Illustrations of these problems and how they can be addressed are given in subsequent chapters.

1.7 More Remarks

Problems with means and the influence of outliers have been known since at least the 19th century. Prior to the year 1960, methods for dealing with these problems were ad hoc compared to the formal mathematical developments related to the ANOVA and least squares regression. What marked the beginning of modern robust methods, resulting in mathematical methods for dealing with robustness issues, was a paper by Tukey (1960) discussing the mixed normal distribution. A few years later, a mathematical foundation for addressing technical issues was developed by a small group of statisticians. Of particular importance is the theory of robustness developed by Huber (1964) and Hampel (1968). These results, plus other statistical tools developed in recent years, and the power of the computer, provide important new methods for comparing groups and studying the association between two or more variables.

1.8 R Software

Software for most of the methods described in this book is not yet available in standard statistical packages. Consequently, a library of over 1700 easy-to-use R functions has been supplied for applying them. The (open source) software R (R Development Core Team, 2010) is free and can be downloaded from www.R-project.org. There are many books that cover the basics of R (e.g., Crawley, 2007; Venables and Smith, 2002; Verzani, 2004; Zuur et al., 2009; also see Becker et al., 1988). The book by Verzani is available on the web at

http://cran.r-project.org/doc/contrib/Verzani-SimpleR.pdf.

A manual for R is also available at

http://www.cran.r-project.org/doc/manuals/R-intro.pdf.

Books that describe S-PLUS (e.g., Becker et al., 1988; Krause and Olson, 2002; Chambers, 1998; Chambers and Hastie, 1992; Fox, 2002; Venables and Ripley, 2000) can be useful when using R.

The R functions written for this book are available in an R package, or they can be downloaded from the author's web page. Seemingly, the easiest way to install the R functions is to go to the web page

Dornsife.usc.edu/cf/labs/wilcox/wilcox-faculty-display.cfm

and download the file Rallfun. Currently, the most recent version is Rallfun-v38. Then use the R command

source('Rallfun-v38')

or

source(file.choose()).

When using the latter command, a window will come up. Click on the file you want to source. Now all of the functions written for this book are part of your version of R until you remove them. The file Rallfun-v38 is also stored on the open science framework at https://osf.io/nvd59/quickfiles.

A second way of gaining access is via the R package WRS (maintained by Felix Schönbrodt). Go to

https://github.com/nicebread/WRS.

You will see a list of R commands for installing WRS. Copy and paste the R commands into R. Then use the R command

```
library(WRS)
```

to gain access to the functions.

A subset of the R package WRS is available in the CRANS R package WRS2 (Mair and Wilcox, 2019). A possible appeal of this package is that it contains help files, and it is easily installed by using the R command

```
install.packages('WRS2').
```

The R command

```
library(WRS2)
```

provides access to the functions, and it lists the functions that are available. This package contains functions for dealing with many commonly encountered situations, but it does not contain many of the functions described and illustrated in this book. For those who use the software Python, some of the methods described in this book can be used via a package at https://alcampopiano.github.io/hypothesize/, which is maintained by Allan Campopiano. At the moment, this package is limited to some ANOVA methods and robust measures of association.

Here is a partial list of the R packages, available from CRANS, that are utilized in this book:

- akima
- class
- cobs
- e1071

- ggplot2
- glmnet
- MASS
- mboost
- mgcv
- mrfDepth
- multicore
- neuralnet
- parallel
- plotrix
- pwr
- quantreg
- randomForest
- robust
- robustbase
- rrcov
- scatterplot3d
- stats
- enetLTS
- hqreg

All of these packages can be installed with the install.packages command (assuming you are connected to the web). For example, the R command

$$\text{install.packages('akima')}$$

will install the R package akima, which is used when creating three-dimensional plots.

1.9 Some Data Management Issues

Some of the R functions written for this book are aimed at manipulating and managing data in a manner that might be helpful, some of which are summarized in this section. Subsequent chapters provide more details about when and how the functions summarized here might be used.

A common situation is where data are stored in columns with one of the columns indicating the group to which a participant belongs and one or more other columns contain the measures of interest. For example, the data for eight participants might be stored as

10 2 64
4 2 47
8 3 59
12 3 61
6 2 73
7 1 56
8 1 78
15 2 63,

where the second column indicates to which group a participant belongs. There are three groups because the numbers in column 2 have one of three distinct values. For illustrative purposes, suppose that for each participant, two measures of reduced stress are recorded in columns 1 and 3. Then two of the participants belong to group 1, on the first measure of reduced stress their scores are 7 and 8, and on the second, their scores are 56 and 78. Some of the R functions written for this book require sorting these data into groups, and then storing them in either a matrix (with columns corresponding to groups) or in list mode. The R function

fac2list(x,g)

is supplied for accomplishing this goal, where x is an R variable, typically, the column of some matrix or a data frame containing the data to be analyzed, and g is an R variable indicating the levels of the groups to be compared. For a one-way ANOVA, g is assumed to be a single column of values. For a two-way ANOVA, g would have two columns, and for a three-way ANOVA, it would have three columns, each column corresponding to a factor. A maximum of four columns is allowed. The built-in R function

split(x,g)

can be used as well. (In some situations, fac2list is a bit more convenient than split.)

■ **Example**

R has a built-in data set, stored in the R variable ChickWeight, which is a matrix containing four columns of data. The first column contains the weight of chicks, column 4 indicates which of four diets was used, and the second column gives the number of days since birth when the measurement was made, which were 0, 2, 4, 6, 8, 10, 12, 14, 16, 18, 20, and 21. So for each chick, measurements were taken on 12 different days. Imagine that the goal is to sort data on weight into four groups based on the four groups indicated in column 4, and that the results are to be stored in list mode. This is accomplished with the R command

z=fac2list(ChickWeight[,1],ChickWeight[,4]).

The data for group 1 are stored in z[[1]], the data for group 2 are stored in z[[2]], and so on. If the levels of the groups are indicated by numeric values, fac2list puts the levels in ascending order. If the levels are indicated by a character string, the levels are put in alphabetical order.

An alternative to the functions fac2list and split is the built-in R function

unstack(x,...).

It uses the formula convention that is probably familiar to readers who already know R. In the last example, the R command would now look like this:

z=unstack(ChickWeight,ChickWeight[,1] $\sim$ ChickWeight[,4]).

The R function

fac2Mlist(x,grp.col,lev.col,pr=TRUE)

is like the R function fac2list; it can be useful when dealing with a multivariate ANOVA (MANOVA) design using the methods in Section 7.10. Roughly, it sorts data into groups based on the data in the column of x indicated by the argument grp.col. See Section 7.10.3 for more details. When dealing with a between-by-between MANOVA design, the function

fac2BBMlist(x,grp.col,lev.col,pr=TRUE)

can be used.

Now consider a between-by-between or a between-by-within ANOVA design. Some of the functions written for this book assume that the data are stored in list mode, or a matrix with columns corresponding to groups, and that the data are arranged in a particular order: The first K groups belong to the first level of the first factor, the next K belong to the second level of the second factor, and so on.

■ Example

For a 2-by-4 design, with the data stored in the R variable x, having list mode, the data are assumed to be arranged as follows:

	Factor B			
Factor	x[[1]]	x[[2]]	x[[3]]	x[[4]]
A	x[[5]]	x[[6]]	x[[7]]	x[[8]]

The R function fac2list can be useful for such situations.

■

Suppose, for example, the following data are stored in the R matrix m having 13 rows and 4 columns:

```
10 2 64 1
 4 2 47 1
 8 3 59 1
12 3 61 2
 6 2 73 2
 7 1 56 2
 8 1 78 2
15 2 63 2
 9 3 71 1
 2 3 81 1
 4 1 68 1
 5 1 53 1
21 3 49 2
```

The goal is to perform a 3-by-2 ANOVA, where the numbers in column 2 indicate the levels of the first factor and the numbers in column 4 indicate the levels of the second. Further assume that the values to be analyzed are stored in column 1. The first row of data indicates that the value 10 belongs to level 2 of the first factor and level 1 of the second. Similarly, the third row indicates that the value 8 belongs to the third level of the first factor and the first level of the second. Chapter 7 describes R functions for comparing the groups. Using these functions requires storing the data in list mode or a matrix, and the function fac2list can help accomplish this goal with the R command

$$dat=fac2list(m[,1],m[,c(2,4)]).$$

The output stored in dat is

```
[[1]]
[1]  4  5

[[2]]
[1]  7  8

[[3]]
[1]  10   4

[[4]]
[1]   6 15

[[5]]
[1]  8  9  2

[[6]]
[1]  12 21.
```

The R variable dat[[1]] contains the data for level 1 of both factors. The R variable dat[[2]] contains the data for level 1 of the first factor and level 2 of the second. The function also prints the values associated with the levels corresponding to each factor. For the situation at hand, it prints

```
[1]  "Group Levels:"
       [,1] [,2]
[1,]    1    1
[2,]    1    2
[3,]    2    1
[4,]    2    2
[5,]    3    1
[6,]    3    2
```

For example, the third row has 2 in the first column and 1 in the second column, meaning that for level 2 of the first factor and level 1 of the second, the data are stored in m[[3]]. It is noted that the data are stored in the form expected by the ANOVA functions covered in Chapter 7. One of these functions is called t2way. In the illustration, the command

$$t2way(dat=fac2list2(m[,1],m[,c(2,4)]),tr=0)$$

would compare means using a heteroscedastic method appropriate for a 3-by-2 ANOVA design, where the outcome measure corresponds to the data in column 1 of the R variable m. To perform a 3-by-2 ANOVA for the data in column 3, first enter the command

$$m=fac2list2(m[,3],m[,c(2,4)])$$

and then

$$t2way(3,2,m,tr=0).$$

A three-way ANOVA can be handled in a similar manner. Variations of some of the R functions written for this book make it possible to avoid using the R function fac2list. They will be described in subsequent chapters.

■ Example

Consider again the example dealing with the R variable ChickWeight, only now the goal is to store the data in list mode. The R command

$$z=fac2list(ChickWeight[,1],ChickWeight[,c(4,2)])$$

accomplishes this goal.

■

Look closely at the argument ChickWeight[,c(4,2)], and note the use of c(4,2). The 2 comes after the 4 because column 2 corresponds to the within group factor, which in this book always corresponds to the second factor. If ChickWeight[,c(2,4)] had been used, functions in this book aimed at a between-by-within design would assume that column 4 corresponds to the within group factor, which is incorrect.

Another goal that is sometimes encountered is splitting a matrix of data into groups based on the values in one of the columns. For example, column 6 might indicate whether participants are male or female, denoted by the values 0 and 1, and it is desired to store the data for females and males in separate R variables. This can be done with the R function

$$matsplit(m,coln=NULL),$$

which sorts the data in the matrix m into separate R variables corresponding to the values indicated by the argument coln. The function is similar to fac2list, only now two or more columns of a matrix can be sorted into groups rather than a single column of data, as is the case when using fac2list. Also, matsplit returns the data stored in a matrix rather than list mode.

The R function

$$mat2grp(m,coln)$$

also splits the data in a matrix into groups based on the values in column coln of the matrix m. Unlike matsplit, mat2grp can handle more than two values. That is, the column of m indicated by the argument coln can have more than two unique values. The results are stored in list mode.

The R function

$$qsplit(x,y,split.val=NULL)$$

splits the data in x into three groups based on a range of values stored in y. The length of y is assumed to be equal to the number of rows in the matrix x. (The argument x can be a vector rather than a matrix.) If split.val=NULL, the function computes the lower and upper quartiles based on the values in y. Then the corresponding rows of data in x that correspond to y values less than or equal to the lower quartile are returned in qsplit$lower. The rows of data for which y has a value between the lower and upper quartiles are returned in qsplit$middle, and the rows for which y has a value greater than or equal to the upper quartile are returned in qsplit$upper. If two values are stored in the argument split.val, they will be used in place of the quartiles.

■ Example

R has a built-in data set stored in the R variable ChickWeight (a matrix with 4 columns) that deals with weight gain over time and based on different diets. The amount of weight gained is stored in column 1. For illustrative purposes, imagine the goal is to separate the data in column 1 into three groups. The first group is to contain those values that are less than or equal to the lower quartile, the next is to contain the values between the lower and upper quartiles, and the third group is to contain the values greater than or equal to the upper quartile. The command

$$qsplit(ChickWeight[,1],ChickWeight[,1])$$

accomplishes this goal.

■

Two other functions are provided for manipulating data stored in a matrix:

- bw2list
- bbw2list

These two functions are useful when dealing with a between-by-within design and a between-by-between-by-within design and will be described and illustrated in Chapter 8.

To illustrate the next R function, consider data reported by Potthoff and Roy (1964) dealing with an orthodontic growth study, where for each of 27 children, the distance between the pituitary and pterygomaxillary fissure was measured at ages 8, 10, 12, and 14 years. The data can be accessed via the R package nlme and are stored in the R variable Orthodont. The first 10 rows of the data are

```
   Distance Age Subject  Sex
1     26.0   8     M01 Male
2     25.0  10     M01 Male
3     29.0  12     M01 Male
4     31.0  14     M01 Male
5     21.5   8     M02 Male
6     22.5  10     M02 Male
7     23.0  12     M02 Male
8     26.5  14     M02 Male
9     23.0   8     M03 Male
10    22.5  10     M03 Male
```

It might be useful to store the data in a matrix, where each row contains the outcome measure of interest, which is distance in the example. For the orthodontic growth study, this means storing the data in a matrix having 27 rows corresponding to the 27 participants, where each row has four columns corresponding to the four times that measures were taken. The R function

$$long2mat(x,Sid.col,dep.col)$$

accomplishes this goal. The argument x is assumed to be a matrix or a data frame. The argument dep.col is assumed to have a single value that indicates which column of x contains the data to be analyzed. The argument Sid.col indicates the column containing a participant's identification. So for the orthodontic growth study, the command m=long2mat(Orthodont,3,1) would create a 27×4 matrix, with the first row containing the values 26, 25, 29, and 31, the measures associated with the first participant.

The R function

$$longcov2mat(x, Sid.col, dep.col)$$

is like the function long2mat, only the argument dep.col can have more than one value and a matrix of covariates is stored in list mode for each of the n participants. Continuing the last example, the command m=long2mat(Orthodont,3,1) would result in m having list mode, m[[1]] would be a 4×1 matrix containing the values for the first participant, m[[2]] would be the values for the second participant, and so on.

Here are a few other R functions that might be useful. The R function

$$m2l(x)$$

stores the data in the J columns of a matrix in list mode having length J. So the command m=m2l(x) would store the data in column one of the matrix x in m[[1]], the data in column two would be stored in m[[2]], and so forth. (The R function matrix2list accomplishes the same goal.) The R function

$$list2mat(x)$$

takes data stored in list mode having length J and stores them in a matrix that has J columns. That is, x[[1]] becomes column 1, x[[2]] becomes column 2, and so on. The R function

$$l2v(x)$$

converts data in list mode into a single vector of values.

Consider the following data:

```
1        1        1        easy     6
1        1        2        easy     3
1        1        3        easy     2
1        1        4        hard     7
1        1        5        hard     4
1        1        6        hard     1
1        2        1        easy     2
1        2        2        easy     2
1        2        3        easy     7
1        2        4        hard     7
```

```
1        2        5        hard     3
1        2        6        hard     2
2        1        1        easy     1
2        1        2        easy     4
2        1        3        easy     4
2        1        4        hard     7
2        1        5        hard     7
2        1        6        hard     6
2        2        1        easy     2
2        2        2        easy     3
2        2        3        easy     1
2        2        4        hard     7
2        2        5        hard     5
2        2        6        hard     5
```

Imagine that column 2 indicates each participant's identification number, columns 1, 3, and 4 indicate categories, and column 5 is some outcome of interest. Further imagine it is desired to compute some measure of location for each category indicated by the values in columns 1 and 4. This can be accomplished with the R function

$$M2m.loc(m, grpc, col.dat, locfun = tmean, ...),$$

where the argument locfun indicates the measure of location that will be used, which defaults to a 20% trimmed mean, grpc indicates the columns of m that indicate the categories (or levels of a factor), and col.dat indicates the column containing the outcome measure of interest. For the situation at hand, assuming the data are stored in the data frame x, the command M2m.loc(x,c(1,4),5,locfun=mean) returns

```
V1   V4        loc
 1 easy 3.666667
 1 hard 4.000000
 2 easy 2.500000
 2 hard 6.166667
```

So, for example, participants who are in both category 1 and category easy, the mean is 3.67.

1.9.1 Eliminating Missing Values

From a statistical point of view, a simple strategy for handling missing values is to simply eliminate them. There are other methods for dealing with missing values (e.g., Little and Rubin, 2002), a few of which are covered in subsequent chapters. Here, it is merely noted that for convenience, when data are stored in a matrix or a data frame, say m, the R function

$$na.omit(m)$$

will eliminate any row having missing values.

1.10 Data Sets

The methods in this book are illustrated with data from a wide range of situations. These data sets can be downloaded from the author's personal web page:

Dornsife.usc.edu/cf/labs/wilcox/wilcox-faculty-display.cfm

CHAPTER 2

A Foundation for Robust Methods

Measures that characterize a distribution, such as measures of location and scale, are said to be *robust* if slight changes in a distribution have a relatively small effect on their value. As indicated in Chapter 1, the population mean and standard deviation, μ and σ, as well as the sample mean and sample standard deviation, $\bar{X}$ and s^2, are not robust. This chapter elaborates on this problem by providing a relatively non-technical description of some of the tools used to judge the robustness of parameters and estimators. Included are some strategies for identifying measures of location and scale that are robust. The emphasis in this chapter is on finding robust analogs of μ and σ, but the results and criteria described here are directly relevant to judging estimators as well, as will become evident. This chapter also introduces some technical tools that are of use in various situations.

This chapter is more technical than the remainder of the book. When analyzing data, it helps to have some understanding of how robustness issues are addressed, and providing a reasonably good explanation requires some theory. Also, many applied researchers, who do not religiously follow developments in mathematical statistics, might still have the impression that robust methods are ad hoc procedures. Accordingly, although the main goal is to make robust methods accessible to applied researchers, it needs to be emphasized that modern robust methods have a solid mathematical foundation. It is stressed, however, that many mathematical details arise that are not discussed here. The goal is to provide an indication of how technical issues are addressed without worrying about the many relevant details. Readers interested in mathematical issues can refer to the excellent books by Huber and Ronchetti (2009) and Hampel et al. (1986). The monograph by Reider (1994) is also of interest. For a book written at an intermediate level of difficulty, see Staudte and Sheather (1990).

2.1 Basic Tools for Judging Robustness

There are three basic tools that are used to establish whether quantities such as measures of location and scale have good properties: qualitative robustness, quantitative robustness, and infinitesimal robustness. This section describes these tools in the context of location measures, but they are also relevant to measures of scale, as will become evident. These tools not only provide formal methods for judging a particular measure, they can also be used to help derive measures that are robust.

Introduction to Robust Estimation and Hypothesis Testing
https://doi.org/10.1016/B978-0-12-820098-8.00008-7

25

Before continuing, it helps to be more formal about what is meant by a measure of location. A quantity that characterizes a distribution, such as the population mean, is said to be a measure of location if it satisfies four conditions, and a fifth is sometimes added. To describe them, let X be a random variable with distribution F, and let $\theta(X)$ be some descriptive measure of F. Then $\theta(X)$ is said to be a measure of location if for any constants a and b,

$$\theta(X + b) = \theta(X) + b, \tag{2.1}$$

$$\theta(-X) = -\theta(X), \tag{2.2}$$

$$X \geq 0 \text{ implies } \theta(X) \geq 0, \tag{2.3}$$

$$\theta(aX) = a\theta(X). \tag{2.4}$$

The first condition is called *location equivariance*. It simply requires that if a constant b is added to every possible value of X, a measure of location should be increased by the same amount. Let $E(X)$ denote the expected value of X. From basic principles, the population mean is location-equivariant. That is, if $\theta(X) = E(X) = \mu$, then $\theta(X + b) = E(X + b) = \mu + b$. The first three conditions, taken together, imply that a measure of location should have a value within the range of possible values of X. The fourth condition is called *scale equivariance*. If the scale by which something is measured is altered by multiplying all possible values of X by a, a measure of location should be altered by the same amount. In essence, results should be independent of the scale of measurement. As a simple example, if the typical height of a man is to be compared to the typical height of a woman, it should not matter whether the comparisons are made in inches or feet.

The fifth condition that is sometimes added was suggested by Bickel and Lehmann (1975). Let $F_x(x) = P(X \leq x)$ and $F_y(x) = P(Y \leq x)$ be the distributions corresponding to the random variables X and Y. Then X is said to be stochastically larger than Y, if for any x, $F_x(x) \leq F_y(x)$ with strict inequality for some x. If all the quantiles of X are greater than the corresponding quantiles of Y, then X is stochastically larger than Y. Bickel and Lehmann argue that if X is stochastically larger than Y, then it should be the case that $\theta(X) \geq \theta(Y)$, if θ is to qualify as a measure of location. The population mean has this property.

2.1.1 Qualitative Robustness

To understand qualitative robustness, it helps to begin by considering any function $f(x)$, not necessarily a probability density function. Suppose it is desired to impose a restriction on this function so that it does not change drastically with small changes in x. One way of doing this is to insist that $f(x)$ be continuous. If, for example, $f(x) = 0$ for $x \leq 1$, but $f(x) = 10,000$

for any $x > 1$, the function is not continuous, and if $x = 1$, an arbitrarily small increase in x results in a large increase in $f(x)$.

A similar idea can be used when judging a measure of location. This is accomplished by viewing parameters as functionals. In the present context, a functional is just a rule that maps every distribution into a real number. For example, the population mean can be written as

$$T(F) = E(X),$$

where the expected value of X depends on F. The role of F becomes more explicit if expectation is written in integral form, in which case, this last equation becomes

$$T(F) = \int x \, dF(x).$$

If X is discrete and the probability function corresponding to $F(x)$ is $f(x)$,

$$T(F) = \sum x f(x),$$

where the summation is over all possible values x of X.

One advantage of viewing parameters as functionals is that the notion of continuity can be extended to them. Thus, if the goal is to have measures of location that are relatively unaffected by small shifts in F, a requirement that can be imposed is that when viewed as a functional, it is continuous. Parameters with this property are said to have *qualitative robustness*.

Let $\hat{F}$ be the usual empirical distribution. That is, for the random sample $X_1, \ldots, X_n$, $\hat{F}(x)$ is just the proportion of X_i values less than or equal to x. An estimate of the functional $T(F)$ is obtained by replacing F with $\hat{F}$. For example, when $T(F) = E(X) = \mu$, replacing F with $\hat{F}$ yields the sample mean, $\bar{X}$. An important point is that qualitative robustness includes the idea that if $\hat{F}$ is close to F, in a sense to be made precise, then $T(\hat{F})$ should be close to $T(F)$. For example, if the empirical distribution represents a close approximation of F, then $\bar{X}$ should be a good approximation of μ, but this is not always the case.

One more introductory remark should be made. From a technical point of view, continuity leads to the issue of how the difference between distributions should be measured. Here, the Kolmogorov distance is used. Other metrics play a role when addressing theoretical issues, but they go beyond the scope of this book. Readers interested in pursuing continuity, as it relates to robustness, can refer to Hampel (1968).

To provide at least the flavor of continuity, let F and G be any two distributions, and let $D(F, G)$ be the Kolmogorov distance between them, which is the maximum value of $|F(x) - G(x)|$, the maximum being taken over all possible values of x. If the maximum does not exist, the supremum or least upper bound is used instead. That is, the Kolmogorov distance is the least upper bound on $|F(x) - G(x)|$ over all possible values of x. More succinctly, $D(F, G) = \sup |F(x) - G(x)|$, where the notation *sup* indicates *supremum*. For readers unfamiliar with the notion of a *least upper bound*, the Kolmogorov distance is the smallest value of A such that $|F(x) - G(x)| \leq A$. Any A satisfying $|F(x) - G(x)| \leq A$ is called an upper bound on $|F(x) - G(x)|$, and the smallest (least) upper bound is the Kolmogorov distance. Note that $|F(x) - G(x)| \leq 1$ for any x, so for any two distributions, the maximum possible value for the Kolmogorov distance is 1. If the distributions are identical, $D(F, G) = 0$.

Now consider any sequence of distributions, $G_n, n = 1, 2, \ldots$. For example, G_n might be the empirical distribution based n observations randomly sampled from some distribution F. Another sequence of distributions is the contaminated normal with $\epsilon = 1/n$. The functional T is said to be continuous at F, if for *any* sequence G_n, such that $D(G_n, F)$ approaches 0 as n gets large, $|T(G_n) - T(F)|$ approaches 0. In particular, the functional evaluated at the empirical distribution, $T(\hat{F})$, should approach $T(F)$, the functional evaluated at the distribution from which observations are being sampled. Under random sampling, the empirical distribution approaches the true distribution as n gets large, and from standard results, the sample mean approaches the population mean as well. However, there are sequences of distributions for which $D(G_n, F)$ approaches 0 for any F, but the mean of the empirical distribution, the sample mean, does not approach the mean of the true distribution, $T(F) = \mu$, as n gets large. Details are given by Staudte and Sheather (1990, p. 66). Thus, for the Kolmogorov metric, $T(F) = E(X)$ is not continuous. That is, if we require a measure of location that has a continuous functional, the population mean, μ, is ruled out.

An example of a continuous functional, that plays a central role in this book, is the γ-trimmed mean, $0 < \gamma \leq 0.5$. A γ-trimmed mean is the mean of a distribution after the distribution has been transformed in a particular way. More specifically, it is trimmed by truncating the distribution at the γ and $1 - \gamma$ quantiles. Note that if a probability density function is trimmed, it no longer qualifies as a probability density function because the area under the curve is no longer equal to 1; it is equal to $1 - 2\gamma$. Consequently, dividing the trimmed probability density function by $1 - 2\gamma$, the resulting function is again a probability density function. Here, two-sided trimming is assumed unless stated otherwise. (Some authors, when referring to a γ-trimmed mean, assume one-sided trimming, but others assume two-sided trimming instead.) In general then, when referring to a trimmed distribution, this means that the probability density function, $f(x)$, is transformed to

$$\frac{1}{1 - 2\gamma} f(x), \ x_\gamma \leq x \leq x_{1-\gamma},$$

where x_γ and $x_{1-\gamma}$ are the γ and $1 - \gamma$ quantiles. In essence, trimming results in focusing on the middle portion of a distribution.

As a simple example, consider a standard normal distribution after it has been trimmed 20% ($\gamma = 0.2$) and rescaled so that the area under the curve is equal to one. The 0.2 and 0.8 quantiles of the standard normal distribution are -0.84 and 0.84, respectively. Thus, the 20% trimmed analog of the standard normal distribution is defined for $-0.84 \le x \le 0.84$. The standard normal probability density function is

$$\frac{1}{\sqrt{2\pi}} \exp(-x^2/2), \quad -\infty \le x \le \infty,$$

so the 20% trimmed analog of the standard normal probability density function is

$$f(x) = \frac{1}{0.6} \frac{1}{\sqrt{2\pi}} \exp(-x^2/2), \quad -0.84 \le x \le 0.84. \tag{2.5}$$

2.1.2 Infinitesimal Robustness

To provide a relatively simple explanation of infinitesimal robustness, it helps to again consider the situation where $f(x)$ is any function, not necessarily a probability density function. Once more consider what restrictions might be imposed so that small changes in x do not result in large changes in $f(x)$. One such condition is that it be differentiable, and that the derivative be bounded. In symbols, if $f'(x)$ is the derivative, it is required that $f'(x) < B$ for some constant B. The function $f(x) = x^2$, for example, does not satisfy this condition because its derivative, $2x$, increases without bound as x gets large.

Analogs of derivatives of functionals exist, and so a natural way of searching for robust measures of location is to focus on functionals that have a bounded derivative. In the statistics literature, the derivative of a functional, $T(F)$, is called the *influence function* of T at F, which was introduced by Hampel (1968, 1974). Roughly, the influence function measures the relative extent a small perturbation in F has on $T(F)$. Put another way, it reflects the (normed) limiting influence of adding one more observation, x, to a very large sample.

To provide a more precise description of the influence function, let Δ_x be a distribution where the value x occurs with probability one. As is fairly evident, if Y has distribution Δ_x, then $P(Y \le y) = 0$, if $y < x$, and the mean of Y is $E(Y) = x$.

Next, consider a mixture of two distributions where an observation is randomly sampled from distribution F with probability $1 - \epsilon$; otherwise sampling is from the distribution Δ_x. That is, with probability ϵ, the observed value is x. The resulting distribution is

$$F_{x,\epsilon} = (1 - \epsilon)F + \epsilon \Delta_x. \tag{2.6}$$

It might help to notice the similarity between $F_{x,\epsilon}$ and the contaminated or mixed normal described in Chapter 1. In the present situation, F is any distribution, including normal distributions as a special case. Also notice the similarity with the influence curve in Chapter 1. Here, interest is in how the value of x affects the value of some functional when x occurs with probability ϵ. For example, if F has mean μ, then $F_{x,\epsilon}$ has mean $(1 - \epsilon)\mu + \epsilon x$, and the difference between the mean of $F_{x,\epsilon}$ and the mean of F is $\epsilon(x - \mu)$.

Notice that when ϵ is small, $F_{x,\epsilon}$ is similar to F, as measured by the Kolmogorov distance function. To see this, first note that if the distributions are evaluated at any value, say y,

$$|F_{x,\epsilon}(y) - F(y)| = |-\epsilon[F(y) - \Delta_x(y)]|.$$

But F and Δ_x are distributions, so $|F(y) - \Delta_x(y)| \leq 1$. Consequently, the Kolmogorov distance between $F_{x,\epsilon}$ and F is at most ϵ. Moreover, $F_{x,\epsilon}$ and F can be made arbitrarily close by choosing ϵ sufficiently small.

The relative influence on $T(F)$ of having the value of x occur with probability ϵ is

$$\frac{T(F_{x,\epsilon}) - T(F)}{\epsilon},$$

and the *influence function* of T at F is

$$IF(x) = \lim \frac{T(F_{x,\epsilon}) - T(F)}{\epsilon}, \tag{2.7}$$

where the limit is taken as ϵ approaches 0 from above. Approximately, $IF(x)$ is the relative influence of x on some measure that characterizes a distribution, $T(F)$, when the probability of observing the value of x is arbitrarily close to 0. $T(F)$ is said to be *B robust*, or to have *infinitesimal robustness*, if $IF(x)$ is bounded. The *gross error sensitivity* of $T(F)$ is $\sup_x |IF(x)|$.

As already indicated, if $T(F) = E(X)$, $T(F_{x,\epsilon}) - T(F) = \epsilon(x - \mu)$, so $(T(F_{x,\epsilon}) - T(F))/\epsilon = x - \mu$. Thus, the influence function of the population mean is

$$IF(x) = x - \mu,$$

which *does not* depend on F. Especially important is that the influence function is unbounded in x, that is, μ does not have infinitesimal robustness, and its gross error sensitivity is ∞.

2.1.3 Quantitative Robustness

The third approach to judging some quantity that characterizes a distribution is the *breakdown point*, which addresses the notion of quantitative robustness. The general idea is to describe quantitatively, the effect a small change in F has on some functional $T(F)$.

Again consider $F_{x,\epsilon} = (1 - \epsilon)F + \epsilon\Delta_x$, which has mean $(1 - \epsilon)\mu + \epsilon x$. Thus, for any $\epsilon > 0$, the mean goes to infinity as x gets large. In particular, even when ϵ is arbitrarily close to 0, in which case, the Kolmogorov distance between $F_{x,\epsilon}$ and F is small, the mean of $F_{x,\epsilon}$ can be made arbitrarily large by increasing x. The minimum value of ϵ for which a functional goes to infinity as x gets large is called the *breakdown point*. When necessary, the minimum value is replaced by the infimum or greatest lower bound. (This definition oversimplifies technical issues, but it suffices for present purposes. See Huber, 1981, Section 1.4 for more details.) In the illustration, any $\epsilon > 0$ causes the mean to go to infinity, so the breakdown point is 0. In contrast, the median of a distribution has a breakdown point of 0.5, and more generally, the γ-trimmed mean, μ_t, has a breakdown point of γ.

When searching for measures of dispersion, the breakdown point turns out to have considerable practical importance. In some cases, the breakdown point is more important than the efficiency of any corresponding estimator. For the moment, it is merely noted that the standard deviation, σ, has a breakdown point of 0, and this renders it unsatisfactory in various situations.

2.2 Some Measures of Location and Their Influence Function

There are many measures of location. (See Andrews et al., 1972.) This section describes some that are particularly important based on what is currently known. (A few additional measures of location are introduced in Chapters 3 and 6.)

2.2.1 Quantiles

It is convenient to begin with quantiles. For any random variable X with distribution F, the qth quantile, x_q say, satisfies $F(x) = P(X \le x_q) = q$, where $0 < q < 1$. For example, if X is standard normal, the 0.8 quantile is $x_{0.8} = 0.84$ and $P(X \le 0.84) = 0.8$.

In the event that there are multiple x values such that $F(x) = q$, the standard convention is to define the qth quantile as the smallest value of x such that $F(x) \ge q$. For completeness, it is sometimes necessary to define the qth quantile as $x_q = \inf\{x : F(x) \ge q\}$, where inf indicates infimum or greatest lower bound, but this is a detail that is not important here.

The qth quantile has location and scale equivariance, and it satisfies the other conditions for a measure location given by Eqs. (2.1)–(2.4), plus the Bickel–Lehmann condition. Insofar as it is desired to have a measure of location that reflects the typical subject under study, the median, $x_{0.5}$, is a natural choice. The breakdown point of the median is 0.5, and more generally, the breakdown point of the qth quantile is $1 - q$ (e.g., Staudte and Sheather, 1990, p. 56).

For some distributions, x_q has qualitative robustness, but for others, including discrete distributions, it does not. In fact, even if x_q has qualitative robustness at F, it is not qualitative robust at $F_{x,\epsilon}$. That is, there are distributions that are arbitrarily close to F for which x_q is not qualitative robust.

Letting $f(x)$ represent the probability density function and assuming $f(x_q) > 0$, and that $f(x_q)$ is continuous at x_q, the influence function of x_q is

$$I F_q(x) = \begin{cases} \frac{q-1}{f(x_q)}, & \text{if } x < x_q, \\ 0, & \text{if } x = x_q, \\ \frac{q}{f(x_q)}, & \text{if } x > x_q. \end{cases} \tag{2.8}$$

This influence function is bounded, so x_q has infinitesimal robustness.

2.2.2 The Winsorized Mean

One problem with the mean is that the tails of a distribution can dominate its value, and this is reflected by an unbounded influence function, a breakdown point of 0, and a lack of qualitative robustness. Put in more practical terms, if a measure of location is intended to reflect what the typical subject is like, the mean can fail because its value can be inordinately influenced by a very small proportion of the subjects who fall in the tails of a distribution. One strategy for dealing with this problem is to give less weight to values in the tails and pay more attention to those near the center. One specific strategy for implementing this idea is to *Winsorize* the distribution.

Let F be any distribution, and let x_γ and $x_{1-\gamma}$ be the γ and $1 - \gamma$ quantiles. Then a γ-Winsorized analog of F is the distribution

$$F_w(x) = \begin{cases} 0, & \text{if } x < x_\gamma, \\ \gamma, & \text{if } x = x_\gamma, \\ F(x), & \text{if } x_\gamma < x < x_{1-\gamma}, \\ 1, & \text{if } x \geq x_{1-\gamma}. \end{cases}$$

In other words, the left tail is pulled in so that the probability of observing the value x_γ is γ, and the probability of observing any value less than x_γ, after Winsorization, is 0. Similarly, the right tail is pulled in so that, after Winsorization, the probability of observing a value

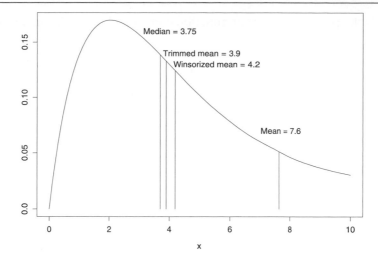

Figure 2.1: Mixed chi-squared distribution.

greater than $x_{1-\gamma}$ is 0. The mean of the Winsorized distribution is

$$\mu_w = \int_{x_\gamma}^{x_{1-\gamma}} x \, dF(x) + \gamma(x_\gamma + x_{1-\gamma}).$$

In essence, the Winsorized mean pays more attention to the central portion of a distribution by transforming the tails. The result is that μ_w can be closer to the central portion of a distribution. It can be shown that μ_w satisfies Eqs. (2.1)–(2.4), so it qualifies as a measure of location, and it also satisfies the Bickel–Lehmann condition.

For the mixed chi-squared distribution described in Chapter 1, the 20% Winsorized mean is approximately $\mu_w = 4.2$, based on simulations with 10,000 replications. Fig. 2.1 shows the position of the Winsorized mean relative to the median, $x_{0.5} = 3.75$, and the mean, $\mu = 7.6$. As is evident, Winsorization results in a measure of location that is closer to the bulk of the distribution. For symmetric distributions, $\mu_w = \mu$.

Like quantiles, there are distributions arbitrarily close to any distribution F for which the Winsorized mean is not qualitative robust. On the positive side, its breakdown point is γ. This suggests choosing $\gamma = 0.5$ to achieve the highest possible breakdown point, but there are some negative consequences, if γ is too far from 0, as will become evident in Chapter 3.

Let

$$C = \mu_w - \frac{\gamma^2}{f(x_\gamma)} - \frac{\gamma^2}{f(x_{1-\gamma})},$$

where again, f is the probability density function corresponding to F. The influence function of the Winsorized mean is

$$IF_w(x) = \begin{cases} x_\gamma - \frac{\gamma}{f(x_\gamma)} - C, & \text{if } x < x_\gamma, \\ x - C, & \text{if } x_\gamma < x < x_{1-\gamma}, \\ x_{1-\gamma} + \frac{\gamma}{f(x_{1-\gamma})} - C, & \text{if } x > x_{1-\gamma}. \end{cases}$$

Notice that the influence function is bounded but not smooth; it has jumps at x_γ and $x_{1-\gamma}$. (Also see Wu and Zuo, 2009.)

2.2.3 The Trimmed Mean

Rather than to Winsorize, another strategy for reducing the effects of the tails of a distribution is to simply remove them, and this is the strategy employed by the trimmed mean. The γ-trimmed mean is

$$\mu_t = \frac{1}{1-2\gamma} \int_{x_\gamma}^{x_{1-\gamma}} x \, dF(x).$$

In words, μ_t is the mean of a distribution after it has been trimmed as described in Section 2.1.1. The trimmed mean is both location- and scale-equivariant; more generally, it satisfies Eqs. (2.1)–(2.4), and it also satisfies the Bickel–Lehmann condition for a measure of location.

The influence function of the trimmed mean is

$$IF_t(x) = \begin{cases} \frac{1}{1-2\gamma}(x_\gamma - \mu_w), & \text{if } x < x_\gamma, \\ \frac{1}{1-2\gamma}(x - \mu_w), & \text{if } x_\gamma \le x \le x_{1-\gamma}, \\ \frac{1}{1-2\gamma}(x_{1-\gamma} - \mu_w), & \text{if } x > x_{1-\gamma}. \end{cases}$$

The influence function is bounded, but it is discontinuous at x_γ and $x_{1-\gamma}$. As already indicated, μ_t is qualitative robust when $\gamma > 0$, and its breakdown point is γ. It can be seen that $E[IF(X)] = 0$.

For the mixed chi-squared distribution described in Chapter 1, the 20% trimmed mean is $\mu_t = 3.9$, and its position relative to the mean, median, and 20% Winsorized mean is shown in Fig. 2.1.

It should be noted that the influence function of the trimmed mean can be derived under very general conditions, which include both symmetric and asymmetric distributions (Huber, 1981). Staudte and Sheather (1990) derive the influence function assuming distributions are symmetric, but their results are easily extended to the asymmetric case.

2.2.4 M-Measures of Location

M-measures of location form a large class of location measures that include the population mean, μ, as a special case. Typically, μ is viewed as $E(X)$, the expected value of the random variable X. However, to gain some insight into the motivation for M-measures of location, it helps to first view μ in a different way.

When searching for a measure of location, one strategy is to use some value, say c, that is in some sense close, on average, to all the possible values of the random variable X. One way of quantifying how close a value c is to all possible values of X is in terms of its expected squared distance. In symbols, $E(X - c)^2$ represents the expected squared distance from c. If c is intended to characterize the typical subject or thing under study, a natural approach is to use the value of c that minimizes $E(X - c)^2$. Viewing $E(X - c)^2$ as a function of c, the value of c minimizing this function is obtained by differentiating, setting the result equal to 0, and solving for c. That is, c is given by the equation

$$E(X - c) = 0, \tag{2.9}$$

so $c = \mu$. In other words, μ is the closest point to all possible values of X in terms of expected squared distance. But μ is not robust, and in the present context, the problem is that $E(X - c)^2$ gives an inordinate amount of weight to values of X that are far from c. Put another way, the function $(x - c)^2$ increases too rapidly as x moves away from c.

The approach just described for deriving a measure of location can be improved by considering a class of functions for measuring the distance from a point, and then searching for a function within this class that has desirable properties. To this end, let $\xi(X - \mu_m)$ be some function that measures the distance from μ_m, and let Ψ be its derivative with respect to μ_m. Attention is restricted to those functions for which $E[\xi(X - \mu_m)]$, viewed as a function of μ_m, has a derivative. In the previous paragraph, $\xi(X - \mu_m) = (X - \mu_m)^2$ and $\Psi(X - \mu_m) = -2(X - \mu_m)$. Then a measure of location that is closest to all possible values of X, as measured by its expected distance, is the value of μ_m that minimizes $E[\xi(X - \mu_m)]$. This means that μ_m is determined by the equation

$$E[\Psi(X - \mu_m)] = 0. \tag{2.10}$$

Typically, the function Ψ is assumed to be odd, meaning that $\Psi(-x) = -\Psi(x)$ for any x. (The reason for this will become clear in Chapter 3.) The value of μ_m that satisfies Eq. (2.10) is called an *M-measure of location*. Obviously, the class of odd functions is too large for practical purposes, but this problem can be corrected, as will be seen. Huber (1981) describes general conditions under which M-measures of location have both quantitative and qualitative robustness.

Table 2.1: Some choices for ξ and Ψ.

Criterion	$\xi(x)$	$\Psi(x)$	Range
Huber	$\frac{1}{2}x^2$	x	$|x| \le K$
	$|x|K - \frac{1}{2}K^2$	$K\,\text{sign}(x)$	$|x| > K$
Andrews	$a[1 - \cos(x/a)]$	$\sin(x/a)$	$|x| \le a\pi$
	$2a$	0	$|x| > a\pi$
Hampel	$\frac{1}{2}x^2$	x	$|x| \le a$
	$a|x| - \frac{1}{2}a^2$	$a\,\text{sign}(x)$	$a < |x| \le b$
	$\frac{a(c|x|-\frac{1}{2}x^2)}{c-b} - \frac{7}{6}a^2$	$\frac{a\,\text{sign}(x)(c-|x|)}{c-b}$	$b < |x| \le c$
	$a(b + c - a)$	0	$|x| > c$
Biweight		$x(1 - x^2)^2$	$|x| < 1$
		0	$|x| \ge 1$

M-measures of location are estimated with M-estimators obtained by replacing F in Eq. (2.10) with the empirical distribution $\hat{F}$. (Details are given in Chapter 3.) It should be remarked that many books and journal articles do not make a distinction between M-measures and M-estimators. Ordinarily, the term M-estimator is used to cover both situations.

Of course, to make progress, criteria are needed for choosing ξ or Ψ. One criterion for a robust measure of location is that its influence function be bounded. It turns out that when μ_m is determined with Eq. (2.10), its influence function has a relatively simple form:

$$IF_m(x) = \frac{\Psi(x - \mu_m)}{E[\Psi'(X - \mu_m)]},$$

where $\Psi'(X - \mu_m)$ is the derivative of Ψ. That is, the influence function is Ψ rescaled by the expected value of its derivative, $E[\Psi'(X - \mu_m)]$. Thus, to obtain a bounded influence function, attention can be restricted to those Ψ that are bounded. From results already given, this rules out the choice $\Psi(X - \mu_m) = X - \mu_m$, which yields $\mu_m = \mu$.

Table 2.1 lists some choices for ξ and Ψ that have been proposed. More recently, Jiang et al. (2019) suggest using $\Psi_{\tau\lambda}(x) = x$, if $|x| \le \tau$ and $\Psi_{\tau\lambda}(x) = x \exp\{-(x^2 - \tau^2)/\lambda\}$, when $|x| > \tau$, where τ and λ are parameters to be determined. The function $\text{sign}(x)$ in Table 2.1 is equal to $-1, 0$, or 1, according to whether x is less than, equal to, or greater than 0. The constants a, b, c, and K can be chosen so that the resulting measure of location has desirable properties. A common strategy is to choose these constants so that when estimating μ_m, the estimator has reasonably high efficiency when sampling from a normal distribution but continues to have high efficiency when sampling from a heavy-tailed distribution instead. For now, these

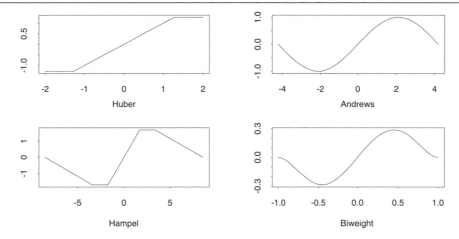

Figure 2.2: Possible choices for influence functions.

constants are left unspecified. As will be seen, further refinements can be made that make it a relatively simple matter to choose Ψ in applied work.

Fig. 2.2 shows a graph of Ψ for the Huber, Andrews, Hampel, and biweight given in Table 2.1. (The biweight also goes by the name of Tukey's bisquare.) Notice that all four graphs are linear, or approximately so, for an interval around 0. This turns out to be desirable when properties of estimators are considered, but the details are postponed for now. Also notice that the biweight and Andrews' Ψ redescend to 0. That is, extreme values are given less weight in determining μ_m, and x values that are extreme enough are ignored.

As a measure of location, μ_m, given by Eq. (2.10), satisfies Eqs. (2.1)–(2.3), but it does not satisfy Eq. (2.4), scale equivariance, for the more interesting choices for Ψ, including those shown in Table 2.1. This problem can be corrected by incorporating a measure of scale into Eq. (2.10) but not just any measure of scale will do. In particular, a measure of scale with a high breakdown point is needed if the M-measure of location is to have a reasonably high breakdown point as well. The standard deviation, σ, has a breakdown point of 0, so some other measure must be used. A method of dealing with scale equivariance is described in Section 2.4.

2.2.5 R-Measures of Location

R-measures of location do not play a role in this book, but for completeness, they are briefly described here. Generally, R-measures of location are derived by inverting tests of hypotheses based on ranks. Let J be some specified function. In functional form, an R-measure of location, μ_r, satisfies

$$\int J\{0.5[q + 1 - F(2\mu_r - x_q)]\}dq = 0.$$

A common choice for J is $J(x) = x - \frac{1}{2}$. This leads to the Hodges–Lehmann estimator, but no details are given here. For symmetric distributions, the Hodges–Lehmann estimator has a well-behaved influence function, and it is bounded and smooth (Staudte and Sheather, 1990). However, for asymmetric distributions, the denominator of the influence function can be very small (Huber, 1981, p. 65), suggesting that practical problems might arise.

Another concern was pointed out by Bickel and Lehmann (1975). For any R-estimator, which estimates an R-measure of location, there are distributions such that the asymptotic efficiency of the R-estimator relative to $\bar{X}$ is zero. Again, it is skewed distributions that create problems. For more on R-estimators, including the Hodges–Lehmann estimator, see Hettmansperger (1984). For a description of situations where R-estimators exhibit practical concerns even when sampling from a symmetric distribution, see Morgenthaler and Tukey (1991).

2.3 Measures of Scale

This section briefly describes some measures of scale that play an important role in robust methods, plus some popular measures of scale that are not robust. (Some additional measures of scale are described in Chapter 3.) As with measures of location, it helps to start with a precise definition of what constitutes a measure of scale.

Any non-negative functional, $\tau(X)$, is said to be a *measure of scale*, if for any constants $a > 0$ and b,

$$\tau(aX) = a\tau(X), \tag{2.11}$$

$$\tau(X + b) = \tau(X), \tag{2.12}$$

$$\tau(X) = \tau(-X). \tag{2.13}$$

The first of these conditions is called *scale equivariance*, the second is called *location invariance*, and the third is *sign invariance*. From basic principles, σ qualifies as a measure of scale.

Suppose X and Y have a symmetric distribution, and that the distribution of $|X|$ is stochastically larger than the distribution of $|Y|$. Bickel and Lehmann (1976) call a measure of scale a *measure of dispersion*, if $\tau(X) \geq \tau(Y)$. It can be seen that σ is a measure of dispersion, but as already mentioned, σ has a breakdown point of 0, and its influence function is unbounded.

Currently, there are two general approaches to measuring scale that are of particular importance: L-measures and M-measures. L-measures are estimated with linear combinations of the order statistics, and M-measures are similar to M-measures of location in the sense that τ is defined by the equation

$$E\left[\chi\left(\frac{X}{\tau}\right)\right] = 0,$$

where χ is some specified function. Typically, χ is an even function, meaning that $\chi(-x) = \chi(x)$.

Mean Deviation from the Mean. A reasonable choice for a measure of scale is

$$\tau(F) = E|X - \mu|.$$

However, its breakdown point is 0, and its influence function is unbounded. On the positive side, the natural estimate of this measure of scale is relatively efficient when sampling from heavy-tailed distributions.

Mean Deviation from the Median. Another popular choice for a measure of scale is

$$\tau(F) = E|X - x_{0.5}|.$$

It might appear that this measure of scale is robust because it uses the median, $x_{0.5}$, but its breakdown point is 0, and its influence function is unbounded.

Median Absolute Deviation. The median absolute deviation, ω, is defined by

$$P(|X - x_{0.5}| \le \omega) = 0.5.$$

In other words, ω is the median of the distribution associated with $|X - x_{0.5}|$, the distance between X and its median. This measure of scale is an M-measure of scale with $\chi(x) = \text{sign}(|x| - 1)$. The breakdown point is 0.5, and this makes it attractive for certain purposes, as will be seen. Its influence function is

$$IF_\omega(x) = \frac{\text{sign}(|x - x_{0.5}| - \omega) - \frac{f(x_{0.5}+\omega)-f(x_{0.5}-\omega)}{f(x_{0.5})}\text{sign}(x - x_{0.5})}{2[f(x_{0.5} + \omega) + f(x_{0.5} - \omega)]}, \qquad (2.14)$$

where $f(x)$ is the probability density function associated with X. Assuming $f(x_{0.5})$ and $2[f(x_{0.5} + \omega) + f(x_{0.5} - \omega)]$ are not equal to 0, IF_ω is defined and bounded. (Alternatives to the median absolute deviation measure of variation were studied by Rousseeuw and Croux, 1993.)

The q Quantile Range. The q quantile range is an L-measure of scale given by

$$\tau(F) = x_{1-q} - x_q, \ 0 < q < 0.5.$$

A special case in common use is the interquartile range, where $q = 0.25$, so τ is the difference between the 0.75 and 0.25 quantiles. Its breakdown point is 0.25. Recalling that the influence functions of $x_{0.75}$ and $x_{0.25}$ are given by Eq. (2.8), the influence function of the interquartile range is $IF_{0.75} - IF_{0.25}$. Letting

$$C = q \left\{ \frac{1}{f(x_q)} + \frac{1}{f(x_{1-q})} \right\},$$

a little algebra shows that the influence function of the q quantile range is

$$IF_{\text{range}} = \begin{cases} \frac{1}{f(x_q)} - C, & \text{if } x < x_q, \\ -C, & \text{if } x_q \le x \le x_{1-q}, \\ \frac{1}{f(x_{1-q})} - C, & \text{if } x > x_{1-q}. \end{cases}$$

The Winsorized Variance. The γ-Winsorized variance is

$$\sigma_w^2 = \int_{x_\gamma}^{x_{1-\gamma}} (x - \mu_w)^2 dF(x) + \gamma[(x_\gamma - \mu_w)^2 + (x_{1-\gamma} - \mu_w)^2].$$

In other words, σ_w^2 is the variance of F after it has been Winsorized. (For a standard normal distribution, and with $\gamma = 0.2$, $\sigma_w^2 = 0.4129$.) It can be shown that σ_w^2 is a measure of scale, it is also a measure of dispersion, and it has a bounded influence function. Welsh and Morrison (1990) report the influence function of a large class of L-measures of scale that contains the Winsorized variance as a special case.

2.4 Scale-Equivariant M-Measures of Location

M-measures of location can be made scale-equivariant by incorporating a measure of scale in the general approach described in Section 2.2.4. That is, rather than determining μ_m with Eq. (2.1), use

$$E\left[\Psi\left(\frac{X - \mu_m}{\tau}\right)\right] = 0, \tag{2.15}$$

where τ is some appropriate measure of scale.

When considering which measure of scale should be used in Eq. (2.15), it helps to notice that τ plays a role in determining whether a value for X is unusually large or small. To illustrate this, consider Huber's Ψ, which, in the present context, is given by

$$\Psi\left(\frac{x - \mu_m}{\tau}\right) = \begin{cases} -K, & \text{if } (x - \mu_m)/\tau < -K, \\ \frac{x - \mu_m}{\tau}, & \text{if } -K \le (x - \mu_m)/\tau \le K, \\ K, & \text{if } (x - \mu_m)/\tau > K. \end{cases}$$

Then according to Ψ, the distance between x and μ_m, $|x - \mu_m|$, is not unusually large or small, if $-K \leq (x - \mu_m)/\tau \leq K$. In this case, the same Ψ used to define the population mean, μ, is being used. If $x - \mu_m > K\tau$, Ψ considers the distance to be relatively large, and the influence of x on μ_m is reduced. Similarly, if $x - \mu_m < -K\tau$, x is considered to be unusually far from μ_m.

For the special case where X is normal and τ is taken to be the standard deviation, σ, x is considered to be unusually large or small, if it is more than K standard deviations from μ. A problem with σ is that its value is inflated by heavy-tailed distributions, and this can mask unusually large or small x values. For example, suppose $K = 1.28$, the 0.9 quantile of the standard normal distribution. Then, if X has a standard normal distribution, $K\sigma = 1.28$, so $x = 3$ is considered to be unusually large by Ψ. Now suppose X has the contaminated normal distribution shown in Fig. 1.1 of Chapter 1. Then $x = 3$ is still fairly far into the right tail, it should be considered unusually large, but now $K\sigma = 1.28 \times 3.3 = 4.224$, so $x = 3$ is not labeled as being unusually large. What is required is a measure of scale that is relatively insensitive to heavy-tailed distributions so that unusual values are not masked. In particular, a measure of scale with a high breakdown point is needed. Among the measures of scale described in Section 2.3, the median absolute deviation, ω, has a breakdown point of 0.5. This is the highest possible breakdown point, and it is higher than any other measure of scale described in Section 2.3. This suggests using ω in Eq. (2.15), and this choice is typically made. There are other considerations when choosing τ in Eq. (2.15), such as efficiency, but ω remains a good choice. M-measures of location, defined by Eq. (2.15), satisfy the four requirements for measures of location given by Eqs. (2.1)–(2.4).

The influence function of M-measures of location, defined by Eq. (2.15), takes on a more complicated form versus the influence function associated with Eq. (2.10). Moreover, it depends on the choice of scale, τ. As an illustration, suppose $\tau = \omega$ is used, where ω is the median absolute deviation measure of scale introduced in Section 2.5. Let $y = (x - \mu_m)/\omega$. Then, if Eq. (2.15) is used to define a measure of location, the influence function is

$$I F_m(x) = \frac{\omega \Psi(y) - I F_\omega(x)\{E[\Psi'(y)y]\}}{E[\Psi'(y)]},$$

where $I F_\omega$ is given by Eq. (2.14). Note that because the influence function of ω, $I F_\omega$, is bounded, $I F_m$ is bounded as well.

The breakdown point depends on the choice for K and the measure of scale, τ. For the common choice $\tau = \omega$, the breakdown point does not depend on K and is equal to 0.5. Despite this, μ_m can have a value that is further from the median than the 20% trimmed mean. For example, with Huber's Ψ and the common choice of $K = 1.28$, $\mu_m = 4.2$, approximately, for the mixed chi-squared distribution in Fig. 2.1. In contrast, $\mu_t = 3.9$, and the median is $x_{0.5} = 3.75$. Lowering K to 1, μ_m drops to 4.0.

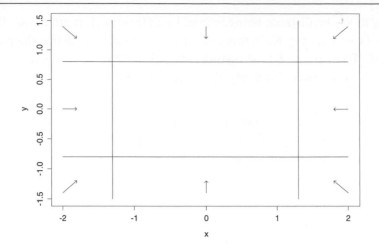

Figure 2.3: Winsorization of a bivariate distribution.

2.5 Winsorized Expected Values

Let $g(X)$ be any function of the continuous random variable X. When working with a single random variable, the γ-*Winsorized expected value* of $g(X)$ is defined to be

$$E_w[g(X)] = \int_{x_\gamma}^{x_{1-\gamma}} g(x)dF(x) + \gamma[g(x_\gamma) + g(x_{1-\gamma})].$$

That is, the expected value of $g(X)$ is defined in the usual way, only with respect to the Winsorized distribution corresponding to F.

Let X and Y be any two continuous random variables with joint distribution F and probability density function $f(x, y)$. What is needed is an analog of Winsorization for any bivariate distribution. Note that any point (x, y) falls in one of nine regions shown in Fig. 2.3, where the corners of the rectangle are determined by the γ and $1 - \gamma$ quantiles of X and Y. That is, the rectangle is given by the four points (x_γ, y_γ), $(x_\gamma, y_{1-\gamma})$, $(x_{1-\gamma}, y_\gamma)$, and $(x_{1-\gamma}, y_{1-\gamma})$. Winsorization of any bivariate distribution consists of pulling in any point outside the rectangle formed by these four points, as indicated by the arrows in Fig. 2.3. For any point inside this rectangle, the Winsorized distribution has probability density function $f(x, y)$. The corners of the rectangle become discrete distributions, even when working with continuous random variables. For example, the point (x_γ, y_γ) has probability $P(X \leq x_\gamma, Y \leq y_\gamma)$. Similarly, the point $(x_\gamma, y_{1-\gamma})$ has probability equal to the probability that $X \leq x_\gamma$ and $Y \geq y_{1-\gamma}$ simultaneously. However, the sides of the rectangle, excluding the four corners, have a continuous distribution when X and Y are continuous.

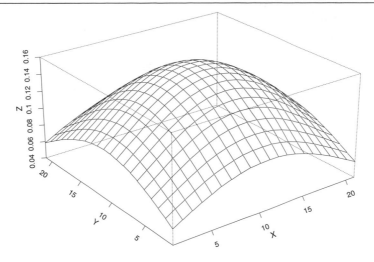

Figure 2.4: Winsorization of a bivariate normal distribution.

Let X and Y be any two random variables with joint distribution F, and let $g(X, Y)$ be any function of X and Y. Following Wilcox (1993b, 1994b), the *Winsorized expected value* of $g(X, Y)$ is defined to be

$$
E_w[g(X, Y)] = \int_{x_\gamma}^{x_{1-\gamma}} \int_{y_\gamma}^{y_{1-\gamma}} g(x, y)dF(x, y)
$$
$$
+ \int_{-\infty}^{x_\gamma} \int_{y_\gamma}^{y_{1-\gamma}} g(x_\gamma, y)dF(x, y) + \int_{-\infty}^{x_\gamma} \int_{-\infty}^{y_\gamma} g(x_\gamma, y_\gamma)dF(x, y)
$$
$$
+ \int_{-\infty}^{x_\gamma} \int_{y_{1-\gamma}}^{\infty} g(x_\gamma, y_{1-\gamma})dF(x, y) + \int_{x_{1-\gamma}}^{\infty} \int_{y_\gamma}^{y_{1-\gamma}} g(x_{1-\gamma}, y)dF(x, y)
$$
$$
+ \int_{x_{1-\gamma}}^{\infty} \int_{-\infty}^{y_\gamma} g(x_{1-\gamma}, y_\gamma)dF(x, y) + \int_{x_{1-\gamma}}^{\infty} \int_{y_{1-\gamma}}^{\infty} g(x_{1-\gamma}, y_{1-\gamma})dF(x, y)
$$
$$
+ \int_{x_\gamma}^{x_{1-\gamma}} \int_{-\infty}^{y_\gamma} g(x, y_\gamma)dF(x, y) + \int_{x_\gamma}^{x_{1-\gamma}} \int_{y_{1-\gamma}}^{\infty} g(x, y_{1-\gamma})dF(x, y).
$$

Fig. 2.4 illustrates the first step when Winsorizing a bivariate distribution. The bivariate distribution of X and Y is trimmed by removing any points outside the rectangle formed by the four points (x_γ, y_γ), $(x_\gamma, y_{1-\gamma})$, $(x_{1-\gamma}, y_\gamma)$, and $(x_{1-\gamma}, y_{1-\gamma})$.

A Winsorized covariance is $\mathrm{COV}_w(X, Y) = E_w[(X - \mu_{wx})(Y - \mu_{wy})]$. If X and Y are independent, $\mathrm{COV}_w(X, Y) = 0$.

The definition of E_w suggests one way of estimating Winsorized parameters. For a random sample, $X_1, \ldots, X_n$, suppose

$$E_w[g(X_1, \ldots, X_n)] = \xi. \tag{2.16}$$

This indicates that ξ be estimated with $\hat{\xi}_w = g(W_1, \ldots, W_n)$, where

$$W_i = \begin{cases} X_{(k+1)}, & \text{if } X_i \leq X_{(k+1)}, \\ X_i, & \text{if } X_{(k+1)} < X_i < X_{(n-k)}, \\ X_{(n-k)}, & \text{if } X_i \geq X_{(n-k)}. \end{cases}$$

$X_{(1)} \leq \cdots \leq X_{(n)}$ are the order statistics, and $k = [\gamma n]$, the greatest integer less than or equal to γn. When Eq. (2.16) holds, $\hat{\xi}_w$ is said to be a Winsorized unbiased estimate of ξ. For example, when dealing with a symmetric distribution, $\bar{W} = \sum W_i / n$ is a Winsorized unbiased estimate of μ_w.

## CHAPTER 3

Estimating Measures of Location and Scale

This chapter describes methods for estimating the measures of location and scale introduced in Chapter 2, and it introduces some additional measures of location and scale that have practical importance. Also, two general approaches to estimating standard errors are described and illustrated. One is based on estimating expressions for the standard errors of estimators, which is perhaps the more common strategy to employ, and the other is based on a so-called bootstrap method. As will be seen, estimating standard errors is often done in a way that is not intuitive nor obvious based on standard statistical training. Another goal is to introduce some outlier detection methods plus some graphical methods for summarizing data that will be used in later chapters.

This chapter is less technical than Chapter 2, but it is important at least to touch on theory so that readers understand why common strategies in applied research turn out to be inappropriate. For example, why is it incorrect to discard outliers among a dependent variable and apply standard techniques using the remaining data? Although this chapter gives the reader some indication of how theoretical problems are addressed, mathematical details are kept to a minimum. Readers interested in a more rigorous description of mathematical issues can refer to Huber and Ronchetti (2009) as well as Hampel et al. (1986). For a book written at an intermediate level of difficulty, see Staudte and Sheather (1990).

3.1 A Bootstrap Estimate of a Standard Error

It is convenient to begin with a description of the most basic bootstrap method for estimating a standard error. Let $\hat{\theta}$ be any estimator based on the random sample $X_1, \ldots, X_n$. The goal is to estimate $\text{VAR}(\hat{\theta})$, the squared standard error of $\hat{\theta}$. The strategy used by the bootstrap method is based on a very simple idea. Temporarily assume that observations are randomly sampled from some *known* distribution, F. Then for a given sample size, n, the sampling distribution of $\hat{\theta}$ could be determined by randomly generating n observations from F, computing $\hat{\theta}$, randomly generating another set of n observations, computing $\hat{\theta}$, and repeating this process many times. Suppose this is done B times, and the resulting values for $\hat{\theta}$ are labeled $\hat{\theta}_1, \ldots, \hat{\theta}_B$. If B is large enough, the values $\hat{\theta}_1, \ldots, \hat{\theta}_B$ provide a good approximation of the distribution of $\hat{\theta}$. In particular, they provide an estimate of the squared standard error of $\hat{\theta}$,

Introduction to Robust Estimation and Hypothesis Testing
https://doi.org/10.1016/B978-0-12-820098-8.00009-9

namely,

$$\frac{1}{B-1} \sum_{b=1}^{B} (\hat{\theta}_b - \bar{\theta})^2,$$

where

$$\bar{\theta} = \frac{1}{B} \sum_{b=1}^{B} \hat{\theta}_b.$$

That is, $\text{VAR}(\hat{\theta})$ is estimated with the sample variance of the values $\hat{\theta}_1, \ldots, \hat{\theta}_B$. If, for example, $\hat{\theta}$ is taken to be the sample mean, $\bar{X}$, the squared standard error would be found to be σ^2/n, approximately, provided B is reasonably large. Of course when working with the mean, it is known that its squared standard error is σ^2/n, so the method just described is unnecessary. The only point is that a reasonable method for estimating the squared standard error of $\hat{\theta}$ has been described.

In practice, F is not known, but it can be estimated with

$$\hat{F}(x) = \frac{\#\{X_i \leq x\}}{n},$$

the proportion of observations less than or equal to x, which provides a non-parametric maximum likelihood estimate of F. The empirical distribution assigns probability $1/n$ to each X_i, so the estimated probability of observing the value X_i is f_i/n, where f_i is the number of times the value X_i occurred among the n observations. All other possible values (values not observed) have an estimated probability of zero. The bootstrap estimate of the standard error is obtained as described in the previous paragraph, except that $\hat{F}$ replaces F. In practical terms, a *bootstrap sample* is obtained by resampling with replacement n observations from $X_1, \ldots, X_n$. This is easily done with the R command

$$\text{sample(x,size=length(x),replace=TRUE)}.$$

To estimate the sampling distribution of $\hat{\theta}$, generate a bootstrap sample from the observations $X_1, \ldots, X_n$ and compute $\hat{\theta}$ based on the obtained bootstrap sample. The result will be labeled $\hat{\theta}^*$ to distinguish it from $\hat{\theta}$, which is based on the observed values $X_1, \ldots, X_n$. Repeat this process B times, yielding $\hat{\theta}_1^*, \ldots, \hat{\theta}_B^*$. These B values provide an estimate of the sampling distribution of $\hat{\theta}$ and, in particular, an estimate of its squared standard error given by

$$S^2 = \frac{1}{B-1} \sum_{b=1}^{B} (\hat{\theta}_b^* - \bar{\theta}^*)^2,$$

where $\bar{\theta}^* = \sum \hat{\theta}_b^*/B$.

How many bootstrap samples should be used? That is, how should B be chosen? This depends, of course, on the goals and criteria that are deemed important. Suppose, for example, estimated standard errors are used to compute a confidence interval. One perspective is to choose B so that the actual probability coverage is reasonably close to the nominal level. Many of the methods in this book are based on this view. However, another approach is to choose B so that if a different collection of bootstrap samples were used, the results would change by a negligible amount. That is, choose B to be sufficiently large so that if the seed in the random number generator is altered, essentially, the same conclusions would be obtained. Booth and Sarkar (1998) derived results on choosing B from this latter point of view, and with the increased speed of computers in recent years, some of the newer methods in this book take this latter view into account.

While the bootstrap estimate of the sampling distribution of a statistic can be argued to be reasonable, the extent to which it has practical value is not immediately clear. The basic bootstrap methods covered in this book are not a panacea for the many problems that confront the applied researcher, as will become evident in subsequent chapters. But with over 1,000 journal articles on the bootstrap, including both theoretical and simulation studies, all indications are that it has great practical value, particularly when working with robust measures of location and scale, as will be seen. Also, there are many proposed ways of possibly improving upon the basic bootstrap methods used in this book, summaries of which are given by Efron and Tibshirani (1993). Some of these look very promising, but the extent to which they have practical value for the problems considered here has not been determined. When testing hypotheses or computing confidence intervals, for some problems, a bootstrap method is the only known method that provides reasonably accurate results.

3.1.1 R Function bootse

As explained in Section 1.7 of Chapter 1, R functions have been written for applying the methods described in this book. The software written for this book is free, and a single command incorporates them into your version of R. Included is the function

$$\text{bootse(x,nboot=1000,est=median),}$$

which can be used to compute a bootstrap estimate of the standard error of virtually any estimator covered in this book. However, an important exception is when dealing with quantile estimators based on only one or two order statistics and there are tied (duplicated) values. The convergence of bootstrap variance estimators of order statistics to the true variance has been demonstrated by Ghosh et al. (1985) and Babu (1986), assuming that sampling is from a continuous distributions, so, in particular, tied values occur with probability zero. However, when tied values can occur, estimates of the standard errors can be highly inaccurate. Some tied values can be tolerated, but at some point, this is no longer the case. Methods for dealing with this issue are described in subsequent sections.

Regarding the R function bootse, the argument x is any R variable containing the data. The argument nboot represents B, the number of bootstrap samples, which defaults to 1,000 if not specified. (As is done with all R functions, optional arguments are indicated by an =, and they default to the value shown. Here, for example, the value of nboot is taken to be 1,000 if no value is specified by the user.) The argument est indicates the estimator for which the standard error is to be computed. If not specified, est defaults to the median. That is, the standard error of the usual sample median will be estimated. So, for example, if data are stored in the R variable blob, the command bootse(blob) will return the estimated standard error of the usual sample median.

3.2 Density Estimators

Before continuing with the main issues covered in this chapter, it helps to first touch on a related problem that plays a role here as well as in subsequent chapters. The problem is estimating $f(x)$, the probability density function, based on a random sample of observations. Such estimators play an explicit role when trying to estimate the standard error of certain location estimators to be described. More generally, density estimators provide a useful perspective when trying to assess how groups differ and by how much.

Generally, kernel density estimators take the form

$$\hat{f}(x) = \frac{1}{nh} \sum_{i=1}^{n} K\left(\frac{x - X_i}{h}\right),$$

where K is some probability density function and h is a constant to be determined. The constant h has been given several names including the span, the window width, the smoothing parameter, and the bandwidth. Some explicit choices for h are discussed later in this section. Often K is taken to be a distribution symmetric about zero, but there are exceptions. A particular choice that has been found to be relatively effective is the Epanechnikov kernel:

$$\begin{aligned} K(t) &= \tfrac{3}{4}(1 - \tfrac{1}{5}t^2)/\sqrt{5}, \quad |t| < \sqrt{5}, \\ &= 0, \qquad\qquad\qquad\quad \text{otherwise.} \end{aligned} \tag{3.1}$$

There is a vast literature on kernel density estimators (Silverman, 1986; Scott, 1992; Wand and Jones, 1995; Simonoff, 1996). (For some more recent results, see, for example, Clements et al., 2003; Devroye and Lugosi, 2001; Messer and Goldstein, 1993; Yang and Marron, 1999; cf. Liu and Brown, 1993; Harpole et al., 2014.) Here, four types of kernel density estimators are summarized for later reference.

3.2.1 Silverman's Rule of Thumb

The first kernel density estimator is based on the Epanechnikov kernel given by Eq. (3.1). Following Silverman (1986), as well as the recommendation made by Venables and Ripley (2002, p. 127), the span is taken to be

$$h = 1.06\min(s, \text{IQR}/1.34)n^{-1/5},$$

where s is the usual sample standard deviation and IQR is some estimate of the interquartile range. That is, IQR estimates the difference between the 0.75 and 0.25 quantiles. (Here, the interquartile range is estimated as described in Section 3.12.5, using what are called the ideal fourths.) This choice for the span has been labeled Silverman's rule of thumb method.

3.2.2 Rosenblatt's Shifted Histogram

The second method, Rosenblatt's shifted histogram estimator, employs results derived by Scott (1979) and Freedman and Diaconis (1981). The computational details are as follows. Set

$$h = \frac{1.2(\text{IQR})}{n^{1/5}}.$$

Let A be the number of observations less than or equal $x + h$. In symbols,

$$A = \#\{X_i \leq x + h\},$$

where the notation $\#\{X_i \leq x + h\}$ indicates the cardinality of the set of observations satisfying $X_i \leq x + h$. Similarly, let

$$B = \#\{X_i < x - h\},$$

the number of observations less than $x - h$. Then the estimate of $f(x)$ is

$$\hat{f}(x) = \frac{A - B}{2nh}.$$

3.2.3 The Expected Frequency Curve

The next estimator, the *expected frequency curve*, is basically a variation of what is called the naive density estimator, and it is related to certain regression smoothers discussed later in this book. It also has similarities to the nearest neighbor method for estimating densities as described in Silverman (1986). The basic idea when estimating $f(x)$, for a given value x, is to use the proportion of observed values among $X_1, \ldots, X_n$ that are "close" to x.

The method begins by computing the median absolute deviation (MAD) statistic, which is just the sample median of the n values $|X_1 - M|, \ldots, |X_n - M|$, where M is the usual sample median described in Section 1.3. (For relevant asymptotic results on MAD, see Hall and Welsh, 1985.) Let MADN=MAD/$z_{0.75}$, where $z_{0.75}$ is the 0.75 quantile of a standard normal distribution. Then x is said to be close to X_i if $|X_i - x|/$MADN $\leq h$, where h again plays the role of a span. Typically, $h = 0.8$ gives good results. (As is evident, there is no particular reason here to use MADN rather than MAD; it is done merely to follow certain conventions covered in Section 3.6, where MAD is introduced in a more formal manner.) Let N_x be the number of observations close to x, in which case, N_x/n estimates the probability that a randomly sampled value is close to x. An estimate of the density at x is

$$\hat{f}(x) = \frac{N_x}{2hn\text{MADN}}.$$

In contrast is the naive density estimator discussed by Silverman (1986, Section 2.3), where essentially MADN is replaced by the value 1. That is, the width of the interval around each point when determining N_x depends in no way on the data, but only on the choice for h.

On rare occasions, data are encountered where MAD is zero. In the event this occurs when computing an expected frequency curve; here, MAD is replaced by IQR (the interquartile range) estimated via the ideal fourths, as described in Section 3.12.5, and MADN is replaced by IQRN, which is IQR divided by $z_{0.75} - z_{0.25}$, where again $z_{0.75}$ and $z_{0.25}$ are the 0.75 and 0.25 quantiles, respectively, of a standard normal distribution. Now an estimate of the density at x is taken to be

$$\hat{f}(x) = \frac{N_x}{2hn\text{IQRN}}.$$

A criticism of the expected frequency curve is that it can miss bimodality when the span is set to $h = 0.8$. This can be corrected by lowering h to say 0.2, but a criticism of routinely using $h = 0.2$ is that it often yields a rather ragged approximation of the true probability density function. With $h = 0.8$ and n small, again, a rather ragged plot can result, but an appealing feature of the method is that it often improves upon the normal kernel in terms of capturing the overall shape of the true distribution.

3.2.4 An Adaptive Kernel Estimator

With large sample sizes, the expected frequency curve typically gives a smooth approximation of the true density, but with small sample sizes, a rather ragged approximation can be obtained. A possible method for smoothing the estimate is to use an adaptive kernel estimate that is known to compete well with other estimators that have been proposed (Silverman,

1986; cf. Politis and Romano, 1997). There are, in fact, many variations of the adaptive kernel estimator, but only one is described here. Following Silverman (1986), let $\tilde{f}(X_i)$ be an initial estimate of $f(X_i)$. Here, $\tilde{f}(X_i)$ is based on the expected frequency curve. Let

$$\log g = \frac{1}{n} \sum \log \tilde{f}(X_i)$$

and

$$\lambda_i = (\tilde{f}(X_i)/g)^{-a},$$

where a is a *sensitivity parameter* satisfying $0 \le a \le 1$. Based on comments by Silverman (1986), $a = 0.5$ is used unless stated otherwise. Then the adaptive kernel estimate of f is taken to be

$$\hat{f}(t) = \frac{1}{n} \sum \frac{1}{h\lambda_i} K\{h^{-1}\lambda_i^{-1}(t - X_i)\},$$

where $K(t)$ is the Epanechnikov kernel given by Eq. (3.1). Following Silverman (1986, pp. 47–48), the span is

$$h = 1.06 \frac{A}{n^{1/5}},$$

where

$$A = \min(s, \text{IQR}/1.34),$$

s is the standard deviation, and IQR is the interquartile range. Again, the interquartile range is estimated as described in Section 3.12.5 (using what are called the ideal fourths).

When using an adaptive kernel estimator, perhaps there are advantages to using some initial estimator other than the expected frequency curve. The relative merits of this possibility have not been explored. One reason for using the expected frequency curve as the preliminary estimate is that it reduces problems due to a restriction in range that are known to be a concern when using the normal kernel as described in Section 3.2.1.

Harpole et al. (2014) compared several kernel density estimators in terms of mean squared error and bias. Their simulation results were based on five distributions. Four of these distributions were based on combinations of normal distributions. The resulting distributions included skewed and bimodal distributions. The fifth distribution was a lognormal distribution. No single estimator dominated, but a rough guide is that for a sample size $n \le 100$, the method in Section 3.2.1 performs relatively well among the situations they considered. For larger sample sizes, the adaptive kernel density estimator in this section performs relatively well. However, a concern with the method in Section 3.2.1 arises when dealing with variables that have a bounded range: The density estimate can indicate positive probabilities well outside the range of possible values as illustrated in Section 3.2.5. The adaptive kernel density estimator in this section avoids this problem. Also see Malec and Schienle (2014).

3.2.5 R Functions skerd, kerSORT, kerden, kdplot, rdplot, akerd, and splot

It is noted that R has a built-in function called density that computes a kernel density estimate based on various choices for K. (This function also contains various options not covered here.) By default, K is taken to be the standard normal density. Here, the R function

$$\text{skerd(x,op=TRUE,kernel='gaussian')}$$

is supplied in the event there is a desire to plot the data based on this collection of estimators. When op=TRUE, the function uses the default density estimator employed by R; otherwise, it uses the method recommended by Venables and Ripley (2002, p. 127). To use the Epanechnikov kernel, set the argument kernel="epanechnikov". For convenience, the function

$$\text{kerSORT(x,xlab=' ',ylab=' ')}$$

is supplied, which defaults to using the Epanechnikov kernel in conjunction with Silverman's rule of thumb rule in Section 3.2.1.

Letting x_q denote the qth quantile, the function

$$\text{kerden(x, q=0.5, xval=0),}$$

written for this book, computes the kernel density estimate of $f(x_q)$ based on the data stored in the R vector x using the Rosenblatt shifted histogram method, described in Section 3.2.2. (Again, see Section 1.7 on how to obtain the functions written for this book.) If unspecified, q defaults to 0.5. The argument xval is ignored unless q=0, in which case, the function estimates $f(x)$ when x is equal to the value specified by the argument xval. The function

$$\text{kdplot(x, rval=15)}$$

plots the estimate of $f(x)$ based on the function kerden, where the argument rval indicates how many quantiles will be used. The default value, 15, means that $f(x)$ is estimated for 15 quantiles evenly spaced between 0.01 and 0.99, and then the function plots the estimates to form an estimate of $f(x)$.

The R function

$$\text{rdplot(x,fr=NA,plotit=TRUE, pts=NA, pyhat=FALSE)}$$

computes the expected frequency curve. The argument fr is the span, h. If not specified, fr=0.8 is used in the univariate case; otherwise, fr=0.6 is used. By default, pts=NA (for not available), in which case, a plot of the estimated density is based on the points $(X_i, \hat{f}(X_i))$, $i = 1, \ldots, n$. If values are stored in pts, the plot is created based on these points. For example, the command rdplot(mydat,pts=c(0,mydat)) will create a plot based on all of the points in my-dat plus the point $(0, \hat{f}(0))$. If pyhat=TRUE, the function returns the $\hat{f}$ values that were computed. So rdplot(mydat,pts=0,pyhat=TRUE) returns $\hat{f}(0)$, and rdplot(mydat,pts=c(1,2),py-hat=TRUE) returns $\hat{f}(1)$ and $\hat{f}(2)$. Setting plotit=FALSE suppresses the plot. (The function can handle multivariate data and produces a plot in the bivariate case. The computational details are outlined in Chapter 6.)

The R function

$$\text{akerd(x,hval=NA,aval=0.5,op=1,fr=0.8,pts=NA,pyhat=FALSE)}$$

applies the adaptive kernel estimate as described in Section 3.2.4, where the argument hval is the span, h, aval is a, the sensitivity parameter, and fr is the span used by the initial estimate based on the expected frequency curve. If the argument op is set to 2, the Epanechnikov kernel is replaced by the normal kernel. The argument hval defaults to NA, meaning that if not specified, the span h is determined as described in Section 3.2.4; otherwise, h is taken to be the value given by hval. Setting pyhat=T, the function returns the $\hat{f}(X_i)$ values. If pts contains values, the function returns $\hat{f}$ for values in pts instead. (The function can be used with multivariate data and produces a plot in the bivariate case.)

For convenience, when working with discrete data, the function

$$\text{splot(x,op=TRUE, xlab='X',ylab='Rel. Freq.')}$$

is supplied, which plots the relative frequencies of all distinct values found in the R variable x. With op=TRUE, a line connecting points marking the relative frequencies is added to the plot.

■ **Example**

Table 3.1 shows data from a study dealing with hangover symptoms for two groups of individuals: sons of alcoholics and a control group. Note that for both groups, zero is the most common value, and it is fairly evident that the data do not have a bell-shaped distribution. The top two panels of Fig. 3.1 show an estimate of the distributions using the R function skerd. (The plots are based on the data for Group 1.) Switching to the R function kerSORT (the method in Section 3.2.1) gives virtually the same results. This illustrates a well-known problem with certain kernel density estimators: A restriction in

Table 3.1:   The effect of alcohol.

Group 1:	0	0	0	0	0	0	0	0	2	2
	3	3	6	9	11	11	11	18	32	41
Group 2:	0	0	0	0	0	0	0	0	0	0
	0	0	0	0	1	2	3	8	12	32

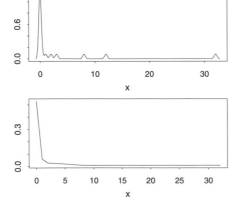
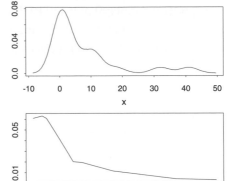

Figure 3.1: An example comparing four plots of data. The upper left panel shows a kernel density estimate using a normal kernel based on the Group 2 data in Table 3.1. The upper right panel is the estimate using the Group 1 data. The bottom left panel used the same data as in the upper left panel, only the adaptive kernel density estimator was used. The lower right panel used the adaptive kernel density estimate with the Group 1 data.

the range of possible values can lead to highly unsatisfactory results. This is clearly the case here because values less than zero are impossible, in contrast to what is suggested, particularly in the top right panel. The bottom two panels are plots of the data using the adaptive kernel density estimator in Section 3.2.4. The method handles the restriction in range reasonably well and provides what seems like a much more satisfactory summary of the data.

■

■ Example

Fig. 3.2 shows the plots created by the R functions just described for $n = 30$ observations randomly sampled from a standard normal distribution. The upper left panel was produced by the function skerd, the upper right panel shows the curve produced by

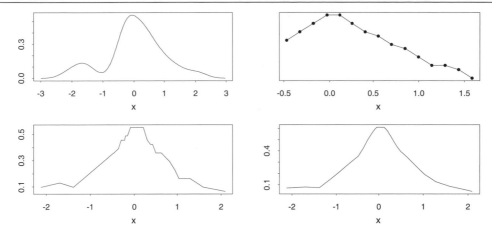

Figure 3.2: Another example comparing four plots of data. The upper left panel shows a kernel density estimate using a normal kernel based on $n = 30$ observations sampled from a standard normal distribution. The upper right panel is the plot (based on the same data) using Rosenblatt's shifted histogram. The lower left panel is the expected frequency curve, and the lower right panel is based on the adaptive kernel estimator.

kdplot (which uses the method in Section 3.2.2), the lower left panel is the expected frequency curve (using the function rdplot), and the lower right panel (created by the function akerd) is based on the adaptive kernel estimator in Section 3.2.4. Plots based on akerd are typically smoother than the plot returned by rdplot, particularly when using small sample sizes.

■

■ Example

To add perspective, 500 observations were generated from the lognormal distribution shown in Fig. 3.3. This particular distribution is defined only for $X \geq 0$. That is, $P(X < 0) = 0$. The upper left panel of Fig. 3.4 shows the plot created by R using the kernel density estimator described in Section 3.2.1. The upper right panel is based on Rosenblatt's shifted histogram (described in the method in Section 3.2.2), the lower left panel is based on the expected frequency curve (using the function kdplot), and the lower right panel is the plot based on the adaptive kernel estimator described in Section 3.2.4. Notice that both rdplot and kdplot do a better job of capturing the shape of the true density. The output from skerd is too Gaussian on the left (for $x \leq 5$), meaning that it resembles a normal curve when it should not, and it performs rather poorly

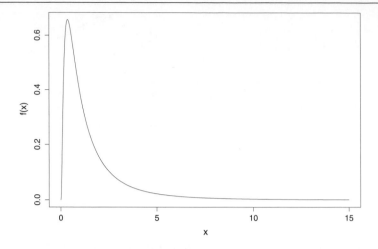

Figure 3.3: A lognormal distribution.

for $X < 0$. The R function kerSORT gives very similar results. A similar problem arises when sampling from an exponential distribution. Setting op=FALSE when using skerd, the restriction in range associated with the lognormal distribution is less of a problem, but the plot becomes rather ragged. Increasing the sample size to $n = 1,000$ and changing the seed in the R random number generator produces results very similar to those in Fig. 3.4. The R function density has optional arguments that replace the normal kernel with other functions; several of these were considered for the situation at hand, but similar results were obtained. So although situations are encountered where the R function skerd produces a smoother, more visually appealing plot versus rdplot, kdplot, and akerd, blind use of this function can be misleading.

■

3.3 The Sample Median and Trimmed Mean

As already indicated, the standard error of the sample mean can be relatively large when sampling from a heavy-tailed distribution, and the sample mean estimates a non-robust measure of location, μ. Trimmed means, which include the usual sample median, represent one way of addressing these problems.

The sample trimmed mean, which estimates the population trimmed μ_t (described in Section 2.2.3), is computed as follows. Let $X_1, \ldots, X_n$ be a random sample, and let $X_{(1)} \leq X_{(2)} \leq \cdots \leq X_{(n)}$ be the observations written in ascending order. The value $X_{(i)}$ is called the ith *order statistic*. Suppose the desired amount of trimming has been chosen to be γ,

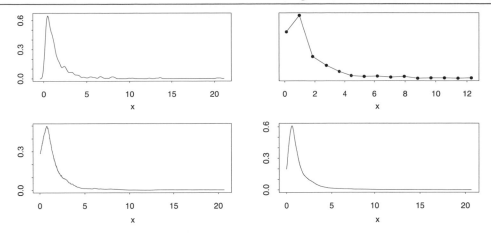

Figure 3.4: Some kernel density estimators can perform poorly when the variable under study is bounded, even with large sample sizes. The upper left panel is a plot based on the function skerd and $n = 500$ randomly sampled observations from the lognormal distribution in Fig. 3.3. The upper right panel is Rosenblatt's shifted histogram (using the function kdplot), the lower left panel shows the plot created by rdplot, and the lower right panel is based on the adaptive kernel estimator.

$0 \le \gamma < 0.5$. Let $g = [\gamma n]$, where $[\gamma n]$ is the value of γn rounded down to the nearest integer. For example, $[10.9] = 10$. The sample trimmed mean is computed by removing the g largest and g smallest observations and averaging the values that remain. In symbols, the sample trimmed mean is

$$\bar{X}_t = \frac{X_{(g+1)} + \cdots + X_{(n-g)}}{n - 2g}. \tag{3.2}$$

In essence, the empirical distribution is trimmed in a manner consistent with how the probability density function was trimmed when defining μ_t. As indicated in Chapter 2, two-sided trimming is assumed unless stated otherwise. (For relevant asymptotic results, see Gribkova, 2017.)

Note that trimmed means contain the usual sample median as a special case, which corresponds to using $g = [(n - 1)/2]$. Miller (1988) noted that the sample median is a biased estimator of the population median and suggested that as a result, it should not be used. However, the sample median is unbiased, in contrast to the sample mean when distributions are skewed (e.g., Rousselet and Wilcox, 2020). On a more fundamental level, Pajari et al. (2019) argue that judging quantile estimators based on bias is inappropriate. They argue that a bin criterion should be used.

The definition of the sample trimmed mean given by Eq. (3.2) is the one most commonly used. However, for completeness, it is noted that the term trimmed mean sometimes refers to a slightly different estimator (e.g., Reed, 1998; cf. Hogg, 1974), namely,

$$\frac{1}{n(1-2\gamma)}\left(\sum_{i=g+1}^{n-g} X_{(i)} + (g - \gamma n)(X_{(g)} + X_{(n-g+1)})\right).$$

Also see Patel et al. (1988) as well as Kim (1992a). Here, however, the definition given by Eq. (3.2) is used exclusively.

For the trimmed mean to have any practical importance, a value for γ must be chosen. One approach is to choose γ so that $\bar{X}_t$ tends to have a relatively small standard error among commonly occurring situations, and a related restriction might be that little accuracy is lost when sampling from a normal distribution. Based on this view, and other criteria to be described, a good choice for general use is $\gamma = 0.2$. If γ is too small, the standard error of the trimmed mean, $\sqrt{\mathrm{VAR}(\bar{X}_t)}$, can be drastically inflated by outliers or sampling from a heavy-tailed distribution. If γ is too large, the standard error can be relatively large compared to the standard error of the sample mean when sampling from a normal distribution. (Some illustrations are given in Section 3.11.) Empirical investigations based on data from actual studies suggest that the optimal amount of trimming, in terms of minimizing the standard error, is usually between 0 and 0.25 (e.g., Hill and Dixon, 1982; Wu, 2002). While $\gamma = 0.1$, for example, might be more optimal than $\gamma = 0.2$ in certain situations, one argument for using $\gamma = 0.2$ is that it can result in a standard error that is much smaller than the standard error associated with $\gamma = 0.1$ or $\gamma = 0$, but the reverse is generally untrue. That is, $\gamma = 0.2$ guards against complete disaster but sacrifices relatively little in situations where $\gamma = 0.1$ and $\gamma = 0$ are more optimal. In some cases, however, more than 20% trimming might be desirable.

Another approach is to determine γ empirically according to some criterion, such as the standard error. That is, estimate the standard error of $\bar{X}_t$ when, for example, $\gamma = 0, 0.1$, and 0.2, and then use the value of γ corresponding to the smallest estimate. These so-called adaptive trimmed means have been studied by Léger and Romano (1990a, 1990b) and Léger et al. (1992). Or one could determine the amount of trimming based on some measure of skewness and heavy-tailedness. For a comparison of such methods, see Reed (1998) as well as Reed and Stark (1996). The properties of this approach, in the context of testing hypotheses and computing confidence intervals, have not been studied to the extent where γ is chosen to be a prespecified constant. In particular, the practical utility of adaptive trimmed means needs further investigation, so they are not discussed here, but further investigation seems warranted. It should be remarked, however, that empirically determining how much trimming to do is fraught with difficulties that are not always obvious. Interested readers can read the discussions of a paper by Hogg (1974), especially the comments by P. Huber. For some results on

Table 3.2: Self-awareness data.

77	87	88	114	151	210	219	246	253	262
296	299	306	376	428	515	666	1,310	2,611	

so-called hinge estimators, regarding control over the probability of a Type I error, see Kesel-man et al. (2007).

■ **Example**

Dana (1990) conducted a study dealing with self-awareness and self-evaluation. One segment of his study measured the time subjects could keep a portion of an apparatus in contact with a specified target. Table 3.2 shows some data for one of the groups. The sample mean and the sample trimmed means with $\gamma = 0.1$ and 0.2 are 448, 343, and 283, respectively. In this particular case, there is an obvious difference between the three measures of location. However, even if they had been nearly equal, this is not necessarily an indication that the sample mean is satisfactory because the standard error of the trimmed mean can be substantially smaller than the standard error of the mean.

■

It might seem that $\gamma = 0.2$ is equivalent to randomly throwing away 40% of the data, but this is not the case. To see why, notice that the order statistics are dependent even though the observations $X_1, \ldots, X_n$ are independent. This result is covered in basic texts on mathematical statistics. For readers unfamiliar with this result, a brief explanation will help shed some light on other practical problems covered in this chapter.

If the random variables X and Y are independent, then the probability function of X is not altered given the value of Y. This means, in particular, that the range of possible values of X cannot depend on the value of Y. Suppose $X_1, \ldots, X_n$ is a random sample, and for the sake of illustration, suppose the value of each random variable can be any of the integers between 1 and 10 inclusive, each value occurring with some positive probability. Then knowing that $X_2 = 3$, say, tells us nothing about the probability of observing a particular value for X_1. However, suppose $X_{(2)} = 3$. Then the smallest value, $X_{(1)}$, cannot be 4. More generally, $X_{(1)}$ cannot be any number greater than 3. In contrast, if we do not know the value of $X_{(2)}$, or any of the other order statistics, $X_{(1)}$ could have any of the values $1, 2, \ldots, 10$, and these values occur with some positive probability. Thus, knowing the value of $X_{(2)}$ alters the probabilities associated with $X_{(1)}$. That is, $X_{(1)}$ and $X_{(2)}$ are dependent, and dependence occurs because knowing the value of $X_{(2)}$ restricts the range of possible values for $X_{(1)}$. More generally, any two order statistics, say $X_{(i)}$ and $X_{(j)}$, $i \neq j$, are dependent.

3.3.1 R Functions mean, tmean, median, and lloc

R has a built-in function that evaluates the trimmed mean. If observations are stored in the vector x, the R command

$$\text{mean(x,trim=0)}$$

computes the γ-trimmed mean where the argument trim determines the amount of trimming. By default, the amount of trimming is 0. For example, mean(x,0.2) returns the 20% trimmed mean. The value 283 is returned for the data in Table 3.2, assuming the data are stored in the R variable x. Because it is common to use 20% trimming, for convenience, the R function

$$\text{tmean(x,tr=0.2)}$$

has been supplied, which computes a 20% trimmed mean by default using the data stored in the R variable x. The amount of trimming can be altered using the argument tr. So tmean(blob) would compute a 20% trimmed mean for the data stored in blob, and tmean(blob,tr=0.3) would use 30% trimming instead. The usual sample median is computed by the built-in R function

$$\text{median(x).}$$

For convenience, the function

$$\text{lloc(x,est=tmean,...)}$$

is supplied for computing a trimmed mean when data are stored in list mode, a data frame, or a matrix. If x is a matrix or data frame, lloc computes the trimmed mean for each column. Other measures of location can be used via the argument est. (For example, est=median will compute the median.) The argument ... means that an optional argument associated with est can be used.

3.3.2 Estimating the Standard Error of the Trimmed Mean

To have practical value when making inferences about μ_t, properties of the sampling distribution of $\bar{X}_t$ need to be determined. This section takes up the problem of estimating $\sqrt{\text{VAR}(\bar{X}_t)}$, the standard error of the sample trimmed mean.

At first glance, the problem might appear to be trivial. The standard error of the sample mean is $\sigma/\sqrt{n}$, which is estimated with $s/\sqrt{n}$, where

$$s^2 = \frac{1}{n-1}\sum(X_i - \bar{X})^2$$

is the usual sample variance. A common mistake in applied work is to estimate the standard error of the trimmed mean by simply computing the sample standard deviation of the untrimmed observations, and then dividing by $\sqrt{n-2g}$, the square root of the number of observations left after trimming. That is, apply the usual estimate of the standard error using the untrimmed values. To see why this simple idea fails, let $X_1, \ldots, X_n$ be any random variables, possibly dependent with unequal variances, and let $a_1, \ldots, a_n$ be any n constants. Then the variance of $\sum a_i X_i$ is

$$\text{VAR}\left(\sum a_i X_i\right) = \sum_{i=1}^{n}\sum_{i=1}^{n} a_i a_j \text{COV}(X_i, X_j), \tag{3.3}$$

where $\text{COV}(X_i, X_j)$ is the covariance between X_i and X_j. That is,

$$\text{COV}(X_i, X_j) = E\{(X_i - \mu_i)(X_j - \mu_j)\},$$

where $\mu_i = E(X_i)$. When $i = j$, $\text{COV}(X_i, X_j) = \sigma_i^2$, the variance of X_i. When the random variables are independent, Eq. (3.3) reduces to

$$\text{VAR}\left(\sum a_i X_i\right) = \sum_{i=1}^{n} a_i^2 \sigma_i^2. \tag{3.4}$$

Under random sampling, in which case, the variance of each of the n random variables has a common value σ^2, the variance of $\bar{X}$ can be seen to be σ^2/n by taking $a_i = 1/n$, $i = 1, \ldots, n$, in Eq. (3.4). The problem with the sample trimmed mean is that it is a linear combination of dependent random variables, namely, a linear combination of the order statistics, so Eq. (3.4) does not apply; Eq. (3.3) must be used instead. For $i \neq j$, there are asymptotic results that can be used to estimate $\text{COV}(X_{(i)}, X_{(j)})$, the covariance between the ith and jth order statistics; this suggests a method for estimating the standard error of a trimmed mean, but a simpler method for estimating $\text{VAR}(\bar{X}_t)$ is typically used and has been found to give good results.

The influence function of the trimmed mean, $IF_t(x)$, introduced in Chapter 2, provides a convenient and useful way of dealing with the dependence among the order statistics. It can be shown that

$$\bar{X}_t = \mu_t + \frac{1}{n}\sum_{i=1}^{n} IF_t(X_i), \tag{3.5}$$

plus a remainder term that goes to zero as n gets large. Moreover, $E(IF_t(X_i)) = 0$. In words, the sample trimmed mean can be written as μ_t plus a sum of independent, identically distributed random variables (assuming random sampling) having mean 0, plus a term that can be ignored, provided n is not too small. The central limit theorem, applied to Eq. (3.5), shows that the distribution of $\bar{X}_t$ approaches a normal distribution as $n \to \infty$. Fortunately, all indications are that the error term can be ignored even when n is as small as 10. Because $IF_t(X)$ has mean 0, Eq. (3.4) can be used to show that

$$\text{VAR}(\bar{X}_t) = \frac{1}{n^2} \sum E\{(IF_t(X_i))^2\}, \tag{3.6}$$

ignoring the error term. From Chapter 2,

$$(1 - 2\gamma)IF_t(X) = \begin{cases} x_\gamma - \mu_w, & \text{if } x < x_\gamma, \\ X - \mu_w, & \text{if } x_\gamma \leq X \leq x_{1-\gamma}, \\ x_{1-\gamma} - \mu_w, & \text{if } x > x_{1-\gamma}, \end{cases}$$

where μ_w is the Winsorized population mean and x_γ is the γ quantile. The main point here is that an estimate of $E\{(IF_t(X_i))^2\}$ yields an estimate of $\text{VAR}(\bar{X}_t)$ via Eq. (3.6). Note that

$$P\left(IF_t(X) = \frac{x_\gamma - \mu_w}{1 - 2\gamma}\right) = \gamma,$$

$$P\left(IF_t(X) = \frac{x_{1-\gamma} - \mu_w}{1 - 2\gamma}\right) = \gamma.$$

The first step in estimating $E\{(IF_t(X_i))^2\}$ is estimating the population-Winsorized mean, μ_w. This is done by Winsorizing the empirical distribution and computing the sample mean of what results. *Winsorization of a random sample* consists of setting

$$W_i = \begin{cases} X_{(g+1)}, & \text{if } X_i \leq X_{(g+1)}, \\ X_i, & \text{if } X_{(g+1)} < X_i < X_{(n-g)}, \\ X_{(n-g)}, & \text{if } X_i \geq X_{(n-g)}. \end{cases} \tag{3.7}$$

The *Winsorized sample mean* is

$$\bar{X}_w = \frac{1}{n} \sum W_i,$$

which estimates μ_w, the population-Winsorized mean introduced in Chapter 2. In words, Winsorization means that the g smallest values are set equal to $X_{(g+1)}$, the smallest value not trimmed, and the g largest values are set equal to $X_{(n-g)}$, the largest value not trimmed. The sample mean of the resulting values is the Winsorized sample mean. (For a detailed study of the sample Winsorized mean when sampling from a skewed distribution, see Rivest, 1994.)

Table 3.3: Winsorized values for the self-awareness data.

114	114	114	114	151	210	219	246	253	262
296	299	306	376	428	515	515	515	515	

Put another way, Winsorization consists of estimating the γ and $1 - \gamma$ quantiles with $X_{(g+1)}$ and $X_{(n-g)}$, respectively, and estimating the population Winsorized distribution, described in Chapter 2, with the resulting W_i values.

Table 3.3 shows the Winsorized values for the data in Table 3.2 when $\gamma = 0.2$, in which case, $g = [0.2(19)] = 3$. Thus, Winsorizing the observations in Table 3.2 consists of replacing the three smallest observations with $X_{(4)} = 114$. Similarly, because $n - g = 19 - 3 = 16$, the three largest observations are replaced by $X_{(16)} = 515$. The sample mean of the values in Table 3.3 is 293, and this is equal to the 20% Winsorized sample mean for the data in Table 3.2.

The expression for the influence function of the trimmed mean involves three unknown quantities: x_γ, $x_{1-\gamma}$, and μ_w. As already indicated, these three unknown quantities are estimated with $X_{(g+1)}$, $X_{(n-g)}$, and $\bar{X}_w$, respectively. A little algebra shows that an estimate of $E[IF_t(X_i)]$ is $(W_i - \bar{W})/(1 - 2\gamma)$, so an estimate of $E\{(IF(X_i))^2\}$ is $(W_i - \bar{W})^2/(1 - 2\gamma)^2$. Referring to Eq. (3.6), the resulting estimate of $\mathrm{VAR}(\bar{X}_t)$ is

$$\frac{1}{n^2(1 - 2\gamma)^2} \sum (W_i - \bar{W})^2.$$

When there is no trimming, this last equation becomes

$$\frac{n-1}{n} \times \frac{s^2}{n},$$

but typically, s^2/n is used instead. Accordingly, to be consistent with how the standard error of the sample mean is usually estimated,

$$\frac{1}{n(n-1)(1 - 2\gamma)^2} \sum (W_i - \bar{W})^2, \tag{3.8}$$

will be used to estimate $\mathrm{VAR}(\bar{X}_t)$.

The quantity

$$s_w^2 = \frac{1}{n-1} \sum (W_i - \bar{W})^2, \tag{3.9}$$

is called the *sample Winsorized variance*. A common way of writing Eq. (3.8) is in terms of the Winsorized variance:

$$\frac{s_w^2}{(1 - 2\gamma)^2 n}. \tag{3.10}$$

Table 3.4:   Summary of how to estimate the standard error of the trimmed mean.

To estimate the standard error of the trimmed mean based on a random sample of n observations, first Winsorize the observations by transforming the ith observation, X_i, to W_i using Eq. (3.7). Compute the sample variance of the W_i values, yielding s_w^2, the Winsorized sample variance. The standard error of the trimmed mean is estimated to be

$$\frac{s_w}{(1 - 2\gamma)\sqrt{n}},$$

where γ is the amount of trimming chosen by the investigator.

In other words, to estimate the squared standard error of the trimmed mean, compute the Winsorized observations W_i using Eq. (3.7), compute the sample variance using the resulting values, and then divide by $(1 - 2\gamma)^2 n$. It can be seen that $E_w\{(IF(X_i))^2\} = \sigma_w^2/(1 - 2\gamma)^2$, and this provides another way of justifying Eq. (3.10) as an estimate of $\text{VAR}(\bar{X}_t)$. Consequently, the standard error of the sample trimmed mean is estimated with

$$\sqrt{\frac{s_w^2}{(1 - 2\gamma)^2 n}} = \frac{s_w}{(1 - 2\gamma)\sqrt{n}}.$$

Table 3.4 summarizes the calculations used to estimate the standard error of the trimmed mean.

■ **Example**

For the data in Table 3.3, the sample variance is 21,551.4, and this is the Winsorized sample variance for the data in Table 3.2. Because 20% trimming was used, $\gamma = 0.2$, and the estimated standard error of the sample trimmed mean is

$$\frac{\sqrt{21,551.4}}{(1 - 2(.2))\sqrt{19}} = 56.1.$$

In contrast, the standard error of the sample mean is $s/\sqrt{n} = 136$, a value that is approximately 2.4 times larger than the standard error of the trimmed mean.

■

As previously noted, it might seem that the standard error of a trimmed mean could be estimated by simply computing the standard error of the mean using the data left after trimming. For example, if $h = n - 2g$ observations are left after trimming, compute the mean and standard deviation, s, based on these h values and estimate the standard error with s/h. In the last example, this yields 30.2. However, this can result in a highly inaccurate estimate of the standard error of a trimmed mean that can differ substantially from a theoretically sound method. For the situation at hand, the theoretically sound method in Table 3.4 yields 56.1, which is nearly twice as large as the technically unsound estimate (cf. Bakker and Wicherts, 2014).

3.3.3 Estimating the Standard Error of the Sample Winsorized Mean

An estimate of the standard error of the sample Winsorized mean, $\bar{X}_w$, can be derived from the influence function of the population Winsorized mean given in Section 2.2.2. Dixon and Tukey (1968) suggest a simpler estimate:

$$\frac{n-1}{n-2g-1} \times \frac{s_w}{\sqrt{n}},$$

where $g = [\gamma n]$ is the number of observations Winsorized in each tail, so $n - 2g$ is the number of observations that are not Winsorized.

3.3.4 R Functions winmean, winvar, winsd, trimse, and winse

Written for this book, the R function

$$\text{winmean(x,tr=0.2)}$$

computes the Winsorized mean. The optional argument tr is the amount of Winsorizing to be used, which defaults to 0.2 if unspecified. (The R function win also computes the Winsorized mean.) For example, the command winmean(dat) computes the 20% Winsorized mean for the data in the R vector dat. The command winmean(x,0.1) computes the 10% Winsorized mean. If there are any missing values (stored as NA in R), the function automatically removes them.

The function

$$\text{winvar(x,tr=0.2)}$$

computes the Winsorized sample variance, s_w^2. The function

$$\text{winsd(x,tr=0.2)}$$

computes the Winsorized standard deviation. Again, tr is the amount of Winsorization, which defaults to 0.2 if unspecified. The function

$$\text{trimse(x,tr=0.2)}$$

estimates the standard error of the trimmed mean and

$$\text{winse(x,tr=0.2)}$$

estimates the standard error of the Winsorized mean. For example, the R command trimse(x,0.1) estimates the standard error of the 10% trimmed mean for the data stored in the vector x, and winvar(x,0.1) computes the Winsorized sample variance using 10% Winsorization. The R command winvar(x) computes s_w^2 using 20% Winsorization.

3.3.5 Estimating the Standard Error of the Sample Median

As noted in Section 3.3, trimmed means contain the usual sample median, M, as a special case where the maximum amount of trimming is used. When using M, and the goal is to estimate its standard error, alternatives to Eq. (3.10) should be used. Many methods have been proposed, comparisons of which were made by Price and Bonett (2001). In terms of hypothesis testing, an effective and fairly simple estimate appears to be one derived by McKean and Schrader (1984). To apply it, compute

$$k = \frac{n+1}{2} - z_{0.995}\sqrt{\frac{n}{4}},$$

where k is rounded to the nearest integer and $z_{0.995}$ is the 0.995 quantile of a standard normal distribution. Put the observed values in ascending order, yielding $X_{(1)} \leq \cdots \leq X_{(n)}$. The McKean–Schrader estimate of the squared standard error of M is

$$\left(\frac{X_{(n-k+1)} - X_{(k)}}{2 z_{0.995}} \right)^2. \tag{3.11}$$

(Price and Bonett, 2001, recommend a slightly more complicated estimator, but when computing a confidence interval for the median, currently, it seems that their method offers little or no advantage.)

It is stressed that when there are tied values, all known methods for estimating the standard error of the sample median, M, can be highly inaccurate. Methods for making inferences about the population median, when there are tied values, have been derived and are described in subsequent sections.

3.3.6 R Function msmedse

The R function

$$msmedse(x)$$

estimates the standard error of the sample median based on Eq. (3.11).

3.4 The Finite Sample Breakdown Point

Before describing additional measures of location, it helps to introduce a technical device for judging any estimator that is being considered. This is the *finite sample breakdown point* of a statistic, which refers to the smallest proportion of observations that, when altered sufficiently, can render the statistic meaningless. More precisely, the finite sample breakdown point of an estimator refers to the smallest proportion of observations that when altered can cause the value of the statistic to be arbitrarily large or small. The finite sample breakdown point of an estimator is a measure of its *resistance* to contamination. For example, if the ith observation among the observations $X_1, \ldots, X_n$ goes to infinity, the sample mean $\bar{X}$ goes to infinity as well. This means that the finite sample breakdown point of the sample mean is only $1/n$. In contrast, the finite sample breakdown point of the γ trimmed mean is γ. For example, if $\gamma = 0.2$, about 20% of the observations can be made arbitrarily large without driving the sample trimmed mean to infinity, but it is possible to alter 21% of the observations so that $\bar{X}_t$ becomes arbitrarily large. Typically, the limiting value of the finite sample breakdown point is equal to the breakdown point, as defined in Chapter 2, of the parameter being estimated. For example, the breakdown point of the population mean, μ, is 0, which equals $1/n$, as n goes to infinity. Similarly, the breakdown point of the trimmed mean is γ.

Two points should be stressed. First, having a high finite sample breakdown point is certainly a step in the right direction when trying to deal with unusual values that have an inordinate influence, but it is no guarantee that an estimator will not be unduly influenced by even a small number of outliers. (Examples will be given when dealing with robust regression estimators.) Second, various refinements regarding the definition of a breakdown point have been proposed (e.g., Genton and Lucas, 2003), but no details are given here.

3.5 Estimating Quantiles

When comparing two or more groups, the most common strategy is to use a single measure of location, and the median or 0.5 quantile is an obvious choice. It can be highly advantageous to compare other quantiles as well, but the motivation for doing this is best explained in Chapter 5. For now, attention is focused on estimating quantiles and the associated standard error.

There are many ways of estimating quantiles, comparisons of which are reported by Parrish (1990), Sheather and Marron (1990), Sfakianakis and Verginis (2008), and Dielman et al. (1994). Here, two are described, and their relative merits are discussed.

For any q, $0 < q < 1$, let x_q be the qth quantile. For a continuous random variable, or a distribution with no flat spots, x_q is defined by the equation $P(X \leq x_q) = q$. This definition is satisfactory in the sense that there is only one value that qualifies as the qth quantile, so there is no ambiguity when referring to x_q. However, for discrete random variables or distributions with flat spots, special methods must be used to avoid having multiple values that qualify as the qth quantile. There are methods for accomplishing this goal, but they are not directly relevant to the topics of central interest in this book, at least based on current technology, so this issue is not discussed.

Setting $m = [qn + 0.5]$, where $[qn + 0.5]$ is the greatest integer less than or equal to $qn + 0.5$, the simplest estimate of x_q is

$$\hat{x}_q = X_{(m)}, \tag{3.12}$$

the mth observation after the data are put in ascending order. For example, if the goal is to estimate the median, then $q = 1/2$, and if $n = 11$, then $m = [11/2 + 0.5] = 6$, and the estimate of $x_{0.5}$ is the usual sample median, M. Of course, if n is even, this estimator does not yield the usual sample median; it is equal to what is sometimes called the *upper empirical cumulative distribution function estimator.*

3.5.1 Estimating the Standard Error of the Sample Quantile

Assuming that observations are randomly sampled from a continuous distribution, and that $f(x_q) > 0$, the influence function of the qth quantile is

$$IF_q(x) = \begin{cases} \frac{q-1}{f(x_q)}, & \text{if } x < x_q, \\ 0, & \text{if } x = x_q, \\ \frac{q}{f(x_q)}, & \text{if } x > x_q, \end{cases} \tag{3.13}$$

and

$$\hat{x}_q = x_q + \frac{1}{n} \sum IF_q(X_i)$$

plus a remainder term that goes to zero as n gets large. That is, the situation is similar to the trimmed mean in the sense that the estimate of the qth quantile can be written as x_q, the population parameter being estimated, plus a sum of independent identically distributed random variables having a mean of zero, plus a term that can be ignored as the sample size gets large. Consequently, the influence function of the qth quantile can be used to determine the (asymptotic) standard error of $\hat{x}_q$. The result is

$$VAR(\hat{x}_q) = \frac{q(1-q)}{n[f(x_q)]^2}. \tag{3.14}$$

For example, when estimating the population median with $\hat{x}_{0.5}$, the variance of $\hat{x}_{0.5}$ is

$$\frac{1}{4n[f(x_{0.5})]^2},$$

so the standard error of $\hat{x}_{0.5}$ is

$$\frac{1}{2\sqrt{n}f(x_{0.5})}.$$

Moreover, for any q between 0 and 1,

$$2\sqrt{n}f(x_q)(\hat{x}_q - x_q)$$

approaches a standard normal distribution as n goes to infinity.

Using Eq. (3.14) to estimate the standard error of $\hat{x}_q$ requires an estimate of $f(x_q)$, the probability density function of X evaluated at x_q, and this can be done using one of the methods described in Section 3.2. It is suggested that the adaptive kernel estimator be used in most cases, but all four kernel density estimators can be used with the software provided in case there are known reasons for preferring one kernel density estimator over another.

■ **Example**

The data in Table 3.2 are used to illustrate how the standard error of $\hat{x}_{0.5}$ can be estimated when using Rosenblatt's shifted histogram estimate of $f(x)$. There are 19 observations, so $[0.25n + 0.5] = 5$, $[0.75n + 0.5] = 14$, and an estimate of the interquartile range is $X_{(14)} - X_{(5)} = 376 - 151 = 225$, so

$$h = \frac{1.2(225)}{19^{1/5}} = 149.8.$$

The sample median is $M = \hat{x}_{0.5} = X_{(10)} = 262$, so $x_{0.5} + h = 411.8$, and the number of observations less than or equal to 411.8 is $A = 14$. The number of observations less than $\hat{x}_{0.5} - h = 112.2$ is $B = 3$, so:

$$\hat{f}(\hat{x}_{0.5}) = \frac{14 - 3}{2(19)(149.8)} = 0.00193.$$

Consequently, an estimate of the standard error of the sample median is

$$\frac{1}{2\sqrt{19}(0.00193)} = 59.4.$$

∎

3.5.2 R Function qse

The R function

$$qse(x,q=0.5,op=3)$$

estimates the standard error of $\hat{x}_q$ using Eq. (3.11). As indicated, the default value for q is 0.5. The argument op determines which density estimator is used to estimate $f(x_q)$. The choices are:

- op=1, Rosenblatt's shifted histograms,
- op=2, expected frequency curve,
- op=3, adaptive kernel method.

For example, storing the data in Table 3.2 in the R vector x, the command qse(x,op=1) returns the value 64.3. In contrast, using op=2 and op=3, the estimates are 58.94 and 47.95, respectively. So the choice of density estimator can make a practical difference.

3.5.3 The Maritz–Jarrett Estimate of the Standard Error of $\hat{x}_q$

Maritz and Jarrett (1978) derived an estimate of the standard error of sample median, which is easily extended to the more general case involving $\hat{x}_q$. That is, when using a single order statistic, its standard error can be estimated using the method outlined here. It is based on the fact that $E(\hat{x}_q)$ and $E(\hat{x}_q^2)$ can be related to a beta distribution. The beta probability density function, when a and b are positive integers, is

$$f(x) = \frac{(a+b+1)!}{a!b!}x^a(1-x)^b, \ 0 \le x \le 1. \tag{3.15}$$

Details about the beta distribution are not important here. Interested readers can refer to Johnson and Kotz (1970, Chapter 24).

As before, let $m = [qn + 0.5]$. Let Y be a random variable having a beta distribution with $a = m - 1$ and $b = n - m$, and let

$$W_i = P\left(\frac{i-1}{n} \le Y \le \frac{i}{n}\right).$$

Many statistical computing packages have functions that evaluate the beta distribution, so evaluating the W_i values is relatively easy to do. In R, there is the function pbeta(x,a,b) that computes $P(Y \le x)$. Thus, W_i can be computed by setting $x = i/n$, $y = (i-1)/n$, in which case, W_i is pbeta(x,m-1,n-m) minus pbeta(y,m-1,n-m).

Let

$$C_k = \sum_{i=1}^{n} W_i X_{(i)}^k.$$

When $k = 1$, C_k is a linear combination of the order statistics. Linear sums of order statistics are called *L-estimators*. Other examples of L-estimators are the trimmed and Winsorized means already discussed. The point here is that C_k can be shown to estimate $E(X_{(m)}^k)$, the kth moment of the mth order statistic. Consequently, the standard error of the mth order statistic, $X_{(m)} = \hat{x}_q$, is estimated with

$$\sqrt{C_2 - C_1^2}.$$

Note that when n is odd, this last equation provides an alternative to the McKean–Schrader estimate of the standard error of M described in Section 3.3.4. Based on limited studies, it seems that when computing confidence intervals or testing hypotheses based on M, the McKean–Schrader estimator is preferable.

3.5.4 R Function mjse

The R function

$$\text{mjse(x,q=0.5)}$$

computes the Maritz–Jarrett estimate of the standard error of $\hat{x}_q = X_{(m)}$, the mth order statistic, where $m = [qn + 0.5]$. If unspecified, q defaults to 0.5. The command mjse(x,0.4), for example, estimates the standard error of $\hat{x}_{0.4} = X_{(m)}$. If the data in Table 3.2 are stored in the R variable xv, and the median is estimated with $X_{(10)}$, the command mjse(xv) reports that the Maritz–Jarrett estimate of the standard error is 45.8. Using instead the method in Section 3.5.1, based on the adaptive kernel density estimator, the estimate is 43.95. Note that both estimates are substantially less than the estimated standard error of the sample mean, which is 136.

All indications are that the Maritz–Jarrett estimator is more accurate than the method based on Eq. (3.11) used in conjunction with Rosenblatt's shifted histogram described in Section 3.2.2. There are some weak indications that the Maritz–Jarrett estimator remains more accurate when Rosenblatt's shifted histogram is replaced by the adaptive kernel estimator, but an extensive study of this issue has not been conducted. Regardless, the kernel density estimator plays a useful role when dealing with M-estimators of location or when summarizing data.

3.5.5 The Harrell–Davis Estimator

A concern when estimating the qth quantile with $\hat{x}_q = X_{(m)}$, $m = [qn + 0.5]$, is that its standard error can be relatively high. The problem is of particular concern when sampling from a light-tailed or normal distribution. A natural strategy for addressing this problem is to use all of the order statistics to estimate x_q, as opposed to a single order statistic, and several methods have been proposed. One such estimator was derived by Harrell and Davis (1982). To compute it, let Y be a random variable having a beta distribution with parameters $a = (n + 1)q$ and $b = (n + 1)(1 - q)$. That is, the probability density function of Y is

$$\frac{\Gamma(a + b)}{\Gamma(a)\Gamma(b)} y^{a-1}(1 - y)^{b-1}.$$

(Γ is the gamma function, the details of which are not important for present purposes.) Let

$$W_i = P\left(\frac{i - 1}{n} \leq Y \leq \frac{i}{n}\right).$$

Then the Harrell–Davis estimate of the qth quantile is

$$\hat{\theta}_q = \sum_{i=1}^{n} W_i X_{(i)}. \tag{3.16}$$

This is another example of an L-estimator. Asymptotic normality of $\hat{\theta}_q$ was established by Yoshizawa et al. (1985) for $q = 0.5$, only.

In some cases, the Harrell–Davis estimator is much more efficient than $\hat{x}_q$, and this can translate into substantial gains in power when testing hypotheses, as illustrated in Chapter 5. This is not to say, however, that the Harrell–Davis estimator always dominates $\hat{x}_q$ in terms of its standard error. In fact, if the tails of a distribution are heavy enough, the standard error of $\hat{x}_q$ can be substantially smaller than the standard error of $\hat{\theta}_q$, as is illustrated later in this chapter.

The main advantage of $\hat{\theta}_q$ is that it guards against extremely poor efficiency under normality, but as the sample size gets large, it seems that this becomes less of an issue (Sheather and Marron, 1990). A criticism of $\hat{\theta}_q$ is that its finite sample breakdown point is only $1/n$ because all of the order statistics have some positive weight. There are kernel density estimators of quantiles, but they are not discussed because they seem to behave in a manner very similar to the Harrell–Davis estimator used here. (For comparisons of various quantile estimators, see Parrish, 1990; Dielman et al., 1994; and Sfakianakis and Verginis, 2008.)

3.5.6 R Functions qest and hd

The R function

$$qest(x,q=0.5)$$

estimates the qth quantile using Eq. (3.12). The R function

$$hd(x,q=0.5)$$

computes $\hat{\theta}_q$, the Harrell–Davis estimate of the qth quantile. Both functions automatically remove missing values (stored as NA). The default value for q is 0.5. Storing the data in Table 3.2 in the R vector x, the command hd(x) returns the value $\hat{\theta}_{0.5} = 271.7$ as the estimate of the median. Similarly, the estimate of the 0.4 quantile is computed with the command hd(x,0.4), and for the data in Table 3.2 it returns the value 236.

3.5.7 A Bootstrap Estimate of the Standard Error of $\hat{\theta}_q$

The influence function of the Harrell–Davis estimator has not been derived, and there is no simple equation giving its standard error. However, its standard error can be obtained using the bootstrap method in Section 3.1. That is, in Section 3.1, simply replace $\hat{\theta}$ with $\hat{\theta}_q$.

3.5.8 R Function hdseb

The R function

$$hdseb(x,q=0.5,nboot=100)$$

computes the bootstrap estimate of the standard error of the Harrell–Davis estimator for the data stored in the vector x. Of course, the R function bootse in Section 3.1.1 could be used as well; the function hdseb is provided merely for convenience. If, for example, the R command bootse(x,nboot=100,est=hd) is used, this yields the same estimate of the standard error returned by hdseb. When using hdseb, the default value for q is 0.5 and the default value for nboot, which represents B, the number of bootstrap samples, is 100. (But the default estimate when using bootse is nboot=1000.) For example, hdseb(x) uses $B = 100$ bootstrap samples to estimate the standard error when estimating the median. For the data in Table 3.2, this function returns the value 50.8. This is a bit smaller than the estimated standard error of the 20% trimmed mean, which is 56.1, it is a bit larger than the Maritz–Jarrett estimate of the standard error of $\hat{x}_{0.5}$, 45.8, and it is substantially smaller than the estimated standard error of the sample mean, 136. With $B = 25$, the estimated standard error of $\hat{\theta}_{0.5}$ drops from 50.8 to 49.4. When using the Harrell–Davis estimator to estimate the qth quantile, $q \neq 0.5$, an estimate of the standard error is obtained with the command hdseb(x,q), and $B = 100$ will be used. The command hdseb(x,0.3,25) uses $B = 25$ bootstrap samples to estimate the standard error when estimating the 0.3 quantile.

3.6 An M-Estimator of Location

The trimmed mean is based on a predetermined amount of trimming. That is, you first specify the amount of trimming that is desired, after which, the sample trimmed mean, $\bar{X}_t$, can be computed. Another approach is to empirically determine the amount of trimming. For example, if sampling is from a light-tailed distribution, or even a normal distribution, it might be desirable to trim very few observations or none at all. If a distribution is skewed to the right, a natural reaction is to trim more observations from the right versus the left tail of the empirical distribution. In essence, this is what the M-estimator of location does. There are, however, some practical difficulties that arise when using M-estimators of location, and in some cases, trimmed means have important advantages. But there are also important advantages to using M-estimators, especially in the context of regression.

Before describing how an M-estimator is computed, it helps to elaborate on the line of reasoning leading to M-estimators (beyond what was covered in Chapter 2) and to comment on some technical issues. Chapter 2 put μ in the context of minimizing the expected squared difference between X and some constant c, and this was used to provide some motivation for the general approach used to define M-measures of location. In particular, setting $c = \mu$ minimizes $E(X - c)^2$. One practical concern was that if a measure of location is defined as the value of c minimizing $E(X - c)^2$, extreme X values can have an inordinately large effect on the resulting value for c. For a skewed distribution, values of X that are extreme and relatively rare can "pull" the value of μ into the tail of the distribution. A method of addressing this concern is to replace $(X - c)^2$ with some other function that gives less weight to extreme values. In terms of estimators of location, a similar problem arises. The sample mean is the value of c minimizing $\sum (X_i - c)^2$. From basic calculus, minimizing this sum turns out to be equivalent to choosing c such that $\sum (X_i - c) = 0$, and the solution is $c = \bar{X}$. The data in Table 3.2 illustrate that the sample mean can be quite far into the tail of a distribution. The sample mean is 448, yet 15 of the 19 observations have values less than 448. In fact, the data suggest that 448 is somewhere near the 0.8 quantile. M-estimators of location address this problem by replacing $(X_i - c)^2$ with some function that gives less weight to extreme X_i values (cf. Martin and Zamar, 1993).

From Chapter 2, an M-measure of location is the value μ_m such that

$$E \left\{ \Psi \left(\frac{X - \mu_m}{\tau} \right) \right\} = 0, \tag{3.17}$$

where τ is some measure of scale and Ψ is an odd function, meaning that $\Psi(-x) = -\Psi(x)$. (This approach to estimating a measure of location dates back to at least Ellis, 1844.) Some choices for Ψ are listed and described in Table 2.1. Once a random sample of observations is available, the M-measure of location is estimated by replacing expected value with summation in Eq. (3.17). That is, an M-estimator of location is the value $\hat{\mu}_m$ such that

$$\sum \Psi \left(\frac{X_i - \hat{\mu}_m}{\tau} \right) = 0. \tag{3.18}$$

If $\Psi\{(X_i - \hat{\mu}_m)/\tau)\} = (X_i - \hat{\mu}_m)/\tau$, $\hat{\mu}_m = \bar{X}$.

There are three immediate problems that must be addressed if M-measures of location are to have any practical value: choosing an appropriate Ψ, choosing an appropriate measure of scale, τ, and finding a method for estimating μ_m once a choice for Ψ and τ has been made.

First, consider the problem of choosing Ψ. There are many possible choices, so criteria are needed for deciding whether a particular choice has any practical value. Depending on the choice for Ψ there can be 0, 1, or multiple solutions to Eq. (3.18), and this helps to limit the range of functions one might use. If there are zero solutions, this approach to estimation has little value, as is evident. If there are multiple solutions, there is the problem of choosing which solution to use in practice. A reasonable suggestion is to use the solution closest to the median, but estimation problems can persist. One of the more important examples of this (see Freedman and Diaconis, 1982) arises when Ψ is taken to be the so-called biweight:

$$\Psi(x) = \begin{cases} x(1 - x^2)^2, & \text{if } |x| < 1, \\ 0, & \text{if } |x| \geq 1. \end{cases} \tag{3.19}$$

All indications are that it is best to limit attention to those Ψ that yield a single solution to Eq. (3.17). This can be done by limiting attention to Ψ that are monotonic increasing.

Insisting on a single solution to Eq. (3.18) provides a criterion for choosing Ψ, but obviously more is needed. To make progress, it helps to replace Eq. (3.18) with an equivalent approach to defining an estimator of location. First note that from basic calculus, defining a measure of location with Eq. (3.18) is equivalent to defining $\hat{\mu}_m$ as the value minimizing

$$\sum \xi \left(\frac{X_i - \hat{\mu}_m}{\tau} \right),$$

(3.20)

where Ψ is the derivative of ξ. Now, if sampling is from a normal distribution, the optimal estimator, in terms of minimum variance, is the sample mean $\bar{X}$, and the sample mean can be viewed as the value minimizing $\sum (X_i - \hat{\mu}_m)^2$. That is, using

$$\xi \left(\frac{X_i - \hat{\mu}_m}{\tau} \right) = (X_i - \hat{\mu}_m)^2$$

(3.21)

yields $\hat{\mu}_m = \bar{X}$, which is optimal under normality. As already indicated, the problem with this function is that it increases too rapidly as the value of X_i moves away from $\hat{\mu}_m$, and this can cause practical problems when sampling from non-normal distributions for which extreme values can occur. But because this choice of ξ is optimal under normality, a natural strategy is to search for some approximation of Eq. (3.21) that gives nearly the same results when sampling from a normal distribution. In particular, consider functions that are identical to Eq. (3.21) provided X_i is not too extreme.

To simplify matters, temporarily consider a standard normal distribution, and take τ to be σ, the standard deviation, which in this case, is 1. Then the optimal choice for ξ is $(x - \hat{\mu}_m)^2$, as already explained. Suppose instead that ξ is taken to be

$$\xi(x - \hat{\mu}_m) = \begin{cases} -2K(x - \hat{\mu}_m), & \text{if } x < -K, \\ (x - \hat{\mu}_m)^2, & \text{if } -K \leq x \leq K, \\ 2K(x - \hat{\mu}_m), & \text{if } x > K, \end{cases}$$

(3.22)

where K is some constant to be determined. Thus, when sampling from a normal distribution, the optimal choice for ξ is being used provided an observation is not too extreme, meaning that its value does not exceed K or is not less than $-K$. If it is extreme, ξ becomes a linear function, rather than a quadratic function; this linear function increases less rapidly than Eq. (3.21), so extreme values are having less of an influence on $\hat{\mu}_m$.

The strategy for choosing ξ, outlined earlier, is illustrated in the left panel of Fig. 3.5, which shows a graph of $\xi(x - \hat{\mu}_m) = (x - \hat{\mu}_m)^2$ when $\hat{\mu}_m = 0$, and this is the optimal choice for

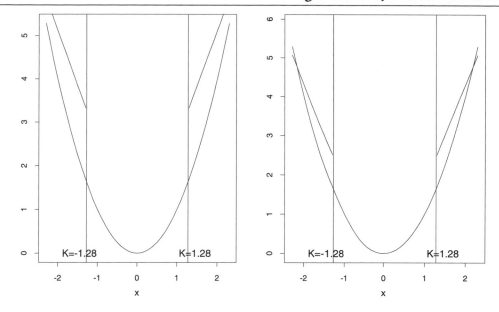

Figure 3.5: An approximation of the optimal function.

ξ when sampling from a standard normal distribution. Also shown is the approximation of the optimal ξ, given by Eq. (3.22), when $K = 1.28$. When $-1.28 \leq x \leq 1.28$, the approximation is exact. When $x < -1.28$ or $x > 1.28$, the straight line above the curve is used to approximate ξ. Because $K = 1.28$ is the 0.9 quantile of a standard normal distribution, there is a 0.8 probability that a randomly sampled observation will have a value between $-K$ and K. Note how Fig. 3.5 suggests that Eq. (3.22) with $K = 1.28$ is a reasonable approximation of $\xi(x - \hat{\mu}_m)^2 = (x - \hat{\mu}_m)^2$.

The left panel of Fig. 3.5 suggests lowering the straight lines to get a better approximation of ξ. The right panel shows what happens when the lines are lowered by $K^2/2$. That is, Eq. (3.22) is replaced by

$$\xi(x - \hat{\mu}_m) = \begin{cases} -2K(x - \hat{\mu}_m) - \frac{K^2}{2}, & \text{if } x < -K, \\ (x - \hat{\mu}_m)^2, & \text{if } -K \leq x \leq K, \\ 2K(x - \hat{\mu}_m) - \frac{K^2}{2}, & \text{if } x > K. \end{cases}$$

However, this modification yields the same equation for determining $\hat{\mu}$, as given by Eq. (3.22) in the next paragraph.

Now, $\hat{\mu}_m$ is the value minimizing Eq. (3.20). Taking the derivative of this equation, with ξ given by Eq. (3.22), and setting the result equal to zero, $\hat{\mu}_m$ is determined by

$$2 \sum \Psi(X_i - \hat{\mu}_m) = 0, \tag{3.23}$$

where:

$$\Psi(x) = \max[-K, \min(K, x)] \tag{3.24}$$

is Huber's Ψ. (For a graph of Huber's Ψ, see Chapter 2.) Of course, the constant 2 in Eq. (3.23) is not relevant to solving for $\hat{\mu}_m$, and typically Eq. (3.23) is simplified to

$$\sum \Psi(X_i - \hat{\mu}_m) = 0. \tag{3.25}$$

There remains the problem of choosing K. One strategy is to choose K so that the large sample (asymptotic) standard error of $\hat{\mu}_m$ is reasonably close to the standard error of the sample mean when sampling from a normal distribution, yet the standard error of $\hat{\mu}_m$ is relatively unaffected when sampling from a heavy-tailed distribution. A common choice is $K = 1.28$, the 0.9 quantile of the standard normal distribution, and this will be used unless stated otherwise. For a more detailed discussion about choosing K, see Huber (1981). In a given situation, some other choice might be more optimal, but $K = 1.28$ guards against relatively large standard errors, while sacrificing very little when sampling from a normal distribution. A more efficacious choice might be made based on knowledge about the distribution being sampled, but the extent to which this strategy can be recommended is unclear.

One more technical issue must be addressed. From Chapter 2, a requirement of a measure of location is that it be scale-equivariant. In the present context, this means that if μ_m is the M-measure of location associated with the random variable X, aX should have $a\mu_m$ as a measure of location for any constant a. If $\hat{\mu}_m$ is estimated with Eq. (3.25), this requirement is not met; Eq. (3.18) must be used instead. Using Eq. (3.18) means, in particular, that a measure of scale, τ, must be chosen. It turns out that the measure of scale need not be efficient in order for $\hat{\mu}_m$ to be efficient. The main concern is that it be reasonably resistant. In particular, it should have a finite sample breakdown point that is reasonably high. A common choice for a measure of scale is the value of ω determined by

$$P(|X - x_{0.5}| < \omega) = \frac{1}{2}, \tag{3.26}$$

where $x_{0.5}$ is the population median. That is, ω is the 0.5 quantile of the distribution of $|X - x_{0.5}|$. If, for example, sampling is from a standard normal distribution, in which case, $x_{0.5} = 0$, ω is determined by

$$P(-\omega \leq Z \leq \omega) = 0.5,$$

where Z has a standard normal distribution. That is, ω is the 0.75 quantile of the standard normal distribution, which is approximately equal to 0.6745.

The standard estimate of ω is the median absolute deviation statistic, MAD, introduced in Section 3.2.3. Its finite sample breakdown point is approximately 0.5. (For more details about the finite sample breakdown point of MAD, see Gather and Hilker, 1997.)

If observations are randomly sampled from a normal distribution, MAD does not estimate σ, the standard deviation; it estimates $z_{0.75}\sigma$, where $z_{0.75}$ is the 0.75 quantile of the standard normal distribution. To put MAD in a more familiar context, it is typically rescaled so that it estimates σ when sampling from a normal distribution. In particular

$$\text{MADN} = \frac{\text{MAD}}{z_{0.75}} \approx \frac{\text{MAD}}{0.6745}$$

is used, and this convention will be followed here. Then for a random sample, Eq. (3.17) says that an M-estimator of location is the value $\hat{\mu}_m$ satisfying

$$\sum \Psi\left(\frac{X_i - \hat{\mu}_m}{MADN}\right) = 0. \tag{3.27}$$

3.6.1 R Function mad

The R function

$$\text{mad(x)}$$

computes MADN. That is, R assumes that MAD is to be rescaled to estimate σ when sampling from a normal distribution. To use R to compute MAD, simply use the command qnorm(0.75)*mad(x). The command qnorm(0.75) returns the 0.75 quantile of a standard normal random variable.

3.6.2 Computing an M-Estimator of Location

Solving Eq. (3.27) for $\hat{\mu}_m$ is usually accomplished with an iterative estimation procedure known as the Newton–Raphson method. It involves the derivative of Ψ, which is given by

$$\Psi'(x) = \begin{cases} 1, & \text{if } -K \leq x \leq K, \\ 0, & \text{otherwise.} \end{cases} \tag{3.28}$$

The steps used to determine $\hat{\mu}_m$ are shown in Table 3.5. Typically, $K = 1.28$ is used, and this choice is assumed henceforth unless stated otherwise.

Table 3.5: How to compute the M-estimator of location $\hat{\mu}_m$.

Set $k = 0$, $\hat{\mu}_k = M$, the sample median, and choose a value for K. A common choice is $K = 1.28$.
Step 1. Let

$$A = \sum \Psi \left(\frac{X_i - \hat{\mu}_k}{\text{MADN}} \right),$$

where Ψ is given by Eq. (3.24).
Step 2. Let

$$B = \sum \Psi' \left(\frac{X_i - \hat{\mu}_k}{\text{MADN}} \right),$$

where Ψ' is the derivative of Ψ given by Eq. (3.28). B is just the number of observations X_i satisfying $-K \le (X_i - \hat{\mu}_k)/\text{MADN} \le K$.
Step 3. Set

$$\hat{\mu}_{k+1} = \hat{\mu}_k + \frac{\text{MADN} \times A}{B}.$$

Step 4. If $|\hat{\mu}_{k+1} - \hat{\mu}_k| < 0.0001$, stop, and set $\hat{\mu}_m = \hat{\mu}_{k+1}$. Otherwise, increment k by one, and repeat steps 1–4.

■ **Example**

For the data in Table 3.2, $\hat{\mu}_0 = M = 262$ and $\text{MADN} = 169$. Table 3.6 shows the resulting values of $\Psi\{(X_i - \mu_0)/\text{MADN}\}$ corresponding to each of the 19 values. The sum of the values in Table 3.6 is $A = 2.05$. The number of Ψ values between -1.28 and 1.28 is $B = 15$, so the first iteration using the steps in Table 3.5 yields

$$\hat{\mu}_1 = 262 + \frac{169 \times 2.05}{15} = 285.1.$$

The iterative estimation process consists of using $\hat{\mu}_1$ to recompute the Ψ values, yielding a new value for A and B, which in turn yields $\hat{\mu}_2$. For the data at hand, it turns out that there is no difference between $\hat{\mu}_2$ and $\hat{\mu}_1$, so the iterative process stops, and $\hat{\mu}_m = 285.1$. If there had been a difference, the Ψ values would be computed again using $\hat{\mu}_2$, and this would continue until $|\hat{\mu}_{k+1} - \hat{\mu}_k| < 0.0001$.

■

When computing the M-estimator of location as described in Table 3.5, the measure of scale, MADN, does not change when iterating. There are also M-estimators where a measure of scale is updated. That is, a measure of scale is simultaneously determined in an iterative fashion. (See Huber, 1981, p. 136.) Currently, it seems that this alternative estimation procedure offers no practical advantage, so it is not discussed. In fact, if a measure of scale is estimated

Table 3.6:   Values of Huber's Ψ for the Self-Awareness Data.

−1.09	−1.04	−1.04	−0.88	−0.66	−0.31	−0.25	−0.09	−0.05	0.0
0.20	0.22	0.26	0.67	0.98	1.28	1.28	1.28	1.28	

simultaneously with a measure of location, using Huber's Ψ, the Bickel–Lehmann condition for a measure of location is no longer satisfied (Bickel and Lehmann, 1975).

Notice that the M-estimator in Table 3.5 empirically determines whether an observation is unusually large or small. In the first step, where $\hat{\mu}_0 = M$, the sample median, X_i is considered unusually small if $(X_i - M)/\text{MADN} < -1.28$, where the typical choice of $K = 1.28$ is being used, and it is unusually large if $(X_i - M)/\text{MADN} > 1.28$. This becomes clearer if the first step in the iterative process is written in a different form. Let i_1 be the number of observations X_i for which $(X_i - M)/\text{MADN} < -1.28$, and let i_2 be the number of observations such that $(X_i - M)/\text{MADN} > 1.28$. Some algebra shows that the value of $\hat{\mu}_1$ in Table 3.5 is

$$\frac{1.28(\text{MADN})(i_2 - i_1) + \sum_{i=i_1+1}^{n-i_2} X_{(i)}}{n - i_1 - i_2}, \tag{3.29}$$

the point being that the sum in this expression is over only those values that are not too large or too small.

To conclude this section, it is noted that there are more formal methods for motivating Huber's Ψ, but no details are given here. Readers interested in technical issues can refer to Huber (1981) or Huber and Ronchetti (2009).

3.6.3 R Function mest

The R function

$$\text{mest(x,bend=1.28)}$$

performs the calculations in Table 3.5. The argument bend corresponds to K in Huber's Ψ and defaults to 1.28 if unspecified. For example, the command mest(x) computes the M-estimator of location for the data in the vector x using $K = 1.28$. The command mestx(x,1.5) uses $K = 1.5$. Increasing K increases efficiency when sampling from a normal distribution, but it increases sensitivity to the tails of the distribution, and efficiency can be lower as well when sampling from a heavy-tailed distribution. To illustrate sensitivity to the tail of a distribution, the mean, median, 20% trimmed mean, M-estimator, with $K = 1.28$, and MOM (described in Section 3.10) are equal to 448, 262, 282.7, 258.1, and 245.4, respectively, for the data in Table 3.2. With $K = 1.2$, the M-estimator is 281.6, and with $K = 1$, the M-estimator is equal to 277.4. Note that MOM has a value less than the median, in contrast to the other location estimators that were used. This illustrates a curious property about the population value of MOM. Suppose a distribution is skewed to the right. Then the population value of MOM can lie between the median and the mode of this distribution, in contrast to any trimmed mean or M-estimator, which typically lie to the right of the median.

3.6.4 Estimating the Standard Error of the M-Estimator

This section describes the first of two methods for estimating the standard error of the M-estimator of location. The method here is based on the influence function of μ_m and the other uses a bootstrap.

As was the case with the trimmed mean and $\hat{x}_q$, the justification for the non-bootstrap estimate of the standard error follows from the result that

$$\hat{\mu}_m = \mu_m + \frac{1}{n} \sum I F_m(X_i),$$

plus a remainder term that goes to zero as n gets large. That is, an M-estimator can be written as its population value plus a sum of independent random variables having mean zero.

The influence function of the M-measure of location has a somewhat complicated form. It depends in part on the measure of scale that is used in Ψ, and here, this is $\omega_N = \omega/0.6745$. The influence function of ω, which is estimated by MAD, is given in Chapter 2. The influence function of ω_N is just the influence function of ω divided by 0.6745. Let

$$A(x) = \text{sign}(|x - \theta| - \omega),$$

where θ is the population median and $\text{sign}(x)$ equals -1, 0, or 1, according to whether x is less than, equal to, or greater than 0. Let

$$B(x) = \text{sign}(x - \theta)$$

and

$$C(x) = A(x) - \frac{B(x)}{f(\theta)}\{f(\theta + \omega) - f(\theta - \omega)\}.$$

The influence function of ω_N is

$$I F_{\omega_N}(x) = \frac{C(x)}{2(0.6745)\{f(\theta + \omega) + f(\theta - \omega)\}}.$$

Estimating $I F_{\omega_N}(X_i)$, the value of the influence function of ω_N at X_i, requires an estimate of the probability density function, and this can be done as described in Section 3.2. Here, the adaptive kernel estimator in Section 3.2.4 will be used unless stated otherwise. Denoting the estimate of the probability density function $f(x)$ with $\hat{f}(x)$ and computing

$$\hat{A}(X_i) = \text{sign}(|X_i - M| - \text{MAD}),$$

$$\hat{B}(X_i) = \text{sign}(X_i - M),$$

$$\hat{C}(X_i) = \hat{A}(X_i) - \frac{\hat{B}(X_i)}{\hat{f}(M)}\{\hat{f}(M + \text{MAD}) - \hat{f}(M - \text{MAD})\},$$

an estimate of $IF_{\omega_N}(X_i)$ is

$$V_i = \frac{\hat{C}(X_i)}{2(.6745)\{\hat{f}(M + \text{MAD}) + \hat{f}(M - \text{MAD})\}}. \tag{3.30}$$

Letting $y = (x - \mu_m)/\omega_N$, the influence function of μ_m is

$$IF_m(x) = \frac{\omega_N \Psi(y) - IF_{\omega_N}(x)\{E(\Psi'(y)y)\}}{E[\Psi'(y)]}. \tag{3.31}$$

Having described how to estimate $IF_{\omega_N}(X_i)$, and because MADN estimates ω_N, all that remains when estimating $IF_m(X_i)$ is estimating $E[\Psi'(Y)Y]$ and $E[\Psi'(Y)]$, where $Y = (X - \mu_m)/\omega_N$. Set

$$Y_i = \frac{X_i - \hat{\mu}_m}{\text{MADN}}$$

and

$$D_i = \begin{cases} 1, & \text{if } |Y_i| \le K, \\ 0, & \text{otherwise.} \end{cases}$$

Then $E[\Psi'(y)]$ is estimated with

$$\bar{D} = \frac{1}{n}\sum D_i.$$

Finally, estimate $E[\Psi'(y)y]$ with

$$\bar{C} = \frac{1}{n}\sum D_i Y_i.$$

The sum in this last equation is just the sum of the Y_i values satisfying $|Y_i| \le K$. The value of the influence function of μ_m, evaluated at X_i, is estimated with

$$U_i = \{(\text{MADN})\Psi(Y_i) - V_i\bar{C}\}/\bar{D}. \tag{3.32}$$

The squared standard error of $\hat{\mu}_m$ can now be estimated from the data. The estimate is

$$\hat{\sigma}_m^2 = \frac{1}{n(n-1)} \sum U_i^2. \tag{3.33}$$

Consequently, the standard error, $\sqrt{VAR(\hat{\mu}_m)}$, is estimated with $\hat{\sigma}_m$. Note that the sum in Eq. (3.33) is divided by $n(n-1)$, not n^2, as indicated by Eq. (3.6). This is done because if no observations are flagged as being unusually large or small by Ψ, $\hat{\mu}_m = \bar{X}$, and Eq. (3.33) reduces to s^2/n, the estimate that is typically used.

The computations just described are straightforward but tedious, so no detailed illustration is given. Interested readers can use the R function mestse, which is described in the next subsection.

3.6.5 R Function mestse

The R function

$$\text{mestse(x,bend=1.28,op=2)}$$

estimates the standard error of the M-estimator using the method just described. The argument bend corresponds to K in Huber's Ψ and defaults to 1.28 if not specified. The argument op indicates which density estimator is used to estimate the influence function. By default (op=2), the adaptive kernel estimator is used; otherwise, Rosenblatt's shifted histogram is employed. If the data in Table 3.2 are stored in the R variable x, the command mestse(x) returns the value 54.1, and this is reasonably close to 56.1, the estimated standard error of the trimmed mean. The 20% trimmed mean and M-estimator have similar influence functions, so reasonably close agreement was expected. A difference between the two estimators is that the M-estimator identifies the four largest values as being unusually large. That is, $(X_i - \hat{\mu}_m)/\text{MADN}$ exceeds 1.28 for the four largest values, while the trimmed mean trims the three largest values only, so the expectation is that $\hat{\mu}_m$ will have a smaller standard. Also, the M-estimator does not identify any of the lower values as being unusual, but the trimmed mean automatically trims three values.

3.6.6 A Bootstrap Estimate of the Standard Error of $\hat{\mu}_m$

The standard error can also be estimated using a bootstrap method. The computational details are essentially the same as those described in Section 3.1. Regarding B, the number of bootstrap samples to be used, $B = 100$ appears to be more than adequate in most situations (Efron, 1987).

The accuracy of the bootstrap method versus the kernel density estimator has not been examined when n is small. Early attempts at comparing the two estimators, via simulations, were complicated by the problem that the kernel density estimator that was used can be undefined because of division by zero. Yet another problem is that the bootstrap method can fail when n is small because a bootstrap sample can yield $\text{MAD} = 0$, in which case, $\hat{\mu}_m^*$ cannot be computed because of division by zero. A few checks were made with $n = 20$ and $B = 1{,}000$ when sampling from a normal or lognormal distribution. Limited results suggest that the bootstrap is more accurate, but a more detailed study is needed to resolve this issue.

3.6.7 R Function mestseb

The R function

$$\text{mestseb(x,nboot=1000,bend=1.28)}$$

computes the bootstrap estimate of the standard error of $\hat{\mu}_m$ for the data stored in the R variable x. The argument nboot is B, the number of bootstrap samples to be used, which defaults to $B = 100$ if unspecified. The default value for bend, which corresponds to K in Huber's Ψ, is 1.28. For example, mestseb(x,50) will compute a bootstrap estimate of the standard error using $B = 50$ bootstrap replications, while mestseb(x) uses $B = 100$. For the data in Table 3.2, mestseb(x) returns the value 53.7, which is in reasonable agreement with 53.2, the estimated standard error using the influence function. The function mestseb sets the seed of the random number generator in R so that results will be duplicated if mestseb is executed a second time with the same data. Otherwise, if mestseb is invoked twice, slightly different results would be obtained because the bootstrap method would use a different sequence of random numbers.

3.7 One-Step M-Estimator

Typically, when computing $\hat{\mu}_m$ with the iterative method in Table 3.5, convergence is obtained after only a few iterations. It turns out that if only a single iteration is used, the resulting estimator has good asymptotic properties (Serfling, 1980). In particular, for large sample sizes, it performs in a manner very similar to the fully iterated M-estimator. An expression for the first iteration was already described, but it is repeated here for convenience, assuming $K = 1.28$. Let i_1 be the number of observations X_i for which $(X_i - M)/\text{MADN} < -1.28$, and let i_2 be the number of observations such that $(X_i - M)/\text{MADN} > 1.28$. The one-step M-estimate of location is

$$\hat{\mu}_{\text{os}} = \frac{1.28(\text{MADN})(i_2 - i_1) + \sum_{i=i_1+1}^{n-i_2} X_{(i)}}{n - i_1 - i_2}. \tag{3.34}$$

While the one-step M-estimator is slightly easier to compute than the fully iterated M-estimator, its influence function has a much more complicated form when distributions are skewed (Huber, 1981, p. 140). In terms of making inferences about the corresponding population parameter, it seems that there are no published results suggesting that the influence function plays a useful role when testing hypotheses or computing confidence intervals. Consequently, details about the influence function are not given here. As for estimating the standard error of the one-step M-estimator, only the bootstrap method will be used, as described in Section 3.1.

3.7.1 R Function onestep

The R function

$$onestep(x, bend=1.28)$$

computes the one-step M-estimator given by Eq. (3.31).

3.8 W-Estimators

W-estimators of location are closely related to M-estimators and usually they give identical results. However, when extending M-estimators to regression, the computational method employed by W-estimators is sometimes used. This method is just another way of solving Eq. (3.27). For completeness, W-estimators are briefly introduced here.

Let

$$w(x) = \frac{\Psi(x)}{x}.$$

In this last equation, Ψ could be any of the functions associated with M-estimators, and the generic measure of scale τ, used to define the general class of M-estimators, could be used. If, for example, τ is estimated with MADN, then $\hat{\mu}_m$ is determined by solving Eq. (3.27), which becomes

$$\sum \left(\frac{X_i - \hat{\mu}_m}{\text{MADN}} \right) w \left(\frac{X_i - \hat{\mu}_m}{\text{MADN}} \right) = 0. \qquad (3.35)$$

Rearranging terms in Eq. (3.35) yields

$$\hat{\mu}_m = \frac{\sum X_i w\{(X_i - \hat{\mu}_m)/\text{MADN}\}}{\sum w\{(X_i - \hat{\mu}_m)/\text{MADN}\}}.$$

This last equation does not yield an immediate value for $\hat{\mu}_m$ because $\hat{\mu}_m$ appears on both sides of the equation. However, it suggests an iterative method for obtaining $\hat{\mu}_m$ that has practical value.

Set $k = 0$, and let $\hat{\mu}_0$ be some initial estimate of $\hat{\mu}_m$. For example, $\hat{\mu}_0$ could be the sample mean. Let

$$U_{ik} = \frac{X_i - \hat{\mu}_k}{\text{MADN}}.$$

Then the iteration formula is

$$\hat{\mu}_{k+1} = \frac{\sum X_i w(U_{ik})}{\sum w(U_{ik})}.$$

That is, given $\hat{\mu}_k$, which is an approximate value for $\hat{\mu}_m$ that solves (3.35), an improved approximation is $\hat{\mu}_{k+1}$. One simply keeps iterating until $|\hat{\mu}_{k+1} - \hat{\mu}_k|$ is small, say less than 0.0001.

The iterative method just described is an example of what is called *iteratively reweighted least squares*. To explain, let $\hat{\mu}$ be any estimate of a measure of location, and recall that $\hat{\mu} = \bar{X}$ is the value that minimizes $\sum(X_i - \hat{\mu})^2$. In the context of regression, minimizing this sum is based on the least squares principle taught in every introductory statistics course. Put another way, the sample mean is the ordinary least squares (OLS) estimator. Weighted least squares, based on *fixed* weights, w_i, determines a measure of location by minimizing $\sum w_i(X_i - \hat{\mu}_m)^2$, and this is done by solving $\sum w_i(X_i - \hat{\mu}_m) = 0$, which yields

$$\hat{\mu} = \frac{\sum w_i X_i}{\sum w_i}.$$

The problem in the present context is that the weights in Eq. (3.33) are not fixed; they depend on the value of $\hat{\mu}_m$, which is not known but updated with each iteration, so this last equation for $\hat{\mu}$ does not apply. Instead, the weights are recomputed according to the value of $\hat{\mu}_k$.

3.8.1 Tau Measure of Location

A variation of the W-estimator just described plays a role in some settings. Called the *tau measure of location*, it is computed as follows: Let

$$W_c(x) = \left(1 - \left(\frac{x}{c}\right)^2\right)^2 I(|x| \leq c),$$

where the indicator function $I(|x| \leq c) = 1$ if $|x| \leq c$; otherwise, $I(|x| \leq c) = 0$. The weights are

$$w_i = W_c \left(\frac{X_i - M}{\text{MAD}} \right),$$

and the resulting measure of location is denoted by

$$\hat{\mu}_\tau = \frac{\sum w_i X_i}{\sum w_i}.$$

Following Maronna and Zamar (2002), $c = 4.5$ is used unless stated otherwise.

3.8.2 R Function tauloc

The R function

$$\text{tauloc(x,cval=4.5)}$$

computes the tau measure of location.

3.8.3 Zuo's Weighted Estimator

Yet another approach to choosing the weights when computing a W-estimator was suggested by Zuo (2010). (It is related to a class of multivariate W-estimators introduced in Section 6.3.7.) Let

$$D_i = 1/(1 + |X_i - M|/\text{MAD}).$$

The weights are taken to be

$$w_i = I_{D_i \geq c} + I_{D_i < c} \frac{e^{-k(1 - D_i^2/c^2)^2} - e^{-k}}{(1 - e^{-k})},$$

where the indicator function $I_{D_i \geq c} = 1$ if $D_i \geq c$; otherwise, $I_{D_i \geq c} = 0$. The constant c satisfies $0 \leq c \leq 1$ and $k > 0$. Zuo suggests using $k = 3$ and $c = 0.2$. The practical advantages of this estimator, relative to the many other robust location estimators that have been studied extensively, are unclear.

3.9 The Hodges–Lehmann Estimator

Chapter 2 mentioned some practical concerns about R-measures of location in general and the Hodges–Lehmann (1963) estimator in particular. But the Hodges–Lehmann estimator plays a fundamental role when applying standard rank-based methods (in particular, the Wilcoxon signed rank test), so for completeness, the details of this estimator are given here.

The *Walsh averages* of n observations refers to all pairwise averages, $(X_i + X_j)/2$, for all $i \le j$. The Hodges–Lehmann estimator is the median of all Walsh averages, namely,

$$\hat{\theta}_{\mathrm{HL}} = \mathrm{med}_{i \le j} \frac{X_i + X_j}{2}.$$

As noted in Section 2.2, there are conditions where this measure of location, as well as R-estimators in general, has good properties. But there are general conditions under which they perform poorly (e.g., Bickel and Lehmann, 1975; Huber, 1981, p. 65; Morgenthaler and Tukey, 1991, p. 15).

3.10 Skipped Estimators

Skipped estimators of location refer to the natural strategy of checking the data for outliers, removing any that are found, and averaging the values that remain. The first skipped estimator appears to be one proposed by Tukey (see Andrews et al., 1972), where checks for outliers were based on a boxplot rule. (Boxplot methods for detecting outliers are described in Section 3.13.) The one-step M-estimator given by Eq. (3.32) is almost a skipped estimator. If we ignore the term $1.28(\mathrm{MADN})(i_2 - i_1)$ in the numerator of Eq. (3.32), an M-estimator removes the value X_i if

$$\frac{|X_i - M|}{\mathrm{MADN}} > 1.28$$

and averages the values that remain. In essence, X_i is declared an outlier if it satisfies this last equation. But based on how the M-estimator is defined, the term $1.28(\mathrm{MADN})(i_2 - i_1)$ arises.

When testing hypotheses, a slight variation of the skipped estimator just described has practical value. This *modified one-step M-estimator* (MOM) simply averages values not declared outliers, but to get reasonably good efficiency under normality, the outlier detection rule used by the one-step M-estimator is modified. Now X_i is declared an outlier if

$$\frac{|X_i - M|}{\mathrm{MADN}} > 2.24$$

(which is a special case of a multivariate outlier detection method derived by Rousseeuw and van Zomeren, 1990). This last expression is known as the *Hampel identifier*, only Hampel used 3.5 rather than 2.24. When using 3.5, this will be called the Hampel version of MOM (HMOM.)

3.10.1 R Functions mom, zwe, and bmean

The R function

$$mom(x,bend=2.24)$$

computes the MOM estimate of location, where the argument bend is the constant used in the Hampel identifier. The R function

$$bmean(x,mbox=TRUE)$$

computes a skipped estimator, where outliers are identified by a boxplot rule covered in Section 3.13. The default value for mbox is TRUE, indicating that Carling's method (described in Section 3.13.3) is used, and mbox=FALSE uses the boxplot rule based on the ideal fourths. The R function

$$zwe(x,k=3, C=0.2)$$

computes Zou's estimator.

3.11 Some Comparisons of the Location Estimators

Illustrations given in the previous sections of this chapter, based on data from actual studies, demonstrate that the estimated standard errors associated with robust estimates of location can be substantially smaller than the standard error of the sample mean. It has been hinted that these robust estimators generally compete well with the sample mean when sampling from a normal distribution, but no details have been given. There is also the concern of how estimators compare under various non-normal distributions, including skewed distributions as a special case. Consequently, this section briefly compares the standard error of the robust estimators to the standard error of the sample mean for a few distributions.

One of the non-normal distributions considered here is the lognormal one shown in Fig. 3.3. The random variable X is said to have a lognormal distribution if the distribution of $Y = \ln(X)$ is normal. It is a skewed distribution for which standard methods for computing confidence intervals for the mean can be unsatisfactory, even with $n = 160$. (Details are given

Table 3.7:   Variances of selected estimators, $n = 10$.

Estimator	Distribution			
	Normal	Lognormal	One-wild	Slash
Mean	0.1000	0.4658	1.0900	∞
$\bar{X}_t$ ($\gamma = 0.1$)	0.1053	0.2238	0.1432	∞
$\bar{X}_t$ ($\gamma = 0.2$)	0.1133	0.1775	0.1433	0.9649
Median	0.1383	0.1727	0.1679	0.7048
$\hat{\mu}_m$ (Huber)	0.1085	0.1976	0.1463	0.9544
$\hat{\theta}_{0.5}$	0.1176	0.1729	0.1482	1.4731
MOM	0.1243	0.2047	0.1409	0.7331
HMOM	0.1092	0.2405	0.1357	1.0272
$\hat{\mu}_\tau$	0.1342	0.3268	2.1610	

in Chapters 4 and 5.) Consequently, there is a general interest in how methods based on alternative measures of location perform when sampling from this particular distribution. The immediate concern is whether robust estimators have relatively small standard errors for this special case.

Table 3.7 shows the variance of several estimators for a few distributions when $n = 10$. (Results for MOM, HMOM, and the Harrell–Davis estimator are based on simulations with 10,000 replications.) As can be seen, the small-sample efficiency of the tau measure of location does not compete well with a 20% trimmed mean, MOM, and the one-step M-estimator. A more detailed study came to the same conclusion (Özdemir and Wilcox, 2012). The distribution *one-wild* refers to sampling from a normal distribution and multiplying one of the observations by 10. Observations are generated from the *slash* distribution by generating an observation from the standard normal distribution and dividing by an independent uniform random variable on the interval (0, 1). Both the one-wild and slash distributions are symmetric distributions with heavier than normal tails. The slash distribution has an extremely heavy tail. In fact, it has infinite variance. The motivation for considering these distributions, particularly the slash distribution, is to see how an estimator performs under extreme conditions. It is unclear how heavy the tails of a distribution might be in practice, so it is of interest to see how an estimator performs for a distribution that represents an extreme departure from normality that is surely unrealistic. If an estimator performs reasonably well under normality and continues to perform well when sampling from a slash distribution, this suggests that it has practical value for any distribution that might arise in practice by providing protection against complete disaster, disaster meaning standard errors that are extremely large compared to some other estimator that might have been used.

The ideal estimator would have a standard error as small as or smaller than any other estimator. None of the estimators in Table 3.7 satisfies this criterion. When sampling from a normal distribution, the sample mean has the lowest standard error, but the improvement over the 10% trimmed mean ($\gamma = 0.1$), the 20% trimmed mean, the Harrell–Davis estimator, and the M-estimator using Huber's Ψ, $\hat{\mu}_m$, is relatively small. Using the sample median is relatively unsatisfactory. For heavy-tailed distributions, the sample mean performs poorly, and its performance can be made as bad as desired by making the tails of the distribution sufficiently heavy. The two estimators that do reasonably well for all of the distributions considered are the 20% trimmed mean and $\hat{\mu}_m$. Note that even the standard error of the Harrell–Davis estimator, $\hat{\theta}_{0.5}$, becomes relatively large when the tails of a distribution are sufficiently heavy. Again, there is the possibility that in practice, $\hat{\theta}_{0.5}$ competes well with the 20% trimmed mean and the M-estimator, but an obvious concern is that exceptions might occur. In situations where interest is specifically directed at quantiles, and in particular, the population median, the choice between the sample median and the Harrell–Davis estimator is unclear. The Harrell–Davis estimator has a relatively small standard error when sampling from a normal distribution, but as the tails of a distribution get heavier, eventually, the sample median performs substantially better. Although MOM has a lower standard error than the median under normality, all other estimators have a lower standard error than MOM for this special case. Switching to the Hampel identifier when using MOM (HMOM), efficiency now competes well with a 20% trimmed mean and the M-estimator based on Huber's Ψ, but for the lognormal distribution, HMOM performs rather poorly, and it is substantially worse than MOM when sampling from the slash distribution. The tau measure of location does not perform all that well, particularly when dealing with the slash distribution. In exploratory studies, one might consider two or more estimators, but in the context of testing hypotheses, particularly in a confirmatory study, some might object to using multiple estimators of location because this will inflate the probability of at least one Type I error. There are methods for adjusting the individual tests so that the probability of at least one Type I error does not exceed a specified value, but such an adjustment might lower power by a substantial amount.

If a skipped estimator is used where outliers are detected via a boxplot rule (described in Section 3.13), good efficiency can be obtained under normality, but situations arise where other estimators offer a distinct advantage. For the situations in Table 3.7, under normality, the variance of this skipped estimator is 0.109 when using Carling's modification of the boxplot method to detect outliers, and switching to the boxplot rule based on the ideal fourths, nearly the same result is obtained. For the lognormal distribution, however, the variances of the skipped estimators are 0.27 and 0.28, approximately, making them the least accurate estimators, on average, excluding the sample mean. They perform the best for the one-wild distribution, but for the slash, they are the least satisfactory excluding the 10% mean and mean.

It cannot be stressed too strongly that no single measure of location always has the lowest standard error. For the data in Table 3.2, the lowest estimated standard error was 45.8, obtained for $\hat{x}_{0.5}$ using the Maritz–Jarrett method. (A bootstrap estimate of the standard error is

42.) The appeal of the 20% trimmed mean and the M-estimator of location is that they guard against relatively large standard errors. Moreover, the potential reduction in the standard error using other estimators is relatively small compared to the possible reduction using the 20% trimmed mean or M-estimator instead.

Based purely on achieving a high breakdown point, the median and an M-estimator (based on Huber's Ψ) are preferable to a 20% trimmed or the Hodges–Lehmann estimator. (For an analysis when sample sizes are very small, see Rousseeuw and Verboven, 2002.) But in terms of achieving accurate probability coverage, methods based on a 20% trimmed mean often are more satisfactory.

3.12 More Measures of Scale

While measures of location are often the focus of attention versus measures of scale, measures of scale are of interest in their own right. Some measures of scale have already been discussed, namely ω estimated by MAD and the Winsorized variance σ_w^2. For results on what is known as the Qn estimator, see Rousseeuw and Croux (1993). (Roughly, it is the 0.25 quantile based on all pairwise absolute differences.) Many additional measures of scale appear in the literature. Two additional measures are described here, which play a role in subsequent chapters.

To begin, it helps to be precise about what is meant by a *scale estimator*. It is any non-negative function, $\hat{\zeta}$, such that for any constants a and b,

$$\hat{\zeta}(a + bX_1, \ldots, a + bX_n) = |b|\hat{\zeta}(X_1, \ldots, X_n). \tag{3.36}$$

From basic principles, the sample standard deviation, s, satisfies this definition. In words, a scale estimator ignores changes in location, and it responds to uniform changes in scale in a manner consistent with what is expected based on standard results related to sample standard deviation, s. In the terminology of Chapter 2, $\hat{\zeta}$ should be location- and scale-equivariant.

A general class of measures of scale that has been found to have practical value stems from the influence function of M-estimators of location when distributions are symmetric. For this special case, the influence function of μ_m takes on a rather simple form:

$$IF_m(X) = \frac{E(\Psi^2(Y))}{\{E(\Psi'(Y))\}^2}, \tag{3.37}$$

$$Y = \frac{X - \mu_m}{K\tau},$$

where τ and Ψ are as in Section 3.6 and K is a constant to be determined. The (asymptotic) variance of $\sqrt{n}\hat{\mu}_m$ is

$$\zeta^2 = \frac{K^2\tau^2 E(\Psi^2(Y))}{\{E(\Psi'(Y))\}^2}, \tag{3.38}$$

and this defines a broad class of measures of scale. (In case it is not obvious, the reason for considering the variance of $\sqrt{n}\hat{\mu}_m$, rather than the variance of $\hat{\mu}_m$, is that the latter goes to 0 as n gets large, and the goal is to define a measure of scale for the distribution under study, not the sampling distribution of the M-estimator that is being used.) Included as a special case among the possible choices for ζ is the usual population standard deviation, σ. To see this, take $\Psi(x) = x$, $K = 1$, and $\tau = \sigma$, in which case, $\zeta = \sigma$.

3.12.1 The Biweight Midvariance

There is the issue of choosing Ψ when defining a measure of scale with Eq. (3.38). For reasons to be described, there is practical interest in choosing Ψ to be the biweight given by Eq. (3.19). The derivative of the biweight is $\Psi'(x) = (1 - x^2)(1 - 5x^2)$ for $|x| < 1$; otherwise, it is equal to 0.

Let K be any constant. For reasons given later, $K = 9$ is a common choice. Also, let τ be ω given by (3.26), which is estimated by MAD. To estimate ζ, set

$$Y_i = \frac{X_i - M}{K \times \text{MAD}},$$

$$a_i = \begin{cases} 1, & \text{if } |Y_i| < 1, \\ 0, & \text{if } |Y_i| \geq 1, \end{cases}$$

in which case, the estimate of ζ is

$$\hat{\zeta}_{bi} = \frac{\sqrt{n}\sqrt{\sum a_i(X_i - M)^2(1 - Y_i^2)^4}}{|\sum a_i(1 - Y_i^2)(1 - 5Y_i^2)|}. \tag{3.39}$$

The quantity $\hat{\zeta}_{bi}^2$ is called a *biweight midvariance*. It appears to have a finite sample breakdown point of approximately 0.5 (Goldberg and Iglewicz, 1992), but a formal proof has not been found.

Explaining the motivation for $\hat{\zeta}_{bi}$ requires some comments on methods for judging estimators of scale. A tempting approach is to compare the standard errors of any two estimators, but this can be unsatisfactory. The reason is that if $\hat{\zeta}$ is a measure of scale, so is $b\hat{\zeta}$, a result that follows from the definition of a measure of scale given by Eq. (3.36). But the variance of $\hat{\zeta}$ is larger than the variance of $b\hat{\zeta}$ if $0 < b < 1$, and, in fact, the variance of $b\hat{\zeta}$ can be made arbitrarily small by choosing b appropriately. What is needed is a measure for comparing scale estimators that is not affected by b. A common method for dealing with this problem (e.g., Lax, 1985; Iglewicz, 1983) is to compare two scale estimators with $\text{VAR}(\ln(\hat{\zeta}))$, the variance of the natural logarithm of the estimators being considered. Note that $\ln(b\hat{\zeta}) = \ln(b) + \ln(\hat{\zeta})$ for any $b > 0$ and scale estimator $\hat{\zeta}$, so $\text{VAR}(\ln(b\hat{\zeta})) = \text{VAR}(\ln(\hat{\zeta}))$. That is, the variance of the logarithm of $\hat{\zeta}$ is not affected by the choice of b.

Another method of comparing scale estimators is in terms of the variance of $\hat{\zeta}/\zeta$. Note that if $\hat{\zeta}$ and ζ are replaced by $b\hat{\zeta}$ and $b\zeta$ for any $b > 0$, the ratio $\hat{\zeta}/\zeta$ remains unchanged, so the problem mentioned in the previous paragraph has been addressed.

Lax (1985) compared over 150 methods of estimating measures of scale, several of which belong to the class of measures defined by Eq. (3.38). Comparisons were made in terms of what is called the *triefficiency* of an estimator. To explain, let $V_{\min}$ be the smallest known value of $\text{VAR}(\ln(\hat{\zeta}))$ among all possible choices for an estimator of scale, $\hat{\zeta}$. Then

$$E = 100 \times \frac{V_{\min}}{\text{VAR}(\ln(\hat{\zeta}))}$$

is a measure of *efficiency*. For some measures of scale, $V_{\min}$ can be determined exactly for a given distribution, while in other situations, $V_{\min}$ is replaced by the smallest variance obtained, via simulations, among the many estimators of scale that are being considered. For example, when sampling from a normal distribution with $n = 20$, the smallest attainable value of $\text{VAR}(\ln(\hat{\zeta}))$ is 0.026, which is attained by s, the sample standard deviation. Thus, for normal distributions, s has efficiency $E = 100$, the best possible value. The main point is that the efficiency of many estimators has been determined for a variety of distributions. Moreover, three distributions have played a major role: normal, one-wild, and slash, already described. The smallest efficiency of an estimator, among these three distributions, is called its *triefficiency*. For example, if $n = 20$, s has efficiency 100, 0, and 0 for these three distributions; the smallest of these three efficiencies is 0, so its triefficiency is 0. Using $K = 9$ in Eq. (3.39) when defining $\hat{\zeta}_{\text{bi}}$, the efficiencies corresponding to these three distributions are 86.7, 85.8, and 86.1, so the triefficiency is 85.8, and this is the highest triefficiency among the measures of scale considered by Lax (cf. Croux, 1994).

3.12.2 R Function bivar

The R function

$$\text{bivar(x)}$$

computes the biweight midvariance using the data stored in the R variable x. For the data in Table 3.2, the function bivar returns the value 25,512.

3.12.3 The Percentage Bend Midvariance and tau Measure of Variation

Two other measures of scale should be mentioned. The first replaces Ψ, the biweight in Eq. (3.38), with Huber's Ψ. Following Shoemaker and Hettmansperger (1982), the particular form of Huber's Ψ used here is

$$\Psi(x) = \max[-1, \min(1, x)]. \tag{3.40}$$

Also, rather than using $\tau = \omega$ when defining Y in Eq. (3.38), a different measure is used instead. Again, let θ be the population median. For any β, $0 < \beta < 0.5$, define ω_β to be the measure of scale determined by

$$P(|X - \theta| < \omega_\beta) = 1 - \beta.$$

Thus, ω_β is the $1 - \beta$ quantile of the distribution of $|X - \theta|$. If X has a standard normal distribution, then ω_β is the $1 - \beta/2$ quantile. For example, if $\beta = 0.1$, $\omega_{0.1} = 1.645$, the 0.95 quantile. Note that when $\beta = 0.5$, ω_β is just the measure of scale ω already discussed and estimated by MAD.

The parameter ω could be rescaled so that it estimates the population standard deviation, σ, when sampling from a normal distribution. That is,

$$\omega_{N,\beta} = \frac{\omega_\beta}{z_{1-\frac{\beta}{2}}},$$

could be used, where $z_{1-\beta/2}$ is the $1 - \beta/2$ quantile of the standard normal distribution. When $\beta = 0.5$, $\omega_{N,\beta}$ is just ω_N, which is estimated by MADN. Shoemaker and Hettmansperger (1982) do not rescale ω, and this convention is followed here. Shoemaker and Hettmansperger choose $\beta = 0.1$, but here $\beta = 0.2$ is used unless stated otherwise. The resulting measure of scale given by Eq. (3.38), with $K = 1$, is called the *percentage bend midvariance* and labeled ζ_{pb}^2. When $\beta = 0.1$, and sampling is from a standard normal distribution, $\zeta_{pb}^2 = 1.03$, while for $\beta = 0.2$, $\zeta_{pb}^2 = 1.05$. A method of estimating ζ_{pb}^2 is shown in Table 3.8.

There are two reasons for including the percentage bend midvariance in this book. First, Bickel and Lehmann (1976) argue that if both X and Y have symmetric distributions about zero and $|X|$ is stochastically larger than $|Y|$, then it should be the case that a measure of scale should be larger for X than it is for Y. That is, if ζ_x is some proposed measure of scale for the random variable X, it should be the case that $\zeta_x > \zeta_y$. Bickel and Lehmann define a

Table 3.8: How to compute the percentage bend midvariance, $\hat{\zeta}_{pb}^2$.

Set $m = [(1 - \beta)n + 0.5]$, the value of $(1 - \beta)n + 0.5$ rounded down to the nearest integer. For good efficiency, under normality, versus the usual sample variance, $\beta = 0.1$ is a good choice, in which case, $m = [0.9n + 0.5]$. For example, if $n = 56$, $m = [0.9 \times 56 + 0.5] = [50.9] = 50$. Let $W_i = |X_i - M|$, $i = 1, \ldots, n$, and let $W_{(1)} \leq \cdots \leq W_{(n)}$ be the W_i values written in ascending order. But a concern with $\beta = 0.1$ is that the breakdown point is a bit low, in which case, something like $\beta = 0.2$ might be preferable. The estimate of ω_β is

$$\hat{\omega}_\beta = W_{(m)},$$

the mth largest of the W_i values. Put another way, $W_{(m)}$ is the estimate of the $1 - \beta$ quantile of the distribution of W.
Next, set

$$Y_i = \frac{X_i - M}{\hat{\omega}_\beta},$$

$$a_i = \begin{cases} 1, & \text{if } |Y_i| < 1, \\ 0, & \text{if } |Y_i| \geq 1, \end{cases}$$

in which case, the estimated percentage bend midvariance is

$$\hat{\zeta}_{pb}^2 = \frac{n\hat{\omega}_\beta^2 \sum \{\Psi(Y_i)\}^2}{(\sum a_i)^2}, \qquad (3.41)$$

where

$$\Psi(x) = \max[-1, \min(1, x)].$$

measure of scale that satisfies this property to be a *measure of dispersion*. The biweight midvariance is not a measure of dispersion (Shoemaker and Hettmansperger, 1982). In contrast, if Huber's Ψ is used in Eq. (3.38) with $K = 1$, the resulting measure of scale, the percentage bend midvariance, is a measure of dispersion. However, if Huber's Ψ is used with $K > 1$, the resulting measure of scale is not a measure of dispersion (Shoemaker and Hettmansperger, 1982). (The Winsorized variance is also a measure of dispersion.) The second reason is that a slight modification of the percentage bend midvariance yields a useful measure of association (a robust analog of Pearson's correlation coefficient) when testing for independence.

The inclusion of the biweight and percentage bend midvariance is motivated by results in Lax (1985). Note that Lax refers to these measures of variation as *A-estimators*, but here, following Shoemaker and Hettmansperger (1984), the terms biweight and percentage bend midvariance are used. More recently, Randal (2008) compared these measures of scale to more recently proposed estimators and again concluded that the biweight and percentage bend midvariances perform relatively well. Two measures of scale not included in the study by Randal are Rocke's (1996) TBS estimator, which is introduced in Section 6.3.3 in the more general setting of multivariate data, and the tau measure of scale described in Yohai and Zamar (1988). Checks on the efficiency of these estimators indicate that under normality, the percentage bend midvariance and the tau measure of variation perform relatively well. For the one-wild distribution and the contaminated normal distribution in Section 1.1, the biweight and percentage bend midvariances are best. But for a sufficiently heavy-tailed distribution (the slash distribution), the tau measure of scale offers an advantage.

Finally, the other measure of variation that should be mentioned is the *tau measure of variation* given by

$$\zeta_\tau^2 = \frac{\mathrm{MAD}^2}{n} \sum \rho_c \left(\frac{X_i - \mu_{tau}}{MAD} \right),$$

where $\rho_c = \min(x^2, c^2)$ and μ_{tau} is the tau measure of location introduced in Section 3.8.1. Following Maronna and Zamar, $c = 3$ is used unless stated otherwise.

3.12.4 R Functions pbvar and tauvar

The R function

$$\mathrm{pbvar(x,beta=0.2)}$$

computes the percentage bend midvariance for the data stored in the vector x. The default value for the argument beta, which is β in Table 3.8, is 0.2. An argument for using beta=0.1 is that the resulting estimator is about 85% as efficient as the sample variance under normality. With beta=0.2, it is only about 67% as efficient, but a concern about beta=0.1 is that the breakdown point is only 0.1. For the data in Table 3.2, the function returns the value 54,422 with beta=0.1, while with beta=0.2, the estimate is 30,681 and with beta=0.5, the estimate is 35,568. In contrast, the biweight midvariance is estimated to be 25,512. In terms of resistance, beta=0.5 is preferable to beta=0.1 or 0.2, but for other goals discussed in subsequent chapters, beta=0.1 or 0.2 might be preferred for general use. The R function

$$\mathrm{tauvar(x,cval=3)}$$

computes the tau measure of variation.

3.12.5 The Interquartile Range

The population interquartile range is the difference between the 0.75 and 0.25 quantiles, $x_{0.75} - x_{0.25}$; it plays a role when dealing with a variety of problems to be described. As previously noted, many quantile estimators have been proposed, so there are many ways in which the interquartile range might be estimated. A simple quantile estimator, $\hat{x}_q$, was described in Section 3.3; this leads to a simple estimate of the interquartile range, but for various purposes, alternative estimates of the interquartile range have been found to be useful. In particular, when checking data for outliers, results in Frigge et al. (1989) suggest using what are called the *ideal fourths* (cf. Carling, 2000; Cleveland, 1985; Hoaglin and Iglewicz, 1987; Hyndman and Fan, 1996).

The computations are as follows. Let $j = [(n/4) + (5/12)]$. That is, j is $(n/4) + (5/12)$ rounded down to the nearest integer. Let

$$h = \frac{n}{4} + \frac{5}{12} - j.$$

Then the estimate of the lower quartile (the 0.25 quantile) is given by

$$q_1 = (1 - h)X_{(j)} + hX_{(j+1)}. \tag{3.42}$$

Letting $k = n - j + 1$, the estimate of the upper quartile, is

$$q_2 = (1 - h)X_{(k)} + hX_{(k-1)}. \tag{3.43}$$

So the estimate of the interquartile range is

$$\text{IQR} = q_2 - q_1.$$

3.12.6 R Functions idealf, idrange, idealfIQR, and quantile

The R function

$$\text{idealf(x)}$$

computes the ideal fourths for the data stored in the R variable x. The R function

$$\text{idrange(x,na.rm=FALSE)}$$

and the R function

$$idealfIQR(x)$$

compute the interquartile range based on the idea fourths. By default, idrange does not remove missing values. By setting na.rm=TRUE when using idrange, missing values are removed. The R function idealfIQR removes missing values automatically.

It is noted that Hyndman and Fan (1996) compared nine quantile estimators, each based on the weighted average of two order statistics. These estimators can be computed with the R function

$$quantile(x,type=7),$$

which uses their seventh estimator. They suggest using their eighth estimator, which can be applied by setting the argument type=8.

3.13 Some Outlier Detection Methods

This section summarizes some outlier detection methods, two of which are variations of so-called boxplot techniques. One of these methods has, in essence, already been described, but it is convenient to include it here along with a description of relevant software. For a method aimed at detecting outliers that influence non-robust statistical methods, see Rosenbusch et al. (2020).

3.13.1 Rules Based on Means and Variances

We begin with a method for detecting outliers that is known to be unsatisfactory, but it is a natural strategy to consider, so its limitations should be made explicit. The rule is to declare X_i an outlier if

$$\frac{|X_i - \bar{X}|}{s} > K,$$

where K is some constant. Basic properties of normal distributions suggest appropriate choices for K. For illustrative purposes, consider $K = 2.24$. (For a modification that is based in part on fitting a g-and-h distribution to the data, see Xu et al. (2014).) A concern about this rule, and indeed any rule based on the sample mean and standard deviation, is that if suffers from *masking*, meaning that the very presence of outliers masks their detection. Outliers affect the sample means, but in a certain sense, they have a bigger impact on the standard deviation.

■ **Example**

Consider the values 2, 2, 3, 3, 3, 4, 4, 4, 100,000, 100,000. Obviously, the value 100,000 is an outlier, and surely any reasonable outlier detection method would flag the value 100,000 as being unusual. But the method just described fails to do so.

■

3.13.2 A Method Based on the Interquartile Range

The standard boxplot approach to detecting outliers is based on the interquartile range. As previously noted, numerous quantile estimators have been proposed, and when checking for outliers, a good method for estimating the quartiles appears to be the ideal fourths, q_1 and q_2, described in Section 3.12. Then a commonly used rule is to declare X_i an outlier if

$$X_i < q_1 - k(q_2 - q_1) \text{ or } X_i > q_2 + k(q_2 - q_1), \tag{3.44}$$

where $k = 1.5$ is used unless stated otherwise.

3.13.3 Carling's Modification

One useful way of characterizing an outlier detection method is with its *outside rate per observation*, p_n, which refers to the expected proportion of observations declared outliers. So, if m represents the number of points declared outliers based on a sample of size n, $p_n = E(m/n)$. A common goal is to have p_n reasonably small, say approximately equal to 0.05, when sampling from a normal distribution. A criticism of the boxplot rule given by Eq. (3.44) is that p_n is somewhat unstable as a function of n; p_n tends to be higher when sample sizes are small. To address this, Carling (2000) suggests declaring X_i an outlier if

$$X_i < M - k(q_2 - q_1) \text{ or } X_i > M + k(q_2 - q_1), \tag{3.45}$$

where M is the usual sample median, q_1 and q_2 are given by Eqs. (3.42) and (3.43), respectively, and

$$k = \frac{17.63n - 23.64}{7.74n - 3.71} \tag{3.46}$$

(cf. Schwertman and de Silva, 2007).

3.13.4 A MAD-Median Rule

Henceforth, the MAD-median rule for detecting outliers will refer to declaring X_i an outlier if

$$\frac{|X_i - M|}{\text{MAD}/0.6745} > K,$$

where K is taken to be $\sqrt{\chi^2_{0.975,1}}$, the square root of the 0.975 quantile of a chi-squared distribution with one degree of freedom, which is approximately 2.24 (cf. Davies and Gather, 1993). This rule is a special case of a multivariate outlier detection method proposed by Rousseeuw and van Zomeren (1990).

Detecting outliers based on MAD and the median has the appeal of being able to handle a large number of outliers because both MAD and the median have the highest possible breakdown, 0.5. We will see, however, that for certain purposes, the choice between a boxplot rule and a MAD-median rule is not always straightforward.

3.13.5 R Functions outbox, out, and boxplot

The R function

$$\text{outbox(x,mbox=FALSE,gval=NA)}$$

checks for outliers using one of two boxplot methods just described. As usual, the argument x is any vector containing data. Using mbox=FALSE results in using the method in Section 3.13.2. Setting mbox=TRUE results in using Carling's modification in Section 3.13.3. The argument gval can be used to alter the constant k. If unspecified, $k = 1.5$ when using the method in Section 3.13.2, and k is given by Eq. (3.46) when using the method in Section 3.13.3.

The R function

$$\text{out(x)}$$

checks for outliers using the MAD-median rule in Section 3.13.4. (This function contains additional arguments that are related to detecting outliers among multivariate data, but the details are postponed for now.)

The built-in R function

$$\text{boxplot(x)}$$

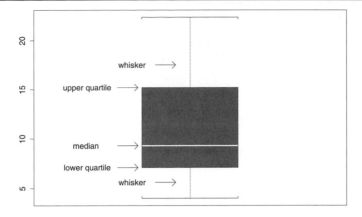

Figure 3.6: An example of a boxplot when there are no outliers.

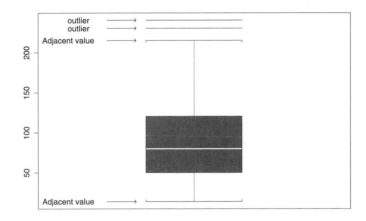

Figure 3.7: An example of a boxplot when there are outliers.

creates the usual graphical version of the boxplot, examples of which are shown in Figs. 3.6 and 3.7, but this function does not use the ideal fourths. Several variations of this method for plotting data have been proposed that were recently summarized by Marmolejo-Ramos and Tian (2010).

It should be noted that the boxplot rule for detecting outliers has been criticized on the grounds that it might declare too many points outliers when there is skewness. More precisely, if a distribution is skewed to the right, among the larger values that are observed, too many might be declared outliers. Hubert and Vandervieren (2008) review the literature and suggest a modification of the boxplot rule that is based in part on a robust measure of skewness, called the *medcouple*, which was introduced by Brys et al. (2004) and is given by

$$MC = \text{med}(h(X_i, X_j)), \ X_i \leq M \leq X_j,$$

where for all $X_i \neq X_j$,

$$h(X_i, X_j) = \frac{(X_j - M) - (M - X_i)}{X_j - X_i}.$$

If $MC > 0$, declare X_i values outside the interval

$$[q_1 - 1.5e^{-4MC}(q_2 - q_1), \, q_2 + 1.5e^{3MC}(q_2 - q_1)]$$

as potential outliers. If $MC < 0$, declare X_i values outside the interval

$$[q_1 - 1.5e^{-3MC}(q_2 - q_1), \, q_2 + 1.5e^{4MC}(q_2 - q_1)]$$

as potential outliers.

There is, however, a feature of this adjusted boxplot rule that should be mentioned. Imagine a distribution that is skewed to the right. Among the larger values, the adjusted boxplot rule might declare fewer points outliers, as intended, but among the lower values, it might declare points outliers that are not flagged as outliers by the boxplot rule. Bruffaerts et al. (2014) raise other concerns about this outlier detection method and suggest a more involved approach. The method suggested by Walker et al. (2018) suffers from similar problems.

3.13.6 R Functions adjboxout and adjbox

The R function

$$\text{adjboxout(x)}$$

applies the adjusted boxplot rule just described. The function returns the values flagged as outliers. It also returns values stored in \$cl and \$cu, which are the lower and upper ends of the interval used to determine whether a value is an outlier. To create a boxplot, the R function

$$\text{adjbox(x)}$$

can be used, which is stored in the R library robustbase.

■ **Example**

Consider the values

12, 33, 47, 55, 85, 87, 87, 96, 97, 99, 113, 118, 128, 138, 165, 202, 213, 218, 275, 653.

Both the boxplot rule and the adjusted boxplot rule declare the value 653 to be an outlier, which certainly seems reasonable based on a casual inspection of the data. But unlike the boxplot rule, the adjusted boxplot rule declares the values 12, 33, and 47 outliers as well.

■

3.14 Exercises

1. Included among the R functions written for this book is the function ghdist(n,g=0,h=0). It generates n observations from a so-called g-and-h distribution, which is described in more detail in Chapter 4. The command ghdist(30,0,0.5) will generate 30 observations from a symmetric, heavy-tailed distribution. Generate 30 observations in this manner and create the density estimates using the functions skerd, kdplot, rdplot, and akerd. Repeat this 20 times, and comment on the pattern of results.

2. In the study by Dana (1990) on self-awareness, described in this chapter (in connection with Table 3.2), a second group of subjects yielded the observations

   59 106 174 207 219 237 313 365 458 497 515 529 557 615 625 645 973 1,065 3,215.

   Compute the sample median, the Harrell–Davis estimate of the median, the M-estimate of location (based on Huber's Ψ), and the 10% and 20% trimmed means. Estimate the standard errors for each location estimator and compare the results.

3. For the data in Exercise 1, compute MADN, the biweight midvariance, and the percentage bend midvariance. Compare the results to those obtained for the data in Table 3.2. What do the results suggest about which group is more dispersed?

4. For the data in Exercise 1, estimate the deciles using the Harrell–Davis estimator. Do the same for the data in Table 3.2. Plot the difference between the deciles as a function of the estimated deciles for the data in Exercise 1. What do the results suggest? Estimate the standard errors associated with each decile estimator.

5. Comment on the strategy of applying the boxplot to the data in Exercise 2, removing any outliers, computing the sample mean for the data that remain, and then estimating the standard error of this sample mean based on the sample variance of the data that remain.

6. Cushny and Peebles (1904) conducted a study on the effects of optical isomers of hyoscyamine hydrobromide in producing sleep. For one of the drugs, the additional hours of sleep for 10 patients were

   $$0.7, -1.6, -0.2, -1.2, -0.1, 3.4, 3.7, 0.8, 0, \text{ and } 2.$$

   Compute the Harrell–Davis estimate of the median, the mean, the 10% and 20% trimmed means, and the M-estimate of location. Compute the corresponding standard errors.

7. Set $X_i = i$, $i = 1, \ldots, 20$, and compute the 20% trimmed mean and the M-estimate of location based on Huber's Ψ. Next set $X_{20} = 200$, and compute both estimates of location. Replace X_{19} with 200, and again, estimate the measures of location. Keep doing this until the upper half of the data is equal to 200. Comment on the resistance of the M-estimator versus 20% trimming.

8. Repeat the previous exercise, only this time, compute the biweight midvariance, the 20% Winsorized variance, and the percentage bend midvariance. Comment on the resistance of these three measures of scale.

9. Set $X_i = i$, $i = 1, \ldots, 20$, and compute the Harrell–Davis estimate of the median. Repeat this, but with X_{20} equal to 1,000 and then 100,000. When $X_{20} = 100,000$, would you expect $\hat{x}_{0.5}$ or the Harrell–Davis estimator to have the smaller standard error? Verify your answer.

10. Argue that if Ψ is taken to be the biweight, it approximates the optimal choice for Ψ under normality when observations are not too extreme.

11. Verify that Eq. (3.29) reduces to s^2/n if no observations are flagged as being unusually large or small by Ψ.

12. Using R, generate $n = 20$ observations from a standard normal distribution, compute the mean, 20% trimmed mean, median, and one-step M-estimate, and repeat this 10,000 times. Compare the results using a boxplot.

13. Using R, generate $n = 20$ observations from the mixed normal distribution in Fig. 1.1. This can be done with the R function cnorm stored in Rallfun. Compute the mean, 20% trimmed mean, median, and one-step M-estimate, and repeat this 10,000 times. Compare the results using a boxplot.

## CHAPTER 4

Inferences in the One-Sample Case

A fundamental problem is testing hypotheses and computing confidence intervals for the measures of location described in Chapters 2 and 3. As will be seen, a method that provides accurate probability coverage for one measure of location can perform poorly with another. That is, the recommended method for computing a confidence interval depends in part on which measure of location is of interest. An appeal of the methods in this chapter is that when computing confidence intervals for robust measures of location, it is possible to get reasonably accurate probability coverage in situations where no known method for the mean gives good results.

4.1 Problems When Working With Means

It helps to first describe problems associated with Student's t. When testing hypotheses or computing confidence intervals for μ, it is assumed that

$$T = \frac{\sqrt{n}(\bar{X} - \mu)}{s} \tag{4.1}$$

has a Student's t distribution with $\nu = n - 1$ degrees of freedom. This implies that $E(T) = 0$, and that T has a symmetric distribution. From basic principles, this assumption is correct when observations are randomly sampled from a normal distribution. However, at least three practical problems can arise. First, there are problems with power and the length of the confidence interval. As indicated in Chapters 1 and 2, the standard error of the sample mean, $\sigma/\sqrt{n}$, becomes inflated when sampling from a heavy-tailed distribution, so power can be poor relative to methods based on other measures of location, and the length of confidence intervals, based on Eq. (4.1), become relatively long—even when σ is known. (For a detailed analysis of how heavy-tailed distributions affect the probability coverage of the t test, see Benjamini, 1983.) Second, the actual probability of a Type I error can be substantially higher or lower than the nominal α level. When sampling from a symmetric distribution, generally the actual level of Student's t test will be less than the nominal level. When sampling from a symmetric, heavy-tailed distribution the actual probability of Type I error can be substantially lower than the nominal α level, and this further contributes to low power and relatively long confidence intervals. From theoretical results reported by Basu and DasGupta (1995), problems with low power can arise even when n is large. When sampling from a skewed distribution with relatively light tails, the actual probability coverage can be substantially less

Introduction to Robust Estimation and Hypothesis Testing
https://doi.org/10.1016/B978-0-12-820098-8.00010-5
Copyright © 2022 Elsevier Inc. All rights reserved.
107

than the nominal $1 - \alpha$ level resulting in inaccurate conclusions and this problem becomes exacerbated as we move toward (skewed) heavy-tailed distributions. Third, when sampling from a skewed distribution, T also has a skewed distribution, it is no longer true that $E(T) = 0$, and the distribution of T can deviate enough from a Student's t distribution so that practical problems arise. These problems can be ignored if the sample size is sufficiently large, but given data it is difficult knowing just how large n has to be.

When sampling from a lognormal distribution, Westfall and Young (1993) indicate that $n > 160$ is required. However, using a more accurate simulation, if a confidence interval is considered to be reasonably accurate when the actual probability coverage is between 0.925 and 0.975, given the goal of computing a 0.95 confidence interval, $n > 130$ is required. As we move away from the lognormal distribution toward skewed distributions where outliers are more common, $n > 300$ might be required. Problems with controlling the probability of a Type I error are particularly serious when testing one-sided hypotheses. And this has practical implications when testing two-sided hypotheses because it means that a biased hypothesis testing method is being used, as will be illustrated. For more detailed illustrations, see Wilcox and Rousselet (2018).

Problems with low power were illustrated in Chapter 1, so further comments are omitted. The second problem, that probability coverage and Type I error probabilities are affected by departures from normality, is illustrated with a class of distributions obtained by transforming a standard normal distribution in a particular way. Of course, the seriousness of a Type I error depends on the situation. Presumably there are instances where an investigator does not want the probability of a Type I error to exceed 0.1; otherwise the common choice of $\alpha = 0.05$ would be replaced by $\alpha = 0.1$ in order to increase power. Bradley (1978) argues that if a researcher makes a distinction between $\alpha = 0.05$ and $\alpha = 0.1$, the actual probability of a Type I error should not exceed 0.075; the idea being that otherwise it is closer to 0.1 than 0.05, and he argues that it should not drop below 0.025.

Suppose Z has a standard normal distribution, and for some constant $h \geq 0$, let

$$X = Z\exp\left(\frac{hZ^2}{2}\right).$$

Then X has what is called an *h distribution*. When $h = 0$, $X = Z$, so X is standard normal. As h gets large, the tails of the distribution of X get heavier, and the distribution is symmetric about 0. (More details about the h distribution are described in Section 4.2.) Further suppose sampling is from an h distribution with $h = 1$, which has very heavy tails. Then with $n = 20$ and $\alpha = 0.05$, the actual probability of a Type I error, when using Student's t to test H_0: $\mu = 0$, is approximately 0.018 (based on simulations with 10,000 replications). Increasing n to 100, the actual probability of a Type I error is approximately 0.019. So even now, Bradley's criterion is not satisfied. A reasonable suggestion for dealing with this problem is to inspect the empirical distribution to determine whether the tails are relatively light. There are various ways this might be done, but there is no known empirical rule that reliably indicates whether the Type I error probability will be substantially lower than the nominal level when attention is restricted to using Student's t test.

To illustrate the third problem, and provide another illustration of the second, consider what happens when sampling from a skewed distribution with relatively light tails. In particular, suppose X has a lognormal distribution, meaning that for some normal random variable, Y, $X = \exp(Y)$. This distribution is light-tailed in the sense that the expected proportion of values declared an outlier, using the MAD-median rule used to define the MOM estimator in Section 3.7, is relatively small. (Gleason, 1993, also argues that a lognormal distribution is light-tailed.)

For convenience, assume Y is standard normal, in which case $E(X) = \sqrt{e}$, where $e = \exp(1) \approx 2.71828$, and the standard deviation is approximately $\sigma = 2.16$. Eq. (4.1) assumes that $T = \sqrt{n}(\bar{X} - \sqrt{e})/s$ has a Student's t distribution with $n - 1$ degrees of freedom. The left panel of Fig. 4.1 shows a (kernel density) estimate of the actual distribution of T when $n = 20$; the symmetric distribution is the distribution of T under normality. As is evident, the actual distribution is skewed to the left, and its mean is not equal to 0. Simulations indicate that $E(T) = -0.54$, approximately. The right panel shows an estimate of the probability density function when $n = 100$. The distribution is more symmetric compared to $n = 20$, but it is clearly skewed to the left.

Let μ_0 be some specified constant. The standard approach to testing H_0: $\mu \leq \mu_0$ is to compute T with $\mu = \mu_0$ and reject H_0 if $T > t_{1-\alpha}$, where $t_{1-\alpha}$ is the $1 - \alpha$ quantile of Student's t distribution with $\nu = n - 1$ degrees of freedom and α is the desired probability of a Type I error. If H_0: $\mu \leq \sqrt{e}$ is tested when X has a lognormal distribution, H_0 should not be rejected, and the probability of a Type I error should be as close as possible to the nominal level, α. If $\alpha = 0.05$ and $n = 20$, the actual probability of a Type I error is approximately 0.008 (Westfall and Young, 1993, p. 40). As indicated in Fig. 4.1, the reason is that T has a distribution that is skewed to the left. In particular, the right tail is much lighter than the one assumed by Student's t distribution, and this results in a Type I error probability that is substantially smaller than the nominal 0.05 level. Simultaneously, the left tail is too thick. The 0.05 quantile of Student's t distribution with 19 degrees of freedom is -1.73. Consequently, when testing H_0: $\mu \geq \sqrt{e}$ at the 0.05 level, the actual probability of rejecting is 0.153. Increasing n to 160, the actual probability of a Type I error is 0.022 and 0.109 for the one-sided hypotheses being considered. And when observations are sampled from a heavier-tailed distribution, control over the probability of a Type I error deteriorates.

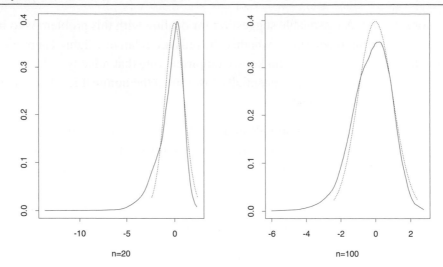

Figure 4.1: Non-normality can seriously affect Student's t. The left panel shows an approximation of the actual distribution of Student's t when sampling from a lognormal distribution and $n = 20$ and the right panel is when $n = 100$.

The results just summarized illustrate an important fact. Even when the sample mean has, to a close approximation, a normal distribution, Student's t test can be unsatisfactory. For example, when sampling from a lognormal distribution, the sample mean has approximately a normal distribution with a sample size of 40. But suppose the goal is to test some two-sided hypothesis so that the Type I error probability is 0.05. To satisfy Bradley's criterion for the situation at hand, a sample size of at least 200 is needed when using Student's t.

Generally, as we move toward a skewed distribution with heavy tails, the problems illustrated by Fig. 4.1 become exacerbated. As an example, suppose sampling is from a squared lognormal distribution that has mean exp(2). (That is, if X has a lognormal distribution, $E(X^2) = \exp(2)$.) Fig. 4.2 shows plots of T values based on sample sizes of 20 and 100. (Again, the symmetric distributions are the distributions of T under normality.)

It is noted that when testing H_0: $\mu < \mu_0$, and when a distribution is skewed to the right, improved control over the probability of a Type I error can be achieved using a method derived by Chen (1995). However, even for this special case, problems with controlling the probability of a Type I error remain in some situations, and power problems plague any method based on means. (A generalization of this method to some robust measure of location might have some practical value, but this has not been established as yet.) Banik and Kibria (2010) compared numerous methods for computing a two-sided confidence interval for the mean. In terms of probability coverage, none of the methods were completely satisfactory when the

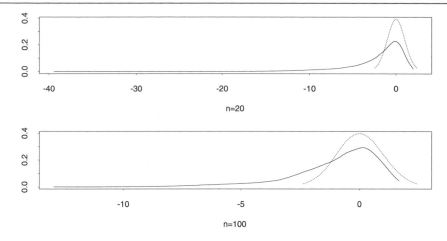

Figure 4.2: The same as Fig. 4.1, only now sampling is from a squared lognormal distribution. This illustrates that as we move toward heavy-tailed distributions, problems with non-normality are exacerbated.

sample size is small. For $n \geq 50$, Chen's method performed reasonably well among the distributions considered, including situations where sampling is from a lognormal distribution. But the lognormal distribution is relatively light-tailed. How well Chen's method performs when sampling from a skewed, heavy-tailed distribution, or even a symmetric, heavy-tailed distribution (such as the mixed normal), appears to be unknown.

4.1.1 P-Values and Testing for Equality: Hypothesis Testing Versus Decision Making

Before continuing, it might help to comment on the common goal of testing for exact equality. Consider, for example, the goal of testing H_0: $\mu_t = \mu_0$, where μ_0 is some specified constant. Tukey (1991) argues that this is nonsensical because surely μ_t differs from μ_0 at some decimal place. Put another way, Tukey is arguing that we know the probability that the null hypothesis is true without any data: zero. Jones and Tukey (2000) suggest dealing with this issue using Tukey's three-decision rule. If the null hypothesis is rejected and $\bar{X}_t < \mu_0$, decide that $\mu_t < \mu_0$. If the null hypothesis is rejected and $\bar{X}_t > \mu_0$, decide that $\mu_t > \mu_0$. If the null hypothesis is not rejected, make no decision. So the goal is not to test for exact equality, but determine the extent to which it is reasonable to make a decision about whether μ_t is greater or less than μ_0. Even if exact equality can be argued to be a reasonable possibility, this perspective on hypothesis testing would seem to be useful.

Put another way, the goal is not to test some hypothesis. Rather, the goal is to determine the extent to which it is reasonable to make a decision about which group has the larger measure of location. Note that from Tukey's perspective, a p-value quantifies the strength of the empirical evidence that a decision can be made. This interpretation of a p-value is consistent with the interpretation developed by R. A. Fisher in the 1920s; see Biau et al. (2010) for more details. But a p-value does not reflect the probability that a correct decision has been made. Also, a p-value close to zero does not provide useful information about rejecting (making a decision) again if the study were replicated; this is a power issue. Moreover, a low p-value does not mean that there is an important or a large difference between the population parameter of interest and its hypothesized value. Similarly, a large p-value is not compelling evidence that there is no important difference. To briefly summarize, a p-value can be argued to provide useful information in the context of Tukey's three-decision rule, but it leaves considerable uncertainty about the true nature of the population under investigation. Wasserstein et al. (2019) provide a more detailed discussion regarding the use of the p-value and Kmetz (2019) provides more details regarding how p-values are misinterpreted.

4.2 The g-and-h Distribution

One of the main goals in this chapter is to recommend certain procedures for computing confidence intervals and testing hypotheses, and to discourage the use of others. These recommendations are based in part on simulations, some of which generate observations from a so-called g-and-h distribution. This section is included for readers interested in the motivation and details of such studies. Readers primarily concerned with how methods are applied, or which methods are recommended, can skip or skim this section.

A basic problem is establishing whether a particular method for computing a confidence interval has probability coverage reasonably close to the nominal $1 - \alpha$ level when the sample size is small or even moderately large. When investigating the effect of non-normality, there is the issue of deciding which non-normal distributions to consider when checking the properties of a particular procedure via simulations. One approach, which provides a partial check on how a method performs, is to consider four types of distributions: normal, symmetric with a heavy tail, asymmetric with a light tail, and asymmetric with a heavy tail. But how heavy-tailed and asymmetric should they be? A natural approach is to use distributions that are similar to those found in applied settings. But coming to terms with what constitutes a reasonable range of values is difficult at best. Several papers have been published with the goal of characterizing the range of heavy-tailedness and skewness that a researcher is likely to encounter (e.g., Pearson and Please, 1975; Sawilowsky and Blair, 1992; Micceri, 1989; Hill and Dixon, 1982; Wilcox, 1990a). The most striking feature of these studies is the extent to which they differ. For example, some papers suggest that distributions are never extremely skewed, while others indicate the exact opposite. In a sexual attitude study by Pedersen et al. (2002), the skewness and kurtosis, based on 105 participants, is 15.9 and 256.3, respectively. In a related study based on 16,288 participants, the 10 variables had estimated skewness that ranged between

Table 4.1: Some properties of the g-and-h distribution.

g	h	κ_1	κ_2	$\hat{\kappa}_1$	$\hat{\kappa}_2$	μ	$\mu_t(20\%)$	μ_m
0.0	0.0	0.00	3.00	0.0	3.00	0.0000	0.0000	0.0000
0.0	0.5	0.00	—	0.0	11,986.2	0.0000	0.0000	0.0000
0.5	0.0	1.75	8.9	1.81	9.7	0.2653	0.0541	0.1047
0.5	0.5	—	—	120.10	18,393.6	0.8033	0.0600	0.0938

52.1 and 115.5, and kurtosis that ranged between 3,290 and 13,357. In a review of 440 large-sample psychological studies, Micceri (1989) reported that 97% (35 of 36 studies) "of those distributions exhibiting kurtosis beyond the double exponential (3.00) also showed extreme or exponential skewness." Moreover, 72% (36 of 50) distributions that exhibited skewness greater than 2 also had tail weights that were heavier than the double exponential.

One way of attempting to span the range of skewness and heavy-tailedness that one might encounter is to run simulations where observations are generated from a g-and-h distribution. An observation X is generated from a g-and-h distribution by first generating Z from a standard normal distribution and then setting

$$X = \frac{\exp(gZ) - 1}{g}\exp(hZ^2/2),$$

where g and h are non-negative constants that can be chosen so that the distribution of X has some characteristic of interest. When $g = 0$ this last equation is taken to be

$$X = Z\exp(hZ^2/2).$$

When $g = h = 0$, $X = Z$, so X has a standard normal distribution. When $g = 0$, X has a symmetric distribution. As h increases, the tails of the distribution get heavier. As g increases, the distribution becomes more skewed. The case $g = 1$ and $h = 0$ corresponds to a lognormal distribution that has been shifted to have a median of zero. Note that within the class of g-and-h distributions, the lognormal is skewed with a relatively light tail. Hoaglin (1985) provides a detailed description of various properties of the g-and-h distribution, but only a few properties are listed here. (For results on estimating g and h based on a random sample, see Xu and Genton, 2015.) Table 4.1 summarizes the skewness and kurtosis values for four selected situations that have been used in published studies and are considered at various points in this book. In Table 4.1, skewness and kurtosis are measured with $\kappa_1 = \mu_{[3]}/\mu_{[2]}^{1.5}$ and $\kappa_2 = \mu_{[4]}/\mu_{[2]}^2$, where $\mu_{[k]} = E(X - \mu)^k$. When $g > 0$ and $h \geq 1/k$, $\mu_{[k]}$ is not defined and the corresponding entry is left blank.

A possible criticism of simulations based on the g-and-h distribution is that observations generated on a computer have skewness and kurtosis that are not always the same as the theoretical values listed in Table 4.1. The reason is that observations generated on a computer come from some bounded interval on the real line, so $\mu_{[k]}$ is finite even when in theory it is not. For this reason, Table 4.1 also reports $\hat{\kappa}_1$ and $\hat{\kappa}_2$, the estimated skewness and kurtosis based on 100,000 observations. (Skewness is not estimated when $g = 0$ because it is known that $\kappa_1 = 0$.) The last three columns of Table 4.1 show the value of μ, the 20% trimmed mean, μ_t, and μ_m, the M-measure of location, which were determined via numerical quadrature. For completeness, it is noted that for the lognormal distribution, $\kappa_1 = 6.2$, $\kappa_2 = 114$, the 20% trimmed mean is $\mu_t = 1.111$, and $\mu_m = 1.1857$. (For additional properties of the g-and-h distribution, see Headrick et al., 2008, as well as the documentation for the R package gk available at https://arxiv.org/pdf/1706.06889.pdf.)

Ideally, a method for computing a confidence interval will have accurate probability coverage when sampling from any of the four g-and-h distributions in Table 4.1. It might be argued that when g or h equals 0.5, the corresponding distribution is unrealistically non-normal. The point is that if a method performs well under seemingly large departures from normality, this offers some reassurance that it will perform well for distributions encountered in practice. Of course, even if a method gives accurate results for the four distributions in Table 4.1, this does not guarantee accurate probability coverage for any distribution that might arise in practice. In most cases, there is no known method for proving that a particular technique always gives good results.

Another possible criticism of the four g-and-h distributions in Table 4.1 is that perhaps the skewed, light-tailed distribution ($g = 0.5$ and $h = 0$) does not represent a large enough departure from normality. In particular, Wilcox (1990a) found that many random variables he surveyed had estimated skewness greater than 3, but the skewness of this particular g-and-h distribution is only 1.8, approximately. For this reason, it might also be important to consider the lognormal distribution when studying the small-sample properties of a particular method.

Table 4.2 shows the estimated probability of a Type I error (based on simulations with 10,000 replications) when using Student's t to test $H_0: \mu = 0$ with $n = 12$ and $\alpha = 0.05$. (The notation $t_{0.025}$ refers to the 0.025 quantile of Student's t distribution.) For example, when sampling from a g-and-h distribution with $g = 0.5$ and $h = 0$, the estimated probability of a Type I error is $0.000 + 0.420 = 0.420$, which is about eight times as large as the nominal level. Put another way, if $H_0: \mu < 0$ is tested with $\alpha = 0.025$, the actual probability of rejecting when $\mu = 0$ is approximately 0.42, over 16 times larger than the nominal level. Note that for fixed g, as the tails get heavier (h increases from 0 to 0.5), the probability of a Type I error decreases. This is not surprising because sampling from a heavy-tailed distribution inflates the standard deviation s, which in turn results in longer confidence intervals. A similar result, but to a lesser extent, is found when using robust measures of location.

Table 4.2:   One-sided Type I error
probabilities when using Student's t, $n = 12$,
$\alpha = 0.025$.

g	h	$P(T > t_{0.975})$	$P(T < t_{0.025})$
0.0	0.0	0.025	0.025
0.0	0.5	0.015	0.016
0.5	0.0	0.000	0.420
0.5	0.5	0.000	0.295

4.2.1 R Functions ghdist, rmul, rngh, rmul.MAR, ghtrim, and gskew

The R function

$$ghdist(n,g=0,h=0)$$

generates n observations from a g-and-h distribution. By default, the R function

$$rmul(n, p=2,cmat=NULL, rho=0, mar.fun=ghdist, OP=FALSE, g=0, h=0,...)$$

generates n vectors of observations from a p-variate normal distribution having a correlation matrix that defaults to the identity matrix. When OP=FALSE, this is done via the R function mvrnorm. This appears to be a bit more satisfactory than the older version of this function, which corresponds to OP=TRUE. The argument cmat, if specified, is taken to be a covariance matrix used to generate the data. Non-normal marginal distributions are generated via the argument mar.fun, which defaults to the R function ghdist, in which case the arguments g and h control skewness and kurtosis. A possible criticism of the R function rmul is that after it transforms the marginal distributions, the correlation among the resulting variables can differ somewhat from the value indicated by the argument rho (Kowalchuk and Headrick, 2010). The R function

$$rngh(n,rho=0,p=2,g=0,h=0,ADJ=TRUE)$$

deals with this issue. It begins by generating data from a multivariate normal distribution but with the correlations adjusted so that after transforming the marginal distributions to a g-and-h distribution, the resulting variables have a common correlation indicated by the argument rho. To reduce execution time in a simulation study, the following approach can be used. Suppose it is desired to generate data that have a correlation equal to 0.5 and that rngh indicates the adjusted value for rho is 0.6. Henceforth, we use the function rmul with the argument rho=0.6.

The R function

$$\text{rmul.MAR(n, p=2, g=rep(0,p), h=rep(0,p), rho=0, cmat=NULL)}$$

is like the R function rmul, only it can be used to generate data where the marginal distributions have different g-and-h distributions.

The function

$$\text{ghtrim(tr=0.2, g = 0, h = 0)}$$

computes the population trimmed mean of a g-and-h distribution. The R function

$$\text{gskew(g)}$$

computes the skewness of a g-and-h distribution when $h = 0$.

4.3 Inferences About the Trimmed, Winsorized Means

When working with the trimmed mean, μ_t, an analog of Eq. (4.1) is

$$T_t = \frac{(1 - 2\gamma)\sqrt{n}(\bar{X}_t - \mu_t)}{s_w}. \tag{4.2}$$

When $\gamma = 0$, $T_t = T$ given by Eq. (4.1). Tukey and McLaughlin (1963) suggest approximating the distribution of T_t with a Student's t distribution having $n - 2g - 1$ degrees of freedom, where, as in Chapter 3, $g = [\gamma n]$ is the integer portion of γn. Then $n - 2g$ is the number of observations left after trimming. The resulting two-sided $1 - \alpha$ confidence interval for μ_t is

$$\bar{X}_t \pm t_{1-\alpha/2}\frac{s_w}{(1 - 2\gamma)\sqrt{n}}, \tag{4.3}$$

where $t_{1-\alpha/2}$ is the $1 - \alpha/2$ quantile of Student's t distribution with $n - 2g - 1$ degrees of freedom. Let μ_0 be some specified constant. Then under the null hypothesis

$$H_0: \mu_t = \mu_0,$$

T_t becomes

$$T_t = \frac{(1 - 2\gamma)\sqrt{n}(\bar{X}_t - \mu_0)}{s_w}, \tag{4.4}$$

and H_0 is rejected if $|T_t| > t_{1-\alpha/2}$. In the context of Tukey's three-decision rule, rejection is taken to mean that a decision is made about where μ_t is less than or greater than μ_0. One-sided tests can be performed in the usual way. In particular, reject H_0: $\mu_t \leq \mu_0$ if $T_t > t_{1-\alpha}$, the $1 - \alpha$ quantile of Student's t distribution with $n - 2g - 1$ degrees of freedom. Similarly, reject H_0: $\mu_t \geq \mu_0$ if $T_t < t_\alpha$.

Based on various criteria, plus a slight variation of the sample trimmed mean used here, Patel et al. (1988) found the Tukey–McLaughlin approximation to be reasonably accurate when sampling from various distributions. They also report that for $\gamma = 0.25$, a better approximation is a Student's t distribution with $n - 2.48g - 0.15$ degrees of freedom. For $n < 18$, they suggest a more refined approximation, but another method, to be described, gives more satisfactory results.

Additional support for using a Student's t distribution with $n - 2g - 1$ degrees of freedom is reported by Wilcox (1994a). Using a Winsorized analog of a Cornish–Fisher expansion of T_t, a correction term for skewness was derived and compared to the correction term used when there is no trimming. As the amount of trimming increases, the magnitude of the correction term decreases, indicating that the probability coverage using Eq. (4.3) should be closer to the nominal level than the probability coverage when $\gamma = 0$. Numerical results for the lognormal distribution indicate that as γ increases, the magnitude of the correction term decreases rapidly up to about $\gamma = 0.2$.

The left panel of Fig. 4.3 shows the probability density function of T_t with 20% trimming when $n = 20$ and sampling is from a lognormal distribution. The symmetric distribution is the assumed distribution of T_t when testing hypotheses. The actual distribution is skewed to the left, but the tail of the distribution is not as heavy as the tail of the distribution shown in Fig. 4.1. The result is that when testing H_0: $\mu_t > 0$, the probability of a Type I error will be greater than the nominal level, but not as much versus no trimming. For example, if $\alpha = 0.025$, the actual probability of a Type I error is approximately 0.062 with 20% trimming versus 0.132 when using the mean to test H_0: $\mu > \sqrt{e}$. The right panel of Fig. 4.3 shows the distribution of T_t when n is increased to 100. Note that the distribution is reasonably symmetric, as is assumed when using T_t, versus the right panel of Fig. 4.1, which is clearly skewed to the left. This illustrates the general expectation that when using the 20% trimmed mean, probability coverage will improve more rapidly as the sample size increases, versus confidence intervals based on means. If a distribution is both skewed and sufficiently heavy-tailed, problems with controlling the probability of a Type I error can persist unless n is fairly large. That is, as the amount of trimming increases, problems with controlling the probability of a Type I error decrease, but even with 20% trimming, not all practical problems are eliminated using the method in this section. Increasing the amount of trimming beyond 20%, such as using a median, could be used, but at the risk of low power if indeed a distribution is normal or

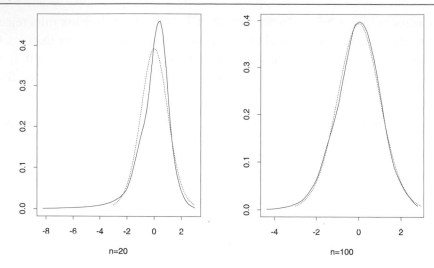

Figure 4.3: The distribution of T_t with 20% trimming when sampling from a lognormal distribution. Compare this to the distribution of t shown in Fig. 4.1.

relatively light-tailed. A better strategy seems to be to use a bootstrap method described in Section 4.4.

It was previously noted that when sampling from a symmetric, heavy-tailed distribution (an h distribution with $h = 1$), the actual probability of a Type I error can be as low as 0.018 when testing H_0: $\mu = 0$ with Student's t test, $n = 20$, and $\alpha = 0.05$. In contrast, with 20% trimming, the probability of a Type I error is approximately 0.033. Generally, if a symmetric distribution is sufficiently heavy-tailed, roughly meaning that the expected proportion of values declared outliers is relatively high, actual Type I error probabilities can drop below the nominal level. In some situations it currently seems that this problem can persist no matter which location estimator is used.

■ Example

Table 4.3 shows the average LSAT scores for the 1973 entering classes of 15 American law schools. (LSAT is a national test for prospective lawyers.) The sample mean is $\bar{X} = 600.3$ with an estimated standard error of 10.8. The 20% trimmed mean is $\bar{X}_t = 596.2$ with an estimated standard error of 14.92, and with $15 - 6 - 1 = 8$ degrees of freedom, the 0.95 confidence interval for μ_t is $(561.8, 630.6)$. In contrast, the 0.95 confidence interval for μ is $(577.1, 623.4)$, assuming T given by Eq. (4.1) does indeed have a Student's t distribution with 14 degrees of freedom. Note that the length of the confidence interval for μ is smaller, and in fact is a subset of the confidence interval

Table 4.3:   Average LSAT scores for 15 law schools.

545	555	558	572	575	576	578	580
594	605	635	651	653	661	666	

Figure 4.4: A boxplot of the LSAT scores.

for μ_t. This might seem to suggest that the sample mean is preferable to the trimmed mean for this particular set of data, but closer examination suggests the opposite conclusion. As already illustrated, if sampling is from a light-tailed, skewed distribution, the actual probability coverage for the sample mean can be smaller than the nominal level. For the situation at hand, the claim that $(577.1, 623.4)$ is a 0.95 confidence interval for the mean might be misleading and overly optimistic. Fig. 4.4 shows a boxplot of the data, which indicates that the central portion of the data is skewed to the right. Moreover, there are no outliers suggesting the possibility that sampling is from a relatively light-tailed distribution. Thus, the actual probability coverage of the confidence interval for the mean might be too low—a longer confidence interval might be needed to achieve 0.95 probability coverage. It is not being suggested, however, that if there had been outliers, there is reason to believe that probability coverage is not too low. For example, boxplots of data generated from a lognormal distribution frequently have values flagged as outliers, and as already noted, sampling from a lognormal distribution can result in a confidence interval for μ that is too short.

As for computing a confidence interval for the population Winsorized mean, results in Dixon and Tukey (1968) suggest using

$$\bar{X}_w \pm t_{1-\alpha/2}\left(\frac{n-1}{n-2g-1}\right)\left(\frac{s_w}{\sqrt{n}}\right),$$

where again the degrees of freedom are $n - 2g - 1$. It appears that the accuracy of this confidence interval, when sampling from a skewed distribution, has not been studied.

4.3.1 Comments on Effect Size and Non-Normal Distributions

A simple way of characterizing the extent μ_t differs from some hypothesized value is with $\mu_t - \mu_0$, which is estimated with $\bar{X}_t - \mu_0$. Another approach is to use

$$d = k \frac{\bar{X}_t - \mu_0}{s_w},$$
(4.5)

where the constant k is chosen so that under normality s_w / k estimates the standard deviation σ. When using a 20% trimmed mean, $k = 0.642$. This is a one-sample version of the robust measure of effect size derived by Algina et al. (2005). For normal distributions, d estimates

$$\delta = \frac{\mu - \mu_0}{\sigma}.$$

For a standard normal distribution, consider $\mu = 0.2$ and 0.5. Then $\delta = 0.2$ and 0.5, respectively. For illustrative purposes assume $\delta = 0.2$ represents a small effect size and $\delta = 0.5$ is a medium effect size. Note that under normality this means that for a small effect size, μ corresponds to the 0.58 quantile of the null distribution. For a medium effect size μ corresponds to the 0.69 quantile. Now focus on a mixed normal and note that when $\mu = 0.5$, $\delta = 0.15$, which would be characterized as being small. However, $\mu = 0.5$ corresponds to the 0.68 quantile, which suggests a medium effect size. That is, a slight departure from normality can have a substantial impact on δ that can mask a relatively large quantile shift in location.

This suggests a non-parametric, quantile shift approach to measuring effect size that plays a role in various situations described in subsequent chapters. Let θ denote any measure of location. Let Z denote a random variable where $\theta = \theta_0$ but otherwise the distribution is unknown. When testing $H_0: \theta = \theta_0$, θ_0 given, a measure of effect size that captures the spirit of δ is $Q = P(Z \leq \theta)$. That is, Q represents the quantile of the null distribution corresponding to θ. If when dealing with a normal distribution, $\delta = 0.1, 0.3$, and 0.5 are considered small, medium, and large effect sizes, this corresponds to $Q = 0.54, 0.62$, and 0.69, respectively. In a similar manner, if $\delta = -0.1, -0.3$, and -0.5 are considered small, medium, and large effect sizes, this corresponds to $Q = 0.46, 0.38$, and 0.031, respectively.

Let $\hat{\theta}$ be some estimate of θ. An estimate of Q when $\theta_0 = 0$ is simply

$$\hat{Q} = \frac{1}{n} \sum I(X_i - \hat{\theta} \leq \hat{\theta}), \tag{4.6}$$

where the indicator function $I(X_i - \hat{\theta} \leq \hat{\theta}) = 1$ if $X_i - \hat{\theta} \leq \hat{\theta}$ and zero otherwise. More generally, the estimate is

$$\hat{Q} = \frac{1}{n} \sum I(X_i - \hat{\theta} + \theta_0 \leq \hat{\theta}). \tag{4.7}$$

4.3.2 R Functions trimci, winci, D.akp.effect.ci, and depQSci

The R function

$$\text{trimci(x,tr=0.2,alpha=0.05, null.value = 0, pr = TRUE)}$$

computes a $1 - \alpha$ confidence interval for μ_t using Eq. (4.3) based on the data stored in the R vector x, where x is any R variable containing data, tr is the desired amount of trimming (the value of γ), and alpha is α. The default amount of trimming is $\gamma = 0.2$ (20%), and the default value for α is 0.05. For example, the command trimci(w,0.1,0.025) returns two values: the lower and upper ends of the 0.975 confidence interval for the 10% trimmed mean using the data stored in w. The command trimci(w) returns a 0.95 confidence interval for the 20% trimmed mean. The function also returns a p-value when testing H_0: $\mu_t = \mu_0$, where the hypothesized value μ_0 is indicated by the argument null.value. Setting the argument pr=FALSE suppresses the message printed by the function.

The R function

$$\text{trimciv2(x,tr=0.2,alpha=0.05,null.value=0,pr=TRUE)}$$

is exactly the same as trimci, only the measure of effect size d given by Eq. (4.5) is reported as well. The R function

$$\text{D.akp.effect.ci(x,null.value=0, tr=0.2, alpha=.05, nboot=500, SEED=TRUE)}$$

also computes the effect size d given by Eq. (4.5) as well as a confidence interval.

The R function

$$\text{winci(x,tr=0.2,alpha=0.05)}$$

computes a confidence interval for the population Winsorized mean.

The R function

depQSci(x,y=NULL, null.value=0, locfun=median, alpha=.05, nboot=500, SEED=TRUE,...)

computes $\hat{Q}$ given by Eq. (4.7). By default, the median is used as described in the previous section, but an alternative measure of location can be used via the argument locfun. A confidence interval is computed as well.

Also see the R function dep.ES.summary.CI in Section 5.9.20.

4.4 Basic Bootstrap Methods

The method used to compute a confidence interval for a trimmed mean, described in Section 4.3, is based on the fundamental strategy developed by Laplace about two centuries ago: When using $\hat{\theta}$ to estimate some parameter of interest, θ, estimate the standard error of $\hat{\theta}$ with say $\hat{\Upsilon}$, and try to approximate the distribution of

$$\frac{\hat{\theta} - \theta}{\hat{\Upsilon}}.$$

Laplace accomplished this by appealing to his central limit theorem, which he publicly announced in 1810. That is, assume this last equation has a standard normal distribution.

An alternative approach is to use some type of bootstrap method. There are many variations; see Efron and Tibshirani (1993), Chernick (1999), Davison and Hinkley (1997), Hall and Hall (1995), Lunneborg (2000), Mooney and Duval (1993), and Shao and Tu (1995). Here attention is focused on two basic types (with some extensions described in subsequent chapters). Alternative methods are not considered because either they currently seem to have no practical advantage for the problems considered here, in terms of controlling the probability of a Type I error or yielding accurate probability coverage, or the practical advantages of these alternative methods have not been adequately investigated when sample sizes are small or moderately large.

4.4.1 The Percentile Bootstrap Method

The first basic version is the so-called percentile bootstrap. It begins by obtaining a *bootstrap sample* of size n. That is, values are obtained by randomly sampling with replacement n values from $X_1, \ldots, X_n$, yielding $X_1^*, \ldots, X_n^*$.

Let $\hat{\theta}^*$ be an estimate of θ based on this bootstrap sample. Of course, a new bootstrap sample can be generated to yield a new bootstrap estimate of θ. Repeating this process B times yields B bootstrap estimates: $\hat{\theta}_1^*, \ldots, \hat{\theta}_B^*$. Let $\ell = \alpha B/2$, rounded to the nearest integer, and let $u = B - \ell$. Letting $\hat{\theta}_{(1)}^* \leq \cdots \leq \hat{\theta}_{(B)}^*$ represent the B bootstrap estimates written in ascending order, an approximate $1 - \alpha$ confidence interval for θ is

$$(\hat{\theta}_{(\ell+1)}^*, \hat{\theta}_{(u)}^*).$$

An outline of the theoretical justification of the method is as follows. Imagine that the goal is to test

$$H_0: \theta = \theta_0,$$

where θ_0 is some given constant. Let $p^* = P(\hat{\theta}^* < \theta_0)$. That is, p^* is the probability that a bootstrap estimate of θ is less than the hypothesized value, θ_0. The value of p^* is not known, but it is readily estimated with

$$\hat{p}^* = \frac{A}{B},$$

where A is the number of bootstrap estimates among $\hat{\theta}_{(1)}^* \leq \cdots \leq \hat{\theta}_{(B)}^*$ that are less than θ_0. Under fairly general conditions, if the null hypothesis is true, the distribution of $\hat{p}^*$ approaches a uniform distribution as n and B get large (e.g., Liu and Singh, 1997; Hall, 1988a, 1988b). This suggests rejecting H_0 when $\hat{p}^* \leq \alpha/2$ or $\hat{p}^* \geq 1 - \alpha/2$. A little algebra shows that this leads to the percentile bootstrap confidence interval described in the previous paragraph. A (generalized) p-value is $2\min(\hat{p}^*, 1 - \hat{p}^*)$.

A practical problem is choosing B. If the goal is to control the probability of a Type I error, $B = 500$ suffices for some problems, even with n very small, but $B = 2000$ or larger might be needed for other situations. And in some instances, such as when making inferences about the population mean, the method performs poorly even when both B and n are fairly large. A rough characterization is that if a location estimator has a low finite sample breakdown point, the percentile method might be unsatisfactory, but with a relatively high finite sample breakdown, it performs reasonably well, even with small sample sizes, and in fact it appears to be the method of choice in many situations. More details are provided as we consider various parameters of interest. Also, when dealing with regression, we will see situations where even with a low finite sample breakdown point, a percentile bootstrap method performs relatively well.

4.4.2 R Functions onesampb and hdpb

The R function

$$onesampb(x, est=onestep, alpha=0.05, nboot=2000, SEED=TRUE, nv=0,$$
$$null.value=NULL,...)$$

can be used to compute a percentile bootstrap $1 - \alpha$ confidence interval when using virtually any estimator available through R. When testing hypotheses, the null value can be indicated via the argument nv, which defaults to zero. (The argument null.value can also be used to indicate the null value.) A p-value, based on the null value, is returned. (Using SEED=TRUE sets the seed of the random number generator so that results are always duplicated when using the same data.) The argument est indicates the estimator to be used, which defaults to the one-step M-estimator. The argument ... can be used to supply values for any additional parameters associated with the estimator indicated by the argument est. For example, to compute a 0.9 confidence interval based on a 10% trimmed means, using 2,000 bootstrap samples, use the command

$$onesampb(x, est=mean, alpha=0.1, nboot=2000, tr=0.1).$$

(When using a trimmed mean, the R function trimpb in Section 4.4.6 can be used as well.) The command

$$onesampb(x, est=pbvar)$$

computes a 0.95 confidence based on the percentage bend midvariance.

The R function

$$hdpb(x, est = hd, tr=0.2, nboot = 2000, SEED = TRUE, nv = 0, q=0.5,...)$$

computes a confidence interval for quantiles using the Harrell–Davis estimator. The argument q indicates the quantile to be used and defaults to the median. This function is essentially the same as onesampb and is supplied merely for convenience.

4.4.3 Bootstrap-t Method

The main alternative to the percentile bootstrap is the *bootstrap-t* method, which also has been called a *percentile-t* technique. When working with means, for example, the strategy is to use the observed data to approximate the distribution of

$$T = \frac{\sqrt{n}(\bar{X} - \mu)}{s}$$

by proceeding as follows:

1. Generate a bootstrap sample $X_1^*, \ldots, X_n^*$.
2. Compute $\bar{X}^*$, s^*, and $T^* = \sqrt{n}(\bar{X}^* - \bar{X})/s^*$ based on the bootstrap sample generated in step 1.
3. Repeat steps 1 and 2 B times, yielding T_b^*, $b = 1, \ldots, B$.

The T_b^* values provide an approximation of the distribution of T and in particular an estimate of the $\alpha/2$ and $1 - \alpha/2$ quantiles.

When testing H_0: $\mu = \mu_0$, there are two variations of the bootstrap-t method that deserve comment. The first is the *equal-tailed* method. Let $T_{(1)}^* \leq \cdots \leq T_{(B)}^*$ be the T_b^* values written in ascending order, let $\ell = \alpha B/2$, rounded to the nearest integer, and let $u = B - \ell$. Then H_0 is rejected if

$$T \leq T_{(\ell)}^* \quad \text{or} \quad T \geq T_{(u)}^*.$$

Rearranging terms, a $1 - \alpha$ confidence interval for μ is

$$\left(\bar{X} - T_{(u)}^* \frac{s}{\sqrt{n}}, \ \bar{X} - T_{(\ell)}^* \frac{s}{\sqrt{n}} \right). \tag{4.8}$$

This last equation might appear to be incorrect because $T_{(u)}^*$, the estimate of the $1 - \alpha/2$ quantile of the distribution of T, is used to compute the lower end of the confidence interval. Simultaneously, $T_{(\ell)}^*$, an estimate of the $\alpha/2$ quantile, is used to compute the upper end of the confidence interval. It can be seen, however, that this last equation follows from the decision rule that rejects H_0: $\mu = \mu_0$ if $T \leq T_{(\ell)}^*$ or $T \geq T_{(u)}^*$. Also, when computing the upper end of the confidence interval, $T_{(\ell)}^*$ will be negative, which is why the term $T_{(\ell)}^* \frac{s}{\sqrt{n}}$ is subtracted from $\bar{X}$.

The second variation of the bootstrap-t method, which yields a *symmetric confidence interval*, uses

$$T^* = \frac{\sqrt{n}|\bar{X}^* - \bar{X}|}{s^*}.$$

Let $c = (1 - \alpha)B$, rounded to the nearest integer. Now a $1 - \alpha$ confidence interval for μ is

$$\bar{X} \pm T^*_{(c)} \frac{s}{\sqrt{n}}.$$

Asymptotic results (Hall, 1988a, 1988b) suggest that it tends to have more accurate probability coverage than the equal-tailed confidence interval, but some small-sample exceptions are noted later.

An interesting theoretical property of the bootstrap-t method is that it is second order correct. Roughly, when using T, as the sample size increases, the discrepancy between the actual probability coverage and the nominal level goes to zero at the rate $1/\sqrt{n}$ as n gets large, meaning that the method is first order correct. But when using the bootstrap-t method, the discrepancy goes to zero at the rate $1/n$. That is, the discrepancy goes to zero faster compared to methods that rely on the central limit theorem.

Again there is the practical issue of choosing B, the number of bootstrap samples. The default choices for B used by the R functions in this book are based on the goal of achieving reasonably good control over the probability of a Type I error. But arguments can be made that perhaps a larger value for B has practical value, the concern being that otherwise there might be some loss of power. Racine and MacKinnon (2007a) discuss this issue at length and propose a method for choosing the number of bootstrap samples. (Also see Jöckel, 1986.) Davidson and MacKinnon (2000) propose a pretest procedure for choosing B. Theoretical results derived by Olive (2010) suggest using $B \geq nlog(n)$.

4.4.4 Bootstrap Methods When Using a Trimmed Mean

As previously indicated, the 20% trimmed mean can be expected to provide better control over the probability of a Type I error and more accurate probability coverage, compared to the mean, in various situations. In some cases, however, even better probability coverage and control of Type I error probabilities might be desired, particularly when the sample size is small. Some type of bootstrap method can make a substantial difference, with the choice of method depending on how much trimming is done.

First it is noted that the bootstrap methods in Sections 4.4.1 and 4.4.2 are readily applied when using a trimmed mean. When using the percentile bootstrap method, generate a bootstrap sample and compute the sample trimmed mean, yielding $\bar{X}^*_{t1}$. Repeat this process B times, yielding $\bar{X}^*_{t1}, \ldots, \bar{X}^*_{tB}$. Then an approximate $1 - \alpha$ confidence interval for μ_t is given by

$$(\bar{X}^*_{t(\ell+1)}, \bar{X}^*_{t(u)}),$$

where again ℓ is $\alpha B/2$ rounded to the nearest integer, $u = B - \ell$, and $\bar{X}^*_{t(1)} \leq \cdots \leq \bar{X}^*_{t(B)}$ are the B bootstrap trimmed means written in ascending order.

The bootstrap-t extends to trimmed means in a straightforward manner as well, and to be sure the details are clear they are summarized in Table 4.4. In the context of testing $H_0 \colon \mu_t = \mu_0$ versus $H_1 \colon \mu_t \neq \mu_0$, reject if $T_t < T^*_{t(\ell)}$ or $T_t > T^*_{t(u)}$, where

$$T^*_t = \frac{(1 - 2\gamma)\sqrt{n}(\bar{X}^*_t - \bar{X}_t)}{s^*_w}. \tag{4.9}$$

As for the symmetric, two-sided confidence interval, now use

$$T^*_t = \frac{(1 - 2\gamma)\sqrt{n}|\bar{X}^*_t - \bar{X}_t|}{s^*_w}, \tag{4.10}$$

in which case a two-sided confidence interval for μ_t is

$$\bar{X}_t \pm T^*_{t(c)} \frac{s_w}{(1 - 2\gamma)\sqrt{n}}. \tag{4.11}$$

The choice between the percentile bootstrap versus the bootstrap-t, based on the criterion of accurate probability coverage, depends on the amount of trimming. With no trimming, all indications are that the bootstrap-t is preferable (e.g., Westfall and Young, 1993). Consequently, early investigations based on means suggested using a bootstrap-t when making inferences about a population trimmed mean, but more recent studies indicate that as the amount of trimming increases, at some point the percentile bootstrap method offers an advantage. In particular, simulation studies indicate that when the amount of trimming is 20%, the percentile

Table 4.4: Summary of the bootstrap-t method for a trimmed mean.

To apply the bootstrap-t (or percentile-t) method when working with a trimmed mean, proceed as follows:
1. Compute the sample trimmed mean, $\bar{X}_t$.
2. Generate a bootstrap sample by randomly sampling with replacement n observations from $X_1, \ldots, X_n$, yielding $X^*_1, \ldots, X^*_n$.
3. When computing an equal-tailed confidence interval, use the bootstrap sample to compute T^*_t given by Eq. (4.9). When computing a symmetric confidence interval, compute T^*_t using Eq. (4.10) instead.
4. Repeat steps 2 and 3, yielding $T^*_{t1}, \ldots, T^*_{tB}$. $B = 599$ appears to suffice in most situations when $n \geq 12$.
5. Put the $T^*_{t1}, \ldots, T^*_{tB}$ values in ascending order, yielding $T^*_{t(1)}, \ldots, T^*_{t(B)}$.
6. Set $\ell = \alpha B/2$, $c = (1 - \alpha)B$, round both ℓ and c to the nearest integer, and let $u = B - \ell$.

The equal-tailed $1 - \alpha$ confidence interval for μ_t is

$$\left(\bar{X}_t - T^*_{t(u)} \frac{s_w}{\sqrt{n}}, \ \bar{X}_t - T^*_{t(\ell)} \frac{s_w}{\sqrt{n}} \right) \tag{4.12}$$

and the symmetric confidence interval is given by Eq. (4.9).

bootstrap confidence interval should be used rather than the bootstrap-t (e.g., Wilcox, 2001a). Perhaps with slightly less trimming the percentile bootstrap continues to give more accurate probability coverage in general, but this issue has not been studied extensively.

One issue is whether Eq. (4.7) yields a confidence interval with reasonably accurate probability coverage when sampling from a light-tailed, skewed distribution. To address this issue, attention is again turned to the lognormal distribution, which has $\mu_t = 1.111$. First consider what happens when the bootstrap-t is not used. With $n = 20$ and $\alpha = 0.025$, the probability of rejecting H_0: $\mu_t > 1.111$ when using Eq. (4.4) is approximately 0.065, about 2.6 times as large as the nominal level. In contrast, the probability of rejecting H_0: $\mu_t < 1.111$ is approximately 0.010. Thus, the probability of rejecting H_0: $\mu_t = 1.111$ when testing at the 0.05 level is approximately $0.065 + 0.010 = 0.075$. If the bootstrap-t method is used instead, with $B = 599$, the one-sided, Type I error probabilities are now 0.035 and 0.020, so the probability of rejecting H_0: $\mu_t = 1.111$ is approximately 0.055 when testing at the 0.05 level. (The reason for using $B = 599$, rather than $B = 600$, stems from results in Hall, 1986, showing that B should be chosen so that α is a multiple of $(B + 1)^{-1}$. On rare occasions this small adjustment improves matters slightly, so it is used here.) As we move toward heavy-tailed distributions, generally the actual probability of a Type I error tends to decrease.

For completeness, when testing a two-sided hypothesis or computing a two-sided confidence interval, asymptotic results reported by Hall (1988a, 1988b) suggest using the bootstrap-t method where

$$T_t^* = \frac{(1 - 2\gamma)\sqrt{n}|\bar{X}_t^* - \bar{X}_t|}{s_w^*}. \tag{4.13}$$

As previously noted,

$$\bar{X}_t \pm T_{t(c)}^* \frac{s_w}{(1 - 2\gamma)\sqrt{n}}, \tag{4.14}$$

where $c = (1 - \alpha)B$, rounded to the nearest integer. This is an example of a *symmetric* two-sided confidence interval. That is, the confidence interval has the form $(\bar{X}_t - \hat{c}, \bar{X}_t + \hat{c})$, where $\hat{c}$ is determined with the goal that the probability coverage be as close as possible to $1 - \alpha$. In contrast, an *equal-tailed* two-sided confidence interval has the form $(\bar{X}_t - \hat{a}, \bar{X}_t + \hat{b})$, where $\hat{a}$ and $\hat{b}$ are determined with the goal that $P(\mu_t < \bar{X}_t - \hat{a}) \approx P(\mu_t > \bar{X}_t + \hat{b}) \approx \alpha/2$. The confidence interval given by Eq. (4.12) is equal-tailed. In terms of testing H_0: $\mu_t = \mu_0$ versus H_1: $\mu_t \neq \mu_0$, Eq. (4.14) is equivalent to rejecting if $T_t < -1 \times T_{t(c)}^*$, or if $T_t > T_{t(c)}^*$. When Eq. (4.14) is applied to the lognormal distribution with $n = 20$, a simulation estimate of the actual probability of a Type I error is 0.0532 versus 0.0537 using (4.12). Thus, in terms of Type I error probabilities, there is little separating these two methods for this special case, but in practice, the choice between these two methods can be important, as will be seen.

Table 4.5: Values of $\hat{\alpha}$ corresponding to three critical values, $n = 12$, $\alpha = 0.025$.

g	h	$P(T_t < -t)$	$P(T_t > t)$	$P(T_t < T^*_{t(\ell)})$	$P(T_t > T^*_{t(u)})$	$P(T_t < -T^*_{t(c)})$	$P(T_t > T^*_{t(c)})$
0.0	0.0	0.031	0.028	0.026	0.030	0.020	0.025
0.0	0.5	0.025	0.022	0.024	0.037	0.012	0.024
0.5	0.0	0.047	0.016	0.030	0.023	0.036	0.017
0.5	0.5	0.040	0.012	0.037	0.028	0.025	0.011

Table 4.5 summarizes the values of $\hat{\alpha}$, an estimate of the probability of a Type I error, when performing one-sided tests with $\alpha = 0.025$, and when the critical value is estimated with one of the three methods described in this section. The first estimate of the critical value is t, the $1 - \alpha/2$ quantile of Student's t distribution with $n - 2g - 1$ degrees of freedom. That is, reject if T_t is less than $-t$ or greater than t depending on the direction of the test. The second estimate of the critical value is $T^*_{t(\ell)}$ or $T^*_{t(u)}$ (again depending on the direction of the test), where $T^*_{t(\ell)}$ and $T^*_{t(u)}$ are determined with the equal-tailed bootstrap-t method. The final method uses $T^*_{t(c)}$ resulting from the symmetric bootstrap-t as used in Eq. (4.14). Estimated Type I error probabilities are reported for the four g-and-h distributions discussed in Section 4.2. For example, when sampling is from a normal distribution ($g = h = 0$), $\alpha = 0.025$, and when H_0 is rejected because $T_t < -t$, the actual probability of rejecting is approximately 0.031. In contrast, when $g = 0.5$ and $h = 0$, the probability of rejecting is estimated to be 0.047, about twice as large as the nominal level. (The estimates in Table 4.5 are based on simulations with 1,000 replications when using one of the bootstrap methods, and 10,000 replications when using Student's t.) If sampling is from a lognormal distribution, not shown in Table 4.5, the estimate increases to 0.066, which is 2.64 times as large as the nominal 0.025 level. For $(g, h) = (0.5, 0.0)$ and $\alpha = 0.05$, the tail probabilities are 0.094 and 0.034.

Note that the choice between Eq. (4.12) and Eq. (4.14), the equal-tailed and symmetric bootstrap methods, is not completely clear based on the results in Table 4.5. An argument for Eq. (4.14) is that the largest estimated probability of a Type I error in Table 4.5, when performing a two-sided test, is $0.036 + 0.017 = 0.053$, while when using Eq. (4.12) the largest estimate is $0.037 + 0.028 = 0.065$. A possible objection to Eq. (4.14) is that in some cases it is too conservative—the tail probability can be less than half the nominal 0.025 level. Also, if one can rule out the possibility that sampling is from a skewed distribution with very heavy tails, Table 4.5 suggests using Eq. (4.12) over Eq. (4.14), at least based on probability coverage.

There are other bootstrap techniques that might have a practical advantage over the bootstrap-t method, but at the moment this does not appear to be the case when γ is close to zero. However, extensive investigations have not been made, so future investigations might alter this view. One approach is to use a bootstrap estimate of the actual probability coverage when using T_t with Student's t distribution and then adjust the α level so that the actual probability coverage is closer to the nominal level (Loh, 1987a). When sampling from a lognormal distribution with $n = 20$, the one-sided tests considered above now have actual Type I error probabilities approximately equal to 0.011 and 0.045, which is a bit worse than the results with the bootstrap-t. Westfall and Young (1993) advocate yet another method that estimates the p-value of T_t. For the situation considered here, simulations (based on 4,000 replications and $B = 1,000$) yield estimates of the Type I error probabilities equal to 0.034 and 0.017. Thus, at least for the lognormal distribution, these two alternative methods appear to have no practical advantage when $\gamma = 0.2$, but of course a more definitive study is needed. Another interesting possibility is the ABC method discussed by Efron and Tibshirani (1993). The appeal of this method is that accurate confidence intervals might be possible with a substantially smaller choice for B, but there are no small-sample results on whether this is the case for the problem at hand. Additional calibration methods are summarized by Efron and Tibshirani (1993).

■ Example

Consider again the law data in Table 4.3, which have $\bar{X}_t = 596.2$ based on 20% trimming. The symmetric bootstrap-t confidence interval, based on Eq. (4.14), is $(541.6, 650.9)$, which was computed with the R function trimcibt described in Section 4.4.6. As previously indicated, the confidence interval for μ_t, based on Student's t distribution and given by Eq. (4.3), is $(561.8, 630.6)$, which is a subset of the interval based on Eq. (4.14). In fact, the length of this confidence is 68.8 versus 109.3 using the bootstrap-t method. The main point here is that the choice of method can make a substantial difference in the length of the confidence interval, the ratio of the lengths being $68.8/109.3 = 0.63$. This might seem to suggest that using Student's t distribution is preferable, because the confidence interval is shorter. However, as previously noted, it appears that sampling is from a light-tailed, skewed distribution, and this is a situation where using Student's t distribution can yield a confidence interval that does not have the nominal probability coverage—the interval can be too short. The 0.95 confidence interval for μ is $(577.1, 623.4)$, which is even shorter and probably very inaccurate in terms of probability coverage. If instead the equal-tailed bootstrap-t method is used, given by (4.12), the resulting 0.95 confidence interval for the 20% trimmed mean is $(523.0, 626.3)$, which is also substantially longer than the confidence interval based on Student's t distribution. To reiterate, all indications are that trimming, versus no trimming, generally improves probability coverage when using Eq. (4.3) and sampling is from a skewed, light-tailed distribution, but the percentile bootstrap or bootstrap-t method can give even better results, at least when n is small.

4.4.5 Singh's Modification

Consider a random sample where say 15% of the observations are outliers. Of course, if a 20% trimmed mean is used, these outliers do not have an undue influence on the estimate as well as the standard error. Note, however, that when generating a bootstrap sample, by chance the number of outliers could exceed 20%, which can result in a relatively long confidence interval. Singh (1998) derived theoretical results showing that this problem can be addressed by Winsorizing the data before taking a bootstrap sample, provided the amount of Winsorizing does not exceed the amount of trimming. So if inferences based on a 20% trimmed mean are to be made, theory allows taking bootstrap samples from the Winsorized data provided the amount of Winsorizing does not exceed 20%. When using a percentile bootstrap method, for example, confidence intervals are computed in the usual way. That is, the only difference from the basic percentile bootstrap method in Section 4.4.1 is that observations are resampled with replacement from the Winsorized data.

Although theory allows the amount of Winsorizing to be as large as the amount of trimming, if we Winsorize as much as we trim, probability coverage can be unsatisfactory, at least with small to moderate sample sizes (Wilcox, 2001a). However, if for example 10% Winsorizing is done when making inferences based on a 20% trimmed mean, good probability coverage is obtained.

Singh's results extend to the bootstrap-t method. But all indications are that achieving accurate probability coverage is difficult. Presumably this problem becomes negligible as the sample size increases, but just how large the sample must be to obtain reasonably accurate probability coverage is unknown.

4.4.6 R Functions trimpb and trimcibt

The R function trimpb (written for this book) computes a 0.95 confidence interval using the percentile bootstrap method. It has the general form

trimpb(x, tr=0.2,alpha=0.05,nboot=2000,WIN=FALSE,plotit=FALSE,win=0.1,pop=1),

where x is any R vector containing data, tr again indicates the amount of trimming, alpha is α, and nboot is B, which defaults to 2,000. The argument WIN controls whether Winsorizing is done. If the argument plotit=TRUE, a plot of the bootstrap trimmed means is created, and the type of plot created is controlled by the argument pop. The choices are:

- pop=1, expected frequency curve,
- pop=2, kernel density estimate (using a normal kernel),

- pop=3, boxplot,
- pop=4, stem-and-leaf,
- pop=5, histogram,
- pop=6, adaptive kernel density estimate.

The function

   trimcibt(x, tr=0.2,alpha=0.05, nboot=599, WIN=FALSE ,plotit=FALSE ,win=0.1,op=1)

performs the bootstrap-t method. Now if plotit=TRUE, a plot of the $T_{t1}^*, \ldots, T_{tB}^*$ values is created based on the adaptive kernel estimator if op=1. If op=2, an expected frequency curve is used.

4.5 Inferences About M-Estimators

A natural way of computing a confidence interval for μ_m, an M-measure of location, is to estimate the standard error of $\hat{\mu}_m$ with $\hat{\sigma}_m$, as described in Chapter 3, and consider intervals having the form $(\hat{\mu}_m - c\hat{\sigma}_m, \hat{\mu}_m + c\hat{\sigma}_m)$ for some appropriate choice for c. This strategy seems to have merit when sampling from a symmetric distribution, but for asymmetric distributions it can be unsatisfactory (Wilcox, 1992). If, for example, c is determined so that the probability coverage is exactly $1 - \alpha$ when sampling from a normal distribution, the same c can yield a confidence interval with probability coverage substantially different from $1 - \alpha$ when sampling from asymmetric distributions instead. Moreover, it is unknown how large n must be so that the resulting confidence interval has probability coverage reasonably close to the nominal level.

One alternative approach is to apply a bootstrap-t method, but simulations do not support this strategy, at least when $n \leq 40$. Could the bootstrap-t method be improved by using something like the adaptive kernel density estimator when estimating the standard of $\hat{\mu}_m$? All indications are that probability coverage remains unsatisfactory. Currently, the most effective method is the percentile bootstrap (but direct comparisons with the method studied by Kuonen, 2005, have not been made).

As before, generate a bootstrap sample by randomly sampling n observations, with replacement, from $X_1, \ldots, X_n$, yielding $X_1^*, \ldots, X_n^*$. Let $\hat{\mu}_m^*$ be the M-estimator of location based on the bootstrap sample just generated. Repeat this process B times, yielding $\hat{\mu}_{m1}^*, \ldots, \hat{\mu}_{mB}^*$. The $1 - \alpha$ confidence interval for μ_m is

$$(\hat{\mu}_{m(\ell+1)}^*, \hat{\mu}_{m(u)}^*), \tag{4.15}$$

Table 4.6: Values of $\hat{\alpha}$ when using Eq. (4.15), $B = 399$, $\alpha = 0.05$, $n = 20$.

g	h	$P(\hat{\mu}^*_{m(\ell)} > 0)$	$P(\hat{\mu}^*_{m(u)} < 0)$
0.0	0.0	0.030	0.034
0.0	0.5	0.029	0.036
0.5	0.0	0.023	0.044
0.5	0.5	0.023	0.042

where $\ell = \alpha B / 2$, rounded to the nearest integer, $u = B - \ell$, and $\hat{\mu}^*_{m(1)} \leq \cdots \leq \hat{\mu}^*_{m(B)}$ are the B bootstrap values written in ascending order.

The percentile bootstrap method appears to give fairly accurate probability coverage when $n \geq 20$ and $B = 399$, but for smaller sample sizes the actual probability coverage can be less than 0.925, with $\alpha = 0.05$. Increasing B to 599 does not appear to improve the situation very much. Another problem is that the iterative method of computing $\hat{\mu}_m$ can break down when applying the bootstrap and n is small. The reason is that if more than half of the observations have a common value, MAD $= 0$, resulting in division by zero when computing $\hat{\mu}_m$. Because the bootstrap is based on sampling with replacement, as n gets small, the probability of getting MAD $= 0$ within the bootstrap sample increases. Of course, problems might also arise in situations where some of the X_i have a common value. Similar problems arise when using $\hat{\mu}_{os}$ instead. This might suggest abandoning the M-estimator, but as noted in Chapter 3, there are situations where it might be preferred over the trimmed mean.

Table 4.6 shows the estimated probability of observing $\hat{\mu}^*_{m(\ell)} > 0$ and the probability of $\hat{\mu}^*_{m(u)} < 0$ when observations are generated from a g-and-h distribution that has been shifted so that $\mu_m = 0$. For example, when sampling from a normal distribution, the probability of a Type I error, when testing H_0: $\mu_m = 0$, is $0.030 + 0.034 = 0.064$. If sampling is from a lognormal distribution, the two tail probabilities are estimated to be 0.019 and 0.050, so the probability of a Type I error when testing H_0: $\mu_m = 0$ is 0.069. Increasing B to 599, the estimated probability of a Type I error is now 0.070. Thus, there is room for improvement, but probability coverage and control over the probability of a Type I error might be deemed adequate in some situations.

Tingley and Field (1990) suggest yet another method for computing confidence intervals based on *exponential tilting* and a *saddlepoint approximation* of a distribution. (Also see Gatto and Ronchetti, 1996, as well as Robinson et al., 2003, for related results.) They illustrate the method when dealing with M-estimators, but their results are quite general and might have practical interest when using other measures of location. While preparing this chapter, the author ran a few simulations to determine how their method performs when working with

M-estimators. When sampling from a standard normal distribution, with $n = 25$ and simulations based on 10,000 replications, the estimated Type I error probability was $\hat{\alpha} = 0.078$ when testing at the $\alpha = 0.05$ level. In contrast, $\hat{\alpha} = 0.064$ when using the percentile bootstrap. Perhaps there are situations where the Tingley–Field method offers a practical advantage, but this has not been established as yet.

From an efficiency point of view, the one-step M-estimator (with Huber's Ψ) given by Eq. (3.25) can be a bit more satisfactory than the modified one-step M-estimator (MOM) in Section 3.10. (With sufficiently heavy tails, MOM can have better efficiency.) With very small sample sizes, it seems that reasonably accurate confidence intervals are easier to obtain when using MOM.

4.5.1 R Functions mestci and momci

The R function onesampb can be used to compute confidence intervals based on MOM or an M-estimator. This function also returns a p-value based on the null value indicated by the argument nv. For convenience, the R function

$$\text{mestci(x,alpha=0.05,nboot=399,bend=1.28,os=FALSE)}$$

is supplied for the special case where the goal is to compute a $1 - \alpha$ confidence interval for μ_m (an M-measure of location based on Huber's Ψ) using the percentile bootstrap method. The default value for alpha (α) is 0.05, nboot is the number of bootstrap samples to be used, which defaults to 399, and bend is the bending constant used in Huber's Ψ, which defaults to 1.28. (See Chapter 3.) The argument os is a logical variable that defaults to FALSE meaning that the fully iterated M-estimator is to be used. Setting os=TRUE causes the one-step M-estimator, $\hat{\mu}_{os}$, to be used. The R function

$$\text{momci(x,alpha=0.05,nboot=2000,bend=2.24,SEED=TRUE,null.value=0)}$$

is supplied for situations where there is specific interest in the modified one-step M-estimator. (The argument bend refers to the constant used by the Hampel identifier described in Section 3.10.) The function also reports a p-value based on the null value indicated by the argument null.value.

■ **Example**

If the law data in Table 4.3 are stored in the R variable x, the command mestci(x) returns a 0.95 confidence interval for μ_m equal to $(573.8, 629.1)$. For these data, the function also prints a warning that because the number of observations is less than 20, division by zero might occur when computing the bootstrap M-estimators, but in this particular case this problem does not arise. Note that the length of the confidence interval is shorter than the length of the confidence interval for the trimmed mean, based on Eq. (4.3), but with such a small sample size, and because sampling appears to be from a light-tailed distribution, the probability coverage of both confidence intervals might be less than 0.95. The command mestci(x,os=TRUE) computes a 0.95 confidence interval using the one-step M-estimator. This yields $(573.8, 629.5)$, which is nearly the same as the 0.95 confidence interval based on $\hat{\mu}_m$.

■

4.6 Confidence Intervals for Quantiles

This section addresses the problem of computing a confidence interval for x_q, the qth quantile. Many strategies are available, but only a few are listed here.

Consider the interval $(X_{(i)}, X_{(j)})$. As a confidence interval for the qth quantile, the exact probability coverage of this interval is

$$\sum_{k=i}^{j-1} \binom{n}{k} q^k (1-q)^{n-k} \tag{4.16}$$

(e.g., Arnold et al., 1992). An issue is whether alternative methods might give shorter confidence intervals, but it seems that among the alternatives listed here, little is known about this possibility. Imagine that a confidence interval is sought that has probability coverage at least 0.95, say. If n is small and fixed, as q goes to zero or one, it becomes impossible to achieve this goal. For example, if $n = 30$ and $q = 0.05$, the highest possible probability coverage is 0.785. So an issue is whether other methods can be found that perform reasonably well in this case.

Next consider techniques based on the Harrell–Davis estimator, $\hat{\theta}_q$. A simple method that seems to be reasonably effective, at least for $\alpha = 0.05$ and $n \geq 20$, is to use the percentile bootstrap. Another approach that appears to be about as effective as the percentile bootstrap is described here. It continues to give good results in situations covered in Chapter 5, where the percentile bootstrap can be unsatisfactory, but there are situations where the reverse is true as well. (For other methods that have been considered, see Wilcox, 1991b.)

Let $\hat{\sigma}_{hd}$ be the bootstrap estimate of the standard error of $\hat{\theta}_q$, which is described in Chapter 3. Here, $B = 100$ bootstrap samples are used to compute $\hat{\sigma}_{hd}$. Temporarily assume that sampling is from a normal distribution and suppose c is determined so that the interval

$$(\hat{\theta}_q - c\hat{\sigma}_{hd}, \hat{\theta}_q + c\hat{\sigma}_{hd}) \tag{4.17}$$

has probability coverage $1 - \alpha$. Then simply continue to use this interval when sampling from non-normal distributions. There is the practical problem that c is not known, but it is easily estimated by running simulations on a computer. Suppose c is to be chosen with the goal of computing a 0.95 confidence interval. For normal distributions, simulations indicate that for n fixed, c does not vary much as a function of the quantile being estimated, provided $n \geq 11$ and attention is restricted to those quantiles between 0.3 and 0.7. For convenience, c was determined for $n = 11, 15, 21, 31, 41, 61, 81, 121$, and 181, and then a regression line was fitted to the resulting pairs of points, yielding

$$\hat{c} = 0.5064n^{-0.25} + 1.96, \tag{4.18}$$

where the exponent, -0.25, was determined using the half-slope ratio of Tukey's resistant regression line. (See, for example, Velleman and Hoaglin, 1981; Wilcox, 1996a.) When dealing with the 0.2 or 0.8 quantile, (4.18) gives reasonably good results for $n > 21$. For $11 \leq n \leq 21$, use

$$\hat{c} = \frac{-6.23}{n} + 5.01.$$

Critical values have not been determined for $n < 11$. For the 0.1 and 0.9 quantiles, use

$$\hat{c} = \frac{36.2}{n} + 1.31$$

when $11 \leq n \leq 41$; otherwise use Eq. (4.18).

As a partial check on the accuracy of the method, it is noted that when observations are generated from a lognormal distribution, the actual probability coverage when working with the median, when $n = 21$ and $\alpha = 0.05$, is approximately 0.959, based on a simulation with 10,000 replications. For the 0.1 and 0.9 quantiles it is 0.974 and 0.928, respectively. However, with $n = 30$ and $q = 0.05$, this method performs poorly. And it might perform poorly when there are tied values. (More details regarding tied values are given in Section 4.6.1 and subsequent chapters.)

Another approach is to use $\hat{x}_q$ to estimate the qth quantile as described in Section 3.5, estimate the standard error of $\hat{x}_q$ with $\hat{\sigma}_{mj}$, the Maritz–Jarrett estimator described in Section 3.4, and then assume that

$$Z = \frac{\hat{x}_q - x_q}{\hat{\sigma}_{mj}}$$

has a standard normal distribution. Then an approximate $1 - \alpha$ confidence interval for the qth quantile is

$$(\hat{x}_q - z_{1-\alpha/2}\hat{\sigma}_{mj}, \; \hat{x}_q + z_{1-\alpha/2}\hat{\sigma}_{mj}), \tag{4.19}$$

Table 4.7: Values of $\hat{\alpha}$ when using (4.19), $n = 13$, and $\alpha = 0.05$.

g	h	$\hat{\alpha}$
0.0	0.0	0.067
0.0	0.5	0.036
0.5	0.0	0.062
0.5	0.5	0.024

where $z_{1-\alpha/2}$ is the $1 - \alpha/2$ quantile of the standard normal distribution. Table 4.7 shows $\hat{\alpha}$, the estimate of one minus the probability coverage, for the four g-and-h distributions discussed in Section 4.2 when $q = 0.5$, $\alpha = 0.05$, and $n = 13$. The estimates are based on simulations with 10,000 replications. When sampling from a lognormal distribution, $\hat{\alpha} = 0.067$.

A variation of this last method is to replace the Maritz–Jarrett estimate of the standard error with an estimate based on Eq. (3.11), which requires an estimate of $f(x_q)$, the probability density function evaluated at x_q. If $f(x_q)$ is estimated with the adaptive kernel method in Section 3.2.4, a relatively accurate 0.95 confidence interval can be obtained with $n = 30$ and $q = 0.05$. In fact, this is the only method known to perform reasonably well for this special case. As we move from normality toward heavy-tailed distributions, this method continues to perform tolerably well up to a point (a g-and-h distribution with $g = h = 0.2$), but eventually it will fail. For example, with $g = h = 0.5$, the probability coverage is approximately 0.92, but increasing n to 40, the probability coverage is approximately 0.95.

4.6.1 Beware of Tied Values When Making Inferences About Quantiles

When making inferences about quantiles, tied values can create serious practical problems when computing confidence intervals and testing hypotheses. For the special case where the goal is to compute a confidence interval for the population median, the method in the next section can be used when tied values occur. But when comparing the median of two or more distributions, techniques based on estimates of the standard error of M, which simultaneously assume M has a normal distribution, can be highly unsatisfactory, even with large sample sizes. The first general problem is getting a reasonably accurate estimate of the standard error. As noted in Section 3.3.4, all known estimates of the standard error of the sample median can be extremely inaccurate. The second difficulty is that the sampling distribution of M can be poorly approximated by a normal distribution, even with a large sample size. Indeed, there are situations where the distribution of M converges to a discrete distribution (e.g., Koenker, 2005, p. 150).

More generally, when using any quantile estimator that is based on only one or two order statistics, methods for estimating the standard error can be highly inaccurate, and asymptotic normality cannot be assumed. With very few tied values, methods based on estimates of the standard error might continue to perform well. But at some point this is no longer true. The safest strategy at the moment is to use a method that does not require an estimate of the standard error. The distribution-free method based on Eq. (4.16) is one possibility. Another is to use a percentile bootstrap method in conjunction with the Harrell–Davis estimator. This approach can yield shorter confidence intervals compared to the method based on Eq. (4.16). But when dealing with quantiles close to zero or one, it is unclear when this approach becomes unsatisfactory.

To underscore why tied values can cause problems when working with the median, and to illustrate a limitation of the central limit theorem, imagine a random sample $X_1, \ldots, X_n$, where each X_i has the binomial probability function

$$\binom{15}{x} 0.3^x 0.7^{15-x}.$$

So, for example, the probability that a randomly sampled participant responds with the value 13 is 0.09156. As is evident, with a sample size of $n = 20$, tied values are guaranteed since there are only 16 possible responses. The left panel of Fig. 4.5 shows a plot of the relative frequencies associated with 5,000 sample medians, with each sample median based on $n = 20$ randomly sampled observations. The plot resembles somewhat a normal curve, but note that only five values for the sample median occur. Now look at the right panel, which was created in the same manner as the left panel, only with a sample size of $n = 100$ for each sample median. Blind reliance on the central limit theorem would suggest that the plot will look more like a normal distribution than the left panel, but clearly this is not the case. Now only three values for the sample median are observed. In practical terms, methods for making inferences about the median, which assume the sample median has a normal distribution, can be disastrous when tied values can occur.

4.6.2 A Modification of the Distribution-Free Method for the Median

When the goal is to compute a confidence interval for the population median, the following method can be used even when there are tied values. Suppose W is a binomial random variable with probability of success $p = 0.5$ and n trials. For any integer k greater than 0 and less than $[n/2]$, let $\gamma_k = P(k \leq W \leq n - k)$, the probability that the number of successes, W, is between k and $n - k$, inclusive. Then a distribution-free γ_k confidence interval for the median is

$$(X_{(k)}, X_{(n-k+1)}).$$

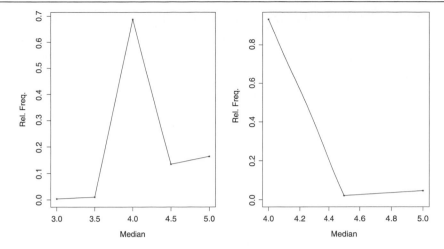

Figure 4.5: When tied values can occur, the sample median might not be asymptotically normal. The left panel shows the sampling distribution of the median with $n = 20$. The right panel is the sampling distribution with $n = 100$.

That is, the probability coverage is exactly γ_k under random sampling (e.g., Hettmansperger and McKean, 1998; also see Yohai and Zamar, 2004). This is just a special case of the first method described in the previous section based on Eq. (4.16). However, when dealing with a discrete distribution, it can be shown by example that the actual probability coverage is greater than γ_k. This is related to how quantiles are defined, which was described in Section 2.2.1.

Because the binomial distribution is discrete, it is not possible, in general, to choose k so that the probability coverage is exactly equal to $1 - \alpha$. For example, if $n = 10$, a 0.891 and 0.978 confidence interval can be computed, but not a 0.95 confidence interval as is often desired. Hettmansperger and Sheather (1986) suggest using the following method for dealing with this issue. First, determine k such that $\gamma_{k+1} < 1 - \alpha < \gamma_k$. Next, compute

$$I = \frac{\gamma_k - (1 - \alpha)}{\gamma_k - \gamma_{k+1}}$$

and

$$\lambda = \frac{(n - k)I}{k + (n - 2k)I}.$$

Then an approximate $1 - \alpha$ confidence interval is

$$(\lambda X_{(k+1)} + (1 - \lambda)X_{(k)}, \lambda X_{(n-k)} + (1 - \lambda)X_{(n-k+1)}). \tag{4.20}$$

Results reported by Sheather and McKean (1987), as well as Hall and Sheather (1988), support the use of this method. However, to guarantee that the probability coverage is at least $1 - \alpha$, use the exact method described in the previous paragraph. See Frey and Zhang (2017) for more details.

4.6.3 R Functions qmjci, hdci, sint, sintv2, qci, qcipb, and qint

The R function

$$\text{qmjci(x,q=0.5,alpha=0.05,op=1)}$$

computes a $1 - \alpha$ confidence interval for the qth quantile using Eq. (4.19) and the data stored in the R vector x. The function returns the lower and upper values of the confidence interval. The default value for q is 0.5 and the default value for alpha (α) is 0.05. (The accuracy of this confidence interval for $q \neq 0.5$ and n small has not been studied.) With op=1, the Maritz–Jarrett estimate of the standard error is used, and with op=2, the McKean–Schrader estimate (described in Section 3.3.4) is used instead. (With op=2, only q=0.5 is allowed.) With op=3, the function estimates the standard error via the adaptive kernel estimate of $f(x_q)$. The function

$$\text{qci(x,q=0.5,alpha=0.05)}$$

returns the same confidence interval as qmjci with op=3 and is provided in case it is convenient.

The R function

$$\text{hdci(x,q=0.5,nboot=100)}$$

computes a 0.95 confidence interval using the Harrell–Davis estimator and Eq. (4.17). As indicated, the default value for q is 0.5, and the default number of bootstrap samples, nboot, is 100. The R function

$$\text{qcipb(x,q=0.5,alpha=0.05,nboot=2000,SEED=TRUE,nv=0)}$$

uses the Harrell–Davis estimator in conjunction with a percentile bootstrap method. Unlike the R function hdci, qcipb is not restricted to 0.95 confidence intervals.

The function

$$\text{sint(x,alpha=0.05)}$$

computes a confidence interval for the median using Eq. (4.20), where α is taken to be 0.05 if not specified. To get a p-value when testing the hypothesis that the population median is equal to some specified value, use the R function

$$\text{sintv2(x,alpha=0.05,nullval=0),}$$

where the null value is specified by the argument nullval, which defaults to 0. Confidence intervals for other quantiles, based on Eq. (4.16) in Section 4.6, which do not use interpolation, are computed by the function

$$\text{qint(x,q=0.5,alpha=0.05).}$$

The exact probability coverage is reported as well.

■ **Example**

Staudte and Sheather (1990) illustrate the use of Eq. (4.20) with data from a study on the lifetimes of EMT6 cells. The values are 10.4, 10.9, 8.8, 7.8, 9.5, 10.4, 8.4, 9.0, 22.2, 8.5, 9.1, 8.9, 10.5, 8.7, 10.4, 9.8, 7.7, 8.2, 10.3, 9.1. Both the sample median, M, and $\hat{x}_{0.5}$ are equal to 9.1. The resulting 0.95 confidence interval reported by sint is $(8.72, 10.38)$. In contrast, the confidence interval based on Eq. (4.19), as computed by the R function qmjci, is $(8.3, 9.9)$. The lengths of the confidence intervals are about the same. The main difference is that the confidence interval based on (4.19) is centered about the sample median, 9.1, while the confidence interval based on Eq. (4.20) is not. The Harrell–Davis estimate of the median is 9.26, and a 0.95 confidence interval based on Eq. (4.17), computed with the R function hdci, is $(8.45, 10.08)$.

■

4.7 Empirical Likelihood

Empirical likelihood methods (Owen, 2001) represent another non-parametric approach for computing a confidence interval for the population mean that should be noted. Asymptotic results suggest that a Bartlett corrected empirical likelihood approach is superior to using a bootstrap-t method (DiCiccio et al., 1991).

The empirical likelihood method can be used to construct a confidence interval for μ, but for simplicity it is described in terms of testing H_0: $\mu = \mu_0$. Consider distributions F_p, $p = (p_1, \ldots, p_n)$ supported on the sample $X_1, \ldots, X_n$, where X_i is assigned mass p_i. For a specified value of μ, the empirical likelihood $L(\mu)$ is defined to be the maximum value of Πp_i over all such distributions that satisfy $\sum X_i p_i = \mu$. Because Πp_i attains its overall maximum when $p_i = n^{-1}$, it follows that the empirical likelihood is maximized when $\mu = \bar{X}$. The empirical likelihood ratio for testing H_0 is

$$W = -2\log\{L(\mu_0)/L(\bar{X})\}.$$

When the null hypothesis is true, W has approximately a chi-squared distribution with 1 degree of freedom. In particular, reject H_0 at the α level if $W \geq c$, where c is the $1 - \alpha$ quantile of a chi-squared distribution with 1 degree of freedom.

Bartlett Corrected Empirical Likelihood

The Bartlett corrected empirical likelihood method is applied as follows. Let $\hat{\mu}_j = n^{-1}\sum(X_i - \bar{X})^j$ and

$$a = \frac{1}{2}\hat{\mu}_4\hat{\mu}_2^{-2} - \frac{1}{3}\hat{\mu}_3^2\hat{\mu}_2^{-3}.$$

Then the null hypothesis is rejected if $W(1 - an^{-1}) \geq c$.

Table 4.8 reports simulation estimates (based on 1,000 replications) of the Type I error probability for the empirical likelihood (EL) method, the Bartlett corrected empirical likelihood (BCEL), the equal-tailed bootstrap-t (BEQ) and the symmetric bootstrap-t (BSYM). The distributions considered are normal, chi-squared with 1 degree of freedom (χ_1^2), a Student's t with 5 degrees of freedom (t_5), a lognormal distribution (LogN), the contaminated normal (cnorm) shown in Fig. 1.1, and some g-and-h distributions. Glen and Zhao (2007) derive theoretical results indicating that the empirical likelihood methods can be unsatisfactory when sampling from contaminated normal. And the results in Table 4.8 illustrate that they can indeed be highly unsatisfactory. (Also see Wilcox, 2010e.)

As previously noted, Bradley (1978) suggests that generally, at a minimum, the actual Type I error probability should be between 0.025 and 0.075 when testing at the 0.05 level. Based on this criterion, none of the methods are satisfactory. However, for skewed distributions for which the median proportion of outliers does not exceed 0.05, the symmetric bootstrap method gives satisfactory results. The symmetric bootstrap method can be too conservative when sampling from a symmetric heavy-tailed distribution, but this might be judged to be less serious than having an actual Type I error greater than 0.075, as is the case when using the empirical likelihood methods. Note that with $n = 20$, the symmetric bootstrap method has a

Table 4.8:   Estimated Type I error probabilities.

n	Distribution	EL	BCEL	BEQ	BSYM
20	Normal	0.074	0.064	0.058	0.045
	χ_1^2	0.117	0.103	0.068	0.080
	t_5	0.075	0.059	0.067	0.036
	LogN	0.137	0.120	0.099	0.104
	Cnorm	0.169	0.138	0.116	0.010
	(g,h)=(0.2,.0)	0.090	0.072	0.083	0.035
	(g,h)=(0.2,0.2)	0.094	0.080	0.083	0.047
	(g,h)=(0.5,0.5)	0.270	0.241	0.231	0.186
50	Normal	0.052	0.050	0.055	0.049
	χ_1^2	0.074	0.069	0.055	0.059
	t_5	0.062	0.058	0.072	0.048
	LogN	0.068	0.062	0.058	0.054
	Cnorm	0.137	0.125	0.145	0.011
	(g,h)=(0,0.2)	0.061	0.057	0.073	0.037
	(g,h)=(0.2,0.2)	0.074	0.066	0.080	0.050
	(g,h)=(0.5,0.5)	0.215	0.203	0.207	0.194

EL=empirical likelihood.
BCEL=Bartlett corrected empirical likelihood.
BEQ=bootstrap-t, equal-tailed.
BSYM=bootstrap-t, symmetric.

Type I error probability of 0.080 when sampling from a chi-squared distribution with 1 degree of freedom. Increasing the sample size to $n = 25$, the estimate drops to 0.065, and for $n = 30$ it is 0.059.

Some additional simulations were run with $n = 100$ and it was found that the empirical likelihood methods continue to perform poorly when sampling from the heavy-tailed distributions considered here. With $n = 200$ they perform well when sampling from the contaminated normal, but estimates exceed 0.15 when sampling from the g-and-h distribution when $g = h = 0.5$.

Results on how to improve the empirical likelihood method, when working with the mean, are reported by Vexler et al. (2009), but control over the Type I error probability remains rather poor when dealing with non-normal distributions. Also see Glen and Zhao (2007). For a review of empirical likelihood methods when dealing with regression, see Chen and Van Keilegom (2009).

As for $n = 50$, the empirical likelihood methods compete better with the bootstrap-t methods, but the symmetric bootstrap-t performs well in situations where the empirical likelihood methods are unsatisfactory based on Bradley's criterion. Again a criticism of the symmetric bootstrap-t is that for a symmetric heavy-tailed distribution (the contaminated normal), the Type I error probability drops below 0.025. But the other three methods have estimates greater than 0.12. So for general use, the symmetric bootstrap-t seems best.

Some additional simulations were run with $n = 100$ and it was found that the empirical likelihood methods continue to perform poorly when sampling from the heavy-tailed distributions considered here. With $n = 200$ they perform well when sampling from the contaminated normal, but estimates exceed 0.15 when sampling from the g-and-h distribution when $g = h = 0.5$. Wang (2016) derived a two-stage method aimed at improving Bartlett corrected empirical likelihood. The extent to which the method performs well for the situations in Table 4.8 has not been determined.

4.8 Inferences About the Probability of Success

Methods designed specifically for comparing discrete distributions will be seen to provide an interesting perspective on robust methods aimed at comparing robust measures of location. Some of these techniques are based in part on methods for making inferences about the probability of success, p, when dealing with a binomial distribution. Numerous methods aimed at computing a confidence interval for p have been proposed. This section describes and comments on some of these techniques. Included is an R function for making inferences about the probability cell probabilities of a multinomial distribution.

Let w denote the number of successes among n trials. A classic approach was derived by Clopper and Pearson (1934). The lower and upper ends of their $1 - \alpha$ confidence interval are $B(\alpha/2; w, n - w + 1)$ and $B(1 - \alpha/2; w + 1, n - w)$, respectively, where $B(q; u, v)$ is the qth quantile of a beta distribution with shape parameters u and v. It guarantees that the actual coverage probability is at least $1 - \alpha$, but in general does not give the shortest-length confidence interval.

Brown et al. (2002) compared various techniques and concluded that the Agresti–Coull method, which stems from Agresti and Coull (1998), performs relatively well. Let

$$\hat{p} = \frac{w}{n}$$

be the proportion of successes among the n observations and let c denote the $1 - \alpha/2$ quantile of a standard normal distribution. Let

$$\tilde{n} = n + c^2,$$

$$\tilde{w} = w + \frac{c^2}{2},$$

and

$$\tilde{p} = \frac{\tilde{w}}{\tilde{n}}.$$

The Agresti–Coull $1 - \alpha$ confidence interval for the probability of success, p, is

$$\tilde{p} \pm c\sqrt{\frac{\tilde{p}(1 - \tilde{p})}{\tilde{n}}}.$$

The Agresti–Coull method is, in essence, a simple approximation of the score method derived by Wilson (1927). Wilson's confidence interval is

$$\left(\hat{p} + c^2/(2n) \pm c\sqrt{[\hat{p}(1 - \hat{p}) + c^2/(4n)]/n} \right)/(1 + c^2/n).$$

There is a substantial literature in support of Wilson's method that is summarized by Zou et al. (2009).

Results reported by Blyth (1986) suggest proceeding as follows when w is equal to 0, 1, $n - 1$, or n.

- If $w = 0$,

$$c_U = 1 - \alpha^{1/n},$$

$$c_L = 0.$$

- If $w = 1$,

$$c_L = 1 - \left(1 - \frac{\alpha}{2}\right)^{1/n},$$

$$c_U = 1 - \left(\frac{\alpha}{2}\right)^{1/n}.$$

- If $w = n - 1$,

$$c_L = \left(\frac{\alpha}{2}\right)^{1/n},$$

$$c_U = \left(1 - \frac{\alpha}{2}\right)^{1/n}.$$

- If $w = n$,

$$c_L = \alpha^{1/n}$$

and

$$c_U = 1.$$

The cases $w = 0$ and $w = n$ can be shown to be the Clopper–Pearson confidence intervals. Otherwise, Blyth recommends a method stemming from Pratt (1968), which is computed as follows. Let

$$A = \left(\frac{w+1}{n-w}\right)^2,$$

$$B = 81(w+1)(n-w) - 9n - 8,$$

$$C = -3c\sqrt{9(w+1)(n-w)(9n+5-c^2)} + n + 1,$$

$$D = 81(w+1)^2 - 9(w+1)(2+c^2) + 1,$$

$$E = 1 + A\left(\frac{B+C}{D}\right)^3,$$

in which case the upper end of the confidence interval is

$$c_U = \frac{1}{E}.$$

As for the lower end, now let

$$A = \left(\frac{w}{n-w-1}\right)^2,$$

$$B = 81(w)(n-w-1) - 9n - 8,$$

$$C = 3c\sqrt{9x(n-w-1)(9n+5-c^2)} + n + 1,$$

$$D = 81w^2 - 9w(2+c^2) + 1,$$

$$E = 1 + A\left(\frac{B+C}{D}\right)^3.$$

The lower end of the confidence interval is

$$c_L = \frac{1}{E}.$$

Kulinskaya et al. (2008, p. 140) derived yet another method for computing a confidence interval. Let $\check{p} = (w+3)/(n+3/4)$,

$$A = \sin\left(\arcsin\left(\sqrt{\check{p}}\right) - \frac{c}{2\sqrt{n}}\right),$$

and

$$B = \sin\left(\arcsin\left(\sqrt{\tilde{p}}\right) + \frac{c}{2\sqrt{n}}\right),$$

where again c is the $1 - \alpha/2$ quantile of a standard normal distribution. Their $1 - \alpha$ confidence interval is (A^2, B^2).

Schilling and Doi (2014) derive a more complex method that guarantees that the actual probability coverage is greater than or equal to the specified level. For example, if the goal is to compute a 0.95 confidence interval, the actual probability that the confidence interval contains the true probability of success, p, is at least 0.95. In addition, the Schilling–Doi method is designed to provide the optimal confidence interval. This means that when computing a $1 - \alpha$ confidence interval, the shortest possible confidence interval is computed that guarantees that the probability coverage is at least $1 - \alpha$. A practical limitation is that execution time quickly becomes prohibitive as the sample size increases. The involved calculations are not described here, but an R function for applying the method is supplied. For yet more methods, see Cao et al. (2020).

For convenience, denote the Agresti–Coull, Pratt, Clopper–Pearson, Kulinskaya et al., Schilling–Doi, and Wilson methods by AC, P, CP, KMS, SD, and WIL, respectively. To provide at least a glimpse of how these methods compare, consider the case where $n = 25$. A simulation was performed aimed at estimating the actual probability that the 0.95 confidence interval does not contain the true probability of success (Wilcox, 2020b). This was done for $p = 0.05$ (0.01) 0.95, with each estimate based on 10,000 replications. Fig. 4.6 summarizes the results when using AC (the dashed line) and CP (the solid line). The top horizontal line corresponds to 0.075 and the lower horizontal line is 0.025. As can be seen, AC performs well for $0.2 \leq p \leq 0.8$, but otherwise it can be highly unsatisfactory. Switching to P or KMS does not improve matters. Consider, for example, $p = 0.15$ and again suppose the goal is to compute a $1 - \alpha = 0.95$ confidence interval. Based on a simulation with 50,000 replications, the actual value of α is 0.112, 0.113, 0.024, 0.113, 0.044, and 0.117 for methods AC, P, CP, KMS, SD, and WIL, respectively. (For method SD, 5,000 replications were used.) So in this particular case, only SD performs reasonably well. Increasing the sample size to $n = 30$ the estimates for AC, P, CP, KMS, and WIL are 0.073, 0.073, 0.024, 0.073, and 0.077, respectively. For $n = 35$ the estimates are 0.053, 0.053, 0.037, 0.053, and 0.053.

4.8.1 R Functions binom.conf.pv and cat.dat.ci

The R function

```
binom.conf.pv(x = sum(y), nn = length(y), y=NULL, method='AC', AUTO=TRUE,
          PVSD=FALSE, alpha=0.05, nullval=0.5, NOTE=TRUE)
```

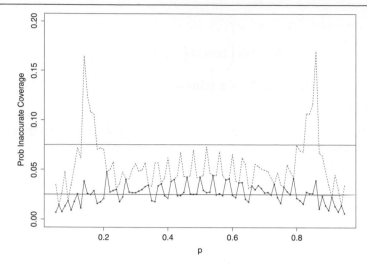

Figure 4.6: The dashed line indicates the probability of an inaccurate confidence interval when using method AC, $n = 25$. The solid dark line is the probability when using CP.

computes a $1 - \alpha$ confidence interval for p using one of six methods and it reports a p-value when testing the hypothesis indicated by the argument nullval. The choices for the argument method are:

- AC: Agresti–Coull,
- P: Pratt,
- CP: Clopper–Pearson,
- KMS: the Kulinskaya et al. method,
- WIL: Wilson,
- SD: Schilling–Doi.

The argument AUTO=TRUE means that if $n < 35$, the Schilling–Doi (SD) method will be used. Otherwise, the argument method determines which method will be used; Agresti–Coull is the default method. If $w = 0, 1, n - 1$, or n, the method recommended by Blyth is used. The argument PVSD=FALSE means that a p-value will not be computed when method='SD', the goal being to avoid high execution time. To get a p-value, set PVSD=TRUE. The R function binom.conf is like the function binom.conf.pv, only no p-value is reported.

When dealing with discrete data where the sample space is relatively small (a multinomial distribution), the R function

$$\text{cat.dat.ci(x,alpha=0.05)}$$

computes a confidence interval for the probability of observing the value x. This is done for every value of x that is observed.

■ **Example**

Suppose there is 1 success in 80 trials. Then binom.conf.pv(1,80) reports that the 0.95 confidence interval for p is (0.00032, 0.0451). The confidence interval based on method='SD' (the Schilling–Doi method) is (0.001, 0.067). If the R object my-dat contains 0, 1, 1, 1, 0, 0, 1, 1, 0, 1; then the command binom.conf.pv(y=my-dat,AUTO=FALSE) returns an estimate of p equal to 0.6 and a 0.95 confidence interval equal to (0.31, 0.83), which is the Agresti–Coull 0.95 confidence interval. The p-value, when testing H_0: $p = 0.5$, is equal to 0.53. The command binom.conf.pv(y=my-dat, method='SD', PVSD=TRUE) returns the Schilling–Doi 0.95 confidence interval, (0.29, 0.85). Now the p-value is 0.76.

■

4.9 Concluding Remarks

To summarize a general result in this chapter, there is a plethora of methods one might use to compute confidence intervals and test hypotheses. Many methods can be eliminated based on published studies, but several possibilities remain. As noted in Chapter 3, there are arguments for preferring trimmed means over M-estimators, and there are arguments for preferring M-estimators instead, so the choice between the two is not particularly obvious. In terms of computing confidence intervals, all indications are that when working with the 20% trimmed, reasonably accurate probability coverage can be obtained over a broader range of situations versus an M-estimator or mean. When working with the 20% trimmed mean and the sample size is small, a percentile bootstrap method is generally preferable to the confidence interval given by Eq. (4.3). As already stressed, the 20% trimmed mean can have a relatively small standard error when sampling from a heavy-tailed distribution, but other criteria can be used to argue for some other measure of location. If the sample size is at least 20, M-estimators appear to be a viable option based on the criterion of accurate probability coverage. An advantage of the modified one-step M-estimator (MOM) is that accurate confidence intervals can be computed with small sample sizes in situations where methods based on M-estimators are not quite satisfactory, and it is flexible about how many observations are trimmed, in contrast to a trimmed mean. From an efficiency point of view, M-estimators based on Huber's Ψ generally have a bit of an advantage over MOM, when sampling from a normal distribution or a distribution where the expected proportion of outliers is less an 0.1. However, as the expected proportion of outliers increases, MOM can have a smaller standard error than the one-step M-estimator (Özdemir and Wilcox, 2012). And in terms of controlling Type I error probabilities, it seems that using MOM in conjunction with a percentile bootstrap method is a bit more satisfactory than using a one-step M-estimator, particularly when dealing with skewed, relatively light-tailed distributions. Inferences about quantiles might appear to be rather uninteresting at this point, but they can be used to address important issues that are ignored by other measures of location, as will be seen in Chapter 5. Put more generally, different methods for summarizing data can reveal important and interesting features that other methods miss.

4.10 Exercises

1. Describe situations where the confidence interval for the mean might be too long or too short. Contrast this with confidence intervals for the 20% trimmed mean and μ_m.

2. Compute a 0.95 confidence interval for the mean, 10% mean, and 20% mean using the data in Table 3.1 of Chapter 3. Examine a boxplot of the data and comment on the accuracy of the confidence interval for the mean. Use both Eq. (4.3) and the bootstrap-t method.

3. Compute a 0.95 confidence interval for the mean, 10% mean, and 20% mean using the lifetime data listed in the example of Section 4.6.2. Use both Eq. (4.3) and the bootstrap-t method.

4. Use the R functions qmjci, hdci, and sint to compute a 0.95 confidence interval for the median based on the LSAT data in Table 4.3. Comment on how these confidence intervals compare to one another.

5. The R function rexp generates data from an exponential distribution. Use R to estimate the probability of getting at least one outlier, based on a boxplot, when sampling from this distribution. Discuss the implications for computing a confidence interval for μ.

6. If the exponential distribution has variance $\mu_{[2]} = \sigma^2$, then $\mu_{[3]} = 2\sigma^3$ and $\mu_{[4]} = 9\sigma^4$. Determine the skewness and kurtosis. What does this suggest about getting an accurate confidence interval for the mean?

7. Do the skewness and kurtosis of the exponential distribution suggest that the bootstrap-t method will provide a more accurate confidence interval for μ_t versus the confidence interval given by Eq. (4.3)?

8. For the exponential distribution, would the sample median be expected to have a relatively high or low standard error? Compare your answer to the estimated standard error obtained with data generated from the exponential distribution.

9. Discuss the relative merits of using the R function sint versus qmjci and hdci.

10. Verify Eq. (4.5) using the decision rule about whether to reject H_0 described in Section 4.4.3.

11. For the LSAT data in Table 4.3, compute a 0.95 bootstrap-t confidence interval for mean using the R function trimcibt with plotit=T. Note that a boxplot finds no outliers. Comment on the plot created by trimcibt in terms of achieving accurate probability coverage

when using Student's t. What does this suggest about the strategy of using Student's t if no outliers are found by a boxplot?

12. Generate 20 observations from a g-and-h distribution with $g = h = 0.5$. (This can be done with the R function ghdist, written for this book.) Examine a boxplot of the data. Repeat this 10 times. Comment on the strategy of examining a boxplot to determine whether the confidence interval for the mean has probability coverage at least as high as the nominal level.

## CHAPTER 5

Comparing Two Groups

A natural approach to comparing two distributions is to focus on a single measure of location. There are two well-known but distinct perspectives regarding how one might proceed. The first is to focus on how the typical measure associated with the first group compares to a typical measure associated with the second group. If, for example, X and Y are two independent random variables with trimmed means μ_{t1} and μ_{t2}, respectively, one might test H_0: $\mu_{t1} = \mu_{t2}$. Another perspective is to focus instead on the trimmed mean of the typical difference. That is, if a single observation X is sampled from the first group, and a single observation is sampled from the second group Y, what is the typical value of $D = X - Y$? For example, one might focus on the trimmed mean of the typical difference, μ_{tD}.

To elaborate on the difference between these views, let $X_1, \ldots, X_n$ and $Y_1, \ldots, Y_m$ be random samples from the first and second group, respectively, and let $\bar{X}_t$ and $\bar{Y}_t$ be the corresponding trimmed means. Then a test of H_0: $\mu_{t1} = \mu_{t2}$ would be based in part on $\bar{X}_t - \bar{Y}_t$, the difference between the sample trimmed means. But from the other perspective, one would compute $D_{ij} = X_i - Y_j$ for all $i = 1, \ldots, n$ and $j = 1, \ldots, m$, and then compute $\bar{D}_t$, the trimmed mean based on all nm D values, which estimates μ_{tD}. With no trimming, $\bar{D}_t = \bar{X}_t - \bar{Y}_t$. But otherwise, they can differ. This latter approach is an integral component of classic and modern rank-based methods, where the median of the D_{ij} values plays a central role, as will be seen in Section 5.7. The point is that, under general conditions, $\mu_{t1} - \mu_{t2} \neq \mu_{tD}$. That is, multiple perspectives might be needed to get a deep sense of how groups compare.

Another complication is that focusing on a single measure of location might not suffice. For example, if one or both distributions are skewed, the difference between the means might be large compared to the difference between the trimmed means, or any other measure of location that might be used. As is evident, the reverse can happen, where the difference between the trimmed means is large and the difference between the means is not. Of course, two or more measures of location might be compared, but this might miss interesting differences and trends among subpopulations of participants.

To elaborate, it helps to consider a simple but unrealistic situation: two normal distributions that have means $\mu_1 = \mu_2$. Then any test of the hypothesis H_0: $\mu_1 = \mu_2$ should not reject. But suppose the variances differ. To be concrete, suppose an experimental method is being compared to a control group, and the control group has variance $\sigma_1^2 = 1$, while the experimental method has $\sigma_2^2 = 0.5$. Then the experimental group is effective in the sense that low scoring

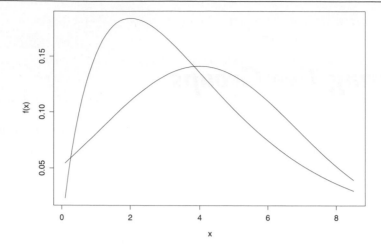

Figure 5.1: Two different distributions with equal means and variances.

participants in the experimental group have higher scores than low scoring participants in the control group. Similarly, the experimental method is detrimental in the sense that high scoring participants in the experimental group tend to score lower than high scoring participants in the control group. That is, different subpopulations of participants respond in different ways to the experimental method. Of course, in this simple example, one could compare the variances of the two groups, but for various reasons to be explained and illustrated, it can be useful to compare the quantiles of the two groups instead.

As another example, consider the two distributions in Fig. 5.1. The distributions differ, the effectiveness of one method over the other depends on which quantiles are compared, yet the distributions have identical means and variances. (The skewed distribution is a chi-squared distribution with 4 degrees of freedom, so, the mean and variance are 4 and 8, respectively, and the symmetric distribution is normal.)

In this book, an experimental method is defined to be *completely effective* compared to a control group if each quantile of the experimental group is greater than the corresponding quantile of the control group. In symbols, if θ_1 and θ_2 are the qth quantiles of the control and experimental group, respectively, the experimental method is said to be *completely effective* if $\theta_2 > \theta_1$ for any q. This implies that the experimental method is stochastically larger than the distribution associated with the control. If $\theta_1 > \theta_2$ for some q, but $\theta_1 < \theta_2$ for others, the experimental method is defined to be *partially effective*. Both of the illustrations just described correspond to situations where an experimental method is only partially effective. There are situations where comparing measures of location and scale can be used to establish whether an experimental method is completely effective (e.g., Wilcox, 1990b), but this requires assumptions that are not always met and cannot always be tested in an effective manner, so, this approach is not pursued here. Note that Student's t test assumes that $\sigma_1 = \sigma_2$, even when $\mu_1 \neq \mu_2$. If this assumption is met, and distributions are normal, then $\mu_1 > \mu_2$ implies that the experimental method is completely effective. The practical concern is that, when these two assumptions are not met, as is commonly the case, such a conclusion can be highly misleading and inaccurate.

5.1 The Shift Function

There are various ways multiple quantiles might be compared. This section describes an approach based on the so-called shift function. The basic idea, which was developed by Doksum (1974, 1977), as well as Doksum and Sievers (1976), is to plot the quantiles of the first group versus the differences between the quantiles. That is, plot θ_1 versus

$$\Delta(\theta_1) = \theta_2 - \theta_1, \tag{5.1}$$

where θ_2 is the qth quantile of the second group. $\Delta(\theta_1)$ is called a *shift function*. It measures how much the first group must be shifted so that it is comparable to the second group at the qth quantile.

The shift function is illustrated with some data from a study by Salk (1973), which are stored in the file salk_dat.txt and can be accessed as described in Section 1.10. The goal was to study weight gain in newborns based on infants who weighed at least 3,500 grams at birth. The experimental group was continuously exposed to the sound of a mother's heartbeat. The data are stored in column 1 of the file salk_dat.txt. For the moment, attention is focused on comparing the deciles rather than all of the quantiles. For the control group (not exposed to the sound of a heartbeat), the Harrell–Davis estimates of the deciles (the 0.1, 0.2, 0.3, 0.4, 0.5, 0.6, 0.7, 0.8, and 0.9 quantiles) are -171.7, -117.4, -83.1, -59.7, -44.4, -32.1, -18.0, 7.5, and 64.9, respectively. For the experimental group, the estimates are -55.7, -28.9, -10.1, 2.8, 12.2, 22.8, 39.0, 61.9, and 102.7, and this yields an estimate of the shift function. For example, an estimate of $\Delta(x_{0.1})$ is $\hat{\Delta}(-171.7) = -55.7 - (-171.7) = 116$. That is, the weight gain among infants at the 0.1 quantile of the experimental group is estimated to be 116 grams higher than that of the infants corresponding to the 0.1 quantile of the control. Fig. 5.2 shows a plot of the estimated deciles for the control group versus $\hat{\Delta}$. Notice the apparent monotonic decreasing relationship between θ_1, weight gain in the control group, versus Δ. The plot suggests that exposure to the sound of a heartbeat is most effective for infants who gain the least amount of weight after birth. As is fairly evident, this type of detailed comparison can be important and useful. If, for example, an experimental method is expensive or invasive, knowing how different subpopulations compare might affect the policy or strategy one is willing to adopt when dealing with a particular problem.

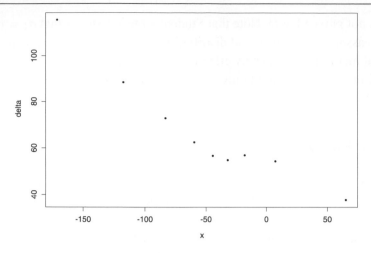

Figure 5.2: The x-axis indicates the deciles for the first group, and the y-axis indicates the difference between the decile of the second group versus that of the first.

Next, attention is turned to the more general setting, where the goal is to compare all of the quantiles rather than just the deciles. Suppose the value x satisfies $P(X \leq x) = q$. As noted in Chapter 3, from a strictly technical point of view, x is not necessarily the qth quantile. If x is the qth quantile, the difference between the two distributions at θ_1 is measured with $\Delta(\theta_1)$ given by Eq. (5.1). If x is not the qth quantile, as might be when sampling from a discrete distribution, the difference between the two distributions is measured with

$$\Delta(x) = \theta_2 - x.$$

Let $X_1, \ldots, X_n$ and $Y_1, \ldots, Y_m$ be random samples from the first and second group, respectively. As usual, let $X_{(1)} \leq \cdots \leq X_{(n)}$ be the order statistics. Following Doksum and Sievers (1976), $\Delta(x)$ can be estimated as follows. Let $\hat{q} = \hat{F}(x)$ be the proportion of observations in the control group that are less than or equal to x. In terms of the order statistics, $\hat{q} = i/n$, where i is the largest integer such that $X_{(i)} \leq x$. Note that x qualifies as a reasonable estimate of θ_1, the qth quantile. Then to estimate $\Delta(x)$, all that is needed is an estimate of the qth quantile of Y. The simplest estimate is $Y_{(\ell)}$, where $\ell = [\hat{q}m + 0.5]$, and as usual the notation $[\hat{q}m + 0.5]$ means to round $\hat{q}m + 0.5$ down to the nearest integer. In other words, use the quantile estimator described in Section 3.4 of Chapter 3. Finally, estimate $\Delta(x)$ with

$$\hat{\Delta}(x) = Y_{(\ell)} - x. \tag{5.2}$$

One can then plot $\hat{\Delta}(x)$ versus x to get an overall sense of how the two groups compare.

■ **Example**

As a simple illustration, suppose it is desired to estimate $\Delta(-160)$ for Salk's weight data. There are $n = 36$ participants in the control group, three participants have values less than or equal to -160, so, $\hat{q} = 3/36$. To estimate the $3/36$ quantile of the experimental group, note that there are $m = 20$ participants, so, $\ell = [(3/36)(20) + 0.5] = 2$. Therefore, the estimate of the $3/36$ quantile is the second smallest value in the experimental group, which is -60. Hence, $\Delta(-160)$ is estimated to be $-60 - (-160) = 100$, suggesting that the typical infant who would lose 160 grams at birth would lose 100 grams less if exposed to the sound of a heartbeat. There are also three values in the control group less than or equal to -180, so, $\Delta(-180)$ is estimated to be $\hat{\Delta}(-180) = -60 - (-180) = 120$. Note that the shift function is just a series of straight lines with jumps at the points $(X_i, \hat{\Delta}(X_i))$, $i = 1, \ldots, n$. (A graphical illustration is given in Fig. 5.3, which is discussed at the end of Section 5.1.4.)

There remains the problem of how to make inferences about $\Delta(x)$. Several approaches are described for comparing independent groups, and then their relative merits are summarized. One approach is based on comparing specified quantiles that are estimated with the Harrell–Davis estimator. Two other methods compute a confidence band for all quantiles. These latter two methods are based on two versions of the Kolmogorov–Smirnov test, which are summarized in the next section of this chapter.

■

5.1.1 The Kolmogorov–Smirnov Test

The Kolmogorov–Smirnov test is designed to test

$$H_0: F(x) = G(x), \text{ all } x, \tag{5.3}$$

versus $H_1: F(x) \neq G(x)$ for at least one x, where F and G are the distributions associated with two independent groups (cf. Li et al., 1996). The Kolmogorov–Smirnov test is of interest because, as will be seen, it yields a distribution-free confidence band for the shift function (cf. Fan, 1996).

Let $X_1, \ldots, X_n$ and $Y_1, \ldots, Y_m$ be random samples from two independent groups. Let $\hat{F}(x)$ and $\hat{G}(x)$ be the usual empirical distribution functions. So, $\hat{F}(x)$ is just the proportion of X_i values less than or equal to x, and $\hat{G}(x)$ is the proportion of Y_i values less than or equal to x. The Kolmogorov–Smirnov statistic for testing Eq. (5.3) is based on $\max|\hat{F}(x) - \hat{G}(x)|$, the maximum being taken over all possible values of x. That is, the test statistic is based on an estimate of the Kolmogorov distance between the two distributions. Let Z_i be the $n + m$

pooled observations. In symbols, $Z_i = X_i$, $i = 1, \ldots, n$, and $Z_{n+i} = Y_i$, $i = 1, \ldots, m$. Then the Kolmogorov–Smirnov test statistic is

$$D = \max |\hat{F}(Z_i) - \hat{G}(Z_i)|, \tag{5.4}$$

the maximum being taken over all $n + m$ values of i. That is, for each i, $i = 1, \ldots, n + m$, compute $|\hat{F}(Z_i) - \hat{G}(Z_i)|$, and set D equal to the largest of these values.

When sampling from continuous distributions, in which case, ties occur with probability zero, percentage points of the null distribution of D can be obtained using a recursive method (Kim and Jennrich, 1973). Table 5.1 outlines the calculations. Section 5.1.2 describes an R function that performs the calculations.

Table 5.1:   Computing the percentage points of the Kolmogorov–Smirnov statistic.

To compute $P(D \le c)$, where D is the Kolmogorov–Smirnov test statistic given by Eq. (5.4), let $C(i, j) = 1$ if

$$\left| \frac{i}{n} - \frac{j}{m} \right| \le c, \tag{5.5}$$

and otherwise, $C(i, j) = 0$, where the possible values of i and j are $i = 0, \ldots, n$ and $j = 0, \ldots, m$. Note that there are $(m + 1)(n + 1)$ possible values of D based on sample sizes of m and n. Let $N(i, j)$ be the number of paths over the lattice

$$\{(i, j) : i = 0, \ldots, n; j = 0, \ldots, m\},$$

from $(0, 0)$ to (i, j), satisfying Eq. (5.5). Because the path to (i, j) must pass through either the point $(i - 1, j)$ or $(i, j - 1)$, $N(i, j)$ is given by the recursion relation

$$N(i, j) = C(i, j)[N(i, j - 1) + N(i - 1, j)],$$

subject to the initial conditions $N(i, j) = C(i, j)$ when $ij = 0$. When ties occur with probability zero and $H_0: F(x) = G(x)$ is true,

$$P(D \le c) = \frac{m! n! N(m, n)}{(n + m)!},$$

where the binomial coefficient, $(n + m)!/(m! n!)$, is the number of paths from $(0, 0)$ to (n, m). When working with the weighted version of the Kolmogorov-Smirnov test, D_w, proceed exactly as before, only set $C(i, j) = 1$ if

$$\sqrt{\frac{mn}{n+m}} \left| \frac{i}{n} - \frac{j}{m} \right| \left[\frac{i+j}{n+m} \left(1 - \frac{i+j}{n+m} \right) \right]^{-1/2} \le c.$$

Then:

$$P(D_w \le c) = \frac{m! n! N(m, n)}{(n + m)!}.$$

Suppose the null hypothesis of identical distributions is true. If H_0 is rejected when $D > c$, and the algorithm in Table 5.1 indicates that $P(D > c) = \alpha$, the probability of a Type I error is exactly α when ties are impossible. Moreover, the Kolmogorov–Smirnov test is distribution-free—the probability of a Type I error is exactly α regardless of which distributions are involved. If there are tied values, the probability of a Type I error is less than α, but an exact p-value can be determined by proceeding as described in Table 5.1, only $C(i, j)$ is also set equal to 1 if $i + j < N$ and $Z_{(i+j)} = Z_{(i+j+1)}$ (Schroër and Trenkler, 1995; cf. Hilton et al., 1994).

There is another version of the Kolmogorov–Smirnov test worth considering that is based on a weighted analog of the Kolmogorov distance between two distributions. Let $N = m + n$, $M = mn/N$, $\lambda = n/N$, and $\hat{H}(x) = \lambda \hat{F}(x) + (1 - \lambda)\hat{G}(x)$. Now, the difference between any two distributions, at the value x, is estimated with

$$\frac{\sqrt{M}|\hat{F}(x) - \hat{G}(x)|}{\sqrt{\hat{H}(x)[1 - \hat{H}(x)]}}. \tag{5.6}$$

Then $H_0: F(x) = G(x)$ can be tested with an estimate of the largest weighted difference over all possible values of x. (Also see Büning, 2001.) The test statistic is

$$D_w = \max \frac{\sqrt{M}|\hat{F}(Z_i) - \hat{G}(Z_i)|}{\sqrt{\hat{H}(Z_i)(1 - \hat{H}(Z_i))}}, \tag{5.7}$$

where the maximum is taken over all values of i, $i = 1, \ldots, N$, subject to $\hat{H}(Z_i)[1 - \hat{H}(Z_i)] > 0$. An exact p-value can be determined as described in Table 5.1. An argument for D_w is that it gives equal weight to each x in the sense that the large sample (asymptotic) variance of Eq. (5.6) is independent of x. Put another way, $|\hat{F}(x) - \hat{G}(x)|$, the estimate of the Kolmogorov distance at x, tends to have a larger variance when x is in the tails of the distributions. Consequently, inferences based on the unweighted Kolmogorov–Smirnov test statistic, D, tend to be more sensitive to differences that occur in the middle portion of the distributions. In contrast, Eq. (5.6) is designed so that its variance remains fairly stable as a function of x. Consequently, D_w is more sensitive than D to differences that occur in the tails.

When using D_w and when both m and n are less than or equal to 100, an approximate 0.05 critical value is

$$\frac{1}{95}\{0.48[\max(n, m) - 5] + 0.44|n - m|\} + 2.58,$$

the approximation being derived from the percentage points of D_w reported by Wilcox (1989). When using D, an approximation of the α critical value, when performing a two-sided test, is

$$\sqrt{-\frac{n+m}{2nm}\log(\alpha/2)}$$

(e.g., Hollander and Wolfe, 1973). This approximate critical value is reported by the R function ks described in the next section of this chapter.

5.1.2 R Functions ks, kssig, kswsig, and kstiesig

The R function

$$ks(x, y, w{=}FALSE, sig{=}TRUE, alpha{=}\ 0.05),$$

written for this book, performs the Kolmogorov–Smirnov test, where x and y are any R vectors containing data. (Again, the R functions written for this book can be obtained as described in Section 1.8.) The default value for w is FALSE, indicating that the unweighted test statistic, D, is to be used. Using w=TRUE results in the weighted test statistic, D_w. The default value for sig is TRUE, meaning that the exact p-value is computed using the method in Table 5.1. If sig=FALSE, ks uses the approximate α critical value, where by default, $\alpha = 0.05$ is used. The function returns the value of D or D_w, the approximate α critical value if sig=FALSE, and the exact p-value if sig=TRUE.

■ Example

If Salk's weight-gain data are stored in the R vectors x and y, the command ks(x,y,sig=FALSE) returns the value $D = 0.48$ and reports that the 0.05 critical value is approximately 0.38. (The value of D is stored in the R variable ks$test, and the approximate critical value is stored in ks$crit.) The command ks(x,y) reports the exact p-value, which is 0.0018 and stored in the R variable ks$p.value. Thus, with $\alpha = 0.05$, one would reject the hypothesis of identical distributions. This leaves open the issue of where the distributions differ and by how much, but this can be addressed with the confidence bands and confidence intervals described in the remaining portion of this section. The command ks(x,y,TRUE,FALSE) reports that $D_w = 3.5$, and that an approximate 0.05 critical value is 2.81. The command ks(x, y, TRUE) computes the p-value when using D_w, which in contrast to D, assumes there are no ties. For the weight-gain data, there are ties, and the function warns that the reported p-value is not exact.

For convenience, the functions kssig, kswsig, and kstiesig are also supplied, which compute exact probabilities for the Kolmogorov–Smirnov statistics. These functions are used by the function ks to determine p-values, so, in general, they are not of direct interest when testing H_0: $F(x) = G(x)$, but they might be useful when dealing with other issues covered in this chapter. The function kssig has the form

$$kssig(n,m,c).$$

Assuming random sampling only, it returns the exact probability of a Type I error when using the critical value c, assuming there are no ties and the sample sizes are n and m. In symbols, it determines $P(D > c)$ when computing D with n and m observations randomly sampled from two independent groups having identical distributions. Continuing the illustration involving the weight-gain data, the R command

$$kssig(length(x),length(y),ks(x,y)\$test)$$

computes the p-value of the unweighted Kolmogorov–Smirnov statistic assuming there are no ties. The result is 0.021. Because there are ties in the pooled data, this is higher than the p-value reported by kstiesig, which takes ties into account.

If there are ties among the pooled observations, the exact p-value can be computed with the R function kstiesig. (This is done automatically when using the function ks.) It has the general form

$$kstiesig(x,y,c)$$

and reports the value of $P(D > c|Z)$, where the vector $Z = (Z_1, \ldots, Z_N)$ is the pooled data. For the weight-gain data, $D = 0.48$, and kstiesig(x,y, 0.48) returns the value 0.0018, the same value returned by the function ks. If there are no ties among the observations, kstiesig returns the same p-value as ks(x,y,sig=TRUE), the p-value associated with the unweighted test statistic, D.

5.1.3 Confidence Bands for the Shift Function

This section describes two methods for computing a simultaneous $1 - \alpha$ level confidence band for $\Delta(x)$. Suppose c is chosen, so that $P(D \leq c) = 1 - \alpha$. As usual, denote the order statistics by $X_{(1)} \leq \cdots \leq X_{(n)}$ and $Y_{(1)} \leq \cdots \leq Y_{(m)}$. For convenience, let $X_0 = -\infty$ and $X_{(n+1)} = \infty$. For any x satisfying $X_{(i)} \leq x < X_{(i+1)}$, let

$$k_* = \left[m \left(\frac{i}{n} - \frac{c}{\sqrt{M}} \right) \right]^+,$$

where $M = mn/(m + n)$ and the notation $[x]^+$ means to round up to the nearest integer. For example, $[5.1]^+ = 6$. Let

$$k^* = \left[m \left(\frac{i}{n} + \frac{c}{\sqrt{M}} \right) \right],$$

where k^* is rounded down to the nearest integer. Then a level $1 - \alpha$ simultaneous, distribution-free confidence band for $\Delta(x)$ $(-\infty < x < \infty)$ is

$$[Y_{(k_*)} - x, Y_{(k^*+1)} - x), \tag{5.8}$$

where $Y_{(k_*)} = -\infty$, if $k_* < 0$ and $Y_{(k^*)} = \infty$, if $k^* \geq m + 1$ (Doksum and Sievers, 1976). That is, with probability $1 - \alpha$, $Y_{(k_*)} - x \leq \Delta(x) < Y_{(k^*+1)} - x$ for all x. The resulting confidence band is called an *S band*.

■ **Example**

Suppose a confidence band for Δ is to be computed for Salk's data. For the sake of illustration, consider computing the confidence band at $x = 77$. Because $n = 36$ and $m = 20$, $M = 12.86$. Note that the value 77 is between $X_{(33)} = 70$ and $X_{(34)} = 100$, so, $i = 33$. From the previous subsection, the 0.05 critical value is approximately $c = 0.38$, so

$$k_* = \left[20 \left(\frac{33}{36} - \frac{0.38}{\sqrt{12.86}} \right) \right]^+ = 17.$$

Similarly, $k^* = 20$. The 17th value in the experimental group, after putting the values in ascending order, is $Y_{(17)} = 75$, $Y_{(20)} = 190$, so, the interval around $\Delta(77)$ is

$$(75 - 77, 190 - 77) = (-2, 113).$$

■

An exact confidence band for Δ, called a *W band*, can also be computed with the weighted Kolmogorov–Smirnov statistic. Let c be chosen so that $P(D_w \leq c) = 1 - \alpha$. This time, for any x satisfying $X_{(i)} \leq x < X_{(i+1)}$, let $u = i/n$ and set

$$h_* = \frac{u + \{c(1 - \lambda)(1 - 2\lambda u) - \sqrt{c^2(1 - \lambda)^2 + 4cu(1 - u)}\}/2}{1 + c(1 - \lambda)^2}$$

and

$$h^* = \frac{u + \{c(1 - \lambda)(1 - 2\lambda u) + \sqrt{c^2(1 - \lambda)^2 + 4cu(1 - u)}\}/2}{1 + c(1 - \lambda)^2}.$$

Set $k_* = [h_*m]^+$ and $k^* = [h^*m]$. In words, k_* is the value of h_*m rounded up to the nearest integer, where m is the number of observations in the second group (associated with Y). The value of k^* is computed in a similar manner, only its value is rounded down. Then the confidence band is again given by Eq. (5.8).

5.1.4 R Functions sband and wband

The R functions sband and wband are provided for determining confidence intervals for $\Delta(x)$ at each of the X_i values in the control group, and they can be used to compute confidence bands as well. The function sband has the general form

$$\text{sband(x,y,crit=1.36*sqrt((length(x)+length(y))/(length(x)*length(y))), flag=TRUE,}$$
$$\text{plotit=TRUE , CI=TRUE, sm=TRUE, op=1).}$$

As usual, x and y are any R vectors containing data. The optional argument crit is the critical value used to compute the simultaneous confidence band, which defaults to the approximate 0.05 critical value if unspecified. The default value for flag is TRUE, meaning that the exact probability of a Type I error will be computed based on the critical value stored in the argument crit. The command sband(x,y, 0.2,TRUE) computes confidence intervals using the critical value 0.2, and it reports the exact probability of a Type I error. The argument plotit defaults to TRUE, meaning that a plot of the shift function will be created. If sm=TRUE, the plot of the shift function is smoothed using a regression smoother called lowess when the argument op is equal to one. (Smoothers are described in Chapter 11.) If op is not equal to one, lowess is replaced by a running interval smoother (described in Chapter 11).

The function returns an n-by-3 matrix in the R variable sband$m. The ith row of the matrix corresponds to the confidence band computed at $\hat{\Delta}(X_{(i)})$, the estimate of the shift function at the ith largest X value. For convenience, the first column of the n-by-3 matrix returned by sband contains $\hat{q} = 1/n_1$, $i = 1, \ldots, n_1$, where n_1 is the sample size for the first group. The values in the second column are the lower ends of the confidence band, while the upper ends are reported in column 3. To suppress the confidence intervals, which might be convenient when the sample size is large, set the argument CI=FALSE. The function returns an indication for which quantiles a significant result is obtained. For positive differences, $qsig.greater indicates which quantiles are significant. For example, if the function returns 0.2 and 0.6, a significant difference greater than zero was obtained when comparing the 0.2 and 0.6 quantiles. If a significant difference less than zero occurs, the corresponding quantiles are stored in $q.sig.less. The function also returns the critical value being used in the R variable sband$crit, the number of significant differences in sband$numsig, and if flag=TRUE, the exact probability coverage is indicated by sband$pc. If flag=FALSE, the value of sband$pc will be NA for not available. The function wband, which is used exactly like sband, computes confidence intervals (W bands) using the weighted Kolmogorov–Smirnov statistic, D_w, instead.

■ Example

Doksum and Sievers (1976) report data from a study designed to assess the effects of ozone on weight gain in rats. The experimental group consisted of 22 70-day-old rats kept in an ozone environment for 7 days. In the control group, 23 rats of the same age were kept in an ozone-free environment. The weight gains, in grams, are stored in the file rat_data.txt. Table 5.2 shows the 23-by-3 matrix reported by the R function sband. The ith row of the matrix reports the confidence interval for Δ at the ith largest value among the X values. If there had been 45 observations in the first group, a 45-by-3 matrix would be reported instead. The function reports that numsig=8, meaning there are eight confidence intervals not containing 0, and from Table 5.2 these are the intervals extending from the second smallest to the ninth smallest value in the control group. For example, the second smallest value in the control group is $X_{(2)} = 13.1$, which corresponds to an estimate of the $q = 2/23 \approx 0.09$ quantile of the control group, and the second row in Table 5.2 (labeled [2,]) indicates that the confidence interval for $\Delta(13.1)$ is (NA, −3.0), which means that the confidence interval is $(-\infty, -3.0)$. The interval does not contain 0, suggesting that rats at the 0.09 quantile of the control group tend to gain more weight compared to the rats at the 0.09 quantile of the experimental method. The next eight confidence intervals do not contain 0 either, but the remaining confidence intervals all contain 0. The function sband indicates that the default critical value corresponds to $\alpha = 0.035$. Thus, there is evidence that rats who ordinarily gain a relatively small amount of weight will gain even less weight in an ozone environment.

The R command sband(x,y,flag=TRUE) reports that when using the default critical value, which is reported to be 0.406, the actual probability coverage is $1 - \alpha = 0.9645$, assuming there are no ties. To find out what happens to $1 - \alpha$ when a critical value of 0.39 is used instead, type the command sband(x,y, 0.39,TRUE). The function reports that now, $1 - \alpha = 0.9638$.

The S band suggests that there might be a more complicated relationship between weight gain and ozone than is suggested by a single measure of location. Fig. 5.3 shows the plot of $\hat{\Delta}(x)$ versus x that is created by sband. (The + along the x-axis marks the position of the median in the first group, and the lower and upper quartiles are indicated by an o.) The solid line,

Table 5.2: Confidence intervals for Δ using the ozone data.

	qhat	lower	upper
[1,]	0.04347826	NA	24.2
[2,]	0.08695652	NA	-3.0
[3,]	0.13043478	NA	-3.3
[4,]	0.17391304	NA	-3.4
[5,]	0.21739130	NA	-3.4
[6,]	0.26086957	NA	-2.8
[7,]	0.30434783	NA	-3.5
[8,]	0.34782609	NA	-3.5
[9,]	0.39130435	NA	-1.4
[10,]	0.43478261	-37.8	6.3
[11,]	0.47826087	-37.1	17.5
[12,]	0.52173913	-35.6	21.4
[13,]	0.56521739	-34.3	30.2
[14,]	0.60869565	-34.9	NA
[15,]	0.65217391	-35.0	NA
[16,]	0.69565217	-19.9	NA
[17,]	0.73913043	-20.0	NA
[18,]	0.78260870	-20.5	NA
[19,]	0.82608696	-20.1	NA
[20,]	0.86956522	-18.4	NA
[21,]	0.91304348	-17.3	NA
[22,]	0.95652174	-24.4	NA
[23,]	1.00000000	-26.7	NA

between the two dotted lines, is a graph of $\hat{\Delta}(x)$ versus x. The dotted lines form the approximate 0.95 confidence band. As previously indicated, the actual probability coverage of the confidence band is 0.9645. Notice that the bottom dotted line starts at $X = 21.9$. This is because for $X \leq 21.9$, the lower end of the confidence band is $-\infty$. Also, the lower dotted line stops at $X = 41$ because $X = 41$ is the largest X value available. Similarly, for $X > 24.4$, the upper end of the confidence band is ∞, and this is why the upper dotted line in Fig. 5.3 stops at $X = 24.4$. An interesting feature about Fig. 5.3 is the suggestion that, as weight gain increases in the control group, ozone has less of an effect. In fact, rats in the control group who would ordinarily have a high weight gain might actually gain more weight when exposed to ozone. However, the confidence band makes it clear that more data are needed to resolve this issue.

5.1.5 Confidence Band for Specified Quantiles

This section describes two alternatives to the shift function. An appeal of the shift function is that assuming random sampling only, the simultaneous probability coverage of the difference

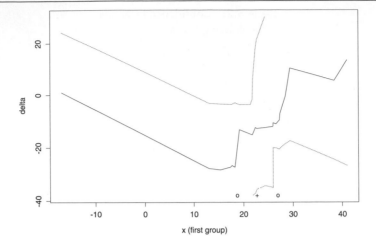

Figure 5.3: The shift function for the ozone data.

between all of the quantiles can be determined. Moreover, power is relatively high when there is interest in quantiles close to the median. But when there is interest in quantiles relatively close to zero or one, power can be relatively poor. Also, tied values might result in relatively poor power as well.

Method Q1

The first alternative method is specifically designed for comparing the deciles. It uses the Harrell–Davis estimator and is designed so that the simultaneous probability coverage of all nine confidence intervals is approximately $1 - \alpha$. One advantage of this approach is that it might have more power than S or W bands when sampling from normal or light-tailed distributions.

Let $\hat{\theta}_{qx}$ be the Harrell–Davis estimate of the qth quantile of the distribution associated with X, $q = 0.1, \ldots, 0.9$. Let $\hat{\theta}_{qy}$ be the corresponding estimate for Y. The goal is to compute confidence intervals for $\theta_{qy} - \theta_{qx}$ such that the simultaneous probability coverage is $1 - \alpha$. A solution that appears to provide reasonably accurate probability coverage for a wide range of distributions begins by computing a bootstrap estimate of the standard errors for $\hat{\theta}_{qx}$ and $\hat{\theta}_{qy}$, as described in Section 3.1. Here, independent bootstrap samples are used for all 18 deciles being estimated. In particular, bootstrap samples are used to estimate the standard error of $\hat{\theta}_{0.1x}$, the Harrell–Davis estimate of the 0.1 quantile corresponding to X, and a different (independent) set of bootstrap samples is used to estimate the standard error of $\hat{\theta}_{0.2x}$. Let $\hat{\sigma}_{qx}^2$ be the bootstrap estimate of the squared standard error of $\hat{\theta}_{qx}$. Then a 0.95 confidence interval for

$\theta_{qx} - \theta_{qy}$ is

$$(\hat{\theta}_{qy} - \hat{\theta}_{qx}) \pm c\sqrt{\hat{\sigma}_{qx}^2 + \hat{\sigma}_{qy}^2}, \tag{5.9}$$

where, when $n = m$,

$$c = \frac{80.1}{n^2} + 2.73. \tag{5.10}$$

The constant c was determined so that the simultaneous probability coverage of all nine differences is approximately 0.95 when sampling from normal distributions. Simulations suggest that when sampling from non-normal distributions, the probability coverage remains fairly close to the nominal 0.95 level (Wilcox, 1995a). For unequal sample sizes, the current strategy for computing the critical value is to set n equal to the smaller of the two sample sizes and use c given by Eq. (5.10). This approach performs reasonably well provided the difference between the sample sizes is not too large, but if the difference is large enough, the actual probability of a Type I error can be substantially smaller than the nominal level, especially when sampling from heavy-tailed distributions. Another approach is to use the nine-variate Studentized maximum modulus distribution to determine c, but Wilcox found this to be less satisfactory. Yet another approach is to use a percentile bootstrap method to determine an appropriate confidence interval, but this is less satisfactory as well.

Method Q2

There are two limitations associated with method Q1. First, it is designed specifically for comparing the deciles. Second, it might be unsatisfactory when there are tied values, meaning that the probability of one or more Type I errors can exceed the nominal level by an unacceptable amount. The mere presence of tied values does not necessarily mean that method Q1 will be unsatisfactory. But to be safe, the method described here currently seems preferable. Perhaps method Q1 has a power advantage when there are no tied values and attention is focused on comparing the deciles, but this issue needs further study.

Method Q2 (Wilcox, 2014) uses a percentile bootstrap method in conjunction with the Harrell–Davis estimator. Details regarding how the percentile bootstrap method is applied are described in a more general context in Section 5.4. Here it is merely noted that method Q2 appears to perform well when comparing the quartiles and both sample sizes are at least 20. When comparing the 0.1 or 0.9 quantiles, both sample sizes should be at least 30. For sample sizes less than 30, Navruz and Özdemir (2018) found that a quantile estimator studied by Sfakianakis and Verginis (2008) performs well when dealing with the lower quantiles, when a distribution is skewed to the right, but not the upper quantiles.

When there are tied values, method Q2 is the only known method that performs reasonably well in terms of controlling the Type I error probability when comparing the lower and upper quartiles. But when there are no tied values, a possible criticism of method Q2 is that the Harrell–Davis estimator has a finite sample breakdown point of only $1/n$. In this case, a quantile estimator based on a single order statistic (the estimator $\hat{x}_q$ in Section 3.5) might be preferred.

5.1.6 R Functions shifthd, qcomhd, qcomhdMC, and q2gci

The R function shifthd, stored in the R package Rallfun, uses method Q1 to compute the 0.95 simultaneous confidence intervals for the difference between the deciles given by Eq. (5.9). The function has the form

$$\text{shifthd(x,y,nboot=200,plotit=TRUE,plotop=FALSE)}.$$

The data corresponding to the two groups are stored in the R vectors x and y, and the default number of bootstrap samples used to compute the standard errors is 200. Thus, the command shifthd(x,y) will use 200 bootstrap samples when computing confidence intervals, while shifthd(x,y,100) will use 100 instead. The function returns a 9-by-3 matrix. The ith row corresponds to the results for the $i/10$ quantile. The first column contains the lower ends of the confidence intervals, the second column contains the upper ends, and the third column contains the estimated difference between the quantiles (the quantile associated with y minus the quantile associated with x). With plotop=FALSE and plotit=TRUE, the function creates a plot where the x-axis contains the estimated quantiles of the first group, as done by sband. With plotop=TRUE, the function plots $q = 0.1, \ldots, 0.9$ versus $(\hat{\theta}_{qy} - \hat{\theta}_{qx})$.

The R function

$$\text{qcomhd(x, y, q = c(0.1, 0.25, 0.5, 0.75, 0.9), nboot = 2000, plotit = TRUE, SEED = TRUE,}$$
$$\text{xlab = 'Group 1', ylab = 'Est.1-Est.2', tr=0.2)}$$

applies method Q2. By default, the 0.1, 0.25, 0.5, 0.75, and 0.9 quantiles are compared, but this can be altered using the argument q. The R function

$$\text{qcomhdMC(x, y, q = c(0.1, 0.25, 0.5, 0.75, 0.9), nboot = 2000, plotit = TRUE,}$$
$$\text{SEED = TRUE, xlab = 'Group 1', ylab = 'Est.1-Est.2', tr=0.2)}$$

is exactly the same as qcomhd, only it takes advantage of a multicore processor if one is available.

The R function

q2gci(x, y, q = c(0.1, 0.25, 0.5, 0.75, 0.9), nboot = 2000, plotit = TRUE, SEED = TRUE,
xlab = 'Group 1', ylab = 'Est.1-Est.2', tr=0.2)

is exactly like the R function qcomhd, only the Harrell–Davis estimator is replaced by the quantile estimator $\hat{X}_q$ in Section 3.5, which is based on a single order statistic.

■ Example

For the ozone data, shifthd returns

```
              lower       upper   Delta.hat
[1,]  -47.75411    1.918352  -22.917880
[2,]  -43.63624   -6.382708  -25.009476
[3,]  -36.04607   -3.237478  -19.641772
[4,]  -29.70039   -0.620098  -15.160245
[5,]  -24.26883   -1.273594  -12.771210
[6,]  -20.71851   -1.740128  -11.229319
[7,]  -24.97728    7.280896   -8.848194
[8,]  -24.93361   19.790053   -2.571780
[9,]  -20.89520   33.838491    6.471643
```

The first row indicates that the confidence interval for $\Delta(x_{0.1})$ is $(-47.75411, 1.918352)$, and that $\hat{\Delta}(x_{0.1})$ is equal to -22.917880. The second row gives the confidence interval for Δ evaluated at the estimated 0.2 quantile of the control group, and so on. The confidence intervals indicate that the weight gain in the two groups differs at the 0.2, 0.3, 0.4, 0.5, and 0.6 quantiles of the control group. Note that, in general, the third column, which reports $\hat{\Delta}(x)$, is increasing. That is, the differences between weight gain are getting smaller, and for the 0.9 quantile, there is the possibility that rats gain more weight in an ozone environment. However, the length of the confidence interval at the 0.9 quantile is too wide to be reasonably sure.

Fig. 5.4 shows the plot created by shifthd for the ozone data. There are nine dots corresponding to the points $(\hat{\theta}_{xq}, \hat{\Delta}(\hat{\theta}_{xq}))$, $q = 0.1, \ldots, 0.9$. That is, the dots are the estimated shift function plotted as a function of the estimated deciles corresponding to the data in the first argument, x. Note that, in general, the dots are monotonic increasing,

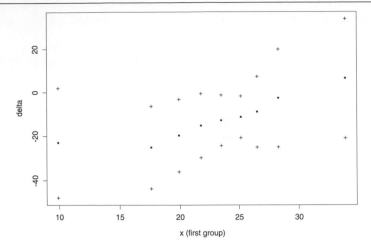

Figure 5.4: The plot created by the function shifthd using the ozone data.

which is consistent with Fig. 5.3. Above and below each dot is a + indicating the ends of the confidence interval.

5.1.7 R Functions g2plot and g5plot

To supplement the shift function, it might help to plot an estimate of the probability density function for the groups under study. The function

$$g5plot(x1,x2, x3=NULL, x4=NULL, x5=NULL, fr=0.8, aval=0.5, xlab='X', ylab='',$$
$$color=rep('black',5)$$

is supplied to help accomplish this goal. So, the function expects at least two vectors of data, and up to five distributions can be plotted. The R function

$$g2plot(x,y,op=4,rval=15,fr=0.8,aval=0.5)$$

is limited to two distributions only, but it contains certain options not available when using g5lot. In particular, the argument op controls the type of graph created. The choices are

- op=1, Rosenblatt shifted histogram,
- op=2, kernel density estimate based on a normal kernel,
- op=3, expected frequency curve,
- op=4, adaptive kernel estimator.

The arguments fr and aval are relevant to the various density estimators as described in Chapter 3.

5.2 Student's T Test

This section reviews some practical concerns about comparing means with Student's t test. From previous chapters, it is evident that Student's t test can have low power under slight departures from normality toward a heavy-tailed distribution. There are some additional issues, however, that help motivate some of the heteroscedastic methods covered in this book.

It is a bit more convenient to switch notation slightly. For two independent groups, let X_{ij}, $i = 1, \ldots, n_j$; $j = 1, 2$, be a random sample of n_j observations from the jth group. Let μ_j and σ_j^2 be the mean and variance associated with the jth group. If the variances have a common value, say $\sigma_1^2 = \sigma_2^2 = \sigma^2$, and if sampling is from normal distributions, then from basic results,

$$T = \frac{\bar{X}_1 - \bar{X}_2 - (\mu_1 - \mu_2)}{\sqrt{\text{MSWG}\left(\frac{1}{n_1} + \frac{1}{n_2}\right)}}, \tag{5.11}$$

has a Student's t distribution with $\nu = n_1 + n_2 - 2$ degrees of freedom, where

$$\text{MSWG} = \frac{(n_1 - 1)s_1^2 + (n_2 - 1)s_2^2}{n_1 + n_2 - 2}$$

is the usual (means squares within groups) estimate of the assumed common variance, σ^2. If the assumptions of normality and equal variances are met, $E(T) = 0$, and the variance of T goes to one as the sample sizes get large. To quickly review, the hypothesis of equal means, $H_0: \mu_1 = \mu_2$, is rejected if $|T| > t$, the $1 - \alpha/2$ quantile of Student's t distribution with $\nu = n_1 + n_2 - 2$ degrees of freedom, and a $1 - \alpha$ confidence interval for $\mu_1 - \mu_2$ is

$$(\bar{X}_1 - \bar{X}_2) \pm t\sqrt{\text{MSWG}\left(\frac{1}{n_1} + \frac{1}{n_2}\right)}. \tag{5.12}$$

Concerns about the ability of Student's t test to control the probability of a Type I error date back to at least Pratt (1964), who established that the level of the test is not preserved if distributions differ in dispersion or shape. If sampling is from normal distributions, the sample sizes are equal, but if the variances are not equal, Eq. (5.12) provides reasonably accurate probability coverage no matter how unequal the variances might be, provided the common sample size is not too small (Ramsey, 1980). For example, if the common sample size is 15 and $\alpha = 0.05$, the actual probability coverage will not drop below 0.94. Put another way, in terms of testing H_0, the actual probability of a Type I error will not exceed 0.06. However, if the sample sizes are equal, but sampling is from non-normal distributions, probability coverage can be unsatisfactory, and if the sample sizes are unequal as well, probability coverage deteriorates even further. Even under normality with unequal sample sizes, there are problems. For example, under normality with $n_1 = 21$, $n_2 = 41$, $\sigma_1 = 4$, $\sigma_2 = 1$, and $\alpha = 0.05$, the actual probability of a Type I error is approximately 0.15. Moreover, Fenstad (1983) argues that $\sigma_1/\sigma_2 = 4$ is not extreme, and various empirical studies support Fenstad's view (e.g., Keselman et al., 1998; Grissom, 2000; Wilcox, 1987a). The illustration just given might appear to conflict with results in Box (1954), but this is not the case. Box's numerical results indicate that under normality, when $1/\sqrt{3} \leq \sigma_1/\sigma_2 \leq \sqrt{3}$, Student's t test provides reasonably good control over the probability of a Type I error, but more recent papers have shown that when $\sigma_1/\sigma_2 > \sqrt{3}$, Student's t test becomes unsatisfactory (e.g., Brown and Forsythe, 1974; Tomarken and Serlin, 1986; Wilcox et al., 1986).

To illustrate what can happen under non-normality, suppose observations for the first group are sampled from a lognormal distribution that has been shifted to have a mean of zero, while the observations from the second group have a normal distribution with mean 0 and standard deviation 0.25. With $n_1 = n_2 = 20$ and $\alpha = 0.025$, the probability of rejecting H_0: $\mu_1 < \mu_2$ is 0.136 (based on simulations with 10,000 replications), while the probability of rejecting H_0: $\mu_1 > \mu_2$ is 0.003. Moreover, Student's t test assumes that $E(T) = 0$, but $E(T) = -0.52$, approximately, again, based on a simulation with 10,000 replications. One implication is that, in addition to yielding a confidence interval with inaccurate probability coverage, the probability of rejecting H_0: $\mu_1 = \mu_2$ with Student's t test has the undesirable property of not being minimized when H_0 is true. That is, Student's t test is biased. If, for example, 0.5 is subtracted from each observation in the second group, the probability of rejecting H_0: $\mu_1 < \mu_2$ drops from 0.136 to 0.083. That is, the mean of the second group has been shifted by a half standard deviation away from the null hypothesis, yet power is less than the probability of rejecting when the null hypothesis is true.

When using Student's t test, poor power properties and inaccurate confidence intervals are to be expected based on results in Chapter 4. To elaborate on why, let $\mu_{[k]} = E(X - \mu)^k$ be the kth moment about the mean of the random variable X. The third moment, $\mu_{[3]}$, reflects skewness, the most common measure being $\kappa_1 = \mu_{[3]}/\mu_{[2]}^{1.5}$. For symmetric distributions, $\kappa_1 = 0$. It can be shown that for two independent random variables, X and Y, having third moments $\mu_{x[3]}$ and $\mu_{y[3]}$, the third moment of $X - Y$ is $\mu_{x[3]} - \mu_{y[3]}$. In other words, if X and Y have equal skewnesses, $X - Y$ has a symmetric distribution. If they have unequal skewnesses, $X - Y$ has a skewed distribution. From Chapter 4, it is known that when X has a skewed distribution and the tails of the distribution are relatively light, the standard confidence interval for μ can have probability coverage that is substantially lower than the nominal level. For symmetric distributions, this problem is much less severe, although probability coverage can be too high when sampling from a heavy-tailed distribution. Consequently, when X_{i1} and X_{i2} have identical distributions, and, in particular, have equal third moments, the third moment of $X_{i1} - X_{i2}$ is zero, suggesting that probability coverage of $\mu_1 - \mu_2$ will not be excessively smaller than the nominal level when using Eq. (5.12). Put another way, if two groups do not differ, $X_{i1} - X_{i2}$ has a symmetric distribution, suggesting that the probability of a Type I error will not exceed the nominal level by too much. (For results supporting this conclusion when dealing with highly discrete data, see Rasch et al., 2007.) However, when distributions differ, and, in particular, have different amounts of skewness, $X_{i1} - X_{i2}$ has a skewed distribution as well, suggesting that probability coverage might be too low, and that T given by Eq. (5.11) does not have a mean of zero as is commonly assumed. This in turn suggests that if groups differ, testing H_0: $\mu_1 = \mu_2$ with Student's t test might result in an undesirable power property—the probability of rejecting might decrease as $\mu_1 - \mu_2$ increases, as was illustrated in the previous paragraph. In fact, if the groups differ, and have unequal variances and differ in skewness, and if the sample sizes differ as well, then confidence intervals based on Eq. (5.12) are not even asymptotically correct. In particular, the variance of T does not go to one as the sample sizes increase (Cressie and Whitford, 1986). In contrast, heteroscedastic methods are asymptotically correct, they give reasonably accurate probability coverage over a wider range of situations than Student's t, so, only heteroscedastic methods are considered in the remainder of this chapter. (For an overview of heteroscedastic methods for means, in the two-sample case, see Sawilowsky, 2002.)

It should be noted, however, that even if two groups have the same amount of skewness, problems with probability coverage and control of Type I error probabilities can arise when distributions differ in scale. This occurs, for example, when sampling from an exponential distribution. Fig. 5.5 shows the probability density function of an exponential distribution, $f(x) = \exp(-x)$. The shape of this distribution is similar to the shape of empirical distributions found in various situations. (An example is given at the end of Section 5.3.3. For examples based on psychometric data, see Sawilowsky and Blair, 1992.) The mean of this distribution is $\mu = 1$, the 20% trimmed mean is $\mu_t = 0.761$, and the M-measure of location (based on Huber's Ψ) is $\mu_m = 0.824$.

Consider two exponential distributions, shifted so that they both have a mean of 0, with the second distribution rescaled so that its standard deviation is four times as large as that of the first. With $n_1 = n_2 = 20$, the probability of a Type I error is 0.082 when testing H_0: $\mu_1 = \mu_2$ at the $\alpha = 0.05$ level. Increasing n_1 to 40, the probability of a Type I error is 0.175, and increasing n_1 to 60, it is 0.245. With $n_1 = n_2 = 40$, it is 0.068.

A natural way of trying to salvage homoscedastic methods is to test for equal variances, and if not significant, assume the variances are equal. Even under normality, this strategy can fail

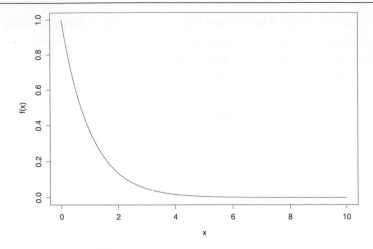

Figure 5.5: An exponential distribution.

because tests for equal variances might not have enough power to detect unequal variances in situations where the assumption should be abandoned, even when the test for equal variances is performed at the 0.25 level (e.g., Hayes and Cai, 2007; Flores and Ocaña, 2019; Markowski and Markowski, 1990; Moser et al., 1989; Wilcox et al., 1986; Zimmerman, 2004). An extensive comparison of methods by Wang et al. (2018) points to a method for testing the homoscedasticity assumption, but situations are still encountered where this approach is unsatisfactory, and there is the added complication that other differences in the distributions, such as differences in skewness, also impact the Type I error probability. Note that from Tukey's perspective, described in Section 4.1.1, the population variances are not equal; they differ at some decimal place. The issue is whether they differ enough to result in inaccurate confidence intervals. The answer is a complicated function of the sample sizes and the nature of the unknown distributions.

5.3 Comparing Medians and Other Trimmed Means

This section considers the problem of computing a $1 - \alpha$ confidence interval for $\mu_{t1} - \mu_{t2}$. Included as a special case is a method for comparing medians, which requires specialized techniques. And there is the related goal of testing

$$H_0: \mu_{t1} = \mu_{t2},$$

the hypothesis that two independent groups have equal trimmed means.

As noted in Section 4.1.1, Tukey (1991) argues that testing for exact equality is nonsensical. For the situation at hand, for example, the argument is that the population means differ at some decimal place. The same argument applies to the situation where trimmed means are being compared. Here, Tukey's three-decision rule (Jones and Tukey, 2000) is used. If the null hypothesis is rejected, a decision is made about which group has the larger population trimmed mean. If the null hypothesis is not rejected, no decision is made. That is, the goal is not to test for equality, but rather to determine which group has the larger trimmed mean. In this context, p-values reflect the strength of the empirical evidence that a decision can be made. But limitations of the p-value summarized in Section 4.1.1 still apply.

Yuen's Method

Yuen (1974) derived a method for comparing trimmed means that is designed to allow unequal Winsorized variances. When there is no trimming ($\gamma = 0$), Yuen's method reduces to Welch's (1938) method for comparing means.

Generalizing the notation of Chapters 3 and 4 in an obvious way, suppose the amount of trimming is γ. For the jth group, let $g_j = [\gamma n_j]$ be the number of observations trimmed from each tail, let $h_j = n_j - 2g_j$ be the number of observations left after trimming, and let s_{wj}^2 be the Winsorized sample variance. From Chapter 3, an estimate of the squared standard error of $\bar{X}_{tj}$ is $s_{wj}^2/\{(1 - 2\gamma)^2 n\}$. However, Yuen estimates the squared standard error with

$$d_j = \frac{(n_j - 1)s_{wj}^2}{h_j(h_j - 1)}. \tag{5.13}$$

It is left as an exercise to verify that both estimates give similar values. In terms of Type I error probabilities and probability coverage, simulations indicate that Yuen's estimate gives slightly better results. Yuen's test statistic is

$$T_y = \frac{\bar{X}_{t1} - \bar{X}_{t2}}{\sqrt{d_1 + d_2}}. \tag{5.14}$$

The null distribution of T_y is approximated with a Student's t distribution with estimated degrees of freedom

$$\hat{v}_y = \frac{(d_1 + d_2)^2}{\frac{d_1^2}{h_1 - 1} + \frac{d_2^2}{h_2 - 1}}.$$

The $1 - \alpha$ confidence interval for $\mu_{t1} - \mu_{t2}$ is

$$(\bar{X}_{t1} - \bar{X}_{t2}) \pm t\sqrt{d_1 + d_2}, \tag{5.15}$$

where t is the $1 - \alpha/2$ quantile of Student's t distribution with $\hat{v}_y$ degrees of freedom. The hypothesis of equal trimmed means is rejected if

$$|T_y| \geq t.$$

(Luh and Guo, 2010, report results on strategies for determining the sample sizes.)

As previously indicated, when two distributions differ, it can be difficult getting a confidence interval for the difference between the means that has probability coverage reasonably close to the nominal level. Theoretical results, supported by simulations, indicate that as the amount of trimming increases from 0 to 20%, Yuen's method yields confidence intervals for $\mu_{t1} - \mu_{t2}$ with probability coverage closer to the nominal level (Wilcox, 1994a). As an illustration, suppose the first group has a normal distribution, and the second group is skewed with $\kappa_1 = 2$ and $n_1 = n_2 = 12$. Wilcox (1994a) reports situations where $H_0: \mu_1 > \mu_2$ is tested with $\alpha = 0.025$, but the actual probability of a Type I error is 0.054. (This result is based on simulations with 100,000 replications.) In contrast, with 20% trimming, the actual probability of a Type I error is 0.022. With $n_1 = 80$ and $n_2 = 20$, the probability of a Type I error can be as high as 0.093—nearly four times higher than the nominal level—when using Welch's test, while with 20% trimming, the actual probability of a Type I error is approximately 0.042 for the same distributions. Of course, by implication, there are some situations where Welch's test will be unsatisfactory when dealing with a two-sided test and $\alpha = 0.05$.

■ **Example**

As another illustration that differences in skewness can make a practical difference, imagine that for the first group, 40 observations are generated from a normal distribution, and for the second group, 20 observations are generated from a lognormal distribution that has been shifted so that it has a mean of zero. When testing at the 0.05 level, the actual level of Welch's test is approximately 0.11. And, if instead, observations for the second group are generated from a g-and-h distribution with $g = h = 0.5$, the actual level is approximately 0.20. Comparing 20% trimmed means instead, the actual levels for these two situations are 0.047 and 0.042, respectively. However, Section 5.3.2 notes that even Yuen's method can be unsatisfactory in terms of Type I errors. An alternative approach to comparing trimmed means is described that gives better results.

■

From Randles and Wolfe (1979, p. 384), the expectation is that the Kolmogorov–Smirnov test will have lower power than Welch's test when sampling from normal distributions with a common variance. More generally, it might seem that when distributions differ in location only, and are symmetric, the Kolmogorov–Smirnov test will have less power than the Yuen–Welch test. Table 5.3 shows the estimated power of these tests for four distributions, $n_1 = n_2 = 25$, and when δ is added to every observation in the first group. The notation KS

Table 5.3: Estimated power, $n_1 = n_2 = 25$, $\alpha = 0.05$.

Distributions	δ	Welch	Yuen ($\gamma = 0.2$)	KS (exact)	KS ($\alpha = 0.052$)
Normal	0.6	0.536	0.464	0.384	0.464
Normal	0.8	0.780	0.721	0.608	0.700
Normal	1.0	0.931	0.890	0.814	0.872
CN1	1.0	0.278	0.784	0.688	0.780
CN2	1.0	0.133	0.771	0.698	0.772
Slash	1.0	0.054	0.274	0.235	0.308

(exact) means that the Kolmogorov–Smirnov critical value was chosen as small as possible with the property that the exact probability of a Type I error will not exceed 0.05. The last column in Table 5.3 shows the power of the Kolmogorov–Smirnov test when the critical value is chosen so that the probability of a Type I error is as close as possible to 0.05. For the situation at hand, the resulting probability of a Type I error is 0.052. The notation CN1 refers to a contaminated normal where, in Eq. (1.1), $\epsilon = 0.1$ and $K = 10$. The notation CN2 refers to a contaminated normal with $K = 20$. As is seen, the exact test does have less power than Welch's test under normality, but the exact test has substantially more power when sampling from a heavy-tailed distribution. Moreover, with $\alpha = 0.052$, the Kolmogorov–Smirnov test has about the same amount of power as Yuen's test with 20% trimming. Another appealing feature of the Kolmogorov–Smirnov test, versus the Yuen–Welch test, is that the Kolmogorov–Smirnov test is sensitive to more features of the distributions. A negative feature of the Kolmogorov–Smirnov test is that when there are tied values among the pooled observations, its power can be relatively low.

Despite the advantages of using a trimmed mean rather than a mean, perhaps it should be stressed that situations occur where Welch's test rejects and Yuen's test does not. This can happen even when there are outliers. For example, if the distributions differ in skewness, the difference between the means can be larger than the difference between the trimmed means to the point that Welch's test will have more power.

Comparing Medians

As the amount of trimming approaches 0.5, Yuen's method breaks down; the method for estimating the standard error becomes highly inaccurate, resulting in inaccurate confidence intervals and poor control over the probability of a Type I error. If there are no tied values in either group, an approach that currently seems to have practical value is as follows. Let M_1 and M_2 be the sample medians corresponding to groups 1 and 2, respectively, and let S_1^2 and S_2^2 be the corresponding McKean–Schrader estimates of the squared standard errors. Then an approximate $1 - \alpha$ confidence interval for the difference between the population medians is

$$(M_1 - M_2) \pm c\sqrt{S_1^2 + S_2^2},$$

where c is the $1 - \alpha/2$ quantile of a standard normal distribution. Alternatively, reject the hypothesis of equal population medians if

$$\frac{|M_1 - M_2|}{\sqrt{S_1^2 + S_2^2}} \geq c.$$

But if there are tied values in either group, control over the probability of a Type I error can be very poor. There are two practical problems, which were noted in Chapter 4. First, with tied values, all known estimators of the standard of the sample median can be highly inaccurate. Second, the sampling distribution of the sample median does not necessarily approach a normal distribution as the sample size gets large. When tied values occur, the only known method for comparing medians that performs well in simulations, in terms of controlling the probability of a Type I error, is the percentile bootstrap method in Section 5.4.2.

5.3.1 R Functions yuen and msmed

The R function

$$\text{yuen(x,y,tr=0.2,alpha= 0.05)}$$

performs the Yuen–Welch method for comparing trimmed means. The default amount of trimming (tr) is 0.2, and the default value for α is 0.05. Thus, the command yuen(x,y) returns a 0.95 confidence interval for the difference between the 20% trimmed means using the data stored in the R vectors x and y. The confidence interval is returned in the R variable yuen$ci. The command yuen(x, y, 0) returns a 0.95 confidence interval for the difference between the means based on Welch's method. The function also returns the value of the test statistic in yuen$teststat, a p-value in yuen$p.value, a $1 - \alpha$ confidence interval in yuen$ci, the estimated degrees of freedom, the estimated difference between the trimmed means, and the estimated standard error.

The R function

$$\text{msmed(x,y,alpha= 0.05)}$$

compares medians using the McKean–Schrader estimates of the squared standard errors.

■ Example

For the ozone data and 20% trimming, the R function yuen indicates that $T_y = 3.4$, the p-value is 0.0037, and the 0.95 confidence interval for $\mu_{t1} - \mu_{t2}$ is $(5.3, 22.85)$. In contrast, with zero trimming (Welch's method), $T_y = 2.46$, the p-value is 0.019, and the 0.95 confidence interval is $(1.96, 20.8)$. Both methods suggest that for the typical rat, weight gain is higher for rats living in an ozone-free environment, but they give a different picture of the extent to which this is true.

■

5.3.2 A Bootstrap-t Method for Comparing Trimmed Means

As previously indicated, when testing hypotheses with the Yuen–Welch method, control of Type I error probabilities is generally better when using 20% trimming versus no trimming at all. However, problems might persist when using 20% trimming, especially when performing a one-sided test and the sample sizes are unequal. For example, if sampling is from exponential distributions with sample sizes of 15 and 30, and the second group has a standard error four times as large as the first, the probability of a Type I error can be twice as large as the nominal level. With $\alpha = 0.025$, $P(T_y < t_{0.025}) = 0.056$, while with $\alpha = 0.05$, the probability is 0.086.

As in the one-sample case, discussed in Chapter 4, there are two fundamental bootstrap methods for dealing with this issue: a bootstrap-t method (sometimes called a percentile-t method) and a percentile bootstrap. The focus in this section is on the bootstrap-t; the percentile bootstrap is discussed in Section 5.4. For the special case, where the goal is to compare trimmed means, the relative merits of these two bootstrap methods are discussed in Section 5.4.2. The bootstrap method advocated by Westfall and Young (1993) has been found to have a practical advantage over the Yuen–Welch method (Wilcox, 1996b), but it seems to have no practical advantage over the bootstrap-t, at least based on extant simulations, so, it is not discussed here.

For the situation at hand, the general strategy of the bootstrap-t method is to estimate the upper and lower critical values of the test statistic, T_y, by running simulations on the available data. This is done by temporarily shifting the two empirical distributions so that they have identical trimmed means, and then generating bootstrap samples to estimate the upper and lower critical values for T_y that would result in a Type I error probability equal to α. Once the critical values are available, a $1 - \alpha$ confidence interval can be computed, as is illustrated later.

One way of describing the bootstrap-t in a more precise manner is as follows. For fixed j, let $X_{1j}^*, \ldots, X_{n_j j}^*$ be a bootstrap sample from the jth group, and set $C_{ij}^* = X_{ij}^* - \bar{X}_{tj}, i =$

$1, \ldots, n_j$. Then C^*_{ij} represents a sample from a distribution that has a trimmed mean of zero. That is, the hypothesis of equal trimmed means is true for the distributions associated with the C^*_{ij} values. Consequently, applying the Yuen–Welch method to the C^*_{ij} values should not result in rejecting the hypothesis of equal trimmed means. Let T^*_y be the value of T_y based on the C^*_{ij} values. To estimate the distribution of T_y when the null hypothesis is true, repeat the process just described B times, each time computing T^*_y based on the resulting C^*_{ij} values. Label the resulting T^*_y values T^*_{yb}, $b = 1, \ldots, B$. Let $T^*_{y(1)} \leq \cdots \leq T^*_{y(B)}$ be the T^*_{yb} values written in ascending order. Set $\ell = \alpha B/2$, round ℓ to the nearest integer, and let $u = B - \ell$. Then an estimate of the lower and upper critical values is $T^*_{y(\ell+1)}$ and $T^*_{y(u)}$. That is, reject H_0: $\mu_{t1} = \mu_{t2}$ if $T_y < T^*_{y(\ell+1)}$ or $T_y > T^*_{y(u)}$. A little algebra shows that a $1 - \alpha$ confidence interval for $\mu_{t1} - \mu_{t2}$ is

$$(\bar{X}_{t1} - \bar{X}_{t2} - T^*_{y(u)}\sqrt{d_1 + d_2}, \bar{X}_{t1} - \bar{X}_{t2} - T^*_{y(\ell+1)}\sqrt{d_1 + d_2}), \qquad (5.16)$$

where d_j, given by Eq. (5.13), is the estimate of the squared standard error of $\bar{X}_{tj}$ used by Yuen. (As in Chapter 4, it might appear that $T^*_{y(u)}$ should be used to compute the upper end of the confidence interval, but this is not the case. Details are relegated to the exercises.) When $\alpha = 0.05$, $B = 599$ appears to suffice in terms of probability coverage, and extant simulations suggest that little is gained using $B = 999$. However, in terms of power, $B = 999$ might make a practical difference. For $\alpha < 0.05$, no recommendations about B can be made for the goal of controlling the Type I error probability.

Table 5.4 provides an equivalent way of describing how to apply the bootstrap-t to the two-sample case, that includes a description of how to compute a symmetric confidence interval instead. The summary in Table 5.4 is very similar to the summary of the one-sample bootstrap-t method given in Table 4.4.

The confidence interval given by Eq. (5.16) is just an extension of the equal-tailed bootstrap-t method described in Chapter 4 to the two-sample case. Chapter 4 noted that there are theoretical results, suggesting that when computing a two-sided confidence interval, a symmetric two-sided confidence interval should be used instead. A symmetric two-sided confidence interval can be obtained for the situation at hand by replacing T_y given by Eq. (5.14) with

$$T_y = \frac{|\bar{X}_{t1} - \bar{X}_{t2}|}{\sqrt{d_1 + d_2}}$$

and letting T^*_y represent the value of T_y based on the bootstrap sample denoted by C^*_{ij}. As before, repeatedly generate bootstrap samples, yielding $T^*_{y1}, \ldots, T^*_{1B}$. Now, however, set $a = (1 - \alpha)B$, rounding to the nearest integer, in which case, the critical value is $c = T^*_{y(a)}$, and the $1 - \alpha$ confidence interval is

$$(\bar{X}_{t1} - \bar{X}_{t1} - c\sqrt{d_1 + d_2}, \bar{X}_{t1} - \bar{X}_{t2} + c\sqrt{d_1 + d_2}). \qquad (5.17)$$

Table 5.4: Summary of the bootstrap-t method for trimmed means.

1. Compute the sample trimmed means, $\bar{X}_{t1}$ and $\bar{X}_{t2}$, and Yuen's estimate of the squared standard errors, d_1 and d_2, given by Eq. (5.13).
2. For the jth group, generate a bootstrap sample by randomly sampling with replacement n_j observations from $X_{1j}, \ldots, X_{nj}$, yielding $X_{1j}^*, \ldots, X_{nj}^*$.
3. Using the bootstrap samples just obtained, compute the sample trimmed means plus Yuen's estimate of the squared standard error, and label the results $\bar{X}_{tj}^*$ and d_j^*, respectively, for the jth group.
4. Compute

$$T_y^* = \frac{(\bar{X}_{t1}^* - \bar{X}_{t2}^*) - (\bar{X}_{t1} - \bar{X}_{t2})}{\sqrt{d_1^* + d_2^*}}.$$

5. Repeat steps 2 through 4 B times, yielding $T_{y1}^*, \ldots, T_{yB}^*$. $B = 599$ appears to suffice in most situations when $\alpha = 0.05$.
6. Put the $T_{y1}^*, \ldots, T_{yB}^*$ values in ascending order, yielding $T_{y(1)}^* \leq \cdots \leq T_{y(B)}^*$. The T_{yb}^* values provide an estimate of the distribution of

$$\frac{(\bar{X}_{t1} - \bar{X}_{t2}) - (\mu_{t1} - \mu_{t2})}{\sqrt{d_1 + d_2}}.$$

7. Set $\ell = \alpha B/2$, rounding to the nearest integer, and let $u = B - \ell$.

The equal-tailed $1 - \alpha$ confidence interval for μ_t is

$$(\bar{X}_{t1} - \bar{X}_{t2} - T_{y(u)}^* \sqrt{d_1 + d_2}, \ \bar{X}_{t1} - \bar{X}_{t2} - T_{y(\ell+1)}^* \sqrt{d_1 + d_2}).$$

($T_{y(\ell)}^*$ will be negative, which is why it is subtracted from $\bar{X}_{t1} - \bar{X}_{t2}$.)
To get a symmetric two-sided confidence interval, replace step 4 with

$$T_y^* = \frac{|(\bar{X}_{t1}^* - \bar{X}_{t2}^*) - (\bar{X}_{t1} - \bar{X}_{t2})|}{\sqrt{d_1^* + d_2^*}}.$$

Set $a = (1 - \alpha)B$, rounding to the nearest integer. The confidence interval for $\mu_{t1} - \mu_{t2}$ is

$$(\bar{X}_{t1} - \bar{X}_{t2}) \pm T_{y(a)}^* \sqrt{d_1 + d_2}.$$

A variation of this approach was derived by Guo and Luh (2000), which is based in part on a transformation stemming from Hall (1992). The basic idea is to transform Yuen's test statistic so that it is approximated reasonably well by a Student's t distribution. Results reported by Keselman et al. (2004) indicate, however, that it is preferable to approximate the null distribution of the test statistic used by Guo and Luh using a bootstrap-t method. Among the situations considered by Keselman et al., a bootstrap-t method was found to perform relatively well when the amount of trimming is set at 10% or 15%. However, with small and unequal sample sizes, situations occur where the method is unsatisfactory when using 10%

trimming. More details are given in Section 5.4.2. In practical terms, it currently seems that a percentile bootstrap method is a bit more satisfactory, and that there can be an advantage in using 20% trimming in terms of controlling the probability of a Type I error. Further evidence for preferring the use of a percentile bootstrap method is reported by Özdemir et al. (2013), who report simulation results when distributions differ in skewness. With no trimming, the bootstrap-t method studied by Keselman et al. can be highly unsatisfactory. For the situations considered by Özdemir et al. (2013), there was little difference between the bootstrap-t and the percentile bootstrap, when using a 20% trimmed mean; the bootstrap-t performed slightly better in terms of controlling the Type I error probability. Further simulations indicate that with 20% trimming, Yuen's method continues to perform reasonably well when both sample sizes are greater than 5. The bootstrap methods described here can break down when either sample size is 6.

5.3.3 R Functions yuenbt and yhbt

The R function

$$yuenbt(x,y,tr=0.2,alpha= 0.05,nboot=599,side=FALSE)$$

computes a $1 - \alpha$ confidence interval for $\mu_{t1} - \mu_{t2}$ using the bootstrap-t method, where the default amount of trimming (tr) is 0.2, the default value for α is 0.05, and the default value for nboot (B) is 599. So far, simulations suggest that in terms of probability coverage, there is little or no advantage to using $B > 599$ when $\alpha = 0.05$. However, there is no recommended choice for B when $\alpha < 0.05$ simply because little is known about how the bootstrap-t performs for this special case. Finally, the default value for side is FALSE, indicating that the equal-tailed two-sided confidence interval is to be used. Using side=TRUE results in the symmetric two-sided confidence interval.

■ **Example**

For the ozone data, yuenbt reports that the 0.95 symmetric two-sided confidence interval for the difference between the trimmed means is $(4.75, 23.4)$. In contrast, the Yuen–Welch method yields a 0.95 confidence interval equal to $(5.3, 22.85)$. The equal-tailed bootstrap-t method yields a 0.95 confidence interval of $(3.78, 21.4)$. The symmetric two-sided confidence interval for the difference between the means is obtained with the command yuenbt(x,y,0,side=TRUE), assuming the data are stored in the R vectors x and y, and the result is $(1.64, 21.2)$. In contrast, yuenbt(x,y,tr=0) yields an equal-tailed confidence interval for the difference between the means: $(1.87, 21.6)$. Note that the lengths of the confidence intervals for the difference between the trimmed means are similar to each other and the length of the confidence interval for the difference between the means, but the next illustration demonstrates that this is not always the case.

■ Example

This example is based on data from a study dealing with the effects of consuming alcohol. (The data are from a portion of a study conducted by M. Earleywine.) The data are stored in the file mitch_dat.txt and can be accessed as indicated in Section 1.10. Two groups of participants reported hangover symptoms the morning after consuming equal amounts of alcohol in a laboratory. Group 1, indicated by a 1 in column 1 of the file mitch_dat.txt, was the control group, and group 2 consisted of sons of alcoholic fathers. Fig. 5.6 shows an adaptive kernel density estimate for the two groups measured at time 1, the second column of the file mitch_dat.txt. Note that the shapes are similar to an exponential distribution, suggesting that confidence intervals for the difference between the means, with probability coverage close to the nominal level, might be difficult to obtain. In fact, even using 20% trimming, the Yuen–Welch method might yield inaccurate probability coverage, as already noted. The main point here is that the length of the confidence intervals based on the Yuen–Welch method can differ substantially from the length of the confidence interval using the bootstrap-t method. The Yuen–Welch method yields a 0.95 confidence interval equal to $(-0.455, 7.788)$, with a p-value of 0.076. In contrast, the equal-tailed bootstrap-t yields a 0.95 confidence interval of $(-4.897, 7.255)$. The ratio of the lengths of the confidence intervals is 0.678. The symmetric bootstrap-t confidence interval is $(-1.357, 8.691)$, and its length divided by the length of the other bootstrap confidence interval is 0.83.

Although 20% trimming performs well under normality in terms of power and efficiency, situations might be encountered where it is desired to use 10% and 15% trimming instead. If this is the case, one strategy is to use the R function

$$\text{yhbt(x, y, tr = 0.15, tr=0.2, nboot = 600, SEED = TRUE, PV=FALSE)},$$

which uses the bootstrap-t version of the test statistic derived by Guo and Luh (2000) that was studied by Keselman et al. (2004). By default, 15% trimming is used. The function returns a confidence interval having probability coverage specified by the argument alpha. A p-value is returned if the argument PV=TRUE, but on occasion, this results in a numerical error causing the function to terminate. And even when not computing a confidence interval, situations are

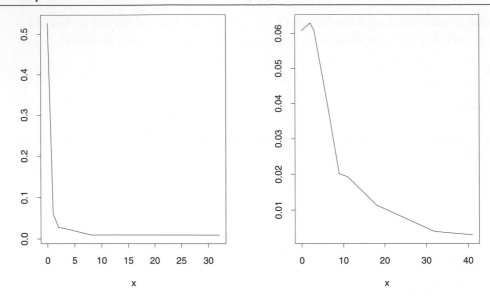

Figure 5.6: Adaptive kernel density estimates for the two groups in the study looking at sons of alcoholic fathers.

encountered where the function is unable to compute a confidence interval. The method is *not* recommended when the goal is to compare means. Another possibility is to use the percentile bootstrap method in Section 5.4.2. Limited studies suggest that even with 10% trimming, there is little or no advantage to using yhbt rather than a percentile bootstrap method. With 20% trimming, results in Özdemir (2013) indicate that the percentile bootstrap method provides better control over the Type I error probability. Moreover, the percentile bootstrap has faster execution time, and computational problems do not arise when computing a p-value or a confidence interval.

5.3.4 Measuring Effect Size

A basic issue is quantifying the extent to which two distributions differ. Peng and Chen (2014) review a collection of methods that might be used. Some general possibilities are:

1. Use the difference between measures of location, such as the difference between the medians. This includes using multiple quantiles, as described in Sections 5.1.3 and 5.1.5.
2. Use a standardized difference. That is, use the difference between two measures of location divided by some measure of variation.
3. Use explanatory power, which has connections to measures of association, such as Pearson's correlation.

4. Use a classification perspective. That is, given a value from one of the groups, what is the probability of correctly deciding whether it came from the first group?
5. Use the probability that a randomly sampled observation from the first group is less than a randomly sampled observation from the second group.
6. Characterize the difference between two measures of location in terms of what is called a quantile shift. This approach has connections to the fifth approach, as will be explained in Section 5.7.2, and a closely related measure provides a useful alternative to standardized differences, as explained later in this section.
7. Plot the data.

Regarding plots, some simple possibilities have already been described: the shift function, plots based on the deciles, boxplots, and plots of the distributions. There are many variations summarized by Rousselet et al. (2017). Also see the R package ggplot2. The remainder of this section focuses on approaches 2–4 listed above. Approaches 5 and 6 are described in Section 5.7.

A Standardized Difference

A simple standardized measure of effect size is

$$\delta = \frac{\mu_1 - \mu_2}{\sigma},$$

where by assumption, $\sigma_1 = \sigma_2 = \sigma$. That is, homoscedasticity is assumed, and the common variance is denoted by σ^2. Cohen (1988) suggests that as a general guide, $\delta = 0.2, 0.5$, and 0.8 correspond to small, medium, and large effect sizes, respectively, and often this suggestion is followed. But what constitutes a large effect size can depend on the situation, and not surprisingly, arguments can be made that other values for δ should be labeled small, medium, and large. The usual estimate of δ, popularly known as Cohen's d, as well as Hedge's g, is

$$d = \frac{\bar{X}_1 - \bar{X}_2}{s},$$

where $s^2 = [(n_1 - 1)s_1^2 + (n_2 - 1)s_2^2]/(n_1 + n_2 - 2)$ estimates the assumed common variance.

There are fundamental concerns regarding this measure of effect size. The first is that δ is based on the mean and variance, which are not robust. Even when dealing with symmetric distributions, heavy-tailed distributions can result in δ being relatively small, when from a graphical perspective, the difference between the two distributions appears to be relatively large. The left panel of Fig. 5.7 shows two normal distributions, both having variance 1, for which $\delta = 0.8$, which is often viewed as a large effect size. But now, look at the right

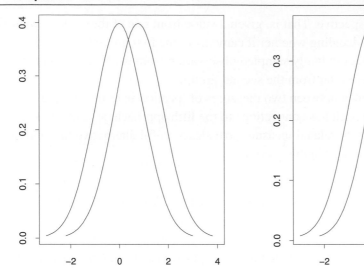

Figure 5.7: The left panel shows two normal distributions for which the measure of effect size δ is 0.8, which is often taken to be a large value. For the right panel, $\delta = 0.24$, despite the similarity to the left panel, illustrating that slight changes in the tails of the distributions can have a major impact on the magnitude of δ.

panel, where again, the difference between the means is 0.8. Despite the similarity with the left panel, $\delta = 0.24$, which is typically considered to be small. The reason δ is substantially smaller in the right panel is that the two distributions are mixed normal distributions, which have variance 10.9. A related concern is that outliers can result in d, the estimate of δ, being relatively small as well, when in fact, for the bulk of the participants, there is a relatively large effect. A second general concern is that δ assumes homoscedasticity.

Algina et al. (2005) suggest using a generalization of δ based on trimmed means and Winsorized variances, namely,

$$\delta_t = \frac{\mu_{t1} - \mu_{t2}}{\sigma_w},\tag{5.18}$$

where it is assumed that the groups have a common (population) Winsorized variance σ_w^2. This will be called method AKP henceforth. The estimate of σ_w^2 is rescaled so that under normality, it estimates the assumed common variance used in the definition of δ. With 20% trimming, this means that the estimate of σ_w^2 is divided by 0.4121. The resulting estimate of δ_t is

$$d_t = 0.642 \frac{\bar{X}_{t1} - \bar{X}_{t2}}{S_w},$$

where

$$S_W^2 = \frac{(n_1 - 1)s_{w1}^2 + (n_2 - 1)s_{w2}^2}{n_1 + n_2 - 2}$$

is the pooled Winsorized variance.

Kulinskaya et al. (2008, p. 177) suggest a measure of effect size that deals with hetero-scedasticity when using means and standard deviations. It takes into account the conclusion by Kulinskaya and Staudte (2006, p. 101) that a natural generalization of δ to the hetero-scedastic case does not appear to be possible without taking into account the relative sample sizes. Let $N = n_1 + n_2$, $q = n_1/N$, and note that the squared standard error of $\bar{X}_1 - \bar{X}_2$ can be written as ς^2/N, where

$$\varsigma^2 = \frac{(1-q)\sigma_1^2 + q\sigma_2^2}{q(1-q)}.$$

They use ς as their measure of scale, in which case, their measure of effect size is

$$\delta_{kms} = \frac{\mu_1 - \mu_2}{\varsigma}, \tag{5.19}$$

which will be called the KMS measure effect size henceforth. Under normality, and when the population variances and sample sizes are equal, $\delta = 2\delta_{kms}$. One way of getting a robust version of δ_{kms} is to replace the means and standard deviations with the trimmed means and Winsorized standard deviations, respectively. As was done by the AKP measure of effect size, the resulting measure of effect size is rescaled, yielding $\delta_{kms.t}$, with the property that $\delta_{kms.t} = 2\delta_t$ under normality and homoscedasticity. The resulting estimator, based on a 20% trimmed mean, is

$$\hat{\delta}_{kms.t} = 0.642\frac{\bar{X}_{t1} - \bar{X}_{t2}}{\hat{\varsigma}_w}, \tag{5.20}$$

where

$$\hat{\varsigma}_w^2 = \frac{(1-q)s_{w1}^2 + qs_{w2}^2}{q(1-q)}.$$

A Quantile Shift Perspective

Momentarily focus on the situation where the first group is viewed as the reference group and the goal is to compare the medians. For example, the first group might be a control group and the second group might be an experimental group. For illustrative purposes, assume that when using Cohen's d, $\Delta = 0.2$, 0.5, and 0.8 correspond to small, medium, and large effect sizes, respectively. Of course, the median for the reference group corresponds to the $q_1 = 0.5$ quantile. The median of the second group corresponds to the q_2 quantile of the reference group. Then q_2 reflects the extent to which the median of the second group is unusual relative to the distribution of the reference group. Momentarily assuming homoscedasticity, standardized measures of effect size are readily interpreted in terms of q_2. Consider, for example, $\Delta = 0.2$. This corresponds to shifting the population median of the reference group from the 0.5 quantile to the 0.58 quantile. In a similar manner, for $\Delta = 0.5$ and 0.8, the population mean is shifted to the 0.69 and 0.79 quantiles, respectively. To the extent to which $\delta = 0.2, 0.5$, and 0.8 correspond to small, medium, and large effect sizes, $q_2 = 0.58, 0.69$, and 0.79 indicate small, medium, and large effect sizes as well. In essence, q_2 is a quantile shift measure of effect size that will be called method QSR henceforth.

Consider any measures of location, θ_1 and θ_2, where θ_1 corresponds to the q_1 quantile of the reference group and θ_2 corresponds to q_2 quantile of the reference group. Let $\delta_q = q_1 - q_2$, which captures the spirit of standardized difference, δ_t, without imposing any parametric family of distributions. Not surprisingly, δ_q can paint a different picture regarding the extent to which two distributions differ. Consider, for example, the situation where both groups have distributions that differ in location only. When dealing with skewed distributions, the interpretation of δ_t, given by Eq. (5.18), can break down. As an illustration, suppose the first group has a lognormal distribution with mean μ_1 and the second group has a lognormal distribution that has been shifted to have mean μ_2. If $\mu_1 = \mu_2$, $\delta_t = \delta_q = 0$. But consider the case where $\delta_t = 0.5$. This corresponds to shifting the mean from about the 0.69 quantile to the 0.29 quantile. So, $\delta_q = -0.4$, suggesting a very large effect size rather than a medium effect size, as suggested by δ_t. Using a 20% trimmed mean for the same distributions and $\delta_t = 0.5$, now, $\delta_q = -0.27$. For $\delta_t = 0.8$, $\delta_q = -0.48$. So, again, from the perspective of δ_q, there is a more pronounced effect size. The reverse can happen; δ_q can indicate an effect size that is less pronounced than indicated by δ_t. For the situation at hand, if $\delta_t = -0.8$, $\delta_q = 0.2$. The extent to which δ_t and δ_q differ is a function, among other things, of the magnitude of $\theta_1 - \theta_2$. When the distributions are similar, δ_t and δ_q will generally have similar values. If the distributions differ substantially, δ_t and δ_q might have similar values, but they can differ substantially, as just illustrated.

The magnitude of δ_q depends on which distribution is used as the reference distribution. In the examples just given, δ_q was defined in terms of the quantiles of the first group. As is evident, if the quantiles of the second group are used instead, this can alter δ_q. A variation of this approach that deals with this issue is described in Section 5.7.2.

Explanatory Power

A robust, heteroscedastic approach to measuring effect size was suggested by Wilcox and Tian (2011), which is based on a generalization of the notion of explanatory power (Doksum and Samarov, 1995). From a regression perspective, if $\hat{Y}$ is the predicted value of Y, given X, the explanatory power is

$$\xi^2 = \frac{\sigma^2(\hat{Y})}{\sigma^2(Y)},$$

the variance of the predicted Y values divided by the variance of the observed Y values. If $\hat{Y}$ is taken to be the usual least squares regression line, then $\xi^2 = \rho^2$, where ρ is Pearson's correlation.

Given that an observation is randomly sampled from the jth group, take $\hat{Y} = \mu_j$, in which case

$$\sigma^2(\hat{Y}) = \sum (\mu_j - \bar{\mu})^2,$$

where $\bar{\mu} = (\mu_1 + \mu_2)/2$. Momentarily assume that equal sample sizes are used. Let $\sigma^2(Y|j)$ be the variance of Y given that an observation is randomly sampled from the jth group, and let $\sigma^2(Y)$ be the unconditional variance of Y. Based on the random sample Y_{ij} ($i = 1, \ldots, n$; $j = 1, 2$), $\sigma^2(Y)$ is estimated with $\hat{\sigma}^2(Y)$, the usual sample variance based on these $2n$ (pooled) observations. So, the estimate of ξ^2 is

$$\hat{\xi}^2 = \frac{\hat{\sigma}^2(\hat{Y})}{\hat{\sigma}^2(Y)}.$$

Now, consider how to estimate ξ^2 when unequal sample sizes are used. First, it is stressed that a key component of the approach used here is defining $\sigma^2(Y)$ in terms of situations where equal sample sizes are used with probability 1. Put another way, $\sigma^2(Y)$ is the estimand associated with the sample variance of the pooled Y_{ij} values when $n_1 = n_2$. Given how $\sigma^2(Y)$ is defined, the problem is finding a reasonable estimate of $\sigma^2(Y)$ when dealing with unequal sample sizes. A simple strategy is to again estimate $\sigma^2(Y)$ with the sample variance based on all $n_1 + n_2$ Y_{ij} values, even when $n_1 \neq n_2$. But this estimation method can be shown to be unsatisfactory: The resulting estimate of ξ^2 can be severely biased. To deal with this, suppose the sample sizes are $n_1 < n_2$ for groups 1 and 2, respectively. If we randomly sample (without replacement) n_1 observations from the second group, we have equal sample sizes from both groups, resulting in a satisfactory estimate of ξ^2. That is, use the estimation method for the equal-sample case, where both groups have sample size n_1. To use all of the data in the second group, repeat this process K times, yielding a series of estimates for ξ^2, which are then averaged to get a final estimate, which we label $\hat{\xi}^2$. The estimate of ξ is just

$$\hat{\xi} = \sqrt{\hat{\xi}^2} \tag{5.21}$$

and is called the *explanatory measure of effect size*.

To get a robust version of ξ^2, simply replace the mean with some robust measure of location, and replace $\sigma^2(Y)$ with some robust measure of variation. Here, unless stated otherwise, a 20% trimmed mean and a 20% Winsorized variance are used, where the Winsorized variance is rescaled to estimate the usual variance, σ^2, when sampling from a normal distribution. For 20% Winsorization, this means that rather than compute the Winsorized variance of the pooled Y_{ij} values with say s_{wy}^2, use $s_{wy}^2/0.4121$. When dealing with medians, one possibility is to proceed as just described but with the Winsorized variance replaced by some other robust measure of variation, such as the percentage bend midvariance. Under normality and homoscedasticity, $\delta = 0.2, 0.5$, and 0.8 roughly correspond to $\xi = 0.15, 0.35$, and 0.50, respectively. If, for example, $\delta = 0.5$ is viewed as a medium effect size, as is often done, this corresponds to $\xi = 0.35$. It is noted that when measuring the strength of the association between two variables via Pearson's correlation, for normal distributions, it has been suggested that $\rho = 0.1, 0.3$, and 0.5 are relatively small, medium, and large values (e.g., Cohen, 1988).

A Classification Perspective

Consider some value x that was randomly generated from one of the two distributions being compared. Another approach to quantifying the extent to which two distributions differ is in terms of the ability to determine whether this value came from the first or second group. There are two components of this approach. The first is choosing a method for making a decision about whether some observed value x came from the first or second group. The second component is finding a reasonably accurate estimate of P_c, the probability of making a correct decision.

Levy (1967) proposed using a classic discriminate analysis method for dealing with the first component, which assumes normality. To deal with non-normality, Wilcox and Muska (1999) used a kernel density estimator. To elaborate, let $\hat{f}_j(x)$ $(j = 1, 2)$ be some kernel density estimate of the distribution associated with the jth group. The decision rule is that x came from the first group if $\hat{f}_1(x) > \hat{f}_2(x)$; otherwise, decide x came from group 2. Other possibilities exist, some of which are mentioned in Chapter 6, but their relative merits for the situation at hand have not been studied. Various strategies for estimating P_c have been compared by Wilcox and Muska (1999). Their results suggest using a variation of the 0.632 bootstrap estimate of P_c, which is described in Section 11.10.3.

A Probabilistic Measure of Effect Size

Yet another useful measure of effect size is

$$p = P(X_1 < X_2),$$

the probability that a randomly sampled observation from the first group is less than a randomly sampled observation from the second group. Estimation of this probability is described in Section 5.7.

5.3.5 R Functions ESfun, akp.effect.ci, KMS.ci, ees.ci, med.effect, qhat, and qshift

The R function ES.summary.CI, described in Section 5.7.3, can be used to compute several measures of effect sizes simultaneously. Confidence intervals are reported as well. This section describes some functions dealing with measures of effect size described in the previous section.

By default, the R function

$$\text{ESfun(x,y,method=c('EP','QS','QStr','AKP','WMW','KMS'), tr=0.2)}$$

estimates explanatory power. Setting the argument method = 'AKP', the effect size δ_t, given by Eq. (5.18), is estimated. (This is done by calling the R function yuenv2.) The function automatically rescales the Winsorized variance so that, based on the amount of trimming used, it estimates the usual variance under normality. To get Cohen's d, set the argument tr=0. The effect size δ_{kms} is computed when method = 'KMS' and method = 'WMW' estimate $p = P(X_{i1} < X_{i2})$. The choices method = 'QS' and 'QStr' refer to the quantile shift measure of effect size described in Section 5.7.2. For method QS, the function returns an estimate of the probability of getting a value less than or equal to the median of the experimental group relative to the distribution of the control group. For example, if the median of an experimental group is 36 and the probability of being less than or equal to 36 in the control group is estimated to be 0.8, the estimate is that 36 is the 0.8 quantile of the control group. Method QStr is the same, only a trimmed mean is used. Using method = 'AKP' is inappropriate when the amount of trimming is close to 0.5 because the Winsorized variance breaks down. When using the median, one possibility is to replace the Winsorized variance with the percentage bend midvariance, and this is done by the R function

$$\text{med.effect(x,y,HD=TRUE,eq.var=FALSE,nboot=100,loc.fun=median,varfun=pbvar).}$$

By default the Harrell–Davis estimator is used. Setting the argument HD=FALSE, the usual sample median is used instead. (Also see Hedges and Olkin, 1985, p. 93.) When $n_1 \neq n_2$, the estimate is based on bootstrap samples of size $\min(n_1, n_2)$. The final estimate is the average of the bootstrap estimates.

The R function

$$\text{akp.effect.ci(x,y, tr = 0.2, nboot = 2000, alpha = 0.05, SEED = TRUE)}$$

computes a confidence interval for δ_t. The R function

$$\text{KMS.ci(x,y, tr=0.2, alpha=0.05, nboot=500, SEED=TRUE)}$$

estimates the KMS measure of effect size.

The R function

$$\text{ees.ci(x,y,SEED=TRUE,nboot=400,tr=0.2,alpha= 0.05)}$$

computes a $1 - \alpha$ confidence interval for $|\xi|$. A percentile bootstrap method is used, but modified so that if the p-value is greater than α when testing H_0: $\mu_{t1} = \mu_{t2}$ with Yuen's method, the lower end of the $1 - \alpha$ confidence interval is set equal to zero. (If the goal is to compute a confidence interval for ξ rather than $|\xi|$, a percentile bootstrap method can be unsatisfactory.)

The R function

$$\text{qhat(x,y,nboot=50,op=2,SEED=TRUE)}$$

estimates P_c, the likelihood of correctly determining that a value x came from the first group based on a kernel density estimate of the distributions. And the R function

$$\text{qshift(x,y,locfun=tmean, ...)}$$

estimates the quantile shift measure of effect size, δ_q, assuming the first group is the reference group. The R function shiftQSci, described in Section 5.7.3, estimates a quantile shift type measure of effect size that does not require designating one of the groups as the reference group.

■ **Example**

A practical issue is the effect of ignoring heteroscedasticity when using δ rather than ξ to measure effect size. That is, can the choice of method alter the extent to which an effect size is deemed to be large? For illustrative purposes, we adopt the convention that $\delta = 0.2$, 0.5, and 0.8 correspond to small, medium, and large effect sizes, respectively. As already noted, under normality and homoscedasticity, these values roughly correspond to $\xi = 0.15$, 0.3, and 0.5. Note that if the group with the larger sample size also has the larger variance, this results in a relatively small value for d. To illustrate how d compares to $\hat{\xi}$, simulations were used to estimate both effect sizes with $n_1 = 80$ and $n_2 = 20$, where the first group has a normal distribution with mean 0.8 and standard deviation 4 and the second group has a standard normal distribution. Based on 1,000 replications, the median value of d was 0.22, which is typically considered to be a small effect size. (The mean value of d was nearly identical to the median.) The median value of $\hat{\xi}$ was 0.40, which suggests a medium effect size. So, even under normality, a heteroscedastic measure of effect size can make a practical difference. If instead, the first group has standard deviation 1 and the second has standard deviation 4, now, the median estimates are 0.42 and 0.32. That is, in contrast to the first situation, the choice between homoscedastic and heteroscedastic measures of effect size makes little difference. If instead, $n_1 = n_2 = 20$, now, the median d value is 0.30, a somewhat small effect size, and the median $\hat{\xi}$ value is 0.34, which suggests a medium effect size instead. The effect of ignoring heteroscedasticity is less of an issue with equal sample sizes, compared to the first situation considered, but it has practical consequences.

■

■ Example

In a study of sexual attitudes, 1,327 males and 2,282 females were asked how many sexual partners they desired over the next 30 years. (The data used in this example, supplied by Lynn Miller, are stored in the file miller.dat and can be downloaded from the author's web page given in Chapter 1.) Welch's test returns a p-value of 0.30, but Yuen's test gives a p-value less than 0.001. Cohen's d is estimated to be 0.049. In contrast, $\hat{\delta}_t = 0.48$, suggesting a medium effect size, and $\hat{\xi} = 0.47$, suggesting a large effect size. However, the usual sample median for both groups was estimated to be one, which results in a measure of effect size of zero, the point being that different measures of effect size can give a decidedly different sense of the extent to which groups differ. That is, multiple measures are needed to understand the extent to which groups differ. One way of getting a more detailed understanding of how the groups differ is to compare the quantiles via the R function qcomhd, which reveals that the largest differences occur in the right tails of the distributions. The difference between the 0.75 quantiles is estimated to be 5.6, and for the 0.90 quantile, the difference is 18.2. That is, males are more likely to give more extreme responses compared to females.

■

5.4 Inferences Based on a Percentile Bootstrap Method

Inferences based on a percentile bootstrap method have been found to be particularly effective when working with a wide range of robust estimators. When comparing two independent groups, the method is applied as follows. First, generate bootstrap samples from each group as described in Table 5.4. Let $\hat{\theta}_j^*$ be the bootstrap estimate of θ_j, where now, θ_j is any parameter of interest associated with the jth group ($j = 1, 2$). Set

$$D^* = \hat{\theta}_1^* - \hat{\theta}_2^*.$$

Repeat this process B times, yielding $D_1^*, \ldots, D_B^*$, let ℓ be $\alpha B/2$, rounded to the nearest integer, and let $u = B - \ell$, in which case, an approximate $1 - \alpha$ confidence interval for $\theta_1 - \theta_2$ is

$$(D_{(\ell+1)}^*, D_{(u)}^*),$$

where $D_{(1)}^* \leq \cdots \leq D_{(B)}^*$.

The theoretical foundation for the method is similar to the theoretical foundation in the one-sample case, described in Chapter 4. Imagine the goal is to test

$$H_0: \theta_1 = \theta_2.$$

For the bootstrap estimates $\hat{\theta}_1^*$ and $\hat{\theta}_2^*$, let

$$p^* = P(\hat{\theta}_1^* > \hat{\theta}_2^*).$$

If the null hypothesis is true, then asymptotically (as both n and B get large), p^* has a uniform distribution. Consequently, reject H_0 if $p^* \leq \alpha/2$ or if $p^* \geq 1 - \alpha/2$. Although p^* is not known, it is readily estimated. Let A be the number of values among $D_1^*, \ldots, D_B^*$ that are greater than zero. Then an estimate of p^* is

$$\hat{p}^* = \frac{A}{B}.$$

For convenience, set

$$\hat{p}_m^* = \min(p^*, 1 - p^*).$$

Then $2\hat{p}_m^*$ is an estimate of what Liu and Singh (1997) call the generalized p-value, and H_0 is rejected if

$$2\hat{p}_m^* \leq \alpha.$$

This last equation leads to the confidence interval given in the previous paragraph.

5.4.1 Comparing M-Estimators

This section comments on the special case, where the goal is to compare M-measures of location based on the one-step M-estimator. Chapter 4 noted that, based on simulations conducted so far, the best approach to computing a confidence interval for μ_m, the M-measure of location, is to use a percentile bootstrap method. When comparing two independent groups using a one-step M-estimator of location, again, a percentile bootstrap method performs fairly well. For example, with sample sizes of 20 or 30, among the situations considered in Özdemir (2013), estimates of the actual Type I error probabilities, when testing at the 0.05 level, ranged between 0.049 and 0.061.

A non-bootstrap confidence interval based on an estimate of the standard error will provide good probability coverage when the sample sizes are sufficiently large, assuming the estimated difference is normally distributed, but it is unknown just how large the sample sizes should be before this approach can be recommended, particularly when distributions are skewed. If both distributions are symmetric, confidence intervals based on estimated standard errors seem to have merit when Student's t distribution is used to determine an appropriate critical value, but there is no good decision rule, based on available empirical data, whether distributions are sufficiently symmetric. (One could test the assumption that distributions are symmetric, but how much power should such a test have to justify the use of a method that assumes symmetric distributions?)

Özdemir (2013) derived an alternative method for which the estimated Type I error probabilities ranged between 0.044 and 0.050 among the situations considered in a simulation study, in contrast to using a percentile bootstrap method where the Type I error probabilities ranged between 0.049 and 0.061. The improved control over the Type I error probability comes at the expense of higher execution time and no confidence interval. An outline of the method is as follows. Let $\hat{\mu}_j$ ($j = 1, 2$) denote the one-step M-estimator for the jth group and denote the estimate of the squared standard error by $\hat{\sigma}^2_{mj}$, which is computed as described in Section 3.6.4. (The R function mestse, described in Section 3.6.5, performs the calculation.) Let

$$w_j = \frac{1/\hat{\sigma}^2_{mj}}{1/\hat{\sigma}^2_{m1} + 1/\hat{\sigma}^2_{m2}},$$

$$\tilde{X} = w_1\hat{\mu}_1 + w_2\hat{\mu}_2,$$

and

$$T_j = \frac{\hat{\mu}_j - \tilde{X}}{\hat{\sigma}_{mj}}.$$

Apply a normalizing transformation, derived by Bailey (1980), to the T_j values:

$$Z_j = \frac{4v_j^2 + 5(2z_{1-\alpha/2}^2 + 3)/24}{4v_j^2 + v_j + (4z_{1-\alpha/2}^2 + 9)/12} v_j^{1/2} \left\{ ln \left(1 + \frac{T_j^2}{v_j} \right) \right\}^{1/2},$$

where $z_{1-\alpha/2}$ is the $1 - \alpha/2$ quantile of a standard normal distribution, $v_j = n_j - i_1 - i_2 - 1$, and i_1 and i_2 are defined as in Section 3.6.2 in connection with Eq. (3.25). The test statistic is

$$D^2 = Z_1^2 + Z_2^2.$$

An α level critical value is determined via a bootstrap-t method as described in Section 5.3.2. Also see method TM in Section 7.6. For a method based on the empirical likelihood technique, see Velina et al. (2019).

5.4.2 Comparing Trimmed Means and Medians

When comparing trimmed means, when the amount of trimming is at least 20%, it currently seems that a percentile bootstrap method is preferable to the bootstrap-t method in Section 5.3.2. With a sufficiently small amount of trimming, a bootstrap-t method provides more accurate results, but there is uncertainty about when this is the case. (Comments on using 10% trimming are given at the end of this section.)

For the special case, where the goal is to compare medians, a slight extension of the percentile bootstrap method is needed in case there are tied values (cf. Freidlin and Gastwirth, 2000). Let M_1^* and M_2^* be the bootstrap sample medians. Let

$$p^* = P(M_1^* > M_2^*) + 0.5P(M_1^* = M_2^*).$$

So, among B bootstrap samples from each group, if A is the number of times $M_1^* > M_2^*$ and C is the number of times $M_1^* = M_2^*$, the estimate of p^* is

$$\hat{p}^* = \frac{A}{B} + 0.5\frac{C}{B}.$$

As usual, the p-value is

$$2\min(\hat{p}^*, 1 - \hat{p}^*).$$

In terms of controlling the Type I error probability, all indications are that this method performs very well regardless of whether tied values occur (Wilcox, 2006c). And in terms of handling tied values, this is the only known method that performs well in simulations.

Section 5.3.2 mentioned a bootstrap-t method (that is performed by the R function yhbt) that is based in part on a test statistic derived by Guo and Luh (2000). As previously noted, Keselman et al. (2004) found that it performs reasonably well in simulations when using 10% and 15% trimming. To extend slightly their results, consider a situation where $n_1 = 40$ observations are sampled from a standard normal distribution. And for the second group, $n_2 = 20$ observations are sampled from a lognormal distribution shifted so that the trimmed mean is zero, after which, the scale is increased by multiplying all observations by 4. When testing at the 0.05 level, when 10% trimming is used, the actual level of the bootstrap-t method is approximately 0.066, compared to 0.050 when using a percentile bootstrap method (based on a simulation with 1,000 replications). Reducing the first sample size to $n_1 = 20$ and the second to $n_2 = 10$, the estimates are now 0.082 and 0.074, respectively. Increasing the amount of trimming to 0.2, again using sample sizes $n_1 = 20$ and $n_2 = 10$, the estimates are 0.081 and 0.063. So, at least in some situations, the percentile bootstrap method has a bit of an advantage when using 10% trimming. And increasing the amount of trimming from 10% to 20% can improve control over the Type I error probability. Results in Özdemir (2013) also indicate that the percentile bootstrap method provides better control over the Type I error probability.

5.4.3 R Functions trimpb2, pb2gen, medpb2, and M2gbt

When comparing independent groups, the R function

$$pb2gen(x,y,alpha= 0.05,nboot=2000,est=onestep,...)$$

can be used to compute a confidence interval for the difference between any two measures of location or scale using the percentile bootstrap method.

As usual, x and y are any R vectors containing data. The default value for α is 0.05, the default for B (nboot) is 2,000. The last argument, est, is any R function that is of interest. The default value for est is onestep, which is the R function described in Chapter 3 for computing a one-step M-estimator. The command pb2gen(dat1,dat2,est=mom), for example, would use the modified one-step M-estimators based on the data stored in the R variables dat1 and dat2.

■ **Example**

For the ozone data, the 0.95 confidence interval for the difference between the M-measures of location returned by pb2gen is (4.8, 21.5). The p-value is 0.006.

■

Medians can be compared via the usual sample median, described in Section 1.3, with the R function pb2gen by setting the argument est=median, and trimmed means can be compared by setting est=tmean. But for convenience, the R function

$$medpb2(x,y,alpha= 0.05,nboot=2000)$$

is supplied, which is designed specifically for comparing medians based on the usual sample median. The R function

$$trimpb2(x,y,alpha= 0.05,nboot=2000)$$

defaults to comparing 20% trimmed means.

The R function

$$M2gbt(x,y,tr=0.2, bend = 1.28, nboot = 599, SEED = TRUE)$$

tests the hypothesis that two independent groups have equal population M-measures of location using the method derived by Özdemir (2013), which was described in Section 5.4.1.

5.5 Comparing Measures of Scale

In some situations, there is interest in comparing measures of scale. Based purely on efficiency, various robust estimators of scale have appeal. First, however, attention is focused on comparing the variances.

5.5.1 Comparing Variances

Consider the goal of testing

$$H_0: \sigma_1^2 = \sigma_2^2, \tag{5.22}$$

the hypothesis that two independent groups have equal variances. Numerous methods have been proposed. Virtually all have been found to be unsatisfactory with small to moderate sample sizes.

A variation of the percentile bootstrap method (Wilcox, 2002) that performs relatively well is performed as follows. Set $n_m = \min(n_1, n_2)$ and for the jth group ($j = 1, 2$), take a bootstrap sample of size n_m. Ordinarily, we take a bootstrap sample of size n_j from the jth group, but when sampling from heavy-tailed distributions and when the sample sizes are unequal, control over the probability of a Type I error can be extremely poor for the situation at hand. Next, for each group, compute the sample variance based on the bootstrap sample, and set D^* equal to the difference between these two values. Repeat this $B = 599$ times, yielding 599 bootstrap values for D, which we label $D_1^*, \ldots, D_{599}^*$. As usual, when writing these values in ascending order, we denote this by $D_{(1)}^* \leq \cdots \leq D_{(B)}^*$. Then an approximate 0.95 confidence interval for the difference between the population variances is

$$(D_{(\ell+1)}^*, D_{(u)}^*), \tag{5.23}$$

where, for $n_m < 40$, $\ell = 6$ and $u = 593$; for $40 \leq n_m < 80$, $\ell = 7$ and $u = 592$; for $80 \leq n_m < 180$, $\ell = 10$ and $u = 589$; for $180 \leq n_m < 250$, $\ell = 13$ and $u = 586$; and for $n_m \geq 250$, $\ell = 15$ and $u = 584$.

The method just described is based on a strategy similar to Gosset's derivation of Student's t: Assume normality, and then make adjustments so that for small sample sizes, accurate probability coverage is obtained. This method appears to perform reasonably well under non-normality, but exceptions can occur when the distributions differ in skewness. What appears to be more satisfactory is to use the method just described, only with $B = 1,000$ and a corresponding adjustment to ℓ and u. Nevertheless, situations can be constructed where the probability coverage is unsatisfactory. An example is when the first group has a standard normal distribution, and the second, a lognormal distribution that has been rescaled to have variance one. In terms of Type I errors, methods based on a robust measure of scale are more reliable. For example, the R function b2ci, described in Section 5.5.4, could be used. But if there is a substantive reason for focusing on the variance, of course robust estimators are no longer appropriate.

The bootstrap method just described is limited to testing at the 0.05 level and does not yield a p-value. A way of comparing variances that yields a p-value, at the expense of no confidence interval, is to use a variation of the heteroscedastic analog of the Morgan–Pitman test in Section 5.9.15. Momentarily assume $n_1 = n_2$. Assuming the data are in random order, one could simply apply the method in Section 5.9.15. Now, suppose that $n_1 < n_2$ and $n_2/n_1 \leq 2$. Then for the first n_1 values in both groups, the method in Section 5.9.15 could be used, yielding a p-value, say p_1. Next, rather than use the first n_1 values in group 2, use the final n_1 values, taken in reverse order, yielding the p-value p_2. Let q_1 and q_2 be adjusted p-values via Hochberg's method in Section 7.4.7, which can be computed with the R function p.adjust. Then a p-value when testing Eq. (5.22) is $\min\{q_1, q_2\}$. This approach is readily extended to situations where $n_2/n_1 > 2$. But again, if the two distributions differ substantially in terms of skewness, this approach can be unsatisfactory. See Section 5.9.15 for details.

When sampling from a distribution that is not too skewed and not very heavy-tailed, the method in Shoemaker (2003) might be used instead. Herbert et al. (2011) derived yet another method for comparing variances. How well it performs under non-normality, including situations where distributions differ in skewness, needs more research. In terms of controlling the probability of a Type I error, any practical advantages the method might have over the modified percentile bootstrap method or the Morgan–Pitman approach have not been determined.

5.5.2 R Functions comvar2 and varcom.IND.MP

The R function

$$\text{comvar2(x,y,nboot=1000,SEED=TRUE)}$$

compares variances using the bootstrap method just described. The method can only be applied with $\alpha = 0.05$; modifications based on other α values have not been derived. The function returns a 0.95 confidence interval for $\sigma_1^2 - \sigma_2^2$ plus an estimate of $\sigma_1^2 - \sigma_2^2$ based on the difference between the sample variances, $s_1^2 - s_2^2$, which is labeled vardif.

The R function

$$\text{varcom.IND.MP(x,y,SEED=TRUE)}$$

performs the heteroscedastic analog of the Morgan–Pitman test. The function first randomly permutes the data to ensure that the data are in random order.

5.5.3 Comparing Biweight Midvariances and Deviations From the Median

For some robust measures of scale, the percentile bootstrap method, described in Section 5.4, has been found to perform well. In particular, Wilcox (1993a) found that it gives good results when working with the biweight midvariance. (Other methods were considered but found to be unsatisfactory, so they are not discussed.) There is some indirect evidence that it will give good results when working with the percentage bend midvariance, but this needs to be checked before it can be recommended. Using the 20% Winsorized variance also performs well.

In some situations, it might be desirable to get a more detailed understanding of how two distributions differ in scale. For example, there might be more deviation in the right tails compared to the left tails. Moreover, the deviation in the more extreme ends of the tails might be relevant, a feature that is missed by robust measures of scatter. One way of proceeding is to first center the data for both groups based on some measure of location. For example, one could shift the data so that each group has a median of zero. Next, compare all of the quantiles using the shift function, in Section 5.1.3, on the centered data. Or a few of the upper and lower quantiles could be compared using method Q2, described in Section 5.1.5. Plotting the distributions of the centered data can be useful as well.

5.5.4 R Functions b2ci, comvar.locdis, and g5.cen.plot

Robust measures of scale are easily compared with the R function pb2gen in Section 5.4.3. For convenience, the function

$$b2ci(x,y,alpha= 0.05,nboot=2000,est=bivar)$$

has been supplied; it defaults to comparing the biweight midvariances. (When using pb2gen, setting est=bivar returns the same results when using the default settings of b2ci.)

The R function

comvar.locdis(x,y,loc.fun=median,CI=FALSE, plotit=TRUE,xlab='First Group', ylab='Est.2 - Est.1', sm=TRUE, QCOM=FALSE, q=c0(.1,0.25,0.75,0.9), MC=FALSE,nboot=2000,...)

centers the data based on the measure of location indicated by the argument loc.fun, which defaults to the median. Then it compares the quantiles using the shift function. To compare specific quantiles using method Q2 from Section 5.1.5, set the argument QCOM=TRUE. By default, the 0.1, 0.25, 0.75, and 0.9 quantiles are compared. The R function

g5.cen.plot(x1, x2, x3 = NULL, x4 = NULL, x5 = NULL, fr = 0.8, aval = 0.5, xlab = 'X', ylab = '', color = rep('black', 5), main = NULL, sub = NULL, loc.fun = median)

plots an estimate of the centered data using an adaptive kernel density estimator. Up to five distributions can be plotted.

■ **Example**

For the ozone data, the 0.95 confidence interval returned by the R function b2ci is $(-538, -49)$, with a p-value of 0.012. In this particular case, the function comvar.locdis does not reject. Using the function g5.cen.plot, the resulting plot suggests that there is more variation in the left tail of the first group, but in the right tail, there is more variation associated with the second group. The details are left as an exercise.

■

5.6 Permutation Tests

This section describes a permutation test for comparing the distributions corresponding to two independent groups, an idea introduced by R. A. Fisher in the 1930s. The method is somewhat similar to bootstrap techniques, but it accomplishes a different goal, as will become evident. There are many extensions and variations of the method about to be described, including a range of techniques aimed at multivariate data (e.g., Good, 2000; Pesarin, 2001; Rizzo and Székely, 2010), but only the basics are included here.

The permutation test in this section can be used with virtually any measure of location or scale, but regardless of which measure of location or scale is used, in essence, the goal is to test the hypothesis that the groups under study have identical distributions. To illustrate the basics, the method is first described using means. The steps are as follows (cf. Chowdhury et al., 2015):

1. Compute $d = \bar{X}_1 - \bar{X}_2$, the difference between the sample means, where the sample sizes are n_1 and n_2.
2. Pool the data.
3. Consider any permutation of the pooled data, compute the sample mean of the first n_1 observations, compute the sample mean using the remaining n_2 observations, and compute the difference between these sample means.
4. Repeat the previous step for all possible permutations of the data, yielding, say, L differences: $\hat{\delta}_1, \ldots, \hat{\delta}_L$.
5. Put these L differences in ascending order, yielding $\hat{\delta}_{(1)} \leq \cdots \leq \hat{\delta}_{(L)}$.
6. Reject the hypothesis of identical distributions if $d < \hat{\delta}_{(\ell+1)}$ or if $d > \hat{\delta}_{(u)}$, where $\ell = \alpha L/2$, rounded to the nearest integer, and $u = L - \ell$.

Although this variation of the permutation test is based on the sample mean, it is known that it does not provide satisfactory inferences about the population means. In particular, it does not control the probability of a Type I error when testing H_0: $\mu_1 = \mu_2$, and it does not yield a satisfactory confidence interval for $\mu_1 - \mu_2$. For example, Boik (1987) establishes that when the goal is to compare means, unequal variances can affect the probability of a Type I error, even under normality, when testing H_0: $\mu_1 = \mu_2$. If the sample means are replaced by the sample variances, it can be seen that differences between the population means can affect the probability of a Type I error even when the population variances are equal. (The details are left as an exercise.) However, the method provides an exact distribution-free method for testing the hypothesis that the distributions are identical. For results on using a permutation test with the mean replaced by a robust estimator, see Lambert (1985). When the goal is to compare medians, again, a permutation test can be unsatisfactory (Romano, 1990). For yet another situation where a permutation is unsatisfactory, see Kaizar et al. (2011). Chung and Romano (2013) summarize general theoretical concerns and limitations. They indicate a variation of the permutation test aimed at addressing the concerns described in their paper, but situations can be constructed where it is unsatisfactory in terms of controlling Type I error probability when using means. For example, consider the situation where both X and Y have lognormal distributions that have been shifted to have a population mean of zero. When Y has a standard deviation four times larger than the standard deviation of X, the actual Type I error probability exceeds 0.11 when $n_1 = n_2 = 50$ and when testing at the 0.05 level. For $n_1 = n_2 = 100$, the actual level is approximately 0.083. Also see Janssen and Pauls (2005). The relative merits of the method are in need of further research.

In practice, particularly with large sample sizes, generating all permutations of the pooled data can be impractical. A method for dealing with this problem is to simply use B random permutations instead. Now, proceed as described above, only L is replaced by B.

5.6.1 R Function permg

The R function

$$permg(x,y,alpha= 0.05,est=mean,nboot=1000)$$

performs the permutation test based on B random permutations of the pooled data. (The argument nboot corresponds to B.) By default, means are used, but any measures of location or scale can be used via the argument est.

5.7 Methods Based on Ranks and the Typical Difference

There is another approach to comparing two independent groups that deserves consideration. Consider two independent variables, X_1 and X_2. Note that in Section 5.3.4, measures of effect size were characterized in terms of measures of location and scale associated with their corresponding distributions. In this section, the focus is on the distribution of $X_1 - X_2$, the typical difference between a randomly sampled observation from the first group and a randomly sampled observation from the second group. As will be made clear, this approach includes as a special case, rank-based or non-parametric methods aimed at making inferences about

$$p = P(X_{i1} < X_{i2}), \tag{5.24}$$

the probability that a randomly sampled observation from the first group is smaller than a randomly sampled observation from the second. When there is no difference between the groups and the distributions are identical, $p = 1/2$. The value of p has a natural interest, and some have argued that in many situations, it is more interesting than the difference between any two measures of location (e.g., Cliff, 1993). Additional arguments for comparing groups based on p can be found in Acion et al. (2006), Kraemer and Kupfer (2006), and Vargha and Delaney (2000). For example, in clinical trials, of interest, is the probability that method A is more effective than method B.

The best-known approach to comparing two independent groups, based on an estimate of p, is the Wilcoxon–Mann–Whitney test. The method might appear to provide a reasonable way of testing

$$H_0: p = 0.5, \qquad (5.25)$$

but a fundamental concern is that when distributions differ, there are general conditions under which the Wilcoxon–Mann–Whitney test uses the wrong standard error. More precisely, the standard error used by the Wilcoxon–Mann–Whitney was derived under the assumption that groups have identical distributions. When the distributions differ, under general conditions, the derivation no longer holds. Modern methods use an estimate of the correct standard error regardless of whether the distributions differ. Pratt (1964) establishes that the Wilcoxon–Mann–Whitney test is biased and documents its inability to control the probability of a Type I error when testing $H_0: p = 0.5$.

Let $\mathcal{D}_{ij} = X_{i1} - X_{j2}, i = 1, \dots n_1, j = 1, \dots, n_2$. That is, $\mathcal{D}_{ij}$ represents all pairwise differences. Often the Wilcoxon–Mann–Whitney test is described as a method for comparing the population medians, but it can be unsatisfactory in this regard (e.g., Fung, 1980; Hettmansperger, 1984). To elaborate, for two independent random variables, let θ_D be the population median associated with $\mathcal{D}$, and let θ_1 and θ_2 denote the population median for the jth group ($j = 1, 2$). It is left as an exercise to show that, under general conditions, $\theta_D \neq \theta_1 - \theta_2$. Although the Wilcoxon–Mann–Whitney test does not provide a direct test of the hypothesis that two groups have equal medians, it is based on an estimate of p, and when $p = 0.5$, $\theta_D = 0$. A method for making inferences about θ_D is described near the end of Section 5.7.1.

Various attempts have been made to improve on the Wilcoxon–Mann–Whitney test, but not all of them are listed here. Interested readers can refer to Baumgartner et al. (1998), Brunner and Munzel (2000), Mee (1990), Ryu and Agresti (2008), Zhou (2008), Fligner and Policello (1981), and Newcombe (2006a, 2006b) plus the references they cite (cf. Neuhäuser, 2003). Ruscio and Mullen (2012) compare 12 methods and conclude that a bias corrected and accelerated (BCa) bootstrap method performs relatively well. This particular bootstrap method is similar to the percentile bootstrap, but rather than compute a confidence interval as described in Section 4.4.1, adjusted quantiles of the bootstrap distribution are used that are based on its bias and its skewness (e.g., Davison and Hinkley, 1997). However, Ruscio and Mullen (2012) note that when testing the hypothesis that $p = 0.5$, other methods tend to perform better. For methods designed specifically for categorical data (having a multinomial distribution), see

Ryu and Agresti (2008). (For a summary of related techniques aimed at testing the hypothesis that two independent groups have identical distributions, see Marozzi and Reiczigel, 2018).

Motivated by results in Neuhäuser et al. (2007) as well as Ruscio and Mullen (2012), the focus here is on the methods derived by Cliff (1996) as well as Brunner and Munzel (2000). Both methods allow tied values, and both use a correct estimate of the standard error even when distributions differ. Both methods can break down as P approaches zero or one. Included is a slight modification of the methods aimed at dealing with this issue. The modification works well when using Cliff's method but not when using the Brunner–Munzel method.

It is noted that situations can be constructed where, with many tied values, Cliff's method seems to be a bit better than the Brunner–Munzel method in terms of guaranteeing an actual Type I error probability less than the nominal α level. When testing at the 0.05 level, Cliff's method seems to do an excellent job of avoiding actual Type I error probabilities less than 0.04. In contrast, the Brunner–Munzel method can have an actual Type I error rate close to 0.07 when tied values are common and sample sizes are small. However, with no tied values and when both samples are greater than or equal to 10, the Brunner–Munzel method can be more satisfactory in terms of controlling the Type I error probability. Also, with very large sample sizes, the Brunner–Munzel method is more convenient from a computational point of view.

Neubert and Brunner (2007) derive a permutation test that provides slightly better control over the Type I error probability, compared to the Brunner–Munzel method, when the sample sizes are small. In their simulations when both sample sizes are equal to 7, or when one sample size is 7 and the other is 15, they found situations where the actual level of the Brunner–Munzel method is as high as 0.059 among the three types of distributions they considered. In contrast, the actual level using the permutation method ranged between 0.053 and 0.054. When using the bootstrap method proposed by Reiczigel et al. (2005), limited results indicate the actual level tends to be lower than the nominal level.

5.7.1 The Cliff and Brunner–Munzel Methods

This section describes two methods for testing H_0: $p = 0.5$. The first was derived by Cliff (1996), and the other was derived by Brunner and Munzel (2000). When tied values are impossible, the basic goal is to make inferences about $p = P(X_1 < X_2)$, the probability that a randomly sampled observation from the first group is less than a randomly sampled observation from the second group. But when tied values can occur, a different formulation is required. Let

$$p_1 = P(X_1 > X_2),$$

$$p_2 = P(X_1 = X_2),$$

and

$$p_3 = P(X_1 < X_2).$$

For convenience, set $P = p_3 + 0.5p_2 = p + 0.5p_2$. The usual generalization to tied values replaces $H_0\colon p = 0.5$ with

$$H_0\colon P = 0.5.$$

So, when tied values occur with probability zero, this hypothesis becomes $H_0\colon p = 0.5$.

Cliff's Method

Cliff prefers a slightly different perspective, namely, testing

$$H_0\colon \delta = p_1 - p_3 = 0.$$

It is readily verified that $\delta = 1 - 2P$.

For the ith observation in group 1 and the hth observation in group 2, let

$$d_{ih} = \begin{cases} -1, & \text{if } X_{i1} < X_{h2}, \\ 0, & \text{if } X_{i1} = X_{h2}, \\ 1, & \text{if } X_{i1} > X_{h2}. \end{cases}$$

An estimate of $\delta = P(X_{i1} > X_{i2}) - P(X_{i1} < X_{i2})$ is

$$\hat{\delta} = \frac{1}{n_1 n_2} \sum_{i=1}^{n_1} \sum_{h=1}^{n_2} d_{ih}, \tag{5.26}$$

the average of the d_{ih} values. Let

$$\bar{d}_{i.} = \frac{1}{n_2} \sum_{h=1}^{n_2} d_{ih},$$

$$\bar{d}_{0.h} = \frac{1}{n_1} \sum_{i=1}^{n_1} d_{ih},$$

$$s_1^2 = \frac{1}{n_1 - 1} \sum_{i=1}^{n_1} (\bar{d}_{i.} - \hat{\delta})^2,$$

$$s_2^2 = \frac{1}{n_2 - 1} \sum_{h=1}^{n_2} (\bar{d}_{0.h} - \hat{\delta})^2,$$

$$\tilde{\sigma}^2 = \frac{1}{n_1 n_2 - 1} \sum \sum (d_{ih} - \hat{\delta})^2.$$

Then

$$\hat{\sigma}^2 = \frac{(n_1 - 1)s_1^2 + (n_2 - 1)s_2^2 + \tilde{\sigma}^2}{n_1 n_2}$$

estimates the squared standard error of $\hat{\delta}$. Let z be the $1 - \alpha/2$ quantile of a standard normal distribution. Rather than use the more obvious confidence interval for δ, Cliff (1996, p. 140) recommends

$$\frac{\hat{\delta} - \hat{\delta}^3 \pm z\hat{\sigma}\sqrt{(1 - \hat{\delta}^2)^2 + z^2\hat{\sigma}^2}}{1 - \hat{\delta}^2 + z^2\hat{\sigma}^2}.$$

Cliff's confidence interval for δ is readily modified to give a confidence for P. Letting

$$C_\ell = \frac{\hat{\delta} - \hat{\delta}^3 - z\hat{\sigma}\sqrt{(1 - \hat{\delta}^2)^2 + z^2\hat{\sigma}^2}}{1 - \hat{\delta}^2 + z^2\hat{\sigma}^2},$$

and

$$C_u = \frac{\hat{\delta} - \hat{\delta}^3 + z\hat{\sigma}\sqrt{(1 - \hat{\delta}^2)^2 + z^2\hat{\sigma}^2}}{1 - \hat{\delta}^2 + z^2\hat{\sigma}^2},$$

a $1 - \alpha$ confidence interval for P is

$$\left(\frac{1 - C_u}{2}, \frac{1 - C_\ell}{2} \right). \tag{5.27}$$

When there is complete separation, in which case, the estimate of P is equal to zero or one, Cliff's method does not yield a confidence interval. The result is that when P is close to zero or one, the resulting confidence interval can be highly inaccurate. This issue is addressed as follows. Suppose there is complete separation, and $n_1 = 1$. Given this one value, inferences about P can be made based on the Clopper–Pearson method in Chapter 4. For $n_1 > 1$, and when there is complete separation, all of the Clopper–Pearson confidence intervals yield the same result for each value in the first group, which is the confidence interval used here. Simulations indicate that the probability coverage is fairly close to the nominal level when P is close to zero or one. Of course, some other approach might prove to be better in some sense, but this remains to be determined.

Brunner–Munzel Method

To describe the Brunner–Munzel method, we begin by providing a formal definition of a midrank. Let

$$c^-(x) = \begin{cases} 0, & x \leq 0, \\ 1, & x > 0, \end{cases}$$

$$c^+(x) = \begin{cases} 0, & x < 0, \\ 1, & x \geq 0, \end{cases}$$

and

$$c(x) = \frac{1}{2}(c^+(x) + c^-(x)).$$

The *midrank* associated with X_i is

$$\frac{1}{2} + \sum_{j=1}^{n} c(X_i - X_j).$$

In essence, midranks are the same as ranks when there are no tied values. If tied values occur, the ranks of tied values are averaged.

■ **Example**

Consider the values

$$7, 7.5, 7.5, 8, 8, 8.5, 9, 11, 11, 11.$$

If there were no tied values, their ranks would be 1, 2, 3, 4, 5, 6, 7, 8, 9, and 10. The midranks are easily determined as follows. Because there are two values equal to 7.5, their ranks are averaged, yielding a rank of 2.5 for each. There are two values equal to 8, their original ranks were 4 and 5, so, their midranks are both 4.5. There are three values equal to 11, and their original ranks are 8, 9, and 10; the average of these ranks is 9, so, their midranks are all equal to 9. The midranks corresponding to all 10 of the original values are

$$1, 2.5, 2.5, 4.5, 4.5, 6, 7, 9, 9, 9.$$

■

To apply the Brunner–Munzel method, first, pool the data and compute midranks. Let $N = n_1 + n_2$ represent the total sample size (the number of observations among the pooled data), and let R_{ij} be the midrank associated with X_{ij} (the ith observation in the jth group) based on the pooled data. Let

$$\bar{R}_j = \frac{1}{n_j} \sum_{i=1}^{n_j} R_{ij}.$$

Compute the midranks for the data in group 1, ignoring group 2, and label the results $V_{11}, \ldots V_{n_1 1}$. Do the same for group 2 (ignoring group 1), and label the midranks $V_{12}, \ldots V_{n_2 2}$. The remaining calculations for testing H_0: $P = 0.5$, or for computing a confidence interval for P, are shown in Table 5.5.

Like Cliff's method, the Brunner–Munzel method can be unsatisfactory when P is close to zero or one. The modification of Cliff's method could of course be used when dealing with the Brunner–Munzel method, but this can be highly unsatisfactory. More precisely, as P approaches zero or one, there are situations where the modified Cliff method performs reasonably, while the Brunner–Munzel method and the modified Brunner–Munzel method perform poorly (Wilcox, 2020b).

Inferences About θ_D

As previously noted, θ_D is the population median of the typical difference between a randomly sampled observation from the first group and a randomly sampled observation from the second group. Let $D_{ik} = X_{i1} - X_{k2}, i = 1, \ldots, n_1, k = 1, \ldots, n_2$. The estimate of θ_D is simply the sample median based on the D_{ik} values. A percentile bootstrap can be used to compute a confidence interval for θ_D. Cliff's confidence interval for P is readily adapted to computing a confidence interval for θ_D as well (Wilcox, 2018c). Briefly, let $\hat{P}(\mathbf{X}_1, \mathbf{X}_2)$ denote the estimate of P based on the random samples $\mathbf{X}_1$ and $\mathbf{X}_2$. Let ω_ℓ be a constant such that $\hat{P}(\mathbf{X}_1 - \omega_\ell, \mathbf{X}_2) = c_\ell$, where c_ℓ is the lower end of the confidence interval for P given by Eq. (5.27). That is, shifting each value in the first group by ω_ℓ results in an estimate of P equal to c_ℓ. In a similar manner, ω_u is chosen so that $\hat{P}(\mathbf{X}_1 - \omega_u, \mathbf{X}_2) = c_u$, where c_u is the upper end of the confidence interval for P given by Eq. (5.27). Let $d_\ell = \hat{\theta}_D(\mathbf{X}_1 - \omega_\ell, \mathbf{X}_2)$ be the estimate of θ_D when the data in the first group are shifted by ω_ℓ, and let $d_u = \hat{\theta}_D(\mathbf{X}_1 - \omega_u, \mathbf{X}_2)$. Then

$$(d_\ell, d_u) \tag{5.28}$$

Table 5.5:   The Brunner–Munzel method for two independent groups.

Compute

$$S_j^2 = \frac{1}{n_j - 1} \sum_{i=1}^{n_j} \left(R_{ij} - V_{ij} - \bar{R}_j + \frac{n_j + 1}{2} \right)^2,$$

$$s_j^2 = \frac{S_j^2}{(N - n_j)^2},$$

$$s_e = \sqrt{N} \sqrt{\frac{s_1^2}{n_1} + \frac{s_2^2}{n_2}},$$

$$U_1 = \left(\frac{S_1^2}{N - n_1} + \frac{S_2^2}{N - n_2} \right)^2,$$

and

$$U_2 = \frac{1}{n_1 - 1} \left(\frac{S_1^2}{N - n_1} \right)^2 + \frac{1}{n_2 - 1} \left(\frac{S_2^2}{N - n_2} \right)^2.$$

The test statistic is

$$W = \frac{\bar{R}_2 - \bar{R}_1}{\sqrt{N} s_e},$$

and the degrees of freedom are

$$\hat{v} = \frac{U_1}{U_2}.$$

Decision Rule: Reject H_0: $P = 0.5$ if $|W| \geq t$, where t is the $1 - \alpha/2$ quantile of a Student's t distribution with $\hat{v}$ degrees of freedom. An estimate of P is

$$\hat{P} = \frac{1}{N} (\bar{R}_2 - \bar{R}_1) + \frac{1}{2}.$$

The estimate of $\delta = p_1 - p_3$ is

$$\hat{\delta} = 1 - 2\hat{P}.$$

An approximate $1 - \alpha$ confidence interval for P is

$$\hat{P} \pm t s_e.$$

is a $1 - \alpha$ confidence interval for θ_D. Generally, Eq. (5.28) performs well and appears to be a bit better than a percentile bootstrap. However, for an extreme shift in location, $d_\ell = d_u$ can result when using Eq. (5.28). The percentile bootstrap method avoids this, but d_ℓ and d_u can be nearly identical.

5.7.2 A Quantile Shift Measure of Effect Size Based on the Typical Difference

Section 5.3.4 described a quantile shift measure of effect size, method QSR, that has connections to a standardized measure of effect size. As indicated, a possible appeal of this approach is that it captures the spirit of a standardized measure without imposing a parametric family of distributions. But a possible concern is that its value depends on which of the two groups is used as the reference group. Presumably, this is not an issue in some situations. For example, a control group might be used as the reference group. But of interest is a method that eliminates this issue. In general, this can be done by focusing on a quantile shift method based on the distribution of $\mathcal{D}$ (Wilcox, 2018c). However, exceptions can occur as noted below in conjunction with the R function shiftQSci.

First note that if two distributions are identical, $\mathcal{D}$ has a symmetric distribution about zero. Let F_0 be the distribution of $\mathcal{D} - \theta_D$, where again, θ_D is the median of $\mathcal{D}$. That is, for the distribution F_0, H_0: $\theta_D = 0$ is true. The idea is to measure effect size based on the extent to which θ_D represents a shift in location to some relatively high or low quantile associated with F_0. More formally, the measure of effect size is taken to be

$$Q = F_0(\theta_D). \tag{5.29}$$

This will be called method QS henceforth. For identical distributions, $\theta_D = 0$ and $Q = 0.5$. If $Q = 0.8$, this corresponds to a shift in location to the 0.8 quantile. Under normality and homoscedasticity, Cohen's d equal to 0.2, 0.5, and 0.8 corresponds approximately to $Q = 0.55$, 0.65, and 0.70, respectively (Wilcox, 2018c). An estimate of Q is

$$\hat{Q} = \frac{1}{n_1 n_2} \sum \sum I(\mathcal{D}_{ik} - \hat{\theta}_D \leq \hat{\theta}_D),$$

where the indicator function $I(\mathcal{D}_{ik} - \hat{\theta}_D \leq \hat{\theta}_D) = 1$ if $\mathcal{D}_{ik} - \hat{\theta}_D \leq \hat{\theta}_D$; otherwise, $I(\mathcal{D}_{ik} - \hat{\theta}_D \leq \hat{\theta}_D) = 0$. Wilcox (2018c) studies various methods for computing a confidence interval for Q; a percentile bootstrap method was found to perform relatively well even for $n_1 = n_2 = 10$.

5.7.3 R Functions cidv2, bmp, wmwloc, wmwpb, loc2plot, shiftQSci, akp.effec.ci, ES.summary, ES.summary.CI, ES.sum.REL.MAG, and loc.dif.summary

The R function

```
cidv2(x,y,alpha= 0.05, plotit=FALSE, pop=0, fr=0.8, rval=15, xlab='',ylab='')
```

performs the modification of Cliff's method for making inferences about $\delta = P(X_{i1} > X_{i2}) - P(X_{i1} < X_{i2})$. The function also reports a confidence interval for $P = p_3 + 0.5p_2$, which is labeled ci.p. The estimate of P is labeled phat. The function cid is the same as cidv2 except that no p-value is reported.

The function

$$bmp(x,y, alpha= 0.05, plotit=TRUE, pop=0, fr=0.8, rval=15, xlab='', ylab='')$$

performs the modified Brunner–Munzel method. It returns a p-value when testing H_0: $P = 0.5$, plus an estimate of P labeled phat, and a confidence interval for P labeled ci.p. (An estimate of $\delta = p_1 - p_3$, labeled d.hat, is returned as well.)

It is noted that the permutation version of the Brunner–Munzel method can be applied via the R function

$$brunnermunzel.permutation.test(y1,y2)$$

assuming that the R package brunnermunzel has been installed. This R package also contains the Brunner–Munzel method. The function wmwRZR, stored in the WRS package, performs the method derived by Reiczigel et al. (2005).

When plotit=TRUE, the R functions bmp and cidv2 create plots based on the $n_1 n_2$ differences, $\mathcal{D}_{ij} = X_{i1} - X_{h2}, i = 1, \ldots, n_1, h = 1, \ldots, n_2$. The distribution of $\mathcal{D}$ will have a median equal to zero if the distributions are identical. For reasons previously mentioned, the R functions cidv2 and bmp can be viewed as methods aimed at testing the hypothesis that the distribution of $\mathcal{D}$ has a median of zero. The argument pop determines the type of plot that will be created. The choices are:

- pop=0, adaptive kernel density estimate,
- pop=1, expected frequency curve,
- pop=2, Rosenblatt's shifted histogram,
- pop=3, boxplot,
- pop=4, stem-and-leaf,
- pop=5, histogram,
- pop=6, kernel density using a normal kernel.

The argument fr is the span when using a kernel density estimator, and rval indicates how many points are used by Rosenblatt's shifted histogram when creating the plot. (See Section 3.2.5.) Labels can be added to the x-axis and y-axis via the arguments xlab and ylab, respectively. Let d_q $(0 < q < 0.5)$ denote the qth quantile of the distribution of $\mathcal{D}$. If the distributions being compared are identical, $d_q + d_{1-q} = 0$. This can be tested with the R function

cbmhd(x,y,qest=hd, alpha=0.05, q=0.25, plotit=FALSE, pop=0, fr=0.8, rval=15, xlab=' ',ylab=' ', nboot=600, SEED=TRUE).

The R function

wmwloc(x,y, na.rm=TRUE, est=median, ...)

computes an estimate of θ_D, the median of $\mathcal{D}$. The median can be replaced by some other measure of location via the argument est. The R function

wmwpb(x, y = NULL, est = median, tr=0.2, nboot = 2000, SEED = TRUE,...)

computes a $1 - \alpha$ confidence interval for θ_D using the percentile bootstrap method. The R function

loc2dif.ci(x,y,est=median,alpha= 0.05,nboot=2000,SEED=TRUE)

computes a $1 - \alpha$ confidence interval for θ_D based on Eq. (5.28). (Under certain circumstances, this function uses a percentile bootstrap instead.) A plot of an estimate of the distribution of D is created by the R function

loc2plot(x,y,plotfun=akerd,xlab='X',ylab='',...).

By default, an adaptive kernel density estimator is used. (The R function wmwplot also computes an estimate of the distribution of D.) The R function

shiftQSci(x,y,locfun = median, alpha = 0.05, nboot = 500, SEED = TRUE, ...)

computes a confidence interval for the quantile shift measure of location, method QS, described in Section 5.7.2. In general, shiftQSci(x,y) + shiftQSci(y,x) = 1, but there are exceptions. This can occur, for example, when there are tied values and the argument locfun = median.

The R function

$$\text{ES.summary}(x,y,\text{NULL.V}=c(0, 0, 0.5, 0.5, 0.5,0), \text{REL.M}=\text{NULL},\text{n.est}=1000000)$$

computes six measures of effect size, which are labeled AKP, the robust analog of Cohen's d given by Eq. (5.18); EP, the explanatory measure of effect size given by Eq. (5.21); QS, the quantile shift given by Eq. (5.29); QStr, the quantile shift measure of effect size given by Eq. (5.29) but with the median replaced by a trimmed mean; $P(X_{i1} < X_{i2})$; and KMS, the measure of effect size given by Eq. (5.19). As previously mentioned, under normality and homoscedasticity, 0.2, 0.5, and 0.8 are sometimes characterized as small, medium, and large effect sizes, respectively, when using Cohen's d. But what is viewed as small, medium, and large can depend on the situation. The argument REL.M can be used to indicate what constitutes a small, medium, and large effect size when using AKP, basically Cohen's d, under normality and homoscedasticity. The corresponding values for the other measures of effect size are estimated based on n.est values sampled from a normal distribution. (This is done by calling the R function ES.sum.REL.MAG.) The R function

$$\text{ES.summary.CI}(x,y, \text{tr}=0.2, \text{SEED}=\text{TRUE}, \text{alpha}=0.05, \text{nboot}=1000,$$
$$\text{NULL.V}=c(0,0,0.5,0.5,0.5,0),\text{REL.M}=\text{NULL},\text{n.est}=1000000)$$

is the same as ES.summary, only confidence intervals are reported for each measure of effect size. (Execution time can be a bit high due to the percentile bootstrap used in conjunction with the explanatory measure of effect size.) The R function

$$\text{loc.dif.summary}(x, y)$$

computes the difference between several measures of location: mean, 20% trimmed mean, median, one-step M-estimator, and θ_D, the median of the typical difference.

■ **Example**

For the hangover data, the Brunner–Munzel method returns a p-value of 0.042, and its 0.95 confidence interval for P is $(0.167, 0.494)$, so, H_0: $P = 0.5$ is rejected at the 0.05 level. Cliff's method also rejects at the 0.05 level, the 0.95 confidence interval for P being $(0.198, 0.490)$.

■

■ Example

For the ozone data, the R function ES.summary.CI returns

```
               Est  NULL   S     M     L    ci.low    ci.up
AKP         1.0533757  0.0  0.20  0.50  0.80  0.2457780  1.9432561
EP          0.7889100  0.0  0.14  0.34  0.52  0.2353012  0.9657369
QS (median) 0.7094862  0.5  0.55  0.64  0.71  0.5948617  0.9288538
QStr        0.7312253  0.5  0.55  0.64  0.71  0.5098814  0.9328063
WMW         0.2391304  0.5  0.45  0.36  0.29  0.1158292  0.4298723
KMS         0.4833453  0.0  0.10  0.25  0.40  0.1128440  0.9014797
```

In this particular case, all six measures indicate a large effect size, but there are situations where some indicate a small effect size and others indicate a large effect size, as illustrated at the end of Section 5.3.5. This can occur simply because they are sensitive to different features of the data.

■ Example

Measures of location provide some sense of how much groups differ, robust measures can provide more power compared with methods based on means, and rank-based methods provide yet another perspective. But sometimes, more might be needed to understand the nature and the extent to which two groups differ, as illustrated here with data dealing with measures of self-regulation for children in grades 6 to 7. The first group consisted of families with both parents, and the second group consisted of children from families with a single parent. The sample sizes are 245 and 230, respectively. Testing at the 0.05 level, no difference between the groups is found based on Student's t test, Welch's heteroscedastic method for means, Yuen's method for trimmed means (in Section 5.3), the bootstrap methods for M-estimators and trimmed means covered in this chapter, and the rank-based methods as well. But is it possible that these methods are missing some true difference? That is, perhaps the distributions differ, but the hypothesis testing methods just listed are insensitive to this difference. The upper left panel of Fig. 5.8 shows the shift function for these two groups. The function sband indicates that from about the 0.2 to 0.3 quantiles, the groups differ, and the 0.47 and 0.48 quantiles differ as well. To add perspective, the upper right plot shows the adaptive kernel density estimates created by the function g2plot. The lower left panel shows a

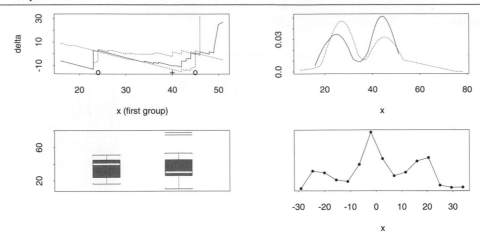

Figure 5.8: Four graphs summarizing how two groups differ based on a measure of self-regulation.

boxplot of the data, and the lower right panel is a Rosenblatt shifted histogram created by the function cid (which performs Cliff's heteroscedastic analog of the Wilcoxon–Mann–Whitney test).

5.8 Comparing Two Independent Binomial and Multinomial Distributions

Many methods have been proposed for comparing two independent binomial distributions, three of which are described here. Santner et al. (2007) compare several methods and recommend a technique derived by Coe and Tamhane (1993). Evidently, there is no R function for performing the complex calculations. Moreover, it is unknown how this method compares to the techniques summarized here. (Also see Berger, 1996.) The first method described here was derived by Storer and Kim (1990), the second was derived by Kulinskaya et al. (2010), and the third is by Zou et al. (2009). For convenience, these methods are labeled SK, KMS, and ZHZ, respectively. Their relative merits are discussed at the end of this section.

Method SK

Let p_j ($j = 1, 2$) be the probability of success associated with the jth group, and let r_j be the number of successes among n_j trials. The goal is to test H_0: $p_1 = p_2$. Or in the context of Tukey's three-decision rule, to what extent is it reasonable to make decisions about which group has the larger probability of success? Note that the possible number of successes in the first group is any integer, x, between 0 and n_1, and for the second group, it is any integer, y, between 0 and n_2. For any x and y, set

$$a_{xy} = 1$$

if

$$\left| \frac{x}{n_1} - \frac{y}{n_2} \right| \geq \left| \frac{r_1}{n_1} - \frac{r_2}{n_2} \right|;$$

otherwise

$$a_{xy} = 0.$$

Let

$$\hat{p} = \frac{r_1 + r_2}{n_1 + n_2}.$$

The test statistic is

$$T = \sum_{x=0}^{n_1} \sum_{y=0}^{n_2} a_{xy} b(x, \, n_1, \, \hat{p}) b(y, \, n_2, \, \hat{p}),$$

where

$$b(x, \, n_1, \, \hat{p}) = \left(\begin{array}{c} n_1 \\ x \end{array} \right) \hat{p}^x (1 - \hat{p})^{n_1 - x},$$

and $b(y, \, n_2, \, \hat{p})$ is defined in an analogous fashion. The null hypothesis is rejected if

$$T \leq \alpha.$$

That is, T is the p-value. The Storer–Kim method does not provide a confidence interval, but it currently seems that, typically, it offers a bit more power compared to method KMS.

Method KMS

The confidence interval for $p_1 - p_2$, derived by Kulinskaya et al. (2010), is

$$\frac{\hat{w}}{u} \sin \left(\arcsin \left[\frac{u \hat{\Delta} + \hat{v}}{\hat{w}} \right] \pm z_{1-\alpha/2} \sqrt{\frac{u}{2 n_1 n_2 / N}} \right) - \frac{\hat{v}}{u},$$

where $z_{1-\alpha/2}$ is the $1 - \alpha/2$ quantile of a standard normal distribution, again, r_1 and r_2 are the observed number of successes, $0 \leq A \leq 1$ is chosen by the user, $u = 2((1 - A)^2 \frac{n_2}{N} + A^2 \frac{n_1}{N})$, $\hat{\Delta} = (r_1 + 0.5)/(n_1 + 1) - (r_2 + 0.5)/(n_2 + 1)$, $\hat{\psi} = A(r_1 + 0.5)/(n_1 + 1) + (1 - A)(r_2 + 0.5)/(n_2 + 1)$, $\hat{v} = (1 - 2\hat{\psi})(A - \frac{n_2}{N})$, and $\hat{w} = \sqrt{2u\hat{\psi}(1 - \hat{\psi}) + \hat{v}^2}$. Here, following the suggestion made by Kulinskaya et al., $A = 0.5$ is used.

Method ZHZ

The method derived by Zou et al. (2009) is applied as follows. Let (ℓ_j, u_j) be a $1 - \alpha$ confidence interval for p_j $(j = 1, 2)$. Here, these confidence intervals are based on the Agresti–Coull method. Zou et al. use the confidence interval derived by Wilson (1927). Compared to using the Agresti–Coull method, simulations indicate Type I error probabilities differ by at most a few units in the fourth decimal place. Then an approximate $1 - \alpha$ confidence interval for $p_1 - p_2$ is (L, U), where

$$L = \hat{p}_1 - \hat{p}_2 - \sqrt{(\hat{p}_1 - \ell_1)^2 + (u_2 - \hat{p}_2)^2}$$

and

$$U = \hat{p}_1 - \hat{p}_2 + \sqrt{(u_1 - \hat{p}_1)^2 + (\hat{p}_2 - \ell_2)^2}.$$

The KMS method appears to be generally superior to the method derived by Agresti and Caffo (2000), as well as a method recommended by Beal (1987). It also competes well with a method derived by Newcombe (1998), which performed well among the methods compared by Brown and Li (2005). When testing at the 0.05 level, there are situations where ZHZ performs well in terms of accurate probability coverage and offers a power advantage over the other methods considered here. However, when either sample size is less than 35, there are situations where it is unsatisfactory. If the goal is to avoid Type I errors well above the nominal level, when testing H_0: $p_1 = p_2$, the results in Wilcox (2019e) suggest using SK or KMS. Method ZHZ can be improved by replacing the Agresti–Coull method with the method derived by Schilling and Doi (2014), but now, it offers no practical advantage over KMS. Method SK might offer a power advantage but at the expense of no confidence interval.

Reiczigel et al. (2008) generalized results derived by Sterne (1954) that yields a minimum volume confidence region for the two probabilities of success. Their method can be used, among other things, to compute a p-value when testing the hypothesis that two probabilities are equal. However, methods SK and ZHZ appear to have a slight edge in terms of power, at least when testing at the 0.05 level. A more systematic study is needed to resolve this issue. The involved computational details are not described, but an R function for computing the confidence region derived by Reiczigel et al. is provided in Section 5.8.1.

5.8.1 R Functions binom2g and bi2CR

The R function

$$binom2g(r1 = sum(elimna(x)), n1 = length(elimna(x)), r2 = sum(elimna(y)),$$
$$n2 = length(elimna(y)), x = NA, y = NA, method = c('KMS', 'ECP' 'SK', 'ZHZ'),$$
$$binCI = acbinomci, alpha = 0.05, null.value = 0, iter=2000)$$

tests H_0: $p_1 = p_2$ using method KMS by default. Setting the argument method='ZHZ', method ZHZ is used and method='SK' is method SK. When the sample sizes are small, the hypothesis is true and the unknown probabilities are close to zero or one, the actual levels of methods KMS, SK, and ZHZ can be well below the nominal level. Method ECP can be more satisfactory. It is essentially method KMS, only a simulation is used to determine a critical p-value based on an estimate of the assumed common probability of success. The argument iter determines how many replications are used in the simulation. When using ZHZ, the argument binCI indicates the function used to compute (ℓ_j, u_j). By default the Agresti–Coull method is used, which was described in Section 4.8. The function can be used either by specifying the number of successes in each group (arguments r1 and r2) and the sample sizes (arguments n1 and n2), or the data can be in the form of two vectors containing 1s and 0s, in which case, you use the arguments x and y.

The R function

$$bi2CR(r1, n1, r2, n2, alpha= 0.05, xlab='p1', ylab='p2')$$

plots the $1 - \alpha$ confidence region for (p_1, p_2) based on the method derived by Reiczigel et al. (2008).

■ **Example**

If for the first group there are 7 successes among 12 observations, and for the second group there are 22 successes among 25 observations, the command binom2g(7,12, 22, method='ZHZ') returns a p-value of 0.047. The command binom2g(7, 12, 22, 25, method='KMS') returns a p-value equal to 0.07, the point being that with a small sample size, the choice of method can matter. The p-value returned by binom2g(7,12, 22, 25, method='SK') is 0.044.

■

5.8.2 Comparing Discrete (Multinomial) Distributions

There are a variety of methods that might be used to compare discrete distributions when the cardinality of the sample space is relatively small. One approach is to test

$$H_0: P(X = x) = P(Y = x) \tag{5.30}$$

for all values x in the sample space. That is, the goal is to test the hypothesis that the distributions are identical. This can be accomplished with a chi-squared test, which can be applied with the R function disc2com. (This function is also stored as disc2.chi.sq.) Assuming that the data are ordinal, two more options are the Kolmogorov–Smirnov test in Section 5.1.1 and the Wilcoxon–Mann–Whitney test, which can be applied with the built-in R function wilcox.test.

As previously noted, a possible criticism of testing the hypothesis that discrete random variables have identical distributions, from the point of view of Tukey's three-decision rule, is that surely $P(X = x)$ and $P(Y = x)$ differ at some decimal place. For each x in the sample space, a more interesting goal is to determine whether a decision can be made regarding whether $P(X = x)$ is greater than or less than $P(Y = x)$. This can be accomplished with the R function binband, described in the next section, which is based on the Storer–Kim method for comparing two independent binomial distributions. Method KMS can be used as well. And of course, there is the issue of whether $P(X = x) - P(Y = x)$ is clinically important.

Note that the shift function in Section 5.1.3 provides an alternative perspective regarding how the distributions differ. Yet another option is method Q2 in Section 5.1.5. The method that provides the most power depends on how the distributions differ, which is unknown.

5.8.3 R Functions binband, splotg5, and cumrelf

The R function

binband(x,y,KMS=FALSE, alpha = 0.05, plotit = TRUE, op = TRUE, xlab = 'X',
ylab = 'Rel. Freq.', method = 'hoch')

tests the hypothesis given by Eq. (5.30). By default, the individual probabilities are compared using the SK method. Setting the argument KMS=TRUE, method KMS is used. If plotit=TRUE, the function plots the relative frequencies for all distinct values found in each of two groups. The argument method indicates which method is used to control FWE. The default is Hochberg's method.

The R function

splotg5(x,y,op=TRUE,xlab='X',ylab='Rel. Freq.')

is supplied in case it is desired to plot the relative frequencies for all distinct values found in two or more groups. The function is limited to a maximum of five groups. With op=TRUE, a line connecting the points corresponding to the relative frequencies is formed.

The R function

$$\text{cumrelf}(x,\text{xlab}='X',\text{ylab}='CUM REL FREQ')$$

plots the cumulative relative frequency of two or more groups. The argument x can be a matrix with columns corresponding to groups, or x can have list mode.

■ **Example**

Consider a study aimed at comparing two methods for reducing shoulder pain after surgery. For the first method, the shoulder pain measures are

2, 4, 4, 2, 2, 2, 4, 3, 2, 4, 2, 3, 2, 4, 3, 2, 2, 3, 5, 5, 2, 2

and for the second method they are

5, 1, 4, 4, 2, 3, 3, 1, 1, 1, 1, 2, 2, 1, 1, 5, 3, 5.

The R function binband returns

Value	p1.est	p2.est	p1-p2	p.value	p.adj
1	0.00000000	0.3888889	−0.38888889	0.00183865	0.009
2	0.50000000	0.1666667	0.33333333	0.02827202	0.113
3	0.18181818	0.1666667	0.01515152	0.94561448	0.946
4	0.22727273	0.1111111	0.11616162	0.36498427	0.946
5	0.09090909	0.1666667	−0.07575758	0.49846008	0.946

For response 1, the p-value is 0.0018, and the Hochberg adjusted p-value is 0.009, indicating that for a familywise error rate of 0.05, a decision would be made that the second group is more likely to respond 1 compared to the first group. In contrast, for the probability of a response 5, the p-value is 0.56. So, the data indicate that the second group is more likely to respond 5, but based on Tukey's three-decision rule, no decision is made about which group is more likely to rate the pain as being 5. In contrast, for both Student's t and Welch's method, the p-value is 0.25. Comparing the medians as well as the 20% trimmed means, again, no difference is found at the 0.05 level. Note that in terms of controlling the familywise error rate, there is no need to first perform a chi-squared test of the hypothesis that the distributions are identical.

■

■ Example

Erceg-Hurn and Steed (2011) investigated the degree to which smokers experience negative emotional responses (such as anger and irritation) upon being exposed to anti-smoking warnings on cigarette packets. Smokers were randomly allocated to view warnings that contained only text, such as "Smoking Kills," or warnings that contained text and graphics, such as pictures of rotting teeth and gangrene. Negative emotional reactions to the warnings were measured on a scale that produced a score between 0 and 16 for each smoker, where larger scores indicate greater levels of negative emotions. (The data are stored on the author's web page in the file smoke.csv.) Testing the 0.05 level, the means and medians are higher for the graphic group. But to get a deeper understanding of how the groups differ, look at Fig. 5.9, which shows plots of the relative frequencies based on the R function splotg5. Note that the plot suggests that the main difference between the groups has to do with the response zero: the proportion of participants in the text only group responding zero is 0.512, compared to 0.192 for the graphics group. Testing the hypothesis that the corresponding probabilities are equal, based on the R function binband, the p-value is less than 0.0001. The probabilities associated with the other possible responses do not differ at the 0.05 level except for the response 16; the p-value is 0.0031. For the graphics group, the probability of responding 16 is 0.096, compared to 0.008 for the text only group. So, a closer look at the data, beyond comparing means and medians, suggests that the main difference between the groups has to do with the likelihood of giving the most extreme responses possible, particularly the response zero.

■

5.9 Comparing Dependent Groups

When comparing dependent groups, there are three distinct approaches that might be used. To be concrete, imagine that n couples, one male and the other female, are randomly sampled who are related in some manner. For illustrative purposes, suppose these couples are married. The first approach is in terms of the marginal distributions. When comparing medians, for example, the goal would be to characterize how the typical male compares to the typical female. A second approach is to use the difference scores $D_i = X_i - Y_i$, $i = 1, \ldots, n$. Now, the goal is to characterize the typical difference between a wife and her husband. This approach is dis-

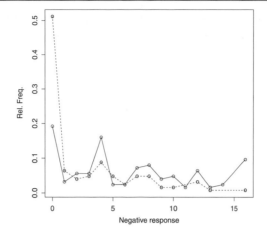

Figure 5.9: Plot created by the function s2plot based on the anti-smoking data.

tinct from the first approach because, under general conditions, the difference between the marginal population medians does not equal the population median of the difference scores. The third perspective is the typical difference between a male and female, which is essentially the approach used in Section 5.7. If, for example, $(X_1, Y_1), \ldots, (X_n, Y_n)$ is a random sample of n pairs of observations, one could use some measure of location based on the n^2 differences $\mathcal{D}_{ij} = X_i - Y_j$ ($i = 1, \ldots, n$, $j = 1, \ldots, n$). Section 5.9.9 discusses this approach when dealing with dependent groups. Methods for dealing with each of these approaches are described in this section. The R function dep.dif.fun in Section 5.9.10 can be used with any of five methods based on difference scores, including the method in Section 5.9.9. For a description of measures of effect size that might be used, see Section 5.9.19.

5.9.1 A Shift Function for Dependent Groups

Lombard (2005) derives an extension of the shift function, described in Section 5.1, to dependent groups, which belongs to the first approach, where the focus is on the marginal distributions (cf. Wilcox, 2006f). Let $X_{(1)} \leq \cdots \leq X_{(n)}$ be the X_i values written in ascending order ($i = 1, \ldots, n$). Let

$$\hat{F}(x) = \frac{1}{n} \sum I(X_i \leq x)$$

be the estimate of $F(x)$, the marginal distribution of X, where the indicator function $I(X_i \leq x) = 1$ if $X_i \leq x$; otherwise, $I(X_i \leq x) = 0$. The estimate of the distribution of Y, $\hat{G}(x)$, is defined in a similar manner. Denote the combined set $\{X_1, \ldots X_n, Y_1, \ldots Y_n\}$, written in ascending order, by $\{Z_{(1)} \leq \cdots \leq Z_{(2n)}\}$. Lombard's (2005) method for computing confidence intervals for the difference between each quantile stems from the test statistic

$$K = (n/2)^{1/2} \max|\hat{F}(Z_i) - \hat{G}(Z_i)|,$$

which can be used to test

$$H_0: F(x) = G(x), \text{ for all } x,$$

versus

$$H_1: F(x) \neq G(x), \text{ for at least one } x.$$

If $x > 0$, let

$$\psi_y(x) = (2\pi x^3)^{-1/2} y \exp(-y^2/2x);$$

otherwise, $\psi_y(x) = 0$. Let R_i be the rank of X_i values among the X values, and let S_i be the rank of Y_i among the Y values. Let f_i be the frequency of occurrence of the value i among the values $\max\{R_1, S_1\}, \ldots, \max\{R_n, S_n\}$. Then the α level critical value used by Lombard is the value c solving

$$\frac{1}{n} \sum_{i=1}^{n} f_i \times \psi_c(i - f_1 - \cdots - f_i) = \alpha. \tag{5.31}$$

Here, the Nelder and Mead (1965) algorithm is used to determine c.

Let $[z]$ denote the integer portion of z, and for $k \geq 0$, let $Y_{(-k)} = X_{(-k)} = -\infty$ and $Y_{(n+1+k)} = X_{(n+1+k)} = \infty$. The quantile matching function q is given by $G(q(x)) = F(x)$. It specifies the functional relationship between the marginal distributions and reflects the difference between quantiles. Lombard's confidence interval for $q(X_{(j)})$, the quantile matching function evaluated at $X_{(j)}$, is

$$(Y_{(j-[(2n)^{1/2}c])}, Y_{(j+[(2n)^{1/2}c])}),$$

which is designed to have, approximately, simultaneous probability coverage $1 - \alpha$. So, when the marginal distributions are identical, the interval

$$(Y_{(j-[(2n)^{1/2}c])} - X_{(j)}, Y_{(j+[(2n)^{1/2}c])} - X_{(j)}) \tag{5.32}$$

should contain zero for any j.

5.9.2 R Function lband

The R function

lband(x,y=NA, alpha= 0.05, plotit=TRUE, sm=TRUE,ylab='delta', xlab='x (first group)')

computes Lombard's shift function for dependent groups. If the argument y=NA, the function assumes the argument x is a matrix with two columns or it has list mode. By default, the shift function is plotted. To avoid the plot, set the argument plotit=FALSE. If the argument sm=TRUE, the plot of shift function is smoothed using lowess.

5.9.3 Comparing Specified Quantiles

Section 5.1.5 noted that when using the shift function for independent groups, power might be relatively low when dealing with quantiles relatively close to zero or one. The same concern is relevant when using the distribution-free method in Section 5.9.1. A way of dealing with this concern is to compare instead a collection of specified quantiles. This section describes three techniques that might be used.

Method D1

The first method is similar to method Q1 in Section 5.1.5. The goal is to test

$$H_0: \theta_{1q} = \theta_{2q}, \tag{5.33}$$

the hypothesis that the qth quantiles of the marginal distributions are equal. Or from the point of view of Tukey's three-decision rule, the goal is to determine whether it is reasonable to make a decision about which distribution has the larger qth quantile.

For convenience, the remainder of this section denotes a random sample of n pairs of observations by $(X_{11}, X_{12}), \ldots, (X_{n1}, X_{n2})$. Let $\hat{\theta}_{jq}$ be the Harrell–Davis estimate of the qth quantile associated with the jth marginal distribution, $q = 0.1(0.1)0.9$. Then $\hat{d}_q = \hat{\theta}_{1q} - \hat{\theta}_{2q}$ estimates the difference between the qth quantiles. For the problem at hand, bootstrap samples are obtained by resampling with replacement n pairs of points. That is, n rows of data are sampled, with replacement, from the n-by-2 matrix

$$\begin{pmatrix} X_{11}, X_{12} \\ \vdots \\ X_{n1}, X_{n2} \end{pmatrix}.$$

This is in contrast to the case of independent groups where bootstrap samples are obtained by resampling from $X_{11}, \ldots, X_{n1}$ and separate (independent) bootstrap samples are obtained by resampling from $X_{12}, \ldots, X_{n2}$.

Let $(X_{11}^*, X_{12}^*), \ldots, (X_{n1}^*, X_{n2}^*)$ be the bootstrap sample obtained by resampling n pairs of points, let $\hat{\theta}_{jq}^*$ be the Harrell–Davis estimate of x_{jq}, the qth quantile of the jth group, based on the values $X_{1j}^*, \ldots, X_{nj}^*$, and let $\hat{d}_q^* = \hat{\theta}_{1q}^* - \hat{\theta}_{2q}^*$. Repeat this bootstrap process B times, yielding $\hat{d}_{q1}^*, \ldots, \hat{d}_{qB}^*$. Then an estimate of the squared standard error of $\hat{d}_q$ is

$$\hat{\sigma}_{dq}^2 = \frac{1}{B-1} \sum_{b=1}^{B} (\hat{d}_{qb}^* - \bar{d}_q)^2,$$

where

$$\bar{d}_q = \frac{1}{B} \sum_{b=1}^{B} \hat{d}_{qb}^*.$$

Setting

$$c = \frac{37}{n^{1.4}} + 2.75,$$

$$(\hat{\theta}_{1q} - \hat{\theta}_{2q}) \pm c\hat{\sigma}_{dq} \tag{5.34}$$

yields a confidence interval for $x_{1q} - x_{2q}$, where c was determined so that the simultaneous probability coverage is approximately 0.95.

Notice that the method uses only one set of B bootstrap samples for all nine quantiles being compared. That is, the same bootstrap samples are used to compute $\hat{\sigma}_{dq}^2$ for each $q = 0.1, \ldots, 0.9$. In contrast, when comparing independent groups, 18 sets of B bootstrap samples are used, one for each of the 18 quantiles being estimated, so, execution time on a computer will be faster when working with dependent groups. The original motivation for using 18 sets of bootstrap values was to approximate the critical value using a nine-variate Studentized maximum modulus distribution. However, the approximation proved to be rather unsatisfactory when sample sizes are small. When working with independent groups, it might be possible to get accurate confidence intervals using only one set of B bootstrap samples, but this has not been investigated.

Method D2

As was the case, when using method Q1 in Section 5.5.1, there is some concern with method D1 when there are tied values; the actual Type I error probability might differ from the nominal level by an unacceptable amount. Two more concerns are that method D1 is designed specifically for comparing all of the deciles. It is not designed to deal with a subset of the deciles or a collection of other quantiles, and it is limited to testing at the 0.05 level. Method D2 (Wilcox and Erceg-Hurn, 2012) is designed to deal with these concerns. But a possible advantage of D1 is that when there are no tied values and the goal is to compare all of the deciles, it might have more power than method D2.

Briefly, method D2 uses a percentile bootstrap method in conjunction with the Harrell–Davis estimator. Bootstrap samples are generated as done by method D1. Again, let $\hat{d}_q^* = \hat{\theta}_{1q}^* - \hat{\theta}_{2q}^*$ be the estimated difference between qth quantiles based on this bootstrap sample. Repeat this process B times, yielding $\hat{d}_{q1}^*, \ldots, \hat{d}_{qB}^*$. Put these B values in ascending order, yielding $\hat{d}_{q(1)}^* \leq \cdots \leq \hat{d}_{q(B)}^*$. Then a $1 - \alpha$ confidence interval for $q_1 - q_2$ and a p-value can be computed as described in Section 5.4. (The probability of one or more Type I errors is controlled using Hochberg's method in Section 7.4.7.)

A variation of this method replaces the Harrell–Davis estimator with an estimate of the qth quantile based on a single order statistic as described in Section 3.5. If there are no tied values and the distributions are sufficiently heavy-tailed, this approach might have more power. But if there are tied values, control over the Type I error probability can be poor and power can be relatively low.

Method D3

Method D3 uses the second general approach, where the focus is on the difference scores. This is in contrast to methods D1 and D2, which focus on the quantiles of the marginal distributions. Consider, for example, a study aimed at assessing the effectiveness of some method for treating depressive symptoms. If the distribution of the difference scores is symmetric about zero, there is a sense in which intervention is relatively ineffective. But if decreases in any symptoms are greater than the typical increases, it might be argued that intervention is worthwhile. For instance, the estimate of the 0.75 quantile of the difference scores might be 5.57 and the estimate of the 0.25 quantile might be -3.19, suggesting that the drop in depression characterized by the 0.75 quantile is larger than the corresponding increase in depression represented by the 0.25 quantile. Let δ_q denote the qth quantile of the difference scores. An issue, then, is whether one can reject H_0: $-\delta_{0.25} = \delta_{0.75}$, which would support the conclusion that there is a sense in which the positive effects of intervention outweigh the negative effects. More generally, there is interest in testing

$$H_0\colon \delta_q + \delta_{1-q} = 0, \; q < 0.5. \tag{5.35}$$

Note that the sign test might also be used in this situation. A possible appeal of method D3 is that it takes into account the magnitude of any increases and decreases.

To test Eq. (5.35), generate a bootstrap sample as done in method D2. Let $\hat{\delta}_q^*$ be the Harrell–Davis estimate of the qth quantile based on this bootstrap sample, and let $\hat{\Delta}^* = \hat{\delta}_q^* + \hat{\delta}_{1-q}^*$ ($q < 0.5$). Repeat this process B times, yielding $\hat{\Delta}_b^*$ ($B = 1, \ldots, B$). Then an approximate $1 - \alpha$ confidence interval for $\theta_q + \theta_{1-q}$ can be computed as described in Section 5.4. A p-value can be computed as well.

5.9.4 R Functions shiftdhd, Dqcomhd, qdec2ci, Dqdif, and difQpci

The R function

shiftdhd(x,y,nboot=200, plotit=TRUE, plotop=FALSE, xlab='x (first group)', ylab='Delta')

computes a confidence interval for the difference between the quantiles, when comparing two dependent random variables, using the method D1 described in the previous subsection. The confidence intervals are designed so that the simultaneous probability coverage is approximately 0.95. As with shifthd, the default number of bootstrap samples (nboot) is $B = 200$, which appears to suffice in terms of controlling the probability of a Type I error. Simulations indicate that this has a practical advantage over $B = 100$ (Wilcox, 1995b). The last argument, plotit, defaults to TRUE, meaning that a plot of the shift function is created. The command shiftdhd(x,y,plotit=FALSE) avoids the plot. By default, plotop=FALSE, meaning that when plotting the results, the x-axis indicates the estimate of the quantiles based on the first group. The y-axis indicates the estimate of $q_1 - q_2$, the estimated difference between the quantiles, and the x-axis indicates the estimates of q_1. Setting plotop=TRUE, the x-axis indicates which quantiles are being estimated. By default, the deciles are compared, so, now, the x-axis would indicate that the plotted confidence intervals correspond to the values 0.1(0.1)0.9.

The R function

Dqcomhd(x,y,q = c(1:9)/10, nboot = 1000, plotit = TRUE, SEED = TRUE, xlab = 'Group 1', ylab = 'Est.1-Est.2', na.rm = TRUE, alpha= 0.05)

compares quantiles using method D2 in the previous section. By default, all of the deciles are compared, which are estimated via the Harrell–Davis estimator. The argument q can be used to specify the quantiles to be compared. The argument na.rm=TRUE means that vectors with missing values will be removed; otherwise, all of the available data are used. The R function qdec2ci(x,y,nboot=500, alpha= 0.05, pr=FALSE, SEED=TRUE, plotit=TRUE)

also applies method D2, but only a single order statistic is used to estimate the quantiles rather than all of the order statistics, as done by the Harrell–Davis estimator.

The R function

Dqdif(x,y=NULL,q=0.25, nboot=1000, plotit=TRUE, xlab='Group 1 - Group 2', SEED=TRUE, alpha= 0.05)

applies method D3. By default, the 0.25 and 0.75 quantiles are compared. That is, $q = 0.25$ is used in the null hypothesis given by Eq. (5.35). A plot of the distribution of D is created as well. Unlike Dqdif, the R function

difQpci(x,y, q=seq(5,40,5)/100, xlab='Quantile', ylab='Group 1 minus Group 2', plotit=TRUE, alpha= 0.05, nboot=1000, SEED=TRUE, LINE=FALSE)

can test Eq. (5.35) for a collection of q values. By default, Eq. (5.35) is tested using $q=$ 0.05(0.05)0.40.

■ Example

An illustration in Section 5.3.3 is based on data from a study on the effects of consuming alcohol. Another portion of the study compared the effects of drinking alcohol for the same participants measured at three different times. The data for the control group measured at times 1 and 3 are stored in columns 2 and 4 of the file mitch_dat.txt. No differences are found between the means or trimmed means, but this leaves open the possibility that one or more quantiles are significantly different. Comparing the deciles with shiftdhd, no significant differences are found with $\alpha = 0.05$. However, there are tied values, which can impact both power and the accuracy of the confidence intervals. Fig. 5.10 shows the results when using the R function Dqcomhd. The x-axis indicates the estimate of the quantiles at time 1 based on the Harrell–Davis estimator. The y-axis corresponds to $\hat{d}_q = \hat{\theta}_{2q} - \hat{\theta}_{1q}$, the estimated difference between the qth quantiles, $\delta_q = \theta_{2q} - \theta_{1q}$. Below and above line a, + marks the ends of the confidence interval for δ_q. The p-values corresponding to the quantiles 0.1(0.1)0.9 are 0.014, 0.016, 0.022, 0.026, 0.052, 0.120, 0.918, and 0.272.

■

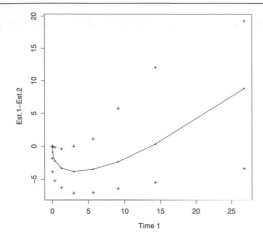

Figure 5.10: Plot created by the function Dqcomhd using a portion of the alcohol data.

5.9.5 Comparing Trimmed Means

As previously noted, when comparing dependent groups, under general conditions, the difference between the marginal population medians does not equal the population median of the difference scores. This result generalizes to the γ-trimmed means, $\gamma > 0$. That is, if $D_i = X_{i1} - X_{i2}$, under general conditions, $\mu_{td} \neq \mu_{t1} - \mu_{t2}$, where μ_{td} is the population trimmed mean corresponding to D. Note that inferences about μ_{td} are readily made using results in Chapter 4. For example, if pairs of observations are stored in the R variables time1 and time2, a confidence interval for μ_{td}, based on 20% trimming, can be computed with the R command trimci(time1-time2), and trimpb(time1-time2) will use a percentile bootstrap method instead. Alternatively, one can use the command onesampb(time1-time2,est=tmean). Of course, in some situations, there is little or no difference between comparing the trimmed means of the marginal distributions versus making inferences about a trimmed mean associated with difference scores. However, situations arise, particularly when comparing multiple groups, where the choice between the two strategies can make a considerable difference, as will be illustrated.

The remainder of this section focuses on a (non-bootstrap) method for comparing the trimmed means associated with the marginal distributions. Suppose $(X_{11}, X_{12}), \ldots, (X_{n1}, X_{n2})$ is a random sample of n pairs of observations from some bivariate distribution. The goal is to compute a confidence interval for $\mu_{t1} - \mu_{t2}$, the difference between the trimmed means. A simple approach is to estimate the squared standard error of $\bar{X}_{t1} - \bar{X}_{t2}$, and then use Student's t distribution with appropriate degrees of freedom to get a confidence interval or test the hypothesis that the trimmed means are equal. This can be done using a simple generalization of Yuen's method.

Before continuing, it is remarked that for the special case, where the goal is to compare the marginal medians, if there are tied values, only one method is known to perform well: a percentile bootstrap method. The R function dmedpb, described in Section 8.3.3, can be used to accomplish this goal. And even with no tied values, the method in this section should not be used because it is based on an unsatisfactory estimate of the standard error.

The process begins by Winsorizing the marginal distributions. In symbols, fix j and let $X_{(1)j} \leq X_{(2)j} \leq \cdots \leq X_{(n)j}$ be the n values in the jth group written in ascending order. Next, set

$$W_{ij} = \begin{cases} X_{(g+1)j}, & \text{if } X_{ij} \leq X_{(g+1)j}, \\ X_{ij}, & \text{if } X_{(g+1)j} < X_{ij} < X_{(n-g)j}, \\ X_{(n-g)j}, & \text{if } X_{ij} \geq X_{(n-g)j}, \end{cases}$$

where, as usual, g is the number of observations trimmed or Winsorized from each end of the distribution corresponding to the jth group. With 20% trimming, $g = [0.2n]$, where $[0.2n]$ means to round $0.2n$ down to the nearest integer. The expression for W_{ij} says that $W_{ij} = X_{ij}$, if X_{ij} has a value between $X_{(g+1)j}$ and $X_{(n-g)j}$. If X_{ij} is less than or equal to $X_{(g+1)j}$, set $W_{ij} = X_{(g+1)j}$, and if X_{ij} is greater than or equal to $X_{(n-g)j}$, set $W_{ij} = X_{(n-g)j}$. Put another way, the observations are Winsorized with the dependent random variables remaining paired together, and this is consistent with the Winsorization of a bivariate distribution described in Chapter 2.

As an illustration, consider the eight pairs of observations

X_{i1}:	18	6	2	12	14	12	8	9
X_{i2}:	11	15	9	12	9	6	7	10

With 20% Winsorization, $g = 1$, so, the smallest observation in each group is pulled up to the next smallest value. Thus, for the first row of data, the value 2 is Winsorized by replacing it with 6. Similarly, the largest value, 18, is replaced by the value 14. For the second row of data, 6 becomes 7 and 15 becomes 12. This yields

W_{i1}:	14	6	6	12	14	12	8	9
W_{i2}:	11	12	9	12	9	7	7	10

The population Winsorized covariance between X_{i1} and X_{i2} is, by definition, $\sigma_{w12} = E_w[(X_{i1} - \mu_{w1})(X_{i2} - \mu_{w2})]$, where E_w indicates the Winsorized expected value as defined in Chapter 2, and μ_{wj} is the population Winsorized mean of the jth group. It follows from the influence function of the trimmed mean that the squared standard error of $\bar{X}_{t1} - \bar{X}_{t2}$ is

$$\frac{1}{(1-2\gamma)^2 n}\{\sigma_{w1}^2 + \sigma_{w2}^2 - 2\sigma_{w12}\},$$

which reduces to a standard result when there is no trimming ($\gamma = 0$). The Winsorized covariance is estimated with the sample covariance between the W_{i1} and W_{i2} values:

$$\frac{1}{n-1} \sum (W_{i1} - \bar{W}_1)(W_{i2} - \bar{W}_2),$$

where $\bar{W}_j$ is the Winsorized mean associated with the jth random variable. Generalizing Yuen's approach in an obvious way, the squared standard error of $\bar{X}_{t1} - \bar{X}_{t2}$ can be estimated with:

$$\frac{1}{h(h-1)} \left\{ \sum (W_{i1} - \bar{W}_1)^2 + \sum (W_{i2} - \bar{W}_2)^2 - 2 \sum (W_{i1} - \bar{W}_2)(W_{i2} - \bar{W}_2) \right\},$$

where $h = n - 2g$ is the effective sample size. Letting

$$d_j = \frac{1}{h(h-1)} \sum (W_{ij} - \bar{W}_j)^2$$

and

$$d_{12} = \frac{1}{h(h-1)} \sum (W_{i1} - \bar{W}_1)(W_{i2} - \bar{W}_2),$$

H_0: $\mu_{t1} = \mu_{t2}$ can be tested with

$$T_y = \frac{\bar{X}_{t1} - \bar{X}_{t2}}{\sqrt{d_1 + d_2 - 2d_{12}}}, \tag{5.36}$$

which is rejected if $|T_y| > t$, the $1 - \alpha$ quantile of Student's t distribution with $h - 1$ degrees of freedom. A $1 - \alpha$ confidence interval for $\mu_{t1} - \mu_{t2}$ is

$$(\bar{X}_{t1} - \bar{X}_{t2}) \pm t\sqrt{d_1 + d_2 - 2d_{12}}.$$

Effect Size

Measures of effect size can be computed via the R function dep.ES.summary.CI, described in Section 5.9.20. It includes a measure based on a trimmed mean. Confidence intervals for the measures of effect size are computed as well.

5.9.6 R Function yuend

The R function

```
yuend(x,y,tr=0.2,alpha= 0.05)
```

computes a confidence interval for $\mu_{t1} - \mu_{t2}$, the difference between the trimmed means corresponding to two dependent groups, using the method described in the previous subsection of this chapter. As usual, the default amount of trimming is tr $= 0.2$, and alpha defaults to 0.05. The resulting confidence interval is returned in the R variable yuend$ci, the p-value is indicated by yuend$p.value, the estimated difference between the trimmed means in yuend$dif, the estimated standard error in yuend$se, the test statistic in yuend$teststat, and the degrees of freedom in yuend$df.

■ Example

As a simple illustration, suppose the cholesterol levels of participants are measured before and after some treatment is administered, yielding

Before:	190, 210, 300, 240, 280, 170, 280, 250, 240, 220,
After:	210, 210, 340, 190, 260, 180, 200, 220, 230, 200.

Storing the before scores in the R vector x and the after scores in y, the command yuend(x,y) returns

```
$ci
[1] -14.20510   67.87177

$p.value
[1] 0.1536335

$est1
[1] 238.5

$est2
[1] 211.6667

$dif
[1] 26.83333

$se
[1] 15.96465

$teststat
```

```
[1] 1.680797

$n
[1] 10

$df
[1] 5
```

Thus, the 0.95 confidence interval for $\mu_{t1} - \mu_{t2}$ is $(-14.23, 67.87)$, and the estimated difference between the trimmed means is 26.83. The p-value is 0.153.

5.9.7 A Bootstrap-t Method for Marginal Trimmed Means

A bootstrap-t method can be used to compute a $1 - \alpha$ confidence interval for $\mu_{t1} - \mu_{t2}$. (Again, when using difference scores, simply proceed as in Chapter 4.) Begin by generating a bootstrap sample as described in Section 5.7.1. That is, n pairs of observations are obtained by randomly sampling with replacement pairs of observations from the observed data. As usual, label the results (X_{i1}^*, X_{i2}^*), $i = 1, \ldots, n$. Now, proceed along the lines in Section 5.3.2. More precisely, set $C_{ij}^* = X_{ij}^* - \bar{X}_{tj}$. Let T_y^* be the value of T_y, given by Eq. (5.36), based on the C_{ij}^* values just computed. Repeat this process B times, yielding T_{yb}^*, $b = 1, \ldots, B$. Let $T_{y(1)}^* \leq \cdots \leq T_{y(B)}^*$ be the T_{yb}^* values written in ascending order. Set $\ell = \alpha B/2$ and $u = (1 - \alpha/2)B$, rounding both to the nearest integer. Then an estimate of the lower and upper critical values is $T_{y(\ell+1)}^*$ and $T_{y(u)}^*$. An equal-tailed $1 - \alpha$ confidence interval for $\mu_{t1} - \mu_{t2}$ is

$$(\bar{X}_{t1} - \bar{X}_{t2} + T_{y(u)}^*\sqrt{d_1 + d_2 - 2d_{12}}, \ \bar{X}_{t1} - \bar{X}_{t2} + T_{y(\ell+1)}^*\sqrt{d_1 + d_2 - 2d_{12}}). \qquad (5.37)$$

To get a symmetric confidence interval, replace T_{yb}^* by its absolute value, set $a = (1 - \alpha)B$, rounding to the nearest integer, in which case, the $1 - \alpha$ confidence interval for $(\mu_{t1} - \mu_{t2})$ is

$$(\bar{X}_{t1} - \bar{X}_{t2}) \pm T_{y(a)}^*\sqrt{d_1 + d_2 - 2d_{12}}.$$

5.9.8 R Function ydbt

The R function

```
ydbt(x,y, tr=0.2, alpha= 0.05, nboot=599, side=FALSE, plotit=FALSE, op=1)
```

computes a bootstrap-t confidence interval for $\mu_{t1} - \mu_{t2}$ when dealing with paired data. As usual, the default amount of trimming is tr=0.2, and α defaults to alpha=0.05. The number of bootstrap samples defaults to nboot=599. Using side=FALSE results in an equal-tailed confidence interval, while side=TRUE returns a symmetric confidence interval instead. Setting the argument plotit to TRUE creates a plot of the bootstrap values, where the type of plot is controlled via the argument op. The possible values for op are 1, 2, and 3, which correspond to the adaptive kernel estimator, the expected frequency curve, and a boxplot, respectively.

5.9.9 Inferences About the Typical Difference

This section deals with the third general approach to comparing dependent groups where the focus is on the typical difference. That is, inferences are made based on $\mathcal{D}_{ik} = X_{i1} - X_{k2}$ ($i = 1, \ldots, n$, $j = 1, \ldots, n$). Let $M_{\mathcal{D}}$ denote the sample median based on all n^2 $\mathcal{D}_{ik}$ values. Of course, some other measure of location could be used, but in terms of efficiency, $M_{\mathcal{D}}$ compares well to smaller amounts of trimming, even under normality (Wilcox, 2006d). Letting $\theta_{\mathcal{D}}$ be the population median associated with $M_{\mathcal{D}}$, a basic percentile bootstrap method has been found to perform well, in terms of controlling the probability of a Type I error, when testing

$$H_0: \theta_{\mathcal{D}} = 0.$$

That is, randomly sample with replacement n pairs of observation, compute $M_{\mathcal{D}}^*$ based on this bootstrap sample, and repeat this process B times. Note that when comparing independent groups, this hypothesis corresponds to $H_0: p = 0.5$, where p is the probabilistic measure of effect size discussed in Section 5.7. That is, the method in this section is a generalization of the Wilcoxon–Mann–Whitney test to dependent groups.

5.9.10 R Functions loc2dif, l2drmci, and dep.dif.fun

The R function

$$\text{loc2dif(x, y = NULL, est = median, na.rm = TRUE, plotit = FALSE,}$$
$$\text{xlab = ' ', ylab = ' ', ...)}$$

computes $M_{\mathcal{D}}$ for the data stored in x (time 1 for example) and y (time 2). If y is not specified, it is assumed x is a matrix with two columns. The argument na.rm=TRUE means that the function will eliminate any pair where one or both values are missing. If it is desired to use all of the available data, set na.rm=FALSE. If the argument plotit=TRUE, the function plots an estimate of the distribution of $\mathcal{D}$. The R function

$$\text{l2drmci(x,y=NA,est=median, alpha=0.05, na.rm=TRUE)}$$

tests H_0: $\theta_{\mathcal{D}} = 0$. The argument na.rm is used as was done with loc2dif.

■ **Example**

Rao (1948) reports data on cork boring weights taken from 28 trees, which are stored in the file cork_data.txt and can be accessed as described in Section 1.10. The borings were taken from the north, east, west, and south sides of each tree. Here, the south and east sides of the trees are compared. The function loc2dif returns 1. That is, among all trees, the median difference between south and east sides is estimated to be 1. In contrast, the median difference for a randomly sampled tree, meaning the median of the difference scores associated with the 28 trees, is 3. So, now, we have information on how the two sides of the same tree compare, which is not the same as the difference among all the trees. Finally, the difference between the marginal medians is 1.5. This tells us something about how the typical weight for the south side of a tree compares to the typical weight of the east side. But it does not provide any direct information regarding the typical difference among all of the trees. The R function l2drmci returns a p-value of 0.336. So, based on the differences among all pairs of trees, it fails to reject the hypothesis H_0: $\theta_{\mathcal{D}} = 0$ at the 0.10 level. However, if the $n = 28$ difference scores are used, the R function sintv2 returns a p-value of 0.088. Now (at the 0.10 level), it rejects and concludes that for a randomly sampled tree, the median difference between the two weights differs from 0, the only point being that different perspectives can alter the p-value substantially.

■

For convenience, the R function

$$\text{dep.dif.fun(x,y,tr=0.2, alpha=0.05, nboot=2000,}$$
$$\text{method=c('TR','TRPB','MED','AD','SIGN'))}$$

can be used to compare dependent groups based on difference scores using any of five methods: trimmed means based on the Tukey–McLaughlin method (TR), trimmed means based on a percentile bootstrap method (TRPB), medians using the method in Section 4.6.2 (MED), the typical difference as described in Section 5.9.9 (AD), and the sign test (SIGN).

5.9.11 *Percentile Bootstrap: Comparing Medians, M-Estimators, and Other Measures of Location and Scale*

The percentile bootstrap method in Chapter 4 is readily extended to comparing various parameters associated with the marginal distributions of two dependent variables. When X_{i1} and X_{i2} are dependent, bootstrap samples are obtained by randomly sampling pairs of observations with replacement. That is, proceed as described in Section 5.9.1, yielding the pairs of bootstrap samples

$$(X_{11}^*, X_{12}^*)$$
$$\vdots$$
$$(X_{n1}^*, X_{n2}^*).$$

Let θ_j be any parameter of interest associated with the jth marginal distribution. Let $\hat{\theta}_j^*$ be the bootstrap estimate of θ_j based on $X_{1j}^*, \ldots, X_{nj}^*$, and let $d^* = \hat{\theta}_1^* - \hat{\theta}_2^*$. Repeat this process B times, yielding $d_1^*, \ldots, d_B^*$, and write these B values in ascending order, yielding $d_{(1)}^* \leq \cdots \leq d_{(B)}^*$, in which case, a $1 - \alpha$ confidence interval for $\theta_1 - \theta_2$ is

$$(d_{(\ell+1)}^*, d_{(u)}^*),$$

where $\ell = \alpha B/2$, rounded to the nearest integer, and $u = B - \ell$.

A (generalized) p-value can be computed as well. Let

$$p^* = P(\hat{\theta}_1^* > \hat{\theta}_2^*) + 0.5P(\hat{\theta}_1^* = \hat{\theta}_2^*).$$

The first probability on the right side of this equation is estimated with the proportion of bootstrap samples, among all B bootstrap samples, for which $\theta_1^* > \theta_2^*$. And the second probability is estimated with the proportion of times $\hat{\theta}_1^* = \hat{\theta}_2^*$. For a two-sided hypothesis, now, reject if $\hat{p}^* \leq \alpha/2$ or if $\hat{p}^* \geq 1 - \alpha/2$. The estimate of the (generalized) p-value is

$$2\min(\hat{p}^*, 1 - \hat{p}^*).$$

The percentile bootstrap method just described can be unsatisfactory when the goal is to make inferences about means and variances, but it appears to be reasonably effective when working with M-estimators, the Harrell–Davis estimate of the median, as well as the biweight midvariance (Wilcox, 1996a). By this is meant that the actual probability of a Type I error will not be much larger than the nominal level.

However, a concern is that with small sample sizes, the actual probability of a Type I error can drop well below the nominal level when working with M-estimators or the modified

M-estimator described in Chapter 3. For example, there are situations where, when testing at the 0.05 level, the actual probability of a Type I error can be less than 0.01. Wilcox and Keselman (2002) found a method that reduces this problem in simulations. Note that if based on the original data, $\hat{\theta}_1 = \hat{\theta}_2$, it should be the case that $\hat{p}^* = 0.5$, but situations arise where this is not the case. The idea is to shift the data so that $\hat{\theta}_1 = \hat{\theta}_2$, compute $\hat{p}^*$ based on the shifted data, and then correct the bootstrap p-value given above. More precisely, let $\hat{q}^*$ be the value of $\hat{p}^*$ based on the shifted data. A so-called *bias-adjusted p-value* is

$$2\min(\hat{p}_a^*, \ 1 - \hat{p}_a^*),$$

where $\hat{p}_a^* = \hat{p}^* - 0.1(\hat{q}^* - 0.5)$. As the sample size increases, $\hat{q}^* - 0.5 \rightarrow 0$, and the adjustment becomes negligible.

Note that the pairs of bootstrap values $(\hat{\theta}_{1b}^*, \hat{\theta}_{2b}^*)$, $b = 1, \ldots, B$, provide an approximate $1 - \alpha$ confidence region for (θ_1, θ_2). The bootstrap method aimed at comparing the measures of location associated with the marginal distributions essentially checks to see how deeply a line through the origin, having slope 1, is nested within the cloud of bootstrap values. Here, the depth of this line is measured by how many bootstrap points lie above it, which corresponds to how often $\hat{\theta}_{1b}^* < \hat{\theta}_{2b}^*$ among the B bootstrap pairs of points.

5.9.12 R Functions two.dep.pb and bootdpci

The R function

```
two.dep.pb(x,y=NULL,alpha=0.05, est=tmean, plotit=FALSE, dif=TRUE, nboot=NA,
          xlab='Group 1',ylab='Group 2',pr=TRUE, SEED=TRUE,...)
```

uses a percentile bootstrap method to compare two dependent groups. The default value for nboot is NA, which, in effect, causes $B = 1,000$ bootstrap samples to be used. By default, a 20% trimmed mean is used. Setting plotit=TRUE, the bootstrap cloud of points is plotted. The argument dif controls whether inferences are made based on difference scores. By default, difference scores are used. To compare measures of location associated with the marginal distributions, set dif=FALSE. The older R function

```
bootdpci(x,y,est=onestep, nboot=NA, alpha= 0.05, plotit=TRUE, dif=TRUE, BA=FALSE,...)
```

can be used as well. It defaults to using a one-step M-estimator based on Huber's Ψ.

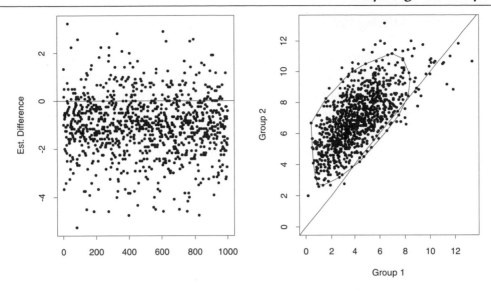

Figure 5.11: Plot of the bootstrap values used by the function bootdpci when analyzing the hangover data. The left panel is for dif=TRUE, meaning that difference scores were analyzed. The right panel is for dif=FALSE, meaning that marginal measures of location are compared.

■ Example

If the hangover data for times 1 and 2 are stored in the R variables t1 and t2, the command bootdpci(t1,t2,est=tmean) computes a 0.95 confidence interval for the 20% trimmed mean associated with the difference scores. The (generalized) p-value is 0.369. The left panel of Fig. 5.11 shows the resulting plot of the bootstrap values. The p-value reflects the proportion of points below the horizontal line at zero. The command bootdpci(x,y,est=tmean,dif=FALSE) compares the marginal distributions instead. Now, the (generalized) p-value is 0.063. The right panel of Fig. 5.11 shows the resulting plot. The p-value reflects the proportion of points below the line having slope 1 and intercept 0.

■

The command bootdpci(x,y,est=winvar,dif=FALSE) would compute a 0.95 confidence interval for the difference between the 20% Winsorized variances. The command bootdpci(x,y) would attempt to compute a 0.95 confidence interval based on a one-step M-estimator, but for the data used here, eventually this results in an error in the R function hpsi. The reason, which was already discussed in Chapter 4, is that with small sample sizes, there is a good chance that an M-estimator based on a bootstrap sample cannot be computed due to division by zero.

5.9.13 Handling Missing Values

Numerous methods have been proposed for handling missing values when dealing with means, none of which are completely satisfactory. A simple approach is the so-called *complete case* method, where any pair of observations is eliminated if one of the values is missing, after which, methods previously described are applied to the data that remain. This section describes three methods for handling missing values when using a robust measure of location that use all of the available data, assuming missing values occur in a manner that does not alter the marginal measures of location. For example, it might be the case that values are missing completely at random (MCAR), meaning that the process resulting in missing values is independent of both the observed and the missing values. A weaker assumption, that allows the analysis to be performed without taking into account the mechanism that creates missing values, is called missing at random (MAR). MAR is taken to mean that, given the observed data, the missingness mechanism does not depend on the unobserved data. (For a description of other mechanisms leading to missing values, see Little and Rubin, 2002.) At the end of this section, comments are made about the relative merits of the methods about to be described.

Method M1

The first method is based on a straightforward generalization of the method in Lin and Stivers (1974), assuming that the goal is to compare the marginal trimmed means, as opposed to the trimmed mean of the difference scores. It is assumed than n pairs of observations are randomly sampled, where both values are available, which is denoted by $(X_1, Y_1), \ldots, (X_n, Y_n)$. The corresponding (marginal) γ-trimmed means are denoted by $\bar{X}_t$ and $\bar{Y}_t$. For the first marginal distribution, additional n_1 observations are sampled for which the corresponding Y value is not observed. These observations are denoted by $X_{n+1}, \ldots, X_{n+n_1}$, and the trimmed mean of these n_1 observations is denoted by $\tilde{X}_t$. Similarly, n_2 observations are sampled for which the corresponding value for the first marginal distribution is not observed and the trimmed mean is denoted by $\tilde{Y}_t$. Let $h_j = [\gamma n_j]$ ($j = 1, 2$), and let $\lambda_j = h/(h + h_j)$, where $h = [\gamma n]$. Then an estimate of the difference between the marginal trimmed means, $\mu_{tD} = \mu_{t1} - \mu_{t2}$, is

$$\hat{\mu}_{tD} = \lambda_1 \bar{X}_t - \lambda_2 \bar{Y}_t + (1 - \lambda_1)\tilde{X}_t - (1 - \lambda_2)\tilde{Y}_t,$$

a linear combination of three independent random variables. The squared standard of $\lambda_1 \bar{X} - \lambda_2 \bar{Y}$ is

$$\sigma_0^2 = \frac{1}{(1 - 2\gamma)^2 n}(\lambda_1^2 \sigma_{wx}^2 + \lambda_2 \sigma_{wy}^2 - 2\lambda_1 \lambda_2 \sigma_{wxy}), \tag{5.38}$$

where σ_{wxy} is the population Winsorized covariance between X and Y. The squared standard error of $(1 - \lambda_1)\tilde{X}$ is

$$\sigma_1^2 = \frac{(1 - \lambda_1)^2 \sigma_{wx}^2}{(1 - 2\gamma)^2 (n + n_1)} \tag{5.39}$$

and the squared standard error of $(1 - \lambda_2)\tilde{Y}$ is

$$\sigma_2^2 = \frac{(1 - \lambda_2)^2 \sigma_{wy}^2}{(1 - 2\gamma)^2 (n + n_2)}. \tag{5.40}$$

So, the squared standard error of $\hat{\mu}_{tD}$ is

$$\tau^2 = \sigma_0^2 + \sigma_1^2 + \sigma_2^2.$$

For convenience, let $N_1 = n + n_1$ and $g_1 = [\gamma N_1]$. The Winsorized values corresponding to $X_1, \ldots, X_{N_1}$ are

$$W_{xi} = \begin{cases} X_{(g_1+1)}, & \text{if } X_i \le X_{(g_1+1)}, \\ X_i, & \text{if } X_{(g_1+1)} < X_i < X_{(N_1-g_1)}, \\ X_{(N_1-g_1)}, & \text{if } X_i \ge X_{(N_1-g_1)}. \end{cases}$$

The (sample) Winsorized mean is

$$\bar{W}_x = \frac{1}{N_1} \sum_{i=1}^{N_1} W_{xi},$$

an estimate of the Winsorized variance, σ_{wx}^2, is

$$s_{wx}^2 = \frac{1}{N_1 - 1} \sum (W_{xi} - \bar{W}_x)^2,$$

and an estimate of σ_{wy}^2 is obtained in a similar fashion. The Winsorized covariance between X and Y is estimated with

$$s_{wxy} = \frac{1}{n-1} \sum_{i-1}^{n} (W_{xi} - \tilde{W}_x)(W_{yi} - \tilde{W}_y),$$

where

$$\tilde{W}_x = \frac{1}{n} \sum_{i=1}^{n} W_{xi}$$

and $\tilde{W}_y$ is defined in a similar manner.

The sample Winsorized variances yield estimates of σ_0^2, σ_1^2, and σ_2^2, say $\hat{\sigma}_0^2$, $\hat{\sigma}_1^2$, and $\hat{\sigma}_2^2$, in which case, an estimate of the squared standard error of $\hat{\mu}_{tD}$ is

$$\hat{\tau}^2 = \hat{\sigma}_0^2 + \hat{\sigma}_1^2 + \hat{\sigma}_2^2.$$

So, a reasonable test statistic for testing the hypothesis of equal (marginal) trimmed means is

$$T = \frac{\hat{\mu}_{tD}}{\hat{\tau}}. \tag{5.41}$$

There remains the problem of approximating the null distribution of T, and here, a basic bootstrap-t method is used. To make sure the details are clear, the method begins by randomly sampling with replacement $N = n + n_1 + n_2$ pairs of observations from $(X_1, Y_1), \ldots, (X_N, Y_N)$, yielding $(X_1^*, Y_1^*), \ldots, (X_N^*, Y_N^*)$. Based on this bootstrap sample, compute the absolute value of the test statistic as just described, and label the result T^*. Repeat this process B times, and put the resulting T^* values in ascending order, yielding $T_{(1)}^* \leq \cdots \leq T_{(B)}^*$. Then an approximate $1 - \alpha$ confidence interval for Δ_t is

$$\hat{\Delta}_t \pm T_{(c)}^* \hat{\tau},$$

where $c = (1 - \alpha)B$ rounded to the nearest integer.

Method M2

Method M2 is based on the usual percentile bootstrap method. For the situation at hand, generate a bootstrap sample using all N pairs of observations, and let $\tilde{D}_t^* = \tilde{X}_t^* - \tilde{Y}_t^*$, where $\tilde{X}_t^*$ is the trimmed mean based on all of the X_i^* values not missing and $\tilde{Y}_t^*$ is computed in a similar manner. Repeat this B times, put the resulting $\tilde{D}_t^*$ values in ascending order, and label the results $\tilde{D}_{t(1)}^* \leq \cdots \leq \tilde{D}_{t(B)}^*$. Then an approximate $1 - \alpha$ confidence interval for μ_{tD} is

$$(\tilde{D}_{t(\ell+1)}^*, \tilde{D}_{t(u)}^*),$$

where $\ell = \alpha B/2$, rounded to the nearest integer, and $u = B - \ell$. A p-value is computed in the usual manner. That is, estimate $p = P(\hat{\mu}_{tD}^* > 0)$ with $\hat{p}$, the proportion of $\tilde{D}_t^*$ values greater than 0. Then a (generalized) p-value is

$$P = 2\min(\hat{p}, 1 - \hat{p}).$$

Method M3

Method M3 is based on θ_D, the median of the distribution of $D = X - Y$. The method begins by forming all pairwise differences among all of the observed X and Y values. That is, for all of the values that are available, compute $D_{ij} = X_i - Y_j$ $(i = 1, \ldots, N_1, \; j = 1, \ldots, N_2)$, resulting in $N_1 \times N_2$ D_{ij} values. Then an estimate of θ_D is obtained by computing the sample median of the D_{ij} values.

Again, a basic percentile bootstrap method is used to make inferences about θ_D. Generate a bootstrap sample as done in method M2, and let $\hat{\theta}_D^*$ be the resulting estimate of θ_D. Repeat this process B times, yielding $\hat{\theta}_{Db}^*$, $b = 1, \ldots, B$. Next, put these B values in ascending order, yielding $\hat{\theta}_{D(1)}^* \leq \cdots \leq \hat{\theta}_{D(B)}^*$, and let ℓ and u be defined as before. Then a $1 - \alpha$ confidence interval for θ_D is

$$(\hat{\theta}_{D(\ell+1)}^*, \; \hat{\theta}_{D(u)}^*).$$

This method can be applied with the R function l2drmci in Section 5.9.10 by setting the argument na.rm=FALSE, meaning that any row of data that has a missing value is not removed.

Comments on Choosing a Method

Based on results in Wilcox (2011b), both method M3 and method M2 coupled with a 20% trimmed mean perform reasonably well in terms of controlling the probability of a Type I error, with M2 having perhaps a slight advantage. Method M1 tends to have an actual type error probability less than the nominal level, again, using a 20% trimmed mean, sometimes substantially so. However, as the amount of trimming decreases, at some point a percentile bootstrap method (method M2) will not perform well, suggesting that eventually, method M1 will be more satisfactory than method M2. But at what point this is the case has not been determined. Not surprisingly, method M1 can be highly unsatisfactory when working with means.

Another general strategy is to impute missing values. As noted in Section 8.2.4, when testing hypotheses, this approach does not work well in general.

5.9.14 R Functions rm2miss and rmmismcp

The R function

```
rmmismcp(x,y = NA, tr=0.2, con = 0, est = tmean, plotit = TRUE, grp = NA, nboot = 500,
         SEED = TRUE, xlab = 'Group 1', ylab = 'Group 2', pr = FALSE, ...)
```

deals with missing values when the goal is to test the hypothesis $H_0: \mu_{t1} = \mu_{t2}$ using method M2, described in the previous section. In particular, rather than using the complete case analysis strategy, it uses all of the available data to compare the marginal trimmed means, assuming any missing values occur at random. With 20% trimming or more, it appears to be one of the better methods for general use when there are missing values. By default, a 20% trimmed mean is used, but other measures of location can be used via the argument est. For example, rmmismcp(x,y,est=onestep) would compare the groups with a one-step M-estimator. The function returns a confidence interval for the difference between the marginal measures of location. If the argument y=NA, it is assumed that the argument x is a matrix with columns corresponding to groups. (The function can handle more than two groups; see Section 8.1.4 for more details.) When there are two groups and the argument plotit=TRUE, a plot of the bootstrap estimates is created.

The R function

$$rm2miss(x,y,tr=0)$$

also tests $H_0: \mu_{t1} = \mu_{t2}$ using method M1.

5.9.15 Comparing Variances and Robust Measures of Scale

This section deals with testing

$$H_0: \sigma_1^2 = \sigma_2^2, \tag{5.42}$$

the hypothesis that two dependent random variables have equal variances. The best-known method is the *Morgan–Pitman* test. Set

$$U_i = X_{i1} - X_{i2}$$

and

$$V_i = X_{i1} + X_{i2}$$

$(i = 1, \ldots, n)$, and let ρ_{uv} be the (population) value of Pearson's correlation between U and V. It can be shown that if H_0 is true, then $\rho_{uv} = 0$, so, a test of the hypothesis of equal variances is obtained by testing

$$H_0: \rho_{uv} = 0. \tag{5.43}$$

The Morgan–Pitman test of Eq. (5.42) is based on the test statistic

$$T_{uv} = r_{uv}\sqrt{\frac{n-2}{1-r_{uv}^2}},$$

where r_{uv} is the usual estimate of Pearson's correlation based on (U_i, V_i) $(i = 1, \ldots, n)$, and T_{uv} is assumed to have a Student's t distribution with $n - 2$ degrees of freedom. However, McCulloch (1987) as well as Mudholkar et al. (2003) show that when sampling from heavy-tailed distributions, T_{uv} does not converge to a standard normal distribution as the sample size increases. For heavy-tailed distributions, the variance of T_{uv} converges to a value that is greater than one, which in turn results in Type I error probabilities greater than the nominal level. Indeed, as the sample size increases, control over the Type I error probability deteriorates. Wilcox (2015a) established that when the null hypothesis of equal variances is true, as we move toward a heavy-tailed distribution, there is heteroscedasticity. That is, T_{uv} is not using a correct estimate of the standard error. There are several methods for testing Eq. (5.43) that allow heteroscedasticity. Simulations indicate that the HC4 method (Section 9.3.14) performs fairly well when there is heteroscedasticity and the marginal distributions have the same amount of skewness.

However, control over the probability of a Type I error can be unsatisfactory when the marginal distributions differ sufficiently in terms of skewness. For example, if X_1 has a standard normal distribution and X_2 has a lognormal distribution, scaled to have variance one, the actual Type I error probability is generally well above the nominal level. The difficulty is that for the bulk of the U and V values, there is a strong positive association even though $\rho_{uv} = 0$. Getting a value for r_{uv} reasonably close to zero is a function in part of getting outliers among the U and V values that are properly placed (Wilcox, 2020a). Exercise 16 at the end of this chapter illustrates the problem. Currently, it is unknown how to improve on this method. In practical terms, if there is specific interest in the variances, the method described here is a relatively good approach among extant techniques. If the distributions are reasonably symmetric and do not differ too much in terms of skewness, the method in this section performs well. For example, if the first distribution is standard normal and the second is a g-and-h distribution with skewness 1.5, $n = 20$ suffices in terms of controlling the Type I error probability at the 0.05 level. If the second distribution has skewness 1.8, $n = 100$ is needed, and if the skewness is 2.7, now, $n = 500$ is required. There are situations where there is strong evidence that the HC4 method performs well when dealing with data from an actual study (e.g., Holladay et al., 2020). But there are situations where all known methods can be unsatisfactory in terms of a Type I error.

Using a robust measure of scale can provide better control over Type I errors, with the understanding that the variances are not being compared in general. Comparing robust measures of scale can be accomplished using a percentile bootstrap method via the R function rmrvar described in the next section. Briefly, independent bootstrap samples are taken for each marginal distribution. That is, a bootstrap sample is taken to estimate the measure of scale associated with the first variable, and then another bootstrap sample is used to estimate the measure of scale associated with the second variable. Otherwise, proceed as described in Section 5.9.11.

Another approach is to compare the quantiles of the marginal distributions after they have been centered to have a common measure of location, as done in Section 5.5.3 for independent variables. The difference in the quantiles in the tails of the distributions provides a more detailed perspective on how the variation of the distributions differ. The method in Section 5.5.3 is readily extended to the situation at hand via the R functions lband and Dqcomhd. The plot created by the R function g5.cen.plot in Section 5.5.4 can be helpful as well for the situation at hand.

5.9.16 R Functions comdvar, rmVARcom, and RMcomvar.locdis

The R function

$$comdvar(x,y,alpha= 0.05)$$

tests the hypothesis of equal variances using the HC4 method described in the previous section. The R function

$$rmVARcom(x, y = NULL, alpha = 0.05, est = bivar, plotit = TRUE, nboot = 500, SEED = TRUE, ...)$$

can be used to compare robust measures of scatter using a percentile bootstrap method. The argument x is assumed to be a vector or a matrix with two columns. The R function

$$RMcomvar.locdis(x,y, loc.fun=median,CI=FALSE,plotit=TRUE,xlab='First Group', ylab='Est.2 - Est.1',ylabQCOM='Est.2 - Est.1', sm=TRUE, QCOM=TRUE,q=c(0.1,0.25,0.75,0.9), MC=FALSE, nboot=2000,...)$$

centers the data, and then compares the quantiles. By default, the data are centered to have a median of zero and the 0.1, 0.25, 0.75, and 0.9 quantiles are compared using the R function Dqcomhd. Setting the argument QCOM=FALSE, the R function lband is used, which provides details about all quantiles using a distribution-free method, at the expense of possibly substantially less power compared to using QCOM=TRUE. Now, the label for the y-axis is given by the argument ylabQCOM.

5.9.17 The Sign Test

For completeness, one can also compare two dependent groups with the sign test. Let $p = P(X_{i1} < X_{i2})$, the probability that for a randomly sampled pair of observations, the first observation is less than the second. Letting $W_i = 1$, if $X_{i1} < X_{i2}$, $w = \sum W_i$ has a binomial distribution with probability of success p, in which case, methods in Section 4.8 can be used. As is evident, the sign test belongs to the second general approach to comparing dependent groups, which is based on difference scores.

5.9.18 R Function signt

For convenience, the R function

$$\text{signt}(x, y = \text{NULL, dif} = \text{NULL, alpha} = 0.05, \text{method} = \text{'AC', AUTO} = \text{TRUE,}$$
$$\text{PVSD=FALSE})$$

is supplied for performing a sign test. If the argument y is not specified, it is assumed that either x is a matrix with two columns corresponding to two dependent groups or x has list mode. The function computes the differences $X_{i1} - X_{i2}$ ($i = 1, \ldots, n$), and then eliminates all differences that are equal to zero, leaving N values. Next, it determines the number of pairs for which $X_{i1} < X_{i2}$, and then it calls the R function binom.conf in Section 4.8.1, where the arguments AUTO, PVSD, and method are explained.

5.9.19 Effect Size for Dependent Groups

The choice for a measure of effect size depends on which of the three general approaches to comparing dependent groups is adopted. The first approach is based on the marginal distributions, but care must be taken regarding the association between the groups. For example, the derivation of method KMS given by Eq. (5.19) assumes there is no association between the two location estimators. The AKP measure of effect size, given by Eq. (5.18), could be used assuming homoscedasticity and ignoring any association.

Another, perhaps more satisfactory approach is to focus on difference scores. Now, the measures of effect size summarized in Section 4.3.1 could be used via the R functions described in Section 4.3.2. And of course $p = P(X_{i1} < X_{i2})$, the probability that for a randomly sampled pair of observations, the first observation is less than the second, can be used as well, which is estimated via the R function sign described in Section 5.9.18.

A third approach is to focus on all pairwise differences $\mathcal{D}_{ij}$, as described in Section 5.7.2. From this perspective, the quantile shift measure of effect size might be used.

5.9.20 R Function dep.ES.summary.CI

The R function

dep.ES.summary.CI(x,y=NULL, tr=0.2, alpha=0.05, REL.MAG=NULL,
SEED=TRUE,nboot=1000)

computes four measures of effect size based on difference scores. The argument x is assumed to be difference scores when the argument y=NULL. If values for y are provided, effect sizes are computed based on x-y. The first effect size is

$$d = k\frac{\bar{X}_t - \mu_0}{s_w},$$ (5.44)

where $\bar{X}_t$ is the trimmed mean of the difference scores and the constant k is chosen so that under normality, s_w/k estimates the standard deviation σ. The next two are quantile shift measures of effect as described in Section 4.3.1: a quantile shift measure based on the median and a quantile shift measure based on a trimmed mean. The default amount of trimming is 0.2. The fourth is the probability that for a randomly sampled pair of observations, the first observation is less than the second, which is labeled SIGN.

What constitutes a large effect size depends on the situation. The default assumption is that $0.1, 0.3$, and 0.5 correspond to small, medium, and large effect sizes, respectively, when using Eq. (5.44). Suppose that under normality, $d = 0.2, 0.5$, and 0.8 are considered small, medium, and large effect sizes. Setting the argument REL.MAG=c(0.2, 0.5, 0.8), the function returns the corresponding values for the other measures of effect size used here. Included are bootstrap confidence intervals for the first three measures. For the sign test, the R function binom.conf is used. See Section 4.8.

5.10 Exercises

1. For Salk's data stored in the file salk_dat.txt, compare the two groups using the weighted Kolmogorov–Smirnov test. Plot the shift function and its 0.95 confidence band. Compare the results with the unweighted test.
2. For the ozone data stored in the file rat_data.txt, compare the two groups using the weighted Kolmogorov–Smirnov test. Plot the shift function and its 0.95 confidence band. Compare the results with the unweighted test.
3. Summarize the relative merits of using the weighted versus unweighted Kolmogorov–Smirnov test. Also discuss the merits of the Kolmogorov–Smirnov test relative to comparing measures of location.

4. Consider two independent groups having identical distributions. Suppose four obser-vations are randomly sampled from the first and three from the second. Determine $P(D = 1)$ and $P(D = 0.75)$, where D is given by Eq. (5.4). Verify your results with the R function kssig.

5. Compare the deciles only, using the Harrell–Davis estimator and Salk's data.

6. Verify that if X and Y are independent, the third moment about the mean of $X - Y$ is $\mu_{x[3]} - \mu_{y[3]}$.

7. Apply the Yuen–Welch method using Salk's data, where the amount of trimming is 0, 0.05, 0.1, and 0.2. Compare the estimated standard errors of the difference between the trimmed means.

8. Compute a confidence interval for p using Salk's data.

9. The example at the end of Section 5.3.3 examined some data from an experiment on the effects of drinking alcohol. Another portion of the study consisted of measuring the ef-fects of alcohol over 3 days of drinking. The scores for the control group, for the first 2 days of drinking, are 4, 47, 35, 4, 4, 0, 58, 0, 12, 4, 26, 19, 6, 10, 1, 22, 54, 15, 4, and 22. The experimental group had scores 2, 0, 7, 0, 4, 2, 9, 0, 2, 22, 0, 3, 0, 0, 47, 26, 2, 0, 7, and 2. Verify that the hypothesis of equal 20% trimmed means is rejected with $\alpha = 0.05$. Next, verify that this hypothesis is not rejected when using the equal-tailed bootstrap-t method, but that it is rejected when using the symmetric percentile-t procedure instead. Comment on these results.

10. Section 5.9.6 used some hypothetical data to illustrate the R function yuend with 20% trimming. Use the function to compare the means. Verify that the estimated standard error of the difference between the sample means is smaller than the standard error of the difference between the 20% trimmed means. Despite this, the p-value is smaller when comparing trimmed means versus means. Why? Make general comments on this result. Next, compute the 20% trimmed mean of the difference scores. That is, set $D_i = X_{i1} - X_{i2}$ and compute the trimmed mean using the D_i values. Compare this to the difference between the trimmed means of the marginal distributions, and make additional comments about comparing dependent groups.

11. The file pyge.dat (see Section 1.8) contains pretest reasoning IQ scores for students in grades 1 and 2 who were assigned to one of three ability tracks. (The data are from Elashoff and Snow, 1970, and originally collected by R. Rosenthal.) The file pygc.dat contains data for a control group. The experimental group consisted of children for whom positive expectancies had been suggested to teachers. Compare the 20% trimmed means of the control group to the experimental group using the function yuen, and verify that the 0.95 confidence interval is $(-7.12, 27.96)$. Thus, you would not reject the hy-pothesis of equal trimmed means. What reasons can be given for not concluding that the two groups have comparable IQ scores?

12. Continuing the last exercise, examine a boxplot of the data. What would you expect to happen if the 0.95 confidence interval is computed using a bootstrap-t method? Verify your answer using the R function yuenbt.

13. The file tumor.dat contains data on the number of days to occurrence of a mammary tumor in 48 rats injected with a carcinogen and subsequently randomized to receive either the treatment or the control. The data were collected by Gail et al. (1980) and represent only a portion of the results they reported. (Here, the data are the number of days to the first tumor. Most rats developed multiple tumors, but these results are not considered here.) Compare the means of the two groups with Welch's method for means, and verify that you reject with $\alpha = 0.05$. Examine a boxplot, and comment on what this suggests about the accuracy of the confidence interval for $\mu_1 - \mu_2$. Verify that you also reject when comparing M-measures of location. What happens when comparing 20% trimmed means or when using the Kolmogorov–Smirnov test?

14. Let $D = X - Y$, let θ_D be the population median associated with D, and let θ_X and θ_Y be the population medians associated with X and Y, respectively. Verify that under general conditions, $\theta_D \neq \theta_X - \theta_Y$.

15. Using R, generate 30 observations from a standard normal distribution and store the values in x. Generate 20 observations from a chi-squared distribution with one degree of freedom, and store them in z. Compute y=4(z-1), so, x and y contain data sampled from distributions having identical means. Apply the permutation test based on means with the function permg. Repeat this 200 times, and determine how often the function rejects. What do the results indicate about controlling the probability of a Type I error with the permutation test when testing the hypothesis of equal means? What does this suggest about computing a confidence interval for the difference between the means based on the permutation test?

16. Section 5.9.15 described a method for comparing the variances of two dependent variables. It was noted that when distributions differ in shape, the method can fail to control the Type I error probability. Execute the following R commands to get a sense of why:

```
set.seed(23)
sdval=sqrt(exp(1))*sqrt(exp(1)-1)
#The standard deviation of a lognormal distribution.
x=rnorm(500)
y=rlnorm(500)/sdval
sd(x)
sd(y) # so x and y have nearly identical sample standard deviations
u=x-y
v=x+y
regplot(u,v,xlab='U',ylab='V')
```

```
# Clearly the bulk of the points have a positive association.
#  Shown is the regression line based on the
#   Theil--Sen estimator in Chapter~10
```

The next command shows the least squares regression line:

```
regplot(u,v,regfun=ols,xlab='U',ylab='V')
 #But the  OLS regression line has a nearly
 #horizontal slope due to the outliers in the upper left corner.
```

For the data generated here, comdvar does not reject at the 0.05 level. But for the two distributions considered here, the actual Type I error probability is 0.145 based on a simulation with 5,000 replications. Getting accurate results depends in part on getting outliers in the upper left corner of the scatterplot.

## CHAPTER 6

Some Multivariate Methods

The goal in this chapter is to discuss some basic problems and issues related to multivariate data and how they might be addressed. Then some inferential methods, based on the concepts introduced in this chapter, are described. As usual, no attempt is made to provide an encyclopedic coverage of all techniques. Indeed, for some problems, many strategies are now available, for some purposes, there are reasons for preferring certain ones over others, but the reality is that more needs to be done in terms of understanding the relative merits of these methods, and it seems that no single technique can be expected to be satisfactory among all situations encountered in practice.

6.1 Generalized Variance

It helps to begin with a brief review of a basic concept from standard multivariate statistical methods. Consider a random sample of n observations from some p-variate distribution and let

$$s_{jk} = \frac{1}{n-1} \sum_{i=1}^{n} (X_{ij} - \bar{X}_j)(X_{ik} - \bar{X}_k)$$

be the usual sample covariance between the jth and kth variables, where $\bar{X}_j = \sum_i X_{ij}/n$ is the sample mean of the jth variable. Letting $\mathbf{S}$ represent the corresponding sample covariance matrix, the generalized sample variance is

$$G = |\mathbf{S}|,$$

the determinant of the covariance matrix. The property of G that will be exploited here is that it is sensitive to outliers. Said another way, to get a value for G that is relatively small requires that all points be tightly clustered together. If even a single point is moved farther away from the others, G will increase.

Although not used directly, it is noted that an R function

$$gvar(m)$$

has been supplied to compute the generalized variance based on the data in m, where m can be any R matrix having n rows and p columns.

Introduction to Robust Estimation and Hypothesis Testing
https://doi.org/10.1016/B978-0-12-820098-8.00012-9
253

6.2 Depth

A general problem that has taken on increased relevance in applied work is measuring or characterizing how deeply a point is located within a cloud of data. Many strategies exist, but the focus here is on methods that have been found to have practical value when estimating location or testing hypotheses. (For a formal description and discussion of properties measures of depth should have, see Zuo and Serfling, 2000a, 2000b. For a general theoretical perspective, see Mizera, 2002.) This is not to suggest that all alternative methods for measuring depth should be eliminated from consideration, but more research is needed to understand their relative merits when dealing with the problems covered in this book.

6.2.1 Mahalanobis Depth

Certainly the best-known approach to measuring depth is based on the Mahalanobis distance. The squared *Mahalanobis distance* between a point $\mathbf{x}$ (a column vector having length p) and the sample mean, $\bar{\mathbf{X}} = (\bar{X}_1, \ldots, \bar{X}_p)'$, is

$$d^2 = (\mathbf{x} - \bar{\mathbf{X}})' \mathbf{S}^{-1} (\mathbf{x} - \bar{\mathbf{X}}). \tag{6.1}$$

A convention is that the deepest points in a cloud of data should have the largest numerical depth. Following Liu and Singh (1997), the *Mahalanobis depth* is taken to be

$$M_D(\mathbf{x}) = [1 + (\mathbf{x} - \bar{\mathbf{X}})' \mathbf{S}^{-1} (\mathbf{x} - \bar{\mathbf{X}})]^{-1}. \tag{6.2}$$

So the closer a point happens to be to the mean, as measured by Mahalanobis distance, the larger is its Mahalanobis depth.

Mahalanobis distance is not robust and is known to be unsatisfactory for certain purposes to be described. Despite this, it has been found to have value for a wide range of hypothesis testing problems, and it has the advantage of being fast and easy to compute with existing software.

6.2.2 Halfspace Depth

Another preliminary is the notion of halfspace depth, an idea originally developed by Tukey (1975); it reflects a generalization of the notion of ranks to multivariate data. In contrast to other strategies, halfspace depth does not use a covariance matrix.

The idea is, perhaps, best conveyed by first focusing on the univariate case, $p = 1$. Given some number x, consider any partitioning of all real numbers into two components: those values below x and those above. All values less than or equal to x form a *closed halfspace*, and all points strictly less than x form an open halfspace. In a similar manner, all points greater than or equal to x form a closed halfspace, and all points greater than x form an open halfspace. In statistical terms, $F(x) = P(X \le x)$ is the probability associated with a closed halfspace formed by all points less than or equal to x. The notation

$$F(x^-) = P(X < x)$$

represents the probability associated with an open halfspace. Tukey's halfspace depth associated with the value x is intended to reflect how deeply x is nested within the distribution $F(x)$. Its formal definition is

$$T_D(x) = \min\{F(x),\ 1 - F(x^-)\}. \tag{6.3}$$

That is, the depth of x is the smallest probability associated with the two closed halfspaces formed by x. The estimate of $T_D(x)$ is obtained simply by replacing F with its usual estimate, $\hat{F}(x)$, the proportion of X_i values less than or equal to x, in which case, $1 - \hat{F}(x^-)$ is the proportion of X_i values greater than or equal to x. So for $p = 1$, Tukey's halfspace depth is estimated to be the smaller of two proportions: the proportion of observed values less than or equal to x and the proportion greater than or equal to x. When F is a continuous distribution, the maximum possible depth associated with any point is 0.5 and corresponds to the population median. However, the maximum possible depth in a sample of observations can exceed 0.5.

■ Example

Consider the values 2, 5, 9, 14, 19, 21, and 33. The proportion of values less than or equal to 2 is $1/7$, and the proportion greater than or equal to 2 is $7/7$, so the halfspace depth of 2 is $1/7$. The halfspace depth of 14 is $4/7$. Note that 14 is the usual sample median, which can be viewed as the average of the points having the largest halfspace depth.

■

■ Example

For the values 2, 5, 9, 14, 19, and 21, both the values 9 and 14 have the highest halfspace depth among the six observations, which is 0.5, and the average of these two points is the usual sample median. The value 1, relative to the sample 2, 5, 9, 14, 19, and 21, has a halfspace depth of zero.

■

Now we generalize to the bivariate case, $p = 2$. For any line, the points on or above this line form a closed halfspace, as do the points on or below the line. Note that for any bivariate distribution, there is a probability associated with the two closed halfspaces formed by any line. For $p = 3$, any plane forms two closed halfspaces: those points on or above the plane, as well as the points on or below the plane, and the notion of a halfspace generalizes in an obvious way for any p.

For the general p-variate case, consider any point $\mathbf{x}$, where again, $\mathbf{x}$ is a column vector having length p, let $\mathcal{H}$ be any closed halfspace containing the point $\mathbf{x}$, and let $P(\mathcal{H})$ be the probability associated with $\mathcal{H}$. That is, $P(\mathcal{H})$ is the probability that an observation occurs in the halfspace $\mathcal{H}$. Then roughly, the halfspace depth of the point $\mathbf{x}$ is the smallest value of $P(\mathcal{H})$ among all halfspaces $\mathcal{H}$ containing $\mathbf{x}$. More formally, Tukey's halfspace depth is

$$T_D = \inf_{\mathcal{H}}\{P(\mathcal{H}) : \mathcal{H} \text{ is a closed halfspace containing } \mathbf{x}\}. \tag{6.4}$$

For $p > 1$, halfspace depth can be defined instead as the least depth of any one-dimensional projection of the data (Donoho and Gasko, 1992). To elaborate, consider any point $\mathbf{x}$ and any p-dimensional (column) vector $\mathbf{u}$ having unit norm. That is, the *Euclidean norm* of $\mathbf{u}$ is $\|\mathbf{u}\| = \sqrt{u_1^2 + \cdots + u_p^2} = 1$. Then a one-dimensional projection of $\mathbf{x}$ is $\mathbf{u}'\mathbf{x}$ (where $\mathbf{u}'$ is the transpose of $\mathbf{u}$). For any projection, meaning any choice for $\mathbf{u}$, depth is defined by Eq. (6.3). In the p-variate case, the depth of a point is defined to be its minimum depth among all possible projections, $\mathbf{u}'\mathbf{X}$. Obviously, from a practical point of view, this does not immediately yield a viable algorithm for computing halfspace depth based on a sample of n observations, but it suggests an approximation that has been found to be relatively effective.

A data set is said to be in *general position* if there are no ties, no more than two points are on any line, no more than three points are in any plane, and so forth. It should be noted that the maximum halfspace depth varies from one situation to the next and in general does not equal 0.5. If the data are in general position, the maximum depth lies roughly between $1/(p + 1)$ and 0.5 (Donoho and Gasko, 1992).

Halfspace depth is metric-free in the following sense. Let $\mathbf{A}$ be any non-singular $p \times p$ matrix, and let $\mathbf{X}$ be any $n \times p$ matrix of n points. Then the halfspace depths of these n points are unaltered under the transformation $\mathbf{XA}$. More formally, halfspace depth is an *affine invariant*.

6.2.3 Computing Halfspace Depth

For $p = 2$ and 3, halfspace depth, relative to $\mathbf{X}_1, \ldots, \mathbf{X}_n$, can be computed exactly (Rousseeuw and Ruts, 1996; Rousseeuw and Struyf, 1998). For $p > 3$, a general framework for computing the exact halfspace depth of a point was developed by Dyckerhoff and Mozharovskyi

(2016). This framework provides a class of algorithms that might be used. (Various strategies for approximating halfspace depth are reviewed in their paper. Also see Struyf and Rousseeuw, 2000 as well as Cuesta-Albertos and Nieto-Reyes, 2008.) Zuo (2019) and Shao and Zuo (2019) deal with the issue of computing halfspace depth in high dimensions. Here, three methods are described for approximating the halfspace depth of a point. When testing hypotheses, the first approximation has been found to have practical value in a variety of situations. The practical benefits of using some alternative method for computing halfspace depth, when testing hypotheses, has not been determined. (For more details about the first two approximations, see Wilcox, 2003a.)

Approximation A1. The first approximation is based on the one-dimensional projection definition of depth. First, an informal description is given, after which, the computational details are provided. The method begins by computing some multivariate measure of location, say $\hat{\boldsymbol{\theta}}$. There are many reasonable choices, and for present purposes, it seems desirable that it be robust. A simple choice is to use the marginal medians, but a possible objection is that they do not satisfy a criterion discussed in Section 6.3. (The marginal medians are not affine equivariant.) To satisfy this criterion, the minimum covariance determinant (MCD) estimator in Section 6.3.2 is used with the understanding that the practical advantages of using some other estimator has received virtually no attention. Given an estimate of the center of the data, consider the line formed by the ith observation, $\mathbf{X}_i$, and the center. For convenience, call this line $\mathcal{L}$. Now, (orthogonally) project all points onto the line $\mathcal{L}$. That is, for every point $\mathbf{X}_j$, $j = 1, \ldots, n$, draw a line through it that is perpendicular to the line $\mathcal{L}$. Where this line intersects $\mathcal{L}$ is the projection of the point. Fig. 6.1 illustrates the process where the point marked by a circle indicates the center of the data, the line going through the center is line $\mathcal{L}$, and the arrow indicates the projection of the point. That is, the arrow points to the point on the line $\mathcal{L}$ that corresponds to the projection. Next, repeat this process for each i, $i = 1, \ldots, n$. So, for each projected point $\mathbf{X}_j$, we have a depth based on the ith line formed by $\mathbf{X}_i$ and $\hat{\boldsymbol{\theta}}$. Call this depth d_{ij}. For fixed j, the halfspace depth of $\mathbf{X}_j$ is approximated by the minimum value among $d_{1j}, \ldots, d_{nj}$.

Now a more precise description of the calculations is given. For any i, $i = 1, \ldots, n$, let

$$\mathbf{U}_i = \mathbf{X}_i - \hat{\boldsymbol{\theta}},$$

$$B_i = \mathbf{U}_i \mathbf{U}_i'$$
$$= \sum_{k=1}^{p} U_{ik}^2,$$

and for any j ($j = 1, \ldots, n$) let

$$W_{ij} = \sum_{k=1}^{p} U_{ik} U_{jk}$$

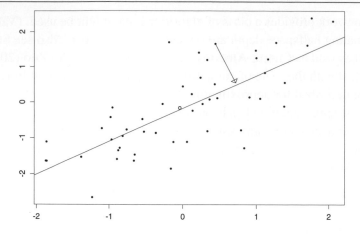

Figure 6.1: An illustration of projecting a point onto a line.

and

$$T_{ij} = \frac{W_{ij}}{B_i}(U_{i1}, \ldots, U_{ip}).\tag{6.5}$$

The distance between $\hat{\theta}$ and the projection of $\mathbf{X}_j$ (when projecting onto the line connecting $\mathbf{X}_i$ and $\hat{\theta}$) is

$$D_{ij} = \text{sign}(W_{ij})\|T_{ij}\|,$$

where $\|T_{ij}\|$ is the Euclidean norm associated with the vector T_{ij}. Let d_{ij} be the depth of $\mathbf{X}_j$ when projecting points onto the line connecting $\mathbf{X}_i$ and $\hat{\theta}$. That is, for fixed i and j, the depth of the projected value of $\mathbf{X}_j$ is

$$d_{ij} = \min(\#\{D_{ij} \leq D_{ik}\}, \#\{D_{ij} \geq D_{ik}\}),$$

where $\#\{D_{ij} \leq D_{ik}\}$ indicates how many D_{ik} $(k = 1, \ldots, n)$ values satisfy $D_{ij} \leq D_{ik}$. Then the depth of $\mathbf{X}_j$ is taken to be

$$L_j = \min d_{ij},$$

the minimum being taken over all $i = 1, \ldots, n$.

Approximation A2. The second approximation of halfspace depth does not use a measure of location, rather, it uses all projections between any two points. That is, method A1 forms n lines, namely, the lines passing through the center of the scatterplot and each of the n points. Method A2 uses $(n^2 - n)/2$ lines, namely, all lines formed by any two (distinct) points. An advantage of method A2 is that, in situations where the exact depth can be determined, it has been found to be more accurate on average than method A1 or the method recommended by Rousseeuw and Struyf (1998); see Wilcox (2003a). Another disadvantage of method A1 is that in some situations to be described, the MCD estimate of location that it uses cannot be computed because the covariance matrix of the data is singular. (Switching to the minimum volume ellipsoid (MVE) estimator described in Section 6.3.1 does not correct this problem.) A possible appeal of method A2 is that depth can still be computed in these situations. A negative feature of A2 is that with n large, execution time can be relatively high.

Although A2 is a more accurate approximation of halfspace depth than A1, perhaps in applied work, this is not a concern. That is, we can think of A1 as a method of defining the depth of a point in a scatterplot, and for practical purposes, maybe any discrepancies between the exact halfspace depth and the depth of a point based on A1 have no negative consequences. This issue has received virtually no attention.

Approximation A3. A third approximation was derived by Rousseeuw and Struyf (1998), but no attempt is made to describe the involved details. A positive feature is that it has low execution time even for fairly large data sets, but it currently seems to be less accurate than the two previous approximations covered in this section. The practical consequences of using method A3 over A2 and A1, when dealing with inferential methods, have not been studied.

6.2.4 R Functions depth2, depth, fdepth, fdepthv2, and unidepth

The R function

$$unidepth(x,pts=NA)$$

is designed for univariate data only and computes the depth of the values in pts relative to the data stored in x. If the argument pts is not specified, the function computes the depth of each value in x relative to all the values.

The function

$$depth(U, V, m)$$

handles bivariate data only and computes the exact halfspace depth of the point (U,V) within the data contained in the matrix m. For example, depth(0,0,m) returns the halfspace depth of (0, 0). The function

$$depth2(m,pts=NA)$$

also handles bivariate data only, but unlike depth, it computes the depth of all the points stored in the matrix pts. If pts is not specified, the exact depths of all the points stored in m are determined. (The function depth is supplied mainly for convenience when computing the halfspace depth of a single point.)

For p-variate data, $p \geq 1$, the function

$$\text{fdepth(m,pts=NA,plotit =TRUE,cop = 2)}$$

computes the approximate halfspace depth of all the points in pts relative to the data in m using method A1. (For $p = 1$, the exact halfspace depth is computed by calling the function unidepth.) If pts is not specified, the function returns the halfspace depth of all points in m. The argument cop indicates the location estimator, $\hat{\theta}$, that will be used by method A1. By default, the MCD estimate is used. Setting cop=3 results in using the marginal medians, and cop=4 uses the MVE estimator (discussed in Section 6.3.1). If simultaneously pts is not specified, m contains two columns of data (so bivariate data are being analyzed), and with plotit=TRUE, a scatterplot of the data is created that marks the center of the data corresponding to the location estimator specified by the argument cop, and it creates what Liu et al. (1999, p. 789) call the 0.5 depth contour. It is a polygon (a convex hull) containing the central half of the data as measured by the depth of the points. The function

$$\text{fdepthv2(m,pts=NA,plotit =TRUE)}$$

uses method A2 to approximate halfspace depth.

6.2.5 Projection Depth

Halfspace depth is defined in terms of the position of the points; it does not use any notion of distance. In various situations, it is convenient and useful to have a measure of depth that captures the spirit of halfspace that includes a measure of distance. There are several ways this might be done. The approximation of halfspace depth, represented by method A1 in Section 6.2.3, is one approach that has been found to have practical value when testing hypotheses. Roughly, for each of the n projections used in method A1, compute the distance between the estimated center $\hat{\theta}$ and the ith point. The *projection distance* associated with the jth point, $\mathbf{X}_j$, is taken to be the largest distance among all n projections after the distances are standardized by dividing by some measure of scale. This can be converted to a measure of projection depth using the same strategy applied to the Mahalanobis distance.

More precisely, compute T_{ij} as given by Eq. (6.5) and let

$$D_{ij} = \|T_{ij}\|.$$

So for the projection of the data onto the line connecting $\mathbf{X}_i$ and $\hat{\boldsymbol{\theta}}$, D_{ij} is the distance between $\mathbf{X}_j$ and $\hat{\boldsymbol{\theta}}$. Now let

$$d_{ij} = \frac{D_{ij}}{q_2 - q_1}, \tag{6.6}$$

where for fixed i, q_2 and q_1 are the ideal fourths based on the values $D_{i1}, \ldots, D_{in}$. The *projection distance* associated with $\mathbf{X}_j$, say $p_d(\mathbf{X}_j)$, is the maximum value of d_{ij}, the maximum being taken over $i = 1, \ldots, n$. To convert to a measure of depth, simply use

$$P_D(\mathbf{X}_j) = \frac{1}{1 + p_d(\mathbf{X}_j)}. \tag{6.7}$$

A variation of the notion of projected distance is discussed by Donoho and Gasko (1992). The main difference is that they use median absolute deviation (MAD) as a measure of scale for the D values in Eq. (6.6) rather than the interquartile range, $q_2 - q_1$. Here, we take this to mean that Eq. (6.6) becomes

$$d_{ij} = \frac{0.6745 D_{ij}}{\text{MAD}}, \tag{6.8}$$

where for fixed i, MAD is based on the values $D_{i1}, \ldots D_{in}$. An obvious advantage of MAD over the interquartile range is that MAD has a higher breakdown point. However, based on other criteria (to be described), the use of the interquartile range has been found to have practical value.

6.2.6 R Functions pdis, pdisMC, and pdepth

The R function

$$\text{pdis(m,pts=m,MM=FALSE,cop=3,dop=1,center=NA)}$$

computes projection distances, p_d, for the data stored in pts relative to the matrix m. So by default, the function returns the distances for each row in m. If MM=FALSE is used, distances are scaled using the interquartile range, and MM=TRUE uses MAD. The argument cop indicates which measure of location will be used. The choices, some of which are described in Section 6.3, are:

- cop=1, Donoho–Gasko median,
- cop=2, MCD estimate of location,
- cop=3, marginal medians,
- cop=4, MVE estimate of location,
- cop=5, OP (skipped) estimator.

If a value is supplied for the argument center (a vector having length p), this value is used as a measure of location and the argument cop is ignored. (The argument dop is relevant when using the Donoho–Gasko analog of the trimmed mean. See Section 6.3.5.) If a multicore processor is available, execution time can be reduced by using the R function

$$\text{pdisMC(m,MM=FALSE,cop=3,dop=1,center=NA)}$$

instead of the function pdis. Note that unlike pdis, there is no argument pts. This function computes only the distances for the data stored in the matrix m.

For convenience, the R function

$$\text{pdepth(m,pts=m,MM=FALSE,cop=3,dop=1,center=NA,MC=FALSE)}$$

computes the projection measure of depth for each row stored in pts, relative to m, based on p_d.

6.2.7 More Measures of Depth

There are many alternatives to the notion of projection depth in Section 6.2.5, some of which are outlined in this section. One approach is to use the Mahalanobis distance but with the mean and usual covariance matrix replaced by robust estimators. Some possibilities are mentioned later in this chapter. This can suffice for certain purposes, but unlike the other measures described here, it uses an ellipsoid to represent the central portion of a data cloud.

Two measures that are similar in spirit to halfspace depth are simplicial depth (Liu, 1990; cf. Serfling and Wang, 2016) and majority depth proposed by K. Singh. (See Liu and Singh, 1997, p. 268.) Both methods are geometric in nature and do not rely on some measure of multivariate scatter. Simplicial depth can be computed via the R function sdepth in the R package mrfDepth. (For some possible concerns about simplicial depth, see Zuo and Serfling, 2000a, 2000b.) Certain methods in this book use, by default, halfspace depth. Any practical advantages simplicial depth might enjoy over halfspace and majority depth have not been discovered as yet, so further details are omitted.

Zuo (2003) studies a notion of projected-based depth that is a broader generalization of other projection-based approaches. Again, let $\mathbf{u}$ be any p-dimensional (column) vector having unit norm, and let

$$O(\mathbf{x}; F) = \sup \frac{|\mathbf{u}'\mathbf{x} - \theta(F_u)|}{\sigma(F_u)}$$

be some measure of outlyingness of the point **x** with respect to the distribution F, where F_u is the distribution of $\mathbf{u}'\mathbf{x}$, $\theta(F_u)$ is some (univariate) measure of location and $\sigma(F_u)$ is some measure of scale associated with F_u, and the supremum is taken over all possible choices for **u** such that $\|u\| = 1$. Then the projection depth of the point **x** is taken to be

$$P_D(\mathbf{x}) = \frac{1}{1 + O(\mathbf{x})}.$$

For results on the exact computation of this notion of depth, in the bivariate case, see Zuo and Lai (2011). A projection median is the point that maximizes $P_D(\mathbf{x})$, which can be computed with the R function dmedian, described in Section 6.3.14. (Also see Liu, 2017.) For results on the uniqueness of this median, see Zuo (2013).

Another approach that has been used in a variety of situations is called the bagdistance of a point (Hubert et al., 2017). Roughly, for the point **x**, consider the ray extending from the halfspace median to **x**. Let $\mathbf{c_x}$ be the point where the ray intersects the boundary of the convex hull containing the deepest half of the data. The bagdistance is the Euclidean distance between **x** and the halfspace median divided by the distance between halfspace median and $\mathbf{c_x}$, say $p_{\text{bag}}(\mathbf{x})$. Bagdepth is $1/(1 + p_{\text{bag}}(\mathbf{x}))$.

Koshevoy and Mosler (1997) proposed what they call zonoid depth, which is based on multivariate trimmed regions that are centered about the mean. It has been found to be useful in certain situations when using a classification method (method C2) described in Section 6.16.

6.2.8 R Functions zdist, zoudepth prodepth, Bagdist, bagdepth, and zonoid

The R function

$$\text{zdist(m,pts=m,zloc=median,zscale=mad)}$$

computes Zuo's notion of projection distance, $O(\mathbf{x}; F)$, for each point stored in the matrix pts, relative to the data stored in the matrix m. The arguments zloc and zscale correspond to $\theta(F_u)$ and $\sigma(F_u)$, respectively. For convenience, the R function

$$\text{zoudepth(m,pts=m,zloc=median,zscale=mad)}$$

computes the projection depth, P_D.

The R function

$$\text{prodepth(x,pts=x,ndir=1000)}$$

computes a projection-type measure of depth for the points in pts, relative to x, via the R package DepthProc. It uses random projections where the number of projections is determined by the argument ndir. An advantage of the function prodepth, compared to zoudepth, is that execution time can be substantially lower.

The R function

$$\text{Bagdist(x,z=NULL)}$$

computes the bagdistance for the points in z via the R package mrfDepth. If z is NULL, the bagdistance for every point in x is computed. The R function

$$\text{bagdepth(x,z=NULL)}$$

computes bagdepth. And

$$\text{zonoid(x,z=NULL)}$$

computes zonoid depth via the R package ddalpha.

6.3 Some Affine Equivariant Estimators

One of the most difficult problems in robust statistics has been the estimation of multivariate shape and location. Many such estimators have been proposed (e.g., Davies, 1987; Donoho, 1982; Kent and Tyler, 1996; Lopuhaä, 1991; Maronna and Zamar, 2002; Rousseeuw, 1984; Stahel, 1981; Tamura and Boos, 1986; Tyler, 1994; Wang and Raftery, 2002; Woodruff and Rocke, 1994; Zhang et al., 2012; cf. Adam and Bejda, 2018). A concern about early attempts, such as multivariate M-estimators as described in Huber (1981), is that when working with p-variate data, typically, they have a breakdown point of at most $1/(p + 1)$. So in high dimensions, a very small fraction of outliers can result in very bad estimates. Several estimators have been proposed that enjoy a high breakdown point. But simultaneously achieving relatively high accuracy, versus the vector of means when sampling from a multivariate normal distribution, has proven to be a very difficult problem.

In Section 2.1, a basic requirement for θ to qualify as a measure location was that it be both scale- and location-equivariant. Moreover, a location estimator should satisfy this property as well. That is, if $T(\mathbf{X}_1, \ldots, \mathbf{X}_n)$ is to qualify as a location estimator, it should be the case that for constants a and b,

$$T(a\mathbf{X}_1 + b, \ldots, a\mathbf{X}_n + b) = aT(\mathbf{X}_1, \ldots, \mathbf{X}_n) + b.$$

So, for example, when transforming from feet to centimeters, the typical value in feet is transformed to the appropriate value in centimeters. In the multivariate case, a generalization of this requirement, called *affine equivariance*, is that for a p-by-p non-singular matrix $\mathbf{A}$ and vector $\mathbf{b}$ having length p,

$$T(\mathbf{X}_1\mathbf{A} + \mathbf{b}, \ldots, \mathbf{X}_n\mathbf{A} + \mathbf{b}) = T(\mathbf{X}_1, \ldots, \mathbf{X}_n)\mathbf{A} + \mathbf{b}, \tag{6.9}$$

where now $\mathbf{X}_1, \ldots, \mathbf{X}_n$ is a sample from a p-variate distribution and each $\mathbf{X}_i$ is a (row) vector having length p. So in particular, the estimate is transformed properly under rotations of the data as well as changes in location and scale. The sample means of the marginal distributions are affine equivariant, but typically, when applying any of the univariate estimators in Chapter 3 to the marginal distributions, an affine equivariant estimator is not obtained.

A measure of scatter, say $\mathbf{V}(\mathbf{X})$, is said to be *affine equivariant* if

$$\mathbf{V}(\mathbf{A}\mathbf{X} + \mathbf{b}) = \mathbf{A}\mathbf{V}(\mathbf{X})\mathbf{A}'. \tag{6.10}$$

The usual sample covariance matrix is affine equivariant but not robust.

From Donoho and Gasko (1992, p. 1811), no affine equivariant estimator can have a breakdown point greater than

$$\frac{n - p + 1}{2n - p + 1}. \tag{6.11}$$

(Also see Lopuhaä and Rousseeuw, 1991.)

The rest of this section describes some of the estimators that have been proposed, and a particular variation of one of these methods is described in Section 6.5.

6.3.1 Minimum Volume Ellipsoid Estimator

One of the earliest affine equivariant estimators to achieve a breakdown point of approximately 0.5 is the MVE estimator, a detailed discussion of which can be found in Rousseeuw and Leroy (1987). Consider any ellipsoid containing half of the data. (An example in the bivariate case is shown in Fig. 6.2.) The basic idea is to search among all such ellipsoids for the one having the smallest volume. Once this subset is found, the mean and covariance matrix of the corresponding points are taken as the estimated measure of location and scatter, respectively. Typically, the covariance matrix is rescaled to obtain consistency at the multivariate normal model (e.g., Marazzi, 1993, p. 254). A practical problem is that it is generally difficult to find the smallest ellipse containing half of the data. That is, in general, the collection of all subsets containing half of the data is so large, determining the subset that has the minimum volume is impractical, so an approximation must be used. Let h be equal to $n/2 + 1$, rounded

down to the nearest integer. An approach to computing the MVE estimator is to randomly select h points, without replacement, from the n points available, compute the volume of the
ellipse containing these points, and then repeat this process many times. The set of points
yielding the smallest volume is taken to be the minimum volume ellipsoid. (For relevant software, see Section 6.4.5.)

6.3.2 The Minimum Covariance Determinant Estimator

An alternative to the MVE estimator, which also has a breakdown point of approximately 0.5,
is the MCD estimator. (For a generalization of this estimator aimed at high-dimensional data,
see Bulut, 2020.) Rather than search for the subset of half the data that has the smallest volume, search for the half that has the smallest generalized variance. (For results on computing
the MCD estimator, see Schnys et al., 2010.) Recall from Section 6.1 that for the determinant
of the covariance matrix (the generalized variance) to be relatively small, it must be the case
that there are no outliers. That is, the data must be tightly clustered together. The MCD estimator searches for the half of the data that is most tightly clustered together among all subsets
containing half of the data, as measured by the generalized variance. Like the MVE estimator,
typically, it is impractical considering all subsets of half the data, so an approximate method
must be used. An algorithm for accomplishing this goal is described in Rousseeuw and van
Driessen (1999); also see Atkinson (1994) as well as Hubert et al. (2012). For asymptotic
results, see Butler et al. (1993). Once an approximation of the subset of half of the data has
been determined that minimizes the generalized variance, compute the usual mean and covariance matrix based on this subset. This yields the MCD estimate of location and scatter.
Bernholdt and Fischer (2004) indicate that this algorithm can provide a poor approximation
of the MCD estimator. Results reported by Hawkins and Olive (2002) also raise concerns
about this estimator. But as a diagnostic tool, MCD seems to have practical value when used
in conjunction with other methods covered in this chapter. For relevant software, see Section 6.4.5.

Herwindiati et al. (2007) suggest a variation of the MCD estimator that searches for the subset
of the data that minimizes the trace of the corresponding covariance matrix rather than the
determinant, which they call the *minimum variance vector* (MVV) method. It has the same
breakdown point as the MCD method and is simpler to compute. Herwindiati et al. suggest
that the method is applicable when dealing with large, high-dimensional data sets. In terms of
identifying outliers, limited results suggest that it performs as well as the MCD estimator, but
further study is needed. Also see Zhang et al. (2012).

A generalization of the MCD estimator, aimed at dealing with missing values, can be applied
via the R function CovNAMcd, which is included in the R package rrcovNA. For details regarding the algorithm that is used, see Todorov et al. (2011). It seems that this method for

handling missing values has practical value when the goal is to detect outliers. When testing hypotheses, some other approach appears to be preferable, at least at the moment. (See Section 8.2 for more comments about this issue.)

Another issue is the random component of the MCD estimator. A criticism is that as a result, it is not permutation-invariant. That is, reordering the variables will impact the estimate of location and scatter. A deterministic version of the MCD (DETMCD) estimator was derived by Hubert et al. (2012) for dealing with this.

6.3.3 S-Estimators and Constrained M-Estimators

One of the earliest treatments of S-estimators can be found in Rousseeuw and Leroy (1987, p. 263). A particular variation of this method that appears to be especially interesting is the translated-biweight S-estimator (TBS) proposed by Rocke (1996). Generally, S-estimators of multivariate location and scatter are values for $\hat{\theta}$ and $\mathbf{S}$ that minimize $|\mathbf{S}|$, the determinant of $\mathbf{S}$, subject to

$$\frac{1}{n}\sum_{i=1}^{n}\xi(((\mathbf{X}_i - \hat{\theta})'\mathbf{S}^{-1}(\mathbf{X}_i - \hat{\theta}))^{1/2}) = b_0, \tag{6.12}$$

where b_0 is some constant and (as in Chapter 2) ξ is a non-decreasing function. Lopuhaä (1989) showed that S-estimators are in the class of M-estimators with standardizing constraints. Rocke (1996) showed that S-estimators can be sensitive to outliers even if the breakdown point is close to 0.5.

Rocke (1996) proposed a modified biweight estimator, which is essentially a constrained M-estimator, where for values of m and c to be determined, the function $\xi(d)$, when $m \leq d \leq m + c$, is

$$\xi(d) = \frac{m^2}{2} - \frac{m^2(m^4 - 5m^2c^2 + 15c^4)}{30c^4} + d^2\left(0.5 + \frac{m^4}{2c^4} - \frac{m^2}{c^2}\right)$$
$$+ d^3\left(\frac{4m}{3c^2} - \frac{4m^3}{3c^4}\right) + d^4\left(\frac{3m^2}{2c^4} - \frac{1}{2c^2}\right) - \frac{4md^5}{5c^4} + \frac{d^6}{6c^4},$$

for $0 \leq d < m$,

$$\xi(d) = \frac{d^2}{2},$$

and for $d > m + c$,

$$\xi(d) = \frac{m^2}{2} + \frac{c(5c + 16m)}{30}.$$

The values for m and c can be chosen to achieve the desired breakdown point and the *asymptotic rejection probability*, roughly referring to the probability that a point will get zero weight when the sample size is large. If the asymptotic rejection probability is to be γ, say, then m and c are determined by

$$E_{\chi_p^2}(\xi(d)) = b_0$$

and

$$m + c = \sqrt{\chi_{p,1-\gamma}^2},$$

where $\chi_{p,1-\gamma}^2$ is the $1 - \gamma$ quantile of a chi-squared distribution with p degrees of freedom. (For a generalized S-estimator, designed to handle missing values, assuming that sampling is from an elliptical distribution, see Danilov et al., 2012.) For some computational concerns regarding S-estimators, see He and Wang (1997, p. 258) as well as Huber and Ronchetti (2009, p. 197).

A related estimator, called DetS, was derived by Hubert et al. (2015a). It improves upon the FastS estimator derived by Salibian-Barrera and Yohai (2006), and when dealing with high-dimensional data, it has relatively low execution time and good robustness properties.

6.3.4 R Functions tbs, DETS, and DETMCD

The R function

$$\text{tbs(m)}$$

computes the TBS measure of location and scatter just outlined using code supplied by David Rocke. The R function

$$\text{DETS(m)}$$

computes the DetS estimator via the R package rrcov. The R function

$$\text{DETMCD(m)}$$

computes the deterministic version of the MCD estimator mentioned in Section 6.3.2.

6.3.5 Donoho–Gasko Generalization of a Trimmed Mean

Another approach to computing an affine equivariant measure of location was suggested and studied by Donoho and Gasko (1992). The basic strategy is to compute the halfspace depth for each of the n points, remove those that are not deeply nested within the cloud of data, and then average those points that remain. The *Donoho–Gasko γ trimmed mean* is the average of all points that are at least γ deep in the sample. That is, points having depth less than γ are trimmed, and the mean of the remaining points is computed. An analog of the median, which has been called *Tukey's median*, is the average of all points having the largest depth (cf. Adrover and Yohai, 2002; Bai and He, 1999; Tyler, 1994). If the maximum depth of $\mathbf{X}_i$, $i = 1, \ldots, n$, is greater than or equal to γ, then the breakdown point of the Donoho–Gasko γ trimmed mean is $\gamma/(1 + \gamma)$. For symmetric distributions, the breakdown point is approximately 0.5, but because the maximum depth among a sample of n points can be approximately $1/(1 + p)$, the breakdown point could be as low as $1/(p + 2)$. If the data are in general position, the breakdown point of Tukey's median is greater than or equal to $1/(p + 1)$. (The influence function, assuming a type of symmetry for the distribution, was derived by Chen and Tyler, 2002.)

■ Example

The data file eeg.txt, which can be accessed as indicated in Section 1.10, contains results reported by Raine et al. (1997), who were interested in comparing EEG measures for murderers versus a control group. For the moment, consider the first two columns of data only. The exact halfspace depths, determined by the function depth2, are

0.1428570 0.1428570 0.3571430 0.2142860 0.2142860 0.2142860 0.0714286
0.2142860 0.1428570 0.2142860 0.0714286 0.0714286 0.0714286 0.0714286

There are five points with depth less than 0.1. Eliminating these points and averaging the values that remain yields the Donoho–Gasko 0.1 trimmed mean, $(-0.042, -0.783)$. The halfspace median corresponds to the deepest point, which is $(0.07, -0.44)$.

■

Another multivariate generalization of a trimmed mean was studied by Liu et al. (1999). Any practical advantages it might have over the Donoho–Gasko γ trimmed mean have not been discovered as yet, so for brevity, no details are given here.

6.3.6 R Functions dmean and dcov

The R function

$$\text{dmean(x,tr=0.2,dop=1,cop=2)}$$

computes the Donoho–Gasko trimmed mean. When the argument tr is set equal to 0.5, it computes Tukey's median, namely, the average of the points having the largest halfspace depth. The argument dop controls how the halfspace depth is approximated. With dop=1, method A1 in Section 6.2.3 is used to approximate halfspace depth when $p > 2$, while dop=2 uses method A2. If $p = 2$, halfspace depth is computed exactly. When using method A1, the center of the scatterplot is determined using the estimator indicated by the argument cop. The choices are:

- cop=2, MCD estimator,
- cop=3, marginal medians,
- cop=4, MVE estimator.

When n is small relative to p, the MCD and MVE estimators cannot be computed, so in these cases, use dop=2. For small sample sizes, execution time is low.

Consider again Zuo's notion of projection depth described in Section 6.2.7. When $\theta(F_u)$ is taken to be the median, and $\sigma(F_u)$ is MAD, and if the average of the deepest points are used as a measure of location, we get another affine equivariant generalization of the median. Comparisons with other affine equivariant median estimators are reported by Hwang et al. (2004) for the bivariate case. They conclude that this estimator and Tukey's median compare well to other estimators they considered.

The R function

$$\text{dcov(x,tr=0.2,dop=1,cop=2)}$$

computes the covariance matrix after the data are trimmed as done by the Donoho–Gasko trimmed mean.

6.3.7 The Stahel–Donoho W-Estimator

Stahel (1981) and Donoho (1982) propose the first multivariate, equivariant estimator of location and scatter that has a high breakdown point. It is a weighted mean and covariance matrix where the weights are a function of how "outlying" a point happens to be. The more outlying a point, the less weight it is given. The notions of Mahalanobis depth, robust analogs of

Mahalanobis depth based perhaps on the MVE or MCD estimators, and halfspace depth are examples of how to measure the outlyingness of a point. Attaching some weight w_i to $\mathbf{X}_i$, which is a function of how outlying the point $\mathbf{X}_i$ happens to be, yields a generalization of W-estimators mentioned in Section 3.8 (cf. Hall and Presnell, 1999). Here, the estimate of location is

$$\hat{\boldsymbol{\theta}} = \frac{\sum_{i=1}^n w_i \mathbf{X}_i}{\sum_{i=1}^n w_i}, \tag{6.13}$$

and the measure of scatter is

$$\mathbf{V} = \frac{\sum_{i=1}^n w_i (\mathbf{X}_i - \hat{\boldsymbol{\theta}})(\mathbf{X}_i - \hat{\boldsymbol{\theta}})'}{\sum_{i=1}^n w_i}. \tag{6.14}$$

The Donoho–Gasko trimmed mean in Section 6.3.4 is a special case where the least deep points get a weight of zero; otherwise, points get a weight of one. Other variations of this approach are based on the multivariate outlier detection methods covered in Section 6.4. For general theoretical results on this approach to estimation, see Tyler (1994). Properties of certain variations were reported by Maronna and Yohai (1995). Also see Arcones et al. (1994), Bai and He (1999), He and Wang (1997), Donoho and Gasko (1992), Gather and Hilker (1997), Zuo (2003), Zuo et al. (2004a), and Zuo et al. (2004b). Gervini (2002) derives the influence function assuming that sampling is from an elliptical distribution. (For an extension of M-estimators to the multivariate case that has a high breakdown point and deals with missing values, see Chen and Victoria-Feser, 2002, as well as Frahm and Jaekel, 2010).

Zuo et al. (2004a) suggest a particular variation of the Donoho–Gasko W-estimator for general use. (Yet another variation was suggested by Van Aelst et al., 2012.) Let P_i be the projection depth of $\mathbf{x}_i$ described at the end of Section 6.2.7. Let C be the median of the P_i values. If $P_i < C$, set

$$w_i = \frac{\exp(-K(1 - P_i/C)^2) - \exp(-K)}{1 - \exp(-K)},$$

otherwise, $w_i = 1$, and the measures of location and scatter are given by Eqs. (6.13) and (6.14), respectively. From Zuo et al., setting the constant $K = 3$ results in good asymptotic efficiency, relative to the sample mean, under normality.

6.3.8 R Function sdwe

The R function

$$\text{sdwe(x,K=3)}$$

computes the Stahel–Donoho W-estimator as suggested by Zuo et al. (2004a, 2004b).

6.3.9 Median Ball Algorithm

This section describes a multivariate measure of location and scatter, introduced by Olive (2004), which is based on what he calls the reweighted median ball algorithm (RMBA). It is an iterative algorithm that begins with two initial estimates of location and scatter. The first, labeled $(T_{0,1}, \mathbf{C}_{0,1})$, is taken to be the usual mean and covariance matrix. The other starting value, $(T_{0,2}, \mathbf{C}_{0,2})$, is the usual mean and covariance based on the $c_n \approx n/2$ cases that are closest to the coordinate-wise median in Euclidean distance. Compute all n Mahalanobis distances $D_i(T_{0,j}, \mathbf{C}_{0,j})$ based on the jth starting value. The next iteration consists of estimating the usual mean and covariance matrix based on the c_n cases corresponding to the smallest distances, yielding $(T_{1,j}, \mathbf{C}_{1,j})$. Repeating this process, based on $D_i(T_{1,j}, \mathbf{C}_{1,j})$, yields an updated measure of location and scatter, $(T_{2,j}, \mathbf{C}_{2,j})$. As done by Olive, unless stated otherwise, it is assumed five iterations are used, yielding $(T_{5,j}, \mathbf{C}_{5,j})$. The RMBA estimator of location, labeled T_A, is taken to be $T_{5,i}$, where $i = 1$, if the determinant $|\mathbf{C}_{5,1}| \leq |\mathbf{C}_{5,2}|$, and otherwise, $i = 2$. And the measure of scatter is

$$\mathbf{C}_{\text{RMBA}} = \frac{\text{MED}(D_i^2(T_A, \mathbf{C}_A))}{\chi_{p,0.5}^2} \mathbf{C}_A.$$

(Also see Olive and Hawkins, 2010.)

6.3.10 R Function rmba

The R function

```
rmba(m,csteps=5)
```

computes the RMBA measure of location and scatter, where the argument csteps controls the number of iterations. (The R code was graciously supplied by David Olive.)

6.3.11 OGK Estimator

Yet another estimator that is sometimes recommended is the orthogonal Gnanadesikan-Kettenring (OGK) estimator, derived by Maronna and Zamar (2002). In its general form, it is applied as follows. Let $\sigma(X)$ and $\mu(X)$ be any measure of dispersion and location, respectively. The method begins with the robust covariance between any two variables, say X and Y, which was proposed by Gnanadesikan and Kettenring (1972):

$$\text{cov}(X, Y) = \frac{1}{4}(\sigma(X + Y)^2 - \sigma(X - Y)^2). \tag{6.15}$$

When $\sigma(X)$ and $\mu(X)$ are the usual standard deviation and mean, respectively, the usual covariance between X and Y results. Here, following Maronna and Zamar, $\sigma(X)$ is taken to be the tau scale of Yohai and Zamar (1988), which was introduced in Section 3.12.3. Using this measure of scale in Eq. (6.15), the resulting measure of covariance will be denoted by $v(X, Y)$.

Following the notation in Maronna and Zamar (2002), let $\mathbf{x}_i$ be the ith row of the $n \times p$ matrix $\mathbf{X}$. Then Maronna and Zamar define a scatter matrix $\mathbf{V}(\mathbf{X})$ and a location vector $\mathbf{t}(\mathbf{X})$ as follows:

1. Let $\mathbf{D} = \mathrm{diag}(\sigma(X_1), ..., \sigma(X_p))$ and $\mathbf{y}_i = \mathbf{D}^{-1}\mathbf{x}_i$, $i = 1, \ldots, n$.
2. Compute $\mathbf{U} = (U_{jk})$ by applying v to the columns of $\mathbf{Y}$. So $U_{jj} = 1$ and for $j \neq k$, $U_{jk} = v(Y_j, Y_k)$.
3. Compute the eigenvalues λ_j and eigenvectors $\mathbf{e}_j$ of $\mathbf{U}$, and let $\mathbf{E}$ be the matrix whose columns are the $\mathbf{e}_j$'s. (So $\mathbf{U} = \mathbf{E}\Lambda\mathbf{E}'$.)
4. Let $\mathbf{A} = \mathbf{DE}$, $\mathbf{z}_i = \mathbf{A}^{-1}\mathbf{x}_i$, in which case

$$\mathbf{V}(\mathbf{X}) = \mathbf{A}\Gamma\mathbf{A}',$$

and

$$\mathbf{t}(\mathbf{X}) = \mathbf{A}\boldsymbol{v},$$

where $\Gamma = \mathrm{diag}(\sigma^2(Z_1), \ldots, \sigma^2(Z_p))$, $\boldsymbol{v} = (\mu(Z_1), \ldots, \mu(Z_p))$, and μ is taken to be the tau measure of location in Section 3.8.1.

Maronna and Zamar (2002) note that the above procedure can be iterated and report results suggesting that a single iteration be used. More precisely, compute $\mathbf{V}$ and $\mathbf{t}$ for $\mathbf{Z}$ (the matrix corresponding to $\mathbf{z}_i$ computed in step 4), and then express them in the original coordinate system, namely, $\mathbf{V}_2 = \mathbf{A}\mathbf{V}(\mathbf{Z})\mathbf{A}'$ and $\mathbf{t}_2(\mathbf{X}) = \mathbf{A}\mathbf{t}(\mathbf{Z})$. Maronna and Zamar show that the estimate can be improved by a reweighting step. Let

$$d_i = \sum_j \left(\frac{z_{ij} - \mu(Z_j)}{\sigma(Z_j)} \right)$$

and $w_i = I(d_i \leq d_0)$, where

$$d_0 = \frac{\chi^2_{p,\beta}\mathrm{med}(d_1, \ldots, d_n)}{\chi^2_{p,0.5}},$$

$\chi^2_{p,\beta}$ is the β quantile of the chi-squared distribution with p degrees of freedom, and "med" denotes the sample median. The measure of location is now estimated to be

$$\mathbf{t}_w = \frac{\sum w_i \mathbf{x}_i}{\sum w_i},$$

and the measure of scatter is

$$\mathbf{V}_w = \frac{\sum w_i (\mathbf{x}_i - \mathbf{t}_w)(\mathbf{x}_i - \mathbf{t}_w)'}{\sum w_i}.$$

A generalization of the OGK estimator, aimed at dealing with missing values, can be applied via the R function CovNAOgk, which is included in the R package rrcovNA. See Todorov and Filzmoser (2010) for details as well as Todorov et al. (2011) regarding the algorithm that is used. Roughly, missing values are imputed. Evidently, there are no published results on how well it performs when testing hypotheses.

6.3.12 R Function ogk

The R function

$$\text{ogk(x,sigmamu=aulc,v=gkcov,n.iter=1,beta=0.9,...)}$$

computes the OGK measure of location and scale.

6.3.13 An M-Estimator

As noted at the beginning of this section, a concern about (affine equivariant) M-estimators is that they have a breakdown point of at most $1/(p+1)$. Also, Devlin et al. (1981, p. 361) found that M-estimators could tolerate even fewer outliers than indicated by this upper bound. Despite this, in situations where p is small, this approach might be deemed satisfactory. For example, Zu and Yuan (2010) suggest an approach to a mediation analysis that is based in part on an M-estimator with Huber weights, which was derived by Maronna (1976). A slight modification of the Zu and Yuan method has been found to perform relatively well in simulations, so for completeness, Maronna's M-estimator is outlined here. (Details of the Zu and Yuan method for performing a mediation analysis are outlined in Section 11.7.2.)

The computation of this estimator is accomplished via an iterative scheme that corresponds to a multivariate version of the W-estimator in Section 3.8. Roughly, an initial estimate of the mean and covariance matrix is computed, which here is taken to be usual mean $\bar{\mathbf{X}}$ vector and covariance matrix $\mathbf{S}$. Based on this initial estimate, squared Mahalanobis distances are computed:

$$d_i^2 = (\mathbf{X}_i - \bar{\mathbf{X}})' \mathbf{S}^{-1} (\mathbf{X}_i - \bar{\mathbf{X}}).$$

Imagine that one wants to downweight a proportion κ of the observations. Let ϱ^2 be the $1 - \kappa$ quantile of a chi-squared distribution with p degrees of freedom. Let $w_i = 1$, if $d_i \leq \varrho$; otherwise, $w_i = \varrho/d_i$. Then an updated estimate of the mean and covariance matrix is given by

$$\bar{\mathbf{X}} = \sum w_i \mathbf{X}_i / n$$

and

$$\mathbf{S} = \frac{1}{\tau n} \sum w_i^2 (\mathbf{X}_i - \bar{\mathbf{X}})(\mathbf{X}_i - \bar{\mathbf{X}})',$$

respectively, where τ is chosen so that $\mathbf{S}$ is an unbiased estimate of the covariance matrix under normality. These updated estimates are used to update the squared Mahalanobis distances, which in turn yields a new updated estimate of the mean and covariance matrix. This process is continued until convergence is achieved.

6.3.14 R Functions MARest and dmedian

The R function

$$\text{MARest(x,kappa=0.1)}$$

computes Maronna's M-estimator of location and scatter, where the argument kappa corresponds to κ in the previous section.

The R function

$$\text{dmedian(x,depfun=pdepth,...)}$$

computes the median of a cloud of data, which is taken to be the deepest point based on some measure of depth. A possible appeal of this median is that when using a projection-type measure of depth, as described, for example, in Sections 6.2.5 and 6.2.7, the median is unique when dealing with continuous variables (Zuo, 2013). By default, the projection measure of depth in Section 6.2.5 is used. Alternative projection-type measures were described in Section 6.2.7 and can be applied with the R functions in Section 6.2.8.

6.4 Multivariate Outlier Detection Methods

An approach to detecting outliers when working with multivariate data is to simply check for outliers among each of the marginal distributions using one of the methods described in Chapter 3. However, this approach can be unsatisfactory: Outliers can be missed because this

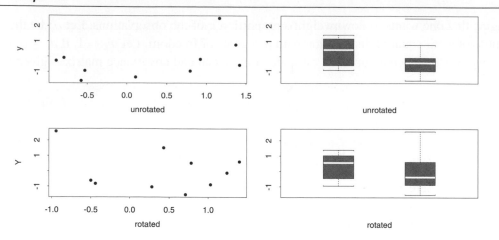

Figure 6.2: The upper right panel shows a boxplot for the X and Y values shown in the upper left panel. The lower right panel shows a boxplot for the X and Y values after the points in the upper left panel are rotated to the position shown in the lower left panel. Now the boxplots find no outliers, in contrast to the unrotated case.

approach does not take into account the overall structure of the data. In particular, any multivariate outlier detection method should be invariant under rotations of the data. Methods based on the marginal distributions do not satisfy this criterion.

To illustrate the problem, consider the observations in the upper left panel of Fig. 6.2. The upper right panel shows a boxplot of both the X and Y values. As indicated, one Y value is flagged as an outlier. It is the point in the upper right corner of the scatterplot. If the points are rotated such that they maintain their relative positions, the outlier should remain an outlier, but for some rotations, this is not the case. For example, if the points are rotated 45 degrees, the scatterplot now appears as shown in the lower left panel with the outlier in the upper left corner. The lower right panel shows a boxplot of the X and Y values after the axes are rotated. Now the boxplots do not indicate any outliers because they fail to take into account the overall structure of the data. What is needed is a method that is invariant under rotations of the data. In addition, any outlier detection method should be invariant under changes in scale. All of the methods in this section satisfy these two criteria.

All but two of the multivariate outlier detection methods described in this section can be used with p-variate data for any $p > 2$. The two exceptions are the relplot and bagplot, which are described in the next section. Early attempts at identifying multivariate outliers, described in Sections 6.4.1, 6.4.3, and 6.4.4, assume that outliers can be identified as points lying outside of an appropriately chosen ellipsoid. A general strategy is to use a robust analog of Mahalanobis distance obtained by replacing the mean and covariance matrix with robust alternatives. These methods ignore any asymmetry in the data cloud. The bagplot in Section 6.4.1

and the methods described in Sections 6.4.7 and 6.4.9 take a more flexible approach to any asymmetry in the data.

The methods listed here are far from exhaustive. For example, Archimbaud et al. (2018) proposed an invariant coordinate selection method. Liebscher et al. (2012) suggest a method that has a connection with self-organizing maps, which are well-known in the field of artificial neural networks. Also see the R package mvoutlier. Perhaps these methods have certain practical advantages over those described here in Sections 6.4.7 and 6.4.9. This remains to be determined. Section 6.4.15 provides a few more comments about the relative merits of the methods covered here.

6.4.1 The Relplot and Bagplot

A relplot is a bivariate generalization of the boxplot derived by Goldberg and Iglewicz (1992). It is based in part on a bivariate generalization of M-estimators covered in Chapter 3 that belongs to the class of W-estimators described in Section 6.3.6. Let X_{ij} ($i = 1, \ldots, n$, $j = 1, 2$) be a random sample from some bivariate distribution. For fixed j, the method begins by computing M_j, MAD_j, and $\hat{\zeta}_j^2$ using the X_{ij} values, where M_j is the sample median, MAD_j is the median absolute deviation statistic, and $\hat{\zeta}_j^2$ is the biweight midvariance described in Section 3.12.1. Let

$$U_{ij} = \frac{X_{ij}}{9MAD_j},$$

and set $a_{ij} = 1$, if $|U_{ij}| < 1$, and otherwise $a_{ij} = 0$. Let

$$T_j = M_j + \frac{\sum a_{ij}(X_{ij} - M_j)(1 - U_{ij}^2)^2}{\sum a_{ij}(1 - U_{ij}^2)^2}.$$

The remaining computational steps are given in Table 6.1, which yield a bivariate measure of location, (T_{b1}, T_{b2}), a robust measure of variance, s_{b1}^2 and s_{b2}^2, and a robust measure of correlation, R_b. These measures of location and scatter can be extended to $p > 2$ variates, but computational problems can arise (Huber, 1981).

The relplot consists of two ellipses. Once the computations in Table 6.1 are complete, the inner ellipse is constructed as follows. Let

$$Z_{ij} = \frac{X_{ij} - T_{bj}}{s_{bj}}$$

and

$$E_i = \sqrt{\frac{Z_{i1}^2 + Z_{i2}^2 - 2R_b Z_{i1} Z_{i2}}{1 - R_b^2}}.$$

Table 6.1: Computing biweight M-estimators of location, scale, and correlation.

Step 1. Compute $Z_{ij} = (X_{ij} - T_j)/\hat{\zeta}_j$.

Step 2. Recompute T_j and $\hat{\zeta}_j^2$ by replacing the X_{ij} values with Z_{ij}, yielding T_{zj} and $\hat{\zeta}_{zj}^2$.

Step 3. Compute

$$E_i^2 = \left(\frac{Z_{i1} - T_{z1}}{\hat{\zeta}_{z1}}\right)^2 + \left(\frac{Z_{i2} - T_{z2}}{\hat{\zeta}_{z2}}\right)^2.$$

Step 4. For some constant C, let

$$W_i = \left(1 - \frac{E_i^2}{C}\right)^2,$$

if $E_i^2 < C$, and otherwise, $W_i = 0$. Goldberg and Iglewicz (1992) recommend $C = 36$ unless more than half of the W_i values are equal to zero, in which case, C can be increased until a minority of the W_i values is equal to zero.

Step 5. Compute

$$T_{bj} = \frac{\sum W_i X_{ij}}{\sum W_i},$$

$$S_{bj}^2 = \frac{\sum W_i (X_{ij} - T_{bj})^2}{\sum W_i},$$

$$R_b = \frac{\sum W_i (X_{i1} - T_{b1})(X_{i2} - T_{b2})}{S_{b1} S_{b2} \sum W_i}.$$

Step 6. Steps 4–8 are iterated. If step 4 has been performed only once, go to step 7; otherwise, let W_{oi} be the weights from the previous iteration, and stop, if $\sum (W_i - W_{oi})^2/(\sum W_i/n)^2 < \epsilon$.

Step 7. Store the current weight, W_i, into W_{oi}.

Step 8. Compute

$$Z_{i1} = \left(\frac{X_{i1} - T_{b1}}{S_{b1}} + \frac{X_{i2} - T_{b2}}{S_{b2}}\right) \frac{1}{\sqrt{2(1 + R_b)}},$$

$$Z_{i2} = \left(\frac{X_{i1} - T_{b1}}{S_{b1}} - \frac{X_{i2} - T_{b2}}{S_{b2}}\right) \frac{1}{\sqrt{2(1 - R_b)}},$$

$$E_i^2 = Z_{i1}^2 + z_{i2}^2.$$

Go back to step 4.

Let E_m be the median of $E_1, \ldots, E_n$, and let $E_{\max}$ be the largest E_i value such that $E_i^2 < DE_m^2$, where D is some constant. Goldberg and Iglewicz recommend $D = 7$, and this value is used here. Let $R_1 = E_m \sqrt{(1 + R_b)/2}$ and $R_2 = E_m \sqrt{(1 - R_b)/2}$. For each υ between 0 and 360, with steps of 2 degrees, compute $\Upsilon_1 = R_1\cos(\upsilon)$, $\Upsilon_2 = R_2\sin(\upsilon)$, $A = T_{b1} + (\Upsilon_1 + \Upsilon_2)s_{b1}$, and $B = T_{b2} + (\Upsilon_1 - \Upsilon_2)s_{b2}$. The values for A and B form the inner ellipse. The outer ellipse is obtained by repeating this process with E_m replaced by $E_{\max}$.

The bagplot was derived by Rousseeuw et al. (1999) and provides a more flexible approach to dealing with any asymmetry in the data cloud. Roughly, using some measure of depth, it determines the convex hull containing the deepest half of the data, called the bag. Then this bag is inflated by a factor of 3 relative to the median, and the data points outside of the inflated bag are flagged as outliers.

6.4.2 R Functions relplot and Bagplot

The R function

$$\text{relplot(x,y,C=36,epsilon=0.0001,plotit=TRUE)}$$

performs the calculations in Table 6.1 and creates a relplot. Here, x and y are any R vectors containing data, C is a constant that defaults to 36 (see step 4 in Table 6.1), and epsilon is ϵ in step 6 of Table 6.1. (The argument epsilon is used to determine whether enough iterations have been performed. Its default value is 0.0001.) The function returns bivariate measures of location in relplot$mest, measures of scale in relplot$mvar, and a measure of correlation in relplot$mrho. The last argument, plotit, defaults to T, for true, meaning that a graph of the bivariate boxplot (the relplot) will be created. To avoid the plot, simply set the last argument, plotit, to FALSE. For example, the command relfun(x, y, plotit=FALSE) will return the measures of location and correlation without creating the plot.

The R function

$$\text{Bagplot(x,plotit=TRUE, colorbag = NULL, colorloop = NULL, colorchull = NULL,}$$
$$\text{databag = TRUE, dataloop = TRUE, plot.fence = FALSE,type='hdepth')}$$

creates a bagplot, where now the argument x is a matrix with two columns. The argument type indicates the measure of depth that is used. The default is halfspace depth. The other options are 'projdepth' for projection depth and 'sprojdepth' for skewness-adjusted projection depth. The choice can make a substantial difference. Given some data, a diagnostic tool for picking the best measure does not appear to exist. To see details about the other argument, use the R commands library(mrfDepth) and ?bagplot.

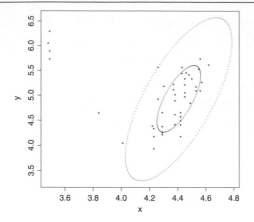

Figure 6.3: A relplot for the star data.

■ **Example**

Rousseeuw and Leroy (1987, p. 27) report the logarithm of the effective temperature at the surface of 47 stars versus the logarithm of its light intensity. Suppose the
(Hertzsprung–Russell) star data are stored in the R variables u and v. Then the R command relplot(u,v) creates the plot shown in Fig. 6.3. The smaller ellipse contains half
of the data. Points outside the larger ellipse are considered to be outliers. The function reports a correlation of 0.7, which is in striking contrast to Pearson's correlation,
$r = -0.21$. Note that six points are flagged as outliers: four in the upper left corner and
two near the outer ellipse. It is left as an exercise to verify that the bagplot flags only
the four points in the upper left corner as outliers when the argument type = 'hdepth' is
used. Using type = 'sprojdepth', several points in the upper right corner are flagged as
outliers in addition to the points in the upper left corner.

■

6.4.3 The MVE Method

A natural way of detecting outliers in p-variate data, $p \geq 2$, is to use the Mahalanobis distance with the usual means and sample covariance matrix replaced by estimators that have a
high breakdown point. One of the earliest such methods is based on the MVE estimators of
location and scale (Rousseeuw and van Zomeren, 1990). Relevant theoretical results are reported by Lopuhaä (1999). Let the column vector $\mathbf{C}$, having length p, be the MVE estimate of
location, and let the p-by-p matrix $\mathbf{M}$ be the corresponding measure of scatter. The distance
of the point $\mathbf{x}_i' = (x_{i1}, \ldots, x_{ip})$ from $\mathbf{C}$ is given by

$$D_i = \sqrt{(\mathbf{x}_i - \mathbf{C})'\mathbf{M}^{-1}(\mathbf{x}_i - \mathbf{C})}. \tag{6.16}$$

If $D_i > \sqrt{\chi^2_{0.975,p}}$, the square root of the 0.975 quantile of a chi-square distribution with p degrees of freedom, then $\mathbf{x}_i$ is declared an outlier. Rousseeuw and van Zomeren recommend this method when there are at least five observations per dimension, meaning that $n/p > 5$. (Cook and Hawkins, 1990, illustrate that problems can arise when $n/p \leq 5$.) A criticism of this method is that it can declare too many points as being extreme (Fung, 1993).

6.4.4 Methods MCD, DETMCD, and DDC

Rather than use the MVE measure of location and scatter to detect outliers, one could, of course, use the MCD estimator instead. That is, in Eq. (6.16), replace $\mathbf{M}$ and $\mathbf{C}$ with the MCD estimates of scatter and location. Or the deterministic version of the MCD (DETMCD) estimator mentioned in Section 6.3.2 could be used.

Cerioli (2010) derived a modification of this method with the goal that under multivariate normality, the probability of declaring one or more points an outlier is equal to some specified value. For additional results on decision rules for declaring a point an outlier, see Cerioli and Farcomeni (2011).

The methods described so far are aimed at detecting rowwise outliers. A cellwise outlier refers to a cell that is an outlier. That is, in the data matrix $\mathbf{X}$, is x_{ij}, the value in the ith row and jth column, unusual in some sense? Cellwise outliers are substantially higher or lower than what could be expected based on the other cells in its column as well as the other cells in its row, taking into account the relations between the columns. For relevant results, see, for example, Van Aelst et al. (2012) and Agostinelli et al. (2015). No details are given here other than to note that an R function for detecting cellwise outliers is available via the R function DDC in the R package cellWise.

6.4.5 R Functions covmve and covmcd

W-estimates of location and scatter, based on the MVE and MCD estimators, can be computed with the R functions

$$\text{covmve(m)}$$

and

$$\text{covmcd(m)}.$$

(See Section 6.5.) In essence, points declared outliers based on the MVE or MCD estimators are removed and the mean and covariance matrix are computed using the data that remain.

The R functions cov.mve and cov.mcd report which subset of half of the data was used to compute the MVE and MCD estimates of location. So it is possible to determine the MVE and MCD estimates of location, if desired.

In some situations, it is convenient to have an R function that returns just the MVE measure of location. Accordingly, the R function

$$mvecen(m)$$

is supplied to accomplish this goal. The R function

$$mcdcen(m)$$

computes the MCD measure of location.

6.4.6 R Functions out and outogk

The R function

$$out(x, cov.fun = cov.mve, plotit = TRUE, xlab = 'X', ylab = 'Y', qval = 0.975, crit = NULL, ...)$$

identifies outliers using the method in Section 6.4.3. The argument cov.fun indicates the measure of location and the covariance matrix that are used, which defaults to the MVE method. The argument x is an n-by-p matrix. The function returns a vector labeled out.id that identifies which rows of data are outliers. And another vector, labeled keep.id, indicates the rows of data that are not declared outliers. In the bivariate case, it creates a scatterplot of the data and marks outliers with an o. To avoid the plot, set the argument plotit=FALSE. Setting cov.fun=covmcd results in replacing the MVE estimator with the MCD estimator. Other options for this argument are ogk, tbs, and rmba, which result in using the estimator OGK, TBS, and median ball algorithm, respectively, but except for tbs, it seems that these options are relatively unsatisfactory. The deterministic version of the MCD (DETMCD) estimator mentioned in Section 6.3.2 could be used by setting the argument cov.mve=DETMCD.

■ **Example**

If the star data in Fig. 6.3 are stored in the matrix stardat, the R command out(stardat) returns the values 7, 9, 11, 14, 20, 30, and 34 in the R variable outmve$out.id. This means, for example, that row 7 of the data in the matrix stardat is an outlier. Six of these points correspond to the outliers in Fig. 6.3 that are to the left of the outer ellipse.

The seventh is the point near the upper middle portion of Fig. 6.3 that lies on or slightly beyond the outer ellipse. (The point is at $x = 4.26$ and $y = 5.57$.)

∎

6.4.7 The MGV Method

An appeal of both the MVE and MCD outlier detection methods is that they are based on high breakdown estimators. That is, they provide an important step toward avoiding *masking*, roughly referring to an inability to detect outliers due to their very presence. (See Section 3.13.1.) But for certain purposes, two alternative methods for detecting outliers have been found to have practical value, one of which is the minimum generalized variance (MGV) method described here. The other is a projection-type method described later in this chapter.

As noted in Chapter 3, the *outside rate per observation* is the expected proportion of points declared outliers. That is, if among n points, A points are declared outliers, the outside rate per observation is $p_n = E(A/n)$. When sampling from multivariate normal distributions, for certain purposes, it is desirable to have p_n reasonably close to zero; a common goal is to have p_n approximately equal to 0.05 (cf. Cerioli, 2010). When all variables are independent, it appears that both the MVE and MCD methods have an outside rate per observation approximately equal to 0.05. But under dependence, this is no longer true; it is higher when using the MCD method. Although the MVE method based on the R function cov.mve appears to have p_n approximately equal to 0.05 under normality, alternative outlier detection methods have been found to have practical advantages for situations to be described.

A multivariate outlier detection method for which p_n is reasonably close to 0.05 under normality, and which has practical value when dealing with problems to be addressed, is the MGV method, which is applied as follows:

1. Initially, all n points are described as belonging to set A.
2. Find the p points that are most centrally located. One possibility is as follows. Let

$$d_i = \sum_{j=1}^{n} \sqrt{\sum_{\ell=1}^{p} \left(\frac{X_{j\ell} - X_{i\ell}}{\text{MAD}_\ell} \right)^2},$$

(6.17)

where MAD_ℓ is the value of MAD based on $X_{1\ell}, \ldots, X_{n\ell}$. The most centrally located points are taken to be the p points having the smallest d_i values. Another possibility, in order to achieve affine equivariance, is to identify the p points having the largest halfspace depth or the largest depth based on the MVE or MCD methods.
3. Remove the p centrally located points from set A, and put them into set B. At this step, the generalized variance of the points in set B is zero.

4. If among the points remaining in set A, the ith point is put in set B, the generalized variance of the points in set B will be changed to some value which is labeled s_{gi}^2. That is, associated with every point remaining in A is the value s_{gi}^2, which is the resulting generalized variance when it, and it only, is placed in set B. Compute s_{gi}^2 for every point in A.

5. Among the s_{gi}^2 values computed in the previous step, permanently remove the point associated with the smallest s_{gi}^2 value from set A, and put it in set B. That is, find the point in set A which is most tightly clustered together with the points in set B. Once this point is identified, permanently remove it from A, and leave it in B henceforth.

6. Repeat steps 4 and 5 until all points are now in set B.

The first p points removed from set A have a generalized variance of zero, which is labeled $s_{g(1)}^2 = \cdots = s_{g(p)}^2 = 0$. When the next point is removed from A and put into B (using steps 4 and 5), the resulting generalized variance of set B is labeled $s_{g(p+1)}^2$, and continuing this process, each point has associated with it some generalized variance when it is put into set B.

Based on the process just described, the ith point has associated with it one of the generalized variances just computed. For example, in the bivariate case, associated with the ith point (X_i, Y_i) is some value $s_{g(j)}^2$, indicating the generalized variance of the set B when the ith point is removed from set A and permanently put in set B. For convenience, this generalized variance associated with the ith point, $s_{g(j)}^2$, is labeled D_i. The p deepest points have D values of zero. Points located at the edges of a scatterplot have the highest D values, meaning that they are relatively far from the center of the cloud of points. Moreover, we can detect outliers simply by applying one of the outlier detection rules in Chapter 3 to the D_i values. Note, however, that we would not declare a point an outlier if D_i is small, only if D_i is large.

In terms of maintaining an outside rate per observation that is stable as a function of n and p, and approximately equal to 0.05 under normality (and when dealing with certain regression problems to be described), a boxplot rule for detecting outliers seems best when $p = 2$, and for $p > 2$, a slight generalization of Carling's modification of the boxplot rule appears to perform well. In particular, if $p = 2$, then declare the ith point an outlier, if

$$D_i > q_2 + 1.5(q_2 - q_1), \tag{6.18}$$

where q_1 and q_2 are the ideal fourths based on the D_i values. For $p > 2$ variables, replace Eq. (6.18) with

$$D_i > M_D + \sqrt{\chi_{0.975,p}^2}(q_2 - q_1), \tag{6.19}$$

where $\sqrt{\chi_{0.975,p}^2}$ is the square root of the 0.975 quantile of a chi-squared distribution with p degrees of freedom and M_D is the usual median of the D_i values. If it is desired to alter the outside rate per observation, this can be done with the R function outmgvad described in Section 6.4.12.

A comment about detecting outliers among the D_i values, using a MAD-median rule, should be made. Using something like the Hampel identifier when detecting outliers has the appeal of using measures of location and scale that have the highest possible breakdown point. When $p = 2$, for example, this means that a point $\mathbf{X}_i$ is declared an outlier, if

$$\frac{|D_i - M_D|}{\text{MAD}_D/.6745} > 2.24, \tag{6.20}$$

where MAD_D is the value of MAD based on the D values. A concern about this approach is that the outside rate per observation is no longer stable as a function of n. This has some negative consequences when addressing problems in subsequent sections. Here, Eq. (6.19) is used because it has been found to avoid these problems and because it has received the most attention so far, but of course, in situations where there are an unusually large number of outliers, using Eq. (6.19) might cause practical problems.

6.4.8 R Function outmgv

The R function

outmgv(x, y = NA, plotit =TRUE, outfun = outbox, se =TRUE, op = 1, cov.fun = rmba, xlab = 'X', ylab = 'Y', SEED =TRUE, ...)

applies the MGV outlier detection method just described. If columns of the input matrix are reordered, this might affect the results due to a rounding error when calling the built-in R function cigen. If the second argument is not specified, it is assumed that x is a matrix with p columns corresponding to the p variables under study, and outmgv checks for outliers for the data stored in x. If the second argument, y, is specified, the function combines the data in x with the data in y and checks for outliers among these $p + 1$ variables. In particular, the data do not have to be stored in a matrix; they can be stored in two vectors (x and y), and the function combines them into a single matrix for you. If plotit=TRUE is used and bivariate data are being studied, a plot of the data will be produced with outliers marked by a circle. The argument outfun can be used to change the outlier detection rule applied to the depths of the points (the D_i values in the previous section). By default, Eq. (6.19) is used. Setting outfun=out, Eq. (6.20) is used. The argument se=TRUE ensures that the results do not change with changes in scale. (The marginal distributions are standardized when calling the R function apgdis.)

6.4.9 A Projection Method

Consider a sample of n points from some p-variate distribution, and consider any projection of the data (as described in Section 6.2.2). A projection-type method for detecting outliers among multivariate data is based on the idea that if a point is an outlier, then it should be an outlier for some projection of the n points. So, if it were possible to consider all possible projections, and if for some projection, a point is an outlier based on the MAD-median rule or the boxplot rule, then the point is declared an outlier. Not all projections can be considered, so the strategy here is to orthogonally project the data onto all n lines formed by the center of the data cloud, as represented by $\hat{\xi}$, and each $\mathbf{X}_i$. It seems natural that $\hat{\xi}$ should have a high breakdown point, and that it should be affine equivariant. Two good choices appear to be the MVE and MCD estimators in Sections 6.3.1 and 6.3.2.

The computational details are as follows. Fix i, and for the point $\mathbf{X}_i$, orthogonally project all n points onto the line connecting $\hat{\xi}$ and $\mathbf{X}_i$, and let D_{ij} be the distance between $\hat{\xi}$ and the projection of $\mathbf{X}_j$. More formally, let

$$\mathbf{A}_i = \mathbf{X}_i - \hat{\xi},$$

$$\mathbf{B}_j = \mathbf{X}_j - \hat{\xi},$$

where both $\mathbf{A}_i$ and $\mathbf{B}_j$ are column vectors having length p, and let

$$\mathbf{C}_j = \frac{\mathbf{A}_i'\mathbf{B}_j}{\mathbf{B}_j'\mathbf{B}_j}\mathbf{B}_j,$$

$j = 1, \ldots, n$. Then when projecting the points onto the line between $\mathbf{X}_i$ and $\hat{\xi}$, the distance of the jth projected point from $\hat{\xi}$ is

$$D_{ij} = \|\mathbf{C}_j\|,$$

where

$$\|\mathbf{C}_j\| = \sqrt{C_{j1}^2 + \cdots + C_{jp}^2}.$$

Here, an extension of Carling's modification of the boxplot rule (similar to the modification used by the MGV method) is used to check for outliers among D_{ij} values. To be certain the computational details are clear, let $\ell = [n/4 + 5/12]$, where $[.]$ is the greatest integer function, and let

$$h = \frac{n}{4} + \frac{5}{12} - \ell.$$

For fixed i, let $D_{i(1)} \leq \cdots \leq D_{i(n)}$ be the n distances written in ascending order. The ideal fourths associated with the D_{ij} values are

$$q_1 = (1 - h)D_{i(h)} + hD_{i(h+1)}$$

and

$$q_2 = (1 - h)D_{i(\ell)} + hD_{i(\ell-1)}.$$

Then the jth point is declared an outlier, if

$$D_{ij} > M_D + \sqrt{\chi^2_{0.975,p}}(q_2 - q_1), \tag{6.21}$$

where M_D is the usual sample median based on the $D_{i1}, \ldots, D_{in}$ values and $\chi^2_{0.95,p}$ is the 0.95 quantile of a chi-squared distribution with p degrees of freedom.

The process just described is for a single projection; for fixed i, points are projected onto the line connecting $\mathbf{X}_i$ to $\hat{\boldsymbol{\xi}}$. Repeating this process for each i, $i = 1, \ldots, n$, a point is declared an outlier, if for any of these projections, it satisfies Eq. (6.21). That is, $\mathbf{X}_j$ is declared an outlier, if for any i, D_{ij} satisfies Eq. (6.21). Note that this outlier detection method approximates an affine equivariant technique for detecting outliers, but it is not itself affine equivariant.

As was the case with the MGV method, a simple and seemingly desirable modification of the method just described is to replace the interquartile range $(q_2 - q_1)$ with the MAD measure of scale based on the values $D_{i1}, \ldots, D_{in}$. So here, MAD is the median of the values

$$|D_{i1} - M_D|, \ldots, |D_{in} - M_D|,$$

which is denoted by MAD_i. Then the jth point is declared an outlier, if for any i

$$D_{ij} > M_D + \sqrt{\chi^2_{0.975,p} \frac{\text{MAD}_i}{0.6745}}. \tag{6.22}$$

Eq. (6.22) represents an approximation of the method given by Eq. (1.3) in Donoho and Gasko (1992). Again, an appealing feature of MAD is that it has a higher finite sample break-down point than the interquartile range. But a negative feature of Eq. (6.22) is that the outside rate per observation appears to be less stable as a function of n. In the bivariate case, for example, it is approximately 0.09 with $n = 10$, and it drops below 0.02 as n increases. For the same situations, the outside rate per observation using Eq. (6.21) ranges, approximately, between 0.043 and 0.038.

Like the MGV method, a change in scale can impact which points are declared outliers. Here, this concern is avoided by standardizing the marginal distributions.

6.4.10 R Functions outpro and out3d

The R function

$$\text{outpro(m,gval=NA,center=NA, plotit=TRUE, op=TRUE, MM=FALSE, cop=3,}$$
$$\text{STAND=TRUE)}$$

checks for outliers using the projection method just described. Here, m is any R variable containing data stored in a matrix (having n rows and p columns). The argument gval can be used to alter the values $\sqrt{\chi^2_{0.95,p}}$ or $\sqrt{\chi^2_{0.975,p}}$ in Eqs. (6.21) and (6.22). These values are replaced by the value stored in gval, if gval is specified. Similarly, the argument center can be used to specify the center of the data cloud, $\hat{\xi}$, that will be used. If not specified, the center is determined by the argument cop. The choices are:

*   cop=1, Donoho–Gasko median,
*   cop=2, MCD,
*   cop=3, median of the marginal distributions,
*   cop=4, MVE.

When working with bivariate data, outpro creates a scatterplot of the data and marks outliers with a circle, and the plot includes a contour indicating the location of the deepest half of the data as measured by projection depth. More precisely, the depth of all points is computed, and among the points not declared outliers, all points having a depth less than or equal to the median depth are indicated. If op=TRUE, the plot creates a 0.5 depth contour based on the data excluding points declared outliers. Setting op=FALSE, the 0.5 depth contour is based on all of the data. If MM=TRUE is used, the interquartile range is replaced by MAD. That is, Eq. (6.22) is used in place of Eq. (6.21). By default, the argument STAND=TRUE means that the marginal distributions are standardized, before checking for outliers, using the median and MAD. (Early versions of the R function outpro used STAND=FALSE.)

The R function outpro is designed so that the expected proportion of points declared outliers is approximately 5% under multivariate normality. If it is desired to alter this rate, this can be done via the R function outproad described in Section 6.4.12.

When working with trivariate data, the R function

$$\text{out3d(x, outfun = outpro, xlab = 'Var 1', ylab = 'Var 2', zlab = 'Var 3', reg.plane =FALSE,}$$
$$\text{regfun =tsreg, COLOR =FALSE)}$$

creates a three-dimensional scatterplot and marks the outliers, identified by the R function outpro, with *. (Setting the argument COLOR=TRUE, outliers are marked with a red circle.)

An alternative outlier detection method can be used via the argument outfun. This function also shows a regression plane when reg.plane=TRUE, assuming the goal is to predict the third variable, given a value for the first two. (That is, column 3 of x is assumed to be the outcome variable, typically labeled y, and columns 1 and 2 contain the predictor variables.) The regression method used is controlled by the argument regfun, which defaults to the Theil–Sen estimator described in Chapter 10.

6.4.11 Outlier Identification in High Dimensions

Filzmoser et al. (2008) note that under normality, if the number of variables is large, the proportion of points declared outliers by the better-known outlier detection methods can be relatively high. This concern applies to all the methods covered here, with the projection method seemingly best at avoiding this problem. But with more than nine variables ($p > 9$), it breaks down as well. Currently, it seems that one of the better ways of dealing with this problem is to use the projection method but with Eq. (6.22) replaced by

$$D_{ij} > M_D + c\frac{\text{MAD}_i}{0.6745},$$

where c is chosen so that the outside rate per observation is approximately equal to some specified value under normality, which is usually taken to be 0.05. Here, the constant c is determined via simulations. That is, n points are generated from a p-variate normal distribution, where all p variables are independent. This process is repeated, say B times, and a value c is determined, so that the expected proportion of points declared outliers is equal to the desired rate. A refinement of this strategy would be to generate data from a multivariate normal distribution that has the same covariance matrix as the data under study. Currently, this does not seem necessary or even desirable, but this issue is in need of further study.

A similar adjustment can be made when using the MGV method to detect outliers, which might be preferred because the MGV method is scale-invariant. Direct comparisons of the performance of the adjusted MGV method and the adjusted projection method have not been made.

It is briefly noted that there is a multivariate outlier detection technique that stems from the notion of geometric quantiles derived by Chaudhuri (1996). Chaouch and Goga (2010) describe an extension of this approach to survey sampling situations. Direct comparisons with the projection method and the MGV method, based on the outside rate per observation, have not been made. Vakili and Schmitt (2014) derive a method that, like the MVE and MCD methods, examines many subsets of the data. How this method compares to the projection method and the MGV method has not been determined.

6.4.12 R Functions outproad and outmgvad

The R function

outproad(m, center = NA, plotit =TRUE, op =TRUE, MM =FALSE, cop = 3, xlab = "VAR 1",
  ylab = "VAR 2", rate = 0.05, iter = 100, ip = 6, pr =TRUE, SEED =TRUE, STAND =TRUE)

is like the R function outpro, only it uses simulations to adjust the decision rule for declaring
a point an outlier as described in the previous section. The argument rate indicates the desired
outside rate per observation under normality. The R function

outmgvad(m, center = NA, plotit =TRUE, op = 1, xlab = "VAR 1", ylab = "VAR 2",
  rate = 0.05, iter = 100, ip = 6, pr =TRUE)

is like the R function outproad, only it is based on the MGV outlier detection technique.

6.4.13 Methods Designed for Functional Data

Functional data analysis refers to methods aimed at analyzing information about curves,
surfaces, or anything else varying over a continuum. In its most general form, each sample
element is a function. The physical continuum over which these functions are defined is often time, but it can be other features, such as spatial location and wavelength. Functional data
methods have been used to analyze human growth curves, weather station temperatures, gene
expression signals, medical images, and human speech. There is a substantial literature dealing with methods aimed at analyzing functional data that goes well beyond the scope of this
book (e.g., Ramsay and Silverman, 2005; Ferraty and Vieu, 2006). The R package fda.usa
(Febrero-Bande and de la Fuente, 2012) can be used to apply a wide range of methods designed specifically for functional data. (Also see Febrero et al., 2008.) The goal in this section
is to describe a method for detecting outlying curves via a functional boxplot derived by Sun
and Genton (2011). Their method is based on the notion of the modified band depth introduced by López-Pintado and Romo (2009). For alternative outlier detection techniques, based
in part on robust principal components, see Hyndman and Shang (2010). Also see Febrero et
al. (2008), Gervini (2012), and Zhang (2013). Also see Ren et al. (2017).

Here, for simplicity, the notion of a band depth is described in the context of how it is computed in practice. López-Pintado and Romo (2009) describe band depth in a broader context
that is not important for present purposes.

Each of n observations is some function $y(t)$ for t in some closed interval $\mathcal{I}$. For example, t
might represent time, and values for the function might range from 0 to 1 minute. Typically,

the function $y(t)$ is not specified. Rather, the values $y(t_1), \ldots, y(t_p)$ are observed, where $t_1 \leq \cdots \leq t_p$, and p can be quite large. So the situation at hand has some connection to the situation in Section 6.4.11.

The band corresponding to two curves, say y_{i_1} and y_{i_2}, is

$$B(y_{i_1}, y_{i_2}) = \{(t, x) : t \in \mathcal{I}, \min(y_{i_1}(t), y_{i_2}(t)) \leq x \leq \max(y_{i_1}(t), y_{i_2}(t))\}.$$

Roughly, $B(y_{i_1}, y_{i_2})$ is the subset of the plane that lies between the two curves y_{i_1} and y_{i_2}. Note that the graph of a function is a subset of the plane that can be denoted by

$$G(y) = \{(t, y(t)) : t \in \mathcal{I}\}.$$

The band depth of some curve y is

$$BD(y) = \binom{n}{2}^{-1} \sum_{i_1 < i_2} G(y) I (G(y) \subseteq B(y_{i_1}, y_{i_2})),$$

where $I(A) = 1$, if A is true; otherwise, $I(A) = 0$. (For measuring depth when dealing with multivariate data, see Claeskens et al., 2014. For an extension of Mahalanobis distance to functional data, see Galeano et al., 2015.) Roughly, the band depth of y is the proportion of bands that contain y. If this proportion is relatively high, y is nested among all the curves in a relatively deep fashion. (There are issues about curves that cross when a band is defined using only two curves as noted by López-Pintado and Romo, 2009. But when using the modified band depth, described next, Sun and Genton, 2011, argue that using only two functions suffices.)

Note that for $t \in \mathcal{I}$, it might be the case that $y(t)$ lies between y_{i_1} and y_{i_2} for some subset of $\mathcal{I}$, but otherwise, it does not. For example, if $\mathcal{I}$ is the closed interval $[0, 1]$, $y(t)$ might have a value between y_{i_1} and y_{i_2} when $t \leq 0.4$, but for $t > 0.4$, this is no longer the case. So 40% of the time, $y(t)$ has a value between $y_{i_1}(t)$ and $y_{i_2}(t)$. The modified band depth takes into account the proportion of times $y(t)$ has a value between y_{i_1} and y_{i_2}.

Let $\lambda(y; y_{i_1}, y_{i_2})$ be the proportion of time that $y(t)$ has a value between $y_{i_1}(t)$ and $y_{i_2}(t)$. The modified band depth is

$$MBD(y) = \binom{n}{2}^{-1} \sum_{i_1 < i_2} \lambda(y; y_{i_1}, y_{i_2}).$$

Let $y_{[1]} \geq \cdots \geq y_{[n]}$ be the curves written in descending order based on their MBD value. So $y_{[1]}$ is the median curve. The sample 50% central region is

$$C_{0.5} = \{(t, y(t)) : \min y_{[r]}(t) \leq y(t) \leq \max y_{[r]}(t)\},$$

where the minimum and maximum are taken over $r = 1, \ldots, m$, and m is the smallest integer not less than $n/2$. So $C_{0.5}$ corresponds to the interquartile range used by a boxplot. Sun and Genton (2011) expand this region by 1.5 to obtain what corresponds to the fences of a boxplot. Any curves outside the fences are flagged as potential outliers.

Rather than use the deepest (median) curve to estimate the typical curve, another strategy is to simply compute some measure of location at each time point. In terms of mean squared error and bias, this approach can result in a substantially more accurate estimate of the typical curve used to generate data in a simulation. See Exercise 18 for more details.

6.4.14 R Functions FBplot, Flplot, medcurve, func.out, spag.plot, funloc, and funlocpb

The R function

FBplot(fit, x = NULL, method ='MBD', depth = NULL, plot =TRUE, prob = 0.5, color = 6, outliercol = 2, barcol = 4, fullout =FALSE, factor = 1.5, xlim = c(1, nrow(fit)), ylim = c(min(fit) - 0.5 * diff(range(fit))),...)

creates a functional boxplot and returns the MBD depth of each curve assuming the R package fda has been installed. The argument fit is a matrix with n rows corresponding to n subjects or curves. The p columns of fit correspond to p measures taken at times $1, \ldots, p$. The argument x indicates the coordinates of curves. By default, x is taken to be $1, \ldots, p$. Any columns with missing values are automatically removed. By default, the median curve is denoted by a black curve. The shaded magenta region corresponds to the interquartile range of a boxplot, and blue lines correspond to the whiskers. Outliers are denoted by dashed lines.

Another option is to simply compute a measure of location associated with each time point, and plot the results. This can be done with the R function

Flplot(x,est=mean,xlab='Time',ylab='Y',plotit=TRUE).

The argument x is assumed to be a matrix with n rows and p columns, and est indicates the measure of location that will be used, which defaults to the mean. The function returns a vector containing the estimates. Simulations suggest that in terms of estimating the true function that generated the data, this approach can be more accurate (based on mean squared error) than using median curve associated with the modified band depth. Note that yet another way of plotting the data is to simply create a boxplot at each time point, which can be done with the R command boxplot(x). The function

$$FQplot(x,est=mean,xlab='Time',ylab='Y',plotit=TRUE)$$

is like the function Flplot, only it includes the upper and lower quartiles.

Color is essential when using the R function FBplot. Otherwise, discerning the regions of the functional boxplot can be difficult. But there is the practical issue that figures with color can be expensive to reproduce in articles. The R function

$$func.out(x,xlab='Time',ylab=' ')$$

is provided in case it helps dealing with this issue. It creates a *spaghetti plot* with solid lines indicating curves that are not flagged as outliers; dashed lines indicate outliers. The argument x is assumed to be an n-by-p matrix (or data frame), where n is the number of subjects for whom there are p values for the function.

Another way of plotting all n curves is with the R function

$$spag.plot(x, regfun =tsreg,xlab = 'Time', ylab = ' ',fit.lin =FALSE, ...).$$

Basically, this function creates the variables needed to compute a *spaghetti plot* via the R function interaction.plot. (There are some additional arguments relevant to the R function interaction.plot that might be of interest. Information about them can be obtained with the R command ?interaction.plot.) If it is desired to use a linear fit, set the argument fit.lin=TRUE, which results in a linear fit that is based on the regression estimator indicated by regfun. This function uses different lines for each curve without indicating which curves are outliers.

The R function

$$medcurve(x)$$

returns the deepest (median) curve. But in terms of estimating the true curve that generated the data, limited simulations indicate that Flplot is more satisfactory based on mean squared error and bias. (See Exercise 18.)

The R function

$$funloc(x,tr=0.2,pts=NULL, npts=25, plotit=TRUE,alpha=0.05, nv=rep(0,ncol(x)),$$
$$xlab='T',ylab='Est.',FBP=TRUE, method='hochberg',COLOR=TRUE)$$

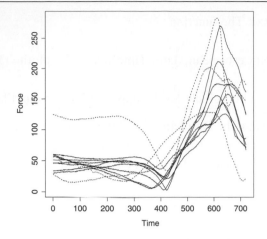

Figure 6.4: Horizontal force in one leg that is produced when swinging a golf club.

computes a trimmed mean at specified time points (columns of the argument x) indicated by the argument pts, as well as a 0.95 confidence interval at each time point. If pts=NULL, the function picks npts time points evenly spaced between the minimum and maximum times. As can be seen, the default is npts=25. The argument nv indicates the null values used when testing hypotheses. Adjusted p-values, designed to control the probability of one or more Type I errors, are reported as well. (See Section 7.4.7.) By default, a functional boxplot is created. If FBP=FALSE and plotit=TRUE, the fur ction plots the trimmed means and $1 - \alpha$ confidence intervals. The R function

funlocpb(x,est=TRUEmean, nboot=2000, SEED=TRUE, pts=NULL,npts=25,plotit=TRUE, alpha=0.05, nv=rep(0,ncol(x)), xlab='T',ylab='Est.',FBP=TRUE, method='hochberg',COLOR=TRUE,...)

is the same as funloc, only a percentile bootstrap method is used.

■ **Example**

This example deals with a study examining the biomechanics used by golfers on a collegiate golf team. A portion of the study focused on the horizontal force produced by one leg when swinging a six iron. (T. Peterson and J. McNitt-Gray generously supplied the data.) Fig. 6.4 shows the plot generated by the R function func.out. As can be seen, 3 of the 11 participants are flagged as outliers.

■

6.4.15 Comments on Choosing an Outlier Detection Method

Choosing an outlier detection method is a non-trivial problem with no single method dominating all others; it seems that several methods deserve serious consideration. In addition to controlling the outside rate per observation, surely a desirable property of any outlier detection method is that it identify points that are truly unusual based on a model that generated the data. Wilcox (2008a) compared several methods, and while no single method was always best, it was found that the MGV and projection methods performed relatively well when the number of variables is $p \leq 9$. Comparisons with the DETMCD, described in Section 6.4.4, have not been made. But as previously noted, with $p > 9$ variables, these two methods break down, in which case, the approach in Section 6.4.11 should be used.

It is worth noting that, *given some data*, the choice of method can matter. To illustrate this point, 20 vectors of observations were generated, where both X and ϵ are independent standard normal variables and $Y = X + \epsilon$. (If of interest, the resulting data are stored on the author's web page in the file Table6_3_dat.txt.) Suppose two additional points are added at $(X, Y) = (2.1, -2.4)$. These two points are clearly unusual compared to the model that generated the data. Using the projection method or the MGV method, the two points $(2.1, -2.4)$ are flagged as outliers, and no other outliers are reported. These two points are declared outliers using the MVE method, but it flags two additional points as outliers. The MCD method finds no outliers.

Gleason (1993) argues that a lognormal distribution is light-tailed. In the univariate case, with $n = 20$, the MAD-median rule given by Eq. (3.45) has an outside rate per observation of approximately 0.13, and a boxplot rule has an outside rate per observation of approximately 0.067. As another illustration that the choice of method can make a difference, consider the case where X and Y are independent, each having a lognormal distribution. For this bivariate case, with $n = 20$, all of the methods considered here have an outside rate above 0.1. For the MVE method, the rate is approximately 0.2. Using instead the projection method and the MGV method, the rates are approximately 0.15 and 0.13, respectively.

Fig. 6.5 shows the plots created by the MGV, MVE, MCD, and the projection methods based on a sample of $n = 50$ pairs of points generated from two independent lognormal distributions. The upper left and lower right panels are based on the projection method and the MGV method, respectively. In this particular instance, they give identical results and flag the fewest points as outliers relative to the other methods used. The upper right panel is the output based on the MVE method, and the lower left panel is based on the MCD method. Although an argument can be made that, in this particular instance, the MVE and MCD methods are less satisfactory, this must be weighed against the ability of the MVE and MCD methods to handle a larger number of outliers. (But if a large number of outliers is suspected, the versions of

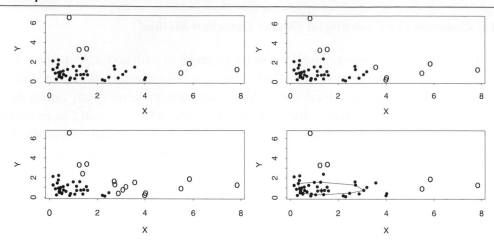

Figure 6.5: Output from four outlier detection methods. The upper left panel used the projection method in Section 6.4.9, the upper right used the MVE method, the lower left is based on the MCD method, and the lower right used the MGV method.

the projection-type method and the MGV method represented by Eqs. (6.20) and (6.19), respectively, might be used.) For more on detecting multivariate outliers, see Kosinski (1999), Liu et al. (1999), Rocke and Woodruff (1996), Peña and Prieto (2001), Poon et al. (2000), Rousseeuw and Leroy (1987), Davies (1987), Fung (1993), and Rousseeuw and van Zomeren (1990).

6.5 A Skipped Estimator of Location and Scatter

Skipped estimators of location and scatter are estimators that search for outliers, discard any that are found, and then compute the mean and usual covariance matrix based on the data that remain. Such estimators are special cases of the W-estimator in Section 6.3.6, where points get a weight of 1 or 0 depending on whether they are declared outliers. Maronna and Yohai (1995) refer to such weight functions as hard rejection weights, as opposed to soft rejection, where the weights gradually descend toward zero as a function of how outlying a point happens to be. When using the outlier projection method in Section 6.4.8, with outliers getting a weight of zero and points getting a weight of one, the corresponding W-estimator will be called the *OP-estimator*. When using the MVE outlier detection method, the skipped estimator will be called the *WMVE estimator*. And when using the MCD outlier detection method, the skipped estimator will be called the *WMCD estimator*.

Note that the methods just described also yield robust analogs of the usual covariance matrix. If outliers are removed via the projection method, and the usual covariance matrix is computed based on the remaining data, this will be called the OP-estimate of scatter.

Table 6.2: Values of R (accuracy), $n = 40$.

h	ρ	$\gamma = 0.10$	$\gamma = 0.15$	$\gamma = 0.20$	DGM	OP	M
0.0	0.0	0.73	0.62	0.50	0.45	0.92	0.81
0.5	0.0	5.99	5.92	5.40	4.11	6.25	8.48
1.0	0.0	4,660.21	5,764.79	5,911.29	4,643.16	5,452.35	10,820.14
0.0	0.7	0.80	0.71	0.61	0.48	0.95	0.44
0.5	0.7	4.74	4.76	4.50	3.20	4.64	5.44
1.0	0.7	1,082.56	1,300.44	1,336.63	1,005.24	1,091.68	1,760.98

To provide at least some sense of how the various location estimators compare, some results on the expected squared standard error are provided when sampling from distributions that are symmetric about zero. More precisely, the measure of accuracy used is

$$R = \frac{\sqrt{E(\sum \bar{X}_j^2)}}{\sqrt{E(\sum \hat{\theta}_j^2)}},$$

where $\hat{\theta}_j^2$ is some competing estimator associated with the jth variable, $j = 1, \ldots, p$. Table 6.2 reports some results for four variations of the Donoho–Gasko trimmed mean, followed by the OP-estimator and the marginal medians. (In Table 6.2, h refers to the type of g-and-h distribution used, as described in Section 4.2, and ρ is the common Pearson correlation among the generated data.) Note that under normality, all four variations of the Donoho–Gasko trimmed mean are the least satisfactory, and method OP performs best among the robust estimators considered. As for the TBS estimator, described in Section 6.3.3, it performs in a manner similar to the Donoho–Gasko trimmed mean with $\gamma = 0.15$ when sampling from a normal distribution. That is, it is less satisfactory than other estimators that might be used. For $h = 0.5$, it performs nearly as well as the skipped estimator (method OP), and for $h = 1$, it is a bit more accurate. As for the WMVE skipped estimator, among the situations considered, it seems to have about the same accuracy as the OP-estimator, with OP offering a slight advantage.

Massé and Plante (2003) report more extensive results on the Donoho–Gasko trimmed mean, plus other estimators not described here. Their results further support the notion that the Donoho–Gasko trimmed mean is relatively inaccurate when sampling from light-tailed distributions. Among the 10 estimators they considered, Massé and Plante (2003) found the spatial median, studied by Haldane (1948) and Brown (1983), to be best. (They did not consider the OP-estimator in their study.) The *spatial median* is the value $\hat{\theta}$ that minimizes

$$\frac{1}{n} \sum \| \hat{\theta} - \mathbf{X}_i \|.$$

It is not affine equivariant, but it is translation equivariant and orthogonally equivariant. One way of computing the spatial median is via the Nelder and Mead (1965) algorithm for minimizing a function. (For related results, see Olsson, 1974, as well as Olsson and Nelson, 1975.) An alternative algorithm for computing the spatial median can be found in Bedall and Zimmermann (1979) as well as Hössjer and Croux (1995). Fritz et al. (2012) compare several algorithms. Ng and Wilcox (2010) compare eight robust estimators for a wide range of situations and conclude that the OP-estimator generally performs best in terms of efficiency, as measured by the generalized variance of the sampling distribution.

6.5.1 R Functions smean, mgvmean, L1medcen, spat, mgvcov, skip, and skipcov

The R function

$$smean(m,cop=3,MM=FALSE,op=1,outfun=outogk,cov.fun=rmba,MC=FALSE,...)$$

computes the OP-estimator of location just described using the data stored in the n-by-p matrix m. The remaining arguments determine which outlier detection method is used. Setting op=1 results in using the projection-type method, and op=2 uses the MGV method. Setting op=3, the method indicated by the argument outfun is used. The initial measure of location used by the outlier detection method is determined by cop, the choices being:

- cop=1, Tukey (halfspace) median,
- cop=2, MCD,
- cop=3, marginal medians.

To take advantage of a multicore processor, with the goal of reducing execution time, set the argument MC=TRUE. If op=3, the outlier detection method indicated by the argument outfun is used.

The R function

$$skipcov(m,cop=6, MM=FALSE,op=1, mgv.op=0, outpro.cop=3)$$

computes the covariance matrix for the data stored in the argument m after outliers are removed. Like the R function smean, op=1 means that a projection method is used to identify outliers. When MM=FALSE, Carling's modification of the boxplot rule is applied to each projection when checking for outliers. When MM=TRUE, a MAD-median rule is used. Setting op=2, the MGV method is used to detect outliers. The argument outpro.cop controls which measure of location is used to compute the projections; see the R function outpro for more details. The R function

$$\text{skip(m,cop=6,MM=FALSE,op=1,mgv.op=0,outpro.cop=3)}$$

returns both the skipped measure of covariance and measure of location. The R function

$$\text{mgvcov(m, MM=FALSE, op=1, cov.fun=rmba)}$$

computes the MGV covariance matrix. For an explanation of the remaining arguments, see the R function outmgv.

The R function

$$\text{spat(m)}$$

computes the spatial median, as does

$$\text{L1medcen(X, tol = 1e-08, maxit = 200, m.init = apply(X, 2, median) trace =FALSE).}$$

The function spat uses the Nelder–Mead algorithm, while L1medcen uses the method described in Hössjer and Croux (1995). These two functions can give slightly different results. Currently, it is unknown why one method might be preferred over the other.

A skipped estimator, with outliers detected via the MGV method, is called the MGV estimator of location and can be computed with the function smean. For convenience, the R function

$$\text{mgvmean(m, op=0, MM=FALSE, outfun=outbox)}$$

is supplied for computing this measure of location. Setting op=0 results in the MGV outlier detection method using pairwise differences when searching for the centrally located points, op=1 uses the MVE method, and op=2 uses MCD. The built-in R function cov.mve, described in Section 6.4.5, is designed to compute the WMVE estimate of location and scatter.

6.6 Robust Generalized Variance

One approach to measuring the overall variation of a cloud of points is with the generalized variance, where the usual covariance matrix is replaced by some robust analog. Based on the criterion of achieving good efficiency, a particular choice for the covariance matrix has been found to be relatively effective when distributions are normal or have moderately heavy tails: the OP-estimator of scatter where Carling's modification of the boxplot rule is applied to each projection of the data. For heavy-tailed distributions, use instead a MAD-median rule (Wilcox, 2006e).

6.6.1 R Function gvarg

The R function

$$gvarg(m, var.fun=cov.mba,...)$$

gives a robust generalized variance for the data stored in the argument m. By default, the RMBA covariance matrix is used because other methods to be described appear to perform reasonably well based on this covariance matrix. The command

$$gvarg(x, skipcov, MM=FALSE)$$

would compute the generalized variance based on the OP-estimate of scatter in conjunction with Carling's modification of the boxplot rule. The command

$$gvarg(x, skipcov, MM=TRUE)$$

would use the MAD-median rule.

6.7 Multivariate Location: Inference in the One-Sample Case

This section describes two methods for making inferences about multivariate measures of location. The first is aimed at the population analog of the OP-estimator. The second is based on an extension of Hotelling's T^2 method to the marginal trimmed means.

6.7.1 Inferences Based on the OP Measure of Location

The immediate goal is to compute a $1 - \alpha$ confidence region for the population measure of location corresponding to the OP-estimator described in the previous section. Alternatively, the method in this section can be used to test the hypothesis that the population measure of location is equal to some specified value.

The basic strategy is to use a general percentile bootstrap method studied by Liu and Singh (1997). Roughly, generate bootstrap estimates, and use the central $1 - \alpha$ bootstrap values as an approximate confidence region. A simple method for determining the central $1 - \alpha$ bootstrap values is to use the Mahalanobis distance. Despite being non-robust, this strategy performs well for a range of situations to be covered. Indeed, for many problems, there is no known reason to prefer another measure of depth, in terms of probability coverage. But for the problem at hand, Mahalanobis depth is unsatisfactory, at least with small to moderate sample sizes; the actual probability coverage can be rather unstable among various distributions, particularly as p gets large. That is, what is needed is a method for which the probability coverage is reasonably close to the nominal level regardless of the distribution associated with the data.

The method begins by generating a bootstrap sample by sampling with replacement n vectors of observations from $\mathbf{X}_1, \ldots, \mathbf{X}_n$, where again, $\mathbf{X}_i$ is a vector having length p. Label the results $\mathbf{X}_1^*, \ldots, \mathbf{X}_n^*$. Compute the OP-estimate of location, yielding $\hat{\boldsymbol{\theta}}^*$. Repeat this B times, yielding $\hat{\boldsymbol{\theta}}_1^*, \ldots, \hat{\boldsymbol{\theta}}_B^*$. Proceeding as in Section 6.2.5, compute the projection distance of each bootstrap estimate, $\hat{\boldsymbol{\theta}}_b^*$, relative to all B bootstrap values, and label the result $d_b^*, b = 1, \ldots, B$. Put these B distances in ascending order, yielding $d_{(1)}^* \leq \cdots \leq d_{(B)}^*$. Set $u = (1 - \alpha)B$, rounding to the nearest integer. A direct application of results in Liu and Singh (1997) indicates that an approximate $1 - \alpha$ confidence region corresponds to the u bootstrap values having the smallest projection distances. As for testing

$$H_0: \boldsymbol{\theta} = \boldsymbol{\theta}_0, \tag{6.23}$$

$\boldsymbol{\theta}_0$ given, let D_0 be the projection distance of $\boldsymbol{\theta}_0$. Set $I_b = 1$, if $D_0 \leq D_b^*$; otherwise, $I_b = 0$. Then the (generalized) p-value is

$$\hat{p} = \frac{1}{B} \sum_{b=1}^{B} I_b,$$

and a direct application of results in Liu and Singh (1997) indicates that H_0 is rejected, if $\hat{p} \leq \alpha$.

However, when testing at the 0.05 level, this method can be unsatisfactory for $n \leq 120$ (and switching to Mahalanobis distance makes matters worse). A better approach is to adjust the decision rule when n is small. In particular, reject if $\hat{p} \leq \alpha_a$, where, for $n \leq 20$, $\alpha_a = 0.02$; for $20 < n \leq 30$, $\alpha_a = 0.025$; for $30 < n \leq 40$, $\alpha_a = 0.03$; for $40 < n \leq 60$, $\alpha_a = 0.035$; for $60 < n \leq 80$, $\alpha_a = 0.04$; for $80 < n \leq 120$, $\alpha_a = 0.045$; and for $n > 120$, use $\alpha_a = 0.05$. Simulations (Wilcox, 2003b) suggest that for $p = 2, \ldots, 8$, reasonably good control over the probability of a Type I error is obtained regardless of the correlations among the p variables under study. That is, the actual probability of a Type I error will not be much larger than the nominal level. However, for $n = 20$, and when sampling from a heavy-tailed distribution, the actual probability of a Type I error can drop below 0.01 when testing at the 0.05 level, so there is room for improvement. Willems et al. (2002) derived a method based on the MCD estimator. How it compares to the method just described is unknown. And there are no simulation results regarding how their method performs when dealing with skewed marginal distributions. Another possible concern, and how it might addressed, is described in Section 6.8.1.

6.7.2 Extension of Hotelling's T^2 to Trimmed Means

Hotelling's T^2 test is a classic method for testing

$$H_0: \boldsymbol{\mu} = \boldsymbol{\mu}_0,$$

where μ represents a vector of p population means and μ_0 is a vector of specified constants. The method is readily generalized to making inferences about the marginal trimmed means via the test statistic

$$T^2 = \frac{h(h-p)}{(n-1)p}(\bar{\mathbf{X}}_t - \boldsymbol{\mu}_0)\mathbf{S}^{-1}(\bar{\mathbf{X}}_t - \boldsymbol{\mu}_0)',$$

where $\mathbf{S}$ is the Winsorized covariance matrix corresponding to the p measures under study, $\bar{\mathbf{X}}_t$ is the vector of marginal trimmed means, and h is the number of observations left after trimming. (The Winsorized covariance for any two variables was described in Section 5.9.13.) When the null hypothesis is true, T^2 has, approximately, an F distribution with degrees of freedom $v_1 = p$ and $v_2 = h - p$. That is, reject at the α level, if

$$T^2 \geq f,$$

where f is the $1 - \alpha$ quantile of an F distribution with $v_1 = p$ and $v_2 = h - p$ degrees of freedom.

6.7.3 R Functions smeancrv2 and hotel1.tr

The R function

```
smeancrv2(m, nullv=rep(0, ncol(m)), nboot=500, plotit=TRUE, MC=FALSE,
          xlab='VAR 1', ylab='VAR 2',STAND=FALSE)
```

tests the hypothesis $H_0: \boldsymbol{\theta} = \boldsymbol{\theta}_0$, where $\boldsymbol{\theta}$ is the population value of the OP-estimator. The null value, $\boldsymbol{\theta}_0$, is specified by the argument nullvec and defaults to a vector of zeros. The argument cop determines the measure of location used by the projection outlier detection method and MM determines the measure of scale that is used; see Section 6.4.10. If m is a matrix having two columns and plotit=TRUE, the function plots the bootstrap values and indicates the approximate 0.95 confidence region. Setting the argument MC=TRUE, a multicore processor can be used to compute the measure of location and the projection distances, which will help

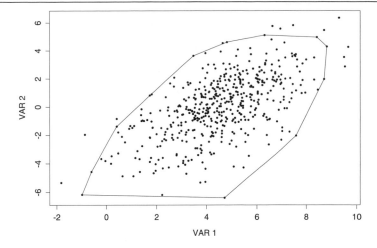

Figure 6.6: The 0.95 confidence region based on the OP-estimate using the cork data, where VAR 1 is the difference between the west and north sides and VAR 2 is the difference between the west and east sides.

reduce execution time. (The function smeancrv2 is the same as the function smeancr, only smeancr does not have an option for using a multicore processor.)

■ Example

This example is based on the cork boring data introduced in Section 5.9.10. For illustrative purposes, suppose the difference scores between the west and north sides of the trees are stored in column one of the R matrix m, and in column two, are the difference scores between the west and east sides. Fig. 6.6 shows the approximate 0.95 confidence region for the typical difference scores based on the OP-estimator and reported by the function smeancr. The (generalized) p-value when testing H_0: $\theta = (0, 0)$ is 0.004, and the 0.05 critical p-value is $\alpha_a = 0.025$, so reject at the 0.05 level.

The R function

$$\text{hotel1.tr(x,null.value=0,tr=0.2)}$$

performs the generalization of Hotelling's T^2 method to trimmed means. The argument null.value can contain a single value, which is taken to mean that all p hypothesized values are equal to the specified value, or the argument null.value can contain p values. The argument x is assumed to be a matrix or data frame. And as usual, tr=0.2 indicates that by default, 20% trimming is used.

6.7.4 Inferences Based on the MGV Estimator

A natural guess is that the inferential method based on the OP-estimator can be used with the MGV estimator as well. It appears, however, that some alternative modification of the bootstrap method is required. For example, if $n = 20$ and $p = 4$, the modified bootstrap method designed for the OP-estimator rejects at the 0.05 level, if $\hat{p}^* \leq 0.02$ and appears to control the probability of a Type I error for a wide range of distributions. However, if this method is applied with the OP-estimator replaced by the MGV estimator, the actual probability of a Type I error exceeds 0.1 when sampling from a normal distribution. To achieve an actual probability of a Type I error approximately equal to 0.05, reject, if $\hat{p}^* \leq 0.006$. Switching to Mahalanobis distance makes matters worse. A better approach is to proceed exactly as was done with the OP-estimator, only use MGV distances when computing the (generalized) p-value.

6.7.5 R Function smgvcr

The R function

    smgvcr(m,nullvec=rep(0,ncol(m)),SEED=TRUE,op=0,nboot=500,plotit=TRUE)

tests the hypothesis $H_0: \theta = \theta_0$, where θ is the population value of the MGV estimator. The null value, θ_0, is specified by the argument nullvec and defaults to a vector of zeros. The argument op determines how the central values are determined when using the MGV outlier detection method; see the function mgvmean.

6.8 The Two-Sample Case

This section deals with the goal of comparing measures of location for the two-sample case. Independent groups are considered first, followed by methods for dependent groups.

6.8.1 Independent Case

This section deals with the goal of testing

$$H_0: \theta_1 = \theta_2 \tag{6.24}$$

for two independent groups, where θ_j ($j = 1, 2$) is some multivariate measure of location. One approach is to use a simple extension of the method in Section 6.7.1. That is, for two independent groups, θ_j represents the population OP measure of location. Now, simply generate bootstrap samples from each group, compute the OP-estimator for each, label the results $\boldsymbol{\theta}_1^*$ and $\boldsymbol{\theta}_2^*$, and set $d^* = \boldsymbol{\theta}_1^* - \boldsymbol{\theta}_2^*$. Repeat this process B times, yielding $d_1^*, \ldots, d_B^*$. Then H_0 is tested by determining how deeply the vector $(0, \ldots, 0)$ is nested within the cloud of d_b^* values, $b = 1, \ldots, B$, again, using the projection depth. If its depth, relative to all B bootstrap estimates, is low, meaning that it is relatively far from the center, then reject. More precisely, let D_b be the OP distance associated with the bth bootstrap sample and let D_0 be the distance associated with $(0, \ldots, 0)$. Set $I_b = 1$, if $D_b > D_0$, and otherwise, $I_b = 0$, in which case, the estimated generalized p-value is

$$\hat{p} = \frac{1}{B} \sum_{b=1}^{B} I_b.$$

Currently, when $\alpha = 0.05$, it is recommended to set $n = \min(n_1, n_2)$ and use α_a as defined in the one-sample case in Section 6.7.1. Checks on this method, when the OP-estimator is replaced by some other estimator, have not been made.

There are two criticisms of this method. First, it is limited to testing at the 0.05 level. Second, as noted in Section 4.3, Tukey (1991) argues that surely any two measures of location differ at some decimal place. A simple way of dealing with both of these issues is to test H_0: $\theta_{1j} = \theta_{2j}$ for each j ($j = 1, \ldots, p$), and then use Tukey's three-decision rule. That is, if the hypothesis is rejected, make a decision about whether $\theta_{1j} < \theta_{2j}$; otherwise, no decision is made. Here, these hypotheses are tested using a percentile bootstrap method. That is, generate bootstrap samples as indicated in Section 6.7.1, compute a measure of location, and repeat this B times. Confidence intervals and p-values are then computed as described in Section 5.4. The main difference from Section 5.4 is that here, θ_{1j} and θ_{2j} are estimated with one of the multivariate location estimators in Section 6.3 or 6.5. Another advantage of this approach, based on simulations, is that it appears to perform well in terms of controlling Type I errors for a range of location estimators, and it is not limited to testing at the 0.05 level. A negative feature is that the actual Type I error probability can drop below 0.025 when testing at the 0.05 level and $n \leq 30$.

Effect Size

One way of measuring effect size is via a classification perspective as described in Section 5.3. When dealing with multivariate data, there are numerous methods that might be used to estimate the probability of correctly determining whether a vector of observations came from the first or second group. Section 6.16 outlines many of the methods that might be used.

The R function manES, described in Section 6.16.2, can be used to estimate the probability of a correct classification.

6.8.2 Comparing Dependent Groups

One way of comparing dependent groups is to focus on the difference scores $D_{ij} = X_{ij} - Y_{ij}$ ($i = 1, \ldots, n$, $j = 1, \ldots, p$) and test Eq. (6.23), where now θ is some population measure of location corresponding to the difference scores D_{ij}. This can be done as described in Section 6.7.1. But again, there is the issue of whether testing for exact equality is meaningful. Moreover, the method in Section 6.7.1 is limited to testing at the 0.05 level. Another way to proceed is to test H_0: $\theta_j = 0$ for each $j = 1, \ldots, p$, where the estimate of θ_j is based on the D_{ij} values in conjunction with one of the estimators in Section 6.3 or 6.5. That is, an estimator is used that takes into account outliers associated with the matrix of difference scores. Here, the null hypothesis is tested using a percentile bootstrap method. If H_0: $\theta_j = 0$ is rejected, again, use Tukey's three-decision rule, and make a decision about whether θ_j is greater than or less than zero. Unlike the method in Section 6.7.1, this approach is not limited to testing at the 0.05 level, and all indications are that it performs well in terms of Type I errors, provided the estimator has a reasonably high breakdown point.

6.8.3 R Functions smean2, mul.loc2g, MUL.ES.sum, Dmul.loc2g, matsplit, and mat2grp

The R function

$$\text{smean2(m1,m2,nullv=rep(0,ncol(m1)), cop=3, MM=FALSE, SEED=NA, nboot=500,}$$
$$\text{plotit=TRUE, MC=FALSE)}$$

tests the hypothesis that two multivariate distributions have the same measure of location based on the skipped (OP) estimator. Here, the data are assumed to be stored in matrices m1 and m2, each having p columns. The argument nullv indicates the null vector and defaults to a vector of zeros. The arguments cop and MM control how outliers are detected when using the projection method; see Section 6.4.10. As usual, to avoid the plot, set plotit=FALSE. To use a multicore processor, set the argument MC=TRUE. The function

$$\text{mul.loc2g(m1,m2,nullv=rep(0,ncol(m1)),locfun=smean, alpha=0.05, SEED=TRUE,}$$
$$\text{nboot=500,...)}$$

tests $H_0: \theta_{1j} = \theta_{2j}$ for each j ($j = 1, \ldots, p$) using a percentile bootstrap method based on the measure of location indicated by the argument locfun. The default estimator is the skipped (OP) estimator that deals with outliers using a projection method. For each variable, the function

$$\text{MUL.ES.sum(x1,x2)}$$

computes several measures of effect size via the R function ES.summary, described in Section 5.7.3.

For dependent groups, the R function

$$\text{Dmul.loc2g(m1, m2 = NULL, nullv = rep(0, ncol(m1)), locfun = smean, alpha = 0.05, SEED}$$
$$\text{= TRUE, nboot = 500, ...)}$$

can be used for $H_0: \theta_j = 0$ ($j = 1, \ldots, p$) based on the difference scores associated with the marginal distributions and the estimator indicated by the argument locfun.

Data Management

The R function

$$\text{matsplit(m,coln)}$$

is supplied in case it helps with data management. It splits the matrix m into two matrices based on the values in the column of m indicated by the argument coln. This column is assumed to have two values only. Results are returned in $m1 and $m2.

The R function

$$\text{mat2grp(m,coln)}$$

also splits the data in a matrix into groups based on the values in column coln of matrix m. Unlike matsplit, mat2grp can handle more than two values (i.e., more than two groups), and it stores the results in list mode.

■ **Example**

Thomson and Randall-Maciver (1905) report four measurements for male Egyptian skulls from five different time periods: 4000 BC, 3300 BC, 1850 BC, 200 BC, and 150 AD. There are 30 skulls from each time period and four measurements: maximal breadth, basibregmatic height, basialveolar length, and nasal height. For illustrative purposes, assume the data are stored in the R variable skull, the four measurements are stored in columns 1–4, and the time period is stored in column 5. Here, the first and last time periods are compared, based on the OP measure of location. First, split the data into five groups based on the time periods. For example, the R command

$$z=mat2grp(skull,5)$$

accomplishes this goal. So, z[[1]] is a matrix containing the data corresponding to the first time period and z[[5]] is a matrix containing the data for the final (fifth) time period, 150 AD. The R command

$$smean2(z[[1]][,1:4],z[[5]][,1:4])$$

compares the two groups based on all four measures. Fig. 6.7 shows the plot created by smean2 when using the first two variables only. The polygon is an approximate 0.95 confidence region for the difference between the measures of location. The p-value is 0.002. (Using all four measures, the p-value is 0.) Using instead the R function mul.loc2g, the p-values corresponding to the four measures are 0.000, 0.012, 0.000, and 0.232.

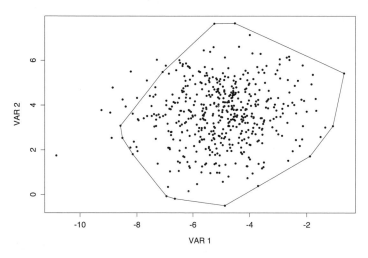

Figure 6.7: Using the first two skull measures for the first and last time periods, the plot shows the 0.95 confidence region for the difference between the OP measures of location.

6.8.4 Comparing Robust Generalized Variances

Robust generalized variances can be compared as well. A percentile bootstrap appears to avoid Type I errors above the nominal level. But situations are encountered where the actual level can be substantially smaller than the nominal level. Corrections are available in some situations (Wilcox, 2006e), which are used by the R function described in the next section, but no details are given here.

6.8.5 R Function gvar2g

The R function

gvar2g(x, y, nboot = 100, DF =TRUE, eop = 1, est = skipcov, tr=0.2, cop = 3, op = 1, MM =FALSE, SEED =TRUE)

compares two independent groups based on a robust version of the generalized variance. By default, the OP covariance matrix is used in conjunction with Carling's modification of the boxplot rule. Setting MM=TRUE, a MAD-median rule is used. If DF=TRUE, and the sample sizes are equal, the function reports an adjusted critical p-value, assuming that the goal is to have a Type I error probability equal to 0.05, the argument est=skipcov, and that other conditions are met. Otherwise, no adjusted critical value is reported. For information about the arguments op, cop, and eop, see the R function skipcov.

6.8.6 Rank-Based Methods

It is noted that there are multivariate rank-based methods for comparing two or more independent groups. Let $F_{jk}(x)$ be the distribution associated with the jth group ($j = 1, \ldots, J$) and kth measure ($k = 1, \ldots, p$). So, for example, $F_{32}(6)$ is the probability that for the third group, the second variable will be less than or equal to 6 for a randomly sampled individual. Munzel and Brunner (2000) derive a test of

$$H_0 \colon F_{1k}(x) = \cdots = F_{Jk}(x) \text{ for all } k = 1, \ldots, p, \tag{6.25}$$

the hypothesis that the J independent groups have identical marginal distributions. (For results regarding how the Munzel–Brunner method compares to several techniques not covered here, see Bathke et al., 2008. A variation of the Munzel–Brunner method can be used in place of the Agresti–Pendergast method, but the relative merits of these two techniques have not been explored.)

Choi and Marden (1997) derive a multivariate analog of the Kruskal–Wallis test. For the jth group and any vector of constants $\mathbf{x} = (x_1, \ldots, x_p)$, let

$$F_j(\mathbf{x}) = P(X_{j1} \leq x_1, \ldots, X_{jK} \leq x_p).$$

So, for example, $F_1(\mathbf{x})$ is the probability that for the first group, the first of the K measures is less than or equal to x_1, the second of the K measures is less than or equal to x_2, and so forth. The null hypothesis is that for any $\mathbf{x}$,

$$H_0: F_1(\mathbf{x}) = \cdots = F_J(\mathbf{x}), \tag{6.26}$$

which is sometimes called the *multivariate hypothesis* to distinguish it from Eq. (6.25), which is called the *marginal hypothesis*. The Choi–Marden method represents an extension of a technique derived by Möttönen and Oja (1995) and is based on a generalization of the notion of a rank to multivariate data, which was also used by Chaudhuri (1996, Section 4). For methods based on spatial ranks, see Paindaveine and Verdebout (2016) and Oja and Randles (2006). Inferential methods based on this approach generally assume distributions differ in location only or that distributions are elliptically symmetric. For relevant R functions, see the R package SpatialNP.

Again, there is the objection to testing for exact equality, which was raised by Tukey (1991) and is discussed in Section 4.1.1. From this point of view, perhaps a more interesting approach is to use the analogs of the Wilcoxon–Mann–Whitney test in Section 5.7.1. For the two-sample case, the goal would be to test $H_0: p_j = 0.5$ for each j ($j = 1, \ldots, p$), where p_j is the probability that for a randomly sampled vector from each group, the jth variable in the first group is less than the jth variable in the second group. If the null hypothesis is rejected, make a decision about whether $p_j < 0.5$. For a global test, see Brunner et al. (2002b).

6.8.7 R Functions mulrank, cmanova, and cidMULT

The R function

$$\text{cidMULT(x1,x2, alpha} = 0.05, \text{BMP} = \text{FALSE)}$$

is for the two-sample case and tests $H_0: p_j = 0.5$ for each j ($j = 1, \ldots, p$) using the methods in Section 5.7.1. By default, Cliff's method is used. Setting BMP=TRUE, the Brunner–Munzel method is used, which can be more computationally convenient when dealing with large sample sizes.

The R function

$$\text{mulrank(J, K, x)}$$

tests (6.25). The data are stored in x, which can be a matrix or have list mode. If x is a matrix, the first K columns correspond to the K measures for group 1, the second K correspond to group 2, and so forth. If stored in list mode, x[[1]], ..., x[[K]] contain the data for group 1, x[[K+1]], ..., x[[2K]] contain the data for group 2, and so on. The R function

$$\text{cmanova(J,K,x)}$$

tests Eq. (6.26).

6.9 Multivariate Density Estimators

This section outlines two multivariate density estimators that will be used when plotting data. The first is based on a simple extension of the expected frequency curve described in Chapter 3, and the other is a multivariate analog of the adaptive kernel density estimator. An extensive discussion of multivariate density estimation goes beyond the scope of this book, but some indication of the method used here, when plotting data, seems warranted.

The strategy behind the expected frequency curve is to determine the proportion of points that are close to $\mathbf{X}_i$. There are various ways this might be done, and here a method based on the MVE covariance matrix is used. Extant results suggest this gives a reasonable first approximation of the shape of a distribution in the bivariate case, but there are many alternative methods for determining which points are close to $\mathbf{X}_i$, and virtually nothing is known about their relative merits for the problem at hand.

Here, the point $\mathbf{X}_{i'}$ is said to be close to $\mathbf{X}_i$, if

$$\sqrt{(\mathbf{X}_{i'} - \mathbf{X}_i)' M^{-1} (\mathbf{X}_{i'} - \mathbf{X}_i)} \leq h,$$

where M is the MVE covariance matrix described in Section 6.3.1 and h is the span. Currently, $h = 0.8$ seems to be a good choice for most situations. Letting N_i represent the number of points close to $\mathbf{X}_i$, $f_i = N_i/n$ estimates the proportion of points close to $\mathbf{X}_i$. In the bivariate case, a plot of the data is created simply by plotting the points $(\mathbf{X}_i, f_i)$.

The expected frequency curve can be used as a first approximation when using an adaptive kernel density estimate. Here, once the expected frequency curve has been computed, the method described by Silverman (1986) is used based on the multivariate Epanechnikov kernel.

An outline of the method is as follows. First, rescale the p marginal distributions. More precisely, let $x_{i\ell} = X_{i\ell}/\min(s_\ell, \mathrm{IQR}_\ell/1.34)$, where s_ℓ and IQR_ℓ are, respectively, the standard deviation and interquartile range based on $X_{1\ell}, \ldots, X_{n\ell}$, $\ell = 1 \ldots, p$. (Here, IQR is computed via the ideal fourths.) If $\mathbf{x}'\mathbf{x} < 1$, the multivariate Epanechnikov kernel is

$$K_e(\mathbf{x}) = \frac{(p+2)(1 - \mathbf{x}'\mathbf{x})}{2c_p};$$

otherwise, $K_e(\mathbf{x}) = 0$. The quantity c_p is the volume of the unit p-sphere: $c_1 = 2$, $c_2 = \pi$, and, for $p > 2$, $c_p = 2\pi c_{p-2}/p$. Similar to Section 3.2, the estimate of the density function is

$$\hat{f}(t) = \frac{1}{n}\sum \frac{1}{h\lambda_i} K\{h^{-1}\lambda_i^{-1}(t - X_i)\},$$

where, following Silverman (1986, p. 86), the span is taken to be

$$h = A(p)n^{-1/(p+4)},$$

$A(1) = 1.77$, $A(2) = 2.78$, and, for $p > 2$,

$$A(p) = \left(\frac{8p(p+2)(p+4)(2\sqrt{\pi})^p}{(2p+1)c_p}\right)^{1/(p+4)}.$$

The quantity λ_i is computed as described in Section 3.2.4, only now the initial estimate of f is based on the multivariate version of the expected frequency curve. The R functions rdplot and akerd, described in Section 3.2.5, perform the calculations.

6.10 A Two-Sample, Projection-Type Extension of the Wilcoxon–Mann–Whitney Test

There are various ways to generalize the Wilcoxon–Mann–Whitney test to the multivariate case, some of which are discussed in Chapter 7. Here, a projection-type extension is described that is based, in part, on the multivariate measures of location covered in this chapter. Consider two independent groups with p measures associated with each. Let $\boldsymbol{\theta}_j$ be any measure of location associated with the jth group ($j = 1, 2$). The basic strategy is to (orthogonally) project the data onto the line connecting the points $\boldsymbol{\theta}_1$ and $\boldsymbol{\theta}_2$, and then consider the proportion of projected points associated with the first group that are "less than" the projected points associated with the second.

To elaborate, let $\mathcal{L}$ represent the line connecting the two measures of location, and let d_j be the Euclidean distance of $\boldsymbol{\theta}_j$ from the origin. For the moment, assume $\boldsymbol{\theta}_1 \neq \boldsymbol{\theta}_2$. Roughly, as we move along $\mathcal{L}$, the positive direction is taken to be the direction from $\boldsymbol{\theta}_1$ toward $\boldsymbol{\theta}_2$, if

$d_1 \leq d_2$; otherwise, the direction is taken to be negative. So if the projection of the point X onto $\mathcal{L}$ corresponds to the point U, the projection of the point Y corresponds to the point V, and moving from U to V corresponds to moving in the positive direction along $\mathcal{L}$, then it is said that X is "less than" Y.

For convenience, distances along the projected line are measured relative to the point midway between $\boldsymbol{\theta}_1$ and $\boldsymbol{\theta}_2$, namely, $(\boldsymbol{\theta}_1 + \boldsymbol{\theta}_2)/2$. That is, the distance of a projected point refers to how far it is from $(\boldsymbol{\theta}_1 + \boldsymbol{\theta}_2)/2$, where the distance is taken to be negative if a projected point lies in the negative direction from $(\boldsymbol{\theta}_1 + \boldsymbol{\theta}_2)/2$. If D_x and D_y are the projected distances associated with two randomly sampled observations, $\mathbf{X}$ and $\mathbf{Y}$, then it is said that $\mathbf{X}$ is "less than," "equal to," or "greater than" $\mathbf{Y}$ according to whether D_x is less than, equal to, or greater than D_y, respectively. In symbols, it is said that $\mathbf{X} \prec \mathbf{Y}$, if $D_x < D_y$, $\mathbf{X} \simeq \mathbf{Y}$, if $D_x = D_y$, and $\mathbf{X} \succ \mathbf{Y}$, if $D_x > D_y$. Extending a standard convention in rank-based methods in an obvious way, to deal with situations where $D_x = D_y$ can occur, let

$$\eta = P(\mathbf{X} \prec \mathbf{Y}) + 0.5 P(\mathbf{X} \simeq \mathbf{Y}).$$

The goal is to estimate η and test

$$H_0: \eta = 0.5. \tag{6.27}$$

First consider estimation. Given an estimated measure of location $\hat{\boldsymbol{\theta}}_j$ for the jth group ($j = 1, 2$), the projected distances are computed as follows. Let $\|\hat{\boldsymbol{\theta}}_j\|$ be the Euclidean norm associated with $\hat{\boldsymbol{\theta}}_j$, and let $S = 1$, if $\|\hat{\boldsymbol{\theta}}_1\| \geq \|\hat{\boldsymbol{\theta}}_2\|$; otherwise, $S = -1$. Let

$$\mathbf{C} = (\hat{\boldsymbol{\theta}}_1 + \hat{\boldsymbol{\theta}}_2)/2,$$

$$\mathbf{B} = S(\hat{\boldsymbol{\theta}}_1 - \hat{\boldsymbol{\theta}}_2),$$

$$A = \|\mathbf{B}\|^2,$$

$$\mathbf{U}_i = \mathbf{X}_i - \mathbf{C},$$

and for any i and $k = 1, \ldots, p$, let

$$W_i = \sum_{k=1}^{p} U_{ik} B_k,$$

$$T_{ik} = \frac{W_i}{A} B_{ik},$$

in which case, the distance associated with the projection of $\mathbf{X}_i$ is

$$D_{xi} = \text{sign}(W_i)\sqrt{\sum_{k=1}^{p} T_{ik}^2},$$

$i = 1, \ldots, m$. The distances associated with the $\mathbf{Y}_i$ values are computed simply by replacing $\mathbf{X}_i$ with $\mathbf{Y}_i$ in the definition of $\mathbf{U}_i$. The resulting distances are denoted by D_{yi}, $i = 1, \ldots, n$.

To estimate η, let

$$V_{ii'} = \text{sign}(D_{xi} - D_{yi'})$$

and

$$\bar{V} = \frac{1}{mn}\sum_{i=1}^{m}\sum_{i'=1}^{n} V_{ii'}.$$

Then extending results in Cliff (1996) in an obvious way,

$$\hat{\eta} = \frac{1 - \bar{V}}{2},$$

is an unbiased estimate of η and takes into account tied values.

When testing Eq. (6.23), Wilcox (2005) found that a basic percentile bootstrap method is unsatisfactory in terms of controlling the probability of a Type I error, but that a slight modification of the method performs reasonably well in simulations. The method begins by subtracting θ_j from every observation in the jth group. In effect, shift the data so that the null hypothesis is true. Now, for each group, generate bootstrap samples from the shifted data and estimate η based on the two bootstrap samples just generated. Label the result $\hat{\eta}^*$. Repeat this B times, yielding $\hat{\eta}_1^*, \ldots, \hat{\eta}_B^*$, and put these B values in ascending order, yielding $\hat{\eta}_{(1)}^* \leq \cdots \leq \hat{\eta}_{(B)}^*$. Then reject H_0, if $\hat{\eta}_{(\ell+1)}^* > \hat{\eta}$, or, if $\hat{\eta}_{(u)}^* < \hat{\eta}$, where $\ell = \alpha B/2$, rounded to the nearest integer, and $u = B - \ell$. Here, $B = 1000$ is assumed unless stated otherwise.

6.10.1 R Functions mulwmw and mulwmwv2

The R function

```
mulwmw(m1,m2,plotit=TRUE,cop=3,alpha=0.05,nboot=1000,pop=4,fr=0.8,pr=FALSE)
```

performs the multivariate extension of the Wilcoxon–Mann–Whitney test just described, where the arguments m1 and m2 are any matrices (having p columns) containing the data for the two groups. The argument pr can be used to track the progress of the bootstrap method used to compute a critical value. If plotit=TRUE, a plot of the projected distances is created, the type of plot being controlled by the argument pop. The choices are

* pop=1, dotplots,
* pop=2, boxplots,
* pop=3, expected frequency curve,
* pop=4, adaptive kernel density estimate.

The argument cop controls which measure of location is used. The choices are:

* cop=1, Donoho–Gasko Median,
* cop=2, MCD estimator,
* cop=3, marginal medians,
* cop=4, OP-estimator.

The R function

  mulwmwv2(m1,m2,plotit=TRUE,cop=3,alpha=0.05,nboot=1000,pop=4,fr=0.8,pr=FALSE)

is the same as mulwmw, only it also reports a robust explanatory measure of effect size, described in Section 5.3.4, based on the projected points.

◼ Example

Fig. 6.8 shows four plots corresponding to the various choices for the argument pop using the skull data used in Fig. 6.6. The upper left panel used pop=1, the upper right panel used pop=2, the lower left used pop=3, and the lower right used pop=4.

◼

6.11 A Relative Depth Analog of the Wilcoxon–Mann–Whitney Test

This section describes another approach to generalizing the Wilcoxon–Mann–Whitney test to the multivariate case. To explain the strategy, first consider the univariate case and let $\mathcal{D} = X - Y$, where X and Y are independent random variables. As explained in Section 5.7, heteroscedastic analogs of the Wilcoxon–Mann–Whitney test are concerned with how deeply zero is nested within the distribution of $\mathcal{D}$. When tied values occur with probability zero, the usual null hypothesis is, in essence, that the depth of zero is equal to the highest possible

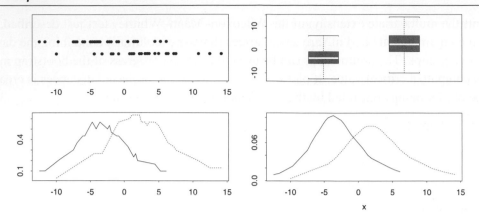

Figure 6.8: An example of the four types of plots created by the function mulwmw.

depth. (That is, the hypothesis is that the median of the distribution of D is zero.) A slightly different formulation, which is useful for present purposes, is to say that the Wilcoxon–Mann–Whitney test is aimed at determining whether the depth of the value zero differs from the maximum possible depth associated with a distribution. A simple way of quantifying this difference is with Q, say, the depth of zero divided by the maximum possible depth, in which case the goal is to test

$$H_0: Q = 1. \tag{6.28}$$

Put a bit more formally, imagine that $\mathbf{X}$ and $\mathbf{Y}$ are independent p-variate random variables and let $\mathcal{D}_j = X_j - Y_j$ be the difference between the jth marginal distributions, $j = 1, \ldots, p$. Let A denote the depth of $\mathbf{0}$ (a vector having length p), relative to the joint distribution of $\mathbf{X} - \mathbf{Y}$, and let B be the maximum possible depth for any point, again, relative to the joint distribution of $\mathbf{X} - \mathbf{Y}$. Then

$$Q = \frac{A}{B}.$$

To estimate Q, let X_{ij} ($i = 1, \ldots, n_1, j = 1, \ldots, p$) and $Y_{i'j}$ ($i' = 1, \ldots, n_2, j = 1, \ldots, p$) be random samples, and for fixed i and i', consider the vector $\mathbf{D}$ formed by the p differences $X_{ij} - Y_{i'j}, j = 1, \ldots, p$. There are $L = n_1 n_2$ such vectors, one for each i and i', which are labeled $\mathbf{D}_\ell, \ell = 1, \ldots, L$. Let P_0 denote the depth of $\mathbf{0}$ relative to the these L vectors, and let P_ℓ be the depth of the ℓth vector, again, relative to the L vectors $\mathbf{D}_\ell, \ell = 1, \ldots, L$. Let $P_m = \max P_\ell$, the maximum taken over $\ell = 1, \ldots, L$. Then an estimate of Q is

$$\hat{Q} = \frac{P_0}{P_m}.$$

Evidently, $\hat{Q}$ is not asymptotically normal when the null hypothesis is true. Note that in this case, Q lies on the boundary of the parameter space. Bootstrap methods have been considered for testing H_0, but their small-sample properties have proven to be difficult to study via simulations because of the high execution time required to compute the necessary depths. Let $N = \min(n_1, n_2)$, and suppose $\alpha = 0.05$. Currently, the only method that has performed well in simulations is to reject, if

$$\hat{Q} \leq c,$$

where for $p = 2$ or 3,

$$c = \max(0.0057N + 0.466, \ 1),$$

for $p = 4$ or 5,

$$c = \max(0.00925N + 0.430, \ 1),$$

for $p = 6$ or 7,

$$c = \max(0.0264N + 0.208, \ 1),$$

for $p = 8$,

$$c = \max(0.0149N + 0.533, \ 1),$$

and for $p > 8$,

$$c = \max(0.04655p + 0.463, \ 1).$$

(See Wilcox, 2005, for more details. Critical values for other choices of α have not been determined.)

6.11.1 R Function mwmw

The R function

```
mwmw(m1,m2,cop=5,pr=TRUE,plotit=TRUE,pop=1,fr=0.8,dop=1,op=1)
```

performs the multivariate extension of the Wilcoxon–Mann–Whitney test just described, where the arguments m1 and m2 are any matrices (having p columns) containing the data for the two groups. The argument cop determines the center of the data that will be used when computing halfspace depth. The choices are:

- cop=1, Donoho–Gasko Median,
- cop=2, MCD estimator,
- cop=3, marginal medians,
- cop=4, MVE estimator,
- cop=5, OP-estimator.

Setting the argument dop=2 causes halfspace depth to be approximated using method A2 in Section 6.2.3; by default, method A1 is used. For bivariate data, a plot is created based on the value of the argument pop, the possible values being 1, 2, and 3, which correspond to a scatterplot, an expected frequency curve, and an adaptive kernel density estimate. The argument fr is the span used by the expected frequency curve. As usual, setting plotit=FALSE avoids the plot. The function returns an estimate of η in the variable phat.

■ Example

The first two skull measures used in the example of Section 6.7.1 are used to illustrate the plot created by the R function mwmw. The plot is shown in Fig. 6.9. The center of the data is marked by an o and based on the OP-estimator, and the null vector is indicated by a +. The function reports that phat is 0.33, which is the estimate of Q. This is less than the critical value 0.62, so reject.

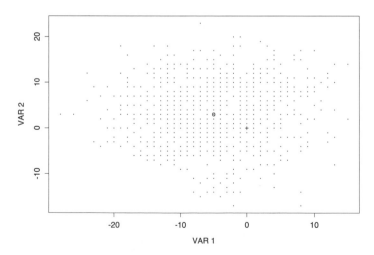

Figure 6.9: An example of the plot created by the function mwmw. The estimated center is marked by an o, and the null center is marked with a +.

6.12 Comparisons Based on Depth

This section describes yet another approach to comparing two independent groups based on multivariate data. The basic idea is that if groups do not differ, the typical depth of the points of the first group, relative to the second, should be the same as the typical depth of the second group, relative to the first. Roughly, the issue is the extent to which the groups are separated as measured by some notion of depth. Here, halfspace depth is used exclusively, simply because this special case has received the most attention from an inferential point of view.

Let $T_D(\mathbf{x}; F)$ represent Tukey's halfspace depth of $\mathbf{x}$ relative to the multivariate distribution F. As usual, let $\mathbf{X}$ and $\mathbf{Y}$ represent independent, p-variate random variables. The corresponding distributions are denoted by F and G. Let

$$R(\mathbf{y}; F) = P_F(T_D(\mathbf{X}; F) \leq T_D(\mathbf{y}; F)).$$

That is, $R(\mathbf{y}; F)$ is the probability that the depth of a randomly sampled $\mathbf{X}$, relative to F, is less than or equal to the depth of some particular point, $\mathbf{y}$, again, relative to F. Said another way, $R(\mathbf{y}; F)$ is the fraction of the F population that is less central than the value $\mathbf{y}$. A *quality index* proposed by Liu and Singh (1993) is

$$Q(F, G) = E_G(R(\mathbf{Y}; F)),$$

the average of all $R(\mathbf{y}; F)$ values with respect to the distribution G. Put another way, for a randomly sampled $\mathbf{X}$ and $\mathbf{Y}$,

$$Q(F, G) = P(D(\mathbf{X}; F) \leq D(\mathbf{Y}; F)),$$

is the probability that the depth of Y is greater than or equal to the depth of X. Liu and Singh show that the range of Q is $[0, 1]$ and when $F = G$, $Q(F, G) = 1/2$. Moreover, when $Q < 1/2$, this reflects a location shift and/or scale increase from F to G. They also develop inferential methods based on Q, where it is assumed that F is some reference distribution. Here, a variation of their method is considered where the goal is to be sensitive to shifts in location. (For relevant asymptotic results, see Zuo and He, 2006. For an alternative method that assumes distributions differ in location only, see Pawar and Shirke, 2019.)

Suppose the sample sizes are m and n for the distributions F and G, respectively. Let

$$\bar{D}_{12} = \frac{1}{m} \sum T_D(\mathbf{X}_i; G_n)$$

be the average depth of the m vectors of observations sampled from F relative to the empirical distribution G_n associated with the second group. If $\bar{D}_{12}$ is relatively small, this can be due to a shift in location or differences in scale. But if

$$\bar{D}_{21} = \frac{1}{n} \sum T_D(\mathbf{Y}_i; F_m)$$

is relatively small as well, this reflects a separation of the two empirical distributions which is roughly associated with a difference in location. (Of course, groups can differ in scale as well when both $\bar{D}_{12}$ and $\bar{D}_{21}$ are small.) So a test of H_0: $F = G$ that is sensitive to shifts in location is one that rejects if

$$\bar{D}_M = \max(\bar{D}_{12},\ \bar{D}_{21})$$

is sufficiently small.

Assuming $m < n$, let $N = (3m + n)/4$. (If $m > n$, $N = (3n + m)/4$.) The only known method that performs well in simulations when testing at the 0.05 level, based on avoiding a Type I error probability greater than the nominal level, is to reject, if $\bar{D}_M \le d_N$, where for $p = 1$,

$$d_N = \frac{-0.4578}{\sqrt{N}} + 0.2536,$$

for $p = 2$,

$$d_N = \frac{-0.3}{\sqrt{N}} + 0.1569,$$

for $p = 3$,

$$d_N = \frac{-0.269}{\sqrt{N}} + 0.0861,$$

for $p = 4$,

$$d_N = \frac{-0.1568}{\sqrt{N}} + 0.0540,$$

for $p = 5$,

$$d_N = \frac{-0.0968}{\sqrt{N}} + 0.0367,$$

for $p = 6$,

$$d_N = \frac{-0.0565}{\sqrt{N}} + 0.0262,$$

for $p = 7$,

$$d_N = \frac{-0.0916}{\sqrt{N}} + 0.0174,$$

and for $p > 8$, $d_N = 0.13$. In terms of Type I errors, the main difficulty is that when sampling from heavy-tailed distributions, the actual Type I error probability can drop well below 0.05

when testing at the 0.05 level (Wilcox, 2003c). (For $p > 8$, as p increases, the actual probability of a Type I error decreases.)

As for a method that is relatively sensitive to differences in scatter, which can be used for p-variate data, first estimate $Q(F, G)$ with

$$\hat{Q}(F, G) = \frac{1}{n} \sum_{i=1}^{n} R(Y_i; F_m). \tag{6.29}$$

(Properties of this estimator are reported by Liu and Singh, 1993.) Similarly, the estimate of $Q(G, F)$ is

$$\hat{Q}(G, F) = \frac{1}{m} \sum_{i=1}^{m} R(X_i; G_n). \tag{6.30}$$

The goal is to test

$$H_0: Q(F, G) = Q(G, F). \tag{6.31}$$

Unlike the method based on $\bar{D}_{12}$ and $\bar{D}_{21}$, a basic percentile bootstrap method performs well in simulations. To begin, generate bootstrap samples from both groups in the usual way and let $\hat{Q}^*(F, G)$ and $\hat{Q}^*(G, F)$ be the resulting bootstrap estimates of $Q(F, G)$ and $Q(G, F)$. Set $D^* = \hat{Q}^*(F, G) - Q^*(G, F)$. Repeat this process B times, yielding D_b^*, $b = 1, \ldots, B$. Put these B values in ascending order, yielding $D_{(1)}^* \leq \cdots \leq D_{(B)}^*$. Then a $1 - \alpha$ confidence interval for $Q(F, G) - Q(G, F)$ is simply $(D_{(\ell+1)}^*, D_{(u)}^*)$, where $\ell = \alpha B/2$, rounded to the nearest integer, and $u = B - \ell$. Of course, reject H_0, if this interval does not contain zero.

6.12.1 R Functions lsqs3 and depthg2

The R function

$$\text{lsqs3(x,y,plotit=TRUE,cop=2)}$$

compares two independent groups based on the statistic $\hat{D}_M$ described in the previous section. For bivariate data, if plotit=TRUE, a scatterplot of the data is produced with the points associated with the second group indicated by a circle. The function

$$\text{depthg2(x,y,alpha=0.05,nboot=500,plotit=TRUE,op=TRUE)}$$

tests Eq. (6.31). If the argument op is set to TRUE, the function prints a message when each bootstrap step is complete.

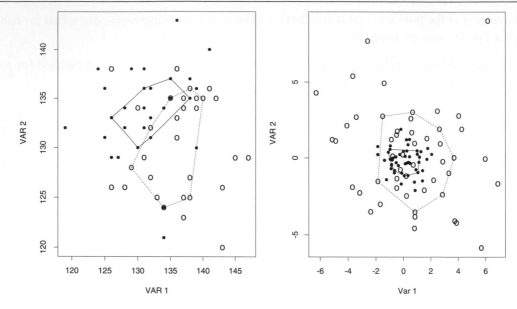

Figure 6.10: The left panel is the plot created by the function lsqs3 using the skull data in Section 6.8.1. The right panel is the plot based on data generated from a bivariate normal distribution, where the marginal distributions have a common mean, but their standard deviations differ. One has a standard deviation of 1 and the other a standard deviation of 3.

■ **Example**

The left panel of Fig. 6.10 shows the plot created by lsqs3 based on the skull data described in Section 6.7.1. (The same plot is created by depthg2.) The function lsqs3 rejects at the 0.05 level, suggesting a shift in location, but depthg2, which is designed to be sensitive to differences in the amount of scatter, does not reject. The right panel shows a scatterplot where both groups have bivariate normal distributions that differ in scale only; the marginal distributions of the first group have standard deviation 1, and for the other group, the marginal distributions have standard deviation 3. (Here, $m = n = 50$.) Now the function lsqs3 finds no difference between the groups, but depthg2 does (at the 0.05 level).

■

■ **Example**

This example is based on data for 24 schizophrenia patients and 18 demographically matched controls. (The data are stored in the file schiz_PP_PM_dat.txt; see Section 1.10.) PP120 is a prepulse inhibition measure taken 120 milliseconds following the onset of an attended stimulus, and PM120 is the prepulse inhibition measure taken 120 milliseconds following the onset of an ignored stimulus. The test statistic returned by lsqs3 is 0.049; the critical value is 0.089, so reject at the 0.05 level. It is left as an exercise to verify that comparing PM120 using means, trimmed means, or an M-estimator, no difference is found at the 0.05 level, but for PP120, the reverse is true. Note, however, that in Fig. 6.11, there is a sense in which PM120 (labeled VAR 2) for the control group lies above and to the left of the points corresponding to the schizophrenia patients. Given PP120, PM120 tends to be greater for the control group.

■

6.13 Comparing Dependent Groups Based on All Pairwise Differences

This section describes an affine invariant method for comparing J dependent groups that is based on a simple extension of the method in Section 6.11. For any $j < m$, let $D_{ijm} = X_{ij} - X_{im}$. Let F be the joint distribution of $\mathbf{D}_{jm}$, and let P be the depth of $\mathbf{0}$ relative to F, divided by the maximum possible depth. Then $0 \le P \le 1$ and the goal is to test

$$H_0 \colon P = 1. \tag{6.32}$$

A simple estimate of P is

$$\hat{P} = \frac{A}{C}, \tag{6.33}$$

where A is the halfspace depth of $\mathbf{0}$ among these n vectors and C is the maximum depth among these n points.

An alternative approach is to determine the halfspace median, which is just the average of the deepest points, and then use the depth of the halfspace median as an estimate of the maximum possible depth. Provided n is not too small, this alternative approach seems to have no practical value, but it can be useful when n is very small. For instance, if $n = 5$, and sampling is from a bivariate normal distribution with a correlation of zero, it is common to have all five depths equal to 0.05, but the depth of the halfspace median is typically close to 0.4.

It should be stressed that affine invariance refers to the D_{ijm} values because it is the depth of these difference scores that are used. It can be seen that the method is not affine invariant in terms of the X_{ij} values.

A technical problem is that, generally, the MCD estimate of location cannot be computed when working with the D_{ijm} values because the corresponding covariance matrix is singular. Consequently, the approximation of halfspace depth with method A1 in Section 6.2.3 is not immediately applicable. To deal with this problem, compute the MCD estimate of location based on the original X_{ij} values, yielding, say $(\hat{\xi}_1, \ldots, \hat{\xi}_J)$, and then approximate the halfspace depth of the D_{ijm} values with method A1 by taking the center of location of D_{ijm} to be $\hat{\theta}_{jm} = \hat{\xi}_j - \hat{\xi}_m$.

An alternative strategy is to use an approximation of halfspace depth that does not require that the D_{ijm} values have a non-singular covariance matrix. This can be accomplished with method A2 in Section 6.2.3.

As was the case in Section 6.10, when the null hypothesis is true, the distribution of $\hat{P}$ is not asymptotically normal. The reason is that for this special case, P lies on the boundary of the parameter space, and so for any general situation where $\hat{P}$ is a consistent estimate of P, it must be that $\hat{P} \leq 1$, with the probability of $\hat{P} = 1$ increasing as the sample sizes get large. For similar reasons, based on theoretical results in Liu and Singh (1997), the expectation is that when H_0 is true, a basic percentile bootstrap method for computing a confidence interval for P will fail, and this has been found to be the case in simulations.

Consider rejecting H_0, if $\hat{P} \leq c$. The following approximations of c appear to perform well:

$J = 2$,	$\hat{c} = -1.46n^{-0.5} + 0.95$,
$J = 3$,	$\hat{c} = -1.71n^{-0.5} + 1.00$,
$J = 4$,	$\hat{c} = -1.77n^{-0.5} + 1.06$,
$J = 5$,	$\hat{c} = -1.76n^{-0.5} + 1.11$,
$J = 6$,	$\hat{c} = -1.62n^{-0.3} + 1.41$,
$J = 7$,	$\hat{c} = -1.71n^{-0.3} + 1.49$,
$J = 8$,	$\hat{c} = -1.38n^{-0.3} + 1.39$.

Note that as $n \to \infty$, $c \to 1$. Moreover, as J increases, c converges to 1 more quickly. So in effect, reject, if $\hat{P} < \min(\hat{c}, 1)$.

The method just described is affine invariant, roughly meaning that it is metric-free. That is, if the $n \times p$ matrix of data is postmultiplied by a non-singular matrix $\mathbf{A}$, $\hat{P}$ is not altered.

6.13.1 R Function dfried

The R function

$$\text{dfried(m,plotit=TRUE,pop=0,fr=0.8,v2=FALSE,op=FALSE)}$$

tests the hypothesis H_0: $P = 1$ as just described. Here, m is any R variable having matrix mode with n rows and p columns. If $p = 2$ and plotit=TRUE, a plot of the difference scores is created with the type of plot controlled by the argument pop. The choices are:

1. pop=0, adaptive kernel density,
2. pop=1, expected frequency curve,
3. pop=2, kernel density estimate using normal kernel,
4. pop=3, R built-in kernel density estimate,
5. pop=4, boxplot.

The argument fr controls the span when using the expected frequency curve. Setting v2=TRUE causes method A2 to be used to approximate halfspace depth, and op=TRUE results in using the depth of Tukey's median as an estimate of the maximum possible halfspace depth.

6.14 Robust Principal Component Analysis

Roughly, principal component analysis (PCA) is aimed at finding m linear combinations of p $(m < p)$ observed variables that explain most of the variability in the data. In the realm of statistical learning, PCA belongs to what are known as unsupervised methods. Roughly, supervised methods involve building a statistical model for predicting or estimating some output based on input variables. With unsupervised statistical learning, there are inputs but no supervising output. A general goal in unsupervised statistical learning is to develop some understanding of the structure of the inputs.

To quickly review the strategy underlying the classic approach to PCA, momentarily consider the situation where $m = 1$. Denoting the data for the jth variable by X_{ij} $(i = 1, \ldots, n, j = 1, \ldots p)$, the goal is to reduce the p variables to a single variable via some linear combination of the p variables, denoted by

$$U_i = \sum_{j=1}^{p} h_j X_{ij},$$

with the constants $h_1, \ldots, h_p$ chosen so as to maximize the variance of the U_i values subject to $\sum h_j^2 = 1$. Put another way, this first principal component is the line that is closest to the data. That is, it is the line for which the sum of squared perpendicular distances between each point and the line is minimized.

Now consider the problem of reducing the p variables down to two variables rather than just one. So for the ith participant, the goal is to compute two linear combinations of the p variables based on two sets of weights:

$$U_{i1} = h_{11} X_{i1} + \cdots + h_{1p} X_{ip},$$

and

$$U_{i2} = h_{21} X_{i1} + \cdots + h_{2p} X_{ip}$$

($i = 1, \ldots, n$), where for fixed k, $\sum h_{jk}^2 = 1$ and the variance of the U_{ik} values is maximized subject to the condition that U_k and U_ℓ have correlation zero, $k \neq \ell$. More generally, m linear combinations are sought that maximize the variance of the marginal distributions with the property that any two linear combinations have zero correlation. This goal is accomplished by taking $\mathbf{h}_1, \ldots \mathbf{h}_p$ to be the eigenvectors of the usual covariance matrix. The columns $U_1, \ldots, U_p$ of the matrix $\mathbf{U}$ are called the *principal components* of $\mathbf{X}$. (For a recent discussion regarding the interpretation of principal components, see Anaya-Izquierdo et al., 2011.) Moreover, the variance of U_k is λ_k, where $\lambda_1 \geq \cdots \geq \lambda_p$ and λ_k is the eigenvalue corresponding to the eigenvector $\mathbf{h}_k$. The U_{ij} are called the *principal component scores*.

But because the usual covariance matrix is not robust, situations are encountered where upon closer scrutiny, the resulting components explain a structure that has been created by a mere one or two outliers (e.g., Huber, 1981, p. 199). This has led to numerous suggestions regarding how the classic PCA method might be made more robust. A simple approach is to replace the covariance matrix with a robust scatter matrix or a robust correlation matrix. Devlin et al. (1981) use an M-estimator with a low breakdown point, so a relatively small number of outliers can cause practical problems. The MVE estimator, as well as the (fast) MCD estimator, might be used, but concerns about these estimators have already been noted. A method based on an S-estimator was studied by Croux and Haesbroeck (2000), and a fast and simple method was proposed by Locantore et al. (1999). Li and Chen (1985) suggest a projection pursuit approach, meaning that directions are sought that maximize or minimize some robust measure of dispersion. (One appealing feature of projection-type methods is that they can be used when the number of variables exceeds the sample size.) Croux and Ruiz-Gazen (2005, Section 5.1) describe an algorithm for implementing the Li and Chen method. (Also see Hubert et al., 2002; Salibián-Barrera et al., 2006.) One negative feature of the Li and Chen method is its computational complexity. Maronna (2005) extends this projection pursuit technique in a manner that improves computational efficiency and statistical performance. Yet another recent suggestion was made by Hubert et al. (2005) that was later refined by Engelen et al. (2005), which is used here. Roughly, the first step is to compute a measure of outlyingness for each of the n points, where n is the sample size. Then for h chosen by the investigator, the h least outlying data points are used to compute a measure of location and scatter, which in turn are used to determine how many components will be retained, as well as the projected data points. Following Engelen et al. (2005), a reweighting step is added based on the orthogonal distances of the observations with respect to the first estimated PCA subspace. It is only at the first stage of the algorithm that the number of points eliminated must be specified via the choice for h. (For some additional results on robust approaches to PCA, see Serneels and Verdonck, 2008; Chen et al., 2009. For a rank-based approach, see Hallin et al., 2014.) For a method that deals with high-dimensional data, see Cevallos-Valdiviezo and Van Aelst (2019).

Generally, the methods just listed are based in part on maximizing some measure of variation associated with the marginal distributions of the p principal components. Another approach is to choose linear combinations (principal components) aimed at maximizing some robust generalized variance associated with the principal component scores (Wilcox, 2008c). That is, take into account the overall structure of the data when measuring variation, in contrast to maximizing the variance of the individual principal component scores. (Details are given in Section 6.14.6.)

There is yet another generalization of PCA that should be mentioned: kernel PCA (Schlölkopf et al., 1998). Roughly, the method first maps the data into a higher-dimensional feature space. (It generalizes regular PCA by replacing the usual inner product with a broader class of functions.) A robust version of kernel PCA has been studied by Debruyne et al. (2010). More information can be found at http://lavaan.org.

6.14.1 R Functions prcomp and regpca

The built-in R function

$$\text{prcomp(x,cor=FALSE)}$$

performs the classic PCA. By default, it uses the covariance matrix rather than the correlation matrix. In case it is useful, the R function

regpca(x, cor =TRUE, loadings =TRUE, SCORES =FALSE, scree =TRUE, xlab = 'Principal Component', ylab = 'Proportion of Variance')

is provided, which performs the classic PCA after first removing any rows of data for which one or more columns have missing values. Unlike prcomp, the function regpca uses the correlation matrix by default. And it creates a scree plot when the argument scree=TRUE, which is a line segment that shows the fraction of the total variance among all m components as a function of the number of components. (The scree plot is illustrated in Section 6.14.8.)

6.14.2 Maronna's Method

This section provides a brief outline of the method proposed by Maronna (2005), which is based in part on an iterative algorithm. Let $\mathbf{x}_i$, $i = 1, \ldots, n$, be a p-dimensional data set, let $q = p - m$, and let $\mathbf{C}$ be an orthonormal $q \times m$ matrix. That is, $\mathbf{CC}' = \mathbf{I}_q$. For some q-vector $\mathbf{a}$, let

$$r_i(\mathbf{C}, \mathbf{a}) = \|\mathbf{Cx}_i - \mathbf{a}\|^2,$$

and let $\sigma(\mathbf{r})$ be a scale statistic, where $\mathbf{r} = (r_1, \ldots, r_n)$. The goal is to determine $\mathbf{C}$ and $\mathbf{a}$ so as to minimize $\sigma(\mathbf{r})$. Maronna considers two choices for $\sigma(\mathbf{r})$: an M-scale and an L-scale. Here, the focus is on the L-scale

$$\sigma(\mathbf{r}) = \sum_{i=1}^{h} r_{(i)},$$

where $r_{(1)} \leq \cdots \leq r_{(h)}$, $h < n$, primarily because it is faster and easier to compute. Following Maronna (2005), h is taken to be the largest integer less than or equal to $(n + p - q + 2)/2$, where $q = p - m$.

6.14.3 The SPCA Method

The *spherical PCA* (SPCA) method was derived by Locantore et al. (1999). Let μ be the L_1 median, which is computed by the R function spat or the R function L1medcen. Let $\mathbf{y}_i = (\mathbf{x}_i - \mu)/\|\mathbf{x}_i - \mu\|$. The procedure consists of using the eigenvectors $\mathbf{b}_1, \ldots, \mathbf{b}_p$ of the covariance matrix of $\mathbf{y}_i$. But the eigenvalues are, in general, not consistent, in which case, they are replaced by

$$\lambda_j = S(\mathbf{b}_j'\mathbf{x}_1, \ldots, \mathbf{b}_j'\mathbf{x}_n)^2,$$

where S is any robust measure of scale. Following Maronna (2005), S is taken to be the MAD statistic. The R package rrcov contains the function PcaLocantore that performs SPCA.

6.14.4 Methods HRVB and MacroPCA

Hubert et al. (2005) suggest a method that combines projection pursuit ideas with robust scatter matrix estimation. An adaptation of this method, called *method HRVB*, was derived by Engelen et al. (2005) and is used here. The computational details are quite involved, so only a brief outline of the method is provided.

The method begins by finding the h least outlying data points. The choice for h is made by the investigator, and Hubert et al. consider choices of the form $h = \max\{[\alpha n], [(n + k_{\max} + 1)/2]\}$, where α is some value between 0.5 and 1, and $k_{\max}$ is the maximum number of components that will be computed; they use $\alpha = 0.75$ and $k_{\max} = 10$, and the same is done here. Next, outlyingness is measured using a maximum standardized distance among the class of all possible projections of the data onto a unidimensional space. Not all projections can be considered, so for n small, they focus on all directions through two points, and for $\binom{n}{2} > 250$, they take at random 250 projections. They then focus on the mean and covariance matrix of the h points that have the smallest distances just computed. The next step computes fast MCD for the projected data resulting from the previous step, which is used to compute a reweighted mean and covariance matrix that increases statistical efficiency. A consistency factor is used to make the estimator unbiased at normal distributions.

Hubert et al. (2019) derive a method that simultaneously deals with three fundamental issues: rowwise outliers, cellwise outliers, and missing values. The computational details are quite involved and not described here. Evidently, the method will be available via the R function MacroPCA in the R package cellWise, but at the moment, this is not the case.

6.14.5 Method OP

Method OP simply removes any outliers detected by the projection approach described in Section 6.4.9. Then classic PCA is applied to the data that remain, and the p-dimensional representation of the data is computed in the usual way.

Croux and Ruiz-Gazen (2005) suggest an algorithm that begins with projections based in part on the L_1 median, but it is evident that their approach differs from method OP. Method OP attempts to eliminate outliers in a manner that takes into account the overall structure of the data. The algorithm used by Croux and Ruiz-Gazen does not do this, but rather searches for projections that maximize a robust measure of scatter applied to the marginal distributions of the scores. Also, the Croux and Ruiz-Gazen (2005) and Hubert et al. (2002) projection algorithms appear to suffer from severe downward bias. It is unknown whether method OP suffers from the same problem.

6.14.6 Method PPCA

Method PPCA is aimed at finding principal components that maximize a robust generalized variance. Let **B** be any $m \times p$ matrix having the property that for any j $(1 \leq j \leq m)$,

$$\sum_{k=1}^{p} b_{jk}^2 = 1$$

and for any $j \neq \ell$,

$$\sum_{k=1}^{p} b_{jk} b_{\ell j} = 0.$$

Given $\mathbf{B}$, the resulting m-dimensional representation of the data is

$$\mathbf{z}_i = \mathbf{B}(\mathbf{x}_i - \boldsymbol{\theta}), \tag{6.34}$$

where $\boldsymbol{\theta}$ is some measure of location. (All of the methods outlined in this section use Eq. (6.34) and differ in how they determine $\mathbf{B}$ and $\boldsymbol{\theta}$.) The $\mathbf{z}_i$ ($i = 1, \ldots, n$) values are the *scores*. (Scores based on the other robust methods in this section are computed in a similar manner.) Let $\hat{\Xi}$ be an estimate of some robust generalized variance based on the $\mathbf{z}_i$ values. Here, the covariance matrix based on the median ball algorithm is used unless stated otherwise, and $\hat{\Xi}$ is taken to be the determinant of this covariance matrix that is computed with the $\mathbf{z}_i$ values.

The goal is to determine the matrix $\mathbf{B}$ that maximizes $\hat{\Xi}$. The method used here begins with an initial estimate of $\mathbf{B}$, say $\mathbf{B}_0$, based on the Hubert et al. (2005) estimator (method HVRB). Then use the Nelder and Mead (1965) algorithm to search for the matrix $\mathbf{B}$ that maximizes $\hat{\Xi}$. (The Nelder–Mead algorithm is applied with the R function nelderv2, which improves on the random search method used by Wilcox, 2008c.)

Regarding the estimation of $\boldsymbol{\theta}$, Wilcox (2008c) considers the (fast) MCD estimator, the L_1 median, Olive's (2004) estimator based on the median ball algorithm, and the mean of the data after points flagged as outliers by the projection method are removed. Simulation results indicate that the choice of location estimator makes little difference when using a random search for the matrix $\mathbf{B}$ that maximizes the generalized variance. However, when using the Nelder–Mead algorithm, Wilcox (2010c) found that the L_1 median, which is computed by the R function spat, performs relatively well, so it is used here.

6.14.7 R Functions outpca, robpca, robpcaS, SPCA, Ppca, and Ppca.summary

The R function

```
outpca(x, cor=TRUE, loadings=TRUE, covlist=NULL, scree=TRUE, SCORES=FALSE,
ADJ=FALSE, scree=TRUE, ALL=TRUE, pval=NULL, cop=3, ADJ=FALSE, SEED=TRUE,
  pr=TRUE,STAND=TRUE, xlab='Principal Component', ylab='Proportion of Variance')
```

eliminates outliers via the projection method and applies classic PCA to the remaining data. The argument pval indicates the number of principal components. Following the convention used by R, the covariance matrix is used by default. To use the correlation matrix, set the argument cor=TRUE. Setting SCORES=TRUE, the principal component scores are returned. If the argument ADJ=TRUE, the R function outproad is used to check for outliers rather than the R function outpro, which is recommended if the number of variables is greater than 9. By default, the argument scree=TRUE, meaning that a scree plot will be created. Another rule that is sometimes used is to retain those components for which the proportion of variance is greater than 0.1. When the proportion is less than 0.1, it has been suggested that the corresponding principal component rarely has much interpretive value.

The function

$$\text{robpcaS(x, pval=ncol(x),SCORES=FALSE)}$$

provides a summary of the results based on the method derived by Hubert et al. (2005), including a scree plot based on a robust measure of variation. The argument pval indicates the number of principal components. A more detailed analysis is performed by the function

$$\text{robpca(x, pval = ncol(x), SEED =TRUE, STAND =TRUE, cst =TRUEmean, varfun = winvar,}$$
$$\text{xlab = 'Principal Component', ylab = 'Proportion of Variance'),}$$

which returns the eigenvalues and other results discussed by Hubert et al. (2005), but these details are not discussed here. A scree plot is created when the argument pval = ncol(x), but not otherwise.

For convenience, the R function

$$\text{SPCA(x, k = 0, kmax = ncol(x), delta = 0.001, na.action = na.fail, scale =FALSE, signflip}$$
$$\text{=TRUE, trace=FALSE, ...)}$$

is provided for applying the SPCA method, described in Section 6.14.3. This function merely eliminates the need to issue the command library(rrcov) when calling the R function PcaLocantore. The argument x is assumed to be an n-by-p matrix. Information about the other arguments can be obtained via the R command ?PcaLocantore, assuming that the R command library(rrcov) has already been issued. The R command screeplot(SPCA(x)) would create a scree plot and summary(SPCA(x)) would return the standard deviations, the proportion of variance, and the cumulative proportions.

The R function

$$Ppca(x, p = ncol(x) - 1, locfun = L1medcen, loc.val = NULL, SCORES =FALSE, gvar.fun = cov.mba, pr =TRUE, SEED =TRUE, gcov = rmba, SCALE =TRUE, ...)$$

applies the method aimed at maximizing a robust generalized variance. This particular function requires the number of principal components to be specified via the argument p, which defaults to $p - 1$. The argument SCALE=TRUE means that the marginal distributions will be standardized based on the measure of location and scale corresponding to the argument gcov, which defaults to the median ball algorithm.

The R function

$$Ppca.summary(x, MC=FALSE, SCALE=TRUE)$$

is designed to deal with the issue of how many components should be used. It calls Ppca using all possible choices for the number of components, computes the resulting generalized standard deviations, and reports their relative size. If access to a multicore processor is available, setting the argument MC=TRUE will reduce execution time. Illustrations in the next section deal with the issue of how many components to use based on the output from the R function Ppca.summary.

6.14.8 Comments on Choosing the Number of Components

First focus on classic PCA. Regarding the choice for p, the number of components to use, a rule that is sometimes used is to retain those components for which the proportion of variance is greater than 0.1. When the proportion is less than 0.1, it has been suggested that the corresponding principal component rarely has much interpretive value. Another way of trying to judge how many principal components to use is by visual inspection of a scree plot, the strategy being to determine where the "elbow" of the curve occurs. This well-known strategy is illustrated with data generated from a multivariate normal distribution with all correlations equal to 0.0 and $n = 200$. The output from the R function regpca is

```
Importance of components:
                          PC1    PC2    PC3    PC4
Standard deviation      1.113  0.963  0.959  0.914
Proportion of Variance  0.316  0.236  0.235  0.213
Cumulative Proportion   0.316  0.552  0.787  1.000
```

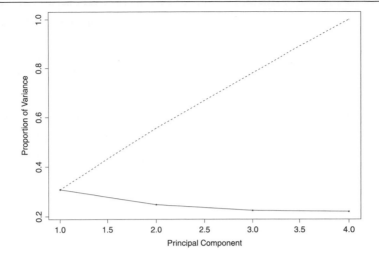

Figure 6.11: The scree plot returned by the R function regpca, where data are multivariate normal with all Pearson correlations equal to zero.

Fig. 6.11 shows the resulting scree plot. The bottom (solid) line shows the variance associated with the principal components. The upper (dashed) line is the cumulative proportion. Note that the lower line is nearly horizontal with no steep declines, suggesting that all four components be used to capture the variability in the data. Also, for each component, the proportion of variance is greater than 0.1.

The output from the function Ppca.summary differs in crucial ways from the other functions described here. To illustrate it, multivariate normal data were generated with all correlations equal to 0.0. The output from Ppca.summary is

```
                   [,1]       [,2]       [,3]       [,4]
Num. of Comp. 1.0000000 2.000000 3.0000000 4.0000000
Gen.Stand.Dev 1.1735029 1.210405 1.0293564 1.0110513
Relative Size 0.9695129 1.000000 0.8504234 0.8353002
```

The second line indicates the (robust) generalized standard deviation given the number of components indicated by the first line. So when using two components, the generalized standard deviation is 1.210405. Note that the generalized standard deviations are not in descending order. Using two components results in the largest generalized standard deviation. But observe that all four generalized standard deviations are approximately equal, which is what we would expect for the situation at hand. The third line of the output is obtained by dividing each value in the second line by the maximum generalized standard deviation. Here, reducing the number of components from four to two does not increase the generalized standard deviation by very much, suggesting that four or maybe three components should be used. Also observe that there is no proportion of variance used here, in contrast to classic PCA. In classic PCA, an issue is how many components must be included to capture a reasonably large proportion of the variance. When using the robust generalized variance, it seems more appropriate to first look at the relative size of the generalized standard deviations using all of the components. If the relative size is small, reduce the number of components. In the example, the relative size using all four components is 0.835, suggesting that perhaps all four components should be used.

Now consider data that were generated from a multivariate normal distribution where all of the correlations are 0.9. Now the output from regpca is

```
Importance of components:
                         PC1     PC2     PC3     PC4
Standard deviation      1.869  0.3444  0.3044  0.2915
Proportion of Variance  0.922  0.0313  0.0244  0.0224
Cumulative Proportion   0.922  0.9531  0.9776  1.0000
```

Note that the first principal component has a much larger standard deviation than the other three principal components. The proportion of variance accounted for by PC1 is 0.922, suggesting that it is sufficient to use the first principal component only to capture the variability in the data. Fig. 6.12 shows the scree plot.

Excluding method PPCA, the robust methods summarized in this section report results similar to the function regpca, only a robust measure of variation, associated with each component, is used. Scree plots can be created as well. However, when using method PPCA, the output from the R function Ppca is interpreted in a different manner. Generally, it is suggested that one first look at the sizes of the generalized standard deviations, relative to the largest generalized standard deviation, starting with $p = m$ components. If the relative size is close to 1, use all m components. If not, consider $p = m - 1$. If the relative size is close to 1, use $p - 1$ components. If not, continue in this manner.

Consider again the multivariate normal data with all correlations equal to 0, which were used to create the scree plot in Fig. 6.11. The output from Ppca.summary is

```
                   [,1]       [,2]       [,3]       [,4]
Num. of Comp.  1.0000000  2.000000  3.0000000  4.0000000
Gen.Stand.Dev  1.1735029  1.210405  1.0293564  1.0110513
Relative Size  0.9695129  1.000000  0.8504234  0.8353002
```

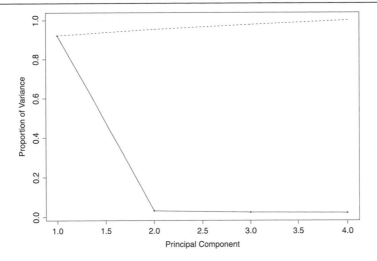

Figure 6.12: The scree plot for multivariate normal data with all Pearson correlations equal to 0.9.

The second line indicates the (robust) generalized standard deviation given the number of components indicated by the first line. So when using two components, the generalized standard deviation is 1.210405. Note that the generalized standard deviations are not in descending order. Using two components results in the largest generalized standard deviation. But observe that all four generalized standard deviations are approximately equal, which is what we would expect for the situation at hand. The third line of the output is obtained by dividing each value in the second line by the maximum generalized standard deviation. Here, reducing the number of components from four to two does not increase the generalized standard deviation by very much, suggesting that four or maybe three components should be used. Also observe that there is no proportion of variance used here, in contrast to classic PCA. In classic PCA, an issue is how many components must be included to capture a reasonably large proportion of the variance. Here, the relative size using all four components is 0.835, suggesting that perhaps all four components should be used.

Consider, again, the data used to create the scree plot in Fig. 6.12. (The data have a multivariate normal distribution with all correlations equal to 0.9.) The output from Ppca.summary is

```
                [,1]      [,2]       [,3]        [,4]
Num. of Comp. 1.000000 2.0000000 3.0000000 4.00000000
Gen.Stand.Dev 2.017774 0.6632588 0.2167982 0.05615346
Relative Size 1.000000 0.3287082 0.1074442 0.02782942
```

As indicated, a single component results in a relatively large generalized standard deviation, suggesting that a single component suffices. The relative sizes corresponding to three and four components are fairly small, suggesting that using three or four components be ruled out. Even with two components, the relative size is fairly small.

■ Example

In an unpublished study by L. Doi, a general goal was to study predictors of reading ability. Here, the focus is on five predictors: two measures of phonological awareness, a measure of speeded naming for digits, a measure of speeded naming for letters, and a measure of the accuracy of identifying lower case letters. Using classic PCA based on the correlation matrix, the R function regpca returns

```
Importance of components:
                      Comp.1 Comp.2 Comp.3 Comp.4 Comp.5
Standard deviation    1.4342 1.0360 0.9791 0.7651 0.57036
Proportion of Variance 0.4114 0.2146 0.1917 0.1170 0.06506
Cumulative Proportion  0.4114 0.6260 0.8178 0.9349 1.00000
```

Note that the proportion of variance exceeds 0.1 with four components or less, which some would take to suggest that four components be used. The R function robpcaS returns

```
                      [,1]    [,2]    [,3]    [,4]    [,5]
Number of Comp.       1.00000 2.00000 3.00000 4.00000 5.000000
Robust Stand Dev      2.23900 1.26512 1.21967 0.97752 0.606995
Proportion Robust var 0.53188 0.16981 0.15783 0.10138 0.039090
Cum. Proportion       0.53188 0.70169 0.85952 0.96090 1.000000
```

which is somewhat similar to the results based on the classic PCA.

However, Ppca.summary returns

```
              [,1]     [,2]      [,3]      [,4]      [,5]
Num. of Comp. 1.000000 2.0000000 3.0000000 4.0000000 5.0000000
Gen.Stand.Dev 1.712513 1.5155318 0.7229315 0.4761138 0.3112773
Relative Size 1.000000 0.8849754 0.4221466 0.2780205 0.1817664
```

The second line shows the robust generalized standard deviations based on the number of components used. Because the relative sizes using three, four, or five components are rather small, the results suggest that two components suffice. (In fairness, it might be argued that a scree plot stemming from classic PCA also suggests that two components be used.)

6.15 Cluster Analysis

Cluster analysis is an exploratory data analysis tool aimed at sorting different objects into groups in a way that the degree of association between two objects is maximal if they belong to the same group, and minimal otherwise. There are many relevant methods that go well beyond the scope of this book (e.g., Everitt et al., 2011). Here, the goal is merely to mention a few R functions that might be useful.

6.15.1 R Functions Kmeans, kmeans.grp, TKmeans, and TKmeans.grp

Cluster analysis can be performed with the R function

$$Kmeans(x,k,xout=FALSE,outfun=outpro).$$

The argument x is a matrix or data frame containing the data, and k indicates the number of clusters to be used. The function calls the built-in R function kmeans, but it automatically removes any rows of data that contain missing values. The R function kmeans uses the k-means method, which partitions the points into k groups such that the sum of squares from points to the assigned cluster centers is minimized. If the argument xout=TRUE, the R function Kmeans removes any points declared outliers via the function specified by the argument outfun. The R function

$$Kmeans.grp(x,k,y,xout=FALSE,outfun=outpro)$$

creates *k* groups based on the data stored in x via the k-means algorithm. It then sorts the data in y into one of these *k* groups and stores the results in an R variable having list mode. For example, based on the command z=kmeans.grp(x,2,y), z[[1]] will contain the data associated with the first cluster and z[[2]] will contain the data associated with the second cluster.

The R function

$$TKmeans(x,k, trim=0.1,scaling=FALSE,runs=100, points=NULL,$$
$$countmode=runs+1,printcrit=FALSE,maxit = 2 * nrow(as.matrix(data)))$$

applies the trimmed k-means method derived by Cuesta-Albertos et al. (1997). It removes any vectors of observations having missing values, and then uses the R function trimkmeans in the R package trimcluster. The R function

$$TKmeans.grp(x,k,xout=FALSE,outfun=out)$$

is like the function Kmeans.grp, only it uses the R function TKmeans to determine the clusters.

6.16 Classification Methods

There is a tremendously vast literature on classification methods that is difficult to cover in an entire book. Indeed, there are multiple strategies with many variations within each resulting in an arguably overwhelming number of choices regarding how to proceed. Suffice it to say that a detailed description of all extant strategies and issues goes well beyond the scope of this book. For a relatively non-technical summary of many of these methods, see, for example, James et al. (2017) or Witten et al. (2017). Also see Venables and Ripley (2002, Chapter 12). Multivariate discriminate analysis is a classic approach, which assumes multivariate normality (e.g., Mardia et al., 1979; Huberty, 1994). Efron (2020) compares and contrasts various methods with least squares and logistic regression. For methods based on functional data, see Krzyśko and Lukasz (2019). For a relevant R package, see Bischl et al. (2016). Lai and McLeod (2020) suggest an approach based on an ensemble quantile classifier. But currently, convenient software for applying their method is not available. Also see Gul et al. (2018).

A highly limited goal in this section is to summarize some classification techniques that have connections to the robust methods in this book, the bulk of which are based on some measure of depth. In some situations, these methods can compete reasonably well with other methods summarized below. But this is not to suggest that they dominate or are to be always preferred over other methods that might be used: No method is always best. Which method performs best depends on the situation. Another goal is to supply a few R functions that make it relatively easy to apply and compare a fairly wide range of techniques.

The most basic goal is to classify things or individuals into one of two categories or groups. For example, the groups might correspond to people who do or do not default on a loan. Given p measures on each individual, how might one make a decision about who will default? Or the issue might be to identify individuals who will develop a particular medical disorder. The goal is to find a rule for identifying the correct group based on p measures. This is done using training data where the correct group is known.

It helps to provide a brief overview of many of the approaches that have been proposed, followed by a description of techniques related to the robust methods in this book. One approach begins with regression trees (e.g., Hastie et al., 2001). The basic strategy is based on splitting the data into J distinct non-overlapping regions. To be a bit more precise, the data are split into high-dimensional rectangles based on the training data. For future data, typically called test data, if the p measures fall in the jth region, the predicted outcome is the most commonly occurring outcome in that region based on the training data. That is, use an estimate of what is generally called Bayes rule, which classifies an object to the class having the highest probability. But trees generally do not perform as well as other methods, and they are not robust. A way of improving them is to aggregate many decision trees using bagging, boosting, and random forests. Bagging is a bootstrap technique, the details of which are described in the next section. (Also see the end of Section 11.5.4.) Basically, for each bootstrap sample, a tree is created, and these trees are combined to create a single prediction model. Random forests are built on bootstrap samples as well, but they are designed to improve on bagging by using a random sample of m predictors, where, typically, m is approximately equal to $\sqrt{p}$. See Calhoun et al. (2020) for recent results on improving on the random forest approach. Boosting (e.g., Breiman, 2001a) is similar, only each tree is grown using information from previously grown trees. Bootstrap samples are not used. For an entire book focused on boosting, see Schapire and Freund (2012). For results on a robust version of boosting, see Li and Bradic (2018). At the moment, there are no convenient R functions that implement their method. A popular variation of boosting is adaboost, which is a weighted version of boosting, where the weights are determined in a sequential manner. Also see Martinez and Gray (2016), as well as the R package gbm.

Another classification strategy is based on neural networks, which are a two-stage regression model, basically a non-linear regression model, the components of which have connections with the logistic function (e.g., Hastie et al., 2001, Section 11.3) or the hyperbolic tangent function. For a method based on the spatial rank function (Möttönen and Oja, 1995), see Makinde and Chakraborty (2018). Logistic regression and non-parametric analogs of logistic regression are two more approaches that might be used. These latter approaches are described in Sections 10.16 and 11.5.8, respectively. One more possibility is to use a quantile classifier. See, for example, Lai and McLeod (2020) plus the references they cite. For an R package that implements this approach, see Hennig and Viroli (2016).

Yet another approach is based on what is called support vector machines (SVMs). Very briefly, imagine that based on the training data, a hyperplane (a line when there are $p = 2$ predictors, a plane when $p = 3$, and generally a flat $p - 1$ surface) can be found that separates the data cloud associated with the two groups. Then this hyperplane could be used to classify any future point for which the correct classification is not known. Rather than use a hyperplane based on the available p measures, SVM transforms the problem into a higher-dimensional space. For example, rather than use the p features $X_1, \ldots, X_p$ to find an appropriate hyper-plane, use instead $X_1, X_1^2, \ldots, X_p, X_p^2$. So now $2p$ features are used rather than the original p measures, resulting in a non-linear hyperplane that is used to classify points. Estimation of the hyperplane can be based on a generalization of the inner product of two vectors called a kernel. The choice for the kernel can make a practical difference.

The remainder of this section describes some methods related to the techniques previously described in this chapter. The only suggestion is that they might prove to have practical value in a given situation.

Method C1

Let $D_j(\mathbf{x})$ be some measure of depth associated with the jth group ($j = 1, 2$) for the point $\mathbf{x}$. Then a natural classification rule is to assign $\mathbf{x}$ to the jth class, if $D_j(\mathbf{x}) = \max(D_1(\mathbf{x}), D_2(\mathbf{x}))$. That is, assign $\mathbf{x}$ to the group for which it has the deepest depth. This basic strategy has been studied by Cui et al. (2008) and Dutta and Ghosh (2012), as well as Ghosh and Chaudhuri (2005).

Method C2

Makinde and Fasoranbaku (2018) proposed a variation of method C1 and showed that in certain situations it competes well with some alternative methods, including SVM. Boosting, bagging, and random forests were among the methods not considered. For a randomly sampled vector $\mathbf{X}$, let $C(a) = P(D(\mathbf{X}) \leq a)$. That is, $C(a)$ represents the cumulative distribution of the depth of a randomly sampled point. Let a_j be the depth of the point $\mathbf{x}$ relative to the jth group. Then assign $\mathbf{x}$ to the first group, if $C(a_1) > C(a_2)$; otherwise, assign it to group 2. They considered several measures of depth. Halfspace depth and zonoid depth (introduced in Section 6.2.7) performed relatively well in the situations they considered.

Method C3

The basic strategy of the next approach stems from Li et al. (2012), who suggest transforming the data to some measure of depth, such as halfspace depth. For two groups, they search for the best separating polynomial based on the transformed data. Hubert et al. (2015b) suggest some modifications of the approach used by Li et al. First, they suggest using a distance measure that is based in part on halfspace depth, but which avoids measures of depth that become zero outside the convex hull of the data. They use bagdistance, outlined at the end of Section 6.2.7. They then use the k-nearest neighbor (kNN) classification rule (Fix and Hodges, 1951) to classify individuals, rather than a separating polynomial, which was found to perform well and is computationally easier to use. Briefly, for each new observation, the kNN rule looks up the k training data points closest to it (typically using Euclidean distance), and then assigns it to the most prevalent group among those neighbors. The value of k is typically chosen by cross-validation to minimize the misclassification rate. Here, the default measure of depth is the projection-based depth computed via the R function prodepth, described in Section 6.2.8, rather than the bagdistance used by Hubert et al. The R function in the next section, which applies the method described here, can be used with any measure of depth. In particular, bagdistance can be used via the R function Bagdist, described in Section 6.2.8. The only reason for choosing prodepth over Bagdist is that the R function prodepth was available prior to the R function Bagdist. The relative merits of these two measures of depth, for the situation at hand, are not known.

Consider the goal of classifying points as belonging to one of $G \geq 2$ groups. For every vector $\mathbf{x}$ in the training data, transform it to the G-variate point

$$(D(\mathbf{x}, P_1), \ldots, D(\mathbf{x}, P_G)),$$

where $D(\mathbf{x}, P_g)$ is some depth measure associated with $\mathbf{x}$ and based on the training data in group P_g ($g = 1, \ldots, G$). Based on these distance measures, classify some future $\mathbf{x}$ based on the kNN rule.

Method C4

Another approach is to project the training data onto the line $\mathcal{L}$ described in Section 6.10. Let d be the projection of $\mathbf{x}$ onto $\mathcal{L}$. Based on the projected data, compute the likelihood of d based on the projected data from group 1 using some kernel density estimator, yielding, say $\hat{\ell}_1$. Here the adaptive kernel density estimator in Section 3.2.4 is used. Do the same for the projected data from group 2, yielding ℓ_2. Classify $\mathbf{x}$ as coming from group 1, if $\hat{\ell}_1 > \hat{\ell}_2$. Otherwise, classify it as coming from group 2. When the distributions for the two groups differ in location only, limited simulations indicate that this approach can compete well with all of the methods described in this section when p is small. It is unknown how well it performs when p is relatively large. Even when p is small, an extensive study of the relative merits of this approach has not been made.

Smoother for Binary Data

One final method is to use the smoother in Section 11.5.8 to estimate the probability that an observation came from the jth group and classify it as coming from the group for which this probability is greater than 0.5. This might perform relatively well when p is reasonably small. For high dimensions, presumably, it can be unsatisfactory due to the curse of dimensionality described in Section 11.5.10. A related approach is to use a logistic regression model described in Section 10.16.

6.16.1 Some Issues Related to Error Rates

Typically, two types of errors are a concern when using any classification technique: false positive (FP) and false negative (FN). Consider, for example, the goal of predicting a foot fracture among college athletes during the next year. The first group does not experience a fracture, but for some individuals, a classification method predicts a fracture. This is a false positive. Similarly, among athletes that develop a fracture, a classification method might predict no fracture for some, which is a false negative. A basic issue is estimating the probability of making an FP or FN error.

Let n_j denote the number of points among the training data that are associated with the jth group. Another issue is dealing with the fact that the extent to which n_1 and n_2 differ can impact the estimate of the decision rule. One approach is to ignore this issue and use one of the classification methods summarized above. Another approach is to estimate the decision rule in a manner that mimics the situation where $n_1 = n_2$. As will be illustrated, the choice between these two approaches depends on the situation.

To elaborate on the impact of $n_1 \neq n_2$, let π_j denote the probability that a randomly sampled point $\mathbf{x}$ belongs to the jth group, and let $f_j(\mathbf{x})$ denote the corresponding probability density function. Let $Y = j$ indicate the event that an observation belongs to the jth group. Bayes' theorem says that

$$P(Y = j | \mathbf{x}) = \frac{\pi_j f_j(\mathbf{x})}{\pi_1 f_1(\mathbf{x}) + \pi_2 f_2(\mathbf{x})}. \tag{6.35}$$

Suppose that $\mathbf{x}$ is classified as coming from the jth group when $P(Y = j | \mathbf{x}) > 0.5$. Further suppose that, based on a random sample of size n, π_j is estimated with n_j/n, the proportion of vectors belonging to the jth group. Note that if the two groups have identical distributions and $n_1 = n_2$, then $p^- = p^+ = 0.5$, where p^- and p^+ denote the probability of FN and FP, respectively. But when $n_1 \neq n_2$, $\mathbf{x}$ is more likely to be classified as coming from the group with the largest sample size. For example, if $n_1 = 900$ and $n_2 = 100$, there is a 0.9 probability that $\mathbf{x}$ will be classified as coming from group 1. And of course, there is the complication that n_j/n might be a meaningless estimate of π_j. For instance, one might have access to $n_2 = 50$ individuals who develop a medical disorder by the age of 25, and a group of $n_1 = 25$ individuals who do not. But in the population as a whole, less than 5% of all individuals will develop the disorder.

Now consider any classification method. If the two groups have identical distributions, one might expect to have $p^- = p^+ = 0.5$. This is exactly what happens when $n_1 = n_2$. But when $n_1 \neq n_2$, this is not what happens when using certain classification techniques. To be concrete, suppose $n_1 = 100$ and $n_2 = 200$. Then there is a tendency for some classification methods to assign future points to group 2 over group 1. That is, there is a kind of classification bias.

When using SVM, for example, and both groups have identical distributions, the likelihood of assigning a future observation to group 2 is about twice as high as assigning it to group 1. Moreover, the tendency to classify a future observation into group 2 can continue to be higher when the distributions differ. Methods for which the two error rates remain about the same when $n_1 \neq n_2$, and the groups have identical distributions, include methods C1 and C2. SVM and C4, and to some extent C3, do not have this property.

Momentarily assume that if $n_1 \neq n_2$, it is desired to have $p^- = p^+ = 0.5$ when the two groups have identical distributions. More broadly, suppose that an adjusted estimate of the classification rule is desired that mimics the situation where $n_1 = n_2$. One strategy is to fill in values for the group with the smallest sample size so that now the sample sizes are equal. The synthetic minority oversampling technique (SMOTE) is a well-known method for implementing this approach (Chawla et al., 2002; Akbani et al., 2004). However, all indications are that an alternative approach is more effective: a balanced version of bootstrap bagging.

For the situation at hand, a balanced version of bootstrap bagging is applied as follows. Take a bootstrap sample of size $n_{\min} = \min(n_1, n_2)$ from the jth group of the training data. So the strategy is in part to mimic the situation where training data consist of the same sample size for each group. Estimate the decision rule, and suppose that for the test data, $g = j$ indicates that $\mathbf{x}$ is classified as coming from group j ($j = 1, 2$). Repeat this process B times, yielding $g_1, \ldots, g_B$. The final decision is to classify $\mathbf{x}$ as coming from group 1, if $f_1 > f_2$, where f_j is the frequency of $g = j$; otherwise, classify $\mathbf{x}$ as coming from group 2. This will be called a balanced bootstrap estimate of the decision rule.

As will be illustrated in the next section, there can be some negative consequences to using a balanced estimate of the decision rule when $n_1 \neq n_2$. This stems from the fact that the relative seriousness of FN versus FP can depend on the situation. For example, a false positive might result in additional medical tests that are relatively inexpensive and non-invasive. But a false negative could result in the further development of a serious medical condition that is life threatening. If training data consist of a preponderance of individuals who do not have the medical disorder under investigation, an adjusted classification rule might make more FN decisions compared to an unadjusted decision rule. That is, it seems prudent to estimate error rates, as described next, using both approaches. Of course, another possibility is to adjust the decision rule. For example, rather than decide $\mathbf{x}$ came from group j, when $P(Y = j|\mathbf{x}) > 0.5$, classify $\mathbf{x}$ as coming from group j, if $P(Y = j|\mathbf{x}) > 0.3$. Generally, bias can be a complex issue that can require close collaboration with individuals who have a deep understanding of relevant clinical issues. An editorial by O'Reilly-Shah et al. (2020) highlights this point when dealing with anesthesia.

Estimating and Comparing Error Rates

A simple method for estimating the probability of making an error is cross-validation. Consider some training data based on a sample of size n_j from the jth group for a total of $n = n_1 + n_2$ vectors of length p. Rather than estimate a classification rule based on all of the training data, imagine that $m_j < n_j$ vectors, randomly sampled without replacement from the training data, are used to determine the classification rule. Then the remaining $n_j - m_j$ vectors can be used as the test data, and because the correct classification is known, one can estimate the probability of FN and FP with $\hat{p}^-$ and $\hat{p}^+$, respectively. Of course, the estimate depends on which vectors are chosen as the training data. To deal with this, imagine that the process just described is repeated D times, yielding $\hat{p}_1^-, \ldots, \hat{p}_D^-$ and $\hat{p}_1^+, \ldots, \hat{p}_D^+$. Then improved estimates are

$$\tilde{p}^- = \frac{1}{D} \sum \hat{p}_d^- \tag{6.36}$$

and

$$\tilde{p}^+ = \frac{1}{D} \sum \hat{p}_d^+. \tag{6.37}$$

As is evident, this process can be applied using two or more methods, which provides some information about their relative merits.

The method just described will be called a bagged estimate of the error rate, which is almost the same as bootstrap bagging. The only difference is that here, the m_j vectors are chosen by random sampling without replacement, while bootstrap samples are generated with replacement. A related strategy that is based on bootstrap samples is the 0.632 estimator described in Section 11.10.3. The relative merits of these two methods, for the situation at hand, appear to be unknown.

Here, two ways of choosing m_j are considered. The first is $m_j = \kappa n_j$, for some κ, $0 < \kappa < 1$. The second is $m_1 = m_2 = \kappa \min(n_1, n_2)$. The former ignores any issues regarding classification bias, when $n_1 \neq n_2$, while the latter deals with it by using training data with equal sample sizes for each of the two groups.

6.16.2 R Functions CLASS.fun, CLASS.bag, class.error.com, and menES

Many of the numerous classification methods previously described can be applied via various R packages. These R packages have very similar features, but they differ enough so as to make comparisons far from convenient. To deal with this, the R function

CLASS.fun(train = NULL, test = NULL, g = NULL, x1 = NULL, x2 = NULL, method = c('KNN', 'DIS', 'DEP', 'SVM', 'RF', 'NN', 'ADA, 'PRO','LSM'), depthfun = prodepth, kernel = 'radial', baselearner = 'btree', ...)

is supplied, making it easy to apply any one of several techniques. There are two ways of entering the data. The first is to store the training and test data in the arguments train and test, followed by a vector, stored in argument g, indicating whether the corresponding training data belong to group 1 or 2. The second is to store the training data for group 1 in argument x1 and the training data for group 2 in argument x2. Now argument g is ignored. The methods available are:

- KNN: method C3,
- DIS: method C1,
- DEP: method C2,
- SVM: support vector machine, requires the R package e1071,
- RF: random forest, requires the R package randomForest,
- NN: neural network, requires the R package neuralnet,
- ADA: adaboost, requires the R package mboost,
- PRO: method C4,
- LSM: a smoother for binary data; setting the argument sm=FALSE, logistic regression is used.

When using SVM, the argument kernel indicates the kernel to be used, which defaults to radial. Other options are linear, polynomial, and sigmoid. The argument baselearner is relevant to adaboost. For information about this argument, use the R commands library(mboost) and ?mboost. The R function

CLASS.BAG(train = NULL, test = NULL, g = NULL, x1 = NULL, x2 = NULL, SEED = TRUE, kernel = "radial", nboot = 100, method = c('KNN', 'SVM', 'DIS', 'DEP', 'PRO', 'NN', 'RF', 'ADA', 'LSM'), depthfun = prodepth, baselearner = 'bbs', sm = TRUE, rule = 0.5, ...)

is like CLASS.fun, only bagging is used. That is, it uses the resampling technique described in the previous section that mimics the situation where $n_1 = n_2$, when in fact, $n_1 \neq n_2$.

The R function

class.error.com(x1=NULL, x2=NULL,train=NULL, g=NULL, method=NULL, pro.p=0.8, nboot=100, EN=FALSE, SEED=TRUE,...)

estimates error rates using cross-validation. By default, error rates are estimated for all nine methods used by CLASS.fun and CLASS.BAG. Execution time when using bagging, method ADA, can be relatively high. To compare a subset of these methods, use the argument method to indicate the methods of interest. For example, method=('KNN','DIS') would compare KNN (method C3) and DIS (method C1). The argument pro.p controls the proportion of the data that will be used as the training data. By default, pro.p=0.8, meaning that when EN=TRUE, $m_1 = m_2 = 0.8 \min(n_1, n_2)$ will be used for the training data. This corresponds to using CLASS.BAG when classifying future observations. The default is EN=FALSE, meaning that $m_j = 0.8n_j$ is used, which corresponds to using CLASS.fun when classifying future observations. The argument nboot corresponds to the value of D in Eqs. (6.36) and (6.37).

The function returns an estimate of the probability of FP (a false positive), which corresponds to classifying a point in the first group, the data stored in x1, as coming from the second group, the data stored in x2. It also returns estimates of the probability of a false negative, the probability that a point in x2 comes from the first group, as well as the probability of making either type of misclassification.

The R function

$$\text{manES(x1, x2, method=NULL, pro.p=0.8, nboot=100, SEED=TRUE,...)}$$

estimates prediction errors via the function class.error.com and reports which method provided the highest estimated probability of a correct decision. The estimate is returned as well. By default, this is done for all methods excluding method ADA. ADA was excluded merely to reduce execution time. It can be included via the argument method. This function might be useful for measuring effect size when comparing two independent groups.

■ Example

To illustrate the R functions when $p = 2$ and the impact of using EN=TRUE versus EN=FALSE, training data were generated for the first group where $n_1 = 50$ observations were generated from a skewed, light-tailed distribution that was shifted to have a median equal to one. For the other class, $n_2 = 100$ observation were generated from a bivariate normal distribution. For both groups, Pearson's correlation is $\rho = 0$. Then some test data were generated for both groups. This was done with the following R commands:

```
set.seed(30)
x1=rmul(50,g=0.5)+1
x2=rmul(100)
```

```
test1=rmul(30,g=0.5)+1
test2=rmul(30)
```

∎

First focus on estimating error rates based on the training data in x1 and x2. The R command class.error.com(x1=x1,x2=x2,EN=TRUE) returns

```
$Error.rates
          KNN        DIS        DEP        SVM         RF         NN        ADA
TE 0.3759375 0.4453125 0.3537500 0.2262500 0.2321875 0.2753125 0.2056250
FP 0.4383333 0.3266667 0.4216667 0.3600000 0.2833333 0.3483333 0.3616667
FN 0.3615385 0.4726923 0.3380769 0.1953846 0.2203846 0.2584615 0.1696154
          PRO        LSM
TE 0.2693750 0.1906250
FP 0.3433333 0.3850000
FN 0.2523077 0.1457692
```

which are estimates based on $m_1 = m_2 = 0.8 \min(n_1, n_2)$ when sampling vectors to serve as the training data. The R command class.error.com(x1=x1,x2=x2,EN=FALSE) returns

```
$Error.rates
         KNN       DIS       DEP       SVM        RF        NN       ADA
TE 0.351875 0.3850000 0.3681250 0.2075000 0.2037500 0.2525000 0.1706250
FP 0.585000 0.3766667 0.3766667 0.5016667 0.3566667 0.4583333 0.3866667
FN 0.212000 0.3900000 0.3630000 0.0310000 0.1120000 0.1290000 0.0410000
         PRO       LSM
TE 0.2856250 0.2006250
FP 0.3433333 0.5216667
FN 0.2510000 0.0080000
```

which are estimates based on unbalanced training data ($m_j = 0.8n_j$). The estimate of the overall error rate corresponds to the first row labeled, TE.

If the goal is to minimize the FN probability, the results suggest using LSM (a smoother designed for binary outcomes) in conjunction with an unbalanced estimate of the classification rule. That is, $m_j = 0.8n_j$ was used when doing cross-validation, and CLASS.fun would be used to classify any future points. This comes at the price of a relatively high rate for FP, but as previously noted, the seriousness of making an FP decision might be considerably less than that of making an FN decision. Here, to minimize the FP rate, the results suggest using RF (random forest) with a balanced estimate of the decision rule, meaning that $m_1 = m_2 = 0.8 \min(n_1, n_2)$. To minimize the overall error rate, ADA (adaboost) appears best in conjunction with a unbalanced estimate.

As a partial check on these results, now consider the test data that were generated. First focus on FN, which refers to the test data stored in test2 as coming from group 1. Using the unadjusted estimate of the classification rule in conjunction with LSM, via the R command CLASS.fun(x1=x1,x2=x2, test=test2, method='LSM'), 25 of the 30 points are correctly classified, 5 are not. Using instead an adjusted estimate of the decision rule with the command CLASS.BAG(x1=x1, x2=x2, test=test2 ,method='LSM'), only 18 points are correctly classified. Using RF, the corresponding correct classifications are 23 and 19. The results suggest that LSM is only slightly better than SVM. For the unadjusted estimate of the classification rule, number of correct classifications is again 25 using SVM. As for FP, classifying data in test1 as coming from group 2, 20 correct decisions are made with LSM and 19 with SVM. But the likelihood that the function class.error.com correctly identifies the method with the lowest error rate is unknown.

Finally, the R package ROCR (Sing et al., 2005) provides access to a collection of other methods for summarizing the performance of classification methods. Included is a function that plots the receiver operating characteristic (ROC) curve. It plots true positive (negative) rates versus false positive (negative) rates as a function of the threshold used to classify points. The R function ROCmul.curve, written for this book, provides a simple way of plotting this curve, via ROCR, for any of the classification methods described here. Multiple curves can be included in the plot.

6.17 Exercises

1. For the EEG data mentioned in Section 6.3.5, compute the MVE, MCD, OP, and the Donoho–Gasko 0.2 trimmed mean for group 1. This corresponds to the first four measures, which are based on murderers.
2. Repeat the last exercise using the data for group 2. These are the measures in the last four columns.
3. For the data used in the last two exercises, check for outliers among the first group using the methods in Section 6.4. Comment on why the number of outliers found differs among the methods.
4. Repeat the last exercise using the data for group 2.
5. Repeat the last two exercises, but now use the cork data mentioned in Section 6.7.3.
6. Suppose that for each row of an n-by-p matrix, its depth is computed relative to all n points in the matrix. What are the possible values that the depths might be?

7. Give a general description of a situation where for $n = 20$, the minimum depth among all points is 3/20.

8. The average LSAT scores (X) for the 1973 entering classes of 15 American law schools and the corresponding grade point averages (Y) are as follows.

   X: 576 635 558 578 666 580 555 661 651 605 653 575 545 572 594
   Y: 3.39 3.30 2.81 3.03 3.44 3.07 3.00 3.43 3.36 3.13 3.12 2.74 2.76 2.88 2.96

   Use a boxplot to determine whether any of the X values are outliers. Do the same for the Y values. Comment on whether this is convincing evidence that there are no outliers. Check for outliers using the MVE, MCD, and projection-type methods described in Section 6.4. Comment on the different results.

9. The MVE method of detecting outliers, described in Section 6.4.3, could be modified by replacing the MVE estimator of location with the Winsorized mean, and replacing the covariances with the Winsorized covariances described in Section 5.9.3. Discuss how this would be done and its relative merits.

10. The file read.dat contains data from a reading study conducted by L. Doi. Columns 4 and 5 contain measures of digit naming speed and letter naming speed. Use both the relplot and the MVE method to identify any outliers. Compare the results and comment on any discrepancies.

11. For the cork boring data mentioned in Section 5.9.10, imagine that the goal is to compare the north, east, and south sides to the west side. How might this be done with the software in Section 6.6.1? Perform the analysis and comment on the results. (The data are stored in the file corkall.dat; see Section 1.10.)

12. For the EEG data used in Exercise 1, compare the two groups with the method in Section 6.7.

13. For the EEG data, compare the two groups with the method in Section 6.9.

14. For the EEG data, compare the two groups with the method in Section 6.10.

15. For the EEG data, compare the two groups with the method in Section 6.11.

16. Argue that when testing Eq. (6.27), this provides a metric-free method for comparing groups based on scatter.

17. The goal is to run simulations to compare the mean squared error and bias of two estimators designed for functional data. Here is some R code for generating data according to a Gaussian process:

```
C=FALSEunction(x, y)exp(-1*abs(x - y))  # Covariances
# Another choice for the covariances:
#  C <- function(x, y) 0.01 * exp(-10000 * x - y)^2
M=FALSEunction(x) sin(x) # The mean for the curves
n=50 # sample size
```

```
p=100 # number of time points
x=r.gauss.pro(n,C=C,M=M,t=x)   # generates data from a Gaussian process
```

Use simulations to compare how well the R functions medcurve and Flplot estimate the typical curve given by M when data are generated from a Gaussian process.

18. This exercise deals with the likelihood of a correct decision when using classification method C3 in Section 6.16. Imagine that training data contain 100 individuals from the first of two groups and 200 from the other. Further imagine that the test data consist of 10 individuals from the first of two groups and 20 from the second. Here is some R code that models this situation when using method KNN or the bagged version of KNN:

set.seed(30)

v1sum=NA

v2sum=NA

for(i in 1:100){

x1=rmul(100)+1

x2=rmul(200)

test=rmul(30)

test[1:10,]=test[1:10]+1

v1=CLASS.fun(x1=x1,x2=x2,test=test,method='DIS')

v2=CLASS.fun(x1=x1,x2=x2,test=test,method='SVM')

v1sum[i]=sum(v1[1:10]==1)

v2sum[i]=sum(v2[1:10]==1)

}.

In terms of correct classification to the first group, the best method is the one for which the mean number of correction classifications is closest to 5. Speculate which method performed best, and check your choice based on v1sum and v2sum. Why was this expected?

## CHAPTER 7

One-Way and Higher Designs for Independent Groups

This chapter describes techniques for testing hypotheses in one-way and higher designs involving independent groups. Included are random effects models plus methods for performing multiple comparisons. This chapter makes no attempt at covering all the designs that are encountered in practice, but it does cover many of the more common designs that are employed. (For comparisons of several non-parametric methods, including some permutation techniques, when dealing with a two-way design, see Harrar et al., 2018.)

In this chapter, only heteroscedastic methods are considered. It might be hoped that as the number of groups increases, problems associated with homoscedastic methods, described in Chapter 5, might be reduced, but there are compelling indications that this is not the case. For example, even under normality with equal sample sizes but unequal variances, problems controlling the probability of a Type I error can arise. With four independent groups, each having 50 observations, the usual analysis of variance F test of

$$H_0: \mu_1 = \mu_2 = \mu_3 = \mu_4$$

can have a Type I error probability approximately equal to 0.09 when testing at the 0.05 level (Wilcox et al., 1986). With unequal sample sizes, the actual probability of a Type I error can exceed 0.3. Under non-normality, control over the probability of a Type I error is even worse. Practical problems with more complicated designs have been found (e.g., Keselman et al., 1995).

It is sometimes suggested that one test the homoscedasticity assumption and, if an appropriate test fails to reject, use a method that assumes equal variances among the groups. However, published papers do not support this strategy (e.g., Hayes and Cai, 2007; Markowski and Markowski, 1990; Moser et al., 1989; Wilcox et al., 1986; Zimmerman, 2004). As noted at the end of Section 5.2 when dealing with the two-sample case, the problem is that tests of the homoscedasticity assumption might not have enough power to detect situations where a violation of the homoscedasticity assumption creates practical problems. Assuming normality, Kim and Cribbie (2017) note that the situation can be improved by testing instead the hypothesis that the variances do not differ substantially.

Introduction to Robust Estimation and Hypothesis Testing
https://doi.org/10.1016/B978-0-12-820098-8.00013-0

One might try to salvage homoscedastic methods by arguing that if the variances are unequal, the means are unequal as well, in which case, a Type I error is not a concern. However, an inability to control the probability of a Type I error often reflects an undesirable power property: The probability of rejecting the null hypothesis is not minimized when the null hypothesis is true. (The hypothesis testing method is biased.) This problem was already pointed out and illustrated in Chapter 5, where shifting one group by a half standard deviation results in a situation where the probability of rejecting is less compared to the situation where H_0 is true. Here, it is merely noted that this problem persists when comparing more than two groups (e.g., Wilcox, 1996a).

Another argument in support of the F test is that it is reasonably good at controlling the probability of a Type I error when distributions are identical. (Tan, 1982, reviews the relevant literature.) That is, it provides a test of the hypothesis that J groups have identical distributions. If the F test is significant, a reasonable argument is that the means differ. However, if the goal is to derive a test that is exclusively sensitive to some measure of location, the F test is unsatisfactory. Even if this problem can be ignored, low power, due to the low efficiency of the sample mean when distributions have heavy tails, remains a concern.

7.1 Trimmed Means and a One-Way Design

From a technical point of view, it is a simple matter to extend the methods in Chapter 5 to situations where the goal is to compare the trimmed means of more than two groups: simply select a heteroscedastic method for means, and then proceed along the lines used to derive the Yuen–Welch test. In essence, replace the sample means by trimmed means, replace estimates of the standard errors with appropriate estimates based on the amount of trimming used, and adjust the degrees of freedom based in part on the number of observations left after trimming. For a one-way design, however, there are many heteroscedastic methods for comparing means, so it is not immediately obvious which to use. Two possibilities are described here, both of which have been examined in simulation studies and found to give relatively good control over the probability of a Type I error. Other methods have been considered by Lix and Keselman (1998) as well as Luh and Guo (1999). For a method based on trimmed means that assumes equal variances, see Lee and Fung (1985). For the special case where the goal is to compare means, Krishnamoorthy et al. (2007) review approaches for handling unequal variances, which are known to be unsatisfactory, and they suggest using instead a parametric bootstrap method that assumes normality. Cribbie et al. (2012) compare this parametric bootstrap method to several other techniques and find it to be unsatisfactory. The actual Type I error probability, when testing at the 0.05 level, can exceed 0.25. A method that performed reasonably well was the Welch-type method for comparing trimmed means, which is described in the next section.

7.1.1 A Welch-Type Procedure and a Robust Measure of Effect Size

The goal is to test

$$H_0: \mu_{t1} = \cdots = \mu_{tJ}, \tag{7.1}$$

where μ_{tj}, $j = 1, \ldots, J$, are the trimmed means corresponding to J independent groups. Table 7.1 describes a method derived by Wilcox (1995e) for testing this hypothesis that reduces to Welch's (1951) adjusted degrees of freedom method for means when there is no trimming. (For a variation of this method, assuming sampling from a symmetric distribution, see Kulinskaya and Dollinger, 2007. When there is no trimming, Celik, 2020, derives a method assuming sampling from a skewed t distribution.) With no trimming, Welch's method can perform poorly in terms of controlling the Type I error probability, even when distributions do not differ in terms of skewness (e.g., Parra-Frutos and Molera, 2019). A criticism of testing (7.1) is that surely the trimmed means differ at some decimal place. As noted in Sections 4.1.1 and 5.3, an argument can be made that a more meaningful goal is to determine whether a decision can be made about whether μ_{tj} is greater or less than μ_{tk} for any $j \neq k$. If this argument is accepted, testing (7.1) can play a useful role via the step-down multiple-comparison procedure described at the end of Section 7.4.1.

Robust, Heteroscedastic Measures of Effect Size

When dealing with more than two groups, one approach to measuring effect size is to use one of the measures in Section 5.3.4 for each pair of groups. This can be done with the R function IND.PAIR.ES, described in Section 7.4.2. If the goal is to use some global measure of effect size, the robust explanatory measure of effect size, described in Section 5.3.4, is readily extended to more than two groups. For simplicity, first consider the situation where means are compared. For equal sample sizes, let $\sigma^2(Y)$ be the estimand corresponding to

$$\hat{\sigma}^2(Y) = \frac{1}{N-1} \sum_{j=1}^{J} \sum_{i=1}^{n} (Y_{ij} - \bar{Y})^2,$$

where $\bar{Y} = \sum \sum Y_{ij}/N$ and $N = nJ$ is the total sample size. Adopting a regression perspective, and given that an observation is randomly sampled from the jth group, the predicted value is μ_j. Let

$$\sigma^2(\hat{Y}) = \frac{1}{J-1} \sum_{j=1}^{J} (\mu_j - \bar{\mu})^2,$$

Table 7.1: Computations for comparing trimmed means.

The goal is to test

$$H_0: \mu_{t1} = \cdots = \mu_{tJ}.$$

For the jth group, let

$$d_j = \frac{(n_j - 1)s_{wj}^2}{h_j \times (h_j - 1)},$$

where h_j is the effective sample size of the jth group (the number of observations left after trimming) and s_{wj}^2 is the Winsorized variance. To test H_0, compute

$$w_j = \frac{1}{d_j},$$

$$U = \sum w_j,$$

$$\tilde{X} = \frac{1}{U} \sum w_j \bar{X}_{tj},$$

$$A = \frac{1}{J - 1} \sum w_j (\bar{X}_{tj} - \tilde{X})^2,$$

$$B = \frac{2(J - 2)}{J^2 - 1} \sum \frac{(1 - \frac{w_j}{U})^2}{h_j - 1},$$

$$F_t = \frac{A}{1 + B}.$$

When the null hypothesis is true, F_t has, approximately, an F distribution with degrees of freedom

$$\nu_1 = J - 1,$$

$$\nu_2 = \left[\frac{3}{J^2 - 1} \sum \frac{(1 - w_j/U)^2}{h_j - 1} \right]^{-1}.$$

where $\bar{\mu} = \sum \mu_j / J$ is the grand mean. The *explanatory measure of effect size* is

$$\xi = \sqrt{\frac{\sigma^2(\hat{Y})}{\sigma^2(Y)}}.$$

Letting

$$\hat{\sigma}^2(\hat{Y}) = \frac{1}{J - 1} \sum_{j=1}^{J} (\bar{Y}_j - \bar{Y})^2,$$

an estimate of ξ is

$$\hat{\xi} = \sqrt{\frac{\hat{\sigma}^2(\hat{Y})}{\hat{\sigma}^2(Y)}}.$$

If the mean and variance are replaced by a trimmed mean and Winsorized variance (scaled to estimate the variance under normality), the resulting estimate of $\hat{\xi}$ can exceed 1 when there are $J > 2$ groups and the amount of trimming is greater than 0.

For unequal sample sizes, let m denote the smallest sample size among the J groups. Randomly sample (without replacement) m observations from each of the groups for which $m < n_j$. Based on the resulting samples of m observations from each group, compute $\hat{\xi}^2$ as just described. Repeat this process K times, yielding a series of estimates for ξ^2, which are then averaged to get a final estimate, which we label $\hat{\xi}^2$. The estimate of ξ is taken to be $\sqrt{\hat{\xi}^2}$.

Another approach is to use the projection distance (as described in Section 6.2.5) of the measures of location from the grand mean. When all J groups have a common sample size, estimating this measures of effect size can be done as described in Section 8.1.1. For unequal sample sizes, again, let m denote the smallest sample size. Next, permute the values in each group and estimate the projection-type measure effect size based on the first m observations from each group. Repeat this B times, and use the mean of these value as the final measure of effect size. For properties of this measure of effect size, see Section 8.1.1.

Yet one more approach is to use a robust analog of the measure of effect size proposed by Kulinskaya and Staudte (2006). Let $q_j = n_j/N$, where $N = \sum n_j$. The measure of effect size is estimated with

$$\frac{J}{2} \sqrt{\sum \frac{(\bar{X}_{tj} - \tilde{X})^2}{s^2_{jwN}}}, \tag{7.2}$$

where $\tilde{X} = \sum q_j \bar{X}_{tj}$ and s^2_{jwN} is the Winsorized variance of the jth group rescaled to estimate the variance when dealing with a normal distribution. This measure of effect size is very similar to the KMS measure of effect size in Chapter 5 when $J = 2$. (The expression under the radical, with no trimming, is the measure of effect size used by Kulinskaya and Staudte when $J > 2$.) Consider the case where $J - 1$ of the groups have a common trimmed mean and the Jth group has a trimmed mean that is larger than the other $J - 1$ trimmed means. If the term $J/2$ is excluded, this measure of effect size decreases as J increases.

7.1.2 R Functions t1way, t1wayv2, t1way.EXES.ci, KS.ANOVA.ES, fac2list, t1wayF, and ESprodis

The R function

$$t1way(x, tr=0.2, grp=NA)$$

performs the calculations in Table 7.1. The data can be stored in any R variable having list mode, or they can be stored in a matrix or a data frame. If x is a matrix or data frame, it is assumed that the columns correspond to groups. So the data for group 1 are stored in column 1, and so on.

It is noted that when using the R package WRS2, rather than WRS or Rallfun, t1way uses the formula convention:

$$t1way(x \sim g, tr=0.2, grp=NA),$$

where g is a factor variable. Also see Villacorta (2017). The R function

$$t1wayv2(x, tr=0.2, grp=NA)$$

is the same as t1way, only the measure of effect size ξ is reported as well. The R function

$$t1way.EXES.ci(x, alpha=0.05, tr=0.2, nboot=500, SEED=TRUE)$$

computes a confidence interval for ξ using a percentile bootstrap method. If t1way does not reject the null hypothesis, the lower end of the confidence interval is set equal to zero.

The R function

$$KS.ANOVA.ES(x, tr=0.2)$$

computes the measure of effect size given by Eq. (7.2). Methods for computing a confidence interval have not yet been studied.

Although familiarity with R is assumed, a brief description of *list mode* is provided in case it helps. List mode is a convenient way of storing data corresponding to several groups under one variable name. It readily accommodates situations where groups have unequal sample sizes. For example, suppose two groups are to be compared, and the data for the two groups are stored in the R vectors x and y having lengths 20 and 30, respectively. The command

$$w=list(x,y)$$

creates a variable, called w, that has list mode. The R variable w[[1]] contains the data stored in x, and w[[2]] contains the data stored in y. Notice the use of the double brackets. In particular, w[[1]] is a vector of observations corresponding to the first group. More generally, in terms of a one-way design, w[[j]] contains the data for the jth group. (For more details about list mode, see the books on R mentioned in Section 1.8.)

The second argument in t1way, tr, indicates the amount of trimming, which defaults to 0.2 (20% trimming). Thus, the command t1way(w) results in a test of the hypothesis that the 20% trimmed means are equal. To compare 10% trimmed means, use the command t1way(w,tr=0.1).

The third argument in t1way, grp, can be used to specify some subset of the populations to be compared. If not specified, all of the groups are used. If, for example, there are four groups, but the goal is to compare groups 1, 2, and 4, ignoring group 3, the command t1way(w,grp=c(1,2,4)) will test the hypothesis H_0: $\mu_{t1} = \mu_{t2} = \mu_{t4}$ using 20% trimmed means. The command t1way(w, 0.1,grp=c(1,4)) would compare the 10% trimmed means of groups 1 and 4.

■ **Example**

Suppose that for three independent groups, the observations are

Group 1:   1, 2, 3, 4, 5, 6, 7, 8, 9, 10,
Group 2:   2, 3, 4, 5, 6, 7, 8, 9, 10, 11,
Group 3:   5, 6, 7, 8, 9, 10, 11, 12, 13, 14.

If the data are stored in the R variable w, in list mode, the command t1way(w,tr=0) tests the hypothesis of equal means. The function returns

```
$TEST:
[1] 4.558442

$nu1:
[1] 2

$nu2:
[1] 18
```

```
$p.value:
[1]   0.02502042
```

In particular, $F_t = 4.56$, with a p-value of 0.025. The command t1way(w,tr=0,c(1,3)) compares the means of groups 1 and 3 and reports a p-value of 0.008. The command t1way(w,grp=c(1,3)) compares the 20% trimmed means of groups 1 and 3 and reports a p-value of 0.039.

■

Note that the third group has the largest sample mean, which is equal to 9.5. Increasing the largest observation in the third group to 40, the sample mean increases to 12.1, suggesting that there is now more evidence that the groups differ, but the p-value *increases* to 0.17. The reason is that the standard error of the sample mean also increases.

Data Management

It is common to have data stored in a matrix or data frame where one of the columns contains the outcome variable of interest and another column indicates the level (group identification) of the factor being studied. Consider, for example, data dealing with plasma retinol, which was downloaded from a web site maintained by Carnegie Mellon University. For illustrative purposes, it is assumed that the data have been stored in the R variable plasma as a data frame. The variable names are:

```
1       AGE: Age (years)
2       SEX: Sex (1=Male, 2=Female).
3       SMOKSTAT: Smoking status (1=Never, 2=Former, 3=Current Smoker)
4       QUETELET: Quetelet (weight/(height^2))
5       VITUSE: Vitamin Use (1=Yes, fairly often, 2=Yes, not often, 3=No)
6       CALORIES: Number of calories consumed per day.
7       FAT: Grams of fat consumed per day.
8       FIBER: Grams of fiber consumed per day.
9       ALCOHOL: Number of alcoholic drinks consumed per week.
10      CHOLESTEROL: Cholesterol consumed (mg per day).
11      BETADIET: Dietary beta-carotene consumed (mcg per day).
12      RETDIET: Dietary retinol consumed (mcg per day)
13      BETAPLASMA: Plasma beta-carotene (ng/ml)
14      RETPLASMA: Plasma Retinol (ng/ml)
```

The first few lines of the data set look like this:

```
64   2   2  21.48380   1  1298.8 57.0   6.3    0.0  170.3 1945 890 200 915
76   2   1  23.87631   1  1032.5 50.1  15.8    0.0   75.8 2653 451 124 727
38   2   2  20.01080   2  2372.3 83.6  19.1   14.1  257.9 6321 660 328 721
40   2   2  25.14062   3  2449.5 97.5  26.5    0.5  332.6 1061 864 153 615
```

Now, imagine that the goal is to compare the three groups based on smoking status, which is indicated in column 3, in terms of plasma beta-carotene, which is stored in column 13. To use the R function t1way, it is necessary to sort the data in column 13 into three groups based on the values stored in column 3. This can be done with the R function

$$\text{fac2list(x,g),}$$

where the argument x is an R variable, usually some column of a matrix or column of a data frame, containing the data to be analyzed (the dependent variable), and g is a column of data indicating the group to which a corresponding value, stored in x, belongs. (When working with a data frame, this latter column of data can be a factor variable.) The output from fac2list is an R variable having list mode. If g contains numeric data, the groups are put in ascending order based on the values in g. If g contains character data, the data are sorted into groups in alphabetical order.

■ **Example**

For the plasma retinol data, imagine the goal is to compare the 20% trimmed means corresponding to the three smoking status groups. The outcome measure of interest is plasma beta-carotene. The groups can be compared using the R commands

$$\text{z=fac2list(plasma[,13],plasma[,3])}$$
$$\text{t1way(z).}$$

The first command sorts the data stored in plasma[,13] into groups based on the values stored in plasma[,3], and it stores the data in the R variable z having list mode. The data stored in plasma[,3] have one of three values: 1, 2, and 3. So z[[1]] contains the data for the first group, z[[2]] contains the data for the second, and z[[3]] the data for the third. If instead, plasma[,3] contained one of three character strings, say "N," "Q," and "S," the data in z would be sorted alphabetically. So now z[[1]] would contain plasma retinol measures for participants designated by "N," z[[2]] would contain plasma retinol measures for participants designated by "Q," and z[[3]] would contain plasma retinol measures for participants designated by "S."

∎

The R function

$$t1wayF(x, fac, tr=0.2, EP = FALSE, pr = TRUE)$$

is like the R function t1way, only x is assumed to be a column of data and fac is a factor variable. That is, this function eliminates the need to use the function fac2list. If EP=TRUE, the function computes the explanatory measure of effect size.

The R function

$$ESprodis(x, est=tmean, REP=100, DIF=FALSE, SEED=TRUE, \ldots)$$

computes a measure of effect size using the projection distance between the measures of location and the grand mean.

∎ **Example**

For the last example, the analysis can be done with the single command

$$t1wayF(plasma[,13], plasma[,3]).$$

∎

7.1.3 A Generalization of Box's Method

For the jth group, again, let h_j be the number of observations left after trimming. Motivated by results in Box (1954) and Rubin (1983), Lix and Keselman (1998) considered testing Eq. (7.1), the hypothesis of equal trimmed means, with

$$F_b = \frac{\sum h_j (\bar{X}_{tj} - \bar{X}_t)^2}{\sum 1 - (h_j/H) S_j^2},$$

where $H = \sum h_j$, $\bar{X}_t = \sum h_j \bar{X}_{tj}/H$, and

$$S_j^2 = \frac{(n_j - 1)s_{wj}^2}{h_j - 1}.$$

When the null hypothesis is true, F_b has, approximately, an F distribution with

$$\hat{v}_1 = \frac{\left(\sum(1 - f_j)S_j^2\right)^2}{\left(\sum S_j^2 f_j\right)^2 + \sum_{j=1}^{J} S_j^4(1 - 2f_j)}$$

and

$$\hat{v}_2 = \frac{\left(\sum_{j=1}^{J}(1 - f_j)S_j^2\right)^2}{\sum_{j=1}^{J} S_j^4(1 - f_j)^2/(h_j - 1)}$$

degrees of freedom, where $f_j = h_j/H$. Currently, it seems that both F_t and F_b give similar protection against Type I errors, with F_b being perhaps slightly better. When there are two groups ($J = 2$), these two methods give exactly the same results. In some situations, F_b has a Type I error probability that exceeds α and is higher than the Type I error probability associated with F_t, but there are situations where the reverse is true. Among the situations considered by Lix and Keselman, F_b is less likely to result in a Type I error probability exceeding 0.075 when testing at the 0.05 level. However, F_b generally has less power.

7.1.4 R Function box1way

The R function

$$\text{box1way(x,tr=0.2,grp=NA)},$$

written for this book, performs the calculations described in the previous subsection. It is used in exactly the same manner as t1way. Thus, the command box1way(w, 0.1,c(1,3)) will test the hypothesis that the 10% trimmed means, associated with the first and third groups, are equal. When comparing only two groups, the R functions box1way, t1way, and yuen all give identical results.

■ **Example**

Suppose the data in Section 7.1.2, used to illustrate the function t1way, are stored in w. Then the command box1way(w) tests the hypothesis of equal 20% trimmed means, and the p-value is reported to be 0.077. Using t1way, the p-value is 0.025.

■

7.1.5 Comparing Medians and Other Quantiles

For the special case, where the goal is to test

$$\theta_1 = \cdots = \theta_J, \tag{7.3}$$

the hypothesis of equal medians, the Yuen–Welch and Box methods for trimmed means are not recommended; an alternative estimate of the standard error is required. Here, two methods are described for testing the hypothesis that J independent groups have a common population median. The first is based on the McKean–Schrader estimate of the standard error, which is used in conjunction with a Welch-type test.

Let M_j be the sample median for the jth group, and let S_j^2 be the McKean–Schrader estimate of the squared standard error of M_j ($j = 1, \ldots, J$). Let

$$w_j = \frac{1}{S_j^2},$$

$$U = \sum w_j,$$

$$\tilde{M} = \frac{1}{U} \sum w_j M_j,$$

$$A = \frac{1}{J-1} \sum w_j (M_j - \tilde{M})^2,$$

$$B = \frac{2(J-2)}{J^2 - 1} \sum \frac{(1 - \frac{w_j}{U})^2}{n_j - 1},$$

$$F_m = \frac{A}{1+B}. \tag{7.4}$$

The hypothesis of equal population medians is rejected, if $F_m \geq f$, the $1 - \alpha$ quantile of an F distribution with $\nu_1 = J - 1$ and $\nu_2 = \infty$ degrees of freedom. However, with small sample sizes, the actual level can be generally less than the nominal level. A better approach is to estimate an appropriate critical value via a simulation assuming normality. That is, randomly sample n_j ($j = 1, \ldots, J$) observations from a standard normal distribution, compute F_m, and repeat this process B times, yielding an approximation of the null distribution of F_m.

It is stressed, however, that all known estimates of the standard error of the sample median can be highly inaccurate when tied values occur, even with large sample sizes. Consequently, the method based on the McKean–Schrader estimator is not recommended when there are tied values. Use instead method LSB, described in Section 7.6, with the M-measures of location replaced by the Harrell–Davis estimator. Using a quantile estimator that uses a weighted average of all the order statistics appears to be crucial (Wilcox, 2015b). For continuous distributions, the method controls the Type I error probability reasonably well with $n \geq 20$. When there are tied values, $n \geq 30$ might be needed. When dealing with the quartiles, $n \geq 50$ is recommended (Wilcox, 2015b).

It is noted that another approach is to perform multiple comparisons using the R function medpb, described in Section 7.4.10, which is designed to control the probability of one or more Type I errors. A possible appeal of this approach is that when dealing with medians, the usual sample median can be used when there are tied values. It has a high breakdown point in contrast to the Harrell–Davis estimator, which has a breakdown point of only $1/n$. However, when dealing with quartiles, using an estimator that gives a positive weight to all of the order statistics, such as the Harrell–Davis estimator, appears to be essential.

7.1.6 R Functions med1way and Qanova

The R function

med1way(x, grp = NA, alpha = 0.05, crit = NA, iter = 5000, SEED = TRUE, pr = TRUE)

tests the hypothesis of equal population medians, assuming there are no tied values based on the test statistic given by Eq. (7.4). The argument x can be a matrix or have list mode. If a matrix, columns correspond to groups. The argument grp can be used to analyze a subset of the groups if desired. By default, all J groups are compared. So the command med1way(disdat,grp=c(1,3,5)) will compare groups 1, 3, and 5 using the data stored in the variable disdat. The argument iter controls how many replications are used in the simulation estimate of the critical value. The function returns the value of the test statistic, F_m, and the p-value.

The R function

Qanova(x, q =0.5, op = 3, nboot = 600, MC = FALSE, SEED = TRUE)

tests the hypothesis of equal population medians using a method that allows tied values. (This function is essentially the same as the function pbadepth in Section 7.6.1, but with the M-estimator replaced by the Harrell–Davis estimator.) The argument q determines the quantile to be used, which defaults to the median. Setting MC=TRUE, the function will take advantage of a multicore processor if one is available. (The argument op indicates how the depth of the null vector, within a bootstrap cloud of points, is measured. The default is to use projection distances.)

7.1.7 Bootstrap-t Methods

Lix and Keselman (1998) found that two other methods for testing hypotheses about trimmed means perform relatively well in terms of controlling the probability of a Type I error, but with small sample sizes and with trimming less than or equal to 20%, none of the methods they considered, including the methods described in this chapter, always guaranteed that the actual probability of a Type I error would be less than 0.075 when testing at the 0.05 level. In some situations the probability of a Type I error exceeds 0.08. It might be argued that this is satisfactory in some situations, but they did not consider situations where distributions differ in skewness. Among the non-normal distributions in their study, they only considered situations where groups have unequal variances. From Chapter 5, if distributions differ in skewness, the expectation is that control over the probability of a Type I error will be worse. Again, one might try to salvage the situation by arguing that if groups have unequal variances or skewness, surely the trimmed means differ, so the probability of a Type I error is not an issue. But as noted in Chapter 5, problems with controlling Type I error probabilities often reflect an unsatisfactory characteristic: Power can go down as the difference between the trimmed means increases, although eventually it will go up. Put another way, the probability of rejecting is not always minimized when the null hypothesis is true.

Chapter 5 describes bootstrap methods for dealing with this problem. Provided the amount of trimming is relatively low (say less than 20%), it currently seems that a bootstrap-t method is relatively effective based on the criterion of controlling the probability of a Type I error. In the present context, a bootstrap-t method refers to any bootstrap technique that is based in part on a test statistic that is a function of estimates of the standard errors of the location estimators being used. With sufficiently large sample sizes, a bootstrap method is not required, but it remains unclear how large the sample sizes must be. This subsection notes that a simple extension of the two-sample bootstrap-t method can be applied to the problem at hand. The strategy is to use the available data to estimate an appropriate critical value when using the Yuen–Welch method to compare trimmed means. Perhaps there is some practical advantage to replacing the Welch-type method with some other procedure, but this remains to be seen. (In the two-sample case, with the amount of trimming less than 20%, a generalization of the Yuen–Welch test seems to have merit; see Othman et al., 2002.)

As was done in Section 5.3.2, the method begins by obtaining a bootstrap sample from each of the J groups: $X_{1j}^*, \ldots, X_{n_j j}^*$. Next, set $C_{ij}^* = X_{ij}^* - \bar{X}_{tj}, i = 1, \ldots, n_j$. Then $C_{1j}^*, \ldots, C_{n_j j}^*$ represents a sample from a distribution that has a trimmed mean of zero, so the hypothesis of equal trimmed means among these J distributions is true. Let F_t^* be the value of F_t (described in Table 7.1), when applied to the C_{ij}^* values. Repeat this process B times, each time, obtaining bootstrap samples and computing F_t using the C_{ij}^* values that result. Label the resulting test statistics $F_{t1}^*, \ldots, F_{tB}^*$. Each time this process is applied, the null hypothesis is true, by construction, so the values $F_{t1}^*, \ldots, F_{tB}^*$ provide an estimate of an appropriate critical value. Letting $F_{t(1)}^* \leq \cdots \leq F_{t(B)}^*$ be the $F_{t1}^*, \ldots, F_{tB}^*$ values written in ascending order, an estimate of the α critical value is $F_{t(m)}^*$, where $u = (1 - \alpha)B$, rounded to the nearest integer. That is, reject the null hypothesis of equal trimmed means if F_t, computed as described in Table 7.1, is greater than or equal to $F_{t(u)}^*$.

A Method Based on Bailey's Transformation

Section 5.4.1 describes a method for comparing two independent groups based on M-estimators. The method is readily generalized to the situation at hand. Simply proceed as done in Section 5.4.1 but with the M-estimator and the estimate of the standard error replaced by a trimmed mean and an estimate of the standard error of the trimmed mean. Next, compute Bailey's transformation Z_j as described in Section 5.4.1, in which case, the test statistic is

$$B_t^2 = \sum Z_j^2. \tag{7.5}$$

A critical value is determined via a bootstrap-t method as outlined above. This approach currently seems best in terms of controlling the Type I error probability (Özdemir et al., 2018).

7.1.8 R Functions t1waybt, btrim, and t1waybtsqrk

The R function

$$\text{t1waybt(x,tr=0.2,alpha= 0.05,grp=NA,nboot=599)}$$

tests the hypothesis of equal trimmed means using the bootstrap-t method. (When using the R package WRS2, t1waybt uses a formula convention.) As with t1way and box1way, the argument x can be any R variable that is a matrix or has list mode. If unspecified, the amount of trimming defaults to tr=0.2, and the argument alpha, corresponding to α, defaults to 0.05. Again, the argument grp can be used to test the hypothesis of equal trimmed means for some subgroup of interest. If unspecified, all J groups are used. The default value for B is nboot=599, which appears to give good results, in terms of controlling the probability of a Type I error, when $\alpha = 0.05$ and $n_j \geq 10$, $j = 1, \ldots, J$. Little is known about how the method performs when $\alpha < 0.05$. Cribbie et al. (2012) found that with 20% trimming, a parametric bootstrap technique performs a bit better than the method in Section 7.1.1. Limited checks indicate that the bootstrap-t method used here is better than the parametric bootstrap method in terms of avoiding Type I error probabilities larger than the nominal level. But extensive comparisons have not been made.

■ **Example**

Again, consider the data in Section 7.1.2. Assuming the data are stored in the variable w, the command t1waybt(w) reports that the 0.05 critical value is 4.97. The value of the test statistic is 2.87, which is the same value reported by the function t1way. The 0.05 critical value used by t1way is 4.1. That is, the bootstrap-t method estimates that t1way is using a critical value that is too small.

■

To add perspective, suppose the data in Section 7.1.2 are shifted so that the trimmed mean for each group is zero. This yields

Group 1: -4.5 -3.5 -2.5 -1.5 -0.5 0.5 1.5 2.5 3.5 4.5
Group 2: -4.5 -3.5 -2.5 -1.5 -0.5 0.5 1.5 2.5 3.5 4.5
Group 3: -4.5 -3.5 -2.5 -1.5 -0.5 0.5 1.5 2.5 3.5 4.5

Now suppose that these values represent the actual distributions associated with the three groups, each value within a group having the same probability of occurring. This is the process used by the bootstrap method. In group 1, for example, there are 10 possible values, every value occurs with equal probability, so the first value, -4.5, occurs with probability 0.1, the second value, -3.5, also occurs with probability 0.1, and so on. As is evident, all three distributions happen to be identical, but in general, this will not be the case. By construction, each of the three distributions has a population trimmed mean equal to zero. Consequently, when F_t is computed using these observations, the probability of rejecting should be α. But if $n_1 = 10$ observations are randomly sampled from the first group, $n_2 = 10$ are randomly sampled from the second, and $n_3 = 10$ from the third, the actual probability of a Type I error is 0.073 when $\alpha = 0.05$, based on a simulation with 1,000 replications. Even without running a simulation, the expectation is that the Type I error probability will be higher than 0.05. The reason is that the bootstrap-t method, when applied to the data in Section 7.1.2, simply performs simulations on the distributions being considered here, and it estimates that the 0.05 critical value is 4.97. But F_t uses a critical value of 4.1 (the 0.95 quantile of an F distribution with 2 and 10 degrees of freedom), which is too small. Put another way, if the bootstrap-t method estimates that the critical value is higher than the critical value used by F_t, in essence, a discrete distribution has been found for which F_t can be expected to have a Type I error probability greater than the nominal level. For the situation at hand, the bootstrap-t estimates the critical value to be 4.97, which corresponds to the 0.968 quantile of an F distribution with 2 and 10 degrees of freedom. Moreover, the discrete distributions being used are estimates of the distributions under study, only shifted, so that the null hypothesis of equal trimmed means is true.

The R function

$$btrim(x,tr=0.2, grp=NA, g=NULL, dp=NULL, nboot=599)$$

is an updated version of t1waybt. In addition to the results reported by t1waybt, the function btrim reports the explanatory measure of effect size. And it has the ability to sort data into groups based on group identification values stored in column g of x, assuming x is a matrix or a data frame. The outcome (dependent variable) of interest is stored in the column indicated by the argument dp. In effect, this eliminates the need to call the function fac2list. For example, btrim(plasma,g=2,dp=4) would sort the data in column 4 of the R variable plasma into groups based on the values stored in column 2.

The R function

$$t1waybtsqrk(x, alpha=0.05, nboot=599, tr=0.2, SEED=TRUE)$$

performs the method based on Bailey's transformation.

7.2 Two-Way Designs and Trimmed Means

This section describes methods for testing hypotheses in a two-way design when working with trimmed means. It is assumed that the reader is familiar with the basic features and terminology of two-way designs, which are covered in numerous books on statistics. To briefly review, the basic goal is to compare groups, taking into account two main factors plus interactions. For example, Steele and Aronson (1995) conduct a study dealing with performance on an aptitude test. They compare test scores of Black and White subjects taking into account how the purpose of the test was presented. The test was presented either as a diagnostic of intellectual ability, as a laboratory tool for studying problem solving, or as both a problem-solving tool and a challenge. This is a 2-by-3 design. The first factor, race, has two levels, while the second factor, type of presentation, has three. As is commonly done, the first factor is generically called Factor A, and the second is called Factor B. The term J-by-K ANOVA refers to a two-way design with Factor A having J levels and Factor B having K levels.

The groups are assumed to be arranged as shown in Table 7.2. Thus, μ_{tjk} is the population trimmed mean associated with the jth level of the first factor and the kth level of the second. Extending standard notation in an obvious way, the grand trimmed mean is the average of all JK trimmed means. In symbols, the grand trimmed mean is

$$\bar{\mu}_t = \frac{1}{JK} \sum_{j=1}^{J} \sum_{k=1}^{K} \mu_{tjk}.$$

Table 7.2:   Trimmed means corresponding to a J-by-K design.

	Factor B				
	μ_{t11}	μ_{t12}	$\cdots$	μ_{t1K}	$\mu_{t1.}$
	μ_{21}	μ_{22}	$\cdots$	μ_{2K}	$\mu_{2.}$
Factor A	$\vdots$	$\vdots$	$\cdots$	$\vdots$	$\vdots$
	μ_{tJ1}	μ_{tJ2}	$\cdots$	μ_{tJK}	$\mu_{tJ.}$
	$\mu_{t.1}$	$\mu_{t.2}$	$\cdots$	$\mu_{t.K}$	

The main effects for Factor A are defined to be

$$\alpha_1 = \mu_{t1.} - \bar{\mu}_t, \cdots, \alpha_J = \mu_{tJ.} - \bar{\mu}_t,$$

where

$$\mu_{tj.} = \frac{1}{K} \sum_{k=1}^{K} \mu_{tjk}.$$

The hypothesis of no main effects for Factor A is

$$H_0: \mu_{t1.} = \cdots = \mu_{tJ.}.$$

When the null hypothesis is true,

$$\alpha_{t1} = \cdots = \alpha_{tJ} = 0,$$

so another common way of writing the null hypothesis is

$$H_0: \sum \alpha_j^2 = 0.$$

Similarly, the levels of Factor B can be compared, ignoring Factor A, by testing

$$H_0: \mu_{t.1} = \mu_{t.2} = \cdots = \mu_{t.K},$$

where

$$\mu_{t.k} = \frac{1}{J} \sum_{j=1}^{J} \mu_{tjk}.$$

The effect size associated with the kth group is written as

$$\beta_k = \mu_{t.k} - \bar{\mu}_t,$$

and often the null hypothesis is written as

$$H_0: \sum \beta_k^2 = 0.$$

The computational steps associated with a two-way design, when testing the hypotheses just listed, are much easier to describe in terms of matrices. Here, a generalization of results in Johansen (1980) is used. The implementation of the method is based on a generalization of the results in Algina and Olejnik (1984).

There are $p = JK$ independent groups with trimmed means $\boldsymbol{\mu}_t = (\mu_{t1}, \ldots, \mu_{tJK})'$. The general strategy for testing main effects and interactions is to test

$$H_0: \mathbf{C}\boldsymbol{\mu}_t = 0, \tag{7.6}$$

where $\mathbf{C}$, which is constructed in a manner to be described, is a k-by-p contrast matrix of rank k, chosen to reflect the hypothesis of interest. For convenience, it is assumed that the sample trimmed means are arranged in a $1 \times p$ matrix

$$\mathbf{X}' = (\bar{X}_{t11} \ldots \bar{X}_{t1K} \ldots, \bar{X}_{tJ1} \ldots \bar{X}_{tJK}),$$

where $\mathbf{X}'$ is the transpose of $\mathbf{X}$.

The construction of the contrast matrix $\mathbf{C}$ is accomplished as follows. For any integer $m \geq 2$, let $\mathbf{C}_m$ be an $(m - 1)$-by-m matrix having the form

$$\begin{pmatrix} 1 & -1 & 0 & 0 & \ldots & 0 \\ 0 & 1 & -1 & 0 & \ldots & 0 \\ & & & \ldots & & \\ 0 & 0 & \ldots & 0 & 1 & -1 \end{pmatrix}.$$

That is, $c_{ii} = 1$ and $c_{i,i+1} = -1$, $i = 1, \ldots, m - 1$. Let $\mathbf{j}'_m$ be a 1-by-m vector of 1s. For example, $\mathbf{j}'_3 = (1, 1, 1)$. The matrix $\mathbf{C}$ for testing main effects and interactions can be constructed with what is called the (right) Kronecker product of matrices, applied to appropriate choices of $\mathbf{C}_m$ and $\mathbf{j}_m$. If $\mathbf{A}$ is any r-by-s matrix and $\mathbf{B}$ is any t-by-u matrix, the Kronecker product of $\mathbf{A}$ and $\mathbf{B}$, written as $\mathbf{A} \otimes \mathbf{B}$, is

$$\begin{pmatrix} a_{11}\mathbf{B} & a_{12}\mathbf{B} & \ldots & a_{1s}\mathbf{B} \\ & \vdots & & \\ a_{r1}\mathbf{B} & a_{r2}\mathbf{B} & \ldots & a_{rs}\mathbf{B} \end{pmatrix}.$$

Table 7.3 shows how to construct the contrast matrix $\mathbf{C}$ for the main effects and interactions in a two-way design. For example, when testing for main effects for Factor A, use $\mathbf{C} = \mathbf{C}_J \otimes \mathbf{j}'_K$.

Table 7.3: How to
construct the
contrast matrix, C,
for a two-way
design.

Effect	C
A	$\mathbf{C}_J \otimes \mathbf{j}'_K$
B	$\mathbf{j}'_J \otimes \mathbf{C}_K$
A × B	$\mathbf{C}_J \otimes \mathbf{C}_K$

Remembering that $p = JK$, the total number of groups, let $\mathbf{V}$ be a p-by-p diagonal matrix with

$$v_{jj} = \frac{(n_j - 1)s_{wj}^2}{h_j(h_j - 1)},$$

$j = 1, \ldots, p$. That is, v_{jj} is Yuen's estimate of the squared standard error of the sample trimmed mean corresponding to the jth group. The test statistic is

$$Q = \bar{\mathbf{X}}'\mathbf{C}'(\mathbf{C}\mathbf{V}\mathbf{C}')^{-1}\mathbf{C}\bar{\mathbf{X}}. \tag{7.7}$$

Let

$$\mathbf{R} = \mathbf{V}\mathbf{C}'(\mathbf{C}\mathbf{V}\mathbf{C}')^{-1}\mathbf{C},$$

and

$$A = \sum_{j=1}^{p} \frac{r_{jj}^2}{h_j - 1},$$

where r_{jj} is the jth diagonal element of $\mathbf{R}$. Asymptotically, a critical value for Q is c, the $1 - \alpha$ quantile of a chi-squared distribution with k degrees of freedom, where, again, k is the rank of $\mathbf{C}$. However, for small sample sizes, an adjusted critical value is needed, which is given by

$$c_{ad} = c + \frac{c}{2k}\left[A\left(1 + \frac{3c}{k + 2}\right)\right].$$

If $Q \geq c_{ad}$, reject H_0.

7.2.1 R Functions t2way and bb.es.main

The R function

$$\text{t2way(J,K,x,grp=c(1:p),tr=0.2,alpha= 0.05)}$$

performs the tests on trimmed means described in the previous section, where J and K denote the number of levels associated with Factors A and B, respectively. When the data are stored in list mode, the first K groups are assumed to be the data for the first level of Factor A, the next K groups are assumed to be data for the second level of Factor A, and so on. In R notation, x[[1]] is assumed to contain the data for level 1 of Factors A and B, x[[2]] is assumed to contain the data for level 1 of Factor A and level 2 of Factor B, and so forth. If, for example, a 2-by-4 design is being used, the data are stored as follows:

	Factor B			
Factor A	x[[1]]	x[[2]]	x[[3]]	x[[4]]
	x[[5]]	x[[6]]	x[[7]]	x[[8]]

For instance, x[[5]] contains the data for the second level of Factor A and the first level of Factor B.

If the data are not stored in the assumed order, grp can be used to correct this problem. Suppose, for example, the data are stored as follows:

	Factor B			
Factor A	x[[2]]	x[[3]]	x[[5]]	x[[8]]
	x[[4]]	x[[1]]	x[[6]]	x[[7]]

That is, the data for level 1 of Factors A and B are stored in the R variable x[[2]], the data for level 1 of A and level 2 of B are stored in x[[3]], and so forth. To use t2way, first enter the R command

$$grp=c(2,3,5,8,4,1,6,7).$$

Then the command t2way(2,4,x,grp=grp) tells the function how the data are ordered. In the example, the first value stored in grp is 2, indicating that x[[2]] contains the data for level 1 of both Factors A and B, the next value is 3, indicating that x[[3]] contains the data for level 1 of A and level 2 of B, while the fifth value is 4, meaning that x[[4]] contains the data for level 2 of Factor A and level 1 of Factor B. As usual, tr indicates the amount of trimming, which defaults to 0.2, and alpha is α, which defaults to 0.05. The function returns the test statistic for Factor A, V_a, in the variable t2way$test.A, and the p-value is returned in t2way$A.p.value. Similarly, the test statistics for Factor B, V_b, and interaction, V_{ab}, are stored in t2way$test.B and t2way$test.AB, with the corresponding p-values stored in t2way$B.p.value and t2way$AB.p.value.

As a more general example, the command

$$t2way(2,3,z,tr=0.11,grp=c(1,3,4,2,5,6),alpha=0.11)$$

would perform the tests for no main effects and no interactions for a 2-by-3 design for the data stored in z, assuming the data for level 1 of Factors A and B are stored in z[[1]], the data for level 1 of A and level 2 of B are stored in z[[3]], and so on. The analysis would be based on 10% trimmed means and $\alpha = 0.11$.

Note that the general form for t2way contains an argument p. It is used by t2way to check whether the total number of groups being passed to the function is equal to JK. If JK is not equal to the number of groups in x, the function prints a warning message. If, however, you want to perform an analysis using some subset of the groups stored in x, this can be done simply by ignoring the warning message. For example, suppose x contains data for 10 groups, but you want to use groups 3, 5, 1, and 9 in a 2-by-2 design. That is, groups 3 and 5 correspond to level 1 of the first factor and levels 1 and 2 of the second. The command

$$t2way(2,2,x,grp=c(3,5,1,9))$$

accomplishes this goal. Note that a value for p is not passed to the function. The only reason p is included in the list of arguments is to satisfy certain requirements of R. The details are not important here and therefore not discussed.

■ Example

Suppose participants are randomly assigned to one of two groups. The first group watches a violent film, and the other watches a non-violent film. Afterwards, suppose aggressive affect is measured, and it is desired to compare both groups, taking gender into account as well. Some hypothetical data are shown in Table 7.4 to illustrate how t2way is used. Suppose the data are stored in the R variable film, having list mode. In particular, assume film[[1]] contains the values for males watching a violent film (the values 8, 7, 5, 6, 10, 14, 2, 3, and 16). The data for males watching a non-violent film are stored in film[[2]], the data for females watching a violent film are stored in film[[3]], and the data for females watching a non-violent film are stored in film[[4]]. Then the command t2way(2,2,film) would perform the appropriate tests for main effects and interactions using 20% trimmed means. If instead, the data for males watching a violent film are stored in film[[2]], and the data for males watching a non-violent film are stored in film[[1]], use the command t2way(2,2,film,grp=c(2,1,3,4)) to compare 20% means, while t2way(2,2,film,tr=0,grp=c(2,1,3,4)) compares means instead.

■

Table 7.4: Hypothetical data on the effect of watching a violent film.

| | Film | |
	Violent	Non-violent
Male	8, 7, 5, 6, 10, 14, 2, 3, 16	2, 4, 6, 7, 11, 12, 12, 3, 4
Female	5, 6, 8, 2, 3, 4, 5, 2	12, 40, 23, 2, 2, 2, 2, 4, 8, 10

The R function

$$bb.es.main(J,K,x,DIF=TRUE,...)$$

computes the explanatory measure of effect size for the main effects.

7.2.2 Comparing Medians

For the special case where the goal is to compare medians, the method in Section 7.2.1 should not be used; the estimate of the standard error performs poorly. One way to proceed is to test global hypotheses as described here, provided there are no tied values. A method that continues to perform well when there are tied values is method LSB in Section 7.6 used in conjunction with the Harrell–Davis estimator. Another approach is to use the multiple-comparison procedure for medians described in Section 7.4.7, which also handles tied values in an effective manner.

Let M_{jk} be the sample median for the jth level of Factor A and the kth level of B, and let n_{jk} and S_{jk}^2 be the corresponding sample size and estimate of the squared standard error of M_{jk}. Here, S_{jk}^2 is the McKean–Schrader estimate. To perform hypotheses dealing with main effects, compute

$$R_j = \sum_{k=1}^{K} M_{jk}, \ W_k = \sum_{j=1}^{J} M_{jk},$$

$$d_{jk} = S_{jk}^2,$$

$$\hat{v}_j = \frac{(\sum_k d_{jk})^2}{\sum_k d_{jk}^2/(n_{jk}-1)}, \ \hat{\omega}_k = \frac{(\sum_j d_{jk})^2}{\sum_j d_{jk}^2/(n_{jk}-1)},$$

$$r_j = \frac{1}{\sum_k d_{jk}}, \ w_k = \frac{1}{\sum_j d_{jk}},$$

$$r_s = \sum_{j=1}^{J} r_j, \ w_s = \sum_{k=1}^{K} w_k,$$

$$\hat{R} = \frac{\sum_j r_j R_j}{r_s}, \quad \hat{W} = \frac{\sum_k w_k W_k}{w_s},$$

$$B_a = \sum_{j=1}^{J} \frac{1}{\hat{v}_j} \left(1 - \frac{r_j}{\sum r_j}\right)^2, \quad B_b = \sum_{k=1}^{K} \frac{1}{\hat{\omega}_k} \left(1 - \frac{w_k}{\sum w_k}\right)^2,$$

$$V_a = \frac{\sum_j r_j (R_j - \hat{R})^2}{(J-1)\left(1 + \frac{2(J-2)B_a}{J^2-1}\right)}, \quad V_b = \frac{\sum_k w_k (W_k - \hat{W})^2}{(K-1)\left(1 + \frac{2(K-2)B_b}{K^2-1}\right)}.$$

The degrees of freedom for Factor A are $v_1 = J - 1$ and $v_2 = \infty$. For Factor B, the degrees of freedom are $v_1 = K - 1$ and $v_2 = \infty$. The hypothesis of no main effect for Factor A is rejected, if $V_a \geq f_{1-\alpha}$, the $1 - \alpha$ quantile of an F distribution with the degrees of freedom for Factor A. Similarly, reject for Factor B, if $V_b \geq f_{1-\alpha}$, with the degrees of freedom for Factor B.

A heteroscedastic test of the hypothesis of no interactions can be performed as follows. Again, let $d_{jk} = S_{jk}^2$ be the McKean–Schrader estimate of the squared standard error of M_{jk}. Let

$$D_{jk} = \frac{1}{d_{jk}},$$

$$D_{.k} = \sum_{j=1}^{J} D_{jk}, \quad D_{j.} = \sum_{k=1}^{K} D_{jk},$$

$$D_{..} = \sum_{j=1}^{J} \sum_{k=1}^{K} D_{jk},$$

$$\tilde{M}_{jk} = \sum_{\ell=1}^{J} \frac{D_{\ell k} M_{\ell k}}{D_{.k}} + \sum_{m=1}^{K} \frac{D_{jm} M_{jm}}{D_{j.}} - \sum_{\ell=1}^{J} \sum_{m=1}^{K} \frac{D_{\ell m} M_{\ell m}}{D_{..}}.$$

The test statistic is

$$V_{ab} = \sum_{j=1}^{J} \sum_{k=1}^{K} D_{jk} (M_{jk} - \tilde{M}_{jk})^2.$$

Let c be the $1 - \alpha$ quantile of a chi-squared distribution with $v = (J-1)(K-1)$ degrees of freedom. Reject the null hypothesis, if $V_{ab} \geq c$.

For small sample sizes, all indications are that an adjusted critical value is preferable. Otherwise, the actual Type I error probability can be well below the nominal level. In particular, compute a critical value using a simulation as done in Section 7.1.5. It is unclear just how large the sample sizes should be before an unadjusted critical value suffices.

7.2.3 R Functions med2way and Q2anova

The computations for comparing medians just described are performed by the R function

$$\text{med2way(J,K,x,grp=c(1:p),p=J*K, ADJ.P.VALUE=TRUE, iter=5000).}$$

If ADJ.P.VALUE=TRUE, it returns a p-value based on a simulation estimate of the null distribution, where the number of replications is specified by the argument iter. The R function

$$\text{Q2anova(J, K, x, nboot = 600, MC = FALSE)}$$

also tests the global hypotheses based on medians. It is based on method LSB in Section 7.6 in conjunction with the Harrell–Davis estimator, which has the advantage of being able to handle tied values.

7.3 Three-Way Designs, Trimmed Means, and Medians

This section extends the hypothesis testing technique in Section 7.2, based on trimmed means, to a three-way design. Again, a generalization of results in Johansen (1980) is used to test global hypotheses. It is assumed that a J-by-K-by-L design is being used, so there are a total of $p = JKL$ independent groups with trimmed means $\boldsymbol{\mu}_t = (\mu_{t1}, \ldots, \mu_{tJKL})'$. The general strategy for testing main effects and interactions is to test

$$H_0 \colon \mathbf{C}\boldsymbol{\mu}_t = 0, \tag{7.8}$$

where $\mathbf{C}$ is constructed in a manner to be described. For convenience, it is assumed that the sample trimmed means are

$$\bar{\mathbf{X}}' = (\bar{X}_{t111}, \ldots \bar{X}_{t11L}, \bar{X}_{t121}, \ldots, \bar{X}_{t12L}, \ldots, \bar{X}_{t1KL}, \ldots, \bar{X}_{tJKL}).$$

That is, the third subscript, which corresponds to the third factor, is incrementing the fastest. Thus, for the first level of the first factor ($J = 1$), the data are arranged as

$$
\begin{array}{ccc}
\bar{X}_{t111} & \cdots & \bar{X}_{t11L} \\
\vdots & \vdots & \vdots \\
\bar{X}_{t1K1} & \cdots & \bar{X}_{t1KL}.
\end{array}
$$

For the second level of the first factor ($J = 2$), the data are arranged as

$$
\begin{array}{ccc}
\bar{X}_{t211} & \cdots & \bar{X}_{t21L} \\
\vdots & \vdots & \vdots \\
\bar{X}_{t2K1} & \cdots & \bar{X}_{t2KL},
\end{array}
$$

Table 7.5: How to construct the contrast matrix, C, for a three-way design.

Effect	C
A	$\mathbf{C}_J \otimes \mathbf{j}'_K \otimes \mathbf{j}'_L$
B	$\mathbf{j}'_J \otimes \mathbf{C}_K \otimes \mathbf{j}'_L$
C	$\mathbf{j}'_J \otimes \mathbf{j}'_K \otimes \mathbf{C}_L$
A × B	$\mathbf{C}_J \otimes \mathbf{C}_K \otimes \mathbf{j}'_L$
A × C	$\mathbf{C}_J \otimes \mathbf{j}'_K \otimes \mathbf{C}_L$
B × C	$\mathbf{j}'_J \otimes \mathbf{C}_K \otimes \mathbf{C}_L$
A × B × C	$\mathbf{C}_J \otimes \mathbf{C}_K \otimes \mathbf{C}_L$

and so on. (The R function fac2list, illustrated in the next section, might help when dealing with data management.)

For any integer $m \geq 2$, again, let $\mathbf{C}_m$ be an $(m-1)$-by-m matrix having the form

$$\begin{pmatrix} 1 & -1 & 0 & 0 & \ldots & 0 \\ 0 & 1 & -1 & 0 & \ldots & 0 \\ & & \ldots & & & \\ 0 & 0 & \ldots & 0 & 1 & -1 \end{pmatrix}.$$

And as in Section 7.2, $\mathbf{j}'_m$ is a 1-by-m vector of 1s. Table 7.5 shows how to construct the contrast matrix $\mathbf{C}$ for the main effects and interactions in a three-way design. For example, when testing for main effects for Factor A, use $\mathbf{C} = \mathbf{C}_J \otimes \mathbf{j}'_K \otimes \mathbf{j}'_L$.

Remembering that $p = JKL$, the total number of groups, let $\mathbf{V}$ be a p-by-p diagonal matrix with

$$v_{jj} = \frac{(n_j - 1)s^2_{wj}}{h_j(h_j - 1)},$$

$j = 1, \ldots, p$. That is, v_{jj} is Yuen's estimate of the squared standard error of the sample trimmed mean corresponding to the jth group. Then:

$$Q = \bar{\mathbf{X}}'\mathbf{C}'(\mathbf{CVC}')^{-1}\mathbf{C}\bar{\mathbf{X}}, \tag{7.9}$$

can be used to test Eq. (7.8). Let

$$\mathbf{R} = \mathbf{VC}'(\mathbf{CVC}')^{-1}\mathbf{C}$$

and:

$$A = \sum_{j=1}^{p} \frac{r^2_{jj}}{h_j - 1},$$

where r_{jj} is the jth diagonal element of **R**. Asymptotically, a critical value for Q is c, the $1 - \alpha$ quantile of a chi-squared distribution with k degrees of freedom. However, for small sample sizes, an adjusted critical value is needed which is given by

$$c_{\text{ad}} = c + \frac{c}{2k}\left[A\left(1 + \frac{3c}{k+2}\right)\right].$$

If $Q > c_{\text{ad}}$, reject H_0.

For the special case where medians are compared, use method LSB in Section 7.6.

7.3.1 R Functions t3way, fac2list, and Q3anova

The R function

$$\text{t3way}(J,K,L,x,\text{tr}=0.2,\text{grp}=c(1:p),\text{alpha}= 0.05,p=J*K*L)$$

tests the hypotheses of no main effects and no interactions in a three-way (J-by-K-by-L) design using the method described in the previous section. Again, x is any R object containing the data, which are assumed to be stored in list mode. The default amount of trimming is tr=0.2, and the default value for alpha is $\alpha = 0.05$. The data are assumed to be arranged such that the first L groups correspond to level 1 of Factors A and B ($J = 1$ and $K = 1$) and the L levels of Factor C. The next L groups correspond to the first level of Factor A, the second level of Factor B, and the L levels of Factor C. If, for example, a 3-by-2-by-4 design is being used, it is assumed that for $J = 1$ (the first level of the first factor), the data are stored in the R variables x[[1]], ..., x[[8]] as follows:

	Factor C			
Factor B	x[[1]]	x[[2]]	x[[3]]	x[[4]]
	x[[5]]	x[[6]]	x[[7]]	x[[8]]

For the second level of the first factor, $J = 2$, it is assumed that the data are stored as

	Factor C			
Factor B	x[[9]]	x[[10]]	x[[11]]	x[[12]]
	x[[13]]	x[[14]]	x[[15]]	x[[16]]

If the data are not stored as assumed by t3way, grp can be used to indicate the proper ordering. As an illustration, consider a 2-by-2-by-4 design and suppose that for $J = 1$, the data are stored as follows:

	Factor C			
Factor B	x[[15]]	x[[8]]	x[[3]]	x[[4]]
	x[[6]]	x[[5]]	x[[7]]	x[[8]]

while for $J = 2$

	Factor C			
Factor B	x[[10]]	x[[9]]	x[[11]]	x[[12]]
	x[[1]]	x[[2]]	x[[13]]	x[[16]]

The R command

$$grp=c(15,8,3,4,6,5,7,8,10,9,11,12,1,2,13,16)$$

followed by the command t3way(2,2,3,x,grp=grp) will test all of the relevant hypotheses at the 0.05 level using 20% trimmed means.

The general form for t3way contains an argument p that is used to check whether $p = JKL$ is equal to the total number of groups contained in x, and it is also used to generate the default value for grp. As far as applications are concerned, this argument can be ignored. (It is necessary only to satisfy certain R requirements that are not relevant here.) If JKL is not equal to the number of groups passed to t3way, the function prints a warning message. If, however, you want to use some subset of the groups in a three-way design, you can do this simply by ignoring the error message and taking care that the proper groups are used in the analysis. In other words, proceed along the lines described in conjunction with t2way. As a simple illustration, if there are 10 groups, but it is desired to use only the first 8 groups in a 2-by-2-by-2 design, the command

$$t3way(2,2,2,x)$$

can be used, assuming the first two groups belong to level 1 of the first two factors and levels 1 and 2 of the third, and so on. If the groups are not in the proper order, grp can be used as already described and illustrated.

■ Example

The example in Section 7.2.1 involved a 2-by-2 design dealing with the effect of watching a violent versus a non-violent film. Extending the illustration, suppose that education is taken into account with one group having a college degree, and the other does not. Some hypothetical data for this 2-by-2-by-2 design are shown in Table 7.6. Suppose the data are stored in the assumed order in the R variable film. Thus, film[[1]] contains the data for level 1 of all three factors (the values 8, 7, 5, 6, 10, 14, 2, 3, and 16), film[[2]] contains the data for no degree, male subjects watching a non-violent film, film[[4]] contains the data for no degree, female subjects watching a non-violent film, and film[[6]] contains the data for male subjects with a degree who watch a

Table 7.6: Hypothetical data on the effect of watching a violent film.

	No degree	
	Violent	**Non-violent**
Male	8, 7, 5, 6, 10, 14, 2, 3, 16	2, 4, 6, 7, 11, 12, 12, 3, 4
Female	5, 6, 8, 2, 3, 4, 5, 2	12, 40, 23, 2, 2, 2, 2, 4

	Degree	
	Violent	**Non-violent**
Male	8, 10, 12, 14, 2, 18, 20	2, 3, 2, 4, 5, 6, 7, 3, 4
Female	4, 5, 6, 7, 6, 5, 4, 7, 8	12, 1, 4, 19, 20, 22, 23, 24, 30

non-violent film. Then the command t3way(2,2,2,film) will test all relevant hypotheses using 20% trimmed means. If it had been the case that the data for no degree, male subjects watching a violent film were stored in film[[2]] and the data for no degree, male subjects watching a non-violent film were stored in film[[1]], but otherwise, the assumed order is correct, the R command t3way(2,2,2,film,grp=c(2,1,3,4,5,6,7,8)) would perform the correct computations.

■

The function returns the various test statistics and corresponding critical values. The value of the test statistic, Q, for main effects for Factor A, is returned in t3way\$Qa, for Factor B, it is returned in t3way\$Qb, and for Factor C, it is returned in t3way\$Qc. The corresponding critical values are returned in t3way\$Qa.crit, t3way\$Qb.crit, and t3way\$Qc.crit. The tests for two-way interactions are stored in t3way\$Qab, t3way\$Qac, and t3way\$Qbc; the critical values are in t3way\$Qab.crit, t3way\$Qac.crit, and t3way\$Qbc.crit; and the test for a three-way interaction is in t3way\$Qabc, with the critical value in t3way\$Qabc.crit.

If data are stored in a matrix, with some of the columns indicating the levels of the factors, it is noted that the function fac2list, described in Section 7.1.2, can be used to store the data in the manner required here. Suppose the data are stored in a matrix, say m, with group numbers for the three factors stored in columns 2, 4, and 6. If, for example, a 2-by-4-by-5 design is being examined, column 2 would contain the group identification numbers for the two levels of the first factor. The values in column 2 might be 1 or 2, or they might be 10 and 16. That is, there are two distinct values only, but they can be any two numbers. If the outcome measures are stored in column 5, the R command

$$\text{dat=fac2list(m[,5],m[,c(2,4,6)])}$$

will store the data in dat, in list mode. If, for example, it is desired to compare 20% trimmed means, this is accomplished with the command

$$\text{t3way(2,4,6,dat)}.$$

When the goal is to compare medians, the R function

$$\text{Q3anova(J, K, L, x, tr=0.2, nboot = 600, MC = FALSE)}$$

can be used even when there are tied values. Another option is to use the R function med3mcp in Section 7.4.10. A possible appeal of this latter function is that it can be used with the usual sample median, which has a higher breakdown point than the Harrell–Davis estimator.

7.4 Multiple Comparisons Based on Medians and Other Trimmed Means

This section summarizes several methods for performing multiple comparisons based on trimmed means, including medians as a special case. Included are methods for testing hypotheses about linear contrasts associated with two-way and three-way designs, which are described and illustrated in Section 7.4.3. The role of linear contrasts is described in many books dealing with the analysis of variance, so, for brevity, details are kept to a minimum. R functions for measuring effect size, based on linear contrasts, are described in Section 7.4.2.

A common goal is to control the probability of at least one Type I error. And a related goal is computing confidence intervals that have some specified simultaneous probability coverage. But another goal that has received increased attention in recent years is to control the *false discovery rate*. (For theoretical results on how this might be done, see Delattre and Roquain, 2015.) To elaborate, when testing C hypotheses, let Q be the proportion of hypotheses that are true and rejected. That is, Q is the proportion of Type I errors among the null hypotheses that are correct. If all hypotheses are false, then $Q = 0$, but otherwise, Q can vary from one experiment to the next. The *false discovery rate* is the expected value of Q.

A common practice is to use multiple-comparison procedures, such as those described in this section, only if a global test, such as those described in Sections 7.1–7.3, rejects. In terms of controlling the probability of one or more Type I errors, the methods in this section do not require that a global test first be performed and rejected. Indeed, based on results reported by Bernhardson (1975), the expectation is that using the multiple-comparison procedures in this chapter, contingent on first rejecting a global hypothesis, would alter their ability to control the probability of at least one Type I error in an unintended way. More precisely, the methods in this section are designed so that the probability of one or more Type I errors is α. If these methods are used contingently on a global test rejecting at the α level, the expectation is that the actual probability of one or more Type I errors will be less than α. In practical terms, a loss in power might result if the multiple-comparison procedures in this chapter are used only if a global test rejects.

7.4.1 Basic Methods Based on Trimmed Means

A relatively simple strategy for performing multiple comparisons and tests about linear contrasts, when comparing trimmed means, is to use an extension of Yuen's method for two groups in conjunction with a simple generalization of Dunnett's (1980) heteroscedastic T3 procedure for means.

Let $\mu_{t1}, \ldots, \mu_{tJ}$ be the trimmed means corresponding to J independent groups. A linear contrast is

$$\Psi = \sum_{j=1}^{J} c_j \mu_{tj},$$

where $c_1, \ldots, c_J$ are specified constants satisfying $\sum c_j = 0$. As a simple illustration, if $c_1 = 1$, $c_2 = -1$, and $c_3 = \cdots = c_J = 0$, then $\Psi = \mu_{t1} - \mu_{t2}$, the difference between the first two trimmed means. Typically, C linear contrasts are of interest, a common goal being to compare all pairs of means. Linear contrasts also play an important role when dealing with two-way and higher designs.

Consider testing

$$H_0 \colon \Psi = 0. \tag{7.10}$$

An extension of the Yuen–Welch method accomplishes this goal. The estimate of Ψ is

$$\hat{\Psi} = \sum_{j=1}^{J} c_j \bar{X}_{tj}.$$

An estimate of the squared standard error of $\hat{\Psi}$ is

$$A = \sum d_j,$$

where

$$d_j = \frac{c_j^2 (n_j - 1) s_{wj}^2}{h_j (h_j - 1)},$$

h_j is the effective sample size of the jth group, and s_{wj}^2 is the Winsorized variance. In other words, estimate the squared standard error of $\bar{X}_t$ as is done in Yuen's method, in which case, an estimate of the squared standard error of $\hat{\Psi}$ is given by A. Let

$$D = \sum \frac{d_j^2}{h_j - 1},$$

set:

$$\hat{v} = \frac{A^2}{D},$$

and let t be the $1 - \alpha/2$ quantile of a Student's t distribution with $\hat{v}$ degrees of freedom. Then an approximate $1 - \alpha$ confidence interval for Ψ is

$$\hat{\Psi} \pm t\sqrt{A}.$$

Let $\Psi_1, \ldots, \Psi_C$ be C linear contrasts of interest, where

$$\Psi_k = \sum_{j=1}^{J} c_{jk}\mu_{tj},$$

and let $\hat{v}_k$ be the estimated degrees of freedom associated with the kth linear contrast, which is computed as described in the previous paragraph. As previously noted, a common goal is to compute a confidence interval for each Ψ_k such that the simultaneous probability coverage is $1 - \alpha$. A related goal is to test H_0: $\Psi_k = 0$, $k = 1, \ldots, C$, such that the *familywise error rate* (FWE), meaning the probability of at least one Type I error among all C tests to be performed, is α. The practical problem is finding a method that adjusts the critical value to achieve this goal. One strategy is to compute confidence intervals having the form

$$\hat{\Psi}_k \pm t_k\sqrt{A_k},$$

where A_k is the estimated squared standard error of $\hat{\Psi}_k$, computed as described when testing Eq. (7.10), and t_k is the $1 - \alpha$ percentage point of the C-variate Studentized maximum modulus distribution with estimated degrees of freedom $\hat{v}_k$. In terms of testing H_0: $\Psi_k = 0$, $k = 1, \ldots, C$, reject H_0: $\Psi_k = 0$, if $|T_k| > t_k$, where

$$T_k = \frac{\hat{\Psi}_k}{\sqrt{A_k}}.$$

(The R software written for this book determines t_k when $\alpha = 0.05$ or 0.01, and $C \leq 28$ using values computed and reported in Wilcox, 1986. For other values of α or $C > 28$, the function determines t_k via simulations with 10,000 replications, where t_k is the $1 - \alpha$ quantile of the distribution of the maximum absolute value of C independent Student t random variables. This value for t_k differs slightly from the $1 - \alpha$ quantile of a Studentized maximum modulus distribution. Bechhofer and Dunnett, 1982, report values up to $C = 32$.) When there is no trimming, the method just described reduces to Dunnett's (1980) T3 procedure for means when all pairwise comparisons are performed.

A Step-Down Multiple-Comparison Procedure

All pairs power refers to the probability of detecting all true differences. For the special case where all pairwise comparisons are to be made, a so-called step-down method might provide higher all pairs power. Motivated by results in Hochberg and Tamhane (1987) as well as Wilcox (1991c), the method is applied as follows:

1. Test the global hypothesis, at the $\alpha_J = \alpha$ level, that all J groups have a common trimmed mean. If H_0 is not rejected, stop, and fail to find any differences among the groups. Otherwise, continue to the next step.
2. For each subset of $J - 1$ groups, test at the $\alpha_{J-1} = \alpha$ level the hypothesis that the $J - 1$ groups have a common trimmed mean. If all such tests are non-significant, stop; otherwise, continue to the next step.
3. For each subset of $J - 2$ groups, test at the $\alpha_{J-2} = 1 - (1-\alpha)^{(J-2)/J}$ level the hypothesis that all $J - 2$ groups have equal trimmed means. If all of these tests are non-significant, stop; otherwise, continue to the next step.
4. Test the hypothesis of equal trimmed means for all subsets of p groups, at the $\alpha_p = 1 - (1-\alpha)^{p/J}$ level, when $p \le J - 2$. If all of these tests are non-significant, stop, and fail to detect any differences among the groups; otherwise, continue to the next step.
5. The final step consists of testing all pairwise comparisons of the groups at the $\alpha_2 = 1 - (1-\alpha)^{2/J}$ level. In this final step, when comparing the jth group to the kth group, either fail to reject, fail to reject by implication from one of the previous steps, or reject. For example, if the hypothesis that groups 1, 2, and 3 have equal trimmed means is not rejected, then, in particular, groups 1 and 2 would not be declared significantly different by implication.

Although this step-down method can increase all pairs power, it should be noted that when comparing means, power can be relatively poor. Consider, for example, four groups, three of which have normal distributions and the third has a heavy-tailed distribution. Even when the first three groups differ substantially, a few outliers in the fourth group can destroy the power of a global test based on means. That is, the first step in the step-down method can fail to reject, in which case, no differences are found. Using a robust measure of location guards against this concern.

7.4.2 R Functions lincon, conCON, IND.PAIR.ES, and stepmcp

The R function

$$\text{lincon(x,con=0,tr=0.2,alpha= 0.05)}$$

tests linear contrasts based on trimmed means. The argument x is assumed to have list mode, tr indicates the amount of trimming, and con is a J-by-C matrix, the kth column containing the contrast coefficients for the kth linear contrast of interest, $k = 1, \ldots, C$. The argument alpha is α, which defaults to 0.05. As usual, x[[1]] contains the data for group 1, x[[2]] the data for group 2, and so on, and the default amount of trimming is tr=0.2, 20%. (The function fac2list, described in Section 1.9, can be used to store the data in list mode when initially the data are stored in a matrix, say m, with 1 or more columns of m containing group identification numbers.) If con is not specified, all pairwise comparisons are performed. The function returns two matrices called test and psihat. If all pairwise comparisons are to be performed, the first two columns of both matrices indicate which groups are being compared. The remaining columns of the first matrix, test, report the test statistic, the 0.95 critical value, the estimated standard error, and the degrees of freedom. If there is interest in using $\alpha = 0.01$, rather than 0.05, these results can be used to determine an appropriate critical value by referring to the table of Studentized maximum modulus distribution previously cited in this section. Columns 3–5 of psihat report $\hat{\Psi}$ and the lower and upper ends of the 0.95 confidence interval. These quantities are found in the columns labeled psihat, ci.lower, and ci.upper, respectively. If specific contrasts are of interest (meaning that a value for con is passed to lincon), the output is the same as just described, only the first two columns of the matrices returned by lincon are replaced by the number of the contrast being examined. That is, the first row of each matrix returned by lincon is numbered 1, meaning that it contains the results for Ψ_1, and so on.

The critical value used by lincon is determined with the goal that the probability of at least one Type I error is less than or equal to 0.05 or 0.01, depending on the argument alpha. A related goal is that the simultaneous probability coverage of all the confidence intervals is greater than or equal to 0.95 or 0.99. This is accomplished by storing exact percentage points of the Studentized maximum modulus distribution in the R functions called smmcrit and smmcrit01 for $C = 2, \ldots, 28$ and selected degrees of freedom. These exact values were determined with the FORTRAN program in Wilcox (1986). For other degrees of freedom, linear interpolation on inverse degrees of freedom is used to determine the 0.95 and 0.99 quantiles. The function assumes that for $v \geq 200$, $v = \infty$. For $C > 28$, or values for α other than 0.05 and 0.01, the R function smmvalv2 is used to compute an approximation of the required percentage point.

The optional argument con is a J-by-C matrix that contains the contrast coefficients to be used. The kth column is assumed to contain the contrast coefficients $c_{1k}, \ldots, c_{Jk}$, which correspond to the kth linear contrast, Ψ_k. As previously indicated, if con is not specified, then all pairwise comparisons are performed. If, for example, the goal is to compare a control group to each of the $J - 1$ other groups, this can be done via the R function

$$\text{conCON(J,conG=1)},$$

where the argument J indicates the number of groups and conG indicates which group is the control. By default, it is assumed that the first group is the control group. The contrast coefficients are returned in $conCON. For example, if data for four groups are stored in the R variable x and the control group corresponds to group 2, the command

$$\text{lincon(x,con=conCON(4,2)\$conCON)}$$

would compare three groups to the control group.

■ **Example**

Suppose all pairwise comparisons are to be performed using the data in Table 7.7. If the data are stored in the R variable x, the command lincon(x) returns

```
$test
       Group Group      test      crit        se         df
[1,]       1     2 0.4151210 3.120264 66.07368 11.374139
[2,]       1     3 0.2590833 3.094621 60.65340 11.900028
[3,]       1     4 0.6099785 3.372408 43.97761  7.892383
[4,]       2     3 0.1708173 3.101733 68.57785 11.749354
[5,]       2     4 0.9975252 3.464224 54.38857  7.177902
[6,]       3     4 0.8926159 3.410628 47.65732  7.578376


$psihat
       Group Group    psihat   ci.lower  ci.upper   p.value
[1,]       1     2 -27.42857 -233.5959 178.7387 0.6857763
[2,]       1     3 -15.71429 -203.4136 171.9850 0.7999983
[3,]       1     4  26.82540 -121.4851 175.1358 0.5590247
[4,]       2     3  11.71429 -200.9959 224.4245 0.8672737
[5,]       2     4  54.25397 -134.1602 242.6682 0.3509425
[6,]       3     4  42.53968 -120.0017 205.0811 0.3995184
```

For example, when comparing groups 1 and 2 with 20% trimmed means, the third and fourth columns stored in $test indicate that the test statistic is 0.415, and the $\alpha = 0.05$ critical value is 3.12. The remaining two columns indicate the estimated standard error

Table 7.7: Data used to illustrate the R function lincon.

Group 1:	119, −53, −77, 32, 194, −34, 48, −73, −69, −95, 175
Group 2:	−25, −22, 158, 208, 245, −70, −95, −68, 161, 28, −73
Group 3:	−95, 438, −72, 290, 3, −86, 136, 43, −27, 76, −79
Group 4:	−37, −88, −23, −50, 45, −36, −79, −86, −66, −73, −11, 16, 0, 47, 218

and degrees of freedom. The results in $psihat indicate that when comparing groups 2 and 3, $\hat{\Psi} = 11.7$, and the 0.95 confidence interval is $(-201, 224)$. The command lincon(x,alpha=0.01) would use $\alpha = 0.01$ instead.

■

■ Example

Again, consider the data in Table 7.7, only now suppose that the fourth group is a control, and it is desired to compare each of the first three groups to the control. That is, the goal is to compare group 1 to group 4, group 2 to group 4, and group 3 to group 4. Then the contrast coefficients for the first linear contrast are $c_{11} = 1$, $c_{21} = c_{31} = 0$, and $c_{41} = -1$, in which case

$$\Psi_1 = 1\mu_{t1} + 0\mu_{t2} + 0\mu_{t3} + (-1)\mu_{t4}$$
$$= \mu_{t1} - \mu_{t4}.$$

In a similar fashion, the contrast coefficients for $\Psi_2 = \mu_{t2} - \mu_{t4}$ are $c_{12} = c_{32} = 0$, $c_{22} = 1$, and $c_{42} = -1$. For $\Psi_3 = \mu_{t3} - \mu_{t4}$, they are $c_{13} = c_{23} = 0$, $c_{33} = 1$, and $c_{43} = -1$.

To use lincon, first store the matrix

$$\begin{pmatrix} 1 & 0 & 0 \\ 0 & 1 & 0 \\ 0 & 0 & 1 \\ -1 & -1 & -1 \end{pmatrix}$$

in any R object. The first column contains the contrast coefficients for the first linear contrast, $(1, 0, 0, -1)$. The second columns contains the contrast coefficients for the second linear contrast, and the third column contains the contrast coefficients associated with Ψ_3. For example, the command

$$\text{MAT=conCON(4,4)\$conCON}$$

stores the contrast coefficients in the R variable MAT. Assuming the data for the four groups are stored in x, the command lincon(x,con=MAT) performs the three comparisons with $\alpha = 0.05$, and the results are

```
$test
         con.num       test       crit       se         df
[1,]           1 0.6099785 2.969545 43.97761 7.892383
[2,]           2 0.9975252 3.040172 54.38857 7.177902
[3,]           3 0.8926159 2.998945 47.65732 7.578376


$psihat
         con.num    psihat   ci.lower ci.upper    p.value
[1,]           1 26.82540 -103.7681 157.4189 0.5590247
[2,]           2 54.25397 -111.0967 219.6046 0.3509425
[3,]           3 42.53968 -100.3820 185.4614 0.3995184


$test:
         con.num       test       crit       se         df
[1,]           1 0.6099785 2.969545 43.97761 7.892383
[2,]           2 0.9975252 3.040172 54.38857 7.177902
[3,]           3 0.8926159 2.998945 47.65732 7.578376


$psihat:
         con.num    psihat   ci.lower ci.upper
[1,]           1 26.82540 -103.7681 157.4189
[2,]           2 54.25397 -111.0967 219.6046
[3,]           3 42.53968 -100.3820 185.4614
```

■

A difference between this output and the output of the previous example is that now the contrasts are numbered under the column labeled con.num. The results for the first linear contrast, $\mu_{t1} - \mu_{t4}$, are stored in the first row of the matrices $test and $psihat. Thus, the test statistic for $H_0: \mu_{t1} = \mu_{t4}$ is 0.61, the critical value is 2.97, the estimate of $\Psi_1 = \mu_{t1} - \mu_{t4}$ is 26.8, and the 0.95 confidence interval is $(-103.8, 157.4)$. Note that the critical values in this example are smaller than those in the previous example. This is because only three contrasts are being tested now, as opposed to six contrasts before.

Effect Size

The R function

$$\text{IND.PAIR.ES(x,con=NULL, fun=ES.summary, tr=0.2,...)}$$

computes measures of effect size. By default, six measures are computed: AKP, EP, and KMS, described in Section 5.3.4, the median and trimmed mean versions of the quantile shift (QS) measure of effect size, described in Section 5.7.2, and the WMW effect size given by Eq. (5.23) in Section 5.7. Again, con is a matrix containing linear contrast coefficients. By default, effect size is estimated for all pairs of groups based on the argument fun. If the argument con is specified, then for each column of con, the data are separated into two groups. The first group is where the contrast coefficient is 1, and the second group is where the contrast coefficient is -1. Then measures of effect size are computed. The argument fun can be any function designed to estimate effect size for two groups where the first two arguments correspond to the groups being compared. To get confidence intervals for the measures of effect size, set fun=ES.summary.CI. Another way to get a single measure of effect size is to use fun=ESfun and include the argument method, the choices for which are summarized in Section 5.3.5. Also see the R function LCES in Section 7.9.3.

■ **Example**

Suppose x is a matrix with four columns. The command

$$\text{IND.PAIR.ES(x,con=conCON(4,2)\$conCON)}$$

would compute measures of effect size when comparing a control group, stored in column 2 of x, to the data for the three other groups stored in columns, 1, 3, and 4. The command IND.PAIR.ES(x,fun=qshift) would compute the shift measure of effect size in Section 5.3.4, where the contrast coefficient equal to 1 indicates the group that is taken to be the reference group. The command IND.PAIR.ES(x,fun=ees.ci) would compute a confidence interval for the explanatory measure of effect size in Section 5.3.4. The command IND.PAIR.ES(x,fun=ESfun,method='AKP') would compute the Algina et al. (2005) robust homoscedastic analog of Cohen's d described in Section 5.3.4. IND.PAIR.ES(x,fun=ES.summary.CI) would compute confidence intervals for the six measures of effect size AKP, EP, KMS, QS (median), QS (trimmed), and WMW.

The R function

$$\text{stepmcp}(x, tr=0.2, alpha=0.05)$$

performs the step-down method for performing all pairwise comparisons of the trimmed means. Each of the global tests is performed via the R function t1way, described in Section 7.1.2, which applies the method in Table 7.1. This function is limited to five groups.

■ **Example**

Here is an example of the output from stepmcp when comparing four groups:

```
        Groups      p-value        p.crit
 [1,]       12    0.003463866    0.02532057
 [2,]       13    0.188209885    0.02532057
 [3,]       14    0.040447312    0.02532057
 [4,]       23    0.278323659    0.02532057
 [5,]       24    0.532323103    0.02532057
 [6,]       34    0.623339080    0.02532057
 [7,]      123    0.012867018    0.05000000
 [8,]      124    0.008151950    0.05000000
 [9,]      134    0.091328319    0.05000000
[10,]      234    0.543503951    0.05000000
[11,]     1234    0.020327162    0.05000000
```

The last line indicates the result when testing the hypothesis that all four groups have equal 20% trimmed means, which is significant at the 0.05 level. Note that when comparing groups 1 and 2, the p-value is less than the critical level, 0.02532, as indicated by line 1. In addition, when testing $H_0: \mu_{t1} = \mu_{t2} = \mu_{t3}$, as well as $H_0: \mu_{t1} = \mu_{t2} = \mu_{t4}$, again, a significant result is obtained. Consequently, reject $H_0: \mu_{t1} = \mu_{t2}$. If, for example, the p-value associated with $H_0: \mu_{t1} = \mu_{t2} = \mu_{t3}$ were 0.06, $H_0: \mu_{t1} = \mu_{t2}$ would not be rejected despite the fact that the p-value associated with $H_0: \mu_{t1} = \mu_{t2}$ is less than the critical p-value. That is, by implication, the hypotheses $H_0: \mu_{t1} = \mu_{t2}$, $H_0: \mu_{t1} = \mu_{t3}$, and $H_0: \mu_{t2} = \mu_{t2}$ would not be rejected regardless of how small the corresponding p-values might be.

■

7.4.3 Multiple Comparisons for Two-Way and Three-Way Designs

Relevant multiple comparisons in a two-way design can be tested using appropriate linear contrasts. Consider, for example, a 3-by-3 design with the trimmed means depicted as follows:

Factor A		Factor B		
		1	2	3
	1	μ_{t1}	μ_{t2}	μ_{t3}
	2	μ_{t4}	μ_{t5}	μ_{t6}
	3	μ_{t7}	μ_{t8}	μ_{t9}

Let

$$\Psi_1 = \mu_{t1} + \mu_{t2} + \mu_{t3} - \mu_{t4} - \mu_{t5} - \mu_{t6},$$

$$\Psi_2 = \mu_{t1} + \mu_{t2} + \mu_{t3} - \mu_{t7} - \mu_{t8} - \mu_{t9},$$

$$\Psi_3 = \mu_{t4} + \mu_{t5} + \mu_{t6} - \mu_{t7} - \mu_{t8} - \mu_{t9}.$$

Then an approach to comparing the main effects for Factor A is to test $H_0: \Psi_\ell = 0$, for $\ell = 1, 2$, and 3. Roughly, the goal is to compare level 1 of Factor A to level 2 of Factor A, then compare levels 1 and 3, and finally, compare levels 2 and 3. Main effects for Factor B, as well as interactions, can be examined in a similar manner. For interactions, this means that for any two levels of Factor A, say j and j' ($j < j'$), and any two levels of Factor B, k and k' ($k < k'$), linear contrast coefficients are generated with the goal of testing

$$H_0: \mu_{tj} - \mu_{tj'} = \mu_{tk} - \mu_{tk'}.$$

For convenience, an R function (described in the next section and called con2way) is provided that generates the contrast coefficients typically used in a two-way design. Three-way designs are handled in a similar manner by generating linear contrast coefficients with the R function con3way, which is also described in the next section.

Consider the goal of making inferences about Factor A, ignoring Factor B. Another approach is to simply pool the data over the levels of Factor B for each level of Factor A. Then pairwise comparisons can be made for all levels of Factor A. In effect, when there are unequal sample sizes, groups with the larger sample sizes can have a more dominant role. Of course, levels of Factor B could be compared in a similar fashion. The R function twoway.pool can be used to pool the data in this manner.

Yet another approach is to first focus on level 1 of Factor A, and perform all pairwise comparisons among the K levels of Factor B. This could be done for each level of Factor A. Of course, the same can be done for each level of Factor B; perform all pairwise comparisons among the *J* levels of Factor A. The R function RCmcp, described in the next section, applies this approach. For measures of effect size, see Section 7.4.13.

7.4.4 R Functions bbmcp, RCmcp, mcp2med, twoway.pool, bbbmcp, mcp3med, con2way, and con3way

The R function

$$bbmcp(J, K, x, tr=0.2, alpha= 0.05, grp=NA, op=FALSE)$$

tests all of the usual pairwise comparisons associated with the levels of each factor, as well as all interactions associated with any two rows and columns, based on trimmed means. (This function is the same as the R function mcp2atm.) It does this by calling the R function

$$con2way(J, K),$$

which generates linear contrast coefficients, and then it calls the R function lincon. If op=FALSE, the $(J^2 - J)/2$ hypotheses associated with Factor A (all pairwise comparisons of the *J* levels) are tested with the probability of one or more Type I errors designated by the argument alpha, which defaults to 0.05. The same is done for Factor B and all relevant interactions. If op=TRUE, the function is designed so that for all comparisons associated with Factor A, Factor B, and all interactions, the probability of one or more Type I errors is alpha. So with op=TRUE, power will be lower because the probability of one or more Type I errors is being controlled for all hypotheses under consideration.

For each level of Factor A, the R function

$$RCmcp(J, K, x, alpha = 0.05, est = tmean, PB = FALSE, SEED = TRUE, pr = TRUE, ...)$$

performs all pairwise comparisons among the levels of Factor B. The results are returned in the R object A having list mode. A[[1]] contains all pairwise comparisons among the *K* levels of B for the first level of A. A[[2]] contains the results for level 2 of Factor A, and so on. Similarly, for each level of Factor B, all pairwise comparisons are performed among the levels of Factor A, and the results are returned in the R object B having list mode. By default, a 20% trimmed mean is used. To compare medians, even when there are tied values, set the argument est=median and PB=TRUE.

For each level of Factor A, the R function

$$\text{twoway.pool(J, K, x)}$$

pools data for all levels of Factor B. The results are returned in the R object A having list mode. And for each level of Factor B, it pools the data for all levels of Factor A and returns the results in the R object B. If, for example, x is a matrix containing data for a two-by-two design, the command lincon(twoway.pool(2,2,x)$A) would pool the data over the levels of Factor B and compare the two levels of Factor A.

For the special case, where the goal is to make inferences about linear contrasts based on medians, the function

$$\text{mcp2med(J,K,x, con=0, alpha= 0.05, grp=NA, op=FALSE)}$$

is supplied, which is based in part on the McKean–Schrader estimate of the standard error of the sample medians. As previously noted, this estimate of the standard error appears to perform reasonably well *with no tied values*, but otherwise, a percentile bootstrap method is recommended for comparing medians, which can be done with the R functions in Section 7.4.7.

■ **Example**

Consider a 2-by-3 design. A portion of the output from the R command con2way(2,3) is

```
$conAB
       [,1] [,2] [,3]
[1,]     1    1    0
[2,]    -1    0    1
[3,]     0   -1   -1
[4,]    -1   -1    0
[5,]     1    0   -1
[6,]     0    1    1
```

The three columns contain the linear contrast coefficients relevant to the three interactions associated with the six groups being compared. Assuming that means are compared, and that they are arranged as indicated in Table 7.2, the first column indicates that the linear contrast of interest is

$$\Psi = \mu_1 - \mu_2 - \mu_4 + \mu_5.$$

The typical goal is to test H_0: $\Psi = 0$, which of course is the same as testing:

$$H_0: \mu_1 - \mu_2 = \mu_4 - \mu_5,$$

the hypothesis of no interaction for levels 1 and 2 of both factors. The second column deals with the interaction associated with levels 1 and 3 of Factor B. And the third column deals with levels 2 and 3. If Factor A had three levels, conAB would have nine columns. The first three would deal with levels 1 and 2 of Factor A, the next three would deal with levels 1 and 3 of Factor A, and the final three would deal with levels 2 and 3. Again, the first three columns of conAB would deal with the three levels of Factor B, namely, levels 1 and 2, levels 1 and 3, and finally, levels 2 and 3.

The R function

$$\text{bbbmcp(J,K,L,tr=0.2,alpha= 0.05,grp=NA,op=FALSE)}$$

is like bbmcp, only it is designed for a three-way design. (This function is the same as the R function mcp3atm.) The linear contrast coefficients are generated by the R function

$$\text{con3way(J,K,L).}$$

So all pairwise comparisons associated with the levels of each factor are performed, as well as all interactions associated with the levels of each factor. For medians, use the R function

$$\text{mcp3med(J,K,L,tr=0.2,alpha= 0.05,grp=NA,op=FALSE).}$$

■ Example

To illustrate the use of the R function con3way when dealing with a three-way interaction, consider a 2-by-2-by-3 design, and focus on the contrast coefficients returned by the command con3way(2,2,3), which are stored in the matrix conABC. The means are assumed to be arranged as described at the beginning of Section 7.3. The first set of contrast coefficients, stored in the first column of conABC, deal with the A-by-B interaction at levels 1 and 2 of Factor C. The second set of contrast coefficients deal with the A-by-B interactions at levels 1 and 3 of Factor C. The contrast coefficients stored in the third column of conABC deal with the A-by-B interactions at levels 2 and 3 of Factor C. For a 2-by-3-by-3 design, there are nine linear contrasts associated with a three-way interaction. The first three deal with the interactions associated with levels 1 and 2 of

Factors A and B, respectively. The first set of linear contrast coefficients (in column 1 of conABC) is relevant to levels 1 and 2 of Factor C, the next is relevant to levels 1 and 3 of Factor C, and the third is relevant to levels 2 and 3 of Factor C. The next three sets of linear contrast coefficients repeat this pattern, only now the focus is on levels 1 and 3 of Factor B (and levels 1 and 2 of Factor A). The final three sets of linear contrast coefficients deal with levels 2 and 3 of Factor B.

■

7.4.5 A Bootstrap-t Procedure

When comparing trimmed means, and when the amount of trimming is relatively small, all indications are that an extension of the bootstrap-t method to multiple comparisons has practical value. So, in particular, when comparing means and when the sample sizes are small, this approach appears to perform relatively well, with the understanding that all methods based on means can be unsatisfactory.

Table 7.8 describes a bootstrap-t method for computing confidence intervals for each of C linear contrasts, Ψ_k, $k = 1, \ldots, C$, such that the simultaneous probability coverage is approximately $1 - \alpha$. The method is essentially the same as the symmetric two-sided confidence interval using the bootstrap-t method described in Table 5.4, only modified, so that for the C linear contrasts, the probability of at least one Type I error is approximately α.

When using the Studentized maximum modulus distribution to compare all pairs of trimmed means, it is known that probability coverage can be more satisfactory when using 20% trimmed means versus no trimming at all. However, concerns persist when any of the sample sizes are small. If the goal is to avoid having the probability of a Type I error excessively higher than α, the bootstrap-t method is a good choice based on extant simulation studies when the amount of trimming is small. A criticism of the bootstrap-t is that when all of the sample sizes are less than or equal to 15, the probability of at least one Type I error can drop below 0.025 when testing at the 0.05 level.

Table 7.9 shows estimates of α (which is one minus the simultaneous probability coverage) when sampling from exponential or lognormal distributions with four groups, $B = 599$, and various configurations of sample sizes, **n**, and standard deviations, σ, when using 20% trimmed means. (Additional simulations are reported by Wilcox, 1996g.) If the sample sizes are large enough, probability coverage will be satisfactory without using the bootstrap-t method, but it is unknown just how large the sample sizes must be. For heavier-tailed distributions, the bootstrap-t method offers less of an advantage, but it is difficult to tell whether it can be safely abandoned simply by looking at the data.

Table 7.8: Bootstrap-t confidence intervals for C linear contrasts.

The goal is to compute confidence intervals for each of C linear contrasts, $\Psi_1, \ldots, \Psi_C$, such that the simultaneous probability coverage is $1 - \alpha$. The kth linear contrast has contrast coefficients $c_{1k}, \ldots, c_{Jk}$.

Step 1. For each of the J groups, generate a bootstrap sample, $X_{ij}^*, i = 1, \ldots, n_j, j = 1, \ldots, J$. For each of the J bootstrap samples, compute the trimmed mean, $\bar{X}_j^*$, and d_j^*, Yuen's estimate of the squared standard error of $\bar{X}_j^*$, $j = 1, \ldots, J$.

Step 2. For the kth linear contrast, compute

$$T_k^* = \frac{|\hat{\Psi}_k^* - \hat{\Psi}_k|}{\sqrt{A_k^*}},$$

where $\hat{\Psi}_k^* = \sum c_{jk} \bar{X}_k^*$ and $A_k^* = \sum c_{jk}^2 d_j^*$.

Step 3. Let

$$T_m^* = \max \{T_1^*, \ldots, T_C^*\}.$$

In words, T_m^* is the maximum of the C values, $T_1^*, \ldots, T_C^*$.

Step 4. Repeat steps 1–3 B times, yielding $T_{mb}^*, b = 1, \ldots, B$.

Let $T_{m(1)}^* \leq \cdots \leq T_{m(B)}^*$ be the T_{mb}^* values written in ascending order, and let $a = (1 - \alpha)B$, rounded to the nearest integer. Then the confidence interval for Ψ_k is

$$\hat{\Psi}_k \pm T_{m(a)}^* \sqrt{A_k},$$

and the simultaneous probability coverage is approximately $1 - \alpha$.

Table 7.9: Simulation estimates of α for some light-tailed distributions.

Distribution	n	σ	Bootstrap	No bootstrap
Exponential	(11,11,11,11)	(1,1,1,1)	0.039	0.060
	(11,11,11,11)	(1,1,1,5)	0.039	0.096
	(15,15,15,15)	(1,1,1,1)	0.050	0.042
	(15,15,15,15)	(1,1,1,5)	0.050	0.083
	(10,15,20,25)	(1,1,1,1)	0.041	0.046
	(10,15,20,25)	(1,1,1,5)	0.040	0.061
	(10,15,20,25)	(5,1,1,1)	0.052	0.098
Lognormal	(11,11,11,11)	(1,1,1,1)	0.015	0.030
	(11,11,11,11)	(1,1,1,5)	0.046	0.091
	(15,15,15,15)	(1,1,1,1)	0.023	0.028
	(15,15,15,15)	(1,1,1,5)	0.052	0.085
	(10,15,20,25)	(1,1,1,1)	0.030	0.033
	(10,15,20,25)	(1,1,1,5)	0.039	0.056
	(10,15,20,25)	(5,1,1,1)	0.063	0.104

7.4.6 R Functions linconbt, bbtrim, and bbbtrim

The R function

$$\text{linconbt}(x, \text{con}=0, \text{tr}=0.2, \text{alpha}=0.05, \text{nboot}=599)$$

is provided for applying the bootstrap-t method when testing d linear contrasts using trimmed means. (The R functions linconb and linconbt are identical.) This function is used exactly like the function lincon in Section 7.4.2, the only difference being the additional argument, nboot, which is used to specify B, the number of bootstrap samples to be used. Again, if con is not specified, all pairwise comparisons are performed. The default value for nboot is 599, which appears to suffice, in terms of controlling the probability of a Type I error, when $\alpha = 0.05$. However, a larger choice for nboot might result in more power. The extent to which accurate probability coverage can be obtained, when $\alpha < 0.05$, is not known.

■ **Example**

Again, consider the data in Table 7.7, and suppose it is desired to compare the first three groups to the fourth, only now a bootstrap-t method is used. The results from linconbt are

```
$psihat
     con.num   psihat   ci.lower ci.upper
[1,]       1 26.82540 -146.2109 199.8617
[2,]       2 54.25397 -159.7458 268.2537
[3,]       3 42.53968 -144.9750 230.0544

$test
     con.num      test       se   p.value
[1,]       1 0.6099785 43.97761 0.5525876
[2,]       2 0.9975252 54.38857 0.3439065
[3,]       3 0.8926159 47.65732 0.4140234

$crit
[1] 3.934646
```

The contrast matrix is also returned in linconb$con. The estimates of Ψ for the three linear contrasts of interest, plus the corresponding standard deviations, are the same as before. However, the confidence intervals are longer using the bootstrap method. The critical value for all three contrasts, used by the first method in this section, is approximately 3, but the bootstrap estimate of the critical value is 3.99. This was expected because the observations in Table 7.7 were generated by first generating values from an exponential distribution, shifting them so that the trimmed mean is equal to zero, and multiplying by 100. That is, observations were generated from a skewed distribution with a relatively light tail. If the observations were generated from a normal distribution instead, the expectation is that there would be little difference between the confidence intervals.

■

For convenience, the R function

$$\text{bbtrim}(J,K,x,tr=0.2,alpha=0.05,nboot=599)$$

is supplied for performing the usual multiple comparisons associated with a J-by-K design using a bootstrap-t method with trimmed means. The function generates the linear contrast coefficients via the R function con2way, and then uses linconbt to test the relevant hypotheses. Multiple comparisons associated with a J-by-K-by-L three-way design are performed with the R function

$$\text{bbbtrim}(J,K,L,x,tr=0.2,alpha=0.05,nboot=599).$$

The contrast coefficients are generated via the R function

$$\text{con3way}(J,K,L).$$

7.4.7 Controlling the Familywise Error Rate: Improvements on the Bonferroni Method

Imagine the goal is to test C hypotheses such that the probability of one or more Type I errors is at most α. A simple way of proceeding is to use the *Bonferroni* method, meaning that each test is performed at the α/C level. However, several improvements on the Bonferroni method have been published that can provide higher power.

Table 7.10: Critical values, d_k, for Rom's method.

k	$\alpha = 0.05$	$\alpha = 0.01$
1	0.05000	0.01000
2	0.02500	0.00500
3	0.01690	0.00334
4	0.01270	0.00251
5	0.01020	0.00201
6	0.00851	0.00167
7	0.00730	0.00143
8	0.00639	0.00126
9	0.00568	0.00112
10	0.00511	0.00101

Rom's Method

Imagine the goal is to test C hypotheses such that the probability of one or more Type I errors is at most α. One way of improving on the Bonferroni method is to use a *sequentially rejective* method derived by Rom (1990), which has been found to have good power relative to several competing techniques. A limitation of Rom's method is that it was derived assuming that independent test statistics are used, but various studies suggest that it continues to control the Type I error reasonably well when the test statistics are dependent.

To apply it, compute a p-value for each of the C tests to be performed, and label them $p_1, \ldots, p_C$. Next, put the p-values in descending order, which are labeled $p_{[1]} \geq p_{[2]} \geq \cdots \geq p_{[C]}$. Proceed as follows:

1. Set k=1.
2. If $p_{[k]} \leq d_k$, where d_k is read from Table 7.10, stop, and reject all C hypotheses; otherwise, go to step 3.
3. Increment k by 1. If $p_{[k]} \leq d_k$, stop, and reject all hypotheses having a significance level less than or equal to d_k
4. If $p_{[k]} > d_k$, repeat step 3.
5. Continue until you reject or all C hypotheses have been tested.

Hochberg's Method

Hochberg's (1988) method for controlling the probability of one or more Type I errors is applied as follows. Again, let $p_1, \ldots, p_C$ be the p-values associated with the C tests, put these p-values in descending order, and label the results $p_{[1]} \geq p_{[2]} \geq \cdots \geq p_{[C]}$. Beginning with $k = 1$ (step 1), reject all hypotheses if

$$p_{[k]} \leq \alpha / k.$$

That is, reject all hypotheses if the largest p-value is less than or equal to α. If $p_{[1]} > \alpha$, proceed as follows:

1. Increment k by 1. If

$$p_{[k]} \leq \frac{\alpha}{k},$$

   stop, and reject all hypotheses having a p-value less than or equal to $p_{[k]}$.
2. If $p_{[k]} > \alpha / k$, repeat step 1.
3. Repeat steps 1 and 2 until you reject or all C hypotheses have been tested.

Rom's method offers a slight advantage over Hochberg's method in terms of power. But Hochberg's method relaxes the assumption that the test statistics are independent. Hochberg's method was derived based on an assumption made by Holm (1979): "all subsets of null hypotheses to be tested could appear as the set of true hypotheses." Suppose the goal is to test $H_0: \mu_1 = \mu_2$, $H_0: \mu_2 = \mu_3$, and $H_0: \mu_1 = \mu_3$. But if the first two hypotheses are true, they imply that the third is true as well. So Holm assumes that one of these hypotheses would not be tested. Note, however, that from the point of view of Tukey's three-decision rule, none of these hypotheses are true. Rather, the goal is to determine which of two groups has the larger mean. Also, suppose $\mu_1 = 2$, $\mu_2 = 4$, and $\mu_2 = 6$. Then it is possible that the first two hypotheses are not rejected, but the third hypothesis is rejected. Also, simulations where the restriction imposed by Holm is ignored indicate that Hochberg's method performs relatively well. See, for example, Section 7.8.5.

Hommel's Method

Yet another method for controlling the probability of one or more Type I errors was derived by Hommel (1988). Let $p_1, \ldots, p_C$ be the p-values associated with C tests, and let

$$j = \max\{i \in \{1, \cdots, C\}: p_{(n-i+k)} > k\alpha / i \text{ for } k = 1, \ldots, i\},$$

where $p_{(1)} \leq \cdots \leq p_{(C)}$ are the p-values written in ascending order. If the maximum does not exist, reject all C hypotheses. Otherwise, reject all hypotheses such that $p_i \leq \alpha / j$. If the Type I error probability is controlled for each of the C tests, then the probability of one or more Type I errors is less than or equal to α. Adjusted p-values based on Hommel's method can be computed via the R function p.adjust, described in Section 7.4.8, which also computes an adjusted p-value based on Hochberg's method. It seems that, typically, there is little or no difference between the adjusted p-values based on the Hochberg and Hommel methods, but situations are encountered where Hommel's method rejects and Hochberg's method does not, and Hommel's method is based on slightly weaker assumptions.

Benjamini–Hochberg Method

Benjamini and Hochberg (1995) proposed a variation of Hochberg's method, where in step 1 of Hochberg's method, $p_{[k]} \leq \alpha/k$ is replaced by

$$p_{[k]} \leq \frac{(C - k + 1)\alpha}{C}$$

(cf. Williams et al., 1999). A criticism of the Benjamini–Hochberg method is that situations can be found where some hypotheses are true, some are false, and the probability of at least one Type I error will exceed α among the hypotheses that are true (Hommel, 1988). In contrast, Hochberg's method does not suffer from this problem (assuming the actual level of each individual test is equal to the nominal level). However, when C hypotheses are tested, let Q be the proportion of hypotheses that are true and rejected. That is, Q is the proportion of Type I errors among the null hypotheses that are correct. The *false discovery rate* is the expected value of Q. That is, if a study is repeated infinitely many times, the false discovery rate is the average proportion of Type I errors among the hypotheses that are true. Benjamini and Hochberg (1995) show that their method ensures that the false discovery rate is less than or equal to α when performing C independent tests.

The k-FWER Procedures

A concern with methods that control the probability of one or more Type I errors is that when many tests are performed, power can be poor because each test is performed at a very low α level. One suggestion for dealing with this issue is to use what is generally known as a k-FWER method. That is, design a method where the probability of k or more Type I errors (the familywise error rate) is equal to some specified value, α. The methods derived by Rom, Hochberg, and Hommel use $k = 1$. For $k > 1$, a variety of methods have been derived, which are reviewed and compared by Keselman et al. (2011). In situations where the nature of the association among the test statistics being used is unknown, their results suggest using a generalization of the method in Holm (1979) that was derived by Lehmann and Romano (2005). Also see Guo et al. (2014).

Let C be the number of tests that are performed and denote the ordered p-values by $p_{(1)} \leq \cdots \leq p_{(k)} \leq \cdots \leq p_{(C)}$, which correspond to hypotheses $H_{(1)}, \ldots, H_{(k)}, \ldots, H_{(C)}$. For specified values for k and α, the generalized Holm procedure is defined stepwise as follows:

Step 0. Set $i = 1$.

Step 1. If $i \leq k$, go to step 2. If $k < i \leq C$, go to step 3. Otherwise, stop, and reject all of the hypotheses.

Step 2. If $p(i) > k\alpha/C$, go to step 4. Otherwise, set $i = i + 1$, and go to step 1.

Step 3. If $p(i) > \frac{k\alpha}{C+k-i}$, go to step 4. Otherwise, set $i = i + 1$, and go to step 1.

Step 4. Reject $H_{(j)}$ for $j < i$, and fail to reject $H_{(j)}$ for $j \geq i$.

If the C tests are independent, the following step-down method, known as the generalized Hochberg method, offers more power:

Step 0. Set $i = C$.

Step 1. If $i > k$, go to step 2. If $1 \leq i \leq k$, go to step 3. Otherwise, stop, and fail to reject all of the hypotheses.

Step 2. If $p(i) \leq \frac{k\alpha}{C+k-i}$, go to step 4. Otherwise, set $i = i - 1$, and go to step 1.

Step 3. If $p(i) \leq k\alpha/C$, go to step 4. Otherwise, set $i = i - 1$, and go to step 1.

Step 4. Reject $H_{(j)}$ for $j \leq i$, and fail to reject $H_{(j)}$ for $j > i$.

(The generalized Hochberg method remains valid if the associations among the m tests satisfy what is known as the MTP$_2$ condition. See Keselman et al., 2011, or Sarkar, 2008, for details.)

7.4.8 R Functions p.adjust and mcpKadjp

The R function

$$p.adjust(p, method = p.adjust.methods, n = length(p))$$

adjusts a collection of p-values based on the Bonferroni method or one of the improvements on the Bonferroni method described in the previous section. (See Wright, 1992, for details about these adjustments.) The available methods include: 'holm', 'hochberg', 'hommel', 'bonferroni', and 'BH', where BH indicates the Benjamini–Hochberg method. The default method is 'holm'.

The R function

$$mcpKadjp(p, k=1, proc = c('Holm'))$$

performs the k-FWER procedures described at the end of the previous section. The argument p contains the p-values to be adjusted. The argument proc indicates which method is to be used, which defaults to Holm. Two other options are: 'Hochberg' and 'BH' (Benjamini–Hochberg).

7.4.9 Percentile Bootstrap Methods for Comparing Medians, Other Trimmed Means, and Quantiles

When the goal is to perform multiple comparisons based on trimmed means, and the amount of trimming is not too small, say at least 20%, a percentile bootstrap method appears to be relatively effective. With 15% trimming, and even 10% trimming, a percentile bootstrap performs reasonably well. For the special case, where the goal is to compare medians or other quantiles, and when tied values occur, it is the only known method that performs well in simulations. In terms of controlling the probability of one or more Type I errors, the methods in Section 7.4.7 can be used.

It is noted that Johnson and Romer (2016) describe an alternative method for comparing the quantiles of multiple groups. Their method uses a bootstrap estimate of the standard errors. So, if tied values can occur, all indications are that a percentile bootstrap method is preferable.

7.4.10 R Functions linconpb, bbmcppb, bbbmcppb, medpb, Qmcp, med2mcp, med3mcp, and q2by2

The R function:

$$linconpb(x, alpha = 0.05, nboot = NA, grp = NA, est = tmean, con = 0, bhop = FALSE, SEED = TRUE, ...),$$

compares measures of location using a percentile bootstrap method. It can be used to compare robust measures of variation as well. For example, setting the argument est=winsd, a 20% Winsorized standard deviation would be used. By default, the function uses a 20% trimmed mean, and all pairwise comparisons are performed. Hochberg's method is used to control the probability of one or more Type I errors. Setting the argument bhop=TRUE, the Benjamini–Hochberg method is used instead. Linear contrasts can be tested by storing linear contrast coefficients in the argument con, which is assumed to be a matrix with rows corresponding to groups. By default, all pairwise comparisons are performed. (The R function tmcppb performs the same calculations as linconpb.)

The R function:

bbmcppb(J, K, x, tr=0.2, JK = J * K, tr=0.2, grp = c(1:JK), nboot = 500, bhop=FALSE,
SEED=TRUE),

performs multiple comparisons for a two-way ANOVA design based on trimmed means. The R function

bbbmcppb(J, K, L, x, tr=0.2, JKL = J * K * L, tr=0.2, grp = c(1:JKL), nboot = 500,
bhop=FALSE, SEED=TRUE)

performs multiple comparisons for a three-way design.

For the special case, where the population medians are to be compared based on the usual sample median (defined in Section 1.3), use the R function

medpb(x, tr=0.2, nboot = NA, grp = NA, est = median, con = 0, bhop=FALSE,
SEED=TRUE).

It performs well when there are tied values, and even with no tied values, it appears to be an excellent choice relative to competing techniques. When dealing with other quantiles, particularly when there are tied values, use the R function

Qmcp(x,q=0.5, con=0,SEED=TRUE,nboot=NA,alpha= 0.05,HOCH=FALSE)),indexR!Qmcp

which is based on the Harrell–Davis estimator. For small sample sizes, using the Bonferroni method to control the probability of one or more Type I errors is better than using Hochberg. How large the sample sizes must be to justify using Hochberg's method is unknown. For convenience, the R functions

med2mcp(,K,x,grp=c(1:p),p=J*K,nboot=NA,alpha=0.05,SEED=TRUE,pr=TRUE,
bhop=FALSE)

and

med3mcp(J,K,L,data, grp=c(1:p), alpha= 0.05, p=J*K*L, nboot=NA,
SEED=TRUE,bhop=FALSE)

are designed to handle two-way and three-way designs, respectively, when dealing with medians. They create the appropriate linear contrast coefficients and call the R function medpb. Other quantiles can be compared via the R function Qmcp; the appropriate linear contrast coefficients can be created with the R functions con2way and con3way.

All of the R functions in this section report p-values for each of the tests that are performed. So adjusted p-values can be computed using the R functions p.adjust and mcpKadjp in Section 7.4.8.

■ **Example**

Consider again, the plasma retinol data described near the end of Section 7.1.2. The goal was to compare the 20% trimmed means corresponding to the three smoking status groups. The outcome measure of interest was plasma beta-carotene, and the three groups being compared correspond to smoking status (1=Never, 2=Former, 3=Current Smoker). Assuming the data are stored in the R object z, as described in Section 7.1.2, consider the two R commands

$$res=lincon(z),$$
$$p.adjust(res\$psihat[,6],method='hommel').$$

A portion of the output from the first R command is:

```
$psihat
      Group Group   psihat     ci.lower ci.upper        p.value
[1,]      1     2 21.89458 -13.3604538 57.14962   0.1386234820
[2,]      1     3 57.78830  21.7636113 93.81300   0.0002009686
[3,]      2     3 35.89372   0.1144305 71.67301   0.0170140002
```

The second command returns adjusted p-values based on Hommel's method. The results are 0.139, 0.0006, and 0.034. So if the desired probability of one or more Type I errors is 0.05, even after adjusting the p-values, group 3 differs significantly from groups 1 and 2. Using Hochberg's method gives exactly the same results.

■

The R function

$$q2by2(x, q = c(0.1, 0.25, 0.5, 0.75, 0.9), nboot = 2000, SEED = TRUE)$$

deals with main effects and interactions when comparing quantiles. The quantiles used are controlled by the argument q. This function is designed for a two-by-two design only.

So the argument x should be a matrix with four columns, or it should have list mode with length 4.

7.4.11 Deciding Which Group Has the Largest Measure of Location

In various situations, a fundamental goal is identifying which of J independent groups has the largest measure of location. Let θ_j denote any measures of location associated with the jth group, and let $\theta_{(1)} \leq \cdots \leq \theta_{(J)}$ denote the measures of location written in ascending order. The goal is to choose a decision rule about which group has the largest measure of location, and then assess in some manner the probability of making a correct decision when a decision is made. This section describes methods for dealing with this issue. Some are based on Yuen's method for comparing trimmed means, and others use a percentile bootstrap method.

Early approaches to this problem were based on means and were aimed at determining sample sizes so that the group with the largest population mean will have the largest sample mean when

$$\theta_{(1)} = \cdots = \theta_{(J-1)} \tag{7.11}$$

and

$$\theta_{(J)} - \theta_{(J-1)} = \delta, \tag{7.12}$$

where $\delta > 0$ is some specified constant (e.g., Bechhofer, 1954; Gibbons et al., 1987; Gupta and Panchapakesan, 1987; Mukhopadhyay and Solanky, 1994). This is known as the indifference zone approach. If $\theta_{(J)} - \theta_{(J-1)} < \delta$, the idea is that it is not important which of these two groups is declared to have the larger measure of location. But if $\theta_{(J)} - \theta_{(J-1)} \geq \delta$, the goal is to be reasonably certain that a correct decision is made. These techniques are generally known as ranking and selection methods. Upper confidence intervals for all distances from the largest measure of location were derived by Hsu (1981), assuming normality and homoscedasticity. This section describes alternative approaches.

Methods RS and MCWB

The first approach is based on comparing trimmed means via Yuen's method. It uses a decision rule based on a multiple-comparison procedure designed to control the FWE rate when in fact all J groups have the same measure of location. That is, $\theta_{(1)} = \cdots = \theta_{(J)}$. Once this decision rule has been derived, the goal is to assess the likelihood of making a decision about which group has the largest measure of location as well as the probability of making a correct decision when a decision is made. Here, this is done in the context of the indifference zone given by Eqs. (7.11) and (7.12).

Let $\hat{\theta}_1, \ldots, \hat{\theta}_J$ denote the estimates of $\theta_1, \ldots, \theta_J$, respectively, and let $\hat{\theta}_{(1)} \leq \cdots \leq \hat{\theta}_{(J)}$ denote the estimates written in ascending order. Let $\theta_{\pi(j)}$ denote the measure of location associated with $\hat{\theta}_{(j)}$. Next, for each j $(j = 1, \ldots, J - 1)$ test

$$H_0: \theta_{\pi(j)} = \theta_{\pi(J)}. \tag{7.13}$$

That is, compare the group with the largest estimate to the remaining $J - 1$ groups. When $j = 1$, for example, the group with the lowest estimate is being compared to the group with the highest estimate. If all $J - 1$ of these hypotheses are rejected, decide that the group with largest estimate has the largest measure of location; otherwise no decision is made. Strictly speaking, the goal is not to test the hypotheses given by Eq. (7.13), but rather, determine the extent to which it is reasonable to make a decision about whether $\theta_{\pi(j)} < \theta_{\pi(J)}$. This is consistent with Tukey's (1991) point that surely $\theta_{\pi(j)}$ and $\theta_{\pi(J)}$ differ at some decimal place.

But there are concerns if FWE is controlled using previously described techniques. Using the Bonferroni method, for example, by testing each hypothesis at the $\alpha/(J - 1)$ level, reasonably good control over FWE is achieved when testing at the 0.05 level and $J = 4$. Estimates of FWE greater than 0.07 were found in simulations, but none were above 0.075. But for $J = 5$, this approach can perform poorly: The actual FWE can exceed 0.08. Controlling FWE with Hochberg's method does not improve matters. The same is true using the methods derived by Hommel and Rom. Controlling FWE in an adequate manner requires a technique designed for the situation at hand.

Momentarily focus on trimmed means, in which case, Eq. (7.13) can be tested with Yuen's method. Let p_j denote the p-value based on Yuen's method when testing Eq. (7.13). Suppose this null hypothesis is rejected if $p_j \leq d_j$. The basic strategy is to first determine d_j, so that the individual tests have a Type I error probability α when dealing with normal distributions and there is heteroscedasticity. Then adjust d_j in a manner that controls FWE, and finally, use this adjustment when dealing with heteroscedasticity and non-normality.

The d_j values are determined as follows. Given J and the corresponding sample sizes, and assuming normality and homoscedasticity, a simulation is used to determine the distribution of the p-values when all $J - 1$ hypotheses are true. More precisely, n_j values are generated from a standard normal distribution for group j, a p-value is computed using Yuen's method, and this process is repeated L times, yielding $p_{\ell j}$ $(\ell = 1, \ldots, L, j = 1, \ldots, J - 1)$. Then for fixed j, d_j is the α sample quantile estimate based on $p_{\ell, j}, \ell = 1, \ldots, L$, the p-values when testing $H_0: \mu_{\pi(j)} = \mu_{\pi(J)}$. That is, d_j is determined so that for each j, the Type I error is approximately equal to α if H_0 is true and rejected when $p_j \leq d_j$. It is informative to note that for $\alpha = 0.05$, $J = 4$ and $n_1 = n_2 = n_3 = n_4 = 20$, (d_1, d_2, d_3) is estimated to be (0.0054, 0.0271, 0.0814) based on a simulation with 10,000 replications. That is, when comparing the two groups having the two largest estimates, for example, if the null hypothesis is rejected when the p-value is less than or equal to 0.0814, the Type I error probability will be 0.05. For $n_1 = n_2 = n_3 = n_4 = 100$, the estimate of (d_1, d_2, d_3) is (0.0053, 0.0295, 0.0922).

There remains the goal of controlling the FWE rate. Here, this is done for the case where all J groups have a common measure of location. Note that the matrix of p-values, $\mathbf{P} = (p_{\ell j})$, having L rows and $J - 1$ columns, provides an estimate of the joint distribution of the p-values when the groups have a common trimmed mean. Based on $\mathbf{P}$, the constant a can be determined so that for $c_j = a d_j$,

$$P(p_1 \leq c_1, \ldots, p_{J-1} \leq c_{J-1}) = \alpha, \tag{7.14}$$

when $\mu_{t1} = \cdots = \mu_{tJ}$. That is, determine a so that the proportion of rows associated with $\mathbf{P}$ that satisfy Eq. (7.14) is equal to α. So if the jth hypothesis is rejected when $p_j \leq c_j$, FWE will be approximately equal to α. Note that $a = 1/(J - 1)$ corresponds to the Bonferroni method. This will be called method RS1-Y henceforth.

When dealing with other robust measures of location, a simple way to proceed is to compute p-values using the percentile bootstrap method; otherwise, proceed exactly as done by method RS1, which will be called method RS1-PB. In particular, use the same c_j values stemming from method Y when using a 20% trimmed mean. All indications are that this approach performs fairly well when using the usual sample median, a one-step M-estimator, and the Harrell–Davis estimate of the median (Wilcox, 2019c).

A limitation of method RS1-Y should be stressed. When dealing with five or more groups, there are situations where the FWE rate is not controlled among the groups having a measure of location equal to $\theta_{(J)}$. Moreover, the confidence intervals resulting from this method do not necessarily have simultaneous probability coverage equal to some specified value. For example, when the first two of six groups both have the largest trimmed mean, $\theta_{(J)}$, which happens to be substantially larger than the remaining four groups, with near certainty, the first two groups would be compared using a critical p-value designed for the situation where all six groups have the same trimmed mean. That is, the test will be performed at a level greater than 0.075 when the goal is to have a Type I error probability of 0.05. The result is that the actual Type I error probability will be larger than 0.075 when in fact they have identical values.

A strategy for dealing with this issue is to test each of the $J - 1$ hypotheses at the p_c level, where p_c is chosen so that the FWE rate is less than or equal to α no matter how many hypotheses are true. This is done here by running a simulation where $\theta_1 = \cdots = \theta_J$, and the goal is to determine the distribution of the minimum p-value among the $J - 1$ tests. Then p_c is taken to be the α quantile of this distribution. This will be called method MCWB. A concern with this approach is that the probability of making a decision can be extremely low compared to using method RS2 described below.

Another possibility is to view methods RS1-Y and RS1-PB in the context of the ranking and selection literature. That is, for the situation at hand, the more relevant issues are the likelihood of being able to make a decision, and the probability of making a correct decision when a decision is made. To provide some perspective, consider the case of $J = 4$ groups that have standard normal distributions. Suppose the first group is shifted to have a measure of location δ. That is, consideration is being given to the indifference zone approach as described by Eqs. (7.11) and (7.12). For a common sample size of 40, the probability of making a decision is 0.017, 0.257, and 0.792 for δ equal to 0.2, 0.5, and 0.8, respectively. (These values were estimated based on the R function PMD.PCD in Section 7.4.12.) Note that the three values for δ correspond to what are sometimes considered small, medium, and large effect sizes based on Cohen's d. The corresponding probabilities of a correct decision (PCDs), given that a decision is made, were estimated to be 0.96, 1, and 1 based on a simulation with 10,000 replications. In reality, the probability of making a decision is as large as or larger than indicated when Eq. (7.12) is true because in all likelihood, $\theta_{(1)} < \cdots < \theta_{(J-1)}$ in contrast to the assumption that Eq. (7.11) is true. For $J = 8$ and $n = 60$, and again considering the indifference zone point of view, the probabilities of making a decision based on δ values of 0.2, 0.5, and 0.8 were estimated to be 0.105, 0.344, and 0.935. The PCD values, given that a decision is made, were estimated to be 0.981, 1, and 1.

The illustrations just described provide a crude way of judging the extent to which specified sample sizes might provide a satisfactory probability of making a decision, and making a correct decision given that a decision is made. A more realistic assessment of the PCD would take into account the fact that there is heteroscedasticity. A way of doing this, given some data, would be to generate data having the same standard deviations as the observed data and to replace δ in Eq. (7.12) with $\delta s_{(J)}$, where $s_{(J)}$ is the sample standard deviation for the group with the largest measure of location. This is the approach used by the R function PMD.PCD in the next section.

As previously indicated, the ranking and selection literature assumes that if $\theta_{(J)} - \theta_{(J-1)} \leq \delta$ for some specified value for δ, a researcher is indifferent to which group is declared to have the largest measure of location. But even when $\delta = 0$, presumably, this is not always the case. For example, the group associated with $\theta_{(J-1)}$ might correspond to some medical procedure that is both expensive and invasive, in contrast to the group associated with $\theta_{(J)}$. Even if a Type I error is committed, and the group associated with $\theta_{(J-1)}$ is declared to have the larger measure of location, presumably, additional measures of effect size would be used to assess the extent to which the group associated with $\theta_{(J-1)}$ is best despite any negative features it might have.

Method RS2-GPB

Another possibility is to focus on a method that ensures that the probability of a correct decision is at least $1 - \alpha$ without specifying an indifference zone. One way of accomplishing this goal is via a percentile bootstrap method. The method begins by estimating some robust measure of location for each group and noting which group has the largest estimate. Suppose group j has the largest estimate. Next, generate a bootstrap sample from each group, compute the bootstrap estimates of location, and note whether the group with the largest bootstrap estimate corresponds to j. Repeat this B times, and let P denote the proportion of times the largest bootstrap estimate corresponds to j. The basic strategy is to determine whether P is sufficiently large to justify making the decision that the largest estimate corresponds to the group with the largest population measure of location. Details about this method are summarized in Wilcox (in press), but no details are described here because Wilcox found that an alternative method, described next, is more satisfactory.

Method RS2-IP

Let $p_1, \ldots, p_{J-1}$ denote the resulting p-values when testing Eq. (7.13). Another way of proceeding is to make a decision if $p_m = \max\{p_1, \ldots, p_{J-1}\} \leq \alpha$. This will be called method RS2-IP, which has a close connection to Hochberg's method. The main difference is that here, only a single decision is made. If the largest p-value exceeds α, no decision is made. Again, the maximum probability of making a mistake occurs when $K = 2$ and is less than or equal to $\alpha/2$. This assumes that when comparing two groups, the method used to test hypotheses ensures that the probability of a Type I error is at most α. When dealing with trimmed means, a simple approach is to use Yuen's method as done by method RS1-Y, but only make a decision if the largest p-value is less than or equal to α. Or a percentile bootstrap method can be used, which has the advantage of not being limited to trimmed means. RS2-IP can have a substantially higher probability of making a decision about which group has the largest measure of location compared to method MCWB. But a possible argument for MCWB is that it might make a decision about some of the groups having a smaller measure of location when RS2-IP makes no decision about which group has the largest measure location. For example, MCWB might indicate that groups 1, 2, and 3 have lower measures of location than group 5. But no decision is made about whether group 4 has a lower measure of location compared to group 5, and RS2-IP fails to make a decision as well. Simulations suggest that RS2-IP is more likely to make a decision compared to RS2-GPB.

7.4.12 R Functions anc.best.PV, anc.bestpb, PMD.PCD, RS.LOC.IZ, best.DO, and bestPB.DO

The R function:

$$\text{anc.best.PV}(x, \text{alpha} = 0.05, \text{tr} = 0.2, \text{iter} = 5000, \text{SEED} = \text{TRUE}),$$

performs method RS1-Y for determining whether a decision can be made about which group has the largest trimmed mean. Included is a p-value related to making a decision about which group has the largest trimmed mean. The function computes the c_j values for $\alpha = 0.001(0.001)0.1(0.01)0.99$, and then determines the smallest α value for which all $J-1$ hypotheses are rejected. If the c_j values are known, they can be specified via the argument p.crit. The function returns confidence intervals based on the values of $c_1, \ldots, c_{J-1}$. However, when there are more than four groups, method RS2-IP might be preferred due to concerns about the Type I error probability previously described. In this case, focus on the p-values for the $J-1$ tests that were performed, and only make a decision if all $J-1$ p-values are less than or equal to α. The R function

$$\text{anc.bestpb}(x, \text{loc.fun} = \text{tmean}, \text{nboot} = 3000, \text{p.crit} = \text{NULL}, \text{alpha} = 0.05, \text{iter} = 5000,$$
$$\text{SEED} = \text{TRUE}, \ldots)$$

performs the bootstrap version of method RS1, RS1-PB.

The R function

$$\text{MCWB}(x, \text{tr}=0.2, \text{alpha}=0.05, \text{SEED}=\text{TRUE}, \text{REPS}=5000,\ldots)$$

performs method MCWB. The critical p-value is determined via a simulation assuming normality. The argument REPS indicates how many replications are used.

For RS1-Y, the R function

$$\text{PMD.PCD}(\text{n}=\text{NULL}, \text{delta}=0.5, \text{x}=\text{NULL}, \text{SIG}=\text{NULL}, \text{alpha}=0.05, \text{p.crit}=\text{NULL}, \text{iter}=5000,$$
$$\text{SEED}=\text{TRUE})$$

uses the indifference zone approach, given by Eqs. (7.11) and (7.12), to determine the probability of making a decision (PMD) about which group has the largest measure of location and the probability of a correct decision (PCD) given that a decision is made. The results provide a reasonable approximation of PMD and PCD when using RS2 instead. The argument n is assumed to be a vector of sample sizes having length J, the number of groups. If data are not available, and say the argument n=c(30, 30, 30, 30), the function estimates both PMD and PCD, assuming normality and homoscedasticity, where the last group is shifted by delta standard deviations. The default value for delta is 0.5. Given some data, it can be passed to the function PMD.PCD via the argument x, in which case the simulation estimates of PMD and PCD are based on normal distributions having standard deviations equal to the sample standard deviations stemming from x. The indifference zone is again used where now the argument delta means δ, taken to be delta $\times s_{(J)}$, where $s_{(J)}$ is the sample standard deviation associated with the group with largest sample trimmed mean.

Suppose that, rather than test hypotheses, it is simply decided that the group with the largest estimate is the group with the largest population measure of location. For a specified collection of sample sizes, the R function

$$\text{RS.LOC.IZ(n,J=NULL, locfun=tmean, delta, iter=10000, SEED=TRUE,...)}$$

estimates the probability of a correct decision, assuming normality, in the context of an indifference zone given by Eqs. (7.11) and (7.12). The argument delta corresponds to δ in Eq. (7.12). Setting delta=0.5, for example, indicates that one of the groups has a measure of location 0.5 standard deviations larger than the other groups. If the argument n contains a single value, the number of groups, J, must be specified. Otherwise, the number of groups is taken to be the length of n.

A non-bootstrap method version of method RS2-IP, which is limited to trimmed means, can be used via the R function

$$\text{best.DO(x, tr = 0.2, ...).}$$

The individual tests are performed with Yuen's method. A bootstrap version of method RS2-IP can be applied via the R function

$$\text{bestPB.DO(x,est=tmean,nboot=NA,SEED=TRUE,pr=TRUE,...).}$$

7.4.13 Determining the Order of the Population Trimmed Means

Another classic problem in the ranking and selection literature is identifying the order of J measures of location. When working with trimmed means, for example, the goal is to identify the group with the smallest population trimmed mean, the group with the next largest population trimmed mean, and so on. The classic ranking and selection method makes a decision based on the sample means. That is, the order of the sample means is taken to be the order of the population means. Assuming normality, the goal is to determine the sample sizes so that the sample means have a reasonably high probability of indicating the proper order. This is done using an indifference zone.

It is noted that the multiple-comparison method in Section 7.4.11 is readily extended to the situation at hand. Now the approach is to test

$$H_0: \theta_{\pi(j)} = \theta_{\pi(j+1)} \tag{7.15}$$

($j = 1, \ldots, J - 1$). If all $J - 1$ of these hypotheses are rejected, decide that the correct order of the population trimmed means has been determined. FWE is controlled in essentially the same manner as done in Section 7.4.11.

7.4.14 R Function ord.loc.PV

The R function

```
ord.loc.PV(x, alpha = 0.05, tr = 0.2, iter = 5000, SEED = TRUE)
```

performs the method described in the previous section.

7.4.15 Judging Sample Sizes

Suppose that all pairs of groups are compared with the R function lincon, described in Section 7.4.1, and that one or more of the hypotheses are not rejected. This might occur because there is little or no difference between the groups. But another possibility is that there is an important difference that was missed. One way of trying to distinguish between these two possibilities is to determine how many observations are needed so that the lengths of the confidence intervals are reasonably short. This can be done with an extension of a two-stage method for means that was developed by Hochberg (1975); see Wilcox (2004a).

Imagine that for all $j < k$, the goal is to compute a confidence interval for $\mu_{tj} - \mu_{tk}$ such that the simultaneous probability coverage is $1 - \alpha$ and the length of each confidence interval is $2m$, where m is the margin of error specified by the researcher. Let h be the $1 - \alpha$ quantile of the range of J independent Student t variates having degrees of freedom $h_1 - 1, \ldots, h_J - 1$, respectively, where $h_j = n_j - 2g_j - 1$ is the number of observations in the jth group left after trimming. Table 7.11 reports the $1 - \alpha$ quantiles of this distribution for selected degrees of freedom, $\alpha = 0.05$ and 0.01, when $h_1 - 1 = \cdots = h_J - 1 = \nu$, say. (For $\nu > 59$, the quantiles can be approximated with the quantiles of a Studentized range statistic with ν degrees of freedom.) For unequal sample sizes, a good choice for the degrees of freedom is:

$$\nu = J \left(\sum \frac{1}{h_j - 1} \right)^{-1}.$$

Let

$$d = \left(\frac{m}{h} \right)^2.$$

The total number of observations needed from the jth group is

$$N_j = \max \left(n_j, \left[\frac{s_{jw}^2}{(1 - 2\gamma)^2 d} \right] + 1 \right). \tag{7.16}$$

So $N_j - n_j$ indicates how many more observations are needed from the jth group.

If the additional $N_j - n_j$ observations can be obtained, confidence intervals are computed as follows. Let $\hat{\mu}_{jt}$ be the trimmed mean associated with the jth group based on all N_j values. For all pairwise comparisons, the confidence interval for $\mu_{jt} - \mu_{kt}$, the difference between the population trimmed means corresponding to groups j and k, is

$$(\hat{\mu}_{jt} - \hat{\mu}_{kt}) \pm hb,$$

where

$$b = \max \left(\frac{s_{jw}}{(1 - 2\gamma)\sqrt{N_j}}, \frac{s_{kw}}{(1 - 2\gamma)\sqrt{N_k}} \right).$$

7.4.16 R Function hochberg

The two-stage method for trimmed means, just described, is performed by the R function

hochberg(x,x2=NA,cil=NA,crit=NA,con=0,tr=0.2,alpha= 0.05,iter=10000).

The first-stage data are assumed to be stored in x, and if the second-stage data are available, they are stored in x2, either in a matrix (having J columns) or in list mode. The argument cil is $2m$, the desired length of the confidence intervals. The argument crit is the $1 - \alpha$ quantile of the range of independent Student t variates, some of which are reported in Table 7.11. If crit is not specified, the appropriate quantile is approximated via simulations with the number of replications controlled by the argument iter. As usual, the default amount of trimming, indicated by the argument tr, is 20%. (The function can also handle linear contrasts specified by the argument con, which is used as described, for example, in Section 7.4.1.)

Table 7.11: Percentage points, h, of the range of J independent t variates.

α	$v=5$	$v=6$	$v=7$	$v=8$	$v=9$	$v=14$	$v=19$	$v=24$	$v=29$	$v=39$	$v=59$
						J=2 groups					
0.05	3.63	3.45	3.33	3.24	3.18	30.01	2.94	2.91	2.89	2.85	2.82
0.01	5.37	4.96	4.73	4.51	4.38	4.11	3.98	3.86	3.83	3.78	3.73
						J=3 groups					
0.05	4.49	4.23	4.07	3.95	3.87	3.65	3.55	3.50	3.46	3.42	3.39
0.01	6.32	5.84	5.48	5.23	5.07	4.69	5.54	4.43	4.36	4.29	4.23
						J=4 groups					
0.05	50.05	4.74	4.54	4.40	4.30	4.03	3.92	3.85	3.81	3.76	3.72
0.01	7.06	6.40	60.01	5.73	5.56	50.05	4.89	4.74	4.71	4.61	4.54
						J=5 groups					
0.05	5.47	5.12	4.89	4.73	4.61	4.31	4.18	4.11	4.06	40.01	3.95
0.01	7.58	6.76	6.35	60.05	5.87	5.33	5.12	50.01	4.93	4.82	4.74
						J=6 groups					
0.05	5.82	5.42	5.17	4.99	4.86	4.52	4.38	4.30	4.25	4.19	4.14
0.01	8.00	7.14	6.70	6.39	6.09	5.53	5.32	5.20	5.12	4.99	4.91
						J=7 groups					
0.05	6.12	5.68	5.40	5.21	5.07	4.70	4.55	4.46	4.41	4.34	4.28
0.01	8.27	7.50	6.92	6.60	6.30	5.72	5.46	5.33	5.25	5.16	50.05
						J=8 groups					
0.05	6.37	5.90	5.60	5.40	5.25	4.86	4.69	4.60	4.54	4.47	4.41
0.01	8.52	7.73	7.14	6.81	6.49	5.89	5.62	5.45	5.36	5.28	5.16
						J=9 groups					
0.05	6.60	6.09	5.78	5.56	5.40	4.99	4.81	4.72	4.66	4.58	4.51
0.01	8.92	7.96	7.35	6.95	6.68	60.01	5.74	5.56	5.47	5.37	5.28
						J=10 groups					
0.05	6.81	6.28	5.94	5.71	5.54	5.10	4.92	4.82	4.76	4.68	4.61
0.01	9.13	8.14	7.51	7.11	6.83	6.10	5.82	5.68	5.59	5.46	5.37

Reprinted, with permission, from R. Wilcox, "A table of percentage points of the range of independent t variables," Technometrics, 1983, 25, 201–204.

7.4.17 Measures of Effect Size: Two-Way and Higher Designs

There are various ways one might characterize the differences among groups when dealing with two-way and higher designs beyond the strategy of reporting measures of location. Some are described here, and other possibilities are described in Section 7.9.2.

Interactions: ME1

First consider a two-way design and focus on Factor A. A simple approach is to ignore the levels of Factor B and use one of the measures of effect size covered in Section 5.3.4. That is, for each level of Factor A, pool the data over the levels of Factor B, in which case any two levels of Factor A can be compared using the R function IND.PAIR.ES in Section 7.4.2. This will be called method ME1. The same can be done for Factor B. An R function (twowayESM) designed to simplify this approach is described in the next section.

Interactions: ME2

Another approach, called ME2, is to first focus on level 1 of Factor A and estimate measures of effect size for all pairs of a group among the K levels of Factor B. This could be done for each level of Factor A. As is evident, the same can be done for each level of Factor B. The R function RCES, described in the next section, applies this approach, which has a close connection with the R function RCmcp in Section 7.4.4.

Interactions: IE1

As for interactions, first consider a two-by-two design and focus on the first level of the first factor, Factor A. A simple approach is to estimate some measure of effect size, say $\hat{\delta}_1$ based on the two groups associated with Factor B. Next, do the same for the second level of Factor A, and then use the difference between the two estimates. This will be called method IE1. For a J-by-K design, this could be done for all relevant tetrad cells.

Another Non-Parametric Approach

Note that measures of effect size in Section 5.7 are based on a non-parametric estimate of the distribution of the difference between two random variables, denoted by $\mathcal{D}_{ij} = X_{i1} - X_{j2}$, $i = 1, \ldots, n_1$, $j = 1, \ldots, n_2$. Yet one more approach for measuring effect size when dealing with linear contrasts is to use a similar approach. That is, use a non-parametric estimate of the distribution of $\Psi = \sum c_j X_j$, in which case, the effect size in Section 4.3.1 could be used, as well as an analog of the Wilcoxon–Mann–Whitney measure of effect size. That is, estimate $P(\Psi < 0)$. Section 7.9.2 outlines how this might be done.

7.4.18 R Functions twowayESM, RCES, interES.2by2, interJK.ESmul, and IND.INT.DIF.ES

For all j and j' such that $1 \leq j < j' \leq J$, the R function

$$\text{twowayESM(J, K, x, fun = ES.summary, ...)}$$

computes measures of effect size for main effects based on method ME1, described in the previous section. The R function

$$\text{RCES(J, K, x, fun=ES.summary, tr=0.2,...)}$$

computes measures of effect size for main effects based on method ME2.

There are three functions that deal with interactions via method IE1. The R function

$$\text{interES.2by2(x, tr=0.2, switch=FALSE)}$$

is limited to a 2-by-2 design. It uses all of the measures of effect size available in the R function ESfun, described in Section 5.3.5. The R function

$$\text{interJK.ESmul(J,K, x, method='QS', tr=0.2, SEED=TRUE)}$$

computes a single measure effect size when dealing with a J-by-K design using the approach based on the argument method, which defaults to a quantile shift measure. Regarding choices for the argument method, see the function ESfun in Section 5.3.5. The R function

$$\text{IND.INT.DIF.ES(J,K,x,...)}$$

computes several measures of effect size simultaneously for all interactions.

7.4.19 Comparing Curves (Functional Data)

The notion of functional data was described in Section 6.4.13. It is noted that several ANOVA-type methods have been developed for functional data (e.g., Cuevas et al., 2004; Górecki and Smaga, 2015; Zhang, 2013). R functions for applying some of these methods can be found in the R package fda, as well as the supplementary material associated with the paper by Górecki and Smaga (2015), and at

$$\text{http://orfe.princeton.edu/~jqfan/papers/pub/hanova.s}$$

Robust methods are in need of further developments. This section describes some additional methods and functions that might help when comparing two independent groups.

The data consist of two matrices, each having p columns corresponding to measures taken at p times, where p is typically large. The time points are denoted by $t_1, \ldots, t_p$. Note that for any column, the two groups can be compared using trimmed means. One option is Yuen's method in Section 5.3, or a percentile bootstrap method can be used.

If the curves are compared at K time points, $1 \le K \le p$, and K is relatively small, controlling FWE, the probability of one or more Type I errors, is relatively straightforward. But as K

increases, little is known about how best to control FWE using some robust technique. To provide at least some perspective, suppose the curve for each group is given by $y(t) = \sin(t)$, $0 \leq t \leq 2\pi$, and that data are generated from a Gaussian process with covariance matrix $C(t_j, t_k) = \exp(-1 * |t_j - t_k|)$. Further assume that the curves are compared at K time points evenly spaced over $p = 100$ time points that are available. A strategy is to choose the number of tests to be performed so as to not lose too much power when adjusting the critical value in some manner that controls FWE, the probability of one or more Type I errors. Simultaneously, it might be desired to choose K reasonably large so as to capture any differences that exist. (Of course, another possibility is that a specific range of times is of particular interest.) Once the K time points are chosen, a simple way of controlling the FWE rate is via Hochberg's method. Another strategy is to use critical values based on the Studentized maximum modulus distribution. But in very limited simulations, when using Yuen's test, both of these approaches were found to be rather unsatisfactory with $K = 25$. A percentile bootstrap method coupled with Hochberg's method gave better results.

To elaborate, consider comparing 20% trimmed means using Yuen's test for each $k = 1, \ldots, K$, where $K = 25$. Simulations indicate that using a Studentized maximum modulus distribution can be unsatisfactory when data are generated with sample sizes $n_1 = n_2 = 60$ and the desired probability of one or more Type I errors, FWE, is 0.05. The actual level exceeds 0.08. Controlling FWE using Hochberg's method gives better results, but it can be rather unsatisfactory when $n = 40$. A percentile bootstrap with $B = 2000$ bootstrap samples performed well with $n_1 = n_2 = 20$, where, again, Hochberg's method is used to control FWE. Using $B = 500$, the actual level can exceed 0.08. So there is weak evidence that a percentile bootstrap will perform relatively well, but clearly, a more definitive simulation study is needed.

7.4.20 R Functions funyuenpb and Flplot2g

The R function

```
funyuenpb(x1,x2,tr=0.2,pts=NULL,npts=25,plotit=TRUE,alpha= 0.05, SEED=TRUE,
nboot=2000, xlab='Time', ylab='Est.dif', FBP=TRUE, method='hochberg', COLOR=TRUE)
```

compares the curves corresponding to two independent groups using the percentile bootstrap method described in the previous section. The argument pts can be used to specify the time points where comparisons are to be made. If pts=NULL, the argument npts indicates how many points will be used, which will be evenly spaced over the p time points available. The function creates two functional boxplots. (Functional boxplots were described in Section 6.4.14.) The plot in the left panel is the functional boxplot for the first group. Also shown is a black line indicating the median curve associated with the second group, the idea being to provide some sense on how the typical curve in group 2 compares to all of the curves in group 1. The right panel contains the functional boxplot of group 2. If the argument FBP=FALSE, the function plots the estimated difference between the trimmed means as well as an approximate confidence band. The R function

$$\text{Flplot2g(x1,x2,est=mean,xlab='Time',ylab='Y',plotit=TRUE)}$$

plots the estimate of the typical curve for each group. It applies a measure of location, indicated by the argument est, to each column of the matrices x1 and x2. If plotit=TRUE, the two resulting curves are plotted. The function also returns the estimates of the typical curves.

7.4.21 Comparing Variances and Robust Measures of Scale

A simple way of comparing variances or robust measures of scale is to use methods in Chapter 5 and control the FWE rate using some improvement on the Bonferroni method. The R functions in the next section can be used to apply this approach.

7.4.22 R Functions comvar.mcp and robVARcom.mcp

The R function

$$\text{comvar.mcp(x,method='hoch',SEED=TRUE,SD=TRUE)}$$

compares variances for all pairs of variables using the extension of the Morgan–Pitman test described in Section 5.5.1. The R function

$$\text{robVARcom.mcp(x,est=winvar, alpha=0.05, nboot=2000, method='hoch', SEED=TRUE)}$$

compares robust measures of scale using a percentile bootstrap method.

7.5 A Random Effects Model for Trimmed Means

This section describes two random effects models based on Winsorization and trimmed means. When there is no trimming, and there is homogeneity of variance, both models reduce to the usual model covered in a basic course on the analysis of variance. The first of the two models is convenient when comparing trimmed means, while the other is convenient when estimating a Winsorized analog of the intraclass correlation coefficient.

It is well known that the standard random effects model provides poor control over the probability of a Type I error when the usual assumptions of normality and homogeneous variances are violated. For example, when testing at the $\alpha = 0.05$ level with four groups and equal sample sizes of 20 in each group, the actual probability of a Type I error can exceed 0.3 (Wilcox, 1994b). A striking feature of the model based on 20% trimming is the extent to which this problem is reduced. Among the various distributions considered by Wilcox (1994b), the highest estimated probability of a Type I error was 0.074.

It is assumed that there is a pool of treatment groups that are of interest, but not all groups can be examined. Instead, J randomly sampled groups are used to make inferences about the pool of treatment groups. Once the J groups are randomly sampled, it is assumed that n_j observations are randomly sampled from the jth group. Let X_{ij} be the ith observation randomly sampled from the jth group, and let μ_{tj} be the population trimmed mean. Generalizing the standard random effects model in a natural way, let $\bar{\mu}_w = E_w(\mu_{tj})$. In words, $\bar{\mu}_w$ is the Winsorized mean for the population of trimmed means being sampled. A generalization of the usual random effects model is

$$X_{ij} = \bar{\mu}_w + b_j + \epsilon_{ij},$$

where $b_j = \mu_{tj} - \bar{\mu}_w$, $E_w(b_j) = 0$, and $E_w(\epsilon_{ij}) = 0$. Let the Winsorized variance of b_j be σ_{wb}^2, and for fixed j, let σ_{wj}^2 be the Winsorized variance of ϵ_{ij}. Also let $\sigma_w^2 = E_w(\sigma_{wj}^2)$, where the Winsorized expectation is taken with respect to a randomly sampled group. When there are no differences among the trimmed means associated with the pool of treatment groups under investigation, $\sigma_{wb}^2 = 0$.

Jeyaratnam and Othman (1985) derived a heteroscedastic method for comparing means in a random effects model, which can be extended to trimmed means (Wilcox, 1994b). The computations are summarized in Table 7.12. Advantages of using trimmed means over means are better control over the probability of a Type I error and the potential of substantially higher power, particularly when distributions have heavier than normal tails. However, there are situations where comparing means might mean more power as well. For example, the variation among the means might be larger than the variation among the trimmed means, so the Jeyaratnam and Othman method might yield more power. This might happen, for example, when distributions are skewed. As usual, the optimal amount of trimming will vary from one situation to the next, but to avoid poor power and undesirable power characteristics (a biased test), 20% trimming is a good choice for general use.

7.5.1 A Winsorized Intraclass Correlation

This subsection describes a Winsorized analog of the usual intraclass correlation. Several estimators of this parameter have been considered (Wilcox, 1994c), one of which is described here.

Table 7.12: Comparing trimmed means in a random effects model.

For the jth group, Winsorize the observations by computing Y_{ij} as described in Section 7.5.1. To test the hypothesis of no differences among the trimmed means, $H_0: \sigma^2_{wb} = 0$, let h_j be the effective sample size of the jth group (the number of observations left after trimming), and compute

$$\bar{Y}_j = \frac{1}{n_j} \sum_{i=1}^{n_j} Y_{ij},$$

$$s^2_{wj} = \frac{1}{n_j - 1} \sum (Y_{ij} - \bar{Y}_j)^2,$$

$$\bar{X}_t = \frac{1}{J} \sum \bar{X}_{tj},$$

$$\text{BSST} = \frac{1}{J-1} \sum_{j=1}^{J} (\bar{X}_{tj} - \bar{X}_t)^2,$$

$$\text{WSSW} = \frac{1}{J} \sum_{j=1}^{J} \sum_{i=1}^{n_j} \frac{(Y_{ij} - \bar{Y}_j)^2}{h_j(h_j - 1)},$$

$$D = \frac{\text{BSST}}{\text{WSSW}}.$$

Let

$$q_j = \frac{(n_j - 1)s^2_{wj}}{J(h_j)(h_j - 1)}.$$

The degrees of freedom are estimated to be

$$\hat{v}_1 = \frac{((J-1)\sum q_j)^2}{(\sum q_j)^2 + (J-2)J \sum q_j^2},$$

$$\hat{v}_2 = \frac{(\sum q_j)^2}{\sum q_j^2/(h_j - 1)}.$$

Reject, if $D > f$, the $1 - \alpha$ quantile of an F distribution with $\hat{v}_1$ and $\hat{v}_2$ degrees of freedom.

It is convenient to switch from the ANOVA model in the previous section to one that is slightly different. As before, the model is written as

$$X_{ij} = \bar{\mu}_w + b_j + \epsilon_{ij},$$

but now, $b_j = \mu_{wj} - \bar{\mu}_w$, the difference between the Winsorized mean of the jth group and $\bar{\mu}_w = E_w(\mu_{wj})$, the Winsorized expected value of μ_{wj} with respect to a randomly sampled group. It can be shown that the Winsorized covariance between any two observations in the jth group, X_{ij} and $X_{i'j}$, $i \neq i'$, is σ_{wb}^2. Let $\bar{\sigma}^2 = E_w(\sigma_{wj}^2)$, the Winsorized expected value of the Winsorized variance associated with a randomly sampled group. Then a heteroscedastic, Winsorized analog of the usual intraclass correlation is

$$\rho_{WI} = \frac{\sigma_{wb}^2}{\sigma_{wb}^2 + \bar{\sigma}^2}.$$

Let s_{wj}^2 be the Winsorized sample variance for the jth group, let $m = [\gamma J]$, where as usual, γ is the amount of Winsorization, let $s_{w(1)}^2 \leq \cdots \leq s_{w(J)}^2$ be the Winsorized sample variances written in ascending order, and let

$$SW = (m+1)s_{w(m+1)}^2 + s_{w(m+2)}^2 + \cdots + s_{w(J-m-1)}^2 + (m+1)s_{w(J-m)}^2,$$

and

$$\bar{s}_w^2 = \frac{SW}{J}.$$

Results in Rao et al. (1981) suggest estimating ρ_{WI} with

$$\hat{\rho}_{\text{WI}} = \frac{\hat{\sigma}_{wb}^2}{\hat{\sigma}_{wb}^2 + \bar{s}_w^2},$$

where

$$\hat{\sigma}_{wb}^2 = \frac{1}{J}\sum \ell_j(\bar{Y}_j - \tilde{Y})^2,$$

$$\ell_j = \frac{n_j}{n_j + 1},$$

and

$$\tilde{Y} = \frac{\sum \ell_j \bar{Y}_j}{\sum \ell_j}.$$

Wilcox (1994c) compares the bias and mean squared error of $\hat{\rho}_{\text{WI}}$ with two alternative estimators and recommends $\hat{\rho}_{\text{WI}}$. When ρ_{WI} is close to zero, some type of bias reduction method might have practical value, but this issue needs further study before a recommendation can be made.

7.5.2 R Function rananova

The R function

$$\text{rananova}(x, tr=0.2, grp=NA)$$

performs the computations for the random effects ANOVA, where x is a data frame or matrix or has list mode, tr is the amount of trimming, which defaults to 0.2, and grp can be used to specify some subset of the groups if desired. If grp is not specified, all groups stored in x are used. The function returns the value of the test statistic, D, which is stored in rananova$teststat, the significance level is stored in rananova$p.value, and an estimate of the Winsorized intraclass correlation, which is computed as described in the previous subsection of this chapter, is returned in the R variable rananova$rho.

■ **Example**

> Assuming the data in Table 7.7 are stored in the R object data, the command rananova(data) returns
>
> ```
> $teststat:
> [1] 0.33194
>
> $df:
> [1]   2.520394 18.693590
>
> $p.value1:
> [1] 0.7687909
>
> $rho:
> [1]   0.0576178
> ```

■

7.6 Bootstrap Global Tests

This section describes bootstrap methods for testing the hypothesis that J independent groups have a common measure of location. The first two can be used with any robust estimator of location. The third is designed specifically for comparing M-measures of location. In more formal terms, the goal is to test

$$H_0: \theta_1 = \cdots = \theta_J, \tag{7.17}$$

where θ_j represents any robust measure of location associated with the jth group.

Method SHB

The first method is based on a test statistic mentioned by Schrader and Hettmansperger (1980) and studied by He et al. (1990). The test statistic is

$$H = \frac{1}{N} \sum n_j (\hat{\theta}_j - \bar{\theta})^2,$$

where $N = \sum n_j$ and

$$\bar{\theta} = \frac{1}{J} \sum \hat{\theta}_j.$$

To determine the critical value, shift the empirical distributions of each group so that the estimated measure of location is zero, generate bootstrap samples from each group in the usual way from each of the shifted distributions, and compute the test statistic based on the bootstrap samples, yielding H^*, say. Repeat this B times, resulting in $H_1^*, \ldots, H_B^*$, and put these B values in order, yielding $H_{(1)}^* \leq \cdots \leq H_{(B)}^*$. Then an estimate of an appropriate critical value is $H_{(u)}^*$, where $u = (1 - \alpha)B$, rounded to the nearest integer, and H_0 is rejected if $H > H_{(u)}^*$. For simulation results on how this method performs, see Wilcox (1993d).

Although this method can be used with any robust estimator, it is not recommended when comparing means or even trimmed means (Özdemir et al., 2018). An exception is situations where the goal is to compare medians, based on the Harrell–Davis estimator, and there are tied values. When using method HSB in conjunction with an M-estimator, Özdemir et al. (2020) found that the actual level is less than 0.06, when testing at the 0.05 level, with a common sample size of 25. The main difficulty is that the actual level can drop below 0.02 in some situations.

Method LSB

The second bootstrap method described here is based on a slight variation of a general approach described by Liu and Singh (1997). Keselman et al. (2002) found that it appears to be better than SHB when working with the modified one-step M-estimator (MOM), described in Section 3.10, and it appears to perform reasonably well when using M-estimators (with Huber's Ψ) and even trimmed means, if the amount of trimming is sufficiently high. Moreover, it performs relatively well when comparing medians via the Harrell–Davis estimator as noted in Section 7.1.5. However, an exception was found by Özdemir et al. (2020) when dealing with the M-estimator based on Huber's Ψ, $J = 6$, samples are drawn from a skewed, light-tailed distribution, and there is heteroscedasticity: The actual level was 0.078 when testing at the 0.05 level.

Let

$$\delta_{jk} = \theta_j - \theta_k,$$

where for convenience, it is assumed that $j < k$. That is, the δ_{jk} values represent all pairwise differences among the J groups. When working with means, for example, δ_{12} is the difference between the means of groups 1 and 2 and δ_{35} is the difference for groups 3 and 5. If all J groups have a common measure of location (i.e., $\theta_1 = \cdots = \theta_J$), then, in particular

$$H_0: \delta_{12} = \delta_{13} = \cdots = \delta_{J-1,J} = 0. \tag{7.18}$$

The total number of δs in Eq. (7.18) is $L = (J^2 - J)/2$.

For each group, generate bootstrap samples from the *original* values. That is, the observations are *not* centered as was done in the previous method. Instead, bootstrap samples are generated from the X_{ij} values. For each group, compute the measure of location of interest based on a bootstrap sample and repeat this B times. The resulting estimates of location are represented by $\hat{\theta}^*_{jb}$ ($j = 1, \ldots, J$, $b = 1, \ldots, B$), and the corresponding estimates of δ are denoted by $\hat{\delta}^*_{jkb}$. (That is, $\hat{\delta}^*_{jkb} = \hat{\theta}^*_{jb} - \hat{\theta}^*_{kb}$.) The general strategy is to determine how deeply $\mathbf{0} = (0, \ldots, 0)$ is nested within the bootstrap values $\hat{\delta}^*_{jkb}$ (where $\mathbf{0}$ is a vector having length L). For the special case, where only two groups are being compared, this is tantamount to determining the proportion of times $\hat{\theta}^*_{1b} > \hat{\theta}^*_{2b}$, among all B bootstrap samples, which is how we proceeded in Chapter 5. But here, we need special techniques for comparing more than two groups.

There remains the problem of measuring how deeply $\mathbf{0}$ is nested within the bootstrap values. Several strategies were described in Chapter 6, but in terms of Type I error probabilities and power, all indications are that, for the situation at hand, the choice of method is irrelevant. However, from a computational point of view, the choice of method can matter, for reasons indicated at the end of this section. For the moment, the focus is on using the Mahalanobis distance.

Let $\hat{\delta}_{jk} = \hat{\theta}_j - \hat{\theta}_k$ be the estimate of δ_{jk} based on the original data, and let $\hat{\delta}^*_{jkb} = \hat{\theta}^*_{jb} - \hat{\theta}^*_{kb}$ based on the bth bootstrap sample ($b = 1, \ldots, B$). (It is assumed that $j < k$.) For notational convenience, we rewrite the $L = (J^2 - J)/2$ differences $\hat{\delta}_{jk}$ as $\hat{\Delta}_1, \ldots, \hat{\Delta}_L$ and the corresponding bootstrap values are denoted by $\hat{\Delta}^*_{\ell b}$ ($\ell = 1, \ldots, L$). Let

$$\bar{\Delta}^*_\ell = \frac{1}{B} \sum_{b=1}^B \hat{\Delta}^*_{\ell b},$$

$$Y_{\ell b} = \hat{\Delta}_{\ell b}^{*} - \bar{\Delta}_{\ell}^{*} + \hat{\Delta}_{\ell},$$

(so the $Y_{\ell b}$ values are the bootstrap values shifted to have mean $\hat{\Delta}_{\ell}$), and let

$$S_{\ell m} = \frac{1}{B-1} \sum_{b=1}^{B} (Y_{\ell b} - \bar{Y}_{\ell})(Y_{mb} - \bar{Y}_{m}),$$

where

$$\bar{Y}_{\ell} = \frac{1}{B} \sum_{b=1}^{B} Y_{\ell b}.$$

(Note that in the bootstrap world, the bootstrap population mean of $\bar{\Delta}_{\ell}^{*}$ is known and is equal to $\hat{\Delta}_{\ell}$.) Next, compute

$$D_b = (\hat{\mathbf{\Delta}}_b^{*} - \hat{\mathbf{\Delta}}) \mathbf{S}^{-1} (\hat{\mathbf{\Delta}}_b^{*} - \hat{\mathbf{\Delta}})',$$

where $\hat{\mathbf{\Delta}}_b^{*} = (\hat{\Delta}_{1b}^{*}, \ldots, \hat{\Delta}_{Lb}^{*})$ and $\hat{\mathbf{\Delta}} = (\hat{\Delta}_1, \ldots, \hat{\Delta}_L)$. D_b measures how closely $\hat{\mathbf{\Delta}}_b^{*}$ is located to $\hat{\mathbf{\Delta}}$. If $\mathbf{0}$ (the null vector) is relatively far from $\hat{\mathbf{\Delta}}$, reject. In particular, put the D_b values in ascending order, yielding $D_{(1)} \leq \cdots \leq D_{(B)}$, and let $u = (1 - \alpha)B$, rounded to the nearest integer. Then reject H_0 if

$$T \geq D_{(u)},$$

where

$$T = (\mathbf{0} - \hat{\mathbf{\Delta}}) \mathbf{S}^{-1} (\mathbf{0} - \hat{\mathbf{\Delta}})'.$$

A p-value can be computed as well and is given by

$$\frac{1}{B} \sum_{b=1}^{B} I_b,$$

where the indicator function $I_b = 1$, if $T < D_b$; otherwise, $I_b = 0$.

Notice that with three groups ($J = 3$), $\theta_1 = \theta_2 = \theta_3$ can be true if and only if $\theta_1 = \theta_2$ and $\theta_2 = \theta_3$. So, in terms of Type I errors, it suffices to test

$$H_0\colon \theta_1 - \theta_2 = \theta_2 - \theta_3 = 0$$

as opposed to testing

$$H_0\colon \theta_1 - \theta_2 = \theta_2 - \theta_3 = \theta_1 - \theta_3 = 0,$$

the hypothesis that all pairwise differences are zero. However, if groups differ, then rearranging the groups could alter the conclusions reached if the first of these hypotheses is tested. For example, if the groups have means 6, 4, and 2, then the difference between groups 1 and 2, as well as between 2 and 3, is 2. But the difference between groups 1 and 3 is 4, so comparing groups 1 and 3 could mean more power. That is, we might not reject when comparing group 1 to 2 and 2 to 3, but we might reject if instead we compare 1 to 3 and 2 to 3. To help avoid different conclusions depending on how the groups are arranged, all pairwise differences among the groups were used. However, a consequence of using all pairwise differences is that situations are encountered where the covariance matrix, used when computing the Mahalanobis distance, is singular. This problem can be avoided by replacing the Mahalanobis distance with the projection distance described in Section 6.2.5.

Method TM

A third approach, limited to the M-estimator based on Huber's Ψ, was derived by Özdemir et al. (2020). Let

$$\omega_j = \frac{\frac{1}{\hat{\sigma}_{mj}^2}}{\frac{1}{\hat{\sigma}_{m1}^2} + \frac{1}{\hat{\sigma}_{m2}^2} + \cdots + \frac{1}{\hat{\sigma}_{mj}^2}}, \tag{7.19}$$

where $\hat{\sigma}_{mj}^2$ is the estimated squared standard error of the sample one-step M-estimator for group j based on Eq. (3.32) in Section 3.6.4. Let $\hat{\theta}^+ = \sum \omega_j \hat{\theta}_j$ and

$$T_j = \frac{\hat{\theta}_j - \hat{\theta}^+}{\hat{\sigma}_{mj}}.$$

The test statistic is

$$TM = \sum_{j=1}^{J} T_j^2. \tag{7.20}$$

A critical value is determined via a bootstrap-t method in essentially the same manner as done by method LSB. Based on the criterion that the actual level should be between 0.04 and 0.06 when testing at the 0.05 level, Özdemir et al. (2020) found that method TM tends to be the best of the methods covered in this section.

7.6.1 R Functions b1way, pbadepth, and boot.TM

The R function b1way performs the first of the percentile bootstrap methods described in the previous subsection. It has the general form

b1way(x,est=onestep,alpha= 0.05,nboot=599),

where x is a matrix (with J columns) or has list mode, alpha defaults to 0.05, and nboot, the value of B, defaults to 599. The argument est is any R function that computes a measure of location. It defaults to onestep, the one-step M-estimator with Huber's Ψ.

The R function

$$\text{pbadepth(x,est=onestep,con=0,alpha= 0.05,nboot=2000,grp=NA,op=1,allp=TRUE,}$$
$$\text{MM=FALSE, MC=FALSE,cop=3,SEED=TRUE,na.rm=FALSE, 0...)}$$

performs the other percentile bootstrap method and uses the one-step M-estimator by default. As usual, the argument ... can be used to reset default settings associated with the estimator being used. The argument allp indicates how the null hypothesis is defined. Setting allp=TRUE, all pairwise differences are used. Setting allp=FALSE, the function tests

$$H_0: \theta_1 - \theta_2 = \theta_2 - \theta_3 = \cdots = \theta_{J-1} - \theta_J = 0.$$

The argument op determines how the depth of a point is measured within a bootstrap cloud. The choices are:

- op=1, Mahalanobis depth,
- op=2, Mahalanobis depth but with the usual covariance matrix replaced by the MCD estimate,
- op=3, projection depth computed via the function pdis (that is, use the measure of depth described in Section 6.2.5).

The default is op=1 in order to reduce execution time. Using op=3 avoids a computational error that can occur when the argument allp=T: In some situations, $\mathbf{S}^{-1}$ (as described in the previous section) cannot be computed. This event appears to be rare with $J \leq 4$, but it can occur. This problem might be avoided by setting allp=FALSE, but in terms of power, this has practical concerns already described. With op=3, a measure of depth is used that does not require inverting a matrix. Given the speed of modern computers, perhaps using op=3 routinely is reasonable. If access to a multicore processor is available, setting the argument MC=TRUE will reduce execution time. (Another option is to use the multiple-comparison procedure in Section 7.6.2, which, again, does not require inverting a matrix.)

The R function

$$\text{boot.TM(x,nboot=599)}$$

performs method TM.

■ **Example**

The command b1way(x,est=hd) would test the hypothesis of equal medians using the Harrell–Davis estimator. For the data in Table 7.7, it reports a test statistic of $H = 329$ and a critical value of 3,627. The command b1way(x) would compare M-measures of location instead.

■

7.6.2 M-Estimators and Multiple Comparisons

This section describes two bootstrap methods for performing multiple comparisons that perform relatively well when comparing M-estimators. The first is based in part on a variation of the bootstrap-t method that uses bootstrap estimates of the standard errors. (For relevant simulation results, see Wilcox, 1993d.) The other does not use estimated standard errors but rather relies on a variation of the percentile bootstrap method.

Variation of a Bootstrap-t Method

For the jth group, generate a bootstrap sample in the usual way and compute $\hat{\mu}^*_{mj}$, the M-estimate of location. Repeat this B times, yielding $\hat{\mu}^*_{mjb}$, $b = 1, \ldots, B$. Let

$$\hat{\tau}^2_j = \frac{1}{B-1} \sum_{b=1}^{B} (\hat{\mu}^*_{mjb} - \bar{\mu}^*)^2,$$

where $\bar{\mu}^* = \sum_b \hat{\mu}^*_{mjb}/B$. Let

$$H^*_{jkb} = \frac{|\hat{\mu}^*_{mjb} - \hat{\mu}^*_{mkb} - (\hat{\mu}_{mj} - \hat{\mu}_{mk})|}{\sqrt{\hat{\tau}^2_j + \hat{\tau}^2_k}}$$

and

$$H^*_b = \max H^*_{jkb},$$

where the maximum is taken over all $j < k$. Put the H^*_b values in order, yielding $H^*_{(1)} \leq \cdots \leq H^*_{(B)}$. Let $u = (1 - \alpha)B$, rounded to the nearest integer. Then a confidence interval for $\mu_{mj} - \mu_{mk}$ is

$$(\hat{\mu}_{mj} - \hat{\mu}_{mk}) \pm H^*_{(u)} \sqrt{\hat{\tau}^2_j + \hat{\tau}^2_k},$$

and the simultaneous probability coverage is approximately $1 - \alpha$. With $\alpha = 0.05$, it seems that $B = 399$ gives fairly good probability coverage when all of the sample sizes are greater than or equal to 21.

The same method appears to perform well when using the Harrell–Davis estimate of the median. The extent to which it performs well when estimating other quantiles has not been determined.

Linear contrasts can be examined using a simple extension of the method for performing all pairwise comparisons. Let

$$\Psi_k = \sum_{j=1}^{J} c_{jk} \mu_{mj},$$

$k = 1, \ldots, C$, be C linear combinations of the M-measures of location. As in Section 7.4, the constants c_{jk} are chosen to reflect linear contrasts that are of interest, and for fixed k, $\sum c_{jk} = 0$. Included as a special case is the situation where all pairwise comparisons are to be performed. As before, generate bootstrap samples, yielding $\hat{\mu}^*_{mjb}$, $b = 1, \ldots, B$, and $\hat{\tau}^2_j$, $j = 1, \ldots, J$. Let

$$\hat{\Psi}_k = \sum_{j=1}^{J} c_{jk} \hat{\mu}_{mj},$$

$$\hat{\Psi}^*_{kb} = \sum_{j=1}^{J} c_{jk} \hat{\mu}^*_{mjb},$$

and

$$H^*_{kb} = \frac{|\hat{\Psi}^*_{kb} - \hat{\Psi}_k|}{\sum c^2_{jk} \hat{\tau}^2_j}.$$

Let

$$H^*_b = \max H^*_{kb},$$

where the maximum is taken over all k, $k = 1, \ldots, C$. Then a confidence interval for Ψ_k is

$$\hat{\Psi}_k \pm H^*_{(u)} \sqrt{\sum c^2_{jk} \hat{\tau}^2_k},$$

where, again, $u = (1 - \alpha)B$, rounded to the nearest integer.

A Percentile Bootstrap Method: Method SR

When comparing modified one-step M-estimators, or M-estimators, if the sample sizes are small, an alternative bootstrap method appears to compete well with the method just described. Imagine that hypotheses for each of C linear contrasts are to be tested. For the cth

Table 7.13: Values of α_c for
$\alpha = 0.05$ and 0.01.

c	$\alpha = 0.05$	$\alpha = 0.01$
1	0.02500	0.00500
2	0.02500	0.00500
3	0.01690	0.00334
4	0.01270	0.00251
5	0.01020	0.00201
6	0.00851	0.00167
7	0.00730	0.00143
8	0.00639	0.00126
9	0.00568	0.00112
10	0.00511	0.00101

hypothesis, let $2\hat{p}_c^*$ be the usual percentile bootstrap estimate of the p-value. Put the $\hat{p}_c^*$ values in *descending* order, yielding $\hat{p}_{[1]}^* \geq \hat{p}_{[2]}^* \geq \cdots \geq \hat{p}_{[C]}^*$. Decisions about the individual hypotheses are made as follows. If $\hat{p}_{[1]}^* \leq \alpha_1$, where α_1 is read from Table 7.13, reject all C of the hypotheses. Put another way, if the largest estimated p-value, $2\hat{p}_{[1]}^*$, is less than or equal to α, reject all C hypotheses. If $\hat{p}_{[1]}^* > \alpha_1$, but $\hat{p}_{[2]}^* \leq \alpha_2$, do not reject the hypothesis associated with $\hat{p}_{[1]}^*$, but the remaining hypotheses are rejected. If $\hat{p}_{[1]}^* > \alpha_1$ and $\hat{p}_{[2]}^* > \alpha_2$, but $\hat{p}_{[3]}^* \leq \alpha_3$, do not reject the hypotheses associated with $\hat{p}_{[1]}^*$ and $\hat{p}_{[2]}^*$, but reject the remaining hypotheses. In general, if $\hat{p}_{[c]}^* \leq \alpha_c$, reject the corresponding hypothesis and all other hypotheses having smaller $\hat{p}_m^*$ values. For other values of α (assuming $c > 1$) or for $c > 10$, use

$$\alpha_c = \frac{\alpha}{c}$$

(which corresponds to a slight modification of a sequentially rejective method derived by Hochberg, 1988). This will be called *method SR*.

Method SR, just described, has the advantage of providing Type I error probabilities close to the nominal level for a fairly wide range of distributions when using a one-step M-estimator or the modified one-step M-estimator. It is *not* recommended, however, when the sample sizes are reasonably large, say greater than about 80, or when using some other location estimator, such as a 20% trimmed mean or median. Method SR does not conform to any known multiple-comparison procedure; it represents a slight modification of a method derived by Rom (1990). But as the sample sizes get large, the actual FWE appears to converge to a value greater than the nominal level. Consequently, if any sample size is greater than 80, use the first of the two methods outlined here, or a percentile bootstrap with Hochberg's method, or perhaps the Benjamini–Hochberg method, which are described in Section 7.4.7.

7.6.3 R Functions linconm and pbmcp

The R function

$$\text{linconm(x,con=0, est=onestep, alpha= 0.05, nboot=500, ...)}$$

computes confidence intervals for measures of location using the first of the methods described in the previous subsection. As usual, x is any R object that is a matrix or has list mode, and con is a J-by-C matrix containing the contrast coefficients of interest. If con is not specified, all pairwise comparisons are performed. The argument est is any estimator, which defaults to the one-step M-estimator if unspecified. Again, alpha is α and defaults to 0.05, and nboot is B, which defaults to 399. The final argument, ..., can be any additional arguments required by the argument est. For example, linconm(w) will use the data stored in the R variable w to compute confidence intervals for all pairwise differences between M-measures of location. The command linconm(w,est=hd) will compute confidence intervals for medians based on the Harrell–Davis estimator, while linconm(w,est=hd,q=0.17) computes confidence intervals for the difference between the 0.7 quantiles, again, using the Harrell–Davis estimator. The command linconm(w,bend=1.1) would use an M-estimator, but the default value for the bending constant in Huber's Ψ would be replaced by 1.1.

The function

$$\text{pbmcp(x, tr=0.2, nboot = NA, grp = NA, est = mom, con = 0, bhop=FALSE, ...)}$$

performs multiple comparisons using method SR described in the previous section. (Method SR should not be used when comparing trimmed means.) By default, all pairwise comparisons are performed, but a collection of linear contrasts can be specified via the argument con, which is used as illustrated in Section 7.4.1. With bhop=F, method SR is used, and setting bhop=T, the Benjamini–Hochberg method is applied instead, which is described in Section 7.4.7.

7.6.4 M-Estimators and the Random Effects Model

Little has been done to generalize the usual random effects model to M-estimators. The approach based on the Winsorized expected value does not readily extend to M-estimators unless restrictive assumptions are made. Bansal and Bhandry (1994) consider M-estimation of the intraclass correlation coefficient, but they assume sampling is from an elliptical and permutationally symmetric probability density function.

7.6.5 Other Methods for One-Way Designs

The methods described in this section are far from exhaustive. For completeness, it is noted that Keselman et al. (2002) compare 56 methods based on means, trimmed means with various amount of trimming, and even asymmetric trimming, and two methods based on MOM. Some of the more successful methods, as measured by the ability to control the probability of a Type I error, were based on trimmed means used in conjunction with transformations studied by Hall (1992) and Johnson (1978). More results supporting the use of these transformations can be found in Guo and Luh (2000) and Luh and Guo (1999).

7.7 M-Measures of Location and a Two-Way Design

As was the case when dealing with one-way designs, comparing M-measures of location in a two-way design requires, at the moment, some type of bootstrap method to control the probability of a Type I error, at least when the sample sizes are small. (There are no results on how large the sample sizes must be to avoid the bootstrap.) The method described here was initially used with M-measures of location, but it can be applied when comparing any measure of location, including trimmed means and the MOM.

Let θ be any measure of location and let

$$\Upsilon_j = \frac{1}{K}(\theta_{j1} + \theta_{12} + \cdots + \theta_{jK})$$

($j = 1, \ldots, J$). So, Υ_j is the average of the K measures of location associated with the jth level of Factor A. The hypothesis of no main effects for Factor A is

$$H_0: \Upsilon_1 = \Upsilon_2 = \cdots = \Upsilon_J.$$

A strategy for testing this hypothesis is to proceed along the lines described in the beginning of Section 7.6. For example, one possibility is to test

$$H_0: \Delta_1 = \cdots = \Delta_{J-1} = 0, \tag{7.21}$$

where

$$\Delta_j = \Upsilon_j - \Upsilon_{j+1},$$

$j = 1, \ldots, J - 1$. Briefly, generate bootstrap samples in the usual manner, yielding $\hat{\Delta}_j^*$, a bootstrap estimate of Δ_j. Then proceed as described in Section 7.6. That is, determine how deeply $\mathbf{0} = (0, \ldots, 0)$ is nested within the bootstrap samples. If $\mathbf{0}$ is relatively far from the center of the bootstrap samples, reject.

For reasons previously indicated, the method just described is satisfactory when dealing with the probability of a Type I error, but when the groups differ, this approach might be unsatisfactory in terms of power depending on the pattern of differences among the Υ_j values. One way of dealing with this issue is to compare all pairs of the Υ_j instead. That is, for every $j < j'$, let

$$\Delta_{jj'} = \Upsilon_j - \Upsilon_{j'},$$

and then test

$$H_0: \Delta_{12} = \Delta_{13} = \cdots = \Delta_{J-1,J} = 0. \tag{7.22}$$

Of course, a similar method can be used when dealing with Factor B.

Now a test of the hypothesis of no interaction is described. For convenience, label the JK measures of location as follows:

	Factor B			
	θ_1	θ_2	$\cdots$	θ_K
Factor A	θ_{K+1}	θ_{K+2}	$\cdots$	θ_{2K}
	$\vdots$	$\vdots$	$\cdots$	$\vdots$
	$\theta_{(J-1)K+1}$	$\theta_{(J-1)K+2}$	$\cdots$	$\theta_{JK}.$

Let $\mathbf{C}_J$ be a $(J-1)$-by-J matrix having the form

$$\begin{pmatrix} 1 & -1 & 0 & 0 & \cdots & 0 \\ 0 & 1 & -1 & 0 & \cdots & 0 \\ & & & \vdots & & \\ 0 & 0 & \cdots & 0 & 1 & -1 \end{pmatrix}.$$

That is, $c_{ii} = 1$ and $c_{i,i+1} = -1$, $i = 1, \ldots, J-1$, and $\mathbf{C}_K$ is defined in a similar fashion. One approach to testing the hypothesis of no interactions is to test

$$H_0: \Psi_1 = \cdots = \Psi_{(J-1)(K-1)} = 0,$$

where

$$\Psi_L = \sum c_{L\ell} \theta_\ell,$$

$L = 1, \ldots, (J-1)(K-1)$, $\ell = 1, \cdots, JK$, and $c_{L\ell}$ is the entry in the Lth row and ℓth column of $\mathbf{C}_J \otimes \mathbf{C}_K$. So, in effect, we have a situation similar to that in Section 7.6. That is, generate bootstrap samples, yielding $\hat{\Psi}_L^*$ values, do this B times, and then determine how deeply $\mathbf{0} = (0, \ldots, 0)$ is nested within these bootstrap samples.

A criticism of this approach is that when groups differ, not all relevant differences are being tested, which might affect power. A strategy for dealing with this problem is setting

$$\Psi_{jj'kk'} = \theta_{jk} - \theta_{jk'} + \theta_{j'k} - \theta_{j'k'},$$

for every $j < j'$ and $k < k'$ and then testing

$$H_0: \Psi_{1212} = \cdots = \Psi_{J-1,J,K-1,K} = 0. \tag{7.23}$$

7.7.1 R Functions pbad2way and mcp2a

The R function

  pbad2way(J,K,x,est=mom,conall=TRUE, alpha= 0.05, nboot=2000,grp = NA, ...)

performs the percentile bootstrap method just described, where J and K indicate the number of levels associated with Factors A and B. The argument conall=TRUE indicates that all possible pairs are to be tested, as described, for example, by Eq. (7.23), and conall=FALSE means that the hypotheses having the form given by Eq. (7.21) will be used instead. The remaining arguments are the same as those used in the function pbadepth, described in Section 7.6.1.

■ **Example**

The data mentioned in Exercise 10, at the end of this chapter, are used to illustrate the R function pbad2way. The study involves a 2-by-2 design with weight gain among rats as the outcome of interest. The factors are source of protein (beef versus cereal) and amount of protein (high versus low). Storing the data in the R object weight, the command

  pbad2way(2,2,weight,est=median)

tests all relevant hypotheses using medians. It is left as an exercise to verify that when using R, the p-values for Factors A and B are 0.28 and 0.059, respectively. The test for no interaction has a p-value of 0.13.

■

For convenience, when working with a two-way design, the function

$$\text{mcp2a(J,K,x,est=mom,con=0,tr=0.2,nboot = NA, grp = NA, 0...)}$$

is supplied for performing all pairwise comparisons for both factors and all interactions. The arguments are generally the same as those used by pbad2way. One difference is that the number of bootstrap samples is determined by the function unless a value for nboot is specified. Another is that, if con=0, all pairwise differences, and all tetrad differences when dealing with interactions, are tested. If a particular set of C linear contrasts is of interest, they can be specified by argument con, a JK-by-C matrix.

7.8 Ranked-Based Methods for a One-Way Design

This section describes some rank-based methods for a one-way design. The techniques in this section are aimed at testing the hypothesis that distributions are identical. As noted in Section 5.3, Tukey (1991) argues that testing for exact equality is non-sensical. For example, surely, the population means differ at some decimal place. From this perspective, the rank-based methods described here determine whether the sample sizes are sufficiently large to detect what is presumably true: The distributions differ in some manner.

The classic method for independent groups is the Kruskal–Wallis test, which is satisfactory, in terms of controlling the probability of a Type I error, when comparing groups having identical distributions. (For a more detailed summary of the theoretical properties of the Kruskal–Wallis test, see Brunner et al., 2019.) But when the distributions differ, under general conditions, an incorrect estimate of the standard errors is being used, which might adversely affect power. A method aimed at improving the Kruskal–Wallis test was derived by Rust and Fligner (1984), assuming that tied values occur with probability zero. The explicit goal stated by Rust and Fligner is to test the hypothesis that J groups have a common median, but under general conditions, it fails to do this in a satisfactory manner. Letting $p_{jk} = P(X_{ij} < X_{ik})$, their technique is appropriate for testing the hypothesis that for all J groups, $p_{jk} = 0.15$. However, their method is based on the assumption that the distributions of the J groups differ in location only. If this assumption is violated, in essence, the Rust–Fligner method is testing the hypothesis that the groups have identical distributions. A possible appeal of their method is that it is asymptotically distribution free under weaker conditions than the Kruskal–Wallis test.

A rank-based method that can handle tied values was derived by Brunner et al. (1997). Extensive comparisons with the Rust–Fligner method, when ties occur with probability zero, have not been made. With small sample sizes, the choice of method might make a practical difference, but a detailed study of when this is the case has yet to be performed. Here, it is merely remarked that situations can be constructed where, with a common sample size of 50, the choice of method makes a practical difference. For example, the Rust–Fligner method can reject at the 0.05 level, even though the Brunner et al. method has a p-value equal to 0.188. Even for normal distributions with unequal variances, the p-values resulting from these two methods can differ substantially.

7.8.1 The Rust–Fligner Method

The basic idea is that, if for any x,

$$H_0: F_1(x) = \cdots = F_J(x)$$

is true, meaning that all J groups have identical distributions, and if ranks are assigned based on the pooled data, then the average ranks among the groups should not differ by too much. Let $V(x) = 1$, if $x \geq 0$; otherwise, $V(x) = 0$. Let

$$R_{ij} = \sum_{\ell=1}^{J} \sum_{m=1}^{n_\ell} V(X_{ij} - X_{m\ell})$$

be the rank of X_{ij} among the pooled observations. Let

$$\bar{R}_{.j} = \sum_i R_{ij}/n_j, \quad U_j = \frac{n_j}{N(N+1)}(\bar{R}_{.j} - \bar{R}),$$

where $\bar{R}$ is the average of all the ranks. Let

$$P_{ij\ell} = \sum_{m=1}^{n_\ell} V(X_{ij} - X_{m\ell}), \quad T_{ij} = \sum_{\ell,\ell \neq j}^{J} P_{ij\ell},$$

where the notation $\sum_{\ell,\ell \neq j}$ means summation over all values of ℓ not equal to j. Let $N = \sum n_j$ be the total number of observations. Compute the matrix $\mathbf{A} = (a_{jk})$, where

$$N^3 a_{jj} = \sum_{m=1}^{n_j} (T_{mj} - \bar{T}_{.j})^2 + \sum_{\ell,\ell \neq j} \sum_{m=1}^{n_\ell} (P_{m\ell j} - \bar{P}_{.\ell j})^2,$$

and for $j \neq k$,

$$N^3 a_{jk} = \sum_{\ell, j \neq \ell \neq k} \sum_m (P_{m\ell j} - \bar{P}_{.\ell j})(P_{m\ell k} - \bar{P}_{.\ell k}) -$$

$$\sum_m (P_{mjk} - \bar{P}_{.jk})(T_{mj} - \bar{T}_{.j}) - \sum_m (P_{mkj} - \bar{P}_{.kj})(T_{mk} - \bar{T}_{.k}),$$

where $\bar{P}_{.j\ell} = \sum_i P_{ij\ell}/n_j$ and $\bar{T}_{.j} = \sum_i T_{ij}/n_j$. Letting $\mathbf{U} = (U_1, \ldots, U_J)$, the test statistic is:

$$Q = N \left(\prod_{j=1}^{J} \frac{n_j - 1}{n_j} \right) \mathbf{U}\mathbf{A}^-\mathbf{U}',$$

where $\mathbf{A}^-$ is any generalized inverse of $\mathbf{A}$. (See, for example, Graybill, 1983, for information about the generalized inverse of a matrix.) When the null hypothesis is true, Q has, approximately, a chi-squared distribution with $J - 1$ degrees of freedom. That is, reject H_0 if Q exceeds the $1 - \alpha$ quantile of a chi-squared distribution having $J - 1$ degrees of freedom.

This method can have relatively good power when sampling from heavy-tailed distributions. Because hypothesis testing methods based on robust measures of location are sensitive to different situations, compared to a method based on the average ranks, the Rust–Fligner method can have more power than methods based on robust measures of location, but there are situations where the reverse is true. That is, in terms of maximizing power, the choice of method depends on how the groups differ, which is not known.

7.8.2 R Function rfanova

The R function

$$\text{rfanova(x,grp)}$$

performs the Rust–Fligner method, where x is any R object that is a matrix (with J columns), or a data frame, or x has list mode. The argument grp indicates which groups are to be used. If grp is unspecified, all J groups are used. (If tied values are detected, the function prints a warning message.) The function returns the value of the test statistic, Q, and the p-value.

■ **Example**

Table 7.6 contains data for eight groups of participants. If the data for the first group are stored in film[[1]], the data for the second group in film[[2]], and so on, the function rfanova reports that the test statistic is $Q = 10.04$ and the p-value is 0.19. The command rfanova(film,grp=c(1,3,4)) would compare groups 1, 3, and 4 only.

■

7.8.3 A Rank-Based Method That Allows Tied Values

Brunner et al. (1997) derive an analog of the Kruskal–Wallis test that allows tied values. Like the Rust–Fligner method, the basic idea is that if

$$H_0: F_1(x) = \cdots = F_J(x)$$

is true, then the average ranks among the groups should not differ by too much. Note that if, for example, there is heteroscedasticity, the null hypothesis is false. Again, pool the data and assign ranks. In the event there are tied values, midranks are used. Let R_{ij} be the resulting rank of X_{ij}. Let

$$\bar{R}_j = \frac{1}{n_j} \sum_{i=1}^{n_j} R_{ij},$$

$$\mathbf{Q} = \frac{1}{N}\left(\bar{R}_1 - \frac{1}{2}, \ldots, \bar{R}_J - \frac{1}{2}\right).$$

The vector $\mathbf{Q}$ contains what are called the *relative effects*. For the jth group, compute

$$s_j^2 = \frac{1}{N^2(n_j - 1)} \sum_{i=1}^{n_j} (R_{ij} - \bar{R}_j)^2,$$

and let

$$\mathbf{V} = N\,\mathrm{diag}\left\{\frac{s_1^2}{n_1}, \ldots, \frac{s_J^2}{n_J}\right\}.$$

Let $\mathbf{I}$ be a J-by-J identity matrix, let $\mathbf{J}$ be a J-by-J matrix of 1s, and set $\mathbf{M} = \mathbf{I} - \frac{1}{J}\mathbf{J}$. (The diagonal entries in $\mathbf{M}$ have a common value, a property required to satisfy certain theoretical restrictions.) The test statistic is

$$F = \frac{N}{\mathrm{tr}(\mathbf{M}_{11}\mathbf{V})}\mathbf{QMQ}', \tag{7.24}$$

where tr indicates trace and $\mathbf{Q}'$ is the transpose of the matrix $\mathbf{Q}$. The null hypothesis is rejected if $F \geq f$, where f is the $1 - \alpha$ quantile of an F distribution with

$$\nu_1 = \frac{M_{11}^2[\mathrm{tr}(\mathbf{V})]^2}{\mathrm{tr}(\mathbf{MVMV})}$$

and

$$\nu_2 = \frac{[\mathrm{tr}(\mathbf{V})]^2}{\mathrm{tr}(\mathbf{V}^2\Lambda)}$$

degrees of freedom and $\Lambda = \mathrm{diag}\{(n_1 - 1)^{-1}, \ldots, (n_J - 1)^{-1}\}$.

7.8.4 R Function bdm

The R function

$$bdm(x)$$

performs the BDM rank-based ANOVA method. Here, x can have list mode or it can be a matrix with columns corresponding to groups. The function returns the value of the test statistic, the degrees of freedom, the vector of relative effects, which is labeled q.hat, and the p-value.

■ **Example**

In schizophrenia research, an issue that has received some attention is whether groups of individuals differ in terms of skin resistance (measured in Ohms). In one such study, the groups of interest were no schizophrenic spectrum disorder, schizotypal or paranoid personality disorder, schizophrenia, predominantly negative symptoms, and schizophrenia, predominantly positive symptoms. Data for a portion of this study are stored in the file skin_dat.txt, which can be retrieved as described in Section 1.10. The function bdm returns a p-value of 0.053. The relative effect sizes (the $\mathbf{Q}$ values) are reported as

```
$output$q.hat:
        [,1]
[1,] 0.4725
[2,] 0.4725
[3,] 0.3550
[4,] 0.7000
```

So, the average of the ranks in group 3 is smallest, and the average is highest for group 4.

■

7.8.5 Inferences About a Probabilistic Measure of Effect Size

Method CHMCP

Consider J independent groups. For groups j and k $(1 \leq J < k \leq J)$, let

$$p_{jk} = P(X_{ij} < X_{ik}) + 0.5P(X_{ij} = X_{ik}).$$

As noted in Section 5.7.1, Cliff's method can be used to test

$$H_0: p_{jk} = 0.5, \tag{7.25}$$

and to compute a confidence interval for p_{jk}. But imagine that it is desired to test Eq. (7.25) for all $j < k$, with the goal that the probability of one or more Type I errors is α. A simple method that seems to be relatively effective is to compute p-values for each test and use Hochberg's method, described in Section 7.4.7, to control the probability of one or more Type I errors. This will be called *method CHMCP*. It has been found to be generally preferable to using a Studentized maximum modulus distribution; see Wilcox (2011c).

Method WMWAOV

It is briefly noted how one might test

$$H_0: p_{12} = p_{13} = \cdots = p_{J-1,J} = 0.5. \tag{7.26}$$

Wilcox (2010d) examines several methods for accomplishing this goal, but only the method that performed well in simulations is described here. A limitation of the method is that it assumes tied values never occur; it can perform poorly, in terms of controlling the probability of a Type I error, when this is not the case.

For the jth and kth groups, let D_{jk} be the distribution of $X_j - X_k$. Recall from Section 5.7.1 that $p_{jk} = 0.5$ corresponds to $\theta_{jk} = 0$, where θ_{jk} is the (population) median of D_{jk}. (As pointed out in Section 5.7.1, under general conditions, $\theta_{jk} \neq \theta_j - \theta_k$.) So testing Eq. (7.26) is tantamount to testing

$$H_0: \theta_{12} = \theta_{13} = \cdots = \theta_{J-1,J} = 0. \tag{7.27}$$

Let X_{ij} ($i = 1, \ldots, n_j$, $j = 1, \ldots, J$) be a random sample of size n_j from the jth group. Generate a bootstrap sample from jth group by randomly sampling with replacement n_j observations from $X_{1j}, \ldots, X_{n_j j}$, which will be labeled $X_{1j}^*, \ldots, X_{n_j j}^*$. Let M_{jk}^*, $j < k$, be the usual sample median based on the $n_j n_k$ differences $X_{ij}^* - X_{\ell k}^*$ ($i = 1, \ldots, n_j$, $\ell = 1, \ldots, n_k$). Repeat this process B times, yielding M_{jkb}^*, $b = 1, \ldots, B$. So M_{jkb}^* represents B vectors, each having length $(J^2 - J)/2$. From Liu and Singh (1997), a p-value for testing Eq. (7.27) can be obtained by measuring how deeply $\mathbf{0} = (0, \ldots, 0)$ is nested within the bootstrap cloud of points. More precisely, let $\mathbf{G_0}$ be the depth of the null vector $(0, \ldots, 0)$ based on the notion of projection distance as described in Section 6.2.5. The projection distance of $\mathbf{M}_b^* = (M_{12b}^*, \ldots M_{J-1,Kb}^*)$ from the center of the bootstrap data cloud is denoted by G_b. Then a p-value is

$$\frac{1}{B} \sum I_b,$$

where the indicator function $I_b = 1$, if $G_0 < G_b$; otherwise, $I_b = 0$. This will be called method WMWAOV. Though seemingly rare, it is possible for method WMWAOV to correctly reject even though method CHMCP finds no differences. Also, when there are tied values, the actual level might be substantially lower than the nominal level. Replacing the median with the Harrell–Davis estimator can help deal with this issue.

Method DBH

When performing all pairwise comparisons, there is a variation of method WMWAOV that should be mentioned. For each pair of groups, apply method WMWAOV and control the probability of one or more Type I errors using Hochberg's method. This will be called method DBH. Note that with only two groups, a p-value can be computed as described, for example, in Section 4.4.1. (If $\hat{p}^*$ is the proportion of M^*_{12b} values less than 0, the p-value is $2\min(\hat{p}^*, 1 - \hat{p}^*)$.) Simulation results indicate that the actual probability of one or more Type I errors will be closer to the nominal level compared to method CHMCP (Wilcox, 2011c). Moreover, method DBH might provide a bit more power. But DBH is not recommended when tied values can occur.

7.8.6 R Functions cidmulv2, wmwaov, and cidM

The R function

$$cidmulv2(x,alpha= 0.05, g=NULL, dp=NULL, CI.FWE=FALSE)$$

tests Eq. (7.25). The output includes a column headed by p.crit, which indicates how small the p-value must be in order to reject using Hochberg's method. If the argument CI.FWE=FALSE, the function returns confidence intervals for each p_{jk} having probability coverage $1 - \alpha$. If CI.FWE=TRUE, the probability coverage corresponds to the "critical" p-value used to make decisions about rejecting Eq. (7.25) based on Hochberg's method. For example, if the goal is to have the probability of one or more Type I errors equal to 0.05, and the second largest p-value is less than or equal to 0.025, Hochberg's method rejects. The confidence interval returned by cidmulv2, for the two groups corresponding to the situation having the second largest p-value, will have probability coverage $1 - 0.025 = 0.1975$. For the next largest p-value, the probability coverage will be $1 - 0.05/3$, and so on. If the argument g is specified, it is assumed that x is a matrix with the dependent variable stored in column dp, and the levels of the factors stored in column g.

The R function

$$wmwaov(x,nboot=500,MC=FALSE, SEED=TRUE, pro.dis=TRUE ,MM=FALSE)$$

performs the WMWAOV method. Setting the argument MC=TRUE, the function takes advantage of a multicore processor, assuming one is available.

Finally, the R function

cidM(x,nboot=1000,alpha= 0.05,MC=FALSE, SEED=TRUE ,g=NULL, dp=NULL)

performs method DBH. (Both wmwaov and cidM check for tied values and print a warning message if any are found.)

7.9 A Rank-Based Method for a Two-Way Design

This section describes a rank-based method for a two-way design derived by Akritas et al. (1997). The basic idea stems from Akritas and Arnold (1994) and is based on the following point of view. For any value x, let

$$\bar{F}_{j.}(x) = \frac{1}{K} \sum_{k=1}^{K} F_{jk}(x)$$

be the average of the distributions among the K levels of Factor B corresponding to the jth level of Factor A. The hypothesis of no main effects for Factor A is

$$H_0 \colon \bar{F}_{1.}(x) = \bar{F}_{2.}(x) = \cdots = \bar{F}_{J.}(x)$$

for any x. Letting

$$\bar{F}_{.k}(x) = \frac{1}{J} \sum_{j=1}^{J} F_{jk}(x)$$

be the average of the distributions for the kth level of Factor B, the hypothesis of no main effects for Factor B is

$$H_0 \colon \bar{F}_{.1}(x) = \bar{F}_{.2}(x) = \cdots = \bar{F}_{.K}(x).$$

As for interactions, first consider a 2-by-2 design. Then no interaction is taken to mean that for any x,

$$F_{11}(x) - F_{12}(x) = F_{21}(x) - F_{22}(x),$$

which has a certain similarity to how no interaction based on means is defined. Here, no interaction in a J-by-K design means that for any two rows and any two columns, there is no interaction as just described. From a technical point of view, a convenient way of stating the hypothesis of no interactions among all JK groups is with

$$H_0\colon F_{jk}(x) - \bar{F}_{j.}(x) - \bar{F}_{.k}(x) + \bar{F}_{..}(x) = 0,$$

for any x, all j ($j = 1, \ldots, J$), and all k ($k = 1, \ldots, K$), where

$$\bar{F}_{..}(x) = \frac{1}{JK} \sum_{j=1}^{J} \sum_{k=1}^{K} F_{jk}(x).$$

Note that the null hypotheses are false when there is heteroscedasticity. Luepsen (2018) demonstrates that for this situation, the Brunner et al. method can have relatively high power compared to competing techniques.

The computations begin by pooling all of the data and assigning ranks. For convenience, let $L = JK$ and let $R_{i\ell}$ be the ranks of the ℓth group ($\ell = 1, \ldots, L$), where the first K groups correspond to the first level of the first factor, the next K correspond to the second level of the first factor, and so on. Let

$$\bar{R}_{\ell} = \frac{1}{n_{\ell}} \sum R_{i\ell}$$

and

$$s_{\ell}^2 = \frac{1}{N^2(n_{\ell} - 1)} \sum (R_{i\ell} - \bar{R}_{\ell})^2,$$

where $N = \sum n_{\ell}$ is the total sample size. Set

$$\mathbf{V} = N \operatorname{diag} \left\{ \frac{s_1^2}{n_1}, \ldots, \frac{s_L^2}{n_L} \right\}.$$

Let $\mathbf{I}_J$ be a J-by-J identity matrix, let $\mathbf{H}_J$ be a J-by-J matrix of 1s, and let

$$\mathbf{P}_J = \mathbf{I}_J - \frac{1}{J} \mathbf{H}_J, \quad \mathbf{M}_A = \mathbf{P}_J \otimes \frac{1}{K} \mathbf{H}_K,$$

$$\mathbf{M}_B = \frac{1}{J} \mathbf{H}_J \otimes \mathbf{P}_K, \quad \mathbf{M}_{AB} = \mathbf{P}_J \otimes \mathbf{P}_K.$$

(The notation $\otimes$ refers to the right Kronecker product.) Let

$$\mathbf{Q} = \frac{1}{N} \left(\bar{R}_1 - \frac{1}{2}, \ldots, \bar{R}_L - \frac{1}{2} \right)$$

be the *relative effects*. The test statistics are

$$F_A = \frac{N}{\text{tr}(\mathbf{M}_{A11}\mathbf{V})}\mathbf{Q}\mathbf{M}_A\mathbf{Q}', \quad F_B = \frac{N}{\text{tr}(\mathbf{M}_{B11}\mathbf{V})}\mathbf{Q}\mathbf{M}_B\mathbf{Q}',$$

$$F_{AB} = \frac{N}{\text{tr}(\mathbf{M}_{AB11}\mathbf{V})}\mathbf{Q}\mathbf{M}_{AB}\mathbf{Q}'.$$

For Factor A, reject, if $F_A \geq f$, where f is the $1 - \alpha$ quantile of an F distribution with degrees of freedom

$$\nu_1 = \frac{M_{A11}^2[\text{tr}(\mathbf{V})]^2}{\text{tr}(\mathbf{M}_A\mathbf{V}\mathbf{M}_A\mathbf{V})}, \quad \nu_2 = \frac{[\text{tr}(\mathbf{V})]^2}{\text{tr}(\mathbf{V}^2\Lambda)},$$

where $\Lambda = \text{diag}\{(n_1 - 1)^{-1}, \ldots, (n_L - 1)^{-1}\}$. Here, M_{A11} is the first diagonal element of the matrix $\mathbf{M}_A$. (By design, all of the diagonal elements of $\mathbf{M}_A$ have a common value.) For Factor B, reject, if $F_B \geq f$, where

$$\nu_1 = \frac{M_{B11}^2[\text{tr}(\mathbf{V})]^2}{\text{tr}(\mathbf{M}_B\mathbf{V}\mathbf{M}_B\mathbf{V})}.$$

(The value for ν_2 remains the same.) As for the hypothesis of no interactions, reject, if $F_{AB} \geq f$, where now

$$\nu_1 = \frac{M_{AB11}^2[\text{tr}(\mathbf{V})]^2}{\text{tr}(\mathbf{M}_{AB}\mathbf{V}\mathbf{M}_{AB}\mathbf{V})},$$

and ν_2 is the same value used to test for main effects.

7.9.1 R Function bdm2way

The R function

$$\text{bdm2way(J, K, x)}$$

performs the two-way ANOVA method described in the previous section.

7.9.2 The Patel–Hoel, De Neve–Thas, and Related Approaches to Interactions

This section describes measures of effect size for interactions that are based on the IE2 approach mentioned in Section 7.4.13. The first is a rank-based approach that was proposed by Patel and Hoel (1973). It can be used even when there are tied values. For a review of related techniques, see Gao and Alvo (2005).

Let $X_1, \ldots, X_4$ be four independent variables associated with the levels of a 2-by-2 design. Temporarily assume ties occur with probability zero, and let

$$p_1 = P(X_1 < X_2)$$

and

$$p_2 = P(X_3 < X_4).$$

The Patel–Hoel definition of no interaction is $p_1 = p_2$. That is, the probability of an observation being smaller under level 1 of Factor B, versus level 2, is the same for both levels of Factor A. Tied values can be handled as described in Section 5.7, in which case, the hypothesis of no interaction is

$$H_0\colon p_1 = p_2.$$

Again, temporarily ignore level 2 of Factor A, and note that the two independent groups corresponding to the two levels of Factor B can be compared in terms of δ as described in Section 5.7.1. Let δ_1 represent δ when focusing on level 1 of Factor A, with level 2 ignored, and let $\hat{\delta}_1$ be the estimate of δ as given by Eq. (5.20). An estimate of the squared standard error of $\hat{\delta}_1$, $\hat{\sigma}_1^2$, is described in Section 5.7.1 as well. Similarly, let δ_2 be the estimate of δ_2 when focusing on level 2 of Factor A, with level 1 ignored, and denote its estimate with $\hat{\delta}_2$. The estimated squared standard error of $\hat{\delta}_2$ is denoted by $\hat{\sigma}_2^2$. It can be seen that the null hypothesis of no interaction just defined corresponds to

$$H_0\colon \Delta = \frac{\delta_2 - \delta_1}{2} = 0.$$

An estimate of $p_1 - p_2$ is

$$\hat{\Delta} = \frac{\hat{\delta}_2 - \hat{\delta}_1}{2},$$

the estimated squared standard error of $\hat{\Delta}$ is

$$S^2 = \frac{1}{4}(\hat{\sigma}_1^2 + \hat{\sigma}_2^2),$$

and a $1 - \alpha$ confidence interval for Δ is

$$\hat{\Delta} \pm z_{1-\alpha/2} S,$$

where $z_{1-\alpha/2}$ is the $1 - \alpha/2$ quantile of a standard normal distribution. The hypothesis of no interaction is rejected if this confidence interval does not contain zero.

For the more general case of a J-by-K design, an analog of Dunnett's T3 method is used to control FWE. When working with levels j and j' of Factor A and levels k and k' of Factor B, we represent the parameter Δ by $\Delta_{jj'kk'}$, its estimate is labeled $\hat{\Delta}_{jj'kk'}$, and the estimated squared standard error is denoted by $S^2_{jj'kk'}$. For every $j < j'$ and $k < k'$, the goal is to test

$$H_0: \Delta_{jj'kk'} = 0.$$

The total number of hypotheses to be tested is

$$C = \frac{J^2 - J}{2} \times \frac{K^2 - K}{2}.$$

The critical value, c, is the $1 - \alpha$ quantile of the C-variate Studentized maximum modulus distribution with degrees of freedom $\nu = \infty$. The confidence interval for $\Delta_{jj'kk'}$ is

$$\hat{\Delta}_{jj'kk'} \pm c S_{jj'kk'},$$

and the hypothesis of no interaction, corresponding to levels j and j' of Factor A and levels k and k' of Factor B, is rejected if this confidence interval does not contain zero. (For relevant simulation results, see Wilcox, 2000.)

De Neve and Thas (2017) suggest using instead

$$P = P(X_1 - X_2 < X_3 - X_4) \tag{7.28}$$

as a measure of effect size. That is, for randomly sampled observations from each of the four groups, focus on the probability that the difference between the first two values is less than the difference between the last two values. They refer to P as the interaction probability of superiority (IPS). An unbiased estimate of P is

$$\hat{P} = \frac{1}{n_1 n_2 n_3 n_4} \sum \sum \sum \sum I(X_{i1} - X_{j2} < X_{k3} - X_{m4}), \tag{7.29}$$

where the indicator function $I(X_{i1} - X_{j2} < X_{k3} - X_{m4}) = 1$ if $X_{i1} - X_{j2} < X_{k3} - X_{m4}$; otherwise, $I(X_{i1} - X_{j2} < X_{k3} - X_{m4}) = 0$. Currently, the best method for computing a confidence interval for P is a percentile bootstrap method where bootstrap samples of size $N = \min\{n_1, n_2, n_3, n_4\}$ are used for all four groups (Multach and Wilcox, 2017).

Generalizing, note that a measure of effect size relevant to interactions could be based on the estimate of the distribution of $X_1 - X_2$ and $X_3 - X_4$. For example, let $D_{1ij} = X_{i1} - X_{j2}$ and $D_{2km} = X_{k3} - X_{m4}$ ($i = 1, \ldots, n_1$, $j = 1, \ldots, n_2$, $k = 1, \ldots, n_3$, $m = 1, \ldots, n_4$). That is, for level 1 of Factor A, estimate the distribution of the typical difference for levels 1 and 2 of Factor B. Do the same for level 2 of Factor A. Then compute a measure of effect size based on the estimate of these two distributions using some method in Section 5.3.4. The R function inter.TDES in the next section deals with this approach.

It is convenient to describe this process in a slightly different fashion. Let $N = \prod n_j$, and note that there are N values for $X_{i1} - X_{j2} - X_{k3} + X_{m4}$ ($i = 1, \ldots, n_1$, $j = 1, \ldots, n_2$, $k = 1, \ldots, n_3$, $m = 1, \ldots, n_4$). For convenience, denote these N values by $W_1, \ldots, W_N$. Then a measure of effect size can be computed based on $W_1, \ldots, W_N$. The estimate of P given by Eq. (7.29) is an estimate of $P(W < 0)$.

The approach just described can be generalized further when dealing with $J \geq 1$ groups given linear contrast coefficients $c_1, \ldots, c_J$. The basic idea is to estimate some measure of effect size based on an estimate of the distribution of $W = \sum c_j X_j$. Two possibilities are the quantile shift measure of effect and the standardized difference as described in Section 4.3.2. Or $P(W < 0)$ could be used, which contains the approach used by De Neve and Thas (2017) as a special case.

When estimating P via Eq. (7.29), the distribution of W was estimated based on $n_1 n_2 n_3 n_4$ W values. In general, when dealing with J groups, mimicking this approach exactly would require computing $\prod n_j$ W values, which can exceed the memory capacity of the typical laptop computer. Here, the strategy for dealing with this issue is as follows. Let $L = \min\{n_1, \ldots, n_J\}$, and for each j ($j = 1, \ldots, J$), let U_{ij} ($i = 1, \ldots, L$) denote L values sampled without replacement from $X_{1j}, \ldots X_{n_j j}$, and let $W_i^* = \sum c_j U_{ij}$. Based on these W^* values, compute a measure of effect size, yielding $\hat{\delta}$. Repeat this process I times, yielding $\hat{\delta}_1, \ldots, \hat{\delta}_I$. The final estimate of the effect size is

$$\bar{\delta} = \frac{1}{I} \sum \hat{\delta}_i. \tag{7.30}$$

7.9.3 R Functions rimul, inter.TDES, LCES, linplot, and plot.inter

The R function

$$\text{rimul(J,K,x,p=J*K,grp=c(1:p), plotit=TRUE ,op=4)}$$

performs the Patel–Hoel test of no interaction. (The argument p=J*K is not important in applied work; it is used to deal with certain conventions in R.) The groups are assumed to be arranged as in Section 7.2.1, and the argument grp is explained in Section 7.2.1 as well. If $J = K = 2$ and plotit=TRUE, the function plots an estimate of the distribution of $X_1 - X_2$ and $X_3 - X_4$ via the function g2plot in Section 5.1.7. The argument op is relevant to g2plot and controls the type of plot that is created.

The R function

$$\text{inter.TDES(x,iter=5,SEED=TRUE,tr=0.2, pr=FALSE, switch=FALSE)}$$

computes six measures of effect size designed for an interaction. It is limited to a 2-by-2 design. Setting switch=TRUE, rows and columns are interchanged. The function estimates the distribution of $X_1 - X_2$ as well as $X_3 - X_4$, and then estimates a measure effect size based on these two distributions. The first, labeled DNT, is the De Neve–Thas measure of effect size given by Eq. (7.28). The others are the same as those computed by the function ESfun, described in Section 5.3.5, which correspond to the measures of effect size in Section 5.3.4. (Method DNT is an analog of the method labeled WMW in Section 5.3.5.) When using explanatory power or a measure based on the trimmed mean, the standard error of the estimate can be relatively high, with a common sample size of 50 and $J > 4$.

The R function

$$\text{LCES(x,con, nreps=200, tr=0.2, SEED=TRUE)}$$

computes $\bar{\theta}$ given by Eq. (7.30) for four measures of effect size: a quantile shift based on the median, a quantile shift based on the trimmed mean, a standardized difference from the null value zero, and an estimate of $P(W < 0)$. The first three correspond to measures of effect size described in Section 4.3.2, where $\hat{\theta}$ is taken to be the median or a 20% trimmed mean. The amount of trimming can be altered via the argument tr. The functions con2way and con3way in Section 7.4.4 can be used to generate the linear contrast coefficients relevant to main effects and interaction when dealing with a two-way or three-way design. For example, if x is a matrix with 12 columns containing data for a 3-by-2-by-2 design, the command

$$\text{LCES(x,con=con3way(3,2,2)\$conA)}$$

would estimate effect sizes for the main effects of Factor A.

The R function

$$\text{linplot(x,con,nreps=200, SEED=TRUE, xlab='DV',ylab=")}$$

plots an estimate of the distribution of $W = \sum c_j X_j$. Here, the argument con must be a single vector of linear contrast coefficients. For convenience, the R function

$$\text{plot.inter(x,nreps=500,SEED=TRUE,xlab='DV',ylab='' ')}$$

is supplied for dealing with De Neve and Thas types of interactions.

7.10 MANOVA Based on Trimmed Means

Multivariate analysis of variance (MANOVA) deals with a generalization of ANOVA to situations where two or more measures are taken on each participant. More formally, consider J independent groups, where, for each participant, p measures are taken. For the jth group, denote the p trimmed means by $\boldsymbol{\mu}_j = (\mu_{tj1}, \ldots, \mu_{tjp})$. The goal is to test

$$H_0: \boldsymbol{\mu}_1 = \cdots = \boldsymbol{\mu}_J. \tag{7.31}$$

One approach is to let $\boldsymbol{\delta}_j = \boldsymbol{\mu}_j - \boldsymbol{\mu}_J$ $(j = 1, \ldots, J-1)$ and test $H_0: \boldsymbol{\delta}_1 = \cdots = \boldsymbol{\delta}_{J-1} = 0$. This could be done with the bootstrap method in Section 6.7.1. Also see Don and Olive (2019). The focus here is on a method that does not require designating a reference group. (For results on a two-way MANOVA that is focused on means, see Zhang, 2011; Vallejo and Ato, 2012. Evidently, extensions of these methods to trimmed means have not been studied.)

Johansen (1980) derives a method for means that allows the covariances associated with the J groups to differ, in contrast to classic methods that assume the J groups have a common covariance matrix. The method represents a heteroscedastic approach to what is called the *general linear model*. Johansen assumes normality, but the method can be extended to trimmed means as described here. For the two-sample case, comparisons with a method derived by Kim (1992b), as well as several other methods, are reported by Wilcox (1995d). Lix et al. (2005) compare several methods based on both means and a 20% trimmed mean, again, for the two-sample case. No single method dominated, and the extent to which the generalization of Johansen's method used here competes well with the methods compared by Lix et al. is unclear.

The version of Johansen's method used here, extended to trimmed means, is applied as follows. For the jth group, there are n_j randomly sampled vectors of observations denoted by $(X_{ij1}, \ldots, X_{ijp})$, $i = 1, \ldots, n_j$. Let $\bar{\mathbf{X}}_j = (\bar{X}_{tj1}, \ldots, \bar{X}_{tjp})$ denote the vector of trimmed means, and let $\mathbf{V}_j$ be the Winsorized covariance matrix. Compute

$$\tilde{R}_j = \frac{n_j - 1}{(n_j - 2g_j)(n_j - 2g_j - 1)} \mathbf{V}_j,$$

where $g_j = \gamma n_j$, rounded down to the nearest integer, and γ is the amount of trimming,

$$\mathbf{W}_j = \tilde{R}_j^{-1},$$

$$\mathbf{W} = \sum \mathbf{W}_j,$$

and

$$A = \frac{1}{2} \sum_{j=1}^{J} [\{tr(\mathbf{I} - \mathbf{W}^{-1}\mathbf{W}_j)\}^2 + tr\{(\mathbf{I} - \mathbf{W}^{-1}\mathbf{W}_j)^2\}]/f_j,$$

where $f_j = n_j - 2g_j - 1$. The estimate of the population trimmed means, assuming H_0 is true, is

$$\hat{\mu}_t = \mathbf{W}^{-1} \sum \mathbf{W}_j \bar{\mathbf{X}}_j.$$

The test statistic is

$$F = \sum_{j=1}^{J} \sum_{k=1}^{p} \sum_{m=1}^{p} w_{mkj}(\bar{X}_{mj} - \hat{\mu}_m)(\bar{X}_{kj} - \hat{\mu}_k), \tag{7.32}$$

where w_{mkj} is the mkth element of $\mathbf{W}_j$, $\bar{X}_{mj}$ is the mth element of $\bar{\mathbf{X}}_j$, and $\hat{\mu}_m$ is the mth element of $\hat{\mu}_t$. Reject the null hypothesis if

$$F \geq c + \frac{c}{2p(J-1)}\left\{A + \frac{3cA}{p(J-1)+2}\right\},$$

where c is the $1 - \alpha$ quantile of a chi-squared distribution with $p(J-1)$ degrees of freedom.

Note that the MANOVA method based on trimmed means uses a measure of location that does not take into account the overall structure of the data. Todorov and Filzmoser (2010) derive a MANOVA method based on the MCD estimator, which does take into account the overall structure, but their method assumes that groups differ in location only.

For the special case, where the goal is to compare two groups only, Yanagihara and Yuan (2005) derive a method for comparing means that compares well to several other heteroscedastic methods, in terms of controlling the probability of a Type I error, when sampling from multivariate normal distributions. Currently, there are no published papers comparing the small-sample properties of the extended Yanagihara and Yuan method to the extension of Johansen's method. (A few simulations were run by the author using a 20% trimmed mean. Situations were found where the extended Yanagihara and Yuan method provides more satisfactory control over the probability of a Type I error, and no situation has been found where the reverse is true, but a more comprehensive study is needed.)

Let

$$T = (\bar{\mathbf{X}}_1 - \bar{\mathbf{X}}_2)'(\tilde{R}_1 + \tilde{R}_2)^{-1}(\bar{\mathbf{X}}_1 - \bar{\mathbf{X}}_2),$$

$$\bar{\mathbf{V}} = \frac{n_2}{n}\mathbf{V}_1 + \frac{n_1}{n}\mathbf{V}_2,$$

where $n = n_1 + n_2$,

$$P_1 = \frac{n_2^2(n-2)}{n^2(n_1-1)}\{\text{tr}(\mathbf{V}_1\bar{\mathbf{V}}^{-1})\}^2 + \frac{n_1^2(n-2)}{n^2(n_2-1)}\{\text{tr}(\mathbf{V}_2\bar{\mathbf{V}}^{-1})\}^2,$$

$$P_2 = \frac{n_2^2(n-2)}{n^2(n_1-1)} \mathrm{tr}(\mathbf{V_1}\bar{\mathbf{V}}^{-1}\mathbf{V_1}\bar{\mathbf{V}}^{-1}) + \frac{n_1^2(n-2)}{n^2(n_2-1)} \mathrm{tr}(\mathbf{V_2}\bar{\mathbf{V}}^{-1}\mathbf{V_2}\bar{\mathbf{V}}^{-1}),$$

and

$$\hat{\nu} = \frac{(h-2-P_1)^2}{(h-2)P_2 - P_1},$$

where $h = h_1 + h_2$ and $h_j = n_j - 2g_j$ $(j = 1, 2)$. The test statistic, based on an extension of the Yanagihara–Yuan method to trimmed means, is

$$T_f = \frac{n-2-P_1}{(n-2)p} T,$$

which has, approximately, an F distribution with p and $\hat{\nu}$ degrees of freedom when the null hypothesis is true. When $p = 1$, this method reduces to Yuen's method described in Section 5.3.

7.10.1 R Functions MULtr.anova, MULAOVp, fac2Mlist, and YYmanova

The R function

$$\mathrm{MULtr.anova}(x, J = NULL, p = NULL, tr=0.2, alpha= 0.05)$$

performs the robust MANOVA method based on the extension of Johansen's method to trimmed means. The argument J defaults to NULL, meaning that x is assumed to have list mode with length J, where x[[j]] contains a matrix with n_j rows and p columns, $j = 1, \ldots, J$. If the arguments J and p are specified, the data can be stored in list mode or a matrix. If stored in list mode, it is assumed that x[[1]] - x[[p]] contain p measures associated with the first group, x[[p+1]] - x[[2p]] contain the p measures for the next group, and so on. If the data are stored in a matrix or data frame, it is assumed that the first p columns of x contain the p measures associated with the first group, the next p columns contain the p measures associated with the second group, and so forth.

The R function

$$\mathrm{MULAOVp}(x, J = NULL, p = NULL, tr=0.2)$$

performs the same robust MANOVA method as the R function MULtr.anova, only it returns a p-value.

The R function

$$\text{fac2Mlist(x,grp.col,lev.col,pr=T)}$$

sorts p-variate data stored in the matrix or data frame x into groups based on the values stored in the column of x indicated by the argument grp.col. The results are stored in list mode in a manner that can be used by MULtr.anov and MULAOVp. For example, the command

$$\text{z=fac2Mlist(plasma,2,c(7:8))}$$

will create groups based on the data in column 2. The result is that z[[1]] will contain the data for the first group stored as a matrix. The first column of this matrix corresponds to data stored in column 7 of the R object plasma, and the second column corresponds to data stored in column 8. Similarly, z[[2]] will contain the data for group 2, and so on.

■ Example

Imagine that two independent groups are to be compared based on measures taken at three different times. One way of comparing the groups is with a robust MANOVA method. If the data for the first group are stored in m1, a matrix having three columns, and if the data for the second group are stored in m2, also having three columns, the analysis can be performed as follows:

```
x=list()
x[[1]]=m1
x[[2]]=m2
MULtr.anova(x).
```

The function returns the test statistic and a critical value.

■ Example

This example is based on data downloaded from Carnegie Mellon University. The first five columns of the data, for the first row of the data, look like this:

```
ID Sex Smoker Opinion Age Order
1   M     N      pos   23    1
```

The next six columns for the first row look like this:

```
U.Trial.1 U.Trial.2 U.Trial.3 S.Trial.1  S.Trial.2 S.Trial.3
    38.4      27.7      25.7      53.1       30.6      30.2
```

These last six columns contain the time participants required to complete a pencil and paper maze test when they were smelling a floral scent and when they were not. The columns headed by U.Trial.1, U.Trial.2, and U.Trial.3 are the times for no scent, which were taken on three different occasions. Here, we compare smokers (Y) and non-smokers (N) based on all three of the no scent measures. So in the notation used here, $J = 2$ and $p = 3$. The first task is storing the data in a manner that can used by the R functions MULtr.anova and MULAOVp. Assuming the data have been stored in the R variable called scent, in a data frame or a matrix, this can be accomplished with the R command

$$z=fac2Mlist(scent,3,c(7:9)).$$

Then the R command

$$MULAOVp(z,2,3)$$

would compare the 20% trimmed means of smokers and non-smokers. The p-value is 0.246.

■

The R function

$$YYmanova(x1,x2,tr=0.2)$$

performs the extension of the Yanagihara–Yuan MANOVA method to trimmed means, which is limited to $J = 2$ groups. The data for the first group are stored in the argument x1, which is assumed to be a matrix with p columns, as is the argument x2, which is assumed to be the data for group 2.

7.10.2 Linear Contrasts

Consider, again, J independent groups where for each participant, p measures are taken. This section deals with the goal of testing a set of linear contrasts in the context of multivariate data. There are two variations. The first uses some marginal measure of location, such as a trimmed mean or M-estimator, and the other uses some multivariate measure of location that takes into account the overall structure of the data, such as those summarized in Section 6.3.

For convenience only, attention is focused on the marginal trimmed means with the understanding that any measure of location can be used. So now we let

$$\Psi = \sum c_j \mu_t,$$

where μ_t is a vector of p trimmed means, and the goal is to test

$$H_0: \Psi = \mathbf{0}.$$

With 20% trimming, currently, a relatively good approach aimed at achieving this goal is to use a percentile bootstrap method. Generate a bootstrap sample by sampling with replacement n_j rows from p-variate data associated with the jth group. Compute the marginal trimmed means, and label the result $\mathbf{X}_t^*$, followed by

$$\hat{\Psi}^* = \sum c_j \bar{\mathbf{X}}_t^*.$$

Repeat B times, yielding $\Psi_1^*, \ldots, \Psi_B^*$. Next, compute the Mahalanobis distance of each Ψ_b^* $(b = 0, \ldots, B)$, say d_b^*, where d_0^* is the distance of the null vector. The center of the bootstrap data cloud is taken to be $\hat{\Psi}$, the estimate of Ψ based on the observed data. And the covariance matrix when computing the Mahalanobis distances is just the sample covariance matrix based on the Ψ_b^* $(b = 0, \ldots, B)$ values. Let P^* be the proportion of d_b values $(b = 0, \ldots, B)$ such that $d_0 \geq d_b$. Then a p-value is $1 - P^*$.

This method is not recommended when using means. It appears to perform well when using a reasonably robust measure of location, but if the breakdown point is close to zero, it can be highly inaccurate.

Limited comparisons suggest that when comparing two groups, the bootstrap method described here performs about as well as the extension of Johansen's method when working with 20% trimmed means in terms of controlling the probability of a Type I error and when sampling from normal distributions. However, the p-values can differ substantially. In simulations, for example, when sampling from normal distributions, Johansen's method can have a substantially larger p-value, but situations where the reverse is true are encountered even though the ability of the two methods to control the Type I error probability is similar. In practical terms, even under normality, the choice of method is not academic in terms of deciding whether to reject the null hypothesis.

Now consider a situation where $p = 3$, $J = 2$, $n_1 = 20$, $n_2 = 40$, the first group has a multivariate normal distribution with common correlation $\rho = 0$, but the other group is generated from a g-and-h distribution with $g = 0.15$, $h = 0$ (a skewed distribution with relatively light tails), and $\rho = 0.16$. Further imagine that for the second group, the marginal distributions are shifted so that they have a trimmed mean of zero. Then the probability of a Type I error when testing at the 0.05 level, and using the bootstrap method described here, is 0.014 (based on 1,000 replications). In contrast, the actual levels using the Yanagihara–Yuan and Johansen methods are 0.053 and 0.045, respectively. A similar result is obtained when the second group now has $g = 0$ and $h = 0.15$ (a symmetric distribution with relatively heavy tails). Of course, this is not convincing evidence that the Yanagihara–Yuan and Johansen methods are generally preferable to the bootstrap method. The only point is that there are situations where indeed they have a practical advantage.

■ Example

For the EEG data mentioned in Section 6.3.5, which can be accessed as indicated in Section 1.10, MULAOVp returns a p-value equal to 0.083. But using the percentile bootstrap method described here, the p-value is 0.789. Situations are encountered, however, where the percentile bootstrap method has a substantially smaller p-value.

■

Note that for $J > 2$ groups, the Yanagihara–Yuan method can be used to perform all pairwise comparisons with the probability of at least one Type I error controlled by Rom's method. Limited simulations suggest that this approach performs relatively well in terms of Type I errors and power, even with small sample sizes. However, the percentile bootstrap method can be used with any robust estimator. Moreover, situations are encountered where both Johansen's method and the Yanagihara–Yuan method cannot be applied because the Winsorized covariance matrix is singular. Because the percentile bootstrap method does not use any covariance matrix, this problem is avoided.

7.10.3 R Functions linconMpb, linconSpb, YYmcp, and fac2BBMlist

The R function

linconMpb(x, tr=0.2, nboot = 1000, grp = NA, est = tmean, con = 0, bhop = FALSE, SEED = TRUE, PDIS = FALSE, J = NULL, p = NULL, 0...)

tests hypotheses, based on linear contrasts, using the percentile bootstrap method described in the previous section, assuming that for each group, some marginal measure of location is used. The argument x is assumed to have list mode, where x[[1]] is a matrix with p columns associated with group 1, x[[2]] is a matrix with p columns associated with group 2, and so on. If x does not have list mode, but rather is a matrix or data frame with the first p columns corresponding to group 1, the next p columns corresponding to group 2, and so forth, then specify how many groups there are via the argument J, or specify how many variables there are via the argument p. By default, all pairwise comparisons are performed based on the marginal 20% trimmed means, but M-estimators, for example, could be used by setting the argument est=onestep. The probability of at least one Type I error is set via the argument alpha and is controlled using Rom's method. As usual, contrast coefficients can be specified via the argument con. Setting the argument PDIS=TRUE, projection distances are used to measure the depth of the null vector in the bootstrap cloud of points.

For each group, it might be desired to use a multivariate measure of location that takes into account the overall structure of the data. That is, use one of the measures of location in Section 6.3. This can be done with a percentile bootstrap method via the R function

$$\text{linconSpb(x,alpha=0.05,nboot=1000, grp=NA,est=smean, con=0,b hop=FALSE,}$$
$$\text{SEED=TRUE, PDIS=FALSE, J=NULL, p=NULL,...).}$$

By default, the OP-estimator of location is used, but this might result in relatively high execution time. (For an alternative approach based on an S-estimator or MM-estimator, see Van Aelst et al., 2012.)

The R function

$$\text{YYmcp(x, tr=0.2, grp = NA, tr=0.2, bhop = FALSE, J = NULL, p = NULL, ...)}$$

performs all pairwise comparisons via the extension of the Yanagihara–Yuan technique used in conjunction with Rom's method for controlling FWE. Setting the argument bhop=TRUE, the Benjamini–Hochberg method is used instead. The arguments J and p are used in the same manner as described in conjunction with the R function linconMpb.

Data Management

The following two R functions might help with data management. The R function

$$\text{fac2BBMlist(x,grp.col,lev.col,pr=TRUE)}$$

is like the function fac2Mlist, only it is designed to handle a between-by-between design. Now the argument grp.col is assumed to contain two values indicating the columns of x that contain the levels of the two factors. The multivariate data are stored in the columns indicated by the argument lev.col. For a J-by-K design, the result is an R variable having list mode with length JK.

■ **Example**

The command

$$z=\text{fac2BBMlist}(\text{plasma},c(2,3),c(7,8))$$

will create groups based on the values in columns 2 and 3 of the R object plasma. In this particular case, there are two levels for the first factor (meaning that column 2 of plasma has two unique values only) and three for the second. The result will be that z[[1]], ..., z[[6]] will each contain a matrix having two columns stemming from the bivariate data in columns 7 and 8 of plasma. Then the commands

$$\text{con}=\text{con2way}(2,3),$$

$$\text{linconMpb}(z,\text{con}=\text{con\$conAB})$$

would test all hypotheses based on the linear contrast coefficients typically used when dealing with interactions.

■

7.11 Nested Designs

Briefly, a two-way nested design refers to a situation where there is a hierarchy among the levels of two factors under study. This is in contrast to a completely crossed design as considered in Section 7.2. For example, a goal might be to compare the efficacy of two medical procedures. The first method is used in K randomly sampled hospitals, with n participants used within each hospital, and the same is done for another K randomly sampled hospitals for the second method. For various reasons, the efficacy of a method might depend on the hospital where it is used. Here, the factor hospital is nested within the two levels corresponding to the medical procedures. (There is no interaction term.) Similarly, the effectiveness of methods for teaching mathematics might depend on the school where they are used. If the goal is to compare J teaching strategies, this might be done based on K randomly sampled schools, with n students within each school being taught based on a particular method. So the factor school is nested within the levels of J methods.

A simple way of dealing with nested designs, in a robust manner that allows heteroscedasticity, is to use the trimmed means from each level of Factor B, which is nested within the levels of Factor A. For the teaching strategies example, methods would be compared based on the trimmed means resulting from each school, where each trimmed mean is based on n participants within each school. For the special case, where means are used, formal statements of this approach are given in Khuri (1992), where heteroscedastic methods are studied. (For the situation where the K levels of the nested factor are fixed, see Guo et al., 2011.)

The goal when dealing with a nested design can be stated in a slightly more formal manner as follows. For the jth level of Factor A, it is assumed that there are K randomly sampled levels of the nested factor. For j fixed, let μ_{tjk} be the population trimmed mean corresponding to level k of the nested Factor B ($j = 1, \ldots, J, k = 1, \ldots, K$). Moreover, with j still fixed, μ_{tjk} is assumed to have some unknown distribution having a trimmed mean denoted by μ_{tj} and variance σ_j^2. There are two goals. The first is to test

$$H_0: \mu_{t1} = \cdots = \mu_{tJ}. \tag{7.33}$$

The second is to perform all pairwise comparisons in a manner that controls the probability of at least one Type I error. That is, the goal is to test

$$H_0: \mu_{tj} = \mu_{tj'}, \tag{7.34}$$

for each $j < j'$, such that the probability of at least one Type I error is approximately equal to α.

Let X_{ijk} be the ith randomly sampled observation from the kth randomly sampled level of Factor B. For fixed j and k, let

$$X_{(1)jk} \leq \cdots \leq X_{(n)jk}$$

be the n observations written in ascending order. For some γ ($0 \leq \gamma < 0.5$), let $g = [\gamma n]$, where $[\gamma n]$ is the value of γn rounded down to the nearest integer. Then the γ sample trimmed mean is

$$\bar{X}_{jk} = \sum_{g+1}^{n-g} X_{(i)jk}.$$

In essence, the unit of analysis becomes the $\bar{X}_{tjk}$ values when applying methods for trimmed means already covered. To elaborate, let $\bar{X}_j$ be the trimmed mean of the values $\bar{X}_{j1}, \ldots, \bar{X}_{jK}$ and let

$$W_{jk} = \begin{cases} \bar{X}_{(g+1)jk}, & \text{if } \bar{X}_{jk} \leq \bar{X}_{(g+1)jk}, \\ \bar{X}_{jk}, & \text{if } \bar{X}_{(g+1)jk} < \bar{X}_{jk} < \bar{X}_{(n-g)jk}, \\ \bar{X}_{(n-g)jk}, & \text{if } \bar{X}_{jk} \geq \bar{X}_{(n-g)jk}. \end{cases}$$

The Winsorized sample mean corresponding to $\bar{X}_{j1}, \ldots, \bar{X}_{jK}$ (j fixed) is

$$\bar{X}_{wj} = \frac{1}{K} \sum_{k=1}^{K} W_{jk}$$

and the Winsorized variance is

$$s_{wj}^2 = \frac{1}{K-1} \sum (W_{jk} - \bar{X}_{wj})^2.$$

Let

$$d_j = \frac{(K-1)s_{wj}^2}{h_j \times (h_j - 1)},$$

where $h_j = K - 2G$, $G = [\gamma K]$,

$$w_j = \frac{1}{d_j},$$

$$U = \sum w_j,$$

$$\tilde{X} = \frac{1}{U} \sum w_j \bar{X}_j,$$

$$A = \frac{1}{J-1} \sum w_j (\bar{X}_j - \tilde{X})^2,$$

$$B = \frac{2(J-2)}{J^2 - 1} \sum \frac{(1 - \frac{w_j}{U})^2}{h_j - 1}.$$

For $J = 2$, a test of Eq. (7.33) can be performed using an analog of Yuen's (1974) method. The test statistic is

$$T_y = \frac{\bar{X}_1 - \bar{X}_2}{\sqrt{d_1 + d_2}}. \tag{7.35}$$

When the null hypothesis is true, T_y has, approximately, a Student's t distribution with degrees of freedom

$$\hat{v}_y = \frac{(d_1 + d_2)^2}{\frac{d_1^2}{h_1 - 1} + \frac{d_2^2}{h_2 - 1}}.$$

For $J \geq 2$, the test statistic

$$F_t = \frac{A}{1 + B}$$

can be used to test Eq. (7.20). When the null hypothesis is true, F_t has, approximately, an F distribution with degrees of freedom

$$\nu_1 = J - 1,$$

$$\nu_2 = \left[\frac{3}{J^2 - 1} \sum \frac{(1 - w_j/U)^2}{h_j - 1} \right]^{-1}.$$

Finally, there is the goal of testing Eq. (7.34) such that the probability of at least one Type I error is approximately equal to α. When comparing groups j and j', reject if $|T_y| \geq c$, where c is the $1 - \alpha$ quantile of a Studentized maximum modulus distribution having degrees of freedom

$$\hat{\nu}_y = \frac{(d_j + d_{j'})^2}{\frac{d_j^2}{h_j - 1} + \frac{d_{j'}^2}{h_{j'} - 1}}.$$

Khuri (1992) derives another approach that compares the means of Factor A based in part on Hotelling's T^2. It is unknown whether an extension of the method to trimmed means, along the lines in Section 6.7.2, has any practical value.

Compared to the method for testing Eq. (7.33), the method aimed at testing Eq. (7.34) has been found to perform well in simulations for a broader range of situations in terms of controlling the probability of a Type I error (Wilcox, 2011a). When $K = 5$ or 6, the method for testing Eq. (7.33) can have an actual Type I error probability exceeding 0.08 when testing at the 0.05 level. Extant results indicate that with $K > 6$, the actual Type I error probability will not exceed 0.075. In contrast, when the method for testing Eq. (7.34) was used, the probability of at least one Type I error never exceeded 0.064. If the distribution of μ_{tjk} (j fixed) is heavy-tailed, both methods can have Type I error probabilities less than 0.025.

7.11.1 R Functions anova.nestA, mcp.nestAP, and anova.nestAP

The R function

$$\text{anova.nestA(x,tr=0.2)}$$

tests the hypothesis given by Eq. (7.33), and the R function

$$\text{mcp.nestA(x,tr=0.2)}$$

tests the hypothesis given by Eq. (7.34). Both of these functions assume that the argument x has list mode with length J. Moreover, x[[j]] $(j = 1, \ldots, J)$ is assumed to contain a matrix with n rows and K columns.

The R function

$$\text{anova.nestAP(x,tr=0.2)}$$

compares the J levels of Factor A after pooling the observations over the levels of the nested factor. The hypothesis that the J levels of Factor A have a common trimmed mean is tested using the method in Section 7.1.1. Multiple comparisons, based on the pooled data, are performed by the R function

$$\text{mcp.nestAP(x,tr=0.2)}.$$

7.12 Methods for Binary Data

For J independent groups and binary data having a binomial distribution, let p_j denote the probability of success associated with the jth group. It is noted that inferences about linear contrasts can be made based on a method derived by Zou et al. (2009). That is, the goal is to test H_0: $\Psi = 0$, where

$$\Psi = \sum c_j p_j, \tag{7.36}$$

where c_j are linear contrast coefficients. Let r_j denote the number of successes associated with the jth group based on a sample size n_j. It is assumed that r_j has a binomial distribution. Let $\hat{p}_j = r_j/n_j$, and let (ℓ_j, u_j) be some appropriate $1 - \alpha$ confidence interval for p_j. Here, the Agresti–Coull confidence interval is used unless stated otherwise. The lower end of the $1 - \alpha$ confidence interval for Ψ is

$$L = \sum c_j \hat{p}_j - \sqrt{\sum [c_j \hat{p}_j - \min(c_j \ell_j, c_j u_j)]^2},$$

and the upper end is

$$U = \sum c_j \hat{p}_j + \sqrt{\sum [c_j \hat{p}_j - \max(c_j \ell_j, c_j u_j)]^2}.$$

Extant simulations indicate that this method performs reasonably well when dealing with $J \geq 3$ groups. Note, however, that it contains method ZHZ in Section 5.8 as a special case, where the goal is to compare two groups only. As explained in Section 5.8, this method can be unsatisfactory when the minimum sample size is less than or equal to 35. That is, when dealing with two groups only, use method KMS or SK, described in Section 5.8. Currently, there are no indications that concerns occur when dealing with $J \geq 3$ groups.

7.12.1 R Functions lincon.bin and binpair

When dealing with binomial distributions, the R function

$$\text{binpair}(r = \text{NULL}, n = \text{NULL}, x = \text{NULL}, \text{method} = \text{'KMS'}, \text{alpha} = 0.05)$$

performs all pairwise comparisons using method KMS, described in Section 5.8, by default. Setting method='SK', method SK would be used. The argument r is a vector containing the number of successes and n contains the corresponding sample sizes. If the data are stored in m, say, where columns correspond to groups, and if the data consist of 0s and 1s, set the argument x=m, in which case, the function computes r and n. The FWE rate is controlled using Hochberg's method in Section 7.4.7.

The R function

$$\text{lincon.bin}(r,n, \text{con=NULL}, \text{alpha=0.05}, \text{null.value=0}, x=\text{NULL},$$
$$\text{method='KMS'}, \text{binCI=acbinomci})$$

can be used to test linear contrasts. As usual, linear contrast coefficients are specified by the argument con. If not specified, all pairwise comparisons are made. When two groups are being compared, the function uses the R function binom2g in Section 5.8.1. So when performing all pairwise comparisons, it gives the same results as binpair. Otherwise, the method in the previous section is used.

7.12.2 Identifying the Group With the Highest Probability of Success

One way of determining whether a decision can be made about which group has the highest probability of success is to mimic method RS1, described in Section 7.4.11. But as noted in Section 7.4.11, there are situations where the FWE rate is not controlled. As was indicated, a way of addressing this concern is to view the problem in terms of an indifference zone. Another approach that avoids having to specify an indifference zone is method RS2-IP, described in Section 7.4.11. For the situation at hand, compare the group that has the largest estimate to each of the other $J - 1$ groups using method KMS. If all $J - 1$ hypotheses are rejected at the α level, decide that the group associated with the largest estimate has the highest probability of success.

7.12.3 R Functions bin.best, bin.best.DO, and bin.PMD.PCD

The R function

$$\text{bin.best}(x, n, p.crit = NULL, alpha = 0.05, iter = 5000, SEED = TRUE)$$

performs the relevant multiple comparisons for identifying the group with the largest probability of success. It is method RS1-Y in Section 7.4.11 adapted to binomial distributions. The arguments x and n are vectors, having length J, containing the number of successes and the sample sizes, respectively.

The R function

$$\text{bin.best.DO}(x,n)$$

uses method RS2-IP. It returns a p-value, which is merely the maximum p-value among the $J - 1$ tests. That is, it is basically method RS2-IP, described in Section 7.4.11. The notion of an indifference zone is not used.

The R function

$$\text{bin.PMD.PCD}(n,p,DO=TRUE,alpha=0.05,p.crit=NULL,iter=5000,SEED=TRUE)$$

returns an estimate of both the PMD (the probability of making a decision) and the PCD (the probability of a correct decision, given that a decision is made)based method RS2-IP when the argument DO=TRUE; otherwise, RS1-Y is used.

7.13 Exercises

1. Describe how M-measures of location might be compared in a two-way design with a percentile bootstrap method. What practical problem might arise when using the bootstrap and sample sizes are small?
2. If data are generated from exponential distributions, what problems would you expect in terms of probability coverage when computing confidence intervals? What problems with power might arise?
3. From well-known results on the random effects model (e.g., Graybill, 1976; Jeyaratnam and Othman, 1985), it follows that

$$\text{BSSW} = \sum \frac{(\bar{Y}_j - \bar{Y})^2}{J - 1}$$

estimates

$$\sigma_{wb}^2 + \sum \frac{\sigma_{wj}^2}{Jn_j},$$

and

$$\text{WSSW} = \sum \sum \frac{(Y_{ij} - \bar{Y}_j)^2}{Jn_j(n_j - 1)}$$

estimates

$$\sum \frac{\sigma_{wj}^2}{Jn_j}.$$

Use these results to derive an alternative estimate of ρ_{WI}.

4. Some psychologists have suggested that teachers' expectancies influence intellectual functioning. The file VIQ.dat contains pretest verbal IQ scores for students in grades 1 and 2 who were assigned to one of three ability tracks. (The data are from Elashoff and Snow, 1970, and originally collected by R. Rosenthal. See Section 1.8 on how to obtain these data.) The experimental group consisted of children for whom positive expectancies had been suggested to teachers. Compare the trimmed means of the control group to the experimental group, taking into account grade and tracking ability. When examining tracking ability, combine ability levels 2 and 3 into one category, so a 2-by-2-by-2 design is used.

5. Using the data in the previous exercise, use the function lincon to compare the experimental group to the control group, taking into account grade and the two tracking abilities. (Again, tracking abilities 2 and 3 are combined.) Comment on whether the results support the conclusion that the experimental and control group have similar trimmed means.

6. Using the data from the previous two exercises, compare the 20% trimmed means of the experimental group to the control, taking into account grade. Also test for no interactions using lincon and linconb. Is there reason to suspect that the confidence interval returned by linconbt will be longer than the confidence interval returned by lincon?

7. Suppose three different drugs are being considered for treating some disorder, and it is desired to check for side effects related to liver damage. Further suppose that the following data are collected on 28 participants.
The values under the columns headed by ID indicate which of the three drugs a subject received. Store these data in an R variable having matrix mode with 28 rows and two columns, with the first column containing the subjects' ID numbers and the second column containing the resulting measure of liver damage. For example, the first subject received the first drug, and liver damage was rated as 92. Compare the groups using t1way.

ID	Damage	ID	Damage	ID	Damage
1	92	2	88	3	110
1	91	2	83	3	112
1	84	2	82	3	101
1	78	2	68	3	119
1	82	2	83	3	89
1	90	2	86	3	99
1	84	2	92	3	108
1	91	2	101	3	107
1	78	2	89		
1	95	3	99		

8. For the data in the previous exercise, compare the groups using both the Rust–Fligner and the Brunner–Dette–Munk method.

9. For the data in the previous two exercises, perform all pairwise comparisons using the Harrell–Davis estimate of the median.

10. Snedecor and Cochran (1967) report weight gains for rats randomly assigned to one of four diets that varied in the amount and source of protein. The data are stored in the file Snedecor_dat.txt, which can be retrieved as described in Section 1.10.
    Verify the results based on the R function pba2way mentioned in the example in Section 7.7.1.

11. Generate data for a 2-by-3 design, and use the function pbad2way. Note the contrast coefficients for interactions. If you again use pbad2way, but with conall=FALSE, what will happen to these contrast coefficients? Describe the relative merits of using conall=TRUE.

12. For the schizophrenia data in Section 7.8.4, compare the groups with t1way and pbadepth.

## CHAPTER 8

Comparing Multiple Dependent Groups

This chapter covers basic methods for comparing dependent groups, including both a between-by-within and a within-by-within design. Three-way designs are covered as well, where one or more factors involve dependent groups.

As noted at the beginning of Section 5.9, when comparing two dependent groups, there are three general approaches that might be used. Methods relevant to all three approaches are described, and comments on their relative merits are provided.

Note that when comparing measures of location associated with the marginal distributions, there are two types of estimators that might be used. The first estimates a measure of location for each marginal distribution, ignoring the other variables under study. That is, for p-variate data X_{ij} ($i = 1, \ldots, n$, $j = 1, \ldots, p$), compute the trimmed mean or some other measure of location using the n values associated with each j. This is in contrast to using a location estimator that takes into account the overall structure of the data when dealing with outliers, such as the OP-estimator in Section 6.5. The bulk of the methods in this chapter are based on the former type of estimator. A multiple comparison procedure that deals with the latter type of estimator is described at the end of Section 8.2.7.

8.1 Comparing Trimmed Means

This section focuses on non-bootstrap methods for testing hypotheses about trimmed means. Methods that use other estimators based on the marginal distributions, such as robust M-estimators, are described in Section 8.2.

8.1.1 Omnibus Test Based on the Trimmed Means of the Marginal Distributions Plus a Measure of Effect Size

For J dependent groups, let μ_{tj} be the population trimmed mean associated with the jth group. That is, μ_{tj} is the trimmed mean associated with the jth marginal distribution. The goal in this section is to test

$$H_0\colon \mu_{t1} = \cdots = \mu_{tJ},$$

the hypothesis that the trimmed means of J dependent groups are equal. A criticism of this hypothesis is that surely the trimmed means differ at some decimal place, but as was the case in Chapter 7, testing this hypothesis might be used in a step-down method, where the ultimate goal is to make decisions about which groups have the larger population trimmed mean. See Romano and Wolfe (2005) for general results on step-down methods.

The method used here is based on a generalization of the Huynh–Feldt method for means, which is designed to handle violations of the sphericity assumption associated with the standard F test. (See Kirk, 1995, for details about sphericity. For simulation results on how the test for trimmed means performs, see Wilcox, 1993c.) The method begins by Winsorizing the values in essentially the same manner as described in Section 5.9.3. That is, fix j, let $X_{(1)j} \leq X_{(2)j} \leq \cdots \leq X_{(n)j}$ be the n values in the jth group written in ascending order, and let

$$Y_{ij} = \begin{cases} X_{(g+1)j}, & \text{if } X_{ij} \leq X_{(g+1)j}, \\ X_{ij}, & \text{if } X_{(g+1)j} < X_{ij} < X_{(n-g)j}, \\ X_{(n-g)j}, & \text{if } X_{ij} \geq X_{(n-g)j}, \end{cases}$$

where g is the number of observations trimmed or Winsorized from each end of the distribution corresponding to the jth group. The test statistic, F, is computed as described in Table 8.1, and Table 8.2 describes how to compute the degrees of freedom.

Effect Size

A simple way of measuring effect size is to use the projection distance (described in Section 6.2.5) between the vector $(\bar{X}_{t1}, \ldots, \bar{X}_{tJ})$ and an estimate of the null vector $(\bar{X}_t, \ldots, \bar{X}_t,)$, where $\bar{X}_t$ is the grand mean in Table 8.1. To provide an indication of its properties, consider the situation where all J groups have a common measure of scale, σ_s, where σ_s is the population measure of scatter estimated by the skipped estimator, the OP-estimator, in Section 6.5. When $\mu_{t1} = \cdots = \mu_{t,J-1}$ and $\mu_{tJ} = \mu_{t1} + \delta\sigma_s$, the distance between $(\bar{X}_{t1}, \ldots, \bar{X}_{tJ})$ and $(\bar{X}_t, \ldots, \bar{X}_t,)$ is approximately δ. For a multivariate normal distribution, $\delta\sigma_s$ corresponds approximately to a shift of δ standard deviations. In general, $\delta\sigma_s$ is less than $\delta\sigma$ because σ_s is a bit smaller than σ. For an R function that measures effect size based on difference scores, see Section 8.3.1.

8.1.2 R Functions rmanova and rmES.pro

The R function

$$\text{rmanova(x,tr=0.2,grp=c(1:length(x)))}$$

tests the hypothesis of equal population trimmed means among J dependent groups using the calculations in Tables 8.1 and 8.2. It is assumed that the first argument contains the data,

Table 8.1: Test statistic for comparing the trimmed means of dependent groups.

Winsorize the observations in the jth group, as described in this section, yielding Y_{ij}. Let $h = n - 2g$ be the effective sample size, where $g = [\gamma n]$, and γ is the amount of trimming. Compute

$$\bar{X}_t = \frac{1}{J} \sum \bar{X}_{tj},$$

$$Q_c = (n - 2g) \sum_{j=1}^{J} (\bar{X}_{tj} - \bar{X}_t)^2,$$

$$Q_e = \sum_{j=1}^{J} \sum_{i=1}^{n} (Y_{ij} - \bar{Y}_{\cdot j} - \bar{Y}_{i \cdot} + \bar{Y}_{\cdot \cdot})^2,$$

where

$$\bar{Y}_{\cdot j} = \frac{1}{n} \sum_{i=1}^{n} Y_{ij},$$

$$\bar{Y}_{i \cdot} = \frac{1}{J} \sum_{j=1}^{J} Y_{ij},$$

$$\bar{Y}_{\cdot \cdot} = \frac{1}{nJ} \sum_{j=1}^{J} \sum_{i=1}^{n} Y_{ij}.$$

The test statistic is

$$F = \frac{R_c}{R_e},$$

where

$$R_c = \frac{Q_c}{J - 1},$$

$$R_e = \frac{Q_e}{(h - 1)(J - 1)}.$$

which can be an n-by-J matrix, the jth column containing the data for the jth group, or an R object having list mode. In the latter case, x[[1]] contains the data for group 1, x[[2]] contains the data for group 2, and so on. As usual, tr indicates the amount of trimming which defaults to 0.2, and grp can be used to compare a subset of the groups. If the argument grp is not specified, the trimmed means of all J groups are compared. If, for example, there are five groups, but the goal is to test H_0: $\mu_{t2} = \mu_{t4} = \mu_{t5}$, the command rmanova(x,grp=c(2,4,5)) accomplishes this goal using 20% trimming.

Table 8.2: How to compute degrees of freedom when comparing trimmed means.

Let

$$v_{jk} = \frac{1}{n-1} \sum_{i=1}^{n} (Y_{ij} - \bar{Y}_{.j})(Y_{ik} - \bar{Y}_{.k})$$

for $j = 1, \ldots, J$ and $k = 1, \ldots, J$, where Y_{ij} is the Winsorized observation corresponding to X_{ij}. When $j = k$, $v_{jk} = s_{wj}^2$, the Winsorized sample variance for the jth group, and when $j \neq k$, v_{jk} is a Winsorized analog of the sample covariance.

Let

$$\bar{v}_{..} = \frac{1}{J^2} \sum_{j=1}^{J} \sum_{k=1}^{J} v_{jk},$$

$$\bar{v}_d = \frac{1}{J} \sum_{j=1}^{J} v_{jj},$$

$$\bar{v}_{j.} = \frac{1}{J} \sum_{k=1}^{J} v_{jk},$$

$$A = \frac{J^2 (\bar{v}_d - \bar{v}_{..})^2}{J - 1},$$

$$B = \sum_{j=1}^{J} \sum_{k=1}^{J} v_{jk}^2 - 2J \sum_{j=1}^{J} \bar{v}_{j.}^2 + J^2 \bar{v}_{..}^2,$$

$$\hat{\epsilon} = \frac{A}{B},$$

$$\tilde{\epsilon} = \frac{n(J-1)\hat{\epsilon} - 2}{(J-1)\{n - 1 - (J-1)\hat{\epsilon}\}}.$$

The degrees of freedom are

$$\nu_1 = (J-1)\tilde{\epsilon},$$

$$\nu_2 = (J-1)(h-1)\tilde{\epsilon},$$

where h is the effective sample size for each group.

The R function

$$\text{rmES.pro(x,est=tmean,...)}$$

computes the measure of effect size described in the previous section.

■ Example

> Section 8.6.2 reports measures of hangover symptoms for participants belonging to one of two groups, with each participant consuming alcohol on three different occasions. For present purposes, focus on group 1 (the control group), with the goal of comparing the responses on the three different occasions. The function rmanova reports a p-value of 0.09.

■

8.1.3 Pairwise Comparisons and Linear Contrasts Based on Trimmed Means

Suppose that for J dependent groups, it is desired to compute a $1 - \alpha$ confidence interval for

$$\mu_{tj} - \mu_{tk},$$

for all $j < k$. That is, the goal is to compare all pairs of trimmed means. One possibility is to compare the jth trimmed mean to the kth trimmed mean using the R function yuend in Chapter 5, and control the familywise error (FWE) rate (the probability of at least one Type I error) with the Bonferroni method. That is, if C tests are to be performed, perform each test at the α/C level. A practical concern with this approach is that the actual probability of at least one Type I error can be considerably less than the nominal level. For example, if $J = 4$, $\alpha = 0.05$, and sampling is from independent normal distributions, the actual probability of at least one Type I error is approximately 0.019 when comparing 20% trimmed means with $n = 15$. If each pair of random variables has correlation 0.1, the probability of at least one Type I error drops to 0.014, and it drops even more as the correlations are increased. Part of the problem is that the individual tests for equal trimmed means tend to have Type I error probabilities less than the nominal level, so performing each test at the α/C level makes matters worse. In fact, even when sampling from heavy-tailed distributions, power can be low compared to using means, even though the sample mean has a much larger standard error (Wilcox, 1997a). One way of improving on this approach is to use results in Rom (1990), Hochberg (1988), or Hommel (1988) to control FWE, which were introduced in Section 7.4.7.

Momentarily consider a single linear contrast

$$\Psi = \sum_{j=1}^{J} c_j \mu_j,$$

where $\sum c_j = 0$, and the goal is to test

$$H_0: \Psi = 0.$$

Or from the perspective of Tukey's three-decision rule, the goal is to characterize the extent to which it is reasonable to make a decision about whether Ψ is less than or greater than zero.

Let Y_{ij} $(i = 1, \ldots, n, j = 1, \ldots, J)$ be the Winsorized values which are computed as described in Section 8.1.1. Let

$$A = \sum_{j=1}^{J} \sum_{k=1}^{J} c_j c_k d_{jk},$$

where

$$d_{jk} = \frac{1}{h(h-1)} \sum_{i=1}^{n} (Y_{ij} - \bar{Y}_j)(Y_{ik} - \bar{Y}_k),$$

and $h = n - 2g$ is the number of observations left in each group after trimming. Let

$$\hat{\Psi} = \sum_{j=1}^{J} c_j \bar{X}_{tj}.$$

The test statistic is

$$T = \frac{\hat{\Psi}}{\sqrt{A}}$$

and the null hypothesis is rejected if $|T| \geq t$, where t is the $1 - \alpha/2$ quantile of a Student's t distribution with $v = h - 1$ degrees of freedom.

When testing C hypotheses, the method derived by Rom (1990) and introduced in Section 7.4.7 appears to be relatively effective at controlling FWE. For the situation at hand, let p_k be the p-value associated with the kth hypothesis, and put these C p-values in descending order, yielding $p_{[1]} \geq \cdots \geq p_{[C]}$. Then:

1. Set $k = 1$.
2. If $p_{[k]} \leq d_k$, where d_k is read from Table 8.3, stop, and reject all C hypotheses; otherwise, go to step 3. (When $k > 10$, then $d_k = \alpha/k$.)
3. Increment k by 1. If $p_{[k]} \leq d_k$, stop, and reject all hypotheses having a p-value less than or equal to d_k.
4. If $P_{[k]} > d_k$, repeat step 3.
5. Continue until you reject or all C hypotheses have been tested.

Note that Table 8.3 is limited to $k \leq 10$ as well as $\alpha = 0.05$ or 0.01. Here, if $k > 10$, or for α values other than 0.05 and 0.01, FWE is controlled with Hochberg's (1988) method. That is, proceed as just indicated, but rather than use d_k read from Table 8.3, use $d_k = \alpha/k$.

Table 8.3: Critical values, d_k, for Rom's method.

k	$\alpha = 0.05$	$\alpha = 0.01$
1	0.05000	0.01000
2	0.02500	0.00500
3	0.01690	0.00334
4	0.01270	0.00251
5	0.01020	0.00201
6	0.00851	0.00167
7	0.00730	0.00143
8	0.00639	0.00126
9	0.00568	0.00112
10	0.00511	0.00101

8.1.4 Linear Contrasts Based on the Marginal Random Variables

An alternative to using linear contrasts based on the trimmed means is to use linear contrasts based on the random variables under study. This approach contains comparisons based on difference scores as a special case. Let

$$D_{ik} = \sum_{j=1}^{J} c_{jk} X_{ij}, \tag{8.1}$$

where for any k $(k = 1, \ldots, C)$, $\sum c_{jk} = 0$, and let μ_{tk} be the population trimmed mean of the distribution from which the random sample $D_{1k}, \ldots, D_{nk}$ was obtained. For example, if $c_{11} = 1$, $c_{21} = -1$, and $c_{31} = \cdots = c_{J1} = 0$, then

$$D_{i1} = X_{i1} - X_{i2},$$

the difference scores for groups 1 and 2, and μ_{t1} is the (population) trimmed mean associated with this difference. Similarly, if $c_{22} = 1$, $c_{32} = -1$, and $c_{12} = c_{41} = \cdots = c_{J1} = 0$, then

$$D_{i2} = X_{i2} - X_{i3}$$

and μ_{t2} is the corresponding (population) trimmed mean. The goal is to test

$$H_0: \mu_{tk} = 0$$

for each $k = 1, \ldots, C$ such that FWE is approximately α. Each hypothesis can be tested using results in Chapter 4. Here, Rom's method, described in Section 8.1.3, is used to control the FWE rate. Measuring effect size can be addressed as described in Sections 4.3.1 and 5.9.19.

It should be noted that the multiple-comparison procedures in this chapter are designed to control the probability of one or more Type I errors. As was the case in Chapter 7, the expectation is that the actual probability of one or more Type I errors will be reduced if the multiple-comparison procedures in this chapter are used contingently on a global test rejecting at the α level. That is, power might be adversely affected (cf. Bernhardson, 1975).

8.1.5 R Functions rmmcp, rmmismcp, trimcimul, wwlin.es, deplin.ES.summary.CI, and boxdif

The R function

$$\text{rmmcp(x,con = 0, tr=0.2, alpha=0.05, dif=TRUE,hoch=TRUE)}$$

performs multiple comparisons among dependent groups using trimmed means and Hochberg's (1988) method for controlling FWE. (Setting the argument hoch=FALSE, Rom's method is used.) By default, difference scores are used. Setting dif=FALSE results in comparing the marginal trimmed means. Again, by default, FWE is controlled with Hochberg's (1988) method. That is, proceed as indicated in Section 8.1.3 but rather than use d_k from Table 8.3, use $d_k = \alpha / k$. The argument con can be used to specify linear contrast coefficients.

When there are values missing at random, method M2, described in Section 5.9.13, can be used to perform multiple comparisons via the R function

$$\text{rmmismcp(x,y = NA, tr=0.2, con = 0, est = tmean, plotit = TRUE, grp = NA, nboot = 500,}$$
$$\text{SEED = TRUE, xlab = 'Group 1', ylab = 'Group 2', pr = FALSE, ...),}$$

which was introduced in Section 5.9.14 and controls the probability of one or more Type I errors using Hochberg's method. By default, 20% trimmed means are used, but other robust estimators can be used via the argument est. When the goal is to make inferences about the median of the difference scores, the R function

$$\text{sintmcp(x, con=0, alpha=0.05)}$$

can be used. Inferences are based on the modification of the distribution-free method described in Section 4.6.2. The R function

$$\text{signmcp(x, con=0, alpha=0.05)}$$

performs the sign test for all pairs of groups when the argument con=0. Letting $W = \sum c_j X_j$, where $c_1, \ldots, c_J$ are linear contrast coefficients, this function can be used to test H_0: $P(W < 0) = 0.5$.

For each of J groups, the R function

$$\text{trimcimul(x, tr = 0.2, alpha=0.05, null.value = 0)}$$

computes a confidence interval for a trimmed mean, and it tests the hypothesis that a trimmed mean is equal to the value indicated by the argument null.value. This function can be used to test hypotheses based on difference scores associated with J dependent groups. If the R variables x1 and x2 are matrices with J columns, and for each j ($j = 1, \ldots, J$), the goal is to test the hypothesis that the differences scores for column j have a population 20% trimmed mean equal to zero, the command trimcimul(x1-x2) accomplishes this goal. If the data are stored in list mode, use the command trimcimul(list.dif(x1,x2)).

The function

$$\text{deplin.ES.summary.CI(x, con=NULL, tr=0.2, REL.MAG=NULL, SEED=TRUE,}$$
$$\text{nboot=1000)}$$

computes measures of effect size based on the values given by Eq. (8.1), where the argument con, a matrix, indicates the linear contrasts of interest. This is done by calling dep.ES.summary.CI for each set of linear contrast coefficients indicated by the argument con. The argument con is used as described in Section 7.4.2. By default, all pairwise difference scores are used. For convenience, the R function

$$\text{wwlin.es (J,K, x, tr = 0.2, REL.MAG = NULL, SEED = TRUE, nboot = 1000)}$$

is provided. It reports measures of effect size for main effects and interactions via the R function deplin.ES.summary.CI. The function

$$\text{boxdif(x,names)}$$

creates boxplots of all pairwise difference scores.

8.1.6 Judging the Sample Size

Let $D_{ijk} = X_{ij} - X_{ik}$, and let μ_{tjk} be a trimmed mean corresponding to D_{ijk}. If when testing H_0: $\mu_{tjk} = 0$ for any $j < k$, a non-significant result is obtained, this might be because the null hypothesis is true, or, of course, a Type II error might have been committed due to a sample size that is too small. Or from the point of view of Tukey's three-decision rule, failing to reject means making no decision about whether μ_{tjk} is greater than or less than zero, which raises the issue of how many more observations are needed so that a decision can be made. A way of dealing with this issue is to use an extension of Stein's (1945) two-stage method for means. Suppose it is desired to have all pairs power greater than or equal to $1 - \beta$ when, for any $j < k$, $\mu_{tjk} = \delta$, that is, the probability of rejecting H_0 for all $j < k$ for which $\mu_{tjk} = \delta$ is to be at least $1 - \beta$. The goal here is to determine whether the sample size used, namely n, is large enough to accomplish this goal, and if not, the goal is to determine how many more observations are needed. The following method performs well in simulations (Wilcox, 2004b).

Let $C = (J^2 - J)/2$ and

$$d = \left(\frac{\delta}{t_\beta - t_{1-\alpha/(2C)}} \right)^2,$$

where $t_{1-\beta}$ is the $1 - \beta$ quantile of Student's t distribution with $v = n - 2g - 1$ degrees of freedom and g is the number of observations trimmed from each tail. (So $n - 2g$ is the number of observations not trimmed.) Let

$$N_{jk} = \max \left(n, \left[\frac{s_{wjk}^2}{(1 - 2\gamma)^2 d} \right] + 1 \right),$$

where s_{wjk}^2 is the Winsorized variance of the D_{ijk} values and the notation $[x]$ refers to the greatest integer less than or equal to x. Then the required sample size in the second stage is

$$N = \max N_{jk},$$

the maximum being taken over all $j < k$. So if $N = n$, the sample size used is judged to be adequate for the specified power requirement.

In the event the additional $N - n$ vectors of observations can be obtained, familiarity with Stein's (1945) original method suggests how H_0 should be tested, but in simulations, a slight modification performs a bit better in terms of power. Let S_{wjk} be the Winsorized variance based on all N of the observations, where the amount of Winsorizing is equal to the amount of trimming. Let $\hat{\mu}_{tjk}$ be the trimmed mean based on all N D_{ijk} differences, and let

$$T_{jk} = \frac{\sqrt{N}(1 - 2\gamma)\hat{\mu}_{tjk}}{S_{wjk}}.$$

Then reject H_0: $\mu_{tjk} = 0$ if $|T_{jk}| \geq t_{1-\alpha/(2C)}$. As would be expected based on Stein's method, the degrees of freedom depend on the initial sample size, n, not the ultimate sample size, N. But contrary to what is expected based on Stein's method, the Winsorized variance when computing T_{jk} is based on all N observations. (All indications are that no adjustment for β is needed when computing d when multiple tests are performed and the goal is to have all pairs power greater than or equal to $1 - \beta$. Also, a variation of the method aimed at comparing the marginal trimmed means has not been investigated.)

8.1.7 R Functions stein1.tr and stein2.tr

Using the method just described, the R function

$$\text{stein1.tr(x,del,alpha= 0.05,pow=0.8,tr=0.2)}$$

determines the required sample size needed to achieve all pairs power equal to the value indicated by the argument pow for a difference specified by the argument del which corresponds to δ. In the event additional data are needed to achieve the desired amount of power, and if these additional observations can be acquired,

$$\text{stein2.tr(x,y,alpha= 0.05,tr=0.2)}$$

tests all pairwise differences. Here, the first-stage data are stored in x (which is a vector or a matrix with J columns) and y contains the second-stage data.

8.1.8 Identifying the Group With the Largest Population Measure of Location

Section 7.4.11 describes two methods for identifying the group with the largest measure of location. But when dealing with dependent variables, an alternative approach is needed that takes into account the correlations among the variables.

As was done in Section 7.4.11, one strategy is to compare the group with the largest estimate to the remaining $J - 1$ groups. That is, the focus is on the marginal trimmed means. The method used here is based in part on comparing the trimmed means using the method in Section 5.9.5. Let $\hat{\theta}_{(1)} \leq \cdots \leq \hat{\theta}_{(J)}$ denote the estimates written in ascending order. Let $\theta_{\pi(j)}$ denote the measure of location associated $\hat{\theta}_{(j)}$, and let p_j denote the p-value associated with the test of

$$H_0: \theta_{\pi(j)} = \theta_{\pi(J)}. \tag{8.2}$$

Suppose this hypothesis is rejected when $p_j \leq c_j$. Mimicking method RS1-Y in Section 7.4.11 in an obvious way, the strategy is to control the FWE rate when $\theta_1 = \cdots = \theta_J$ by determining the values $c_1, \ldots, c_{J-1}$ such that

$$P(p_1 \leq c_1, \ldots, p_{J-1} \leq c_{J-1}) = \alpha. \tag{8.3}$$

This is done by first generating data from a multivariate normal distribution having a mean of zero and a covariance matrix equal to the estimate of the Winsorized covariance matrix based on the observed data. Then Eq. (8.2) is tested for each j ($j = 1, \ldots, J - 1$), yielding $J - 1$ p-values. Finally, proceed as described in Section 7.4.11 by determining adjusted p-values that control the FWE rate when the J groups have a common measure of location. When using 20% trimming, this approach works well in terms of controlling the FWE rate in situations where the Bonferroni and related methods are unsatisfactory (Wilcox, 2019d).

But as was the case in Section 7.4.11, situations can be constructed where a Type I error probability exceeds the nominal level. As done in Section 7.4.11, a possible way of dealing with this issue is to view the goal in terms of an indifference zone. From this perspective, the issue is whether the probability of a correct decision, given that a decision is made, is reasonably high.

Another method described in Section 7.4.11, method RS2-IP, focuses on ensuring a correct decision is made with probability at least $1 - \alpha$. This is done without specifying an indifference zone. Method RS2-GPB is readily extended to the situation at hand, but currently analogs of RS2-IP appear to be a more satisfactory approach. That is, test Eq. (8.2) for each j, and make a decision if all of the $J - 1$ p-values are less than or equal to α. A non-bootstrap method can be used via the R function rmmcp. The R function rmanc.best.DO, described in Section 8.1.10, is provided in order to simplify this approach. When using a percentile bootstrap method, see the R function rmbestPB.DO.

When the correlations are sufficiently high, extant simulations indicate that the percentile bootstrap method provides an advantage in terms of the probability of making a decision. When all of the correlations are zero, the non-bootstrap method is more likely to make a decision.

8.1.9 Identifying the Variable With the Smallest Robust Measure of Variation

Situations arise where the goal is to identify which of J dependent variables has the smallest measure of scale. When dealing with variances, the ranking and selection literature, mentioned in Section 7.4.11, approaches this problem assuming normality, and then the goal is to determine the sample size needed so that the variable with lowest population variance has the smallest sample variance (e.g., Gibbons et al., 1987). This is done in the context of an in-difference zone. The approach used here is simple: Use a robust measure of scale and mimic the bootstrap methods described in Section 8.1.8.

8.1.10 R Functions comdvar.mcp, ID.sm.varPB, rmbestVAR.DO, rmanc.best.PV, RM.PMD.PCD, rmanc.best.PB, RMPB.PMD.PCD, and rmanc.best.DO

For J dependent variables, the R function

$$\text{comdvar.mcp(x,method='hoch')}$$

performs all pairwise comparisons of the variances using the HC4 version of the Morgan–Pitman test described in Section 5.9.15. The R function

$$\text{rmVARcom.mcp(x,est=winvar, alpha=0.05, nboot=500, method='hoch', SEED=TRUE)}$$

deals with robust measures of variation using a percentile bootstrap method.

The R function

$$\text{ID.sm.varPB(x,var.fun=winvar, nboot=500, iter=200, NARM=FALSE,}$$
$$\text{na.rm=TRUE,SEED=TRUE)}$$

performs the method described in Section 8.1.9. It uses the RS2-GPB approach described in Section 8.1.8. Setting PV=TRUE, a p-value is returned at the expense of high execution time. The method might continue to perform well with other robust estimators when PV=TRUE, but this remains to be determined. The RS2-IP approach can be applied with the R function

$$\text{rmbestVAR.DO(x, est = winvar, nboot = NA, SEED = TRUE, pr = TRUE, ...)}$$

Measure of Location: Analogs of Method RS1

The next sets of functions deal with analogs of method RS1, described in Section 8.1.8. The R function

$$\text{rmanc.best.PV(x, alpha = 0.05, tr = 0.2, iter = 5000, SEED = TRUE)}$$

performs the analog of method RS1-Y for identifying the group with the largest trimmed mean. The R function

RM.PMD.PCD(x, tr=0.2, delta=0.5, alpha=0.05, p.crit=NULL, iter=5000, nboot=500,
SEED=TRUE)

computes an estimate of the probability of making a decision in the context of an indifference
zone based on the data stored in x, which can be a matrix or a data frame. (See Section 7.4.11
for a discussion of an indifference zone.) The argument iter (in both of these functions) indicates the number of replications used to estimate the critical p-values. The argument nboot in
the function RM.PMD.PCD indicates how many replications are used to estimate the probability of making a correct decision (PMD). The function returns an estimate of PMD as well
as the probability of a correct decision (PCD) given that a decision is made. This is done via a
simulation where data are generated from a multivariate normal distribution having the same
covariance matrix as the data stored in x. The means are all equal to zero, except the first,
which is equal to delta$\times s_{(J)}$, where $s_{(J)}$ is the sample standard deviation of the group estimated to have the largest trimmed mean. The default value for the argument delta is 0.5. The
R function

$$\text{rmanc.bestPB}(x, \text{alpha} = 0.05, \text{tr} = 0.2, \text{iter} = 5000, \text{SEED} = \text{TRUE},...)$$

performs the percentile bootstrap GPB method in Section 8.1.8. If it rejects, make a decision
about which group has the largest marginal trimmed mean.

The R function

RMPB.PMD.PCD(x,est=tmean, delta=0.5, alpha=0.05, iter=500, nboot=1000,
SEED=TRUE,...)

estimates the probability of making a decision, and making a correct decision when a decision
is made, in the context of an indifference zone in conjunction with the percentile bootstrap
method.

Measure of Location: Analogs of Method RS2

The R function

$$\text{rmanc.best.DO}(x,\text{tr}=0.2,...)$$

performs the non-bootstrap version of method RS2-IP. The R function

rmbestPB.DO(x,est=tmean,nboot=NA,SEED=TRUE,pr=FALSE,...)

is a percentile bootstrap method for implementing the RS2-IP approach.

8.2 Bootstrap Methods Based on Marginal Distributions

This section focuses on bootstrap methods aimed at making inferences about measures of location associated with the marginal distributions. (Section 8.3 takes up measures of location associated with difference scores.) As in previous chapters, two general types of bootstrap methods appear to deserve serious consideration in applied work. (As usual, this is not intended to suggest that all other variations of the bootstrap have no practical value for the problems considered here, only that based on extant studies, the methods covered here seem to perform relatively well.) The first type uses estimated standard errors and reflects extensions of the bootstrap-t methods described in Chapter 5; they are useful when comparing trimmed means. The other is an extension of the percentile bootstrap method, where estimated standard errors do not play a direct role. When comparing robust M-measures of location, this latter approach is the only known way of controlling the probability of a Type I error for a fairly wide range of distributions.

8.2.1 Comparing Trimmed Means

Let μ_{tj} be the population trimmed mean associated with the jth marginal distribution, and consider the goal of testing

$$H_0: \mu_{t1} = \cdots = \mu_{tJ}.$$

An extension of the bootstrap-t method to this problem is straightforward. Set

$$C_{ij} = X_{ij} - \bar{X}_{tj}$$

with the goal of estimating an appropriate critical value, based on the test statistic F in Table 8.1, when the null hypothesis is true. The remaining steps are as follows:

1. Generate a bootstrap sample by randomly sampling, with replacement, n rows of data from the matrix

$$\begin{pmatrix} C_{11}, \ldots, C_{1J} \\ \vdots \\ C_{n1}, \ldots, C_{nJ} \end{pmatrix},$$

   yielding

$$\begin{pmatrix} C_{11}^*, \ldots, C_{1J}^* \\ \vdots \\ C_{n1}^*, \ldots, C_{nJ}^* \end{pmatrix}.$$

2. Compute the test statistic F in Table 8.1 based on the C_{ij}^* values generated in step 1, and label the result F^*.
3. Repeat steps 1 and 2 B times, and label the results $F_1^*, \ldots, F_B^*$.
4. Put these B values in ascending order, and label the results $F_{(1)}^* \leq \cdots \leq F_{(B)}^*$.

The critical value is estimated to be $F_{(u)}^*$, where $u = (1 - \alpha)B$ is rounded to the nearest integer. That is, reject the hypothesis of equal trimmed means if

$$F \geq F_{(u)}^*,$$

where F is the statistic given in Table 8.1 based on the X_{ij} values.

8.2.2 R Function rmanovab

The R function

$$\text{rmanovab(x, tr=0.2, alpha=0.05, grp = 0, nboot = 599)}$$

performs the bootstrap-t method just described.

8.2.3 Multiple Comparisons Based on Trimmed Means

This section describes bootstrap methods for performing multiple comparisons based on trimmed means. First, consider the goal of performing all pairwise comparisons. That is, the goal is to test

$$H_0: \mu_{tj} = \mu_{tk}$$

for all $j < k$. A bootstrap-t method is applied as follows. Generate bootstrap samples as was done in Section 8.2.1, yielding

$$\begin{pmatrix} C_{11}^*, \ldots, C_{1J}^* \\ \vdots \\ C_{n1}^*, \ldots, C_{nJ}^* \end{pmatrix}.$$

For every $j < k$, compute the test statistic T_y, given by Eq. (5.21), using the values in the jth and kth columns of the matrix just computed. That is, perform the test for trimmed means corresponding to two dependent groups using the data $C_{1j}^*, \ldots, C_{nj}^*$ and $C_{1k}^*, \ldots, C_{nk}^*$. Label the resulting test statistic T_{yjk}^*. Repeat this process B times, yielding $T_{yjk1}^*, \ldots, T_{yjkB}^*$. Because these test statistics are based on data generated from a distribution for which the trimmed means are equal, they can be used to estimate an appropriate critical value. In particular, for each b, set

$$T_b^* = \max|T_{yjkb}^*|,$$

the maximum being taken over all $j < k$. Let $T_{(1)}^* \leq \cdots \leq T_{(B)}^*$ be the T_b^* values written in ascending order, and let $u = (1 - \alpha)B$, rounded to the nearest integer. Then $H_0: \mu_{tj} = \mu_{tk}$ is rejected, if $T_{yjk} > T_{(u)}^*$. That is, for the jth and kth groups, test the hypothesis of equal trimmed means using the method in Section 5.9.3, only the critical value is $T_{(u)}^*$, which was determined so that the probability of a least one Type I error is approximately equal to α. Alternatively, the confidence interval for $\mu_{tj} - \mu_{tk}$ is

$$(\bar{X}_{tj} - \bar{X}_{tk}) \pm T_{(u)}^* \sqrt{d_j + d_k - 2d_{jk}},$$

where $\sqrt{d_j + d_k - 2d_{jk}}$ is the estimate of the standard error of $\bar{X}_{tj} - \bar{X}_{tk}$, which is computed as described in Section 5.9.3. The simultaneous probability coverage is approximately $1 - \alpha$. Probability coverage appears to be reasonably good with n as small as 15 when using 20% trimming with $J = 4$, $\alpha = 0.05$, $B = 599$ (Wilcox, 1997a). When there is no trimming, probability coverage can be poor, and no method can be recommended. Also, the power of the bootstrap method, with 20% trimmed means, compares well to an approach based on means and the Bonferroni inequality.

The method is easily extended to situations where the goal is to test C linear contrasts, $\Psi_1, \ldots, \Psi_C$, where

$$\Psi_k = \sum c_{jk} \mu_{tj},$$

and c_{jk} ($j = 1, \ldots, J, k = 1, \ldots, C$) are constants chosen to reflect some hypothesis of interest. As before, Ψ_k is estimated with $\hat{\Psi}_k = \sum c_{jk} \bar{X}_{tj}$, but now the squared standard error is estimated with

$$A_k = \sum_{j=1}^{J} \sum_{\ell=1}^{J} c_{jk} c_{\ell k} d_{j\ell},$$

where

$$d_{jk} = \frac{1}{h(h-1)} \sum (Y_{ij} - \bar{Y}_j)(Y_{ik} - \bar{Y}_k)$$

and Y_{ij} are the Winsorized observations for the jth group. (When $j = k$, $d_{jk} = d_j^2$.)

To compute a $1 - \alpha$ confidence interval for Ψ_k, generate a bootstrap sample, yielding C_{ij}^*, and let

$$T_{yk}^* = \frac{\hat{\Psi}_k^*}{\sqrt{A_k^*}},$$

where $\hat{\Psi}_k^*$ and A_k^* are computed with the bootstrap observations. Repeat this bootstrap process B times, yielding T_{ykb}^*, $b = 1, \ldots B$. For each b, let $T_b^* = \max |T_{ykb}^*|$, the maximum being taken over $k = 1, \ldots, C$. Put the T_b^* values in order, yielding $T_{(1)}^* \leq \cdots \leq T_{(B)}^*$, in which case, an appropriate critical value is estimated to be $T_{(u)}^*$, where $u = (1 - \alpha)B$, rounded to the nearest integer. Then an approximate $1 - \alpha$ confidence interval for Ψ_k is

$$\hat{\Psi}_k \pm T_{(u)}^* \sqrt{A_k}.$$

8.2.4 R Functions pairdepb and bptd

The R function

$$\text{pairdepb(x,tr=0.2,alpha= 0.05,grp=0,nboot=599)}$$

performs all pairwise comparisons among J dependent groups using the bootstrap method just described. The argument x can be an n-by-J matrix of data, or it can be an R variable having list mode. In the latter case, x[[1]] contains the data for group 1, x[[2]] contains the data for group 2, and so on. The argument tr indicates the amount of trimming, which, if unspecified, defaults to 0.2. The value for α defaults to alpha=0.05, and B defaults to nboot=599. The argument grp can be used to test the hypothesis of equal trimmed means using a subset of the groups. If missing values are detected, they are eliminated via the function elimna, described in Section 1.9.1.

■ **Example**

For the alcohol data reported in Section 8.6.2, suppose it is desired to perform all pairwise comparisons using the time 1, time 2, and time 3 data for the control group. The R function pairdepb returns

```
$test:
       Group Group       test        se
[1,]       1     2  -2.115985  1.693459
[2,]       1     3  -2.021208  1.484261
[3,]       2     3   0.327121  1.783234


$psihat:
       Group Group     psihat   ci.lower    ci.upper
[1,]       1     2  -3.5833333  -7.194598  0.02793158
[2,]       1     3  -3.0000000  -6.165155  0.16515457
[3,]       2     3   0.5833333  -3.219376  4.38604218
```

```
$crit:
[1] 2.132479
```

Thus, none of the pairwise differences is significantly different at the 0.05 level.

∎

Assuming the data are stored in the R variable dat, the command pairdepb(dat,grp=c(1,3)) would compare groups 1 and 3, ignoring group 2. It is left as an exercise to show that if the data are stored in list mode, the command ydbt(dat[[1]],dat[[3]]) returns the same confidence interval.

The function

$$bptd(x,tr=0,alpha= 0.05,con=0,nboot=599)$$

computes confidence intervals for each of C linear contrasts, Ψ_k, $k = 1, \ldots, C$, such that the simultaneous probability coverage is approximately $1 - \alpha$. The only difference between bptd and pairedpb is that bptd can handle a set of specified linear contrasts via the argument con. The argument con is a J-by-C matrix containing the contrast coefficients. The kth column of con contains the contrast coefficients corresponding to Ψ_k. If con is not specified, all pairwise comparisons are performed. So for this special case, pairdepb and bptd always produce the same results.

■ **Example**

If there are three dependent groups, con is a 3-by-1 matrix with the values 1, −1, and 0, and the data are stored in the R variable xv, the command bptd(xv,con=con) will compute a confidence interval for $\Psi = \mu_{t1} - \mu_{t2}$, the difference between the 20% trimmed means corresponding to the first two groups. If xv has list mode, the command ydbt(xv[[1]],xv[[2]]) returns the same confidence interval. (The function ydbt is described in Section 5.9.8.)

∎

8.2.5 Percentile Bootstrap Methods

This section describes two types of percentile bootstrap methods that can be used to compare J dependent groups based on any measure of location, θ, associated with the marginal distributions. Included as special cases are M-measures of location and trimmed means. The goal is to test

$$H_0: \theta_1 = \cdots = \theta_J. \qquad (8.4)$$

Method RMPB3

The first method uses the test statistic

$$Q = \sum (\hat{\theta}_j - \bar{\theta})^2,$$

where $\bar{\theta} = \sum \hat{\theta}_j / J$. An appropriate critical value is estimated using an approach similar to the bootstrap-t technique. First, set $C_{ij} = X_{ij} - \hat{\theta}_j$. That is, shift the empirical distributions so that the null hypothesis is true. Next, a bootstrap sample is obtained by resampling, with replacement, as described in step 1 in Section 8.2.1. As usual, label the results

$$\begin{pmatrix} C_{11}^*, \ldots, C_{1J}^* \\ \vdots \\ C_{n1}^*, \ldots, C_{nJ}^* \end{pmatrix}.$$

For the jth column of the bootstrap data just generated, compute the measure location that is of interest and label it $\hat{\theta}_j^*$. Compute

$$Q^* = \sum (\hat{\theta}_j^* - \bar{\theta}^*)^2,$$

where $\bar{\theta}^* = \sum \hat{\theta}_j^* / J$, and repeat this process B times, yielding $Q_1^*, \ldots, Q_B^*$. Put these B values in ascending order, yielding $Q_{(1)}^* \leq \cdots \leq Q_{(B)}^*$. Then reject the hypothesis of equal measures of location if $Q > Q_{(u)}^*$, where, again, $u = (1 - \alpha)B$ is rounded to the nearest integer.

Note that if any of the rows of data has a missing value, a simple strategy is to simply remove such rows and test the hypothesis of equal measures of location using the remaining data to test Eq. (8.1) with the method in this section. An alternative approach is to impute the missing values. Several strategies have been proposed regarding how missing values might be imputed in a robust manner (e.g., Vanden Branden and Verboven, 2009; Danilov et al., 2012). There are indications that these imputation methods are useful when checking for outliers. But currently, all indications are that they are rather unsatisfactory when testing Eq. (8.1). Note that the method in this section could be used in conjunction with the MCD estimator. Moreover, missing values can be computed via the R package rrcovNA. But even with no missing values, when testing at the 0.05 level, the actual level is approximately 0.001 when $n = 40$, $p = 4$, and all correlations are equal to zero. With missing values, matters get worse. Missing values can be imputed when using the OGK estimator. Now, with no missing values, the actual level is approximately 0.009. With 10 missing values, the actual level drops to 0.005. Perhaps there are situations where these and related methods perform well in terms of Type I errors, but this remains to be determined. (There are other techniques that have not been studied.) Currently, there is only one method that performs well in simulations in terms of controlling the Type I error probability, given the goal of using all of the available data: Proceed as described in this section using all of the data in each column to estimate the trimmed mean rather than simply removing any row that has missing values.

Method RMPB4

If the null hypothesis is true, then all J groups have a common measure of location, θ. The next method estimates this common measure of location, and then checks to see how deeply it is nested within the bootstrap values obtained when resampling from the original values. That is, in contrast to method RMPB3, the data are not centered, and bootstrap samples are obtained by resampling rows of data from

$$\begin{pmatrix} X_{11}, \dots, X_{1J} \\ \vdots \\ X_{n1}, \dots, X_{nJ} \end{pmatrix},$$

yielding

$$\begin{pmatrix} X_{11}^*, \dots, X_{1J}^* \\ \vdots \\ X_{n1}^*, \dots, X_{nJ}^* \end{pmatrix}.$$

For the jth group (or column of bootstrap values), compute $\hat{\theta}_j^*$. Repeating this process B times yields $\hat{\theta}_{jb}^*$, $(j = 1, \dots, J, \ b = 1, \dots, B)$. The remaining calculations are performed as outlined in Table 8.4.

For completeness, yet another approach to comparing dependent groups is to use a *mixed linear model* in conjunction with the regression MM-estimator introduced in Chapter 10. Heritier et al. (2009, Section 4.5) summarize the relevant details and computations. The mixed linear model has the form

$$Y = \mathbf{X}\alpha + \sum Z_j \beta_j + \epsilon,$$

where Y is a vector of N measurements, $\mathbf{X}$ is an $n \times q$ design matrix for the fixed effects, $\mathbf{Z}_j$ is an $N \times q_j$ design matrix for the random effects β_j, and ϵ is an N-vector of independent residual errors. The classic version of this model assumes that both β_j and ϵ have

Table 8.4:   Repeated measures ANOVA based on the depth of the grand mean.

Goal: Test the hypothesis

$$H_0 \colon \theta_1 = \cdots = \theta_J.$$

1.   Compute

$$S_{jk} = \frac{1}{B-1} \sum_{b=1}^{B} (\hat{\theta}_{jb}^* - \bar{\theta}_j^*)(\hat{\theta}_{kb}^* - \bar{\theta}_k^*),$$

where

$$\bar{\theta}_j^* = \frac{1}{B} \sum_{b=1}^{B} \hat{\theta}_{jb}^*.$$

(The quantity S_{jk} is the sample covariance of the bootstrap values corresponding to the jth and kth groups.)

2.   Let

$$\hat{\theta}_b^* = (\hat{\theta}_{1b}^*, \ldots, \hat{\theta}_{Jb}^*)$$

and compute

$$d_b = (\hat{\theta}_b^* - \hat{\theta}) \mathbf{S}^{-1} (\hat{\theta}_b^* - \hat{\theta})',$$

where $\mathbf{S}$ is the matrix corresponding to S_{jk}, $\hat{\theta} = (\hat{\theta}_1, \ldots, \hat{\theta}_J)$, $\hat{\theta}_j$ is the estimate of θ based on the original data for the jth group (the X_{ij} values, $i = 1, \ldots, n$), and $\hat{\theta}_b = (\hat{\theta}_{1b}, \ldots, \hat{\theta}_{Jb})$. The value of d_b measures how far away the bth bootstrap vector of location estimators is from $\hat{\theta}$, which is roughly the center of all B bootstrap values.

3.   Put the d_b values in ascending order: $d_{(1)} \leq \cdots \leq d_{(B)}$.

4.   Let $w_j = 1/S_{jj}$, $W_j = w_j / \sum w_j$, $\hat{\theta}_G = (\bar{\theta}, \ldots, \bar{\theta})$, where $\bar{\theta} = \sum W_j \hat{\theta}_j$, and compute

$$D = (\hat{\theta}_G - \hat{\theta}) \mathbf{S}^{-1} (\hat{\theta}_G - \hat{\theta})'.$$

D measures how far away the estimated common value is from the observed measures of location (based on the original data). The original version of this method used $W_j = 1$, which is a bit less satisfactory in terms of controlling the Type I error probability.

5.   Reject if $D \geq d_{(u)}$, where $u = (1 - \alpha)B$, rounded to the nearest integer.

multivariate normal distributions (Laird and Ware, 1982). Copt and Heritier (2007) derive a (non-bootstrap) method for testing hypotheses that is based in part on an appropriate estimate of the standard errors. Also see Koller (2013). For an R package designed for a robust mixed linear model, see Koller (2016). However, a general pattern regarding M-estimators seems to be that non-bootstrap methods that use a test statistic based on an estimate of the standard error can perform poorly in terms of Type I errors and probability coverage when dealing with skewed distributions. Perhaps the MM-estimator, in the context of the mixed linear model, is an exception, but apparently this has not been investigated.

Missing Values

If there are missing values, both methods RMPB3 and RMPB4 are readily modified so that they use all of the available data. When computing a measure of location associated with the jth group, simply compute the measure of location using all of the data, even if there are missing values in the other groups. Assuming missing values occur at random, method RMPB3 has been found to perform well in simulations when using a 20% trimmed mean; method RMPB4 did not perform well (Ma and Wilcox, 2013).

8.2.6 R Functions bd1way and ddep

The R functions

$$bd1way(x, est = tmean, nboot = 599, tr=0.2, misran=FALSE)$$

performs method RMPB3, described in the previous section, and

$$ddep(x, tr=0.2, est = onestep, grp = NA, nboot = 500, WT=TRUE, ...)$$

performs method RMPB4, described in Table 8.4. (The argument WT=FALSE results in using the original version of method RMPB4 by setting $W_j = 1$ in Table 8.4. Starting with version 29 of the Rallfun file described in Section 1.8, bd1way uses by default a 20% trimmed mean, but any other estimator can be used via the argument est. Earlier versions of this function used est=onestep by default.) The argument misran=FALSE means that missing values are handled using casewise deletion. Setting misran=TRUE, all of the data are used, assuming missing values occur at random. As usual, x is any R variable that is a matrix or has list mode, and nboot is B, the number of bootstrap samples that will be used. When there are values missing at random, method M2 in Section 5.9.13 can be used to perform multiple comparisons via the R function rmmismcp in Section 5.9.14. By default, 20% trimmed means are used, but other robust estimators can be used via the argument est.

■ Example

Section 6.7.3 mentioned data dealing with the weight of cork borings taken from the north, east, south, and west sides of 28 trees. Assuming the data are stored in the R matrix cork, the command bd1way(cork) returns a p-value equal to 0.10. If we compare groups using modified one-step M-estimator (MOM) in conjunction with method RMPB4, the p-value is 0.385. (Compare this result to the example in Section 8.3.1.) ∎

■ Example

Again, consider the hangover data used to illustrate rmanova in Section 8.1.2. (The data are listed in Section 8.6.2.) Comparing M-measures of location results in an error because there are too many tied values, resulting in MAD=0 within the bootstrap. Assuming the data are stored in x, the command bd1way(x,est=hd) compares medians based on the Harrell–Davis estimator. The function reports that $Q = 9.96$, with a 0.05 critical value of 6.3, so the null hypothesis is rejected at the 0.05 level. ∎

8.2.7 Multiple Comparisons Using M-Estimators or Skipped Estimators

Next, consider C linear contrasts involving M-measures of location where the kth linear contrast is

$$\Psi_k = \sum_{j=1}^{J} c_{jk}\mu_{mj},$$

and, as usual, the c_{jk} values are constants that reflect linear combinations of the M-measures of location that are of interest, and for fixed k, $\sum c_{jk} = 0$. The goal is to compute a confidence interval for Ψ_k, $k = 1, \ldots, C$, such that the simultaneous probability coverage is approximately $1 - \alpha$. Alternatively, test H_0: $\Psi_k = 0$, with the goal that the probability of at least one Type I error is α.

First, set $C_{ij} = X_{ij} - \hat{\mu}_{mj}$. Next, obtain a bootstrap sample by sampling, with replacement, n rows of data from the matrix C_{ij}. Label the bootstrap values C_{ij}^*. Use the n values in the jth column of C_{ij}^* to compute $\hat{\mu}_{mj}^*$, $j = 1, \ldots, J$. Repeat this process B times, yielding $\hat{\mu}_{mjb}^*$, $b = 1, \ldots, B$. Next, compute the J-by-J covariance matrix associated with the $\hat{\mu}_{mjb}^*$ values. That is, compute

$$\hat{\tau}_{jk} = \frac{1}{B-1}\sum_{b=1}^{B}(\hat{\mu}_{mjb}^* - \bar{\mu}_j^*)(\hat{\mu}_{mkb}^* - \bar{\mu}_k^*),$$

where $\bar{\mu}_j^* = \sum \hat{\mu}_{mjb}^* / B$. Let

$$\hat{\Psi}_k = \sum_{j=1}^{J} c_{jk} \hat{\mu}_{mj},$$

$$\hat{\Psi}_{kb}^* = \sum_{j=1}^{J} c_{jk} \hat{\mu}_{mjb}^*,$$

$$S_k^2 = \sum_j \sum_\ell c_{jk} c_{\ell k} \hat{\tau}_{j\ell},$$

$$T_{kb}^* = \frac{\hat{\Psi}_{kb}^*}{S_k},$$

and

$$T_b^* = \max |T_{kb}^*|,$$

the maximum being taken over $k = 1, \ldots, C$. Then a confidence interval for Ψ_k is

$$\hat{\Psi}_k \pm T_{(u)}^* S_k,$$

where, as usual, $u = (1 - \alpha)B$, rounded to the nearest integer, and $T_{(1)}^* \leq \cdots \leq T_{(B)}^*$ are the T_b^* values written in ascending order. The simultaneous probability coverage is approximately $1 - \alpha$, but for $n \leq 21$ and $B = 399$, the actual probability coverage might be unsatisfactory. Under normality, for example, there are situations where the probability of at least one Type I error exceeds 0.08, with $J = 4$, $\alpha = 0.05$, and $n = 21$, and where all pairwise comparisons are performed. Increasing $B = 599$ does not correct this problem. It seems that $n > 30$ is required if the probability of at least one Type I error is not to exceed 0.075 when testing at the 0.05 level (Wilcox, 1997a).

An alternative approach, which appears to have some practical advantages over the method just described, is to use a simple extension of the percentile bootstrap method in Section 5.9.11. Let $\hat{p}_k^*$ be the proportion of times $\hat{\Psi}_{kb}^* > 0$ among the B bootstrap samples. Then a (generalized) p-value for H_0: $\Psi_k = 0$ is $2\min(\hat{p}_k^*, 1 - \hat{p}_k^*)$. When using M-estimators or MOM, however, a bias-adjusted estimate of the p-value appears to be beneficial; see Section 5.9.11. (With trimmed means, this bias adjustment appears to be unnecessary; see Wilcox and Keselman, 2002). FWE can be controlled with method SR outlined in Section 7.6.2. Again, for large sample sizes, say greater than 80, Hochberg's method (mentioned in Section 8.1.3) appears to be preferable.

Note that all of the methods described so far are based on measures of location that do not take into account the overall structure of the data when dealing with outliers. The skipped estimators in Section 6.5 do take the overall structure of the data into account, and situations might be encountered where this makes a practical difference. A basic percentile bootstrap method can be used to test H_0: $\Psi_k = 0$ and appears to control the probability of a Type I error reasonably well when using the OP-estimator in Section 6.5.

8.2.8 R Functions lindm and mcpOV

The R function

$$\text{lindm}(x,con=0,est=onestep,grp=0,alpha= 0.05,nboot=399,...)$$

computes confidence intervals for C linear contrasts based on measures of location corresponding to J dependent groups using the first method described in Section 8.2.7. The argument x contains the data and can be any n-by-J matrix, or it can have list mode. In the latter case, x[[1]] contains the data for group 1, x[[2]] the data for group 2, and so on. The optional argument con is a J-by-C matrix containing the contrast coefficients. If not specified, all pairwise comparisons are performed. The argument est is any statistic of interest. If unspecified, a one-step M-estimator is used. The argument grp can be used to select a subset of the groups for analysis. As usual, alpha is α and defaults to 0.05, and nboot is B, which defaults to 399. The argument ... is any additional arguments that are relevant to the function est.

■ **Example**

Again, consider the hangover data (reported in Section 8.6.2), where two groups of participants are measured at three different times. Suppose the first row of data is stored in DAT[[1]], the second row in DAT[[2]], the third in DAT[[3]], and so forth, but it is desired to perform all pairwise comparisons using only the group 1 data at times 1, 2, and 3. Then the command lindm(DAT,grp=c(1:3)) attempts to perform the comparisons, but eventually the function terminates with the error message "missing values in x not allowed." This error arises because there are so many tied values, bootstrap samples yield $\text{MAD} = 0$, which in turn makes it impossible to compute $\hat{\mu}^*$. This error can also arise when there are no tied values but one or more of the sample sizes are smaller than 20.

■

The command lindm(DAT,est=hd,gpr=c(1:3)) compares medians instead, using the Harrell–Davis estimator, and returns

```
         con.num        psihat  ci.lower   ci.upper         se
[1,]            1   -3.90507026 -7.880827 0.07068639  2.001571
[2,]            2   -3.82383677 -9.730700 2.08302610  2.973775
[3,]            3    0.08123349 -6.239784 6.40225079  3.182279


$crit:
[1] 1.986318


$con:
     [,1] [,2] [,3]
[1,]    1    1    0
[2,]   -1    0    1
[3,]    0   -1   -1
```

Because the argument con was not specified, the function creates its own set of linear contrasts, assuming all pairwise comparisons are to be performed. The resulting contrast coefficients are returned in the R variable $con. Thus, the first column, containing the values 1, -1, and 0, indicates that the first contrast corresponds to the difference between the medians for times 1 and 2. The results in the first row of $con.num indicate that the estimated difference between these medians is -3.91, and the confidence interval is $(-7.9, 0.07)$. In a similar fashion, the estimated difference between the medians at times 1 and 3 is -3.82, and for time 2 versus time 3, the estimate is 0.08. The command lindm(DAT,est=hd,gpr=c(1:3),q=0.4) would compare the 0.4 quantiles.

The R function

$$\text{mcpOV}(x, \text{alpha}= 0.05, \text{nboot=NA}, \text{grp=NA}, \text{est=smean}, \text{con=0}, \text{bhop=FALSE}, \text{SEED=TRUE},...)$$

is like the R function lindm, only it is designed to handle skipped estimators that take into account the overall structure of the data when checking for outliers. By default, it uses the OP-estimator, which is based on the projection method for detecting outliers.

8.2.9 Comparing Robust Measures of Scatter

Rather than comparing robust measures of location, situations arise where the goal is to compare robust measures of variation associated with the marginal distributions. A basic percentile bootstrap method can be used.

8.2.10 R Function rmrvar

The R function

> rmrvar(x, y = NA, alpha = 0.05, con = 0, est = pbvar, plotit = FALSE, grp = NA, hoch = TRUE, nboot = NA, xlab = 'Group 1', ylab = 'Group 2', pr = TRUE, SEED = TRUE, ...)

can be used to compare robust measures of variation associated with the marginal distributions. It is an extension of the R function rmVARcom, described in Section 5.9.16, that can be used to test hypotheses about linear contrasts. By default, all pairwise comparisons are made.

8.3 Bootstrap Methods Based on Difference Scores and a Measure of Effect Size

Section 8.1.4 describes a method for performing multiple comparisons based on linear combinations of random variables that contains comparing difference scores as a special case. This section describes a bootstrap method for testing the global hypothesis that all of the

$$L = \frac{J^2 - J}{2}$$

differences scores $D_{i\ell}$ ($i = 1, \ldots, n$, $\ell = 1, \ldots, L$) have a common measure of location.

■ **Example**

For four groups ($J = 4$), and each i ($i = 1, \ldots, n$), there are $L = 6$ differences given by

$$D_{i1} = X_{i1} - X_{i2},$$
$$D_{i2} = X_{i1} - X_{i3},$$
$$D_{i3} = X_{i1} - X_{i4},$$
$$D_{i4} = X_{i2} - X_{i3},$$
$$D_{i5} = X_{i2} - X_{i4},$$
$$D_{i6} = X_{i3} - X_{i4}.$$

The goal is to test

$$H_0: \theta_1 = \cdots = \theta_L = 0, \tag{8.5}$$

where θ_ℓ is the population measure of location associated with the ℓth set of difference scores, $D_{i\ell}$ ($i = 1, \ldots, n$). To test H_0 given by Eq. (8.5), resample vectors of D values, but unlike the bootstrap-t, observations are not centered. That is, a bootstrap sample now consists of resampling with replacement n rows from the matrix

$$\begin{pmatrix} D_{11}, \ldots, D_{1L} \\ \vdots \\ D_{n1}, \ldots, D_{nL} \end{pmatrix},$$

yielding

$$\begin{pmatrix} D_{11}^*, \ldots, D_{1L}^* \\ \vdots \\ D_{n1}^*, \ldots, D_{nL}^* \end{pmatrix}.$$

For each of the L columns of the D^* matrix, compute whatever measure of location is of interest, and for the ℓth column, label the result $\hat{\theta}_\ell^*$ ($\ell = 1, \ldots, L$). Next, repeat this B times, yielding $\hat{\theta}_{\ell b}^*$, $b = 1, \ldots, B$, and then determine how deeply the vector $\mathbf{0} = (0, \ldots, 0)$, having length L, is nested within the bootstrap values $\hat{\theta}_{\ell b}^*$. For two groups, this is tantamount to determining how many bootstrap values are greater than zero, which leads to the (generalized) p-value described in Section 5.4. The computational details when dealing with more than two groups are relegated to Table 8.5.

Effect Size

One way of measuring effect size is to use the basic approach in Section 8.1.1. For the situation at hand, the strategy is to measure the projection distance of the null vector $(0, \ldots, 0)$ to the estimate of the difference scores $\theta_1, \ldots, \theta_L$. This measure of effect size has, approximately, the same properties as the method in Section 8.1.1.

8.3.1 R Functions rmdzero and rmES.dif.pro

The R function

$$\text{rmdzero(x,est = mom, grp = NA, nboot = NA, ...)}$$

Table 8.5: Repeated measures ANOVA based on difference scores.

Goal: Test the hypothesis given by Eq. (8.5).
1. Let $\hat{\theta}_\ell$ be the estimate of θ_ℓ. Compute bootstrap estimates as described in Section 8.3, and label them $\hat{\theta}_{\ell b}^*$, $\ell = 1, \ldots, L$, $b = 1, \ldots, B$.
2. Compute the L-by-L matrix

$$S_{\ell\ell'} = \frac{1}{B-1} \sum_{b=1}^{B} (\hat{\theta}_{\ell b}^* - \hat{\theta}_\ell)(\hat{\theta}_{\ell' b}^* - \hat{\theta}_{\ell'}).$$

Readers familiar with multivariate statistical methods might notice that $S_{\ell\ell'}$ uses $\hat{\theta}_\ell$ (the estimate of θ_ℓ based on the original difference values), rather than the seemingly more natural $\bar{\theta}_\ell^*$, where

$$\bar{\theta}_\ell^* = \frac{1}{B} \sum_{b=1}^{B} \hat{\theta}_{\ell b}^*.$$

If $\bar{\theta}_\ell^*$ is used, unsatisfactory control over the probability of a Type I error can result.
3. Let $\hat{\theta} = (\hat{\theta}_1, \ldots, \hat{\theta}_L)$, $\hat{\theta}_b^* = (\hat{\theta}_{1b}^*, \ldots, \hat{\theta}_{Lb}^*)$, and compute

$$d_b = (\hat{\theta}_b^* - \hat{\theta}) S^{-1} (\hat{\theta}_b^* - \hat{\theta})',$$

where S is the matrix corresponding to $S_{\ell\ell'}$.
4. Put the d_b values in ascending order: $d_{(1)} \leq \cdots \leq d_{(B)}$.
5. Let

$$\mathbf{0} = (0, \ldots, 0)$$

having length L.
6. Compute

$$D = (\mathbf{0} - \hat{\theta}) S^{-1} (\mathbf{0} - \hat{\theta})'.$$

D measures how far away the null hypothesis is from the observed measures of location (based on the original data). In effect, D measures how deeply $\mathbf{0}$ is nested within the cloud of bootstrap values.
7. Reject if $D \geq d_{(u)}$, where $u = (1 - \alpha)B$, rounded to the nearest integer.

performs the test on difference scores described in Table 8.5. The R function

$$\text{rmES.dif.pro(x, est = tmean, ...)}$$

returns a measure of effect size based on projection distances.

■ **Example**

For the cork data, rmdzero returns a p-value of 0.034, so, in particular, reject with $\alpha = 0.05$. That is, conclude that the typical difference score is not equal to zero for all pairs of groups. This result is in sharp contrast to comparing marginal measures of location based on a robust M-estimator or MOM and the method in Table 8.4; see the example in Section 8.2.6. The measure of effect size returned by rmES.dif.pro is 0.81, which can be argued to be relatively large based on the discussion of measures of effect size in Chapter 5.

■

8.3.2 Multiple Comparisons

Multiple comparisons based on a percentile bootstrap method and difference scores can be addressed as follows. First, generate a bootstrap sample as described at the beginning of this section, yielding $D_{i\ell}^*$, $\ell = 1, \ldots, L$. When all pairwise differences are to be tested, $L = (J^2 - J)/2$, and $\ell = 1$ corresponds to comparing group 1 to group 2, $\ell = 2$ to comparing group 1 to group 3, and so on. Let $\hat{p}_\ell^*$ be the proportion of times among B bootstrap resamples that $D_{i\ell}^* > 0$. As usual, let

$$\hat{p}_{m\ell}^* = \min(\hat{p}_\ell^*, \, 1 - \hat{p}_\ell^*),$$

in which case, $2\hat{p}_{m\ell}^*$ is the estimated (generalized) p-value for the ℓth comparison.

One approach to controlling FWE is to put the p-values in descending order and to make decisions about which hypotheses are to be rejected using method SR outlined in Section 7.6.2. That is, once $\hat{p}_{mc}^*$ are computed, reject the hypothesis corresponding to $\hat{p}_{mc}^*$, if $\hat{p}_{mc}^* \leq \alpha_c$, where α_c is read from Table 7.13.

As for linear contrasts, consider any specific linear contrast with contrast coefficients $c_1, \ldots, c_J$, set

$$D_i = \sum c_j X_{ij},$$

and let θ_d be some (population) measure of location associated with this sum. Then $H_0: \theta_d = 0$ can be tested by generating a bootstrap sample from the D_i values, repeating this B times, and computing $\hat{p}^*$, the proportion of bootstrap estimates that are greater than zero, in which case, $2\min(\hat{p}^*, \, 1 - \hat{p}^*)$ is the estimated significance level. Then FWE can by controlled in the manner just outlined. Moreover, as noted in Section 8.1.4, measures of effect size can be addressed as described in Sections 4.3.1 and 5.9.17.

Another option is to use a measure of location associated with the marginal distributions. That is, test hypotheses having the form $H_0: \sum c_j\theta_j = 0$, where $\theta_1, \ldots, \theta_J$ are measures of location associated with the J marginal distributions.

When comparing groups using MOM or M-estimators, at the moment, it seems that the method based on difference scores often provides the best power versus testing hypotheses based on measures of location associated with the marginal distributions. Both approaches do an excellent job of avoiding Type I error probabilities greater than the nominal α level. But when testing hypotheses about measures of location associated with the marginal distributions, the actual Type I error probability can drop well below the nominal level in situations where the method based on difference scores avoids this problem. This suggests that the method based on difference scores will have more power, and indeed, there are situations where this is the case even when the two methods have comparable Type I error probabilities. It is stressed, however, that a comparison of these methods, in terms of power, needs further study. Also, the bias-adjusted critical value mentioned in Section 5.9.7 appears to help increase power.

While comparing marginal measures of location based on MOM or an M-estimator seems to result in relatively low power, there is weak evidence that comparing marginal measures of location based on the OP-estimator, via a percentile bootstrap method as mentioned at the end of Section 8.2.7, performs relatively well. But the extent to which this is true needs additional study.

When dealing with trimmed means, including medians as a special case, again, the percentile bootstrap method just described can be used and appears to be a relatively good choice provided the amount of trimming is not too small. With a small amount of trimming, use a bootstrap-t method instead. Controlling the probability of at least one Type I error can be done with Hochberg's method.

Note that when working with the median of difference scores, another approach is to use the method in Section 4.6.2 and control the probability of one or more Type I errors using some improvement on the Bonferroni method. Here, Rom's or Hochberg's method is used.

8.3.3 R Functions rmmcppb, wmcppb, dmedpb, lindepbt, and qdmcpdif

The R function

rmmcppb(x,y = NA, tr=0.2, con = 0, est = mom, plotit = TRUE, dif = TRUE, grp = NA, nboot = NA, BA=FALSE,hoch=FALSE,...)

performs multiple comparisons among dependent groups using the percentile bootstrap methods just described. The argument dif defaults to TRUE, indicating that difference scores will be used, in which case, Hochberg's method is used to control FWE. If dif=FALSE, measures of location associated with the marginal distributions are used instead. If dif=FALSE and BA=TRUE, the bias-adjusted estimate of the generalized p-value (described in Section 5.9.7) is applied; using BA=TRUE (when dif=FALSE) is recommended when comparing groups with M-estimators and MOM, but it is not necessary when comparing 20% trimmed means (Wilcox and Keselman, 2002). If the goal is to compare groups using M-estimators or the MOM and hoch=FALSE, then FWE is controlled using method SR in Section 7.6.2 if the sample size is less than 80; otherwise, Hochberg's method is used as described in Section 8.1.3. If hoch=TRUE, Hochberg's method is used regardless of the sample size. If no value for con is specified, then all pairwise differences will be tested. As usual, if the goal is to test hypotheses other than all pairwise comparisons, con can be used to specify the linear contrast coefficients.

When comparing trimmed means, it appears that Hochberg's method is preferable to method SR in terms of controlling the probability of at least one Type I error. For convenience, the R function

wmcppb(x, tr=0.2, con = 0, est = tmean, plotit = TRUE, dif = TRUE, grp = NA, nboot = NA, BA=FALSE,hoch=TRUE,...)

is supplied. It is the same as the R function rmmcppb, only it defaults to comparing 20% trimmed means, and by default, it uses Hochberg's method rather than method SR. (The R function dtrimpb is the same as the function wmmcppb.)

The R function

dmedpb(x,y=NA,alpha= 0.05,con=0,est=median, plotit=TRUE, dif=FALSE, grp=NA, hoch=TRUE, nboot=NA,xlab='Group 1', ylab='Group 2', pr=TRUE, SEED=TRUE, BA=FALSE,...)

is similar to the R function rmmcppb, only it defaults to comparing marginal medians, and it is designed to handle tied values. Difference scores are used when the argument dif=TRUE, but it is unknown what advantage this approach has, if any, over the R function sintmcp, described in Section 8.1.5. Hochberg's method is used to control FWE. With a small sample size, say less than 30, setting the argument BA=TRUE seems advisable, meaning that the p-value is adjusted as described in Section 5.9.11 (Wilcox, 2006b). Note that by default, the argument dif=FALSE, indicating that the marginal medians are compared. To use difference scores, set dif=TRUE.

The R function

lindepbt(x, con = NULL, tr=0.2, alpha=0.05, nboot=599, dif=TRUE, SEED=TRUE) performs multiple comparisons based on trimmed means using a bootstrap-t method. When the amount of trimming is small, a bootstrap-t method is preferable to a percentile bootstrap method. With 20% trimming, a percentile bootstrap is generally more satisfactory. There are some indications that this remains the case with 10% trimming but an extensive simulation study has not yet been performed. The function reports critical p-values based on Rom's method for controlling the probability of one or more Type I errors. The function returns confidence intervals, but they are not adjusted so that the simultaneous probability coverage is $1 - \alpha$. Rather, each confidence interval is designed to have probability coverage $1 - \alpha$.

■ Example

For the cork boring data, the R function wmcppb (with the argument dif=FALSE as well as dif=TRUE) finds no significant results when the probability of at least one Type I error is taken to be 0.05. But the R function mcpOV (in Section 8.2.8), which compares marginal measures of location via the OP-estimator, finds three significant results, the only point being that the choice of method can make a practical difference. Again, it is stressed that little is known about the extent to which the OP-estimator might have higher power compared to the many other methods that might be used to compare dependent groups.

■

The R function

$$qdmcpdif(x, con = 0, tr=0.2)$$

can be used to perform multiple tests based on difference scores using the method in Section 4.6.2. Testing the hypothesis that other linear contrasts have a population median equal to zero can be done via the argument con.

8.3.4 Measuring Effect Size: R Function DEP.PAIR.ES

Section 5.9.20 described an R function for measuring effect size based on difference scores. The R function

$$DEP.PAIR.ES(x, tr = 0.2, OPT = TRUE, REL.MAG = NULL, SEED = TRUE)$$

is provided for measuring effect size for each pair of variables stored in the argument x, which is assumed to be a matrix or have list mode. The argument REL.MAG can be used to indicate three values: what constitutes a low, medium, or large effect size. If not specified, the default values are 0.1, 0.3, and 0.5, respectively.

8.3.5 Comparing Multinomial Cell Probabilities

Consider a multinomial distribution with cell probabilities $p_1, \ldots, p_K$. Let (l_1, u_1) and (l_2, u_2) denote a $1 - \alpha$ confidence interval for p_j and p_k, respectively, which can be computed as described in Section 4.8. From basic principles, the covariance between $\hat{p}_j$ and $\hat{p}_k$ is

$$\tau_{jk} = -\sqrt{\frac{p_j p_k}{(1 - p_j)(1 - p_k)}},$$

which here is estimated with

$$\hat{\tau}_{jk} = -\sqrt{\frac{\hat{p}_j \hat{p}_k}{(1 - \hat{p}_j)(1 - \hat{p}_k)}}.$$

Then, based on a simple modification of results by Zou (2007), an approximate $1 - \alpha$ confidence interval for $p_j - p_k$ is

$$(L, U), \tag{8.6}$$

where

$$L = \hat{p}_j - \hat{p}_k - \sqrt{(\hat{p}_j - l_1)^2 + (u_2 - \hat{p}_j)^2 - 2\hat{\tau}_{jk}(\hat{p}_j - l_1)(u_2 - \hat{p}_k)}$$

and

$$U = \hat{p}_j - \hat{p}_k + \sqrt{(u_1 - \hat{p}_j^2 + (\hat{p}_j - l_2)^2 - 2\hat{\tau}_{jk}(u_1 - \hat{p}_j)(\hat{p}_k - l_2)}.$$

Note that this includes the special case where the off-diagonal cell probabilities of a two-by-two contingency table are to be compared. When the sample size is small, this confidence interval can be more satisfactory than a well-known technique (e.g., Wilcox, 2017d, Section 15.3) not described here, which is related to McNemar's test. For other approaches to this problem, see Lloyd (1999). Also, method RS2-IP, described in Section 7.4.11, is readily adapted to determining which cell of a multinomial distribution has the largest or smallest probability.

8.3.6 R Functions cell.com, cell.com.pv, and best.cell.DO

The R function

```
cell.com(x,i=1,j=2, alpha=0.05, AUTO=TRUE, method='AC')
```

compares cell probabilities of a multinomial distribution using the method in the previous section. Arguments i and j indicate which cells are compared. AUTO=TRUE means that when $n \leq 35$, the Schilling–Doi method is used to compute confidence intervals for p_i and p_j. The R function

$$\text{cell.com.pv(x,i=1,j=2,method='AC')}$$

computes a p-value. For determining whether it is reasonable to make a decision about which cell has the largest or smallest probability, use the R function

$$\text{best.cell.DO(x, FREQ = NULL, LARGE = TRUE).}$$

■ **Example**

Consider the R command x=c(rep(1,5),rep(2,6),rep(3,24)). The R command best.cell.DO(x) returns a p-value equal to 0.015.

■

8.4 Comments on Which Method to Use

No single method in this chapter dominates based on various criteria used to compare hypothesis testing techniques. However, a few comments can be made about their relative merits that might be useful. First, the expectation is that in many situations where groups differ, all methods based on means perform poorly, in terms of power, relative to approaches based on some robust measure of location, such as MOM or a 20% trimmed mean. Currently, with a sample size as small as 21, the bootstrap-t method described in Section 8.2.2, which is performed by the R function rmanovab, appears to provide excellent control over the probability of a Type I error when used in conjunction with 20% trimmed means. Its power compares reasonably well to most other methods that could be used, but as noted in previous chapters, different methods are sensitive to different features of the data, and arguments for some other measure of location, such as an M-estimator, have been made.

The percentile bootstrap methods in Section 8.2.5 also do an excellent job of avoiding Type I errors greater than the nominal level, but there are indications that when using method RMPB3, when the sample size is small, the actual probability of a Type I error can be substantially less than α, suggesting that some other method might provide better power. Nevertheless, if there is specific interest in comparing M-estimators associated with the marginal distributions, it is suggested that method RMPB3 be used when the sample size is greater than 20. Also, it can be used to compare groups based on MOM, but with very small sample sizes, power might be inadequate relative to other techniques that could be used.

Given the goal of testing some omnibus hypothesis, currently, among the techniques covered in this chapter, it seems that the two best methods for controlling Type I error probabilities and simultaneously providing relatively high power are the bootstrap-t method based on 20% trimmed means and the percentile bootstrap method in Table 8.5, which is based in part on difference scores; the computations are performed by the R function rmdzero. (But also consider the multiple-comparison procedure described in Section 8.1.4.) With near certainty, situations arise where some other technique is more optimal, but typically, the improvement is small. But again, comparing groups with MOM is not the same as comparing means, trimmed means, or M-estimators, and certainly there will be situations where some other estimator has higher power than any method based on MOM or a 20% trimmed mean. If the goal is to maximize power, several methods are contenders for routine use. With sufficiently large sample sizes, trimmed means can be compared without resorting to the bootstrap-t method, but it remains unclear just how large the sample size must be. Roughly, as the amount of trimming increases from 0% to 20%, the smaller the sample size must be to control the probability of a Type I error without resorting to a bootstrap technique. But if too much trimming is done, power might be relatively low. When comparing medians, a percentile bootstrap method is recommended when dealing with tied values.

As for the issue of whether to use difference scores rather than robust measures of location based on the marginal distributions, each approach provides a different perspective on how groups differ, and they can give different results regarding whether groups are significantly different. There is some evidence that difference scores typically provide more power and better control over the probability of a Type I error, but situations are encountered where the reverse is true. A more detailed study is needed to resolve this issue.

As previously mentioned, method RMPB4 performed by the R function ddep, described in Section 8.2.6, is very conservative in terms of Type I errors, meaning that when testing at the 0.05 level, say, often the actual probability of a Type I error will be less than or equal to α and typically smaller than that of any other method described in this chapter. So a concern is that power might be low relative to the many other methods that might be used.

Regarding methods designed for performing multiple comparisons in a manner that controls the probability of at least one Type I error, currently, it seems that using the R function wmcppb, which compares groups based on trimmed means via a percentile bootstrap method, is a good choice, particularly in terms of maximizing power. But again, there are exceptions. Perhaps the method in Section 8.2.7, which is based on the OP-estimator and performed by the R function mcpOV, generally competes well with the R function wmcppb, but little is known about the extent to which this is true. As in previous chapters, in terms of controlling the probability of at least one Type I error, the method used by wmcppb and mcpOV does not assume or require that one first test and reject the global hypothesis that all groups have identical population trimmed means. It is not necessary, for example, that the function rmanovab (described in Section 8.2.2) returns a significant result before using the function wmcppb.

8.5 Some Rank-Based Methods

This section describes two rank-based methods for testing

$$H_0: F_1(x) = \cdots = F_J(x),$$

the hypothesis that J dependent groups have identical marginal distributions. Friedman's test is the best-known test of this hypothesis, but no details are given here. The first of the two methods was derived by Agresti and Pendergast (1986) and has higher power than Friedman's test when sampling from normal distributions, and their test can be expected to have good power when sampling from heavy-tailed distributions, so it is included here. The other method stems from Brunner et al. (2002a, Section 7.2.2), which appears to have an advantage over the Agresti–Pendergast method in terms of power (Tian and Wilcox, 2009).

Method AP

The calculations for the Agresti–Pendergast method are as follows. Pool all the observations and assign ranks. Let R_{ij} be the resulting rank of the ith observation in the jth group. Compute

$$\bar{R}_j = \frac{1}{n} \sum_{i=1}^{n} R_{ij},$$

$$s_{jk} = \frac{1}{n - J + 1} \sum_{i=1}^{n} (R_{ij} - \bar{R}_j)(R_{ik} - \bar{R}_k).$$

Let the vector $\boldsymbol{R}'$ be defined by

$$\mathbf{R}' = (\bar{R}_1, \ldots, \bar{R}_J),$$

and let $\mathbf{C}$ be the $(J - 1)$-by-J matrix given by

$$\begin{pmatrix} 1 & -1 & 0 & \ldots & 0 & 0 \\ 0 & 1 & -1 & \ldots & 0 & 0 \\ \cdot & \cdot & \cdot & \cdot & \cdot & \cdot \\ 0 & 0 & 0 & \ldots & 1 & -1 \end{pmatrix}.$$

The test statistic is

$$F = \frac{n}{J-1}(\mathbf{CR})'(\mathbf{CSC}')^{-1}\mathbf{CR},$$

where

$$\mathbf{S} = (s_{jk}).$$

The degrees of freedom are $\nu_1 = J - 1$ and $\nu_2 = (J-1)(n-1)$, and you reject, if $F > f_{1-\alpha}$, the $1 - \alpha$ quantile of an F distribution with ν_1 and ν_2 degrees of freedom.

Method BPRM

Let R_{ij} be defined as in method AP, and let $\mathbf{R}_i = (R_{i1}, \ldots, R_{iJ})'$ be the vector of ranks for the ith participant, where $(R_{i1}, \ldots, R_{iJ})'$ is the transpose of $(R_{i1}, \ldots, R_{iJ})$. Let

$$\bar{\mathbf{R}} = \frac{1}{n}\sum_{i=1}^{n}\mathbf{R}_i,$$

be the vector of ranked means, let

$$\bar{R}_{.j} = \frac{1}{n}\sum_{i=1}^{n}R_{ij},$$

denote the mean of the ranks for group j, and let

$$\mathbf{V} = \frac{1}{N^2(n-1)}\sum_{i=1}^{n}(\mathbf{R}_i - \bar{\mathbf{R}})(\mathbf{R}_i - \bar{\mathbf{R}})'.$$

The test statistic is

$$F = \frac{n}{N^2\text{tr}(\mathbf{PV})}\sum_{j=1}^{J}\left(\bar{R}_{.j} - \frac{N+1}{2}\right)^2, \tag{8.7}$$

where

$$\mathbf{P} = \mathbf{I} - \frac{1}{J}\mathbf{J},$$

$\mathbf{J}$ is a $J \times J$ matrix of all 1s, and $\mathbf{I}$ is the identity matrix.

Decision Rule

Reject the hypothesis of identical distributions, if

$$F \geq f,$$

where f is the $1 - \alpha$ quantile of an F distribution with degrees of freedom

$$\nu_1 = \frac{[\text{tr}(\mathbf{PV})]^2}{\text{tr}(\mathbf{PVPV})}$$

and $\nu_2 = \infty$. Note that based on the test statistic, a crude description of method BPRM is that it is designed to be sensitive to differences among the average ranks.

8.5.1 R Functions apanova and bprm

The R function

$$\text{apanova(x,grp=0)}$$

performs the Agresti–Pendergast test of equal marginal distributions. As usual, x can have list mode or be an *n*-by-*J* matrix, and the argument grp can be used to specify some subset of the groups. If grp is unspecified, all *J* groups are used. The function returns the value of the test statistic, the degrees of freedom, and the p-value. For example, the command apanova(dat,grp=c(1,2,4)) would compare groups 1, 2, and 4 using the data in the R variable dat.

The R function

$$\text{bprm(x)}$$

performs method BPRM; it returns a p-value.

8.6 Between-by-Within and Within-by-Within Designs

This section describes some methods for testing hypotheses in a between-by-within (or split-plot) design. That is, a J-by-K ANOVA design is being considered, where the J levels of the first factor correspond to independent groups (between subjects), and the K levels of the second factor are dependent (within subjects). Within-by-within designs are covered as well.

8.6.1 *Analyzing a Between-by-Within Design Based on Trimmed Means*

We begin with a between-by-within design. For the jth level of Factor A, let Σ_j be the K-by-K population Winsorized covariance matrix for the K dependent random variables associated with the second factor. The better-known methods for analyzing a split-plot design are based on the assumption that $\Sigma_1 = \cdots = \Sigma_J$, but violating this assumption can result in problems controlling the probability of a Type I error. Keselman et al. (1995) found that a method derived by Johansen (1980) that does not assume there is a common covariance matrix gives better results, so a generalization of Johansen's method to trimmed means is described here. (For related results, see Keselman et al., 1993, 1999; Livacic-Rojas et al., 2010.)

As in the case of a two-way design for independent groups, it is easier to describe the method in terms of matrices. Main effects and interactions are examined by testing:

$$H_0: \mathbf{C}\boldsymbol{\mu}_t = \mathbf{0},$$

where $\mathbf{C}$ is a k-by-JK contrast matrix, having rank k that reflects the null hypothesis of interest. Let $\mathbf{C}_m$ and $\mathbf{j}'$ be defined as in Section 7.3. Then for Factor A, $\mathbf{C} = \mathbf{C}_J \otimes \mathbf{j}'_K$ and $k = J - 1$. For Factor B, $\mathbf{C} = \mathbf{j}'_J \otimes \mathbf{C}_K$ and $k = K - 1$, and the test for no interactions uses $\mathbf{C} = \mathbf{C}_J \otimes \mathbf{C}_K$.

For every level of Factor A, there are K dependent random variables, and each pair of these dependent random variables has a Winsorized covariance that must be estimated. In symbols, let X_{ijk} be the ith observation randomly sampled from the jth level of Factor A and the kth level of Factor B. For fixed j, the Winsorized covariance between the mth and ℓth levels of Factor B is estimated with

$$s_{jm\ell} = \frac{1}{n_j - 1} \sum_{i=1}^{n_j} (Y_{ijm} - \bar{Y}_{.jm})(Y_{ij\ell} - \bar{Y}_{.j\ell}),$$

where

$$Y_{ijk} = \begin{cases} X_{(g+1),jk}, & \text{if } X_{ijk} \leq X_{(g+1),jk}, \\ X_{ijk}, & \text{if } X_{(g+1),jk} < X_{ij} < X_{(n-g),jk}, \\ X_{(n-g),jk}, & \text{if } X_{ijk} \geq X_{(n-g),jk}, \end{cases}$$

and

$$\bar{Y}_{.jm} = \frac{1}{n} \sum_{i=1}^{n} Y_{ijm}.$$

For fixed j, let $\mathbf{S}_j = (s_{jm\ell})$. That is, $\mathbf{S}_j$ estimates Σ_j, the K-by-K Winsorized covariance matrix for the jth level of Factor A. Let

$$\mathbf{V}_j = \frac{(n_j - 1)\mathbf{S}_j}{h_j(h_j - 1)}, \ j = 1, \ldots, J,$$

and let $\mathbf{V} = \text{diag}(\mathbf{V}_1, \ldots, \mathbf{V}_J)$ be a block diagonal matrix. The test statistic is

$$Q = \bar{\mathbf{X}}'\mathbf{C}'(\mathbf{CVC}')^{-1}\mathbf{C}\bar{\mathbf{X}}, \qquad (8.8)$$

where $\bar{\mathbf{X}}' = (\bar{X}_{t11}, \ldots, \bar{X}_{tJK})$. Let $\mathbf{I}_{K \times K}$ be a K-by-K identity matrix, and let $\mathbf{Q}_j$ be a JK-by-JK block diagonal matrix (consisting of J blocks, each block being a K-by-K matrix), where the tth block ($t = 1, \ldots, J$) along the diagonal of $\mathbf{Q}_j$ is $\mathbf{I}_{K \times K}$, if $t = j$, and all other elements are zero. (For example, if $J = 3$ and $K = 4$, then $\mathbf{Q}_1$ is a 12-by-12 matrix block diagonal matrix, where the first block is a 4-by-4 identity matrix, and all other elements are zero. As for $\mathbf{Q}_2$, the second block is an identity matrix, and all other elements are zero.) Compute

$$A = \frac{1}{2}\sum_j^J [\text{tr}(\{\mathbf{VC}'(\mathbf{CVC}')^{-1}\mathbf{CQ}_j\}^2) + \{\text{tr}(\mathbf{VC}'(\mathbf{CVC}')^{-1}\mathbf{CQ}_j)\}^2]/(h_j - 1),$$

where tr indicates trace, and let

$$c = k + 2A - \frac{6A}{k+2}.$$

When the null hypothesis is true, Q/c has, approximately, an F distribution with $\nu_1 = k$ and $\nu_2 = k(k+2)/(3A)$ degrees of freedom, so reject, if $Q/c > f_{1-\alpha}$, the $1 - \alpha$ quantile. Recent simulation results reported by Livacic-Rojas et al. (2010) indicate that this method performs relatively well, in terms of controlling the probability of a Type I error, when comparing means. However, their results do not consider the effect of having different amounts of skewness.

■ **Example**

Consider a 2-by-3 design, where for the first level of Factor A, observations are generated from a multivariate normal distribution with all correlations equal to zero. For the second level of Factor A, the marginal distributions are lognormal that have been shifted to have mean zero. Further suppose that the covariance matrix for the second level is three times larger that the covariance matrix for the first level. If the sample sizes are $n_1 = n_2 = 30$, and the hypothesis of no main effects for Factor A, based on means, is tested at the 0.05 level, the actual level is approximately 0.088. If the sample sizes are $n_1 = 40$ and $n_2 = 70$, and for the first level of Factor A the marginal distributions are g-and-h distributions with $g = h = 0.5$, the probability of a Type I error, again, testing at the 0.05 level, is approximately 0.188. Comparing 20% trimmed means instead, the actual Type I error probability is approximately 0.035.

■

8.6.2 R Functions bwtrim, bw.es.main, and tsplit

The R function

$$bwtrim(J,K,x,tr=0.2,grp=c(1:p),p=J*K)$$

tests the hypotheses of no main effects and no interactions in a between-by-within (split-plot) design, where J is the number of independent groups, K is the number of dependent groups, and the argument x contains the data stored in list mode, a matrix, or a data frame. The optional argument, tr, indicates the amount of trimming, which defaults to 0.2 if unspecified. (The R function tsplit is identical to bwtrim.) The R function

$$bw.es.main(J,K,x,DIF=TRUE,...)$$

computes measures of effect size for the main effects. For Factor A, the explanatory measure of effect size is used. For Factor B, the measure of effect described in Section 8.3 is used if the argument DIF=TRUE. Otherwise, it uses the measure of effect size described in Section 8.1.1.

The groups are assumed to be ordered as described in Section 7.2.1. If the data are not stored in the proper order, grp can be used to indicate how they are stored. For example, if a 2-by-2 design is being used, the R command

$$bwtrim(2,2,x,grp=c(3,1,2,4))$$

indicates that the data for the first level of both factors are stored in x[[3]], the data for level 1 of Factor A and level 2 of Factor B are in x[[1]], and so forth. If x is a matrix or a data frame, the first K columns correspond to the first level of Factor A and the K levels of Factor B, the next K columns correspond to the second level of Factor A, and so on.

■ Example

In a study on the effect of consuming alcohol, hangover symptoms were measured for two independent groups, with each subject consuming alcohol and being measured on three different occasions. One group (group 2) consisted of sons of alcoholics and the other was a control group. Here, $J = 2$ and $K = 3$. The results were as follows.

Group 1, time 1:	0 32 9 0 2 0 41 0 0 0 6 18 3 3 0 11 11 2 0 11
Group 1, time 2:	4 15 26 4 2 0 17 0 12 4 20 1 3 7 1 11 43 13 4 11
Group 1, time 3:	0 25 10 11 2 0 17 0 3 6 16 9 1 4 0 14 7 5 11 14
Group 2, time 1:	0 0 0 0 0 0 0 1 8 0 3 0 0 32 12 2 0 0 0
Group 2, time 2:	2 0 7 0 4 2 9 0 1 14 0 0 0 0 15 14 0 0 7 2
Group 2, time 3:	1 0 3 0 3 0 15 0 6 10 1 1 0 2 24 42 0 0 0 2

Suppose the first row of data is stored in DAT[[1]], the second row in DAT[[2]], the third in DAT[[3]], the fourth in DAT[[4]], the fifth in DAT[[5]], and the sixth in DAT[[6]]. That is, the data are stored as assumed by the function bwtrim. Then the command bwtrim(2,3,DAT,tr=0) will compare the means and return:

```
$Qa
[1] 3.277001

$Qa.siglevel
           [,1]
[1,]   0.07825593

$Qb
[1] 0.8808645

$Qb.p.value
           [,1]
[1,] 0.4250273

$Qab
[1] 1.050766

$Qab.p.value
           [,1]
[1,] 0.3623739
```

The test statistic for Factor A is $Qa = 3.28$, and the corresponding p-value is 0.078. For Factor B, the p-value is 0.425, and for the interaction, it is 0.362.

■

This section described one way of comparing independent groups in a between-by-within subjects design. Another approach is simply to compare the J independent groups for each level of Factor B. That is, do not sum or average the data over the levels of Factor B as was done here. So the goal is to test

$$H_0:\ \mu_{t1k} = \cdots = \mu_{tJk},$$

for each $k = 1, \ldots, K$. The next example illustrates that in applied work, the choice between these two methods can make a practical difference. (Yet another strategy for comparing the levels of Factor A is to apply the robust MANOVA method in Section 7.10.)

■ Example

Section 7.8.4 reports data from a study comparing schizophrenics to a control group based on a measure taken at two different times. Analyzing the data with the function bwtrim, no main effects for Factor A are found, the p-value being 0.245. So no difference between the schizophrenics and the control group was detected at the 0.05 level. But if the groups are compared using the first measurement only, using the function yuen (described in Chapter 5), the p-value is 0.012. For the second measurement, ignoring the first, the p-value is 0.89. (Recall that in Chapter 6, using some multivariate methods, again, a difference between the schizophrenics and the control group was found.)

8.6.3 Data Management: R Function bw2list

Imagine a situation where data are stored in a matrix or a data frame, say x, with one column indicating the levels of the between factor, but the K levels of the within group factor are stored in K columns of the matrix x. In order to use the R function bwtrim, it is necessary to store the data in the format that is allowed. The R function

bw2list(x, grp.col, lev.col)

is provided to help accomplish this goal. The argument grp.col indicates the column containing information about the levels of the independent groups. The values in this column can be numeric or character data. And the argument lev.col indicates the K columns where the within group data are stored. The function returns the data stored in list mode, which can then be used by bwtrim as well as other functions aimed at dealing with a between-by-within design. The function will store the data sorted in ascending (or alphabetical) order based on the values found in the column of x indicated by the argument grp.col. The next example illustrates this feature.

■ **Example**

Imagine that three medications are being investigated regarding their effectiveness to lower cholesterol, and that column 3 of matrix m indicates which medication a participant received. Moreover, columns 5 and 8 contain the participants' cholesterol levels at times 1 and 2, respectively. The R command

$$z=bw2list(m,3,c(5,8))$$

will store the data in z in list mode. If column 3 contains the character values 'P', 'CH', and 'BN', then z[[1]] and z[[2]] will contain the data for times 1 and 2, respectively, corresponding to level 'BN' of Factor A, z[[3]] and z[[4]] will contain the data for level 'CH', and z[[5]] and z[[6]] will contain the data for level 'P'. The R command

$$bwtrim(3,2,z)$$

will compare the groups based on 20% trimmed means.

■

8.6.4 Bootstrap-t Method for a Between-by-Within Design

To apply a bootstrap-t method, when working with trimmed means and dealing with a between-by-within design, first, center the data in the usual way. In the present context, this means you compute

$$C_{ijk} = X_{ijk} - \bar{X}_{tjk},$$

$i = 1, \ldots, n_j, j = 1, \ldots, J, k = 1, \ldots, K$. That is, for the group corresponding to the jth level of Factor A and the kth level of Factor B, subtract the corresponding trimmed mean from each of the observations. Next, for each j, generate a bootstrap sample based on the C_{ijk} values by resampling with replacement n_j vectors of observations from the n_j rows of data corresponding to level j of Factor A. That is, for each level of Factor A, you have an n_j-by-K matrix of data, and you generate a bootstrap sample from this matrix of data as described in Section 8.2.5, where for fixed j, resampling is based on the C_{ijk} values. Label the resulting bootstrap samples C_{ijk}^*. Compute the test statistic Q based on the C_{ijk}^* values as described in Section 8.6.1, and label the result Q^*. Repeat this B times, yielding $Q_1^*, \ldots, Q_B^*$, and then put these B values in ascending order, yielding $Q_{(1)}^* \leq \cdots \leq Q_{(B)}^*$. Next, compute Q using the original data (the X_{ijk} values), and reject, if $Q \geq Q_{(c)}^*$, where $c = (1 - \alpha)$ rounded to the nearest integer.

A crude rule that seems to apply to a wide variety of situations is: The more the distributions (associated with groups) differ, the more beneficial it is to use some type of bootstrap method, at least when sample sizes are small. Keselman et al. (2000) compare the bootstrap method just described to the non-bootstrap method for a split-plot design covered in Section 8.6.1. For the situations they examine, this rule does not apply; it was found that the bootstrap offers little or no advantage. Their study includes situations where the correlations (or covariances) among the dependent groups differ across the independent groups being compared. However, the more complicated the design, the more difficult it becomes to consider all the factors that might influence the operating characteristics of a particular method. One limitation of their study was that the differences among the covariances were taken to be relatively small. Another issue that has not been addressed is how the bootstrap method performs when distributions differ in skewness. Having differences in skewness is known to be important when dealing with the simple problem of comparing two groups only. There is no reason to assume that this problem diminishes as the number of groups increases, and indeed, there are reasons to suspect that it becomes a more serious problem. So currently, it seems that if groups do not differ in any manner or the distributions differ slightly, it makes little difference whether you use a bootstrap versus a non-bootstrap method for comparing trimmed means. However, if distributions differ in shape, there is indirect evidence that a bootstrap method might offer an advantage when using a split-plot design, but the extent to which this is true is not well understood.

8.6.5 R Functions bwtrimbt and tsplitbt

The R function

$$tsplitbt(J,K,x,tr=0.2,alpha=0.05,JK=J*K,grp=c(1:JK),nboot=599)$$

performs a bootstrap-t method for a split-plot design as just described. The data are assumed to be arranged as indicated in conjunction with the R function tsplit (as described in Section 8.6.2), and the arguments J, K, tr, and alpha have the same meanings as before. The argument JK can be ignored, and grp can be used to rearrange the data if they are not stored as expected by the function. (For an R function that might help when dealing with organizing the data in a manner that is accepted by tsplitbt, see Section 8.6.3.)

The R function

$$bwtrimbt(J,K,x,tr=0.2,JK=J*K,grp=c(1:JK),nboot=599)$$

is the same as tsplitbt, only bwtrimbt reports p-values rather than α level critical values.

8.6.6 Percentile Bootstrap Methods for a Between-by-Within Design

Comparing groups based on MOMs, medians, and M-estimators in a between-by-within design is possible using extensions of percentile bootstrap methods already described. And they provide yet another way of comparing trimmed means.

Again consider a two-way design where Factor A consists of J independent groups and Factor B corresponds to K dependent groups. First, consider the dependent groups. One approach to comparing these K groups, ignoring Factor A, is to simply form difference scores, and then apply the method in Section 8.3. More precisely, imagine you observe X_{ijk} ($i = 1, \ldots, n_j$, $j = 1, \ldots, J$, $k = 1, \ldots, K$). That is, X_{ijk} is the ith observation in level j of Factor A and level k of Factor B. Note that if we ignore the levels of Factor A, we can write the data as Y_{ik}, $i = 1, \ldots, N$, $k = 1, \ldots, K$, where $N = \sum n_j$. Now, consider levels k and k' of Factor B ($k < k'$), set

$$D_{ikk'} = Y_{ik} - Y_{ik'},$$

and let $\theta_{kk'}$ be some measure of location associated with $D_{ikk'}$. Then the levels of Factor B can be compared, ignoring Factor A, by testing

$$\theta_{12} = \cdots = \theta_{k-1,k} = 0 \tag{8.9}$$

using the method in Section 8.3. In words, the null hypothesis is that the typical difference score between any two levels of Factor B, ignoring Factor A, is zero.

As for Factor A, ignoring Factor B, one approach is as follows. Momentarily, focus on the first level of Factor B, and note that the levels of Factor A can be described as in Chapter 7. That is, the null hypothesis of no differences among the levels of Factor A is

$$H_0\colon \theta_{11} = \theta_{21} = \cdots = \theta_{J1},$$

where, of course, these J groups are independent, and a percentile bootstrap method can be used. More generally, for any level of Factor B, say the kth, the hypothesis of no main effects is

$$H_0\colon \theta_{1k} = \theta_{2k} = \cdots = \theta_{Jk}$$

($k = 1, \ldots, K$), and the goal is to determine whether these K hypotheses are simultaneously true. Here, we take this to mean that we want to test

$$H_0\colon \theta_{11} - \theta_{21} = \cdots \theta_{J-1,1} - \theta_{J1} = \cdots = \theta_{J-1,K} - \theta_{JK} = 0. \tag{8.10}$$

In this last equation, there are $C = K(J^2 - J)/2$ differences, all of which are hypothesized to be equal to zero. Proceeding along the lines in Chapter 7, for each level of Factor A, generate bootstrap samples as is appropriate for K dependent groups. Label the C differences based on the observed data as $\delta_1, \ldots, \delta_C$, and then denote bootstrap estimates by $\hat{\delta}_c^*$ ($c = 1, \ldots, C$). For example, $\hat{\delta}_1^* = \theta_{11}^* - \theta_{21}^*$. Then we test Eq. (8.10) by determining how deeply the vector $(0, \ldots, 0)$, having length C, is nested within the B bootstrap values, which is done as described in Table 8.5. However, a criticism of this method is that control over the probability of a Type I error can be unsatisfactory (it can exceed 0.075 when testing at the 0.05 level) when the sample size is small.

For Factor A, an alternative approach, which seems more satisfactory in terms of Type I errors, is to base the analysis on the average measures of location across the K levels of Factor B. In symbols, let

$$\bar{\theta}_{j.} = \frac{1}{K} \sum_{k=1}^{K} \theta_{jk},$$

in which case, the goal is to test

$$H_0: \bar{\theta}_{1.} = \cdots = \bar{\theta}_{J.}. \tag{8.11}$$

Again, for each level of Factor A, generate B samples for the K dependent groups as described in Section 8.2.5 in conjunction with method RMPB4. Let $\bar{\theta}_{j.}^*$ be the bootstrap estimate for the jth level of Factor A. For levels j and j' of Factor A, $j < j'$, set $\delta_{jj'}^* = \bar{\theta}_{j.}^* - \bar{\theta}_{j'.}^*$. Then you determine how deeply $\mathbf{0}$, having length $(J^2 - J)/2$, is nested within the B bootstrap values for $\delta_{jj'}^*$ using the method described in Table 8.5. When dealing with Factor A, this approach seems to be more satisfactory than the strategy described in the previous paragraph.

As for interactions, again, there are several approaches one might adopt. Here, an approach based on difference scores among the dependent groups is used. To explain, first consider a 2-by-2 design, and for the first level of Factor A, let $D_{i1} = X_{i11} - X_{i12}, i = 1, \ldots, n_1$. Similarly, for level 2 of Factor A, let $D_{i2} = X_{i21} - X_{i22}, i = 1, \ldots, n_2$, and let θ_{d1} and θ_{d2} be the population measure of location corresponding to the D_{i1} and D_{i1} values, respectively. Then the hypothesis of no interaction is taken to be

$$H_0: \theta_{d1} - \theta_{d2} = 0. \tag{8.12}$$

Again, the basic strategy for testing hypotheses is generating bootstrap estimates and determining how deeply 0 is embedded in the B values that result. For the more general case of a J-by-K design, there are a total of

$$C = \frac{J^2 - J}{2} \times \frac{K^2 - K}{2}$$

equalities, one for each pairwise difference among the levels of Factor B and any two levels of Factor A.

8.6.7 R Functions sppba, sppbb, and sppbi

The R function

$$\text{sppba(J,K,x,est=tmean,grp = c(1:JK), avg=TRUE, nboot=500, SEED = TRUE, MC=FALSE,}$$
$$\text{MDIS=TRUE,...)}$$

tests hypotheses for Factor A as described in the previous section. By default, a 20% trimmed mean is used. Setting the argument est=onestep, a one-step M-estimator would be used instead. The argument avg=TRUE indicates that the averages of the measures of location (the $\bar{\theta}_{j.}$ values) will be used. That is, $H_0\colon \bar{\theta}_{1.} = \cdots = \bar{\theta}_{J.}$ is tested. Otherwise, the hypothesis given by Eq. (8.10) is tested. By default, the argument MDIS=TRUE, meaning that the depths of the points in the bootstrap cloud are based on the Mahalanobis distance. Otherwise, a projection distance is used, which was described in Section 6.2.5. If MDIS=FALSE and MC=TRUE, a multicore processor will be used if one is available. The remaining arguments have their usual meaning.

The R function

$$\text{sppbb(J,K,x,est=tmean,grp = c(1:JK),nboot=500, SEED = TRUE,...)}$$

tests the hypothesis of no main effects for Factor B (as described in the previous section) and

$$\text{sppbi(J,K,x,est=tmean,grp = c(1:JK),nboot=500, SEED = TRUE,...)}$$

tests the hypothesis of no interactions.

Note: The three functions just described use a 20% trimmed mean by default. When using the R package WRS2, the choices for the measure of location are limited to the median, the one-step M-estimator, and MOM. The default is MOM. Moreover, the WRS2 version does not set the seed of the random number generator. In contrast, the functions described here set the seed unless SEED=FALSE.

■ Example

We examine once more the EEG measures for murderers versus a control group mentioned in Section 6.3.5, only now, we use the data for all four sites in the brain where measures were taken. If we label the typical measures for the control group as $\theta_{11}, \ldots, \theta_{14}$ and the typical measures for the murderers as $\theta_{21}, \ldots, \theta_{24}$, we have a 2-by-4, between-by-within design, and a possible approach to comparing the groups is testing

$$H_0\colon \theta_{11} - \theta_{21} = \theta_{12} - \theta_{22} = \theta_{13} - \theta_{23} = \theta_{14} - \theta_{24} = 0.$$

This can be done with the R function sppba, with the argument avg set to FALSE. If the data are stored in a matrix called eeg having eight columns, with the first four corresponding to the control group, then the command sppba(2,4,eeg,avg=FALSE) performs the calculations based on a 20% trimmed mean and returns a p-value equal to 0.116. An alternative approach is to average the value of trimmed means over the four brain sites for each group, and then compare these averages. That is, test $H_0: \bar{\theta}_{1.} = \bar{\theta}_{2.}$, where $\bar{\theta}_{j.} = \sum \theta_{jk}/4$. This can be done with the command sppba(2,4,eeg,avg=TRUE). Now, the p-value is 0.444, illustrating that the p-value can vary tremendously depending on how groups are compared.

∎

8.6.8 Multiple Comparisons

When dealing with multiple comparisons associated with a between-by-within design, there are several approaches that might be taken that answer different questions. This section outlines some of the possibilities.

Method BWMCP

Focusing on trimmed means, multiple comparisons, when dealing with a between-by-within design, can be tested using linear contrasts, which are created in the same manner as outlined in Section 7.4.3. Consider any linear contrast Ψ, with the understanding that multiple linear contrasts are generally of interest. The goal is to test

$$H_0: \Psi = 0.$$

In general, there are C linear contrasts of interest, and often it is desired to have the probability of one or more Type I errors equal to some specified value, α. Two methods for accomplishing this goal seem to be relatively effective: a bootstrap-t method and a percentile bootstrap method.

For convenience, denote the $L = JK$ trimmed means as $\bar{X}_{t1}, \ldots, \bar{X}_{tL}$. The estimate of

$$\Psi = \sum c_\ell \mu_{t\ell},$$

is:

$$\hat{\Psi} = \sum c_\ell \bar{X}_{t\ell},$$

where $c_1, \ldots, c_L$ are the linear contrast coefficients.

To test hypotheses using a bootstrap-t method, first note that the variances and covariances among the sample trimmed means can be estimated using results in Section 5.9.5. (Of course, when two sample trimmed means are independent, their covariance is taken to be zero.) Let **S** denote this L-by-L covariance matrix. (So the diagonal elements are the estimated squared standard errors.) Let **C** be a column matrix having length L that contains the contrast coefficients. Then the squared standard error of $\hat{\Psi}$ is estimated with

$$s_{\hat{\Psi}}^2 = \mathbf{C}'\mathbf{S}\mathbf{C}$$

and an appropriate test statistic is

$$W = \frac{|\hat{\Psi}|}{s_{\hat{\Psi}}}.$$

A bootstrap-t method is used to estimate the null distribution of W. First, compute

$$Y_{ijk} = X_{ijk} - \bar{X}_{tjk},$$

where $\bar{X}_{tjk}$ is the trimmed mean corresponding to level j of Factor A and level k of Factor B. Next, take bootstrap samples based on the Y_{ijk} values. So for level j of Factor A, n_j rows of data are sampled with replacement. Based on this bootstrap sample, compute the test statistic W, which is labeled W^*. Repeat this process B times, yielding $W_1^*, \ldots, W_B^*$. Let $c = (1 - \alpha)B$, rounded to the nearest integer. Then a $1 - \alpha$ confidence interval for Ψ is

$$\hat{\Psi} \pm W_{(c)}^* \frac{s_{\hat{\Psi}}}{\sqrt{n}}.$$

Alternatively, reject the null hypothesis, if $W \geq W_{(c)}^*$.

Section 4.4.3 made a distinction between a symmetric bootstrap-t confidence interval and an equal-tailed confidence interval. Here, a symmetric confidence interval is used. For the situation at hand, there are no results on whether an equal-tailed confidence interval ever offers a practical advantage.

A percentile bootstrap method can be applied as well. As usual, no standard errors are used. For each hypothesis to be tested, corresponding to some linear contrast, a p-value can be computed as indicated at the end of Section 8.2.7.

Method BWAMCP: Comparing Levels of Factor A for Each Level of Factor B

To provide more detail about how groups differ, another strategy is to focus on a particular level of Factor B and perform all pairwise comparisons among the levels of Factor A. Of course, this can be done for each level of Factor B. Effect sizes stemming from Chapter 5 can be estimated with the R function bw.es.A, described in Section 8.6.9.

■ **Example**

Consider again, a 3-by-2 design where the means are arranged as follows:

		Factor B	
		1	2
Factor A	1	μ_1	μ_2
	2	μ_3	μ_4
	3	μ_5	μ_6

For level 1 of Factor B, method BWAMCP would test H_0: $\mu_1 = \mu_3$, H_0: $\mu_1 = \mu_5$, and H_0: $\mu_3 = \mu_5$. For level 2 of Factor B, the goal is to test H_0: $\mu_2 = \mu_4$, H_0: $\mu_2 = \mu_6$, and H_0: $\mu_4 = \mu_6$. These hypotheses can be tested by creating the appropriate linear contrasts and using the R function lincon, which can be done with the R function bwamcp described in the next section.

■

Method BWBMCP: Dealing With Factor B

When dealing with Factor B, there are four variations of method BWMCP that might be used, which are described here, under the appellation method BWBMCP. The first two variations ignore the levels of Factor A and test hypotheses based on the trimmed means. The first variation uses difference scores, and the second uses the marginal trimmed means. Both of these variations begin by pooling the data over the levels of Factor A. In essence, Factor A is ignored. The other two variations do not pool the data over the levels of Factor A but rather perform an analysis based on difference scores or the marginal trimmed means for each level of Factor A. In more formal terms, consider the jth level of Factor A. Then there are $(K^2 - K)/2$ pairs of groups that can be compared. If for each of the J levels of Factor A, all pairwise comparisons are performed, the total number of comparisons is $J(K^2 - K)/2$. As for effect sizes, one possibility is to use the methods based on difference scores mentioned in Section 5.9.19, which are computed by the R function bwbmcp, described in the next section.

■ **Example**

Consider a 2-by-2 design, where the first level of Factor A has 10 pairs of observations and the second has 15. So we have a total of 25 pairs of observations, with the first 10 corresponding to level 1 of Factor A. When analyzing Factor B, pooling the data means the goal is to compare either the difference scores corresponding to all 25 pairs of observations or the marginal trimmed means, again based on all 25 observations. Not pooling means that for level 1 of Factor A either test hypotheses based on difference scores or compare the marginal trimmed means. And the same would be done for level 2 of Factor A.

∎

Method BWIMCP: Interactions

As for interactions, we focus on a 2-by-2 design, with the understanding that the same analysis can be done for any two levels of Factor A and any two levels of Factor B. Rather than define interactions as done when using Method BWMCP, difference scores might be used instead. To elaborate, consider the first level of Factor A. There are n_1 pairs of observations corresponding to the two levels of Factor B. Form the difference scores, which for level j of Factor A are denoted by D_{ij} ($i = 1, \ldots, n_j$), and let μ_{tj} be the population trimmed means associated with these difference scores. Then one way of stating the hypothesis of no interaction is

$$H_0: \mu_{t1} = \mu_{t2}.$$

In words, the hypothesis of no interaction corresponds to the trimmed means of the difference scores associated with level 1 of Factor A being equal to the trimmed means of the difference scores associated with level 2 of Factor A. When either factor has more than two levels, a possible goal is to test all similar hypotheses (associated with any two levels of Factor A and Factor B) in a manner that controls FWE, which might be done using Rom's method or Hochberg's method.

Methods SPMCPA, SPMCPB, and SPMCPI

If it is desired to compare groups using a percentile bootstrap method, which appears to be the best method when comparing groups based on an M-estimator or MOM, analogs of methods BWAMCP, BWBMCP, and BWIMCP can be used, which are called methods SPMCPA, SPMCPB, and SPMCPI, respectively.

8.6.9 R Functions bwmcp, bwmcppb, bwmcppb.adj, bwamcp, bw.es.A, bw.es.B, bwbmcp, bw.es.I, bwimcp, bwimcpES, spmcpa, spmcpb, spmcpbA, and spmcpi

The R function

bwmcp(J, K, x, tr=0.2,con=0, nboot=599)

performs method BWMCP described in the previous section. By default, it creates all relevant linear contrasts for main effects and interactions by calling the R function con2way. The function returns three sets of results corresponding to Factor A, Factor B, and all interactions. The critical value reported for each of the three sets of tests is designed to control the probability of at least one Type I error.

The R function

$$\text{bwmcppb}(J, K, x, \text{est=tmean}, \text{tr=0.2}, \text{nboot = 500}, \text{bhop = FALSE},...)$$

tests the same hypotheses as done by the R function bwmcp, only a percentile bootstrap method is used. This function reports critical p-values aimed at controlling the probability of one or more Type I errors via Hochberg's method when bhop=FALSE. Otherwise, the Benjamini–Hochberg method is used. In contrast to spmcpa, spmcpb, and spmcpi, the function bwmcppb does not have an option about pooling the data. To get adjusted p-values rather than adjusted critical p-values, use the function

$$\text{bwmcppb.adj}(J, K, x, \text{est=tmean}, JK = J * K, \text{method='hoch'}, \text{alpha = 0.05}, \text{grp =c(1:JK)},$$
$$\text{nboot = 500}, \text{SEED = TRUE},...).$$

Hochberg's method is used by default via the R function p.adjust, but other methods can be used via the argument method.

The R function

$$\text{bwamcp}(J, K, x, \text{tr=0.2}, \text{alpha=0.05})$$

performs multiple comparisons associated with Factor A using method BWAMCP, described in the previous section. More precisely, for each level of Factor B, do all pairwise comparisons for the levels of Factor A. The function creates the appropriate set of linear contrasts and calls the R function lincon. Effect sizes are estimated by the R function

$$\text{bw.es.A}(J, K, x, \text{tr=0.2}, \text{alpha=0.05}, \text{pr=TRUE},...).$$

For each level of Factor B, effect sizes are computed for each pair of groups corresponding to the levels of Factor A. Note: To change the default values for what is considered a small, medium, and large effect size, include the argument REL.M. For example, REL.M=c(0.1, 0.3, 0.5) would indicate small, medium, and large effect sizes, respectively.

The R function:

$$\text{spmcpa(J,K,x,est=tmean,JK=J*K, grp=c(1:JK),con=0, avg=FALSE, alpha= 0.05, nboot=NA,}$$
$$\text{pr=TRUE,...)}$$

is like the R function bwamcp, only a percentile bootstrap method is used. If the argument avg=TRUE, the function uses the average of the marginal trimmed means for each level of Factor A.

The R function

$$\text{bwbmcp(J, K, x, tr=0.2, con = 0, alpha=0.05, tr=0.2, dif = TRUE, pool = FALSE, hoch =}$$
$$\text{TRUE, pr = TRUE)}$$

uses method BWBMCP to compare the levels of Factor B. If the argument pool=TRUE, the function pools the data over the levels of Factor A, and then calls the function rmmcp. That is, the levels of Factor A are ignored. If the argument dif=FALSE, the marginal trimmed means are compared instead. By default, pool=FALSE, meaning that

$$H_0: \mu_{tjk} = \mu_{tjk'},$$

is tested for all $k < k'$ and $j = 1, \ldots, J$. For each level of Factor A, the function uses data associated with the levels of Factor B and tests hypotheses via the R function rmmcp. "Critical p-values" are reported in the column headed by p.crit. That is, p.crit indicates how small the p-value must be so that the FWE rate is (approximately) equal to some specified α value. The function also reports four measures of effect size based on difference scores, which were noted in Section 5.9.19. The R function

$$\text{bw.es.B(J, K, x, tr = 0.2, POOL = FALSE, OPT = FALSE, CI = FALSE, SEED = TRUE,}$$
$$\text{REL.MAG = NULL)}$$

computes the same measures of effect size for Factor B via the R function DEP.PAIR.ES. Default values for what constitute small, medium, and large effect sizes can be altered via the argument REL.MAG. By default, this is done for each level of Factor A. The results for level j of Factor A are returned in A[[j]], which contains effect sizes for all pairs of groups associated with Factor B. To get confidence intervals, set the argument CI=TRUE. The R function

$$\text{spmcpb(J, K, x, est = tmean, JK = J * K, grp = c(1:JK), tr=0.2, dif = TRUE, nboot = NA,}$$
$$\text{SEED = TRUE, ...)}$$

is like bwbmcp, only a percentile bootstrap method is used based on the pooled data. That is, the levels of Factor A are ignored. To perform all pairwise comparisons among the levels of Factor B for each level of A, use the R function

$$\text{spmcpbA(J, K, x, est = tmean, JK = J * K, grp = c(1:JK), tr=0.2, dif = TRUE, nboot = NA, SEED = TRUE, ...).}$$

As for interactions, the R function

$$\text{bwimcp(J, K, x, tr=0.2)}$$

compares trimmed means using a non-bootstrap method. The probability of one or more Type I errors is controlled via Hochberg's method. The R function

$$\text{bwimcpES(J, K, x, tr=0.2, CI=FALSE, alpha=0.05)}$$

is the same as bwimcp, only it also reports six measures of effect size. The effect sizes are computed via the R function

$$\text{bw.es.I(J, K, x, tr = 0.2, OPT = FALSE, SEED = TRUE, CI = FALSE, alpha = 0.05, REL.MAG = NULL).}$$

For any two levels of Factor A and Factor B, difference scores are computed for the dependent groups and measures of effect size based on these differences are computed via the function ES.summary, described in Section 5.7.3.

Consider, for example, a 2-by-3 design. A portion of the output looks like this:

```
$output
     A A B B     psihat     p.value      p.crit
[1,] 1 2 1 2   0.1859648  0.3658556  0.02500000
[2,] 1 2 1 3  -0.1513233  0.4500393  0.05000000
[3,] 1 2 2 3  -0.2249659  0.3020950  0.01666667
```

Look at row 1. The results deal with the difference scores associated with levels 1 and 2 of the between factor in conjunction with levels 1 and 2 of Factor B. That is, for level 1 of Factor A, let D_{i1} $(i = 1, \ldots, n_1)$ denote the difference scores based on levels 1 and 2 of Factor B. In a similar manner, for level 2 of Factor A, let D_{i2} $(i = 1, \ldots, n_2)$ denote the difference scores, again, based on levels 1 and 2 of Factor B. The difference between the two trimmed means based on these difference scores is reported under the column headed by psihat. Below this output are estimates of effect size. For example,

```
$Effect.Sizes[[1]]
               Est NULL   S    M    L
AKP         0.13781996  0.0 0.20 0.50 0.80
EP          0.09588576  0.0 0.14 0.34 0.52
QS (median) 0.52810000  0.5 0.55 0.64 0.71
QStr        0.52910000  0.5 0.55 0.64 0.71
WMW         0.47250000  0.5 0.45 0.36 0.29
KMS         0.06440789  0.0 0.10 0.25 0.40
```

are the estimated effect sizes corresponding to the first row of the results labeled $output. Effect.Sizes[[2]] contains the estimates for the second row. Setting the argument CI=TRUE, confidence intervals for these measures of effect size are computed as well.

The R function

$$\text{spmcpi(J,K,x,est=tmean,JK=J*K,grp=c(1:JK),alpha= 0.05,nboot=NA,}$$
$$\text{SEED=TRUE,SR=FALSE,pr=TRUE,...)}$$

uses a percentile bootstrap technique instead. If the argument SR=TRUE and the argument est is equal to either onestep or mom, the probability of one or more Type I errors is controlled with method SR; otherwise Hochberg's method is used. The R function

$$\text{bwiJ2plot(J, K, x, fr=0.8,aval = 0.5, xlab = 'X', ylab = ' ', color = rep('black', 5),}$$
$$\text{BOX=FALSE)}$$

plots the distribution of the difference scores assuming there are two levels for Factor B (meaning that K=2 is required), and that the number of independent groups is $J \leq 5$. Setting the argument BOX=TRUE, boxplots are created instead.

8.6.10 Within-by-Within Designs

The methods for dealing with a between-by-within design are readily extended to a within-by-within design. That is, all JK groups being compared are dependent. For example, the method in Section 8.6.1 can be modified to handle this situation by taking **V** in Eq. (8.5) to be the Winsorized variance-covariance of all JK variables under study. (That is, for a between-by-within design, **V** was a block diagonal matrix, but for the situation at hand, generally, this is no longer the case.) A similar extension can be used when dealing with linear contrasts. Note that when dealing with linear contrasts, again, there are two basic goals that might be of interest. The first is to test hypotheses about linear contrasts stated in terms of the measures of location associated with the marginal distributions. Section 8.1.3 provides explicit details when dealing with trimmed means that can be used to analyze a within-by-within design. The second strategy is to use the approach described in Section 8.1.4. As for effect sizes, one approach is to proceed as described in Section 8.1.4. R functions specifically designed for within-by-within designs are described in the next section.

8.6.11 R Functions wwtrim, wwtrimbt, wwmcp, wwmcppb, wwmcpbt, ww.es, wwmed, and dlinplot

The R function

$$\text{wwtrim}(J, K, x, \text{grp} = c(1{:}p), p = J * K, \text{tr}=0.2)$$

tests for main effects and interactions in a within-by-within design using a modification of the method for trimmed means described in Section 8.6.1. (The modification simply takes into account the possibility that all JK variables might be dependent.) The R function

$$\text{wwtrimbt}(J, K, x, \text{tr}=0.2, \text{JKL} = J * K, \text{grp} = c(1{:}\text{JK}), \text{nboot} = 599, \text{SEED} = \text{TRUE}, ...)$$

is the same as the R function wwtrim, only a bootstrap-t method is used. The R function

$$\text{wwmcp}(J,K,x,\text{tr}=0.2,\text{alpha}= 0.05,\text{dif}=\text{TRUE})$$

performs multiple comparisons relevant to both main effects and interactions. (The function creates the appropriate linear contrasts, and then uses the R function rmmcp.) By default, linear contrasts are created along the lines described in Section 8.1.4. To use linear contrasts based on the marginal trimmed means, set the argument dif=FALSE. The R function

$$\text{wwmcppb}(J,K,x, \text{tr}=0.2, \text{con} = 0,\text{est}=\text{tmean}, \text{plotit} = \text{FALSE}, \text{dif} = \text{TRUE}, \text{grp} = \text{NA}, \text{nboot} = \text{NA}, \text{BA} = \text{TRUE}, \text{hoch} = \text{TRUE}, \text{xlab} = \text{`Group 1'}, \text{ylab} = \text{`Group 2'}, \text{pr} = \text{TRUE}, \text{SEED} = \text{TRUE}, ...)$$

is like the R function wwmcp, only a percentile bootstrap method is used. It defaults to using a 20% trimmed mean, but other measures of location can be used via the argument est. (When using an M-estimator, setting the argument hoch=FALSE is suggested.) This function (using default settings) appears to be a relatively good choice, particularly when dealing with a small sample size. When the amount of trimming is small, use the R function

$$\text{wwmcpbt(J,K,x, tr=0.2, alpha=0.05, dif=TRUE, nboot = 599),}$$

which uses a bootstrap-t method. The R function

$$\text{ww.es(J,K,x, CI=FALSE, tr=0.2, OPT=FALSE, SEED=TRUE, REL.MAG=NULL)}$$

computes measures of effect size via the R function DEP.PAIR.ES.

Another approach is to focus on the medians of difference scores. More broadly, inferences can be made based on the distribution of $W = \sum c_{jk} X_{jk}$, where $c_1, \ldots, c_{JK}$ are linear contrast coefficients and X_{jk} is some random variable associated with level j of Factor A and the level k of Factor B. This can be done based in part on the distribution-free method described in Section 4.6.2. The R function

$$\text{wwmed(J,K,x,alpha=0.05)}$$

uses this approach. Note that when the marginal distributions are identical, W has a symmetric distribution about zero. The R function

$$\text{dlinplot(x, con, nreps = 200, SEED = TRUE, xlab = 'DV', ylab =' ')}$$

plots an estimate of the distribution of W and tests the hypothesis that the median of W is zero. Only a single set of contrast coefficients can be used, in contrast to wwmed, which can handle multiple sets of linear contrasts.

8.6.12 A Rank-Based Approach

This section describes a rank-based approach to a split-plot (or between-by-within subjects) design taken from Brunner et al. (2002a, Chapter 8). There are other rank-based approaches (e.g., Beasley, 2000; Beasley and Zumbo, 2003), but it seems that the practical merits of these competing methods, versus the method described here, have not been explored. For a method based on a wild bootstrap technique, see Friedrich et al. (2017).

Main effects for Factor A are expressed in terms of

$$\bar{F}_{j.}(x) = \frac{1}{K} \sum_{k=1}^{K} F_{jk}(x),$$

the average of the distributions among the K levels of Factor B corresponding to the jth level of Factor A. The hypothesis of no main effects for Factor A is

$$H_0: \bar{F}_{1.}(x) = \bar{F}_{2.}(x) = \cdots = \bar{F}_{J.}(x)$$

for any x. Letting

$$\bar{F}_{.k}(x) = \frac{1}{J} \sum_{j=1}^{J} F_{jk}(x)$$

be the average of the distributions for the kth level of Factor B, the hypothesis of no main effects for Factor B is

$$H_0: \bar{F}_{.1}(x) = \bar{F}_{.2}(x) = \cdots = \bar{F}_{.K}(x).$$

As for interactions, first, consider a 2-by-2 design. Then no interaction is taken to mean that for any x,

$$F_{11}(x) - F_{12}(x) = F_{21}(x) - F_{22}(x).$$

More generally, the hypothesis of no interactions among all JK groups is

$$H_0: F_{jk}(x) - \bar{F}_{j.}(x) - \bar{F}_{.k}(x) + \bar{F}_{..}(x) = 0,$$

for any x, all j $(j = 1, \ldots, J)$, and all k $(k = 1, \ldots, K)$, where

$$\bar{F}_{..}(x) = \frac{1}{JK} \sum_{j=1}^{J} \sum_{k=1}^{K} F_{jk}(x).$$

As usual, let X_{ijk} represent the ith observation for level j of Factor A and level k of Factor B. Here, $i = 1, \ldots, n_j$. That is, the jth level of Factor A has n_j vectors of observations, each vector containing K values. So for the jth level of Factor A, there are a total of $n_j K$ observations, and among all the groups, the total number of observations is denoted by N. So the total number of vectors among the J groups is $n = \sum n_j$, and the total number of observations is $N = K \sum n_j = Kn$.

Pool all N observations and assign ranks. As usual, midranks are used if there are tied values. Let R_{ijk} represent the rank associated with X_{ijk}. Let

$$\bar{R}_{.jk} = \frac{1}{n_j} \sum_{i=1}^{n_j} R_{ijk},$$

$$\bar{R}_{\cdot j\cdot} = \frac{1}{K} \sum_{k=1}^{K} \bar{R}_{\cdot jk},$$

$$\bar{R}_{ij\cdot} = \frac{1}{K} \sum_{k=1}^{K} R_{ijk},$$

$$\hat{\sigma}_j^2 = \frac{1}{n_j - 1} \sum_{i=1}^{n_j} (\bar{R}_{ij\cdot} - \bar{R}_{\cdot j\cdot})^2,$$

$$S = \sum_{j=1}^{J} \frac{\hat{\sigma}_j^2}{n_j},$$

$$U = \sum_{j=1}^{J} \left(\frac{\hat{\sigma}_j^2}{n_j}\right)^2,$$

$$D = \sum_{j=1}^{J} \frac{1}{n_j - 1} \left(\frac{\hat{\sigma}_j^2}{n_j}\right)^2.$$

Factor A: The test statistic is

$$F_A = \frac{J}{(J-1)S} \sum_{j=1}^{J} (\bar{R}_{\cdot j\cdot} - \bar{R}_{\cdots})^2,$$

where $\bar{R}_{\cdots} = \sum \bar{R}_{\cdot j\cdot}/J$. The degrees of freedom are

$$\nu_1 = \frac{(J-1)^2}{1 + J(J-2)U/S^2}$$

and

$$\nu_2 = \frac{S^2}{D}.$$

Reject, if $F_A \geq f$, where f is the $1 - \alpha$ quantile of an F distribution, with ν_1 and ν_2 degrees of freedom.

Factor B: Let

$$\mathbf{R}_{ij} = (R_{ij1}, \ldots, R_{ijK})',$$

$$\bar{\mathbf{R}}_{.j} = \frac{1}{n_j} \sum_{i=1}^{n_j} \mathbf{R}_{ij}, \quad \bar{\mathbf{R}}_{..} = \frac{1}{J} \sum_{j=1}^{J} \bar{\mathbf{R}}_{.j},$$

$n = \sum n_j$ (so $N = nK$),

$$\mathbf{V}_j = \frac{n}{N^2 n_j (n_j - 1)} \sum_{i=1}^{n_j} (\mathbf{R}_{ij} - \bar{\mathbf{R}}_{.j})(\mathbf{R}_{ij} - \bar{\mathbf{R}}_{.j})'.$$

So $\mathbf{V}_j$ is a K-by-K matrix of covariances based on the ranks. Let

$$\mathbf{S} = \frac{1}{J^2} \sum_{j=1}^{J} \mathbf{V}_j$$

and let $\mathbf{P}_K$ be defined as in Section 7.9. The test statistic is

$$F_B = \frac{n}{N^2 \mathrm{tr}(\mathbf{P}_K \mathbf{S})} \sum_{k=1}^{K} (\bar{R}_{..k} - \bar{R}_{...})^2.$$

The degrees of freedom are

$$\nu_1 = \frac{(\mathrm{tr}(\mathbf{P}_K \mathbf{S}))^2}{\mathrm{tr}(\mathbf{P}_K \mathbf{S} \mathbf{P}_K \mathbf{S})}, \quad \nu_2 = \infty,$$

and H_0 is rejected, if $F_B \geq f$, where f is the $1 - \alpha$ quantile of an F distribution, with ν_1 and ν_2 degrees of freedom.

Interactions: Let $\mathbf{V}$ be the block diagonal matrix based on the matrices $\mathbf{V}_j$, $j = 1, \ldots, J$. Letting $\mathbf{M}_{AB}$ be defined as in Section 7.9, the test statistic is

$$F_{AB} = \frac{n}{N^2 \mathrm{tr}(\mathbf{M}_{AB} \mathbf{V})} \sum_{j=1}^{J} \sum_{k=1}^{K} (\bar{R}_{.jk} - \bar{R}_{.j.} - \bar{R}_{..k} + \bar{R}_{...})^2.$$

The degrees of freedom are

$$\nu_1 = \frac{(\mathrm{tr}(\mathbf{M}_{AB} \mathbf{V}))^2}{\mathrm{tr}(\mathbf{M}_{AB} \mathbf{V} \mathbf{M}_{AB} \mathbf{V})}, \quad \nu_2 = \infty.$$

Reject, if $F_A \geq f$ (or if $F_{AB} \geq f$), where f is the $1 - \alpha$ quantile of an F distribution, with ν_1 and ν_2 degrees of freedom.

8.6.13 R Function bwrank

The R function

$$\text{bwrank}(J,K,x)$$

performs a between-by-within ANOVA based on ranks using the method just described. In addition to testing hypotheses as just indicated, the function returns the average ranks $(\bar{R}_{.jk})$ associated with all JK groups as well as the relative effects, $(\bar{R}_{.jk} - 0.5)/N$.

■ **Example**

Lumley (1996) reports data on shoulder pain after surgery; the data are from a study by Jorgensen et al. (1995). The data can be accessed as described in Section 1.10. Two treatment methods were used and measures of pain, based on a five-point scale, were taken at three different times. The output from bwrank is

```
$test.A:
[1] 12.87017

$sig.A:
[1]   0.001043705

$test.B:
[1] 0.4604075

$sig.B:
[1] 0.5759393

$test.AB:
[1] 8.621151

$sig.AB:
[1]   0.0007548441

$avg.ranks:
          [,1]      [,2]      [,3]
[1,] 58.29545 48.40909 39.45455
[2,] 66.70455 82.36364 83.04545
```

```
$rel.effects:
          [,1]        [,2]        [,3]
[1,] 0.4698817  0.3895048  0.3167036
[2,] 0.5382483  0.6655580  0.6711013
```

So at approximately the 0.001 level, treatment methods are significantly different and there is a significant interaction. That is, there is evidence that the distributions differ. Note that the average ranks and relative effects suggest that a disordinal interaction might exist. In particular, for group 1 (the active treatment group), time 1 has higher average ranks versus time 2, and the reverse is true for the second group. However, the Wilcoxon signed rank test fails to reject at the 0.05 level when comparing time 1 to time 2 for both groups. When comparing time 1 versus time 3 for the first group, again, using the Wilcoxon signed rank test, we reject at the 0.05 level, but a non-significant result is obtained for group 2. So, again, a disordinal interaction appears to be a possibility, but the empirical evidence is not compelling. Note that because the outcome measures are limited to one of five values, another approach is to compare the two groups at each time using the function binband, described in Section 5.8.3. It is left as an exercise to verify that at time 1, binband finds no significant difference, and that a plot of the distributions suggests that they are similar. At time 3, responses 1 and 3 are significant at the 0.05 level.

■

■ Example

Section 6.11 illustrates a method for comparing multivariate data corresponding to two independent groups based on the extent to which points from one group are nested within the other. For the schizophrenia data mentioned in Section 6.12.1, it was found that schizophrenics differed from the control group; also see Fig. 6.10. If the two groups are compared based on the OP-estimator (using the function smean2), again, the two groups are found to differ. Comparing the groups with the method for means and trimmed means described in this section, no difference between the schizophrenics and the control group is found at the 0.05 level. Using the rank-based method in this section, again, no difference is found. (The p-value is 0.11.) The only point is that how we compare groups can make a practical difference about the conclusions reached.

■

8.6.14 Rank-Based Multiple Comparisons

Multiple comparisons based on the rank-based methods covered here can be performed using simple combinations of methods already considered. When dealing with Factor A, for example, one can simply compare level j to level j', ignoring the other levels. When comparing all pairs of groups, FWE can be controlled with Rom's method or the Benjamini–Hochberg technique. Factor B and the collection of all interactions (corresponding to any two rows and any two columns) can be handled in a similar manner.

8.6.15 R Function bwrmcp

The R function

$$\text{bwrmcp(J,K,x,grp=NA,alpha= 0.05,bhop=FALSE)}$$

performs all pairwise multiple comparisons using the method of Section 8.6.14, with the FWE rate controlled using Rom's method or the Benjamini–Hochberg method. For example, when dealing with Factor A, the function simply compares level j to level j', ignoring the other levels. All pairwise comparisons among the J levels of Factor A are performed, and the same is done for Factor B and all relevant interactions.

8.6.16 Multiple Comparisons When Using a Patel–Hoel Approach to Interactions

Rather than compare distributions when dealing with an interaction in a between-by-within design, as done by the rank-based method in Section 8.6.12, one could use a simple analog of the Patel–Hoel approach instead. First, consider a two-by-two design, and focus on level 1 of Factor A. Because the two levels of Factor B are dependent, they can be compared with the sign test. In essence, inferences are being made about p_1, the probability that for a randomly sampled pair of observations, the observation from level 1 of Factor B is less than the corresponding observation from level 2. Of course, for level 2 of Factor A, again, the two levels of Factor B can be compared with the sign test. Now, we let p_2 be the probability that for a randomly sampled pair of observations, the observation from level 1 of Factor B is less than the corresponding observation from level 2. Then no interaction can be defined as $p_1 = p_2$.

The hypothesis of no interaction,

$$H_0\text{: } p_1 = p_2,$$

is just the hypothesis that two independent binomials have equal probabilities of success, which can be tested using one of the methods described in Section 5.8. Here, FWE is controlled with Hochberg's method, described in Section 7.4.7.

Tied values are handled in the same manner as with the signed rank test: Pairs of observations with identical values are simply discarded. So among the remaining observations, for every pair of observations, the observation from level 1 of Factor B, for example, is either less than or greater than the corresponding value from level 2.

A variation of the approach in this section is where, for level 1 of Factor B, p_1 is the probability that an observation from level 1 of Factor A is less than an observation from level 2. Similarly, p_2 is now defined in terms of the two levels of Factor A when working with level 2 of Factor B. However, the details of how to implement this approach have not been studied.

8.6.17 R Function BWPHmcp

The Patel–Hoel interaction, described in the previous section, can be computed with the R function

$$\text{BWPHmcp(J,K,x, method='KMS').}$$

See Section 5.8 regarding the choices for the argument method.

8.7 Three-Way Designs

Generally, two-way designs can be extended to a three-way design, where one or more factors involve dependent groups. This section outlines some of the methods that might be used. It is stressed, however, that simulation studies reporting the relative merits of the methods considered are extremely limited.

8.7.1 Global Tests Based on Trimmed Means

The method in Section 8.6.1 is readily extended to a three-way design. Note that matrix $\mathbf{V}$ used in Eq. (8.4) reflects the variances and covariances among the trimmed means, where the covariances are taken to be zero if the groups are independent. Here, $\mathbf{V}$ is computed in a similar manner. That is, for a J-by-K-by-L design, $\mathbf{V}$ is a JKL square matrix that contains the squared standard errors and covariances among the sample trimmed means, with independent trimmed means having a covariance of zero. Once $\mathbf{V}$ is available, compute the test statistic Q given by Eq. (8.4), where now matrix $\mathbf{C}$ is computed as described in Table 7.5.

When dealing with situations where one or more factors involve dependent groups, comments should be made regarding the approximation of the null distribution using an F distribution. Unlike the method in Section 8.6.1, where the second degrees of freedom, v_2, is estimated based on the data, the strategy here is to simply set $v_2 = 999$. The reason is that even with $v_2 = 999$, the actual level of the method can drop well below the nominal level when the sample size is small and 20% trimmed means are used. For example, when dealing with a between-by-between-by-within design, under normality with all correlations equal to zero and $J = 2$, $K = L = 3$, and $n = 25$, the actual Type I error probability is approximately 0.003 when dealing with the A-by-C interaction and testing at the 0.05 level. Increasing n to 50, now the actual level is approximately 0.013, for $n = 100$, it is 0.037, and for $n = 900$, it is 0.04. For the main effect associated with Factor A, the estimated level is 0.050, with $n = 25$ and 0.052, with $n = 900$. A bootstrap-t method appears to suffer from the same problem, but an extensive study of this issue has not been conducted. A similar problem occurs when dealing with a between-by-within-by-within design. Again, $v_2 = 999$ is used, but even with $n = 100$, the actual level can drop as low as 0.025 under normality. Using instead a bootstrap-t method (via the R function bbwtrimbt), with $n = 25$, the actual level was estimated to be 0.067.

Evidently, there are no published studies comparing methods, in terms of Type I errors, when dealing with three-way designs with one or more within group factors. Very limited results suggest that perhaps a better approach, compared to the methods described here, is to use the R functions in Section 8.7.5. They test hypotheses about all of the usual linear contrasts associated with a three-way design using a percentile bootstrap method in conjunction with a trimmed mean. In terms of controlling the probability of one or more Type I errors, performing one of the global tests described here is not required when using the percentile bootstrap methods via the R functions in Section 8.7.5. Moreover, limited results suggest that control over the Type I error probability is more satisfactory when the amount of trimming is 20%. For the situation considered here, where $n = 25$ and the actual Type I error is approximately 0.003, the probability of one or more Type I errors was estimated to be 0.039 when using a percentile bootstrap method via the R function bbwmcppb, based on a simulation with 1,000 replications. (And execution time can be substantially less when using a percentile bootstrap method rather than the bootstrap-t method.) Using instead the R function bwwmcppb, the probability of one or more Type I errors was estimated to be 0.049. But again, a more comprehensive study is needed.

8.7.2 R Functions bbwtrim, bwwtrim, wwwtrim, bbwtrimbt, bwwtrimbt, wwwtrimbt, and wwwmed

The R function

$$bbwtrim(J,K,L,x,grp=c(1:p),tr=0.2)$$

tests all omnibus main effects and interactions associated with a between-by-between-by-within design. The data are assumed to be stored as described in Section 7.3.1. For a between-by-within-by-within design use

$$\text{bwwtrim(J,K,L,x,grp=c(1:p),tr=0.2).}$$

And for a within-by-within-by-within design use

$$\text{wwwtrim(J,K,L,x,grp=c(1:p),tr=0.2).}$$

The R functions

$$\text{bbwtrimbt(J,K,L,x,grp=c(1:p),tr=0.2, nboot = 599, SEED = TRUE),}$$

$$\text{bwwtrim(J,K,L,x,grp=c(1:p),tr=0.2, nboot = 599, SEED = TRUE),}$$

and

$$\text{wwwtrimbt(J,K,L,x,grp=c(1:p),tr=0.2, nboot = 599, SEED = TRUE)}$$

are the same as the functions bbwtrim, bwwtrim, and wwwtrim, respectively, only a bootstrap-t method is used.

The R function

$$\text{wwwmed(J,K,L,x, alpha=0.05)}$$

is like the function wwmed, described in Section 8.6.11, which is based on medians, only adapted to a within-by-within-by-within design.

8.7.3 Data Management: R Functions bw2list and bbw2list

For a between-by-within-by-within design, the R function

$$\text{bw2list(x, grp.col, lev.col),}$$

which was introduced in Section 8.6.3, can be used when dealing with data that are stored in a matrix or a data frame, with one column indicating the levels of the independent groups and other columns containing data corresponding to within group levels. For example, setting the argument grp.col=c(5) would indicate that the levels for Factor A are stored in column 5 and lev.col=c(3,9,10,12) indicates that the within levels data are stored in columns 3, 9, 10, and 12. Note that it must be the case that KL is equal to the number of values stored in lev.col. So lev.col=c(3,9,10,12) would be appropriate if the within factors have two levels each, with the data for two levels of Factor C being stored in columns 10 and 12.

The R function

$$bbw2list(x, grp.col, lev.col)$$

deals with a between-by-between-by-within design and assumes that the argument grp.col contains two values that indicate the columns of x that indicate the levels of Factors A and B. Now, the argument lev.col indicates the columns containing the within data.

■ **Example**

Imagine that for a between-by-between-by-within design, column 14 of the R variable dis contains values indicating the levels of Factor A, column 10 has values that contain the levels of Factor B, and columns 2, 4, and 9 contain the outcome values at times 1, 2, and 3, respectively. Then

$$z=bbw2list(dis, grp.col=c(14,10), lev.col=c(2,4,9))$$

would store the data in list mode in z, after which the command

$$bbwtrim(3,4,3,z)$$

would test the usual hypotheses, assuming that Factors A and B have three and four levels, respectively. The values in columns 14 and 10 would be sorted in ascending order, or in alphabetical order if the values in these columns are character data.

■

8.7.4 Multiple Comparisons

Multiple comparisons in a three-way design can be performed using a straightforward extension of methods described in previous sections. The R function con3way, described in Section 7.4.4, can be used to generate the linear contrast coefficients that are often used. Here, when computing A in Section 8.1.3, we set $d_{jk} = 0$ whenever j and k correspond to independent groups. Otherwise, this term is computed as described in Section 8.1.3. The next two sections summarize some R functions aimed at facilitating the analysis.

8.7.5 R Functions wwwmcp, bbwmcp, bwwmcp, bbwmcppb, bwwmcppb, wwwmcppb, and wwwmcppbtr

When dealing with a within-by-within-by-within design, a non-bootstrap method can be used to test the hypotheses associated with all of the linear contrasts generated by the R function con3way. This can be done with the R function

$$\text{wwwmcp}(J, K, L, x, tr=0.2, alpha=0.05, dif = TRUE, grp = NA).$$

(That is, it uses the R function con3way to generate the linear contrast coefficients, and then it tests the corresponding hypotheses. The function rm3mcp performs the same calculations as wwwmcp.) When dealing with designs where there are both between and within factors, use a bootstrap method via one of the R functions described in the next section. Another approach is to use the R function rmmcp, described in Section 8.1.5, in conjunction with the R function con3way. (For an illustration of how to interpret three-way interactions based on the contrast coefficients returned by con3way, see the example at the end of Section 7.4.4.)

Bootstrap-t Methods

The R function

$$\text{bbwmcp}(J, K, L, x, tr=0.2, JKL = J * K * L, con = 0, tr=0.2, grp = c(1:JKL), nboot = 599, SEED = TRUE, ...)$$

performs all multiple comparisons associated with main effects and interactions using a bootstrap-t method in conjunction with trimmed means when analyzing a between-by-between-by-within design. The function uses con3way to generate all of the relevant linear contrasts, and then uses the function lindep to test the hypotheses. The critical value is designed to control the probability of at least one Type I error among all the linear contrasts associated with Factor A. The same is done for Factor B and Factor C.

The R function

$$\text{bwwmcp}(J, K, L, x, tr=0.2, JKL = J * K * L, con = 0, tr=0.2, grp = c(1:JKL), nboot = 599, SEED = TRUE, ...)$$

handles a between-by-within-by-within design.

Percentile Bootstrap Methods

For a between-by-between-by-within design, the R function

$$\text{bbwmcppb(J, K, L, x, tr=0.2, JKL = J * K * L, con = 0, tr=0.2, grp = c(1:JKL), nboot = 599,}$$
$$\text{SEED = TRUE, ...)}$$

tests hypotheses using a percentile bootstrap method. As for a between-by-within-by-within and a within-by-within-by-within design, use the functions

$$\text{bwwmcppb(J, K, L, x, est=tmean, JKL = J * K * L, con = 0, tr=0.2, grp = c(1:JKL),}$$
$$\text{nboot = 599, SEED = TRUE, ...)}$$

and

$$\text{wwwmcppb(J, K, L, x, est=tmean, JKL = J * K * L, con = 0, tr=0.2, grp = c(1:JKL),}$$
$$\text{nboot = 599, SEED = TRUE, ...),}$$

respectively.

The function

$$\text{wwwmcppbtr(JJ,K,L, x,tr=0.2, alpha=0.05, dif=TRUE, op=FALSE, grp=NA, nboot=2000,}$$
$$\text{SEED=TRUE,pr=TRUE)}$$

is like wwwmcppb, only it is designed for trimmed means only, and by default, it tests H_0: $\theta_d = 0$ as described in Section 8.3.2, the hypothesis that a linear combination of the random variables has a trimmed mean equal to zero. To test hypotheses about the marginal trimmed means, set the argument dif=FALSE.

8.8 Exercises

1. Section 8.6.2 reports data on hangover symptoms. For group 2, use the R function rmanova to compare the trimmed means corresponding to times 1, 2, and 3.
2. For the data used in Exercise 1, compute confidence intervals for all pairs of trimmed means using the R function pairdepb.
3. Analyze the data for the control group mentioned in Section 6.12.1 using the methods in Sections 8.1 and 8.2. The data are stored in the file schiz2.data, which can be accessed as described in Section 1.10. Compare and contrast the results.

4. Repeat Exercise 3 using the rank-based method in Section 8.5. How do the results compare to using a measure of location?
5. Repeat Exercises 3 and 4 using the data for the murderers in Table 6.1.
6. Analyze the data for murderers in Section 6.3.5 using the methods in Sections 8.6.1 and 8.6.4.

## CHAPTER 9

Correlation and Tests of Independence

There are many approaches to finding robust measures of correlation and covariance (e.g., Ammann, 1993; Davies, 1987; Devlin et al., 1981; Goldberg and Iglewicz, 1992; Hampel et al., 1986, Chapter 5; Huber, 1981, Chapter 8; Li and Chen, 1985; Lopuhaä, 1989; Maronna, 1976; Mosteller and Tukey, 1977, p. 211; Wang and Raftery, 2002; Wilcox, 1993b), but no attempt is made to give a detailed description of all the strategies that have been proposed. Some of these measures are difficult to compute, others are not always equal to zero under independence, and from a technical point of view, some do not have all the properties one might want. One of the main goals in this chapter is to describe some tests of zero correlation that have practical value relative to the standard test based on the usual product moment correlation, r. Some alternative methods for testing the hypothesis of independence that are not based on some type of correlation coefficient are described as well. (For a collection of alternative methods for detecting dependence, see Kallenberg and Ledwina, 1999.) For additional ways of measuring the strength of an association, see the description of explanatory power in Section 11.9.

9.1 Problems With Pearson's Correlation

The most common measure of covariance between any two random variables, X and Y, is

$$\mathrm{COV}(X, Y) = \sigma_{xy}$$
$$= E\{(X - \mu_x)(Y - \mu_y)\}.$$

Pearson's correlation is

$$\rho = \frac{\sigma_{xy}}{\sigma_x \sigma_y},$$

which is sometimes called the product moment correlation. A practical concern with ρ is that it is not robust. If one of the marginal distributions is altered slightly, as measured by the Kolmogorov distance, but the other marginal distribution is left unaltered, the magnitude of ρ can be changed substantially. More formally, the influence function of Pearson's correlation is

$$IF(x, y) = xy - \left(\frac{x^2 + y^2}{2}\right)\rho,$$

Introduction to Robust Estimation and Hypothesis Testing
https://doi.org/10.1016/B978-0-12-820098-8.00015-4

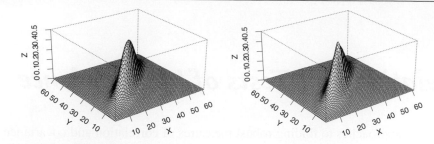

Figure 9.1: Pearson's correlation for the bivariate normal distribution shown on the left panel is 0.8. In the right panel, x has a contaminated normal distribution and Pearson's correlation is 0.2.

which is unbounded (Devlin et al., 1981). That is, Pearson's correlation does not have infinitesimal robustness.

The left panel of Fig. 9.1 shows a bivariate normal distribution, with $\rho = 0.8$. Suppose the marginal distribution of X is replaced by a contaminated normal given by Eq. (1.1), with $\epsilon = 0.9$ and $K = 10$. The right panel of Fig. 9.1 shows the resulting joint distribution. As is evident, there is little visible difference between these two distributions, but in the right panel of Fig. 9.1, $\rho = 0.2$. Put another way, even with an infinitely large sample size, the usual estimate of ρ, given by r in the next paragraph, can be misleading.

Let $(X_1, Y_1), \ldots, (X_n, Y_n)$ be a random sample from some bivariate distribution. The usual estimate of ρ is

$$r = \frac{\sum (X_i - \bar{X})(Y_i - \bar{Y})}{\sqrt{\sum (X_i - \bar{X})^2 \sum (Y_i - \bar{Y})^2}}.$$

A practical concern with r is that it is not resistant—a single unusual point can dominate its value. Moreover, as we move toward a skewed, heavy-tailed distribution, at some point, the sampling distribution of r becomes very unstable, even with a very large sample size.

■ **Example**

Data were generated using the rmul function in Section 4.2.1, with $g = 0.5$, $h = 0.2$, and $\rho = 0.8$, r was computed based on a sample size of one million, and this was repeated 100 times. The interquartile range was 0.009, and the range was 0.034. With $g = h = 0.5$, the interquartile range was 0.11 and the range was 0.50.

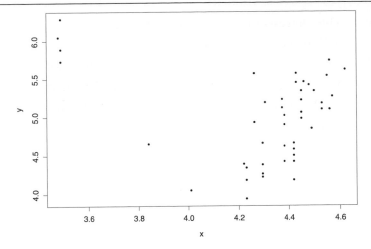

Figure 9.2: Scatterplot of the star data.

■ Example

Fig. 9.2 shows a scatterplot of data on the logarithm of the effective temperature at the surface of 47 stars versus the logarithm of its light intensity. (The data are reported in Rousseeuw and Leroy, 1987, p. 27.) The scatterplot suggests that, in general, there is a positive association between temperature and light, yet, $r = -0.21$. The reason is that the four points in the upper left corner of Fig. 9.2 (which are giant red stars) are outliers that dominate the value of r. (Two additional points are flagged as outliers by the R function out.)

■

From basic principles, if X and Y are independent, then $\rho = 0$. The best-known test of

$$H_0: \rho = 0 \qquad (9.1)$$

is based on the test statistic

$$T = r\sqrt{\frac{n-2}{1-r^2}}. \qquad (9.2)$$

If H_0 is true, T has a Student's t distribution, with $v = n - 2$ degrees of freedom, if at least one of the marginal distributions is normal (e.g., Muirhead, 1982, p. 146) and simultaneously X and Y are independent. In particular, reject $H_0: \rho = 0$, if $|T| > t_{1-\alpha/2}$, the $1 - \alpha/2$ quantile of a Student's t distribution, with $n - 2$ degrees of freedom. When X and Y are independent, there are general conditions under which $E(r) = 0$ and $E(r^2) = 1/(n - 1)$ (Huber, 1981, p. 204). (All that is required is that the distribution of X or Y be invariant under permutations of the components.) This suggests that the test of independence, based on T, will be reasonably robust in terms of Type I errors, and this seems to be the case for a variety of situations (Kowalski, 1972; Srivastava and Awan, 1984). Bishara and Hittner (2012), for example, report good control over the Type I error probability when X and Y are independent and when testing at the 0.05 level. The primary exception was when both X and Y have mixed normal distribution, in which case, the actual level was estimated to be between 0.066 and 0.068 for sample sizes ranging between 10 and 160. However, problems arise in at least three situations: when $\rho = 0$, but X and Y are dependent (e.g., Edgell and Noon, 1984), when performing one-sided tests (Blair and Lawson, 1982), and when considering the more general goal of testing for independence among all pairs of p random variables. Section 5.9.15 describes yet another situation where (9.2) is true, but the T test is highly unsatisfactory because X and Y are dependent.

There is also the problem of computing a confidence interval for ρ. Many methods have been proposed, but simulations do not support their use, at least for small to moderate sample sizes, and it is unknown just how large of a sample size is needed before any particular method can be expected to give good probability coverage (Wilcox, 1991a). A modified percentile bootstrap method appears to perform reasonably well in terms of probability coverage, provided ρ is reasonably close to zero. But as ρ gets close to 1, it begins to break down (Wilcox and Muska, 2001). Many books recommend *Fisher's r-to-Z transformation* when computing confidence intervals, but under general conditions, it is not even asymptotically correct when sampling from non-normal distributions (Duncan and Layard, 1973).

9.1.1 Features of Data That Affect r and T

There are several features of data that affect the magnitude of Pearson's correlation, as well as the magnitude of T, given by Eq. (9.2). These features are important when interpreting robust correlation coefficients, so they are described here.

Five features of data that affect r are:

1. Outliers,
2. The magnitude of the slope around which points are clustered (e.g., Barrett, 1974; Loh, 1987b) (put another way, rotating points can raise or lower r),
3. Curvature,

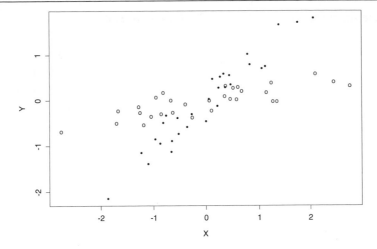

Figure 9.3: An illustration that rotating points can alter Pearson's correlation. The dots have correlation $r = 0.964$. The points marked by a circle are the dots rotated by 35 degrees; these rotated points have $r = 0.81$.

4. The magnitude of the residuals,

5. Restriction of range.

The effect of outliers has already been illustrated. The effect of curvature seems fairly evident, as does the magnitude of the residuals. It is well known that restricting the range of X or Y can lower r, and the star data in Fig. 9.2 illustrate that a restriction in range can increase r as well. The effect of rotating points is illustrated by Fig. 9.3. Thirty points were generated for X from a standard normal distribution, ϵ was taken to have a normal distribution with mean zero and standard deviation 0.25, and then $Y = X + \epsilon$ was computed. The least squares estimate of the slope is 1.00, and it was found that $r = 0.964$. Then the points were rotated clockwise by 35 degrees. The rotated points are indicated by the circles in Fig. 9.3. Now, the least squares estimate of the slope is 0.19, and $r = 0.81$. Rotating the points by 40 degrees, instead, $r = 0.61$. Continuing to rotate the points in the same direction, the correlation will decrease until both r and the least squares estimate of the slope are zero.

Cohen (1988) suggests that $\rho = 0.1, 0.3$, and 0.5 be viewed as small, medium, and large values, respectively. Gignac and Szodorai (2016) find that among 708 meta-analyses, the 25th, 50th, and 75th percentiles corresponded to $r = 0.10$, $r = 0.20$, and $r = 0.30$, and they suggest that these values be considered small, medium, and large instead. The lack of robustness associated with Pearson's correlation makes it more difficult to determine empirically what might be considered small, medium, and large values.

9.1.2 Heteroscedasticity and the Classic Test That $\rho = 0$

Now, consider testing H_0: $\rho = 0$ based on the test statistic T given by Eq. (9.2). As just pointed out, five features of data affect the magnitude of r, and hence T. There is, in fact, a sixth feature that affects T even when $\rho = 0$: heteroscedasticity. In regression, homoscedasticity refers to a situation where the conditional variance of Y, given X, does not depend on X. That is, $\text{VAR}(Y|X) = \sigma^2$. Heteroscedasticity refers to a situation where the conditional variance of Y varies with X. Independence implies homoscedasticity, but $\rho = 0$ does not necessarily mean there is homoscedasticity. The reason heteroscedasticity is relevant when using Eq. (9.2) is that the derivation of the test statistic, T, is based on the assumption that X and Y are independent. Even if $\rho = 0$, when there is heteroscedasticity, the wrong standard error is being used by T.

To illustrate what can happen, imagine that both X and Y have normal distributions, with both means equal to zero. Further assume that both X and Y have variance 1 unless $|X| > 0.5$, in which case, Y has standard deviation $|X|$. So there is dependence, but $\rho = 0$. With $n = 20$ and testing at the $\alpha = 0.05$ level with T, the actual probability of a Type I error is 0.140. With $n = 40$, it is 0.171, and for $n = 200$, it is 0.182. Even though $\rho = 0$, the probability of rejecting is increasing as n gets large because the wrong standard error is being used. So when H_0 is rejected, it is reasonable to conclude dependence, but the nature of the dependence (the reason why H_0 was rejected) is unclear. Put another way, the test statistic T, given by Eq. (9.2), tests the hypothesis that two random variables are independent, rather than the hypothesis that $\rho = 0$. (Section 9.3.13 describes two methods aimed at dealing with this problem.)

9.2 Two Types of Robust Correlations

Here, robust analogs of Pearson's correlation are classified into one of two types: those that protect against outliers among the marginal distributions without taking into account the overall structure of the data, and those that take into account the overall structure of the data when dealing with outliers. In terms of developing tests of the hypothesis of independence between two random variables, it is a bit easier working with the first type. Recently, however, some progress has been made when working with the second. For convenience, the first type will be called a *type M correlation* and the second will be called *type O*.

9.3 Some Type M Measures of Correlation

This section describes four type M correlations and how they can be used to test the hypothesis of independence.

Table 9.1: Computing the percentage bend correlation.

The goal is to estimate the percentage bend correlation, ρ_{pb}, based on the random sample $(X_1, Y_1), \ldots, (X_n, Y_n)$. For the observations $X_1, \ldots, X_n$, let M_x be the sample median. Select a value for β, $0 \leq \beta \leq 0.5$. Compute

$$W_i = |X_i - M_x|,$$

$$m = [(1 - \beta)n].$$

Note that $[(1 - \beta)n]$ is $(1 - \beta)n$ rounded down to the nearest integer. Let $W_{(1)} \leq \cdots \leq W_{(n)}$ be the W_i values written in ascending order. Set

$$\hat{\omega}_x = W_{(m)}.$$

For example, if the observations are 4, 2, 7, 9, and 13, then the sample median is $M_x = 7$, so $W_1 = |4 - 7| = 3$, $W_2 = |2 - 7| = 5$, $W_3 = |7 - 7| = 0$, $W_4 = 2$, and $W_5 = 6$; so $W_{(1)} = 0$, $W_{(2)} = 2$, $W_{(3)} = 3$, $W_{(4)} = 5$, and $W_{(5)} = 6$. If $\beta = 0.1$, $m = [.9(5)] = 4$ and $\hat{\omega} = W_{(4)} = 5$.
Let i_1 be the number of X_i values such that $(X_i - M_x)/\hat{\omega}_x < -1$. Let i_2 be the number of X_i values such that $(X_i - M_x)/\hat{\omega}_x > 1$. Compute

$$S_x = \sum_{i=i_1+1}^{n-i_2} X_{(i)},$$

$$\hat{\phi}_x = \frac{\hat{\omega}_x(i_2 - i_1) + S_x}{n - i_1 - i_2}.$$

Set $U_i = (X_i - \hat{\phi}_x)/\hat{\omega}_x$. Repeat these computations for the Y_i values, yielding $V_i = (Y_i - \hat{\phi}_y)/\hat{\omega}_y$. Let

$$\Psi(x) = \max[-1, \min(1, x)].$$

Set $A_i = \Psi(U_i)$ and $B_i = \Psi(V_i)$. The percentage bend correlation between X and Y is estimated to be

$$r_{pb} = \frac{\sum A_i B_i}{\sqrt{\sum A_i^2 \sum B_i^2}}.$$

9.3.1 The Percentage Bend Correlation

The first type M correlation that has proven to be relatively successful, in terms of controlling Type I error probabilities when testing the hypothesis of independence, is the so-called percentage bend correlation. It is estimated with r_{pb} using the computations described in Table 9.1. Table 9.2 describes how the population parameter corresponding to r_{pb}, ρ_{pb}, is defined. When X and Y are independent, $\rho_{pb} = 0$. Under normality, ρ and ρ_{pb} have very similar values, but ρ_{pb} is more robust, and their population values can differ substantially, even when there is little apparent difference between the bivariate distributions (Wilcox, 1994d).

Table 9.2: Definition of the population percentage bend correlation.

Let

$$\Psi(x) = \max[-1, \min(1, x)],$$

which is a special case of Huber's Ψ. Let θ_x and θ_y be the population medians corresponding to the random variables X and Y, and let ω_x be defined by the equation

$$P(|X - \theta_x| < \omega_x) = 1 - \beta.$$

Shoemaker and Hettmansperger (1982) use $\beta = 0.1$, but the resulting breakdown point might be too low in some cases. The percentage bend measure of location, corresponding to X, is the quantity ϕ_{pbx} such that

$$E[\Psi(U)] = 0,$$

where

$$U = \frac{X - \phi_{pbx}}{\omega_x}.$$

In terms of Chapter 2, ϕ_{pbx} is an M-measure of location for the particular form of Huber's Ψ being used here. Let

$$V = \frac{Y - \phi_{pby}}{\omega_y}.$$

Then the percentage bend correlation between X and Y is

$$\rho_{pb} = \frac{E\{\Psi(U)\Psi(V)\}}{\sqrt{E\{\Psi^2(U)\}E\{\Psi^2(V)\}}}.$$

Under independence, $\rho_{pb} = 0$ and $-1 \le \rho_{pb} \le 1$.

Perhaps it should be emphasized that r_{pb} is not intended as an estimate of ρ. Rather, the goal is to estimate a measure of the strength of the association, ρ_{pb}, that is not overly sensitive to slight changes in the distributions. There might be situations where ρ is of interest despite its lack of robustness, in which case, r_{pb} has little or no value. The situation is similar to finding a robust measure of location. If there is direct interest in the population mean μ, the 20% sample trimmed mean does not estimate μ when distributions are skewed and would not be used. The problem is that μ is not robust, in which case, some other measure of location might be of interest, such as a 20% trimmed mean. In a similar fashion, assuming a linear association, ρ_{pb} provides a robust measure of the strength of the association between two random variables that is designed so that its value is not overly sensitive to a relatively small proportion of the population under study.

Note that the definition of ρ_{pb} depends in part on a measure of scale, ω_x, which is a generalization of MAD. A technical point of some interest is that ω_x is a measure of dispersion.

(See Section 2.3.) If in the definition of ρ_{pb}, $\Psi(x) = \max[-1, \min(1, x)]$ is replaced by $\Psi(x) = \max[-K, \min(K, x)]$ for some $K > 1$, ω_x is no longer a measure of dispersion (Shoemaker and Hettmansperger, 1982).

9.3.2 A Test of Independence Based on ρ_{pb}

When X and Y are independent, $\rho_{pb} = 0$. To test the hypothesis $H_0\colon \rho_{pb} = 0$, assuming independence, compute

$$T_{pb} = r_{pb}\sqrt{\frac{n-2}{1 - r_{pb}^2}}, \qquad (9.3)$$

and reject H_0, if $|T_{pb}| > t_{1-\alpha}$, the $1 - \alpha$ quantile of a Student's t distribution, with $v = n - 2$ degrees of freedom. All indications are that this test provides reasonably good control over the probability of a Type I error for a broader range of situations than the test based on r (Wilcox, 1994d).

The breakdown point of the percentage bend correlation is at most β. However, if $\beta = 0.5$ is used, the power of the test for independence, based on T_{pb}, can be substantially less than the test based on r when sampling from a bivariate normal distribution (as will be illustrated by results in Table 9.5). Here, the default value for β is 0.2. In exploratory studies, several values might be considered.

Like the conventional T test of $H_0\colon \rho = 0$, the method just described for testing $H_0\colon \rho_{pb} = 0$ is sensitive to heteroscedasticity. That is, even when $H_0\colon \rho_{pb} = 0$ is true, if there is heteroscedasticity, the wrong standard error is being used, and the probability of rejecting can increase with the sample size. For a test of $H_0\colon \rho_{pb} = 0$ that is designed to be insensitive to heteroscedasticity, see Section 9.3.13.

9.3.3 R Function pbcor

The R function

$$\text{pbcor(x,y,beta= 0.2)}$$

estimates the percentage bend correlation for the data stored in any two vectors. If unspecified, the argument beta, the value for β when computing the measure of scale $W_{(m)}$, defaults to 0.2. The function returns the value of r_{pb} in pbcor\$cor, the value of the test statistic, T_{pb}, in pbcor\$test, and the p-value in pbcor\$p.value. It is noted that the function pbcor automatically removes any pair of observations for which one or both values are missing.

■ **Example**

The example in Section 8.6.2 of Chapter 8 reports the results of drinking alcohol for two groups of subjects measured at three different times. Consider the measures at times 1 and 2 for the control group. Then $r = 0.37$, and $H_0: \rho = 0$ is not rejected with $\alpha = 0.05$. However, with $\beta = 0.1$ and $r_{pb} = 0.5$, $H_0: \rho_{pb} = 0$ is rejected, the p-value being 0.024.

■

■ **Example**

Consider the star data in Fig. 9.2. As previously noted, $r = -0.21$. In contrast, $r_{pb} = 0.06$, with $\beta = 0.1$. Increasing β to 0.2, $r_{pb} = 0.26$, and the p-value is 0.07. For $\beta = 0.3$, $r_{pb} = 0.3$, with a p-value of 0.04, and for $\beta = 0.5$, $r_{pb} = 0.328$.

■

9.3.4 A Test of Zero Correlation Among p Random Variables

Consider a random sample of n vectors from some p-variate distribution, $X_{i1}, \ldots, X_{ip}$, $i = 1, \ldots, n$. Let ρ_{pbjk} be the percentage bend correlation between the jth and kth random variables, $1 \le j < k \le p$. This section considers the problem of testing

$$H_0: \rho_{pbjk} = 0, \text{ for all } j < k.$$

Put another way, the hypothesis is that the matrix of percentage bend correlations among all p random variables is equal to the identity matrix.

Currently, the best method for testing this hypothesis, in terms of controlling the probability of a Type I error under independence, begins by computing

$$c_{jk} = \sqrt{(n - 2.5) \times \ln\left(1 + \frac{T_{pbjk}^2}{n - 2}\right)},$$

where T_{pbjk} is the statistic given by Eq. (9.3) for testing independence between the jth and kth random variables and ln indicates the natural logarithm. Let

$$z_{jk} = c_{jk} + \frac{c_{jk}^3 + 3c_{jk}}{b} - \frac{4c_{jk}^7 + 33c_{jk}^5 + 240c_{jk}^3 + 855c_{jk}}{10b^2 + 8bc_{jk}^4 + 1000b}.$$

Table 9.3: Estimated Type I error probabilities, $\alpha = 0.05$.

			$p = 4$			$p = 10$		
g	h	n	H_r	GR	H	H_r	GR	H
0.0	0.0	10	0.050	0.070	0.053	0.054	—	0.056
		20	0.049	0.022	0.053	0.048	—	0.046
0.5	0.0	10	0.055	0.076	0.050	0.062	—	0.055
		20	0.058	0.025	0.053	0.059	—	0.051
1.0	0.0	10	0.092	0.111	0.054	0.126	—	0.062
		20	0.091	0.055	0.054	0.126	—	0.055
0.0	0.5	10	0.082	0.106	0.050	0.124	—	0.052
		20	0.099	0.062	0.050	0.152	—	0.054
0.5	0.5	10	0.097	0.118	0.051	0.157	—	0.053
		20	0.097	0.118	0.051	0.185	—	0.053
1.0	0.5	10	0.130	0.158	0.053	0.244	—	0.059
		20	0.135	0.105	0.053	0.269	—	0.055

When H_0 is true,

$$H = \sum_{j<k} z_{jk}^2,$$

has, approximately, a chi-squared distribution, with $p(p-1)/2$ degrees of freedom. Consequently, reject H_0, if $H > \chi_{1-\alpha}^2$, the $1 - \alpha$ quantile.

When the percentage bend correlation is replaced by Pearson's correlation, r, in the hypothesis testing procedure just described, the resulting test statistic will be labeled H_r. Gupta and Rathie (1983) suggest yet another test of the hypothesis that all pairs of random variables have zero correlations, again, using r. Table 9.3 reports the estimated probability of a Type I error for $p = 4$ and 10, and various g-and-h distributions, when using H, H_r, or the Gupta–Rathie (GR) method and when $n = 10$ and 20. For $p = 10$ and $n \leq 20$, the GR method cannot always be computed, and the corresponding entry in Table 9.3 is left blank. As is evident, the test based on the percentage bend correlation is easily the most satisfactory, with the estimated probability of a Type I error (based on simulations with 10,000 replications) ranging between 0.046 and 0.062. If the usual correlation, r, is used instead, the probability of a Type I error can exceed 0.2, and when using method GR, it can exceed 0.15.

9.3.5 R Function pball

The R function

$$\text{pball(m,beta= 0.2)}$$

computes the percentage bend correlation for all pairs of random variables, and it tests the hypothesis that all of the correlations are equal to zero. Here, m is an n-by-p matrix of data. If the data are not stored in a matrix, the function prints an error message and terminates. Again, beta, which is β in Table 9.2, defaults to 0.2. The function returns a p-by-p matrix of correlations in pball$pbcorm, another matrix indicating the p-values for the hypotheses that each correlation is zero, plus the test statistic H and its corresponding p-value.

■ **Example**

Again, consider the alcohol data in Section 8.6.2, where measures of the effect of drinking alcohol are taken at three different times. If the data for the control group are stored in the R matrix amat, the command pball(amat) returns

```
$pbcorm:
          [,1]      [,2]      [,3]
[1,] 1.0000000 0.5028002 0.7152667
[2,] 0.5028002 1.0000000 0.5946712
[3,] 0.7152667 0.5946712 1.0000000


$p.value:
               [,1]          [,2]          [,3]
[1,]            NA 0.023847557 0.0003925285
[2,]   0.0238475571          NA 0.0056840954
[3,]   0.0003925285 0.005684095          NA


$H:
[1] 5.478301e+192


$H.p.value:
[1] 0
```

For example, the correlation between variables 1 and 2 is 0.5, between 1 and 3, it is 0.72, and between 2 and 3, it is 0.59. The corresponding p-values are 0.024, 0.0004, and 0.0057. The test statistic, H, for testing the hypothesis that all three correlations are equal to zero, has a p-value approximately equal to 0. To use $\beta = 0.1$ instead, type the command pball(amat,.1).

Table 9.4: Estimated Type I error probabilities, $\alpha = 0.05$.

		$T_{pb.1}$		$T_{pb.5}$		T		T_w	
g	*h*	*n = 10*	*n = 20*	*n = 10*	*n = 20*	*n = 10*	*n = 20*	*n = 10*	*n = 20*
0.0	0.0	0.050	0.050	0.053	0.049	.049	0.049	0.040	0.045
0.0	0.2	0.050	0.049	0.054	0.049	0.054	0.052	0.043	0.045
0.0	0.5	0.047	0.048	0.053	0.049	0.062	0.067	0.038	0.044
0.5	0.0	0.050	0.049	0.053	0.050	0.039	0.050	0.037	0.043
0.5	0.2	0.048	0.048	0.053	0.050	0.055	0.053	0.044	0.044
0.5	0.5	0.047	0.047	0.053	0.049	0.064	0.065	0.037	0.043
1.0	0.0	0.045	0.048	0.053	0.050	0.054	0.055	0.037	0.043
1.0	0.2	0.046	0.047	0.053	0.050	0.062	0.053	0.041	0.058
1.0	0.5	0.045	0.046	0.053	0.050	0.070	0.065	0.035	0.044
1.0	1.0	0.044	0.045	0.052	0.050	0.081	0.071	0.034	0.043

9.3.6 The Winsorized Correlation

Another (type M) robust analog of ρ is the Winsorized correlation, which is estimated as follows. Based on the random sample $(X_{11}, X_{12}), \dots, (X_{n1}, X_{n2})$, first Winsorize the observations by computing the Y_{ij} values as described in Section 8.1.1. Then compute Pearson's correlation based on the Y_{ij} values. That is, the sample Winsorized correlation is

$$r_w = \frac{\sum (Y_{i1} - \bar{Y}_1)(Y_{i2} - \bar{Y}_2)}{\sqrt{\sum (Y_{i1} - \bar{Y}_1)^2 \sum (Y_{i2} - \bar{Y}_2)^2}}.$$

Here, 20% Winsorization is assumed unless stated otherwise.

Let ρ_w be the population analog of r_w. To test H_0: $\rho_w = 0$, compute

$$T_w = r_w \sqrt{\frac{n-2}{1 - r_w^2}},$$

and reject, if $|T_w| > t_{1-\alpha/2}$, the $1 - \alpha/2$ quantile of a Student's t distribution, with $\nu = h - 2$ degrees of freedom, where h, the effective sample size, is the number of pairs of observations not Winsorized. (Equivalently, $h = n - 2g$, $g = [\gamma n]$, is the number of observations left after trimming.) Unless stated otherwise, $\gamma = 0.2$ is assumed. In terms of Type I error probabilities when testing the hypothesis of zero correlation, the Winsorized correlation appears to compete well with the test based on r under independence, but the percentage bend correlation is better still, at least when $\beta = 0.1$. Like all of the hypothesis testing methods in this section, T_w is sensitive to heteroscedasticity, so a more accurate description of the test statistic T_w is that it tests the hypothesis that two random variables are independent. As for power, the best method among Pearson's correlation, the Winsorized correlation, and the percentage bend correlation depend in part on the values of ρ, ρ_w, and ρ_{pb}, which are unknown. Perhaps there are situations where using ρ_w will result in more power. This depends in part on how much ρ_w differs from ρ_{pb}.

Table 9.4 compares Type I error probabilities when testing for independence using T, T_w, and T_{pb}. The notation $T_{pb.1}$ means that $\beta = 0.1$ is used, and $T_{pb.5}$ means $\beta = 0.5$. (The first two columns in Table 9.4 indicate the g-and-h distribution associated with the marginal distributions.) As can be seen, the test based on r is the least stable in terms of Type I errors.

9.3.7 R Function wincor

The R function

$$\text{wincor}(x, y = \text{NULL}, tr = 0.2)$$

estimates the Winsorized correlation. When the argument y=NULL, the argument x is assumed to be a matrix with $p \geq 2$ columns. The default amount of Winsorization, tr, is 0.2. The function returns the Winsorized correlation, r_w, the Winsorized covariance, and the corresponding p-values.

Section 9.3.4 describes a method for testing the hypothesis that all percentage bend correlations, among all pairs of random variables, are equal to zero. The method is easily extended to test the hypothesis that all Winsorized correlations are equal to zero, but there are no simulation results on how well this approach performs in terms of Type I errors, so it is not recommended at this time.

■ **Example**

For the alcohol data in Section 8.6.2, used to illustrate the R function pball, wincor returns

```
$wcor:
          [,1]       [,2]       [,3]
[1,] 1.0000000 0.5134198 0.6957740
[2,] 0.5134198 1.0000000 0.6267765
[3,] 0.6957740 0.6267765 1.0000000

$wcov:
          [,1]       [,2]       [,3]
```

```
[1,] 44.77895 24.12632 27.68421
[2,] 24.12632 49.31316 26.17105
[3,] 27.68421 26.17105 35.35526

$p.value:
               [,1]          [,2]          [,3]
[1,]             NA 0.023645294  0.001061593
[2,]  0.023645294           NA  0.004205145
[3,]  0.001061593 0.004205145           NA
```

Thus, the estimated correlation between variables 1 and 2 is 0.51, the covariance is 24.1, and the p-value, when testing the hypothesis of independence via the Winsorized correlation, is 0.024. In this particular case, the results are very similar to those obtained with the percentage bend correlation.

9.3.8 The Biweight Midcovariance and Correlation

It should be noted that the percentage bend covariance and correlation are a special case of a larger family of measures of association. Let Ψ be any odd function, such as those summarized in Section 2.2.4. Let μ_x be any measure of location for the random variable X, let τ_x be some measure of scale, let K be some constant, and let $U = (X - \mu_x)/(K\tau_x)$ and $V = (Y - \mu_y)/(K\tau_y)$. Then a measure of covariance between X and Y is

$$\gamma_{xy} = \frac{nK^2\tau_x\tau_y E\{\Psi(U)\Psi(V)\}}{E\{\Psi'(U)\Psi'(V)\}}$$

and a measure of correlation is $\gamma_{xy}/\sqrt{\gamma_{xx}\gamma_{yy}}$. If μ_x and μ_y are measures of location such that $E\{\Psi(U)\} = E\{\Psi(V)\} = 0$, then $\gamma_{xy} = 0$ when X and Y are independent.

Among the many choices for Ψ and K, the so-called biweight midcovariance has played a role in a regression method covered in Chapter 10, so for completeness, an estimate of this parameter is described here. It is based on $K = 9$ and the biweight function described in Table 2.1 of Chapter 2. Let $(X_1, Y_1), \ldots, (X_n, Y_n)$ be a random sample from some bivariate distribution. Let

$$U_i = \frac{X_i - M_x}{9 \times \text{MAD}_x},$$

where M_x and MAD_x are the median and the value of MAD for the X values. Similarly, let

$$V_i = \frac{Y_i - M_y}{9 \times \text{MAD}_y}.$$

Set $a_i = 1$, if $-1 \le U_i \le 1$; otherwise, $a_i = 0$. Similarly, set $b_i = 1$, if $-1 \le V_i \le 1$; otherwise, $b_i = 0$. The sample biweight midcovariance between X and Y is

$$s_{bxy} = \frac{n \sum a_i (X_i - M_x)(1 - U_i^2)^2 b_i (Y_i - M_y)(1 - V_i^2)^2}{(\sum a_i (1 - U_i^2)(1 - 5U_i^2))(\sum b_i (1 - V_i^2)(1 - 5V_i^2))}.$$

The statistic s_{bxx} is the biweight midvariance mentioned in Chapter 3, and

$$r_b = \frac{s_{bxy}}{\sqrt{s_{bxx} s_{byy}}}$$

is an estimate of what is called the **biweight midcorrelation** between X and Y. The main reasons for considering this measure of covariance are that it is relatively easy to compute and it appears to have a breakdown point of 0.5, but a formal proof has not been found.

9.3.9 R Functions bicov and bicovm

The R function

$$\text{bicov(x,y)}$$

computes the biweight midcovariance between two random variables. The R function COR.PAIR, described in Section 9.3.12, also computes this correlation when the argument method='BIC'. The function bicovm computes the biweight midcovariance and midcorrelation for all pairs of p random variables stored in some R variable, m, which can be either an n-by-p matrix or a variable having list mode. It has the form

$$\text{bicovm(m).}$$

■ Example

For the star data in Fig. 9.2, bicovm reports that the biweight midcorrelation is 0.6. In contrast, the highest percentage bend correlation, among the choices 0.1, 0.2, 0.3, 0.4, and 0.5 for β, is 0.33, and a similar result is obtained when Winsorizing instead. As previously noted, these data have several outliers. The main point here is that r_{pb} and r_w offer more resistance than r, but they can differ substantially from other resistant estimators.

■

9.3.10 Kendall's tau

A well-known type M correlation is Kendall's tau. For completeness, it is briefly described here.

Consider two pairs of observations, (X_1, Y_1) and (X_2, Y_2). For convenience, assume tied values never occur, and that $X_1 < X_2$. Then these two pairs of observations are said to be concordant, if $Y_1 < Y_2$; otherwise, they are discordant. For n pairs of points, let $K_{ij} = 1$, if the ith and jth points are concordant, and if they are discordant, $K_{ij} = -1$. Then Kendall's tau is given by

$$\hat{\tau} = \frac{2}{n(n-1)} \sum_{i<j} K_{ij}. \tag{9.4}$$

Under independence, the population value of $\hat{\tau}$, τ, is zero. The usual test of $H_0: \tau = 0$ is to reject if

$$|Z| \geq z_{1-\frac{\alpha}{2}},$$

where

$$Z = \frac{\hat{\tau}}{\sigma_\tau},$$

and

$$\sigma_\tau^2 = \frac{2(2n+5)}{9n(n-1)}.$$

It is left as an exercise to show that heteroscedasticity affects the probability of rejecting when H_0 is true.

If X and Y have the bivariate distribution H, the influence function of Kendall's tau is

$$IF(x, y) = 2(2P_H[(X - x)(Y - y) > 0] - 1 - \tau)$$

(Croux and Dehon, 2010). So, a positive feature of Kendall's tau is that it has infinitesimal robustness. (Its influence function is bounded.) However, although Kendall's tau provides protection against outliers among the X values ignoring Y, or among the Y values ignoring X, it can be seen that outliers can substantially alter its value. Even two points, properly placed, can alter Kendall's tau substantially (e.g., Wilcox, 2017b, Section 8.7.1) (Details are relegated to the exercises.)

9.3.11 Spearman's rho

Assign ranks to the X values ignoring Y, and assign ranks to the Y values ignoring X. Then Spearman's rho, r_s, is just Pearson's correlation based on the resulting ranks. Like all of the correlations in this section, it provides protection against outliers among the X values, ignoring Y, as well as outliers among the Y values, ignoring X, but outliers properly placed can alter its value substantially. Letting ρ_s be the population value of Spearman's rho, the influence function of ρ_s is

$$IF(x, y) = -3\rho_s - 9 + 12\{F(x)G(y) + E(F(X)I(Y \geq y)) + E(G(Y)I(X \geq x))\},$$

where F and G are the marginal distributions of X and Y, respectively, and I is the indicator function (Croux and Dehon, 2010). In terms of asymptotic efficiency and other robustness considerations, results in Croux and Dehon (2010) indicate that Kendall's tau is preferable to Spearman's rho.

When X and Y are independent, $\rho_s = 0$. The usual test of H_0: $\rho_s = 0$ is to reject, if $|T| \geq t$, where t is the $1 - \alpha/2$ quantile of a Student's t distribution, with $\nu = n - 2$ degrees of freedom, and

$$T = \frac{r_s \sqrt{n - 2}}{\sqrt{1 - r_s^2}}.$$

Like all of the hypothesis testing methods in this section, heteroscedasticity affects the probability of rejecting, even when H_0 is true.

Table 9.5 shows estimated power when testing the hypothesis of zero correlation and $\rho = 0.5$. Included is the power of the test based on Kendall's tau, under the column headed "Kend.", and Spearman's rho under the column "Spear". The main point is that different methods can have more or less power than other methods, one reason being that the parameters being estimated can differ, so it is difficult to select a single method for general use based on the criterion of high power. Like Kendall's tau, two points, properly placed, can alter substantially Spearman's rho.

9.3.12 R Functions tau, spear, cor, taureg, COR.ROB, and COR.PAIR

The function

$$\text{tau}(x,y,\text{alpha}= 0.05)$$

computes Kendall's tau, and

$$\text{spear}(x,y)$$

Table 9.5: Estimated power, $n = 20$, $\rho = 0.5$.

g	h	T_w	T	$T_{pb.1}$	$T_{pb0.5}$	Kend.	Spear.
0.0	0.0	0.562	0.637	0.620	0.473	0.551	0.568
0.0	0.2	0.589	0.633	0.638	0.512	0.594	0.597
0.0	0.5	0.614	0.603	0.658	0.552	0.644	0.626
0.0	1.0	0.617	0.573	0.658	0.588	0.692	0.659
0.5	0.0	0.588	0.620	0.629	0.499	0.602	0.602
0.5	0.2	0.602	0.608	0.643	0.525	0.624	0.615
0.5	0.5	0.614	0.591	0.653	0.558	0.656	0.638
0.5	1.0	0.611	0.565	0.608	0.591	0.698	0.664
1.0	0.0	0.621	0.597	0.650	0.537	0.668	0.641
1.0	0.2	0.620	0.585	0.652	0.550	0.667	0.644
1.0	0.5	0.619	0.571	0.652	0.569	0.683	0.652
1.0	1.0	0.611	0.559	0.653	0.595	0.709	0.669

computes Spearman's rho. The built-in R function

cor(x, y = NULL, use = 'everything', method = c('pearson', 'kendall', 'spearman'))

can be used to compute Kendall's tau by setting the argument method='kendall' and Spearman's rho by setting method='spearman'. The R functions tau and spear automatically test the hypothesis of a zero correlation. The R function cor does not. For convenience, the function

taureg(m,y,corfun=tau)

computes the p correlations between every variable in the matrix m, having p columns, and the variable y. The argument corfun can be any function that computes a correlation between two variables only and returns the value in corfun$cor along with the p-value in corfun$p.value. By default, Kendall's tau is used.

For convenience, the R function

COR.ROB(x,method='WIN',tr=.2)

can be used to compute any of several correlation matrices, where x is a matrix with p columns. The default is the Winsorized correlation. The others are the percentage bend (PB), the skipped correlation (skip), the minimum volume ellipsoid (mve), the minimum covariance determinant (mcd), Kendall's tau (Ken), Spearman's rho (Spear), and the biweight correlation (BIC). The skipped correlation defaults to using Pearson's correlation after outliers are removed. The command COR.ROB(x,method='skip',corfun=wincor) would use a Winsorized correlation instead. For bivariate data, the R function

$$\text{COR.pair(x,y,method='WIN',tr=.2)}$$

returns the correlation designated by method.

9.3.13 Heteroscedastic Tests of Zero Correlation

The tests of independence based on type M correlations, previously covered, are sensitive to heteroscedasticity. That is, even when these correlations are equal to zero, if there is heteroscedasticity, the probability of rejecting can increase as the sample size gets large. So when rejecting, it is reasonable to conclude that the variables under study are dependent, but the reason for rejecting might be due more to heteroscedasticity than to the correlation differing from zero. To test the hypothesis that a (type M) correlation is equal to zero in a manner that is insensitive to heteroscedasticity, a percentile bootstrap method can be used.

When using robust correlations with a reasonably high breakdown point, all indications are that a basic percentile bootstrap performs well in terms of controlling the probability of a Type I error. For a random sample $(X_1, Y_1), \ldots, (X_n, Y_n)$, generate a bootstrap sample by resampling with replacement n pairs of points, yielding $(X_1^*, Y_1^*), \ldots, (X_n^*, Y_n^*)$. Compute any of the robust estimators described in this section, and label the result r^*. Repeat this B times, yielding $r_1^*, \ldots, r_B^*$. Let $\ell = \alpha B/2$, rounded to the nearest integer, and let $u = B - \ell$. Then reject the hypothesis of a zero correlation, if $r_{(\ell+1)}^* > 0$ or $r_{(u)}^* < 0$, where $r_{(1)}^* \le \cdots \le r_{(B)}^*$ are the values $r_1^*, \ldots, r_B^*$ written in ascending order. Note that the method is readily extended to situations where the focus is on the association between p independent (explanatory) variables $X_1, \ldots, X_p$ and some dependent variable Y.

For the special case where r is Pearson's correlation, Bishara et al. (2017) compare a wide range of methods aimed at computing a 0.95 confidence interval and conclude that no method is completely satisfactory. A few of the better methods are described here.

The first method is based on a modified percentile bootstrap method (Wilcox and Muska, 2001). When $B = 599$, an approximate 0.95 confidence interval for ρ is

$$(r_{(a)}^*, r_{(c)}^*),$$

where, again, for $n < 40$, $a = 7$ and $c = 593$; for $40 \le n < 80$, $a = 8$ and $c = 592$; for $80 \le n < 180$, $a = 11$ and $c = 588$; for $180 \le n < 250$, $a = 14$ and $c = 585$; while for $n \ge 250$, $a = 15$ and $c = 584$. As usual, if this interval does not contain zero, reject H_0: $\rho = 0$. But this method is limited to 0.95 confidence intervals.

Another approach stems from the heteroscedastic method for computing a confidence interval for the usual least squares regression slope based on the HC4 estimate of the standard error described in Section 10.1.1. First, standardize both X and Y. That is, compute $Z_{xi} = (X_i - \bar{X})/s_x$ and $Z_{yi} = (Y_i - \bar{Y})/s_y$ $(i = 1, \ldots, n)$. Based on these standardized variables, compute V, the HC4 estimate of the squared standard error described in Section 10.1.1. The test statistic is $r/\sqrt{V}$. The null distribution when testing H_0: $\rho = 0$ is taken to be a Student's t distribution, with $n - 2$ degrees of freedom.

Bishara et al. (2017) found that both the HC4 method and the modified percentile method perform relatively well when $\rho = 0.25$ or 0.5 in the sense that both are relatively good at ensuring confidence intervals that have probability coverage at least as large as the nominal 0.95 level. However, when using the HC4 method, the probability coverage can be much higher than intended when $|\rho|$ is relatively high even when n is fairly large. For example, when $\rho = 0.8$, the probability coverage can be well above 0.975, even with $n = 500$. This occurs, for example, when the marginal distributions have a g-and-h distribution, with $g = 1$ and $h = 0$, which is a lognormal distribution shifted to have a median of zero. But when $\rho = 0$, it performs well. This suggests that HC4 might not be asymptotically correct when $\rho \neq 0$.

Bishara et al. (2017) found that a bootstrap with bias correction and acceleration (BCa) performed relatively well in their simulations. But when dealing with a g-and-h distribution, with $g = 1$ and all $h = 0$, the actual probability coverage drops below 0.925, when $n = 500$ and $\rho = 0.5$.

An approach not considered by Bishara et al. (2017) is to use a bootstrap-t method based on the HC4 estimate of the standard error, V. More precisely, first compute r. Next, take a bootstrap sample from the standardized variables (Z_{xi}, Z_{yi}). That is, pairs of standardized variables are chosen at random with replacement. Compute $U^* = (r^* - r)/\sqrt{V^*}$, where r^* and V^* are the values of r and V based on the bootstrap sample. Repeat this B times, yielding U_b^* $(b = 1, \ldots, B)$. Put these values in ascending order, yielding $U_{(1)}^* \leq \cdots \leq U_{(B)}^*$, and let $\ell = \alpha B/2$, rounded to the nearest integer. Let $u = B - \ell + 1$. Then a $1 - \alpha$ confidence interval for ρ is $(r - U_{(u)}^* V, r + U_{(\ell+1)}^* V)$. This approach can yield accurate confidence intervals in situations where the HC4 and BCa methods perform poorly. Extant simulations suggest that this approach performs relatively well, in general, including situations dealing with heavy-tailed distributions.

Hu et al. (2020) derive a confidence interval for Pearson's correlation using an empirical likelihood approach. However, their simulation results are limited to light-tailed distributions. It is unknown how this approach compares to the bootstrap version of the HC4 method or modified percentile bootstrap technique.

For completeness, DiCiccio and Romano (2017) describe a permutation method for testing H_0: $\rho = 0$. Let

$$m_{u,v} = \frac{1}{n} \sum (X_i - \bar{X})^u (Y_i - \bar{Y})^v$$

and:

$$\hat{\tau} = \sqrt{\frac{m_{2,2}}{m_{2,0}m_{0,2}}}.$$

Their test statistic is $S = \sqrt{nr}/\hat{\tau}$. Permute the Y values, and denote the value of the test statistic by S^*. Repeat this for all possible permutations. The null hypothesis is rejected at the α level, if S is less than the $\alpha/2$ quantile or greater than the $1 - \alpha/2$ quantile of the S^* values. (In practice, a large subset of all permutations is used.) Reasons for preferring this method over the other methods mentioned here are unclear.

Of course, none of the methods just described deal with the fact that ρ is not robust. Currently, the bootstrap-t method based on the HC4 estimator appears to perform relatively well. But because the estimate of Pearson's correlation is sensitive to outliers, there are situations where it can be unsatisfactory.

9.3.14 R Functions corb, corregci, pcorb, pcorhc4, and rhohc4bt

The R function

$$corb(x,y,corfun=pbcor,nboot=599,...)$$

tests the hypothesis of a zero correlation using the heteroscedastic bootstrap method just described. By default, it uses the percentage bend correlation, but any correlation can be specified by the argument corfun. For example, the command corb(x,y,corfun=wincor,tr= 0.25) will use a 25% Winsorized correlation. For the situation where the focus is on a single independent variable Y and p independent variables $X_1, \ldots, X_p$, the R function

$$corregci(x,y,corfun=wincor,nboot=599,...)$$

can be used. More formally, letting ϱ_j denote any robust correlation between Y and the jth independent variable, the function tests

$$H_0: \varrho_j = 0$$

for each j ($j = 1, \ldots, p$). The argument x is assumed to be a matrix with p columns. The familywise error (FWE) rate, meaning the probability of making one or more Type I errors, is controlled by Hochberg's method in Section 7.4.7.

When working with Pearson's correlation, the function

$$pcorb(x,y)$$

applies the modified percentile bootstrap method described in the previous section. The R function

$$\text{pcorhc4(x,y,alpha= 0.05)}$$

applies the HC4 method and

$$\text{rhohc4bt(x,y,nboot=2999, alpha= 0.05, SEED=TRUE)}$$

computes a confidence interval for ρ using a bootstrap-t version of the HC4 method. It returns a p-value by finding the smallest value for alpha=0.01(.01)0.99 for which the null hypothesis of a zero correlation is rejected. Consequently, the smallest possible p-value is 0.01.

9.4 Some Type O Correlations

Type M correlations have the property that two properly placed outliers can substantially alter their value. (Illustrations are relegated to the exercises at the end of this chapter.) Type O correlations are an attempt to correct this problem. Section 6.2 described various measures that reflect how deeply a point is nested within a cloud of data, where the measures of depth take into account the overall structure of the data. Roughly, *type O correlations* are correlations that possibly downweight or eliminate one or more points that have low measures of depth. In essence, they are simple extensions of W-estimators described in Section 6.3.6. Included among this class of correlations coefficients are so-called *skipped correlations*, which remove any points flagged as outliers, and then compute some correlation coefficient with the data that remain.

9.4.1 MVE and MCD Correlations

An example of a type O correlation has, in essence, already been described in connection with the MCD and MVE estimators of scatter described in Section 6.3. These measures search for the central half of the data, and then use this half of the data to estimate location and scatter. As is evident, the covariance associated with this central half of the data readily yields a correlation coefficient. For example, simply compute Pearson's correlation based on the central half of the data.

9.4.2 Skipped Measures of Correlation

Skipped correlations are obtained by checking for any outliers using one of the methods described in Section 6.4, removing them, and applying some correlation coefficient to the remaining data. An example is to remove outliers using the MVE or MCD methods and compute Pearson's correlation after outliers are removed. It is noted that when using the R functions cov.mve and cov.mcd, already described, setting the optional argument cor to TRUE causes this correlation to be reported. For example, when using cov.mve, the command:

$$\text{cov.mve(m,cor=TRUE)}$$

accomplishes this goal.

9.4.3 The OP Correlation

In recent years, the term skipped correlation has been used to signify a particular type O correlation coefficient that corresponds to the OP correlation described here (e.g., Pernet et al., 2013; Wilcox, 2015d). The OP correlation coefficient begins by eliminating any outliers using the projection method in Section 6.4.9. Then it computes some correlation coefficient with the data that remain. Pearson's correlation is assumed unless stated otherwise.

Imagine that data are randomly sampled from some bivariate normal distribution. If the goal is to use a skipped correlation coefficient that gives a reasonably accurate estimate of Pearson's correlation, ρ, relative to r, then the OP-estimator is the only skipped estimator known to be reasonably satisfactory.

Let r_p represent the skipped correlation coefficient, and let m be the number of pairs of points left after outliers are removed. A seemingly simple method for testing the hypothesis of independence is to apply the usual T test for Pearson's correlation but with r replaced by r_p and n replaced by m. But this simple solution fails because it does not take into account the dependence among the points remaining after outliers are removed. If this problem is ignored, unsatisfactory control over the probability of a Type I error results (Wilcox, 2010d).

Let

$$T_p = r_p \sqrt{\frac{n-2}{1-r_p^2}}$$

and suppose the hypothesis of independence is rejected at the $\alpha = 0.05$ level, if $|T_p| \geq c$, where

$$c = \frac{6.947}{n} + 2.3197.$$

The critical value c was determined via simulations under normality by determining an appropriate critical value for n ranging between 10 and 200, and then a least squares regression line was fit to the data. For non-normal distributions, when testing at the 0.05 level, all indications are that this hypothesis testing method has an actual Type I error probability reasonably close to the nominal level.

If the goal is to test the hypothesis that two random variables are independent, inferences based on T_p are reasonable. But if the goal is to test the hypothesis that the population OP measure of association is zero, this is no longer the case because it does not deal with heteroscedasticity. Another negative feature is that it is limited to testing at the 0.05 level. These two limitations can be addressed by using a percentile bootstrap method (Wilcox, 2015d). That is, proceed as described in Section 9.3.13, only use the OP correlation coefficient rather than a type M correlation. For $n < 120$, this approach can be improved via method H1 described in the next section.

9.4.4 Inferences Based on Multiple Skipped Correlations

This section considers the goal of extending the hypothesis testing method based on T_p in the previous section to situations where there are $p \geq 2$ random variables. That is, there are two or more pairs of variables and the goal is to control FWE, the familywise error rate. One approach is to focus on homoscedastic methods and test the hypothesis that each pair of variables is independent. Another goal is to test the hypothesis that the skipped correlations are zero in a manner that allows heteroscedasticity. Or in the context of Tukey's decision rule, to what extent is it reasonable to decide whether the skipped correlation is less than or greater than zero? Let $\hat{\tau}_{jk}$ be a skipped correlation between variables j and k, $j < k$. Wilcox (2003d) proposes a homoscedastic based on

$$T_{jk} = \hat{\tau}_{jk} \sqrt{\frac{n-2}{1 - \hat{\tau}_{jk}^2}}.$$

The strategy for controlling FWE is to find approximate estimates of the quantiles of

$$T_{\max} = \max |T_{jk}|. \tag{9.5}$$

The method can be applied via the R function mscor. However, this method is limited to using Spearman's correlation after outliers are removed, it is limited to testing at the 0.05 level, and it does not deal with heteroscedasticity.

Method IND

Improved methods were derived by Wilcox et al. (2018). First, consider the goal of testing for independence using a method that is sensitive to both heteroscedasticity and the value of τ_{jk}. That is, even when $\tau_{jk} = 0$, if there is heteroscedasticity, there is dependence, and the goal is to reject, if this is the case. The strategy is to generate bootstrap samples from each marginal distribution in a manner for which there is no association. Next, compute $T_{\max}$ based on these bootstrap samples, yielding, say T^*. This process is repeated B times, which can be used to estimate the $1 - \alpha$ quantile of the distribution of $T_{\max}$, $T^*_{(c)}$, when the null hypothesis is true for all $j < k$. Then for any $j < k$, reject the hypothesis of independence, if $|T_{jk}| \geq T^*_{(c)}$.

Method ECP

Now consider the goal of testing

$$H_0: \tau_{jk} = 0 \tag{9.6}$$

for every $j < k$ in a manner that deals with heteroscedasticity. As noted at the end of Section 9.4.3, a percentile bootstrap method has been found to perform reasonably well for any j and k. Simulations indicate that the actual level is primarily a function of the sample size, not the distribution from which observations are generated. This suggests a simple modification for dealing with each $j < k$: Use simulations to estimate a critical p-value, p_α, such that FWE is equal to α under normality and homoscedasticity. To elaborate, for any $j < k$, let p_{jk} be the percentile bootstrap p-value when testing Eq. (9.6). Then proceed as follows:

1. Generate n observations from a p-variate normal distribution for which the covariance matrix is equal to the identity matrix.
2. For the data generated in step 1, compute the p-value for each of the $C = (p^2 - p)/2$ hypotheses using a percentile bootstrap method. Let V denote the minimum p-value among the C p-values just computed.
3. Repeat steps 1 and 2 D times, yielding the minimum p-values $U_1, \ldots, U_D$.
4. Let $\hat{p}_\alpha$ be an estimate of the α quantile of the distribution of U. Here, the Harrell–Davis estimator is used.
5. Reject any hypothesis for which $p_{jk} \leq \hat{p}_\alpha$.

Method H1

But a practical concern with the approach just described is that execution time can be quite high. Wilcox et al. (2018) suggest dealing with this issue in the following manner. Let V_n denote the vector of p-values corresponding to n and stemming from the simulation just described. Then given V_n, an adjusted p-value can be computed, which is simply the value q such that $\hat{\theta}_q(V_n) = p_{jk}$, where $\hat{\theta}_q(V_n)$ is some estimate of the qth quantile of the values in V_n.

For example, if the bootstrap p-value is 0.08 and V_n indicates that the level of the test is 0.05 when $p_\alpha = 0.08$ is used, then the adjusted p-value is 0.05. To reduce execution time, vectors of p-values were computed for selected sample sizes and stored in an R function. Here, adjusted p-values are computed in the following manner: Use V_{30} when $20 \leq n \leq 40$, use V_{60} when $41 < n \leq 70$, use V_{80} when $71 < n \leq 100$, and use V_{100} when $101 < n \leq 120$. For $n > 120$, no adjustment is made. FWE is controlled using Hochberg's method unless stated otherwise. Note that the method is readily extended to the situation where the goal is to test the hypothesis of a zero correlation between each of p independent variables and a single dependent variable, as might be done when dealing with regression.

9.4.5 R Functions scor, scorall, scorci, mscorpb, mscorci, mscorciH, scorreg, scorregci, and scorregciH

The R function

$$\text{scor(x,y=NA,corfun=pcor, gval=NA, plotit=TRUE, cop=3, op=TRUE, MC=FALSE)}$$

computes a skipped correlation (the OP correlation coefficient) for bivariate data. If the argument y is not specified, it is assumed that x is an n-by-2 matrix. The argument corfun controls which correlation is computed after outliers are removed; by default, Pearson's correlation is used. The arguments gval, cop, and op are relevant to the projection outlier detection method; see Section 6.4.10. If MC=TRUE, the multicore version of the function outpro is used to detect outliers. To compute a skipped correlation for each pair of variables, when dealing with p-variate data, use the R function

$$\text{scorall(x,corfun=pcor,...).}$$

The R function

$$\text{scorci(x, y, nboot = 1000, alpha = 0.05, V2 = TRUE, SEED = TRUE, plotit = TRUE,}$$
$$\text{STAND = TRUE, corfun = pcor, cop = 3, ...)}$$

computes a $1 - \alpha$ confidence interval for the skipped correlation using a percentile bootstrap method. It deals with two variables only. That is, the argument x is assumed to be a vector. The argument V2=TRUE means that method H1 is used; otherwise, a standard percentile bootstrap is used. The R function

$$\text{scorciMC(x, y, nboot = 1000, alpha= 0.05, SEED = TRUE, plotit = TRUE, corfun = pcor,}$$
$$\text{cop = 3, ...)}$$

is the same as scorci when V2=FALSE, only it takes advantage of a multicore processor if one is available, assuming the R package parallel has been installed.

The R function

$$\text{mscorpb}(x, \text{corfun} = \text{pcor}, \text{nboot} = 500, \text{alpha} = 0.05, \text{SEED} = \text{TRUE}, \text{outfun} = \text{outpro},$$
$$\text{pr} = \text{TRUE})$$

tests the hypothesis of independence using method IND in the previous section. This is done for all pairs of variables stored in the argument x, which is assumed to be a matrix with p columns.

The R function

$$\text{mscorciH}(x, \text{nboot} = 1000, \text{alpha} = 0.05, \text{SEED} = \text{TRUE}, \text{method} = \text{'hoch'}, \text{corfun} = \text{pcor},$$
$$\text{outfun} = \text{outpro}, \text{crit.pv} = \text{NULL}, \text{ALL} = \text{TRUE}, \text{MC} = \text{TRUE}, \text{pr} = \text{TRUE})$$

tests Eq. (9.6) via method H1, which allows heteroscedasticity. All indications are that it controls FWE about as well as method ECP. The argument ALL=TRUE means that the outlier method indicated by the argument outpro is applied to the matrix containing all p variables. ALL=FALSE means that when computing a skipped correlation between variables j and k, the remaining $p - 2$ variables are ignored when checking for outliers. If it is desired to use method ECP, this can be done via the R function

$$\text{mscorci}(x, y = \text{NULL}, \text{nboot} = 1000, \text{alpha} = c(0.05, 0.025, 0.01), \text{SEED} = \text{TRUE},$$
$$\text{STAND} = \text{TRUE}, \text{corfun} = \text{pcor}, \text{outfun} = \text{outpro}, \text{crit.pv} = \text{NULL}, \text{pvals} = \text{NULL},$$
$$\text{hoch} = \text{FALSE}, \text{iter} = 500, \text{pval.SEED} = \text{TRUE}, \text{pr} = \text{TRUE}).$$

Setting the argument hoch=TRUE, p-values are not adjusted as done by method H1 or ECP.

When dealing with regression, an issue is which of the p independent variables has a non-zero correlation with the dependent variable. If the goal is merely to estimate the p correlations, the R function

$$\text{scorreg}(x, y, \text{corfun} = \text{pcor}, \text{cop} = 3, \text{MM} = \text{FALSE}, \text{gval} = \text{NA}, \text{outfun} = \text{outpro}, \text{alpha} = 0.05,$$
$$\text{MC} = \text{NULL}, \text{SEED} = \text{TRUE}, \text{ALL} = \text{TRUE})$$

can be used. The argument ALL is used in the same manner as indicated by the previous R function. The R function scorregciH(x, y, nboot = 1000, alpha = 0.05, SEED = TRUE, corfun = pcor, outfun = outpro, crit.pv = NULL, ALL = TRUE, MC = TRUE, pvals = NULL, iter = 500, pval.SEED = TRUE, pr = TRUE)

applies an analog of method H1.

9.5 A Test of Independence Sensitive to Curvature

Let $(\mathbf{X_1}, Y_1), \ldots, (\mathbf{X}_n, Y_n)$ be a random sample of n pairs of points, where $\mathbf{X}$ is a vector having length p. The goal in this section is to test the hypothesis that $\mathbf{X}$ and Y are independent in a manner that is sensitive to curvature as well as any linear association that might exist (cf. Chaudhuri and Hu, 2019).

Method INDT

The first method described here stems from general theoretical results derived by Stute et al. (1998). It does not assume or require homoscedasticity and is based in part on what is called a wild bootstrap method. Stute et al. establish that other types of bootstrap methods are not suitable when using the test statistic to be described. Essentially, the method in this section is designed to test the hypothesis that the regression surface for predicting Y, given $\mathbf{X}$, is a horizontal plane. Let $E(Y|\mathbf{X})$ represent the conditional mean of Y given $\mathbf{X}$. The goal is to test

$$H_0\colon E(Y|\mathbf{X}) = \mu_y.$$

That is, the conditional mean of Y, given $\mathbf{X}$, does not depend on $\mathbf{X}$.

The test statistic is computed as follows. Let $\bar{Y}$ be the mean based on $Y_1, \ldots, Y_n$. (Using a trimmed mean or some other robust estimator can result in poor control over the probability of a Type I error when Y has a sufficiently skewed distribution.) Fix j, and set $I_i = 1$, if $\mathbf{X}_i \leq \mathbf{X}_j$; otherwise, $I_i = 0$. The notation $\mathbf{X}_i \leq \mathbf{X}_j$ means that for every k, $k = 1, \ldots, p$, $X_{ik} \leq X_{jk}$. Let

$$
\begin{aligned}
R_j &= \tfrac{1}{\sqrt{n}} \sum I_i(Y_i - \bar{Y}) \\
&= \tfrac{1}{\sqrt{n}} \sum I_i r_i,
\end{aligned}
\tag{9.7}
$$

where:

$$r_i = Y_i - \bar{Y}.$$

The test statistic is the maximum absolute value of all the R_j values. That is, the test statistic is

$$D = \max|R_j|. \tag{9.8}$$

An appropriate critical value is estimated with the *wild bootstrap method* as follows. Generate $U_1, \ldots, U_n$ from a uniform distribution, and set

$$V_i = \sqrt{12}(U_i - 0.5),$$

$$r_i^* = r_i V_i,$$

and

$$Y_i^* = \bar{Y}_t + r_i^*.$$

Then based on the n pairs of points $(\mathbf{X}_1, Y_1^*), \ldots, (\mathbf{X}_n, Y_n^*)$, compute the test statistic as described in the previous paragraph, and label it D^*. Repeat this process B times, and label the resulting (bootstrap) test statistics $D_1^*, \ldots, D_B^*$. Finally, put these B values in ascending order, which are labeled $D_{(1)}^* \leq \cdots \leq D_{(B)}^*$. Then the critical value is $D_{(u)}^*$, where $u = (1 - \alpha)B$, rounded to the nearest integer. That is, reject, if

$$D \geq D_{(u)}^*.$$

An alternative test statistic has been studied where D is replaced by

$$W = \frac{1}{n}(R_1^2 + \cdots + R_n^2). \tag{9.9}$$

The critical value is determined in a similar manner as before. First, generate a wild bootstrap sample and compute W, yielding W^*. Repeating this B times, and reject, if

$$W \geq W_{(u)}^*,$$

where, again, $u = (1 - \alpha)B$, rounded to the nearest integer, and $W_{(1)}^* \leq \cdots \leq W_{(B)}^*$ are the B W^* values written in ascending order. The test statistic D is called the *Kolmogorov–Smirnov* test statistic, and W is called the *Cramér–von Mises* test statistic. The choice between these two test statistics is not clearcut. For $p = 1$, currently, it seems that there is little separating them in terms of controlling Type I errors. The extent to which this remains true when $p > 1$ appears to have received little or no attention.

Method MEDIND

A seemingly natural way of generalizing the method to a robust measure of location is to replace $\bar{Y}$ with say the median or 20% trimmed mean. But when the distribution of Y is sufficiently skewed, control over the probability of a Type I error can be highly unsatisfactory. There is, however, an alternative method that can be used with the median, which is based on a modification of a method derived by He and Zhu (2003); see Wilcox (2008e) for details.

Let **x** be the $n \times (p + 1)$ matrix, with the first column containing all 1s and the remaining p columns are the columns of **X**. Following He and Zhu (2003), it is assumed that the design has been normalized so that $n^{-1} \sum \mathbf{x}_j \mathbf{x}'_j - I = o(1)$. Let $r_i = Y_i - \hat{Y}_\gamma$, where $\hat{Y}_\gamma$ is some estimate of the γth quantile of Y. Currently, simulation results on how well the method controls the probability of a Type I error are limited to the quartiles. For the 0.5 quantile, $\hat{Y}_{0.5}$ is taken to be the usual sample median. Here, the lower and upper quartiles are estimated via the ideal fourths. Let

$$\mathbf{W}_i = n^{-1/2} \sum_{k=1}^{n} \psi(r_k) \mathbf{x}_k I(\mathbf{x}_k \leq \mathbf{x}_i),$$

where $\psi(r) = \gamma I(r > 0) + (\gamma - 1) I(r < 0)$. For fixed j, let U_{ij} be the ranks of the n values in the jth column of **x**, $j = 2, \ldots, q$. Let $F_i = \max U_{ij}$, the maximum being taken over $j = 2, \ldots, q$. If $\mathbf{x}_k \leq \mathbf{x}_i$, then $F_k \leq F_i$. The test statistic is D_n, the largest eigenvalue of

$$\mathbf{Z} = \frac{1}{n} \sum \mathbf{W}_i \mathbf{W}'_i.$$

The strategy for determining an appropriate critical value is to temporarily assume normality, use simulations to approximate the $1 - \alpha$ quantile of the null distribution, say c, and then reject the null hypothesis, if $T_n \geq c$, even when sampling from a non-normal distribution.

An advantage of the methods just described is that they are sensitive to a variety of ways two or more variables might be dependent. But a limitation is that when they reject, it is unclear why. That is, these tests do not provide any information about the nature of the association.

9.5.1 R Functions indt, indtall, and medind

The R function

$$\text{indt(x,y,nboot=500,tr= 0.2,flag=1)}$$

tests the hypothesis of independence using method INDT. As usual, x and y are R variables containing data, tr indicates the amount of trimming used when computing the Cramér–von Mises test statistic, and nboot is B. Here, x can be a single variable or a matrix having n rows and p columns. The argument flag indicates which test statistic will be used:

- flag=1 means the Kolmogorov–Smirnov test statistic, D, is used,
- flag=2 means the Cramér–von Mises test statistic, W, is used,
- flag=3 means both test statistics are computed.

■ **Example**

Sockett et al. (1987) report data from a study dealing with diabetes in children. One of the variables was the age of a child at diagnosis, and another was a measure called base deficit. Using the conventional (Student's t) test of H_0: $\rho = 0$ based on Pearson's correlation fails to reject at the 0.05 level. (The p-value is 0.135.) We, again, fail to reject with a skipped correlation, a 20% Winsorized correlation, and a percentage bend correlation. Using the bootstrap methods for Pearson's correlation or the percentage bend correlation, again, no association is detected. But the function indt rejects at the 0.05 level. A possible explanation is that (based on methods covered in Chapter 11), the regression line between these two variables appears to have some curvature, meaning that a linear model appears to be inappropriate, which masks a true association when attention is restricted to one of the correlation coefficients covered in this chapter.

■

The function

$$indtall(x,y=NA,nboot=500,tr=0.2)$$

performs all pairwise tests of independence for the variables in the matrix x, if y=NA. If data are found in y, then the function performs p tests of independence between each of the p variables in x and y. The current version computes only the Kolmogorov–Smirnov test statistic. Each test is performed at the level indicated by the argument alpha.

The function

$$medind(x, y, qval = 0.5, nboot = 1000, SEED = TRUE, tr= 0.2, pr = TRUE, xout = FALSE,$$
$$outfun = out, ...)$$

tests the hypothesis of independence using method MEDIND. The function contains critical values for a range of situations. If a critical value is not available, one is determined via simulations with the number of replications determined by the argument nboot.

9.6 Comparing Correlations: Independent Case

This section deals with comparing correlations associated with two independent groups. Comparing dependent correlations is covered in Section 11.10.1.

9.6.1 Comparing Pearson Correlations

Numerous methods have been proposed for testing the hypothesis that two Pearson correlations are equal. More formally, the goal is to test

$$H_0\colon \rho_1 = \rho_2, \tag{9.10}$$

where r_1 and r_2, the estimates of ρ_1 and ρ_2, respectively, are independent. A comparison of various methods (Wilcox, 2009c) indicates that a modified percentile bootstrap method performs relatively well, in terms of controlling the probability of a Type I error, when the sample sizes are small. Let $N = n_1 + n_2$ be the total number of pairs of observations. For the jth group ($j = 1, 2$), generate a bootstrap sample of n_j pairs of observations. Let r_1^* and r_2^* represent the resulting correlation coefficients, and set

$$D^* = r_1^* - r_2^*.$$

Repeat this process 599 times, yielding $D_1^*, \ldots, D_{599}^*$. Then a 0.95 confidence interval for the difference between the population correlation coefficients, $\rho_1 - \rho_2$, is

$$(D_{(\ell)}^*, D_{(u)}^*),$$

where $\ell = 7$ and $u = 593$, if $N < 40$; $\ell = 8$ and $u = 592$, if $40 \le N < 80$; $\ell = 11$ and $u = 588$, if $80 \le N < 180$; $\ell = 14$ and $u = 585$, if $180 \le N < 250$; and $\ell = 15$ and $u = 584$, if $N \ge 250$.

An alternative approach is to use the HC4 method for estimating an approximation of the squared standard error of r_1 and r_2, say V_1 and V_2, respectively. Then a test statistic is simply

$$T = \frac{r_1 - r_2}{\sqrt{V_1 + V_2}}$$

with the null distribution taken to be a Student's t distribution, with $n_1 + n_2 - 4$ degrees of freedom. A positive feature is that the actual Type I error rate tends to be lower than the nominal level. Another advantage of the HC4 method is that it is not restricted to testing at the 0.05 level. A disadvantage is that if the null hypothesis is true, the actual level can be substantially lower than the nominal level. As the common correlation increases, this problem is exacerbated. Using a bootstrap-t method to estimate the null distribution of T improves matters substantially. (For results on a permutation method, see Sakaori, 2002.)

9.6.2 Comparing Robust Correlations

When comparing robust correlations, all indications are that a basic percentile bootstrap method performs reasonably well. That is, no modification, as described in the previous section, is necessary. For methods where the goal is to compare measures of association among variables that are possibly dependent, see Sections 11.10.1 and 11.10.2.

9.6.3 R Functions twopcor, tworhobt, and twocor

The R function

$$twopcor(x1, y1, x2, y2, SEED = T)$$

computes a confidence interval for $\rho_1 - \rho_2$ using the modified bootstrap method just described. The R function

$$tworhobt(x1,y1,x2,y2, alpha=0.05, nboot=499)$$

tests the hypothesis of equal Pearson correlations using a bootstrap-t method in conjunction with the HC4 method.

The R function

$$twocor(x1, y1, x2, y2, corfun = pbcor, nboot = 599, tr= 0.2, SEED = TRUE, ...)$$

tests the hypothesis that two robust correlation coefficients are equal. The choice of correlation is indicated by the argument corfun, which defaults to the percentage bend correlation. The function returns a $1 - \alpha$ confidence interval and a p-value.

9.7 Exercises

1. Generate 20 observations from a standard normal distribution, and store them in the R variable ep. Repeat this, and store the values in x. Compute y=x+ep, and compute Kendall's tau. Generally, what happens if two pairs of points are added at $(2.1, -2.4)$? Does this have a large impact on tau? What would you expect to happen to the p-value when testing $H_0: \tau = 0$?
2. Repeat Exercise 1 with Spearman's rho, the percentage bend correlation, and the Winsorized correlation.
3. Demonstrate that heteroscedasticity affects the probability of a Type I error when testing the hypothesis of a zero correlation based on any type M correlation and non-bootstrap method covered in this chapter.
4. Use the function cov.mve(m,cor=TRUE) to compute the MVE correlation for the star data in Fig. 9.2. Compare the results to the Winsorized, percentage bend, skipped, and biweight correlations, as well as the M-estimate of correlation returned by the R function relfun.

5. Using the group 1 alcohol data in Section 8.6.2, compute the MVE estimate of correlation, and compare the results to the biweight midcorrelation, the percentage bend correlation using $\beta = 0.1, 0.2, 0.3, 0.4$, and 0.5, the Winsorized correlation using $\gamma = 0.1$ and 0.2, and the skipped correlation.

6. Repeat the previous problem using the data for group 2.

7. The method for detecting outliers, described in Section 6.4.3, could be modified by replacing the MVE estimator with the Winsorized mean and covariance matrix. Discuss how this would be done and its relative merits.

8. Use the data in the file read.dat and test for independence using the data in columns 2, 3, and 10 and the R function pball. Try $\beta = 0.1, 0.3$, and 0.5. Comment on any discrepancies.

9. Examine the variables in the last exercise using the R function mscor.

10. For the data used in the last two exercises, test the hypothesis of independence using the function indt. Why might indt find an association not detected by any of the correlations covered in this chapter?

11. For the data in the file read.dat, test for independence using the data in columns 4 and 5 and $\beta = 0.1$.

12. The definition of the percentage bend correlation coefficient, ρ_{pb}, involves a measure of scale, ω_x, that is estimated with $\hat{\omega} = W_{(m)}$, where $W_i = |X_i - M_x|$ and $m = [(1 - \beta)n]$, where $0 \le \beta \le 0.5$. Note that this measure of scale is defined even when $0.5 < \beta < 1$, provided that $m > 0$. Argue that the finite sample breakdown point of this estimator is maximized when $\beta = 0.5$.

13. If in the definition of the biweight midcovariance, the median is replaced by the biweight measure of location, the biweight midcovariance is equal to zero under independence. Describe some negative consequences of replacing the median with the biweight measure of location.

14. Let X be a standard normal random variable, and suppose Y is a contaminated normal with probability density function given by Eq. (1.1). Let $Q = \rho X + \sqrt{1 - \rho^2} Y$, $-1 \le \rho \le 1$. Verify that the correlation between X and Q is

$$\frac{\rho}{\sqrt{\rho^2 + (1 - \rho^2)(1 - \epsilon + \epsilon K^2)}}.$$

Examine how the correlation changes as K gets large with $\epsilon = 0.1$. What does this illustrate about the robustness of ρ?

## CHAPTER 10

Robust Regression

Suppose $(y_i, x_{i1}, \ldots, x_{ip})$, $i = 1, \ldots, n$, are n vectors of observations randomly sampled from some $(p + 1)$-variate distribution, where $(x_{i1}, \ldots, x_{ip})$ is a vector of predictor values. When dealing with regression, the remaining three chapters write random variables as lower case Roman letters. In some situations the predictors are fixed, known constants, but this distinction is not particularly relevant for most of the results reported here. (Exceptions are noted when necessary.) A general goal is understanding how y is related to the p predictors, which includes finding a method of estimating a conditional measure of location associated with y given $(x_{i1}, \ldots, x_{ip})$, and there is now a vast arsenal of regression methods that might be used. Even when attention is restricted to robust methods, all relevant techniques would easily take up an entire book. In order to reduce the number of techniques to a reasonable size, attention is focused on estimation and hypothesis testing methods that perform reasonably well in simulation studies, in terms of efficiency and probability coverage, particularly when there is a heteroscedastic error term. For more information about robust regression, see Belsley et al. (1980), Birkes and Dodge (1993), Carroll and Ruppert (1988), Cook and Weisberg (1992), Fox (1999), Hampel et al. (1986), Hettmansperger (1984), Hettmansperger and McKean (1998), Huber (1981), Li (1985), Montgomery et al. (2012), Rousseeuw and Leroy (1987), Staudte and Sheather (1990), and Maronna et al. (2006). For methods dealing with functional data, see Gervini and Yohai (2002).

Robust methods that have practical value when the error term is homoscedastic, but are unsatisfactory when the error term is heteroscedastic, are not described or only briefly mentioned. A natural suggestion for trying to salvage homoscedastic methods is to test the hypothesis that there is, indeed, homoscedasticity. But this approach has been found to be unsatisfactory for reasons reviewed in Section 10.1.5.

In regression, the most common assumption is that

$$y_i = \beta_0 + \beta_1 x_{i1} + \cdots + \beta_p x_{ip} + \epsilon_i, \tag{10.1}$$

where $\beta_0, \ldots, \beta_p$ are unknown parameters, $i = 1, \ldots, n$, and ϵ_i are independent random variables with $E(\epsilon_i) = 0$, $\text{VAR}(\epsilon_i) = \sigma^2$. This model implies that the conditional mean of y_i, given $(x_{i1}, \ldots, x_{ip})$, is $\beta_0 + \sum \beta_k x_{ik}$, a linear combination of the predictors. Eq. (10.1) is a *homoscedastic* model, meaning that ϵ_i has a common variance. If the error term, ϵ_i, has variance σ_i^2 and $\sigma_i^2 \neq \sigma_j^2$, for some $i \neq j$, the model is said to be *heteroscedastic*. Even when ϵ has a normal distribution, heteroscedasticity can result in relatively low efficiency when using the conventional (ordinary least squares [OLS]) estimator, meaning that the estimator of β can have a relatively large standard error. (For results on dealing with a non-normal error term, assuming homoscedasticity, see Ivokić et al., 2020.) Also, probability coverage can be poor when computing confidence intervals, as will be illustrated. Dealing with these two problems is one of the major goals in this chapter. Other general goals are dealing with outliers and achieving high efficiency when sampling from heavy-tailed distributions.

A point worth stressing is that the blind acceptance of the linear model given by Eq. (10.1) should be avoided (e.g., Breiman, 2001b). Also see Raper (2020). Smoothers, discussed in Chapter 11, provide a useful perspective on this issue. But for the moment, concerns about using a linear model are ignored.

Before continuing, some comments about notation might be useful. At times, standard vector and matrix notation will be used. In particular, let

$$\mathbf{x}_i = (x_{i1}, \ldots, x_{ip})$$

and

$$\beta = (\beta_1, \ldots, \beta_p)'$$
$$= \begin{pmatrix} \beta_1 \\ \vdots \\ \beta_p \end{pmatrix}.$$

Then

$$\mathbf{x}_i \beta = \beta_1 x_{i1} + \cdots + \beta_p x_{ip}$$

and Eq. (10.1) becomes

$$y_i = \beta_0 + \mathbf{x}_i \beta + \epsilon_i.$$

This chapter begins with a summary of practical problems associated with least squares regression. Then various robust estimators are described and some comments are made about their relative merits. Inferential techniques, based on the robust regression estimators introduced in this chapter, are described and illustrated in Chapter 11.

10.1 Problems With Ordinary Least Squares

Let b_j be any estimate of β_j, $j = 0, 1, \ldots, p$, and let

$$\hat{y}_i = b_0 + b_1 x_{i1} + \cdots + b_p x_{ip}.$$

From basic principles, the OLS estimator arises without making any distributional assumptions. The estimates are the b_j values ($j = 0, \ldots, p$) that minimize

$$\sum (y_i - \hat{y}_i)^2,$$

the sum of the squared residuals. In order to test hypotheses or compute confidence intervals, typically the homoscedastic model given by Eq. (10.1) is assumed with the additional assumption that ϵ has a normal distribution with mean zero. Even when ϵ is normal, but heteroscedastic, problems with computing confidence intervals arise.

Consider, for example, simple regression where there is only one predictor ($p = 1$). Then the OLS estimate of the slope is

$$\hat{\beta}_1 = \frac{\sum (x_{i1} - \bar{x}_1)(y_i - \bar{y})}{\sum (x_{i1} - \bar{x}_1)^2},$$

where $\bar{x}_1 = \sum x_{i1}/n$ and $\bar{y} = \sum y_i/n$, and the OLS estimate of β_0 is

$$\hat{\beta}_0 = \bar{y} - \hat{\beta}_1 \bar{x}_1.$$

Suppose

$$y_i = \beta_0 + \beta_1 x_{i1} + \lambda(x_{i1})\epsilon_i, \tag{10.2}$$

where $\mathrm{VAR}(\epsilon_i) = \sigma^2$ and λ is some unknown function of x_{i1} used to model heteroscedasticity. That is, the error term is now $\lambda(x_{i1})\epsilon_i$, and its variance varies with x_{i1}, so it is heteroscedastic. The usual homoscedastic model corresponds to $\lambda(x_{i1}) \equiv 1$. To illustrate the effect of heterogeneity when computing confidence intervals, suppose both x_{i1} and ϵ_i have standard normal distributions, $\beta_1 = 1$, $\beta_0 = 0$, and $\lambda(x_{i1}) = \sqrt{|x_{i1}|}$. Thus, the error term, $\lambda(x_{i1})\epsilon_i$, has a relatively small variance when x_{i1} is close to zero, and the variance increases as x_{i1} moves away from zero. The standard $1 - \alpha$ confidence interval for the slope, β_1, is

$$\hat{\beta}_1 \pm t_{1-\alpha/2}\sqrt{\frac{\hat{\sigma}^2}{\sum (x_{i1} - \bar{x}_1)^2}},$$

where $t_{1-\alpha/2}$ is the $1 - \alpha/2$ quantile of a Student's t distribution with $n - 2$ degrees of freedom,

$$\hat{\sigma}^2 = \sum \frac{r_i^2}{n - 2},$$

and $r_i = y_i - \hat{\beta}_1 x_{i1} - \hat{\beta}_0$ are the residuals. When the error term is homoscedastic and normal, the probability coverage is exactly $1 - \alpha$.

Let $\hat{\alpha}$ be a simulation estimate of one minus the probability coverage when computing a $1 - \alpha$ confidence interval for β_1. When using the conventional confidence interval for β_1, with $n = 20$, $\lambda(x_{i1}) = \sqrt{|x_{i1}|}$, and $\alpha = 0.05$, $\hat{\alpha} = 0.135$ based on a simulation with 1,000 replications. If $\lambda(x_{i1}) = |x_{i1}|$, the estimate increases to 0.214. If instead x_{i1} has a g-and-h distribution (described in Section 4.2) with $g = 0$ and $h = 0.5$ (a symmetric, heavy-tailed distribution), $\hat{\alpha}$ increases to 0.52. Put another way, when testing $H_0: \beta_1 = 1$ with $\alpha = 0.05$, the actual probability of a Type I error can be more than 10 times the nominal level. A similar problem arises when testing $H_0: \beta_0 = 0$. A natural strategy is to test the assumptions of normality and homoscedasticity, but as already noted, such tests might not have enough power to detect a situation where these assumptions yield poor probability coverage. (Section 10.1.5 provides more details about testing the homoscedasticity assumption.)

Another problem with OLS is that it can be highly inefficient, and this can result in relatively low power. As will be illustrated, this problem arises even when ϵ_i is normal but $\lambda(x_{i1})$ is not equal to one. That is, the error term has a normal distribution but is heteroscedastic. Low efficiency also arises when the error term is homoscedastic but has a heavy-tailed distribution. As an illustration, again consider simple regression. When the error term is homoscedastic, $\hat{\beta}_1$, the OLS estimate of β_1, has variance

$$\frac{\sigma^2}{\sum(x_{i1} - \bar{x}_1)^2}.$$

But from results described and illustrated in Chapters 1, 2, and 3, σ^2, the variance of the error term, becomes inflated if sampling is from a heavy-tailed distribution. That is, slight departures from normality, as measured by the Kolmogorov distance function, result in large increases in the standard error of $\hat{\beta}_1$. (This problem is well known and discussed by Hampel, 1973; Schrader and Hettmansperger, 1980; He et al., 1990, among others.) Note, however, that if x_{i1} is sampled from a heavy-tailed distribution, this inflates the expected value of $\sum(x_{i1} - \bar{x}_1)^2$ relative to sampling from a normal distribution, in which case the standard error of $\hat{\beta}_1$ tends to be smaller versus the situation where x_{i1} is normal. In fact, a single outlier among the x_{i1} values inflates $\sum(x_{i1} - \bar{x}_1)^2$, causing the estimate of the squared standard error to decrease. Consequently, there is interest in searching for methods that have good efficiency when ϵ has a heavy-tailed distribution and maintains relatively high efficiency when the x_i values are randomly sampled from a heavy-tailed distribution as well.

One strategy is to check for outliers, remove any that are found, and proceed with standard OLS methods using the data that remain. (For recent results on detecting influential points, when using OLS, see Roberts et al., 2015.) Removing points associated with outliers among the independent variable is innocuous when testing hypotheses. But removing outliers among the dependent variable and using standard methods for testing hypotheses result in inaccurate estimates of the standard errors. A bootstrap estimate of the standard error could be used instead, but little is known about the effectiveness of this approach. Also, simply removing outliers does not always deal effectively with low efficiency due to a heteroscedastic error term.

Yet another practical problem is that OLS has a breakdown point of only $1/n$. That is, a single point, properly placed, can cause the OLS estimator to have virtually any value. Not only do unusual y values cause problems, outlying x values, called *leverage points*, can have an inordinate influence on the estimated slopes and intercepts.

It should be noted that two types of leverage points play a role in regression: good and bad. Roughly, leverage points are good or bad depending on whether they are reasonably consistent with the true regression line. A *regression outlier* is a point with a relatively large residual. A *bad leverage point* is a leverage point that is also a regression outlier. A *good leverage point* is a leverage point that is not a regression outlier. Leverage points can reduce the standard error of the OLS estimator, but a bad leverage point can result in a poor fit to the bulk of the data.

Despite the many concerns associated with OLS, it is not being suggested that it be completely abandoned. Removing bad leverage points might result in a satisfactory fit to the bulk of the points. Also, in some situations there might be interest in the conditional mean of y given $\mathbf{x}$, even though the population mean is not robust.

10.1.1 Computing Confidence Intervals Under Heteroscedasticity

When using the OLS estimator, various methods have been proposed for computing confidence intervals for regression parameters when the error term is heteroscedastic. One strategy when dealing with heteroscedasticity is to transform the data (e.g., Carroll and Ruppert, 1988). The focus here is on methods that appear to perform well without relying on any transformation. Perhaps situations arise where transformations have practical value relative to the methods described here, but it seems that this issue has not been investigated. The methods described here compete well with homoscedastic methods when indeed the error term is homoscedastic. Attention is restricted to the seemingly better methods followed by comments regarding their relative merits.

Wilcox (1996c) found that for the special case $p = 1$ (one predictor only), only one method performed well among the situations he considered. For $p > 1$ a slight modification is recommended (Wilcox, 2003f) when the goal is to have simultaneous probability coverage equal to $1 - \alpha$ for all p slope parameters. The method begins by sampling, with replacement, n vectors of observations from $(y_i, \mathbf{x}_i)$, $i = 1, \ldots, n$. Put another way, a bootstrap sample is obtained by randomly sampling, with replacement, n rows of data from the n-by-$(p + 1)$ matrix

$$\begin{pmatrix} y_1, x_{11}, \ldots, x_{1p} \\ y_2, x_{21}, \ldots, x_{2p} \\ \vdots \\ y_n, x_{n1}, \ldots, x_{np} \end{pmatrix}.$$

The resulting n vectors of observations are labeled

$$(y_1^*, x_{11}^*, \ldots, x_{1p}^*), \ldots, (y_n^*, x_{n1}^*, \ldots, x_{np}^*).$$

Let $\hat{\beta}_j^*$ be the OLS estimate of β_j, the jth slope parameter, $j = 1, \ldots, p$, based on the bootstrap sample just obtained. Repeat this bootstrap process B times, yielding $\hat{\beta}_{j1}^*, \hat{\beta}_{j2}^*, \ldots, \hat{\beta}_{jB}^*$. Let $\hat{\beta}_{j(1)}^* \le \hat{\beta}_{j(2)}^* \le \cdots \le \hat{\beta}_{j(B)}^*$ be the B bootstrap estimates written in ascending order.

For the special case $p = 1$, a slight modification of the standard percentile bootstrap method is used. When $B = 599$, the 0.95 confidence interval for β_1 is

$$(\hat{\beta}_{1(a+1)}^*, \hat{\beta}_{1(c)}^*),$$

where for $n < 40$, $a = 6$ and $c = 593$; for $40 \le n < 80$, $a = 7$ and $c = 592$; for $80 \le n < 180$, $a = 10$ and $c = 589$; for $180 \le n < 250$, $a = 13$ and $c = 586$; while for $n \ge 250$, $a = 15$ and $c = 584$. Note that this method becomes the standard percentile bootstrap procedure when $n \ge 250$. If, for example, $n = 20$, the lower end of the 0.95 confidence interval is given by $\hat{\beta}_{1(7)}^*$. A confidence interval for the intercept is computed in the same manner, but currently it seems best to use $a = 15$ and $c = 584$ for any n. That is, use the usual percentile bootstrap confidence interval. From results described in Chapter 4, there are situations where the confidence interval for the intercept can be expected to have unsatisfactory probability coverage. In essence, the situation reduces to computing a percentile bootstrap confidence interval for the mean when the slope parameters are all equal to zero. A criticism of this method is that it is limited to $\alpha = 0.05$.

As for $p > 1$, if the goal is to achieve simultaneous probability coverage equal to $1 - \alpha$, it currently appears that the best approach is to use a standard percentile bootstrap method in conjunction with the Bonferroni inequality (Wilcox, 2003f). So, set

$$\ell = \frac{\alpha B}{2p},$$

round ℓ to the nearest integer, and let $u = B - \ell$, in which case the confidence interval for the jth predictor ($j = 1, \ldots, p$) is

$$(\hat{\beta}_{j(\ell+1)}^*, \hat{\beta}_{j(u)}^*).$$

For $p = 1$, the bootstrap confidence interval just described is based on the strategy of finding a method that gives good results under normality and homoscedasticity, and then using this method when there is heteroscedasticity or sampling is from a non-normal distribution. Simulations were then used to see whether the method continues to perform well when sampling from non-normal distributions, or when there is heteroscedasticity. Relative to other methods that have been proposed, the modified bootstrap procedure has a clear advantage. This result is somewhat unexpected because in general, when working with non-robust measures of location and scale, this strategy performs rather poorly. If, for example, the percentile bootstrap method is adjusted so that the resulting confidence interval for the mean has probability coverage close to the nominal level when sampling from a normal distribution, probability coverage can be poor when sampling from non-normal distributions instead.

Another strategy is to obtain bootstrap samples by resampling residuals, as opposed to vectors of observations as is done here. When dealing with heteroscedasticity, theoretical results do not support this approach (Wu, 1986), and simulations indicate that unsatisfactory probability coverage can result. Of course, one could check for homoscedasticity in an attempt to justify resampling residuals, but there is no known way of being reasonably certain that the error term is sufficiently homoscedastic. Again, any test of the assumption of homoscedasticity might not have enough power to detect heteroscedasticity in situations where the assumption should be discarded.

Nanayakkara and Cressie (1991) derive another method for computing a confidence interval for the regression parameters when the error term is heteroscedastic. When the x_{i1} values are fixed and evenly spaced, their method appears to give good probability coverage, but otherwise probability coverage can be unsatisfactory.

Long and Ervin (2000) compare several non-bootstrap methods for dealing with heteroscedasticity and recommend one particular method for general use, which is based on the HC3 estimate of the standard errors. The HC3 estimator is

$$\text{HC3} = (\mathbf{X}'\mathbf{X})^{-1}\mathbf{X}'\text{diag}\left[\frac{r_i^2}{(1-h_{ii})^2}\right]\mathbf{X}(\mathbf{X}'\mathbf{X})^{-1},$$

where r_i $(i = 1, \ldots, n)$ are the usual residuals,

$$h_{ii} = \mathbf{x}_i(\mathbf{X}'\mathbf{X})^{-1}\mathbf{x}_i',$$

and

$$\mathbf{X} = \begin{pmatrix} 1 & x_{11} \cdots & x_{1p} \\ 1 & x_{21} \cdots & x_{2p} \\ \vdots & \vdots & \vdots \\ 1 & x_{n1} \cdots & x_{np} \end{pmatrix},$$

and $\mathbf{x}_i$ is the ith row of $\mathbf{X}$ (e.g., MacKinnon and White, 1985). If $b_0, \ldots, b_p$ are the least squares estimates of the $p + 1$ parameters, the diagonal elements of the matrix HC3 represent the estimated squared standard errors. So if S_j^2 $(j = 0, \ldots, p)$ is the jth diagonal element of HC3, the $1 - \alpha$ confidence interval for β_j is taken to be

$$b_j \pm t S_j,$$

where t is the $1 - \alpha/2$ quantile of a Student's t distribution with $v = n - p - 1$ degrees of freedom. But it is unknown how large n must be to ensure reasonably accurate confidence intervals. For a single predictor, it is known that $n = 60$ might not suffice (Wilcox, 2001b). Cribari-Neto and Lima (2014) describe situations where the actual level exceeds the nominal level when using a quasi-t test, even when the error term has a normal distribution.

Godfrey (2006) suggests an alternative to the HC3 estimator, the HC4 estimator. His results suggest that it is better for general use. But when testing hypotheses, a possible concern is that the use of HC4 can result in the actual Type I error probability being substantially less than the nominal level. Ng and Wilcox (2009) describe situations where the reverse problem occurs. Of particular concern are situations where the error term has a skewed, heavy-tailed distribution. Cribari-Neto and Lima (2014) suggest alternatives to the HC3 and HC4 estimators, one of which appears to perform well when testing the hypothesis of a zero slope based on a quasi-t test.

Let h_{ii} be defined as done when using the HC3 estimator. Let $\bar{h} = \sum h_{ii}/n$, $e_{ii} = h_{ii}/\bar{h}$, and $d_{ii} = \min(4, e_{ii})$. The HC4 estimator is

$$S = (\mathbf{X}'\mathbf{X})^{-1}\mathbf{X}'\mathrm{diag}\left[\frac{r_i^2}{(1 - h_{ii})^{d_{ii}}}\right]\mathbf{X}(\mathbf{X}'\mathbf{X})^{-1}.$$

The diagonal elements of matrix $\mathbf{S}$, which we denote by $S_0^2, S_1^2, \ldots, S_p^2$, are the estimated squared standard errors of $b_0, b_1, \ldots b_p$, respectively. Following Ng and Wilcox (2009), the $1 - \alpha$ confidence interval for β_j is taken to be

$$b_j \pm t S_j,$$

where t is the $1 - \alpha/2$ quantile of a Student's t distribution with $v = n - p - 1$ degrees of freedom. (Cribari-Neto et al., 2007, suggest an alternative to the HC4 estimator, but for the situation at hand it seems to offer no practical advantage; see Ng, 2009b.)

Wald-type Statistics Used in Conjunction with a Wild Bootstrap

Method HC4WB-D

Two wild bootstrap methods should be mentioned, both of which are based in part on what is called a Wald-type statistic. The first, which is labeled the HC4WB-D and based in part on the Rademacher function, is performed as follows:

1. Again let b_j be the OLS estimate of β_j, and compute S_j, the HC4 estimate of the standard error.
2. Compute the Wald test statistic

$$W = (b_j - 0)S_j^{-1}(b_j - 0).$$

3. Generate $D_1, \ldots, D_n$ from a two-point (lattice) distribution. That is,

$$D_i = \begin{cases} -1, & \text{with probability 0.5,} \\ 1, & \text{with probability 0.5.} \end{cases}$$

A bootstrap sample $(y_i^*, \mathbf{x}_i)$, assuming the null hypothesis is true, is given by $y_i^* = \bar{y} + D_i r_i, i = 1, \ldots, n$, where $\hat{\beta} = (\hat{\beta}_1, \ldots, \hat{\beta}_p)'$ are the OLS estimates under the assumption that the null hypothesis is true.

4. Compute the OLS estimate (b_j^*) based on this bootstrap sample as well as the HC4 estimate of the standard error (S_j^*). Compute the Wald test statistic

$$W^* = (b_j^* - 0)S^{*-1}(b_j^* - 0)$$

based on the bootstrap sample.

5. Repeat steps 2–4 B times, yielding $W_b^*, b = 1, \ldots, B$.
6. A p-value for H_0: $\beta_j = 0$ is given by

$$p = \frac{\#\{W_b^* \geq W\}}{B}.$$

Reject H_0 if $p \leq \alpha$.

Method HC4WB-C

Method HC4WB-C is exactly like method HC4WB-D, only now

$$D_i = \sqrt{12}(U_i - 0.5),$$

where U has a uniform distribution over the unit interval.

In contrast to Godfrey (2006), the more extensive simulations by Ng and Wilcox (2009) indicate that these wild bootstrap methods do not have a striking advantage over the non-bootstrap

HC4 method in terms of achieving a Type I error probability reasonably close to the nominal level. However, in terms of minimizing the variability of the Type I error probabilities among the situations that were considered, HC4WB-C was found to be best when testing at the 0.05 level. Although the methods based on the HC4 estimator perform relatively well, there are situations where all methods based on the HC4 estimator fail to control the Type I error probability in a reasonably accurate manner, even with $n = 100$. Problems occur when dealing with skewed distributions with very heavy tails and simultaneously there is a seemingly extreme amount of heteroscedasticity.

10.1.2 An Omnibus Test

Rather than test hypotheses about the individual parameters, a common goal is to test

$$H_0: \beta_1 = \cdots = \beta_p = 0,$$

the hypothesis that all p slope parameters are zero. When using the OLS estimator and $p > 1$, it seems that no method has been found to be effective, in terms of controlling the probability of a Type I error, when there is heteroscedasticity, non-normality, or both. Mammen (1993) studies a method based in part on a wild bootstrap technique. The author ran some simulations as a partial check on this approach and found that it generally performed reasonable well with $n = 30$ and $p = 4$. However, a situation was found where, with a heteroscedastic error term, the actual probability of a Type I error was estimated to be 0.29 when testing at the 0.05 level. In fairness, perhaps for nearly all practical situations, the method performs reasonably well, but resolving this issue is difficult at best.

With the understanding that, when dealing with least squares regression, no single method is always satisfactory in terms of controlling the probability of a Type I error, two methods currently seem best for general use. Both are based on a more general form of the HC4 estimate of the standard error, which was derived by Cribari-Neto (2004).

Let $\mathbf{V}$ be the HC4 estimate of the variances and covariances of $\mathbf{b} = (b_1, \ldots, b_p)'$, the least squares estimate of the slope parameters. A test statistic for testing the hypothesis that all slopes are zero is

$$W = n\mathbf{b}'\mathbf{V}\mathbf{b},$$

which has, approximately, a chi-squared distribution with p degrees of freedom. However, for $p > 1$, this method is unsatisfactory in terms of controlling the probability of a Type I error.

An alternative approach is to use a wild bootstrap method. That is, generate wild bootstrap values y_i^*, as is done in conjunction with the HC4WB-C and HC4WB-D methods. Based on this bootstrap sample, compute the test statistic W, yielding W^*. Repeat this B times, yielding $W_1^*, \ldots, W_B^*$. A p-value is given by

$$\frac{1}{B} \sum I_i,$$

where the indicator function $I_i = 1$ if $W_i \leq W$; otherwise $I_i = 0$.

It is noted that when dealing with a single independent variable, a homoscedastic confidence band for the regression line can be computed with the built-in R function ols.pred.ci. For a heteroscedastic method that can be used with both OLS and robust regression estimators, see the R functions regYci and regYband in Section 11.1.13.

10.1.3 R Functions lsfitci, olshc4, hc4test, and hc4wtest

The R function

lsfitci(x,y,nboot = 599, tr=0.2, SEED=TRUE, xout = FALSE, outfun = out)

is supplied for computing 0.95 confidence intervals for regression parameters, based on the OLS estimator, using the percentile bootstrap method described in Section 10.1.1. As usual, x is an n-by-p matrix of predictors. In contrast to the other R functions in this section, this function is designed for $\alpha = 0.05$ only. Setting the argument xout=TRUE, leverage points are identified with the method indicated by the argument outfun and then they are removed.

The R function

olshc4(x,y,alpha=0.05,xout=FALSE,outfun=out,HC3=FALSE)

computes $1 - \alpha$ confidence intervals for each of the $p + 1$ parameters using the HC4 estimator, and p-values are returned as well. By default, 0.95 confidence intervals are returned. Setting the argument alpha equal to 0.1, for example, will result in 0.9 confidence intervals. Leverage points are removed if the argument xout=TRUE using the R function specified by the argument outfun, which defaults to the projection method in Section 6.4.9. Setting HC3=TRUE results in using the HC3 estimator rather than HC4.

The function

hc4test(x, y, pval = c(1:ncol(x)), xout = FALSE, outfun = outpro, pr = TRUE, plotit = FALSE, xlab = 'X', ylab = 'Y', ...)

tests the hypothesis that all slope parameters are equal to zero. The argument pval controls which independent variables will be included in the model. By default, all are included. With a sufficiently large sample size, this method will perform well in terms of controlling the probability of a Type I error. But it is unclear just how large the sample size needs to be. With a small to moderate sample size all indications are that it is safer to use the R function

hc4wtest(x, y, nboot = 500, SEED=TRUE, RAD = TRUE, xout = FALSE, outfun = outpro, ...),

which uses a wild bootstrap method. When the argument RAD=TRUE, method HC4WB-D is used. Otherwise method HC4WB-C is used.

■ Example

Assuming both x and ϵ have standard normal distributions, 30 pairs of observations were generated according to the model $y = (|x|+1)\epsilon$. The standard F test for H_0: $\beta_1 = 0$ was applied, and this process was repeated 1,000 times. Testing at the 0.05 level, the proportion of Type I errors was 0.144. So the standard F test correctly detects an association about 14% of the time, but simultaneously provides an inaccurate assessment of β_1. This again illustrates that under heteroscedasticity, the standard F test does not control the probability of a Type I error. Using instead the R function olshc4, the proportion of rejections was 0.6, which is reasonably close to the nominal 0.05 level.

■

■ Example

Cohen et al. (1993) report data on the number of hours, y, needed to splice x pairs of wires for a particular type of telephone cable. The left panel of Fig. 10.1 shows a scatterplot of the data. (The data are stored in the file splice.dat, which can be obtained as described in Section 1.8 of Chapter 1.) Note that the data appear to be heteroscedastic. The usual 0.95 confidence interval (multiplied by 1,000 for convenience), based on the assumption of normality and homoscedasticity, is (1.68, 2.44). If the y values are stored in the R vector yvec and the x values are stored in the R variable splice, the command lsfitci(splice,yvec) reports that the 0.95 bootstrap confidence interval is (1.64, 2.57).

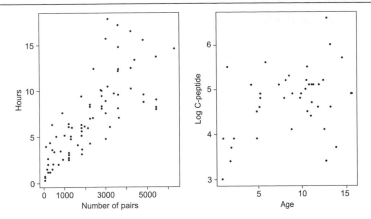

Figure 10.1: Scatterplots of two real data sets that appear to be heteroscedastic.

The ratio of the lengths is $(2.44 - 1.68)/(2.57 - 1.64) = 0.82$. It might be argued that the lengths are reasonably similar. However, the probability coverage of the usual method can be less than the nominal level; it is unclear whether this problem can be ignored for the data being examined, and all indications are that the bootstrap method provides better probability coverage under heteroscedasticity. Consequently, using the bootstrap confidence interval seems more satisfactory.

■ **Example**

Sockett et al. (1987) collected data with the goal of understanding how various factors are related to the patterns of residual insulin secretion in children. (The data can be found in the file diabetes.dat.) One of the response measurements is the logarithm of C-peptide concentration (pmol/ml) at diagnosis, and one of the predictors considered is age. The right panel of Fig. 10.1 shows a scatterplot of the data. The scatterplot suggests that the error term is heteroscedastic, with the smallest variance near age 7. (Results in Chapter 11 lend support for this speculation.) The 0.95 confidence interval for the slope, using the standard OLS method, is $(0.0042, 0.263)$, the estimate of the slope being 0.15. In contrast, lsfitci returns a 0.95 confidence interval of $(-0.00098, 0.029)$, and the ratio of the lengths is $(0.0263 - 0.0042)/(0.029 + 0.00098) = 0.74$. Again there is concern that the standard confidence interval is too short and that its actual probability coverage is less than the nominal level. Note that the standard confidence interval rejects $H_0\colon \beta_1 = 0$, but lsfitci does not.

10.1.4 Comments on Comparing Means via Dummy Coding

A well-known approach to comparing the means of multiple groups is via least squares regression coupled with dummy coding (e.g., Montgomery et al., 2012). Yet another way of dealing with heteroscedasticity when comparing means, beyond the methods in Chapter 7, is to use dummy coding in conjunction with the HC4 estimator. But results in Ng (2009b) do not support this approach. Control over the probability of a Type I error can be unsatisfactory.

10.1.5 Salvaging the Homoscedasticity Assumption

One strategy for trying to salvage the homoscedasticity assumption, when using classic inferential methods associated with the least squares regression estimator, is to simply test the hypothesis that there is homoscedasticity. Most methods for testing this hypothesis have been found to be unsatisfactory in terms of controlling the probability of a Type I error (Lyon and Tsai, 1996). That is, there is a high probability of rejecting the hypothesis that there is homoscedasticity when indeed there is homoscedasticity. Two methods that have been found to perform well in simulations are described in Section 11.4. However, imagine that the classic homoscedastic methods are used if these methods fail to reject the hypothesis that the error term is homoscedastic. A basic issue is whether these methods have enough power to detect situations where there is heteroscedasticity that invalidates the homoscedastic methods that are typically used. With $n \leq 100$, Ng and Wilcox (2011) found that the answer is no. The usual homoscedastic methods can still have actual Type I error probabilities well above the nominal level. Presumably with a large enough sample size, methods for testing the homoscedasticity assumption will have adequate power, but just how large the sample must be is unknown. To the extent it is desired to make inferences about the regression parameters, without being sensitive to heteroscedasticity, all indications are that it is best to abandon homoscedastic methods and always use a heteroscedastic method, which are described in Chapters 11 and 12.

10.2 The Theil–Sen Estimator

This section describes an estimator first proposed by Theil (1950) and later extended by Sen (1968) that is restricted to the case of a single predictor ($p = 1$). Then various extensions to $p > 1$ are discussed.

Temporarily focusing on $p = 1$, one view of the Theil–Sen estimator is that it attempts to find a value for the slope that makes Kendall's correlation tau, between $y_i - b_1 x_i$ and x_i, (approximately) equal to zero. This can be seen to be tantamount to the following method. For any

$i < i'$, for which $x_i \neq x_{i'}$, let

$$S_{ii'} = \frac{y_i - y_{i'}}{x_i - x_{i'}}.$$

The Theil–Sen estimate of the slope is b_{1ts}, the median of all the slopes represented by $S_{ii'}$. Two strategies for estimating the intercept have been proposed. The first is

$$M_y - b_1 M_x,$$

where M_y and M_x are the usual sample medians of the y and x values, respectively. The second estimates the intercept with the median of $y_1 - b_1 x_1, \ldots, y_n - b_1 x_n$. This latter method is used unless stated otherwise.

Sen (1968) derives the asymptotic standard error of the slope estimator, but it plays no role here and is therefore is not reported. Dietz (1987) shows that the Theil–Sen estimator has an asymptotic breakdown point of 0.293. For results on its small-sample efficiency, see Dietz (1989), Talwar (1991), and Wilcox (1998a, 1998b). Sievers (1978) and Scholz (1978) propose a generalization of the Theil–Sen estimator by attaching weights to the pairwise slopes, $S_{ii'}$, but in terms of efficiency it seems to offer little or no advantage, and in some cases its bias is considerably larger, so it is not described here. For variations of the Theil–Sen estimator, see Luh and Guo (2000) as well as Moses and Klockars (2012). Peng ct al. (2008) describe situations where the Theil–Sen estimator is not asymptotically normal and they demonstrate that it can be superefficient when the error term is discontinuous.

It is noted that the Theil–Sen estimator has close similarities to a regression estimator studied by Maronna and Yohai (1993, Section 3.3). This alternative estimator belongs to what are called projection estimators; see Section 10.13.11. Martin et al. (1989) prove that this alternative estimate of the slope is minimax in the class of all regression-equivariant estimates if the error term has a symmetric and unimodal distribution.

A possible concern about the Theil–Sen estimator is that tied values among the dependent variable might negatively impact its efficiency. That is, the magnitude of the standard error of the Theil–Sen estimator might be greater than some alternative estimator that might be used, which in turn might result in relatively low power. A way of improving the efficiency of the Theil–Sen estimator is to replace the usual sample median with the Harrell–Davis estimator (Wilcox and Clark, 2013). So now the slope is estimated via the Harrell–Davis estimator applied to the $S_{ii'}$ values and the intercept is taken to be the median of $y_1 - b_1 x_1, \ldots, y_n - b_1 x_n$ based on the Harrell–Davis estimator. An alternative strategy is to estimate the intercept with

$$\hat{\theta}_y - b_1 \hat{\theta}_x,$$

where $\hat{\theta}_y$ is the Harrell–Davis estimate based on $y_1, \ldots, y_n$ and $\hat{\theta}_x$ is the Harrell–Davis estimate based on $x_1, \ldots, x_n$. But to be consistent with how $p > 1$ independent variables are handled, the former method is used unless stated otherwise.

There are at least three general ways the Theil–Sen estimator might be extended to two or more predictors (cf. Hussain and Sprent, 1983). The first, which will be called method *TS*, is to apply the back-fitting, Gauss–Seidel method as described in Hastie and Tibshirani (1990, pp. 106–108). (For a general description of the Gauss–Seidel method and its properties, see, for example, Dahlquist and Björck, 1974; Golub and van Loan, 1983.) For the situation at hand, the method is applied as follows:

1. Set $k = 0$ and choose an initial estimate for β_j, say $b_j^{(0)}$, $j = 0, \ldots, p$. Here, the initial estimate is taken to be the Theil–Sen estimate of the slope based on the jth regressor only. That is, simply ignore the other predictors to obtain an initial estimate of β_j. The initial estimate of β_0 is the median of

$$y_i - \sum_{j=1}^{p} b_j^{(0)} x_{ij}, \ i = 1, \ldots, n.$$

2. Increment k and take the kth estimate of β_j ($j = 1, \ldots, p$) to be $b_j^{(k)}$, the Theil–Sen estimate of the slope based on the regression estimate of x_{ij} with

$$r_i = y_i - b_0^{(k-1)} - \sum_{\ell=1,\ell \neq j}^{p} b_\ell^{(k-1)} x_{i\ell}.$$

The updated estimate of the intercept, $b_0^{(k)}$, is the median of

$$y_i - \sum_{j=1}^{p} b_j^{(k)} x_{ij}, \ i = 1, \ldots, n.$$

3. Repeat step 2 until convergence.

The second general approach toward extending the Theil–Sen estimator to multiple predictors is based on so-called *elemental subsets*, an idea that appears to have been first suggested in unpublished work by Oja and Niimimaa; see Rousseeuw and Leroy (1987, p. 146, cf. Hawkins and Olive, 2002). In a regression data set, an elemental subset consists of the minimum number of cases required to estimate the unknown parameters of a regression model. With p regressors, the Oja and Niimimaa extension is to use all $N = n!/((p+1)!(n-p-1)!)$ elemental subsets. That is, for each elemental subset, estimate the slope parameters using OLS. At this point, a simple strategy is to use the median of the N resulting estimates. That is, letting $\hat{\beta}_{ji}$ be the estimate of β_j based on the ith elemental subset, $i = 1, \ldots, N$, the final estimate of β_j would be the median of these N values. One practical problem is that the number of elemental subsets increases rapidly with n and p. For example, with $n = 100$ and $p = 4$, the number of elemental subsets is 106,657,320. Also note that intuitively, some elemental subsets will yield a highly inaccurate estimate of the slopes. An alternative strategy is to use $(n^2 - n)/2$ randomly sampled elemental subsets, the same number of elemental subsets used when $p = 1$. In terms of efficiency, results in Wilcox (1998b) support this approach over using all N elemental subsets instead. For convenience, this method is labeled *TSG*. Note that the back-fitting, Gauss–Seidel method eliminates the random component associated with the method just described, and Gauss–Seidel also offers faster execution time.

Let $\hat{\tau}_j$ be Kendall's tau between the jth predictor, x_j, and $y - b_1 x_1 - \cdots - b_p x_p$. A third approach, when generalizing the Theil–Sen estimator to $p > 1$ predictors, is to determine $b_1, \ldots, b_p$ so that $\sum |\hat{\tau}_j|$ is approximately equal to zero. Note that this approach can be used to generalize the Theil–Sen estimator by replacing Kendall's tau with any reasonable correlation coefficient. The relative merits of this third way of extending the Theil–Sen estimator to multiple predictors have not been explored.

Results regarding the small-sample efficiency of the Gauss–Seidel method versus using randomly sampled elemental subsets are reported in Wilcox (2004c). The choice of method can make a practical difference, but currently there is no compelling reason to prefer one method over the other based solely on efficiency.

A criticism of method TSG is that as p increases, its finite sample breakdown point decreases (Rousseeuw and Leroy, 1987, p. 148). Another possible concern is that the marginal medians are location equivariant but not affine equivariant. (See Eq. (6.9) for a definition of affine equivariance when referring to a multivariate location estimator.) A regression estimator T is *affine equivariant* if for any non-singular matrix $\mathbf{A}$,

$$T(\mathbf{x_i A}, \ y_i; i = 1, \ldots, n) = \mathbf{A}^{-1} T(\mathbf{x_i}, \ y_i; i = 1, \ldots, n).$$

Because the marginal medians are not affine equivariant, TSG is not affine equivariant either. Yet one more criticism is that with only $(n^2 - n)/2$ randomly sampled elemental subsets, if n is small, rather unstable results can be obtained, meaning that if a different set of $(n^2 - n)/2$ elemental subsets is used, the estimates can change substantially. If, for example, $n = 20$ and $p = 2$, only 190 resamples are used from among the 1,140 elemental subsets. (Of course, when n is small, it is a simple matter to increase the number of sampled elemental subsets, but just how many additional samples should be taken has not been investigated.)

A regression estimator T is *regression-equivariant* if for any vector $\mathbf{v}$,

$$T(\mathbf{x_i}, \ y_i + \mathbf{x_i v}; i = 1, \ldots, n) = \mathbf{T}(\mathbf{x_i}, \ y_i; i = 1, \ldots, n) + \mathbf{v}.$$

And T is said to be *scale-equivariant* if

$$T(\mathbf{x_i}, cy_i; i = 1, \ldots, n) = \mathbf{c}T(\mathbf{x_i}, y_i; i = 1, \ldots, n).$$

It is noted that method TS also fails to achieve affine equivariance, but it does achieve regression equivariance and scale equivariance.

10.2.1 R Functions tsreg, tshdreg, correg, regplot, and regp2plot

The R function

$$\text{tsreg(x,y,xout=FALSE,outfun=outpro,iter=1,varfun=pbvar,}$$
$$\text{corfun=pbcor,plotit=FALSE,WARN=TRUE, OPT=FALSE, xlab='X',ylab='Y',...)}$$

computes the Theil–Sen regression estimator just described. When OPT=FALSE, the intercept is estimated with the median of $y_1 - b_1 x_1, \ldots, y_n - b_1 x_n$. With OPT=TRUE, the estimate is taken to be $M_y - b_1 M_x$ when there is a single independent variable. When $p > 1$, the default maximum number of iterations when using the Gauss–Seidel method, indicated by the argument iter, is 1. As usual, setting the argument xout=TRUE will eliminate leverage points using the outlier detection method indicated by the argument outfun. If there is a single covariate, plotit=TRUE will create a scatterplot as well as a plot of the estimated regression line.

The R function

$$\text{tshdreg(x,y,HD=TRUE, xout=FALSE, outfun=out, iter=1, varfun=pbvar, corfun=pbcor,}$$
$$\text{plotit=FALSE, tol=0.0001, RES=FALSE, xlab='X', ylab='Y',...)}$$

applies the Theil–Sen estimator but with the usual sample median replaced by the Harrell–Davis estimator.

The function

$$\text{correg(x,y,corfun=tau)}$$

computes the Theil–Sen estimate by determining $b_1, \ldots, b_p$ so that $\sum |\hat{\tau}_j|$ is approximately equal to zero, where $\hat{\tau}_j$ is Kendall's tau between the jth predictor, x_j, and $y - b_1 x_1 - \cdots - b_p x_p$. It generalizes the function tsreg by allowing Kendall's tau to be replaced by some other correlation. For example, correg(x,y,corfun=pbcor) would use the percentage bend correlation coefficient.

When dealing with a single predictor, only two R commands are needed to create a scatterplot that includes a regression line. (The two R functions are plot and abline.) To make this task even easier, the R function

regplot(x,y,regfun=tsreg,xlab='X',ylab='Y', xout=FALSE, outfun=out, theta=50,
phi=25,ticktype='simple',...)

can be used to plot a regression line when there is a single independent variable, and a regression plane when dealing with two independent variables. With two independent variables, the arguments theta and phi can be used to rotate or tilt the plot. Setting ticktype='detail', values along the three axes will be included in the plot. By default the Theil–Sen estimator is used. Other regression estimators can be specified via the argument regfun, which assumes that the estimated slope and intercept, returned by the R function specified by the argument regfun, are stored in $coef.

When there are two independent variables, the R function

regp2plot(x,y, xout=FALSE, outfun=out, xlab='Var 1',ylab='Var 2',zlab='Var 3',
regfun=tsreg, COLOR=FALSE, tick.marks=TRUE,...)

is another way of plotting the regression surface. Setting the argument COLOR=TRUE, the regression plane is indicated by the color blue. (Yet one more option is the R function out3d in Section 6.4.10.)

10.3 Least Median of Squares

The *least median of squares* (LMS) regression estimator appears to have been first proposed by Hampel (1975) and further developed by Rousseeuw (1984). That is, the regression estimates are taken to be the values that minimize

$$\text{MED}(r_1^2, \ldots, r_n^2),$$

the median of the squared residuals. (Also see Davies, 1993; Hawkins and Simonoff, 1993; Rousseeuw and Leroy, 1987; Bertsimas and Mazumder, 2014.) It was the first equivariant estimator to attain a breakdown point of approximately 0.5, but its efficiency relative to the OLS estimator is 0. (And its rate of convergence is $n^{-1/3}$ rather than the usual rate of $n^{-1/2}$.) Despite these negative properties, the LMS estimator is often suggested as a diagnostic tool or a preliminary fit to data.

10.3.1 R Function lmsreg

The built-in R function

lmsreg(x,y)

computes the LMS regression estimator.

10.4 Least Trimmed Squares Estimator

Rousseeuw's (1984) *least trimmed squares* (LTS) estimator is based on minimizing

$$\sum_{i=1}^{h} r_{(i)}^2,$$

where $r_{(1)}^2 \leq \cdots \leq r_{(h)}^2$ are the squared residuals written in ascending order. (For recent results on computing estimates of the parameters, see Mount et al., 2016.) With $h = [n/2] + 1$, the same breakdown point as LMS is achieved. However, $h = [n/2] + [(p+1)/2]$ is often used to maintain regression equivariance. LTS has a relatively low asymptotic efficiency (Croux et al., 1994), but it seems to have practical value. For example, it plays a role in the asymptotically efficient M-estimator described in Section 10.9.

10.4.1 R Function ltsreg

The R function

$$\text{ltsreg(x,y,xout=FALSE,outfun=outpro,tr} = 0.2, \text{plotit} = \text{FALSE, ...)}$$

computes the LTS regression estimate. The amount of trimming is controlled by the argument tr.

10.5 Least Trimmed Absolute Value Estimator

A close variation of the LTS estimator is the least trimmed absolute (LTA) value estimator. Now the strategy is to choose the intercept and slope so as to minimize

$$\sum_{i=1}^{h} |r|_{(i)}, \tag{10.3}$$

where $|r|_{(i)}$ is the ith smallest absolute residual and h is defined as in Section 10.4. (For results on the LTA estimator, see Hawkins and Olive, 1999. For asymptotic results, see Tableman, 1994. For asymptotic results when $h = n$ and the error term is homoscedastic, see Knight, 1998.) Like LTS, the LTA estimator can have a much smaller standard error than the least squares estimator, but its improvement over the LTS estimator seems to be marginal at best (Wilcox, 2001b).

10.5.1 R Function ltareg

The R function

$$\text{ltareg(x,y,tr=0.2,h=NA)}$$

computes the LTA estimate using the Nelder–Mead method for minimizing a function (which was mentioned in Chapter 6 in connection with the spatial median). If no value for h is specified, the value for h is taken to be $n-[\text{tr}(n)]$, where tr defaults to 0.2.

10.6 M-Estimators

Regression M-estimators represent a generalization of the location M-estimator described in Chapter 3. There are, in fact, many variations of M-estimators when dealing with regression, some of which appear to be particularly important in applied work. We begin, however, with some of the early and fairly simple methods. They represent an important improvement on OLS, but by today's standards they are relatively unsatisfactory. Nevertheless, some of these early methods have become fairly well known, so they are included here for completeness.

Generally, M-estimators of location can be extended to regression estimators by choosing a function ξ and then estimating β_j, $(j = 0, \ldots, p)$ with the b_j values that minimize

$$\sum \xi(r_i).$$

Typically ξ is some symmetric function chosen so that it has desirable properties plus a unique minimum at zero. Table 2.1 lists some choices.

As explained in Chapter 2, a measure of scale needs to be used with M-estimators of location so that they are scale-equivariant, and a measure of scale is needed for the more general situation considered here. In the present context, this means that if the y values are multiplied by some constant, c, the estimated slope parameters should be multiplied by the same value. For example, if the estimated slope is 0.304 and $y_1, \ldots, y_n$ are multiplied by 10, then the estimated slope should become 3.04.

Letting τ be any measure of scale, M-estimators minimize

$$\sum \xi\left(\frac{r_i}{\tau}\right).$$

Following Hill and Holland (1977), τ is estimated with

$$\hat{\tau} = \frac{\text{median of the largest } n - p - 1 \text{ of the } |r_i|}{0.6745}. \tag{10.4}$$

Note that $\hat{\tau}$ is resistant, which is needed so that the M-estimator is resistant against unusual y values.

Let Ψ be the derivative of ξ, some of which are given in Table 2.1. Here the focus of attention is on Huber's Ψ given by

$$\Psi(x) = \max[-K, \min(K, x)].$$

For the problem at hand, $K = 2\sqrt{(p+1)/n}$ is used following the suggestion of Belsley et al. (1980). M-estimators solve the system of $p+1$ equations

$$\sum_{i=1}^{n} x_{ij} w_i r_i = 0, \ j = 0, \dots, p, \tag{10.5}$$

where $x_{i0} = 1$ and

$$w_i = \begin{cases} \frac{\Psi(r_i/\hat{\tau})}{r_i/\hat{\tau}}, & \text{if } y_i \neq \hat{y}_i, \\ 1, & \text{if } y_i = \hat{y}_i. \end{cases}$$

As was the case when dealing with M-estimators of location, there is no explicit equation that gives the estimate of the regression parameters; an iterative estimation method must be used instead.

Eq. (10.5) represents a problem in weighted least squares. That is, Eq. (10.5) is equivalent to estimating the parameters by minimizing

$$\sum_i w_i r_i^2.$$

As with W-estimators of location, described in Chapter 3, a technical problem here is that the weights, w_i, depend on unknown parameters. The iterative method used with the W-estimator suggests an iterative estimation procedure for the problem at hand, but the details are postponed until Section 10.8.

10.7 The Hat Matrix

The M-estimator described in Section 10.6 provides resistance against unusual or outlying y values, but a criticism is that it is not resistant against leverage points. In fact the breakdown point is only $1/n$. That is, a single unusual point can completely dominate the estimate of the parameters. Moreover, its influence function is unbounded. An early approach to these problems is based in part on the so-called hat matrix, which is described here. As will become evident, the hat matrix yields an M-estimator that has certain practical advantages—it competes very well with OLS in terms of efficiency and computing accurate confidence intervals under heteroscedasticity and non-normality—but there are situations where it is not as resistant as one might want. In particular, its breakdown point is only $2/n$, meaning that it can handle a single outlier, but two outliers might destroy it. But methods based on the hat matrix have become increasingly well known, so a description of some of them seems in order.

It is a bit easier to convey the idea of the hat matrix in terms of simple regression, so this is done first, after which attention is turned to the more general case where the number of predictors is $p \geq 1$. One strategy for determining whether the point (y_i, x_i) is having an inordinate effect on $\hat{\beta}_1$ and $\hat{\beta}_0$, the OLS estimates of the slope and intercept, is to consider how much the estimates change when this point is eliminated. It turns out that this strategy provides a method for judging whether x_i is a leverage point.

Let

$$h_j = \frac{1}{n} + \frac{(x_j - \bar{x})^2}{\sum (x_i - \bar{x})^2},$$ (10.6)

$j = 1, \ldots, n$. Let $\hat{\beta}_1(i)$, read beta hat sub 1 not i, be the OLS estimate of the slope when the ith pair of observations is removed from the data. Let

$$\hat{y}_i = \hat{\beta}_0 + \hat{\beta}_1 x_i,$$

$$A_j = \frac{\sum x_i^2}{n \sum (x_i - \bar{x})^2} - \frac{x_j \bar{x}}{\sum (x_i - \bar{x})^2},$$

and

$$B_j = \frac{x_j - \bar{x}}{\sum (x_i - \bar{x})^2}.$$

Then the change from $\hat{\beta}_1$ which is based on all of the data, versus the situation where the ith pair of values is removed, can be shown to be

$$\hat{\beta}_1 - \hat{\beta}_1(i) = B_i \frac{r_i}{1 - h_i},$$

the change in the intercept is

$$\hat{\beta}_0 - \hat{\beta}_0(i) = A_i \frac{r_i}{1 - h_i},$$

and the change in the predicted value of y_i, based on x_i, is

$$\hat{y}_i - \hat{y}_i(i) = \frac{h_i}{1 - h_i} r_i.$$

In particular, the bigger h_i happens to be, the more impact there is on the slope, the intercept, and the predicted value of y. Moreover, the change in $\hat{y}_i$ depends only on the residual, r_i, and h_i. But h_i reflects the amount x_i differs from the typical predictor value, $\bar{x}$, because the numerator of the second fraction in Eq. (10.6) is $(x_i - \bar{x})^2$. That is, h_i provides a measure of how much x_i influences the estimated regression equation, which is related to how far x_i is from $\bar{x}$, so h_i provides a method for judging whether x_i is unusually large or small relative to all the predictor values being used.

The result just described can be extended to the more general case where $p \geq 1$. Let

$$\mathbf{X} = \begin{pmatrix} 1 & x_{11} \cdots & x_{1p} \\ 1 & x_{21} \cdots & x_{2p} \\ \vdots & \vdots & \vdots \\ 1 & x_{n1} \cdots & x_{np} \end{pmatrix}.$$

Let $\hat{\boldsymbol{\beta}}$ be the vector of OLS estimates, and let $\hat{\boldsymbol{\beta}}(i)$ be the estimate when $(y_i, \mathbf{x}_i)$ is removed. Then the change in the estimates is

$$\hat{\boldsymbol{\beta}} - \hat{\boldsymbol{\beta}}(i) = (\mathbf{X}'\mathbf{X})^{-1}\mathbf{x}_i' \frac{r_i}{1 - h_{ii}},$$

where h_{ii} is the ith diagonal element of

$$\mathbf{H} = \mathbf{X}(\mathbf{X}'\mathbf{X})^{-1}\mathbf{X}'.$$

The matrix $\mathbf{H}$ is called the *hat matrix* because the vector of predicted y values is given by

$$\hat{\mathbf{y}} = \mathbf{H}\mathbf{y}.$$

In particular,

$$\hat{y}_i = \sum_{j=1}^{n} h_{ij} y_i.$$

In words, the predicted value of y, based on $\mathbf{x}_i$, is obtained by multiplying each element in the ith row of the hat matrix by y_i and adding the results. Furthermore, when $(y_i, \mathbf{x}_i)$ is removed, the vector of predicted values, $\hat{\mathbf{y}}$, is changed by

$$\hat{\mathbf{y}} - \hat{\mathbf{y}}(i) = \mathbf{X}(\mathbf{X}'\mathbf{X})^{-1}(1, x_{i1}, \ldots, x_{ip})' \frac{r_i}{1 - h_{ii}},$$

and the change in the ith predicted value, $\hat{y}_i$, is

$$\hat{y}_i - \hat{y}_i(i) = \frac{h_{ii}}{1 - h_{ii}} r_i.$$

The main point here is that the bigger h_{ii} happens to be, the more impact it has on the OLS estimator and the predicted values. In terms of identifying leverage points, Hoaglin and Welsch (1978) suggest regarding h_{ii} as being large if it exceeds $2(p+1)/n$. (For more information regarding the hat matrix and the derivation of relevant results, see Li, 1985; Huber, 1981; Belsley et al., 1980; Cook and Weisberg, 1992; and Staudte and Sheather, 1990. The R function hat computes h_{ii}, $i = 1, \ldots, n$.)

■ Example

Consider the observations

x: 1 2 3 3 4 4 15 5 6 7,
y: 21 19 23 20 25 30 40 35 30 26.

The h_i values are 0.214, 0.164, 0.129, 0.129, 0.107, 0.107, 0.814, 0.100, 0.107, and 0.129, and $2(p+1)/n = 2(1+1)/10 = 0.4$. Because $h_7 = 0.814 > 0.4$, $x_7 = 15$ would be flagged as a leverage point. The OLS estimate of the slope is 1.43, but if the seventh point is eliminated, the estimate increases to 1.88.

■

There is an interesting connection between the hat matrix and the conditions under which the OLS estimator is asymptotically normal. In particular, a necessary condition for asymptotic normality is that $h_{ii} \to 0$ as $n \to \infty$ (Huber, 1981). A related result turns out to be relevant in the search for an appropriate M-estimator.

10.8 Generalized M-Estimators

This section takes up the problem of finding a regression estimator that guards against leverage points. A natural strategy is to attach some weight, w_i, to $\mathbf{x}_i$ with the idea that the more outlying or unusual $\mathbf{x}_i$ happens to be, relative to all the $\mathbf{x}_i$ values available, the less weight it is given. In general, it is natural to look for some function of the predictor values that reflects in some sense the extent to which the point $\mathbf{x}_i$ influences the OLS estimator. From the previous section, h_{ii} is one way to measure how unusual $\mathbf{x}_i$ happens to be. Its value satisfies $0 < h_{ii} \leq 1$. If $h_{ii} > h_{jj}$, this suggests that $\mathbf{x}_i$ is having a larger impact on the OLS estimator versus $\mathbf{x}_j$.

However, there is a concern. Consider, for example, the simple regression model where $h_{ii} = 1/n + (x_i - \bar{x})^2 / \sum(x_i - \bar{x})^2$. The problem is that $(x_i - \bar{x})^2$ is not a robust measure of the extent to which $\mathbf{x}_i$ is an outlier. Practical problems do indeed arise, but there are some advantages to incorporating the leverage points in an estimation procedure, as will be seen.

One way of attaching a weight to $\mathbf{x}_i$ is with $w_i = \sqrt{1 - h_{ii}}$, which satisfies the requirement that if $h_{ii} > h_{jj}$, $w_i < w_j$. In other words, high leverage points get a relatively low weight. One specific possibility is to estimate the regression parameters as those values solving the $p + 1$ equations

$$\sum_{i=1}^{n} w_i \Psi(r_i/\hat{\tau}) x_{ij} = 0, \tag{10.7}$$

where again $x_{i0} = 1$ and $j = 0, \ldots, p$. Eq. (10.7) is generally referred to as an M-estimator using Mallows weights, which derive their name from Mallows (1975).

Schweppe (see Hill, 1977) takes this one step further with the goal of getting a more efficient estimator. The basic idea is to give more weight to the residual r_i if $\mathbf{x}_i$ has a relatively small weight, w_i. (Recall that outlying x values reduce the standard error of the OLS estimator.) Put another way, Mallows weights can result in a loss of efficiency if there are any outlying $\mathbf{x}$ values, and Schweppe's approach is an attempt at dealing with this problem by dividing r_i by w_i (Krasker and Welsch, 1982). The resulting M-estimator is now the solution to $p + 1$ equations

$$\sum_{i=1}^{n} w_i \Psi(r_i/(w_i\hat{\tau})) x_{ij} = 0, \tag{10.8}$$

$j = 0, \ldots, p$. (For some technical details related to the choice of Ψ and w_i, see Hampel, 1968; Krasker, 1980; and Krasker and Welsch, 1982.) Hill (1977) compares the efficiency of the Mallows and Schweppe estimators to several others and finds that they dominate, with the Schweppe method having an advantage. Solving Eq. (10.8), which includes a choice for the measure of scale, τ, is based on a simple iterative procedure described in Table 10.1. For convenience, the estimator will be labeled $\hat{\beta}_m$.

An even more general framework is to consider $w_i = u(\mathbf{x}_i)$, where $u(\mathbf{x}_i)$ is some function of $\mathbf{x}_i$, chosen to supply some desirable property such as high efficiency. Two choices for $u(\mathbf{x}_i)$, discussed by Markatou and Hettmansperger (1990), are $\sqrt{1 - h_i}$ and $(1 - h_i)/\sqrt{h_i}$. The first choice, already mentioned, is due to Schweppe and was introduced in Handschin et al. (1975). The second choice is due to Welsch (1980).

In the context of testing hypotheses, and assuming ϵ has a symmetric distribution, Markatou and Hettmansperger (1990) recommend $w_i = (1 - h_i)/\sqrt{h_i}$. However, when ϵ has an asymmetric distribution, results in Carroll and Welsh (1988) indicate using $w_i = \sqrt{1 - h_i}$ or Mallows weights. The reason is that otherwise, under general conditions, the estimate of β is not consistent, meaning that it does not converge to the correct value as the sample size gets large. In particular, if Eq. (10.8) is written in the more general form

$$\sum w_i \Psi(r_i/(u(\mathbf{x}_i)\hat{\tau})) x_{ij} = 0,$$

Table 10.1: Iteratively reweighted least squares for M regression, $\hat{\beta}_m$.

To compute an M regression estimator with Schweppe weights, begin by setting $k = 0$ and computing the OLS estimate of the intercept and slope parameters, $\hat{\beta}_{0k}, \ldots, \hat{\beta}_{pk}$. Proceed as follows:

1. Compute the residuals, $r_{i,k} = y_i - \hat{\beta}_{0k} - \hat{\beta}_{1k}x_{i1} - \cdots - \hat{\beta}_{pk}x_{ip}$, let M_k be equal to the median of the largest $n - p$ of the $|r_{i,k}|$, $\hat{\tau}_k = 1.48 M_k$, and let $e_{i,k} = r_{i,k}/\hat{\tau}_k$.
2. Form weights,

$$w_{i,k} = \frac{\sqrt{1 - h_{ii}}}{e_{i,k}} \Psi\left(\frac{e_{i,k}}{\sqrt{1 - h_{ii}}}\right),$$

where

$$\Psi(x) = \max[-K, \min(K, x)]$$

is Huber's Ψ with $K = 2\sqrt{(p+1)/n}$.
3. Use these weights to obtain weighted least squares estimates, $\hat{\beta}_{0,k+1}, \ldots, \hat{\beta}_{p,k+1}$. Increase k by 1.
4. Repeat steps 1–3 until convergence. That is, iterate until the change in the estimated parameters is small.

Carroll and Welsh (1988) show that if $u(\mathbf{x}_i) = 1$, which corresponds to using Mallows weights, the estimate is consistent, but it is not if $u(\mathbf{x}_i) \neq 1$. Thus, this suggests using $w_i = \sqrt{1 - h_i}$ versus $w_i = (1 - h_i)/\sqrt{h_i}$ because for almost all random sequences, $h_i \to 0$ as $n \to \infty$, for any i. In the one-predictor case, it is easy to see that $h_i \to 0$ if x_i are bounded. In fact, as previously indicated, $h_i \to 0$ is a necessary condition for the least squares estimator to be asymptotically normal. (An example where h_i does not converge to zero is $x_i = 2^i$, as noted by Staudte and Sheather, 1990.) For completeness, it is pointed out that there are also *Mallows-type estimators* where weights are given by the leverage points. In particular, perform weighted least squares with weights

$$w_i = \min\left\{1, \left(\frac{b}{h_i}\right)^{j/2}\right\},$$

where $b = h_{(mn)}$, $h_{(r)}$ is the rth ordered leverage value, and j and m are specified; see Hamilton (1992) as well as McKean et al. (1993). The choices $j = 1, 2$, and 4 are sometimes labeled GMM1, GMM2, and GMM4 estimators. Little or nothing is known about how these estimators perform under heteroscedasticity, so they are not discussed further.

To provide at least some indication of the efficiency of $\hat{\beta}_m$, the M-estimator with Schweppe weights, suppose $n = 20$ observations are randomly sampled from the model $y_i = x_i + \lambda(x_i)\epsilon_i$, where both x_i and ϵ_i have standard normal distributions. First consider $\lambda(x_i) \equiv 1$, which corresponds to the usual homoscedastic model, and suppose efficiency is measured with R, the estimated standard error of the OLS estimator divided by the estimated standard error of the M-estimator. Then $R < 1$ indicates that OLS is more efficient, and $R > 1$ indicates that the reverse is true. For the situation at hand, a simulation estimate of R, based on 1,000 replications, is 0.89, so OLS gives better results. If instead $\lambda(x) = |x|$, meaning that the variance of y increases as x moves away from its mean, zero, $R = 1.09$. For $\lambda(x) = x^2$, $R = 1.9$, and for $\lambda(x) = 1 + 2/(|x| + 1)$, $R = 910$, meaning that the OLS estimator is highly unsatisfactory. In the latter case, the variance of y, given x, is relatively large when x is close to its mean.

Suppose instead ϵ has a symmetric heavy-tailed distribution (a g-and-h distribution with $g = 0$ and $h = 0.5$). Then the estimated efficiencies, R, for the four λ functions considered here, are 3.02, 3.87, 6.0, and 226. Thus, for these four situations, OLS performs poorly, particularly for the last situation considered.

Wilcox (1996d) reports additional values of R, again based on 1,000 replications, but using a different set of random numbers (a different seed in the random number generator), yielding $R = 721$ versus 226 for $\lambda(x) = 1 + 2/(|x| + 1)$ and the distributions considered in the previous paragraph. Even if the expected value of R is 10, surely OLS is unsatisfactory. In particular, a large value for R reflects that OLS can yield a wildly inaccurate estimate of the slope when there is a heteroscedastic error term. For example, among the 1,000 replications reported in Wilcox (1996d), where $\beta_1 = 1$ and $R = 721$, there is one case where the OLS estimate of the slope is $-3,140$.

If x has a heavy-tailed distribution, this will favor OLS if ϵ is normal and homoscedastic, but when ϵ is heteroscedastic, OLS can be less efficient than $\hat{\beta}_m$. Suppose for example that x has a g-and-h distribution with $g = 0$ and $h = 0.5$. Then estimates of R, for the four choices for λ considered here, are 0.89, 1.45, 5.48, and 36. Evidently, there are situations where OLS is slightly more efficient than $\hat{\beta}_m$, but there are situations where $\hat{\beta}_m$ is substantially more efficient than OLS, so $\hat{\beta}_m$ appears to have considerable practical value.

10.8.1 R Function bmreg

The R function

```
bmreg(x,y,iter=20,bend=2*sqrt((ncol(x)+1)/nrow(x)),xout=FALSE,outfun=outpro,...)
```

computes the bounded influence M regression with Huber's Ψ and Schweppe weights using the iterative estimation procedure described in Table 10.1. The argument x is any n-by-p matrix of predictors. If there is only one predictor, x can be stored in an R vector as opposed to an R variable having matrix mode. The argument iter controls the maximum number of iterations allowed, which defaults to 20 if unspecified. The argument bend is K in Huber's

Ψ, which defaults to $2\sqrt{(p+1)/n}$. The estimate of the regression parameters is returned in bmreg$coef. The function also returns the residuals in bmreg$residuals, and the final weights (the w_i values) are returned in bmreg$w.

■ Example

The file read.dat contains data from a reading study conducted by Doi. (See Section 1.8 for instructions on how to obtain these data.) One of the goals was to predict WWISST2, a word identification score (stored in column 8). One of the predictors is TAAST1, a measure of phonological awareness (stored in column 2). The OLS estimates of the slope and intercept are 1.72 and 73.56, respectively. The corresponding estimates returned by bmreg are 1.38 and 80.28.

■

■ Example

Although $\hat{\beta}_m$ offers more resistance than OLS, it is important to keep in mind that $\hat{\beta}_m$ might not be resistant enough. The star data in Fig. 6.3 illustrate that problems might arise. The OLS estimates are $\hat{\beta}_1 = -0.41$ and $\hat{\beta}_0 = 6.79$. In contrast, if the data are stored in the R variables x and y, bmreg(x,y) returns $\hat{\beta}_1 = -0.1$ and $\hat{\beta}_0 = 5.53$. As is evident from Fig. 6.3, $\hat{\beta}_m$ does a poor job of capturing the relationship among the majority of points, although it is less influenced by the outliers than is OLS. As noted in Chapter 6, there are several outliers, and there is the practical problem that more than one outlier can cause $\hat{\beta}_m$ to be misleading.

■

10.9 The Coakley–Hettmansperger and Yohai Estimators

The M-estimator $\hat{\beta}_m$, described in the previous section, has a bounded influence function, but a criticism is that its finite sample breakdown point is only $2/n$. That is, it can handle one outlier, but two outliers might destroy it. Coakley and Hettmansperger (1993) derive an M-estimator that has a breakdown point nearly equal to 0.5, a bounded influence function, and high asymptotic efficiency for the normal model. Their strategy is to start with the LTS estimator and then adjust it, the adjustment being a function of empirically determined weights. More formally, letting $\hat{\boldsymbol{\beta}}_0$ (a vector having length $p+1$) be the LTS estimator, their estimator is

$$\hat{\boldsymbol{\beta}}_{\text{ch}} = \hat{\boldsymbol{\beta}}_0 + (\mathbf{X}'\mathbf{B}\mathbf{X})^{-1}\mathbf{X}'\mathbf{W}\Psi(r_i/(w_i\hat{\tau}))\hat{\tau},$$

where $\mathbf{W} = \text{diag}(w_i)$,

$$\mathbf{B} = \text{diag}(\Psi'(r_i/\hat{\tau} w_i)),$$

$\mathbf{X}$ is the n-by-$(p+1)$ design matrix (described in Section 10.7), and $\Psi'(x)$ is the derivative of Huber's Ψ. They suggest using $K = 1.345$ in Huber's Ψ if uncertain about which value to use, so this value is assumed here unless stated otherwise. As an estimate of scale, they use $\hat{\tau} = 1.4826(1 + 5/(n - p)) \times \text{med}\{|r_i|\}$. The weight given to $\mathbf{x_i}$, the ith row of predictor values, is

$$w_i = \min\{1, [b/(\mathbf{x_i} - \mathbf{m_x})'\mathbf{C}^{-1}(\mathbf{x_i} - \mathbf{m_x})]^{a/2}\},$$

where the quantities $\mathbf{m_x}$ and $\mathbf{C}$ are the minimum volume ellipsoid (MVE) estimators of location and covariance associated with the predictors. (See Section 6.3.1.) These estimators have a breakdown point approximately equal to 0.5. When computing w_i, Coakley and Hettmansperger suggest setting b equal to the 0.95 quantile of a chi-squared distribution with p degrees of freedom and using $a = 2$.

The Coakley–Hettmansperger regression method has high asymptotic efficiency when the error term, ϵ, has a normal distribution. That is, as the sample size gets large, the standard error of the OLS estimator will not be substantially smaller than the standard error of $\hat{\beta}_{ch}$. To provide some indication of how the standard error of $\hat{\boldsymbol{\beta}}_{ch}$ compares to the standard error of OLS when n is small, consider $p = 1$, $a = 2$, $K = 1.345$, and suppose both x and ϵ have standard normal distributions. When there is homoscedasticity, R, the standard error of OLS divided by the standard error of $\hat{\boldsymbol{\beta}}_{ch}$, is 0.63 based on a simulation with 1,000 replications. If x has a symmetric, heavy-tailed distribution instead (a g-and-h distribution with $g = 0$ and $h = 0.5$), then $R = 0.31$. In contrast, if $\hat{\beta}_m$ is used, which is computed as described in Table 10.1, the ratio is 0.89 for both of the situations considered here. Of course, the lower efficiency of $\hat{\beta}_{ch}$ must be weighed against the lower breakdown point of $\hat{\beta}_m$. Moreover, if x has a light-tailed distribution, and the distribution of ϵ is heavy-tailed, $\hat{\beta}_{ch}$ offers an advantage over OLS. For example, if x is standard normal, but ϵ has a g-and-h distribution with $g = 0$ and $h = 0.5$, $R = 1.71$. However, replacing $\hat{\beta}_{ch}$ with $\hat{\beta}_m$, $R = 3$.

It might be thought that the choice for the bending constant in Huber's Ψ, $K = 1.345$, is partly responsible for the low efficiency of $\hat{\boldsymbol{\beta}}_{ch}$. If $\hat{\boldsymbol{\beta}}_{ch}$ is modified by increasing the bending constant to $K = 2$, $R = 0.62$ when both x and ϵ are standard normal. If x has the symmetric, heavy-tailed distribution considered in the previous paragraph, again $R = 0.31$. It appears that $\hat{\boldsymbol{\beta}}_{ch}$ is inefficient because of its reliance on LTS regression and because it does not take sufficient advantage of good leverage points.

10.9.1 MM-Estimator

Yet another robust regression estimator that should be mentioned is the MM-estimator derived by Yohai (1987), which has certain similarities to the generalized M-estimators in Section 10.8. It has the highest possible breakdown point, 0.5, and high efficiency under normality. The parameters are estimated by solving an equation similar to Eq. (10.8), with Ψ taken to be some redescending function. A popular choice is Tukey's biweight, given in Table 2.1, which will be used here unless stated otherwise. That is, the regression parameters are estimated by determining the solution to $p + 1$ equations

$$\sum_{i=1}^{n} \Psi(r_i/\hat{\tau})x_{ij} = 0, \tag{10.9}$$

$j = 0, \ldots, p$, where again $\hat{\tau}$ is a robust measure of variation based on the residuals and

$$\Psi(r_i; c) = \frac{r_i}{\hat{\tau}} \left(\left(\frac{r_i}{c\hat{\tau}} \right)^2 - 1 \right)^2 \text{ if } |r_i/\hat{\tau}| \leq c;$$

otherwise $\Psi(r_i; c) = 0$. The choice $c = 4.685$ leads to an MM-estimator with 95% efficiency compared to the least squares estimator and is the default value used here. The ratio p/n is relevant to the efficiency of this estimator; see Maronna and Yohai (2010) for details. In addition to having excellent theoretical properties, the small-sample efficiency of the MM-estimator appears to compare well with other robust estimators, but like several other robust estimators it can be sensitive to what is called contamination bias, as described in Section 10.14.1. Also, situations are encountered where the iterative estimation scheme used to compute the MM-estimator does not converge.

The R function lmrob, which can be accessed via the R package robustbase, can be used to compute confidence intervals and test hypotheses when using the MM-estimator. The function computes estimates of the standard errors and assumes the null distribution of the test statistics has, approximately, a Student's t distribution with $n - p - 1$ degrees of freedom. However, even under normality and homoscedasticity, control over the probability of a Type I error is poor when n is small. When testing at the 0.05 level, the actual level exceeds 0.05 by a substantial amount. With $n = 100$ it performs reasonably well, still assuming normality and homoscedasticity. Evidently there are no results regarding how well the method performs under non-normality and heteroscedasticity. (For an extension of this estimator, see Koller and Stahel, 2011.) If there is interest in testing hypotheses based on the MM-estimator, currently the best approach appears to be the percentile bootstrap methods in Sections 11.1.1 and 11.1.3.

When dealing with multivariate data, Section 6.4.4 makes a distinction between rowwise outliers and cellwise outliers. This means that single entries (cells) in the data matrix are considered as potential outliers and not necessarily a whole row. This distinction arises when dealing with regression as well. Filzmoser et al. (2020) derive a variation of the M regression methods aimed at dealing with this situation. Briefly, cellwise outliers are detected as described by Debruyne et al. (2019), which are then downweighted. Filzmoser et al. focus on MM regression as a starting point, which plays a role in detecting cellwise outliers. R code for applying their method is available in the R package crmReg at github.com/SebastiaanHoppner/CRM. For completeness, it is noted that there are additional M regression methods not covered in this chapter (e.g., Jurečková and Portnoy, 1987).

10.9.2 R Functions chreg and MMreg

The R function

    chreg(x, y, bend = 1.345, SEED=TRUE, xout = FALSE, outfun = outpro, pr = TRUE, ...)

computes the Coakley–Hettmansperger regression estimator. As with all regression functions, the argument x can be a vector or an n-by-p matrix of predictor values. The argument bend is the value of K used in Huber's Ψ.

The R function

    MMreg(x, y, RES = FALSE, xout = FALSE, outfun = outpro, varfun = pbvar, corfun = pbcor,
                                        ...)

computes Yohai's MM-estimator. Setting RES=TRUE, the residuals are returned.

■ Example

If the star data in Fig. 6.3 are stored in the R variables starx and stary, the command

                        chreg(starx,stary)

returns an estimate of the slope equal to 4.0. The R function MMreg estimates the slope to be 2.25. In contrast, the OLS estimate is -0.41, and the M regression estimate, based on the method in Table 10.1, yields $\hat{\beta}_m = -0.1$.

■

10.10 Skipped Estimators

Skipped regression estimators generally refer to the strategy of checking for outliers using one of the multivariate outlier detection methods in Chapter 6, discarding any that are found, and applying some estimator to the data that remain. A natural issue is whether the OLS estimator might be used after outliers are removed, but checks on this approach by the author found that the small-sample efficiency of this method is rather poor, compared to other estimators, when the error term is normal and homoscedastic.

Consider the goal of achieving good small-sample efficiency, relative to OLS, when the error term is both normal and homoscedastic, and $\mathbf{x}_i'$ is multivariate normal, while simultaneously providing protection against outliers. Then a relatively effective method is to first check for outliers among the $(p + 1)$-variate data $(\mathbf{x}_i', y_i)$ $(i = 1, \ldots, n)$ using the MGV or projection method (described in Sections 6.4.7 and 6.4.9), remove any outliers that are found, and then apply the Theil–Sen estimator to the data that remain. Replacing the Theil–Sen estimator with one of the M-estimators previously described has been found to be rather unsatisfactory. (Checks on using the MM-estimator have not been made.) Generally, when using the MGV outlier detection in conjunction with any regression estimator, this will be called an MGV estimator. Here, it is assumed that the Theil–Sen estimator is used unless stated otherwise. When using the projection method for detecting outliers, this will be called an OP-estimator, and again the Theil–Sen estimator is assumed.

10.10.1 R Functions mgvreg and opreg

The R function

mgvreg(x,y,regfun=tsreg,cov.fun=rmba,se=TRUE,varfun=pbvar,corfun=pbcor, SEED=TRUE)

computes a skipped regression estimator where outliers are identified and removed using the MGV outlier detection method in Section 6.4.7. Once outliers are eliminated, the regression estimator indicated by the argument regfun is applied and defaults to the Theil–Sen estimator.

The R function

opreg(x,y,regfun=tsreg, cop=3,MC=FALSE, varfun=pbvar, corfun=pbcor)

is like mgvreg, only the projection detection method in Section 6.4.9 is used to detect outliers. The argument cop determines the measure of location used when checking for outliers. (See Section 6.4.10 for details about the argument cop.) Unlike mgvreg, opreg can take advantage of a multicore processor by setting the argument MC=TRUE. (The arguments varfun and pbcor deal with estimating the strength of the association as described in Section 11.9.)

10.11 Deepest Regression Line

Rousseeuw and Hubert (1999) derive a method of fitting a line to data that searches for the deepest line embedded within a scatterplot. First consider simple regression. Let b_1 and b_0 be any choice for the slope and intercept, respectively, and let r_i $(i = 1, \ldots, n)$ be the corresponding residuals. The candidate fit, (b_0, b_1), is called a non-fit if a partition of the x values can be found such that all of the residuals for the lower x values are negative (positive), but for all of the higher x values the residuals are positive (negative). So, for example, if all of the points lie above a particular straight line, in which case all of the residuals are positive, this line is called a non-fit. More formally, a candidate fit is called a non-fit if and only if a value for v can be found such that

$$r_i < 0 \quad \text{for all } x_i < v$$

and

$$r_i > 0 \quad \text{for all } x_i > v$$

or

$$r_i > 0 \quad \text{for all } x_i < v$$

and

$$r_i < 0 \quad \text{for all } x_i > v.$$

The regression depth of a fit (b_1, b_0), relative to $(x_1, y_1), \ldots, (x_n, y_n)$, is the smallest number of observations that need to be removed to make (b_1, b_0) a non-fit. The deepest regression estimator corresponds to the values of b_1 and b_0 that maximize regression depth. (See Hubert and Rousseeuw, 1998, for a variation of this method.) Bai and He (1999) derive the limiting distribution of this estimator. When the $\mathbf{x}_i$ are distinct, the breakdown point is approximately 0.33.

The idea can be extended to multiple predictors. Let $r_i(b_0, \ldots, b_p)$ be the ith residual based on the candidate fit $b_0, \ldots, b_p$. This candidate fit is called a non-fit if there exists a hyperplane V in $\mathbf{x}$ space such that no $\mathbf{x}_i$ belongs to V and such that $r_i(b_0, \ldots, b_p) > 0$ for all $\mathbf{x}_i$ in one of the open halfspaces corresponding to V and $r_i(b_0, \ldots, b_p) < 0$ in the other open halfspace. Regression depth is defined as in the $p = 1$ case.

10.11.1 R Functions rdepth.orig, Rdepth, mdepreg.orig, and mdepreg

The R function

$$\text{rdepth.orig(d,x,y)}$$

computes the depth of a regression line within a cloud of points for the case of a single independent variable. The argument d should have two values: the intercept followed by the slope. The function

$$\text{Rdepth(x, y, z = NULL, ndir = NULL)}$$

computes regression depth for one or more independent variables via the function rdepth in the R package mfrDepth. The argument z is matrix with $p + 1$ columns containing rowwise the hyperplanes for which to compute the regression depth. The first column should contain the intercepts. If z is not specified, it is set equal to x. The R function

$$\text{mdepreg.orig(x,y)}$$

computes the deepest regression line. For $p > 1$ predictors it uses an approximation of the depth of a hyperplane. This function matches the function mdepreg in the 4th edition of this book. The function

$$\text{mdepreg(x, y, xout = FALSE, outfun = out, ...)}$$

computes the deepest regression line by calling the function rdepthmedian in the R package mfrDepth. It follows more closely the algorithm in Rousseeuw and Hubert (1999). The relative merits of mdepreg versus mdepreg.orig are currently unknown.

10.12 A Criticism of Methods With a High Breakdown Point

It seems that no single method is free from criticism, and regression methods that have a high breakdown point are no exception. A potential problem with these methods is that standard diagnostic tools for detecting curvature, by examining the residuals, might fail (Cook et al., 1992; McKean et al., 1993). Thus, it is recommended that if a regression method with a high breakdown point is used, possible problems with curvature be examined using some alternative technique. Some possible ways of checking for curvature, beyond the standard methods covered in an introductory regression course, are described in Chapter 11.

10.13 Some Additional Estimators

Some additional regression estimators should be outlined. Although some of these estimators have theoretical properties that do not compete well with some of the estimators already described, they might have practical value. For instance, some of the estimators listed here have been suggested as an initial estimate that is refined in some manner or is used as a preliminary screening device for detecting regression outliers. Graphical checks suggest that these estimators sometimes provide a more reasonable summary of the data versus other estimators covered in the previous section. Also, it is not being suggested that by listing an estimator in this section, it necessarily should be excluded from consideration in applied work. In some cases, certain comparisons with estimators already covered, such as efficiency under heteroscedasticity, have not been explored. (For an extension to the so-called general linear model, see Cantoni and Ronchetti, 2001.)

10.13.1 S-Estimators and τ-Estimators

S-estimators of regression parameters, proposed by Rousseeuw and Yohai (1984), search for the slope and intercept values that minimize some measure of scale associated with the residuals. Least squares, for example, minimizes the variance of the residuals and is a special case of S-estimators. The hope is that by replacing the variance with some measure of scale that is relatively insensitive to outliers, we will obtain estimates of the slope and intercept that are relatively insensitive to outliers as well. As noted in Chapter 3, there are many measures of scale. The main point is that if, for example, we use the percentage bend midvariance (described in Section 3.12.3), situations arise where the resulting estimate of the slope and intercept has advantages over other regression estimators we might use. This is not to say that other measures of scale never provide a more satisfactory estimate of the regression parameters. But for general use, it currently seems that the percentage bend midvariance is a good choice. For relevant asymptotic results, see Davies (1990). Hössjer (1992) shows that S-estimators cannot achieve simultaneously both a high breakdown point and high efficiency under the normal model. Also, Davies (1993) reports results on the inherent instability of S-estimators. Despite this, it may have practical value as preliminary fit to data; see Section 10.13.3.

Here a simple approximation of the S-estimator is used. (There are other ways of computing S-estimators, e.g., Croux et al., 1994; Ferretti et al., 1999; perhaps they have practical advantages, but it seems that this possibility has not been explored.) Let

$$R_i = y_i - b_1 x_{1i} - \cdots - b_p x_{ip},$$

and use the Nelder–Mead method (mentioned in Chapter 6) to find the values $b_1, \ldots, b_p$ that minimize S, some measure of scale based on the values $R_1, \ldots, R_n$. The intercept is taken to be

$$b_0 = M_y - b_1 M_1 - \cdots - b_p M_p,$$

where M_j and M_y are the medians of the x_{ij} ($i = 1, \ldots, n$) and y values, respectively. This will be called method *SNM*. (Again, for details motivating the Nelder–Mead method, see Olsson and Nelson, 1975.)

A related approach in the one-predictor case, called the *STS* estimator, is to compute the slope between points j and j', $S_{jj'}$, and take the estimate of the slope to be the value of $S_{jj'}$ that minimizes some measure of scale applied to the values $y_1 - S_{jj'}x_1, \ldots, y_n - S_{jj'}x_n$ values. Here, the back-fitting method is used to handle multiple predictors.

For completeness, τ-estimators proposed by Yohai and Zamar (1988) generalize S-estimators by using a broader class of scale estimates. Gervini and Yohai (2002, p. 584) note that tuning these estimators for high efficiency will result in an increase in bias. An extension of τ-estimators is the class of generalized τ-estimators proposed by Ferretti et al. (1999). Briefly, residuals are weighted, with high leverage values resulting in small weights, and a measure of scale based on these weighted residuals is used to judge a fit to data. For results on computing τ-estimators, see Flores (2010).

10.13.2 R Functions snmreg and stsreg

The R function

$$\text{snmreg(x,y)}$$

computes the SNM estimate as just described. The measure of scale, S, is taken to be the percentage bend midvariance. (When using the Nelder–Mead method, the initial estimate of the parameters is based on the Coakley–Hettmansperger estimator.) The R function

$$\text{stsreg(x,y,sc=pbvar)}$$

computes the STS estimator, where sc is the measure of scale to be used. By default, the percentage bend midvariance is used.

10.13.3 E-Type Skipped Estimators

Skipped estimators remove any outliers among the cloud of data $(x_{i1}, \ldots, x_{ip}, y_i)$, $i = 1, \ldots, n$, and then fit a regression line to the data that remain. E-type skipped estimators (where E stands for error term) look for outliers among the residuals based on some preliminary fit, remove (or downweight) the corresponding points, and then compute a new fit to the data. Rousseeuw and Leroy (1987) suggest using LMS to obtain an initial fit, remove any points for which the corresponding standardized residuals are large, and then apply least squares to the data that remain. But He and Portnoy (1992) show that the asymptotic efficiency is 0.

Another E-type skipped estimator is to apply one of the outlier detection methods in Chapter 3 to the residuals. For example, first fit a line to the data using the STS estimator described in Section 10.13.1. Let r_i $(i = 1, \ldots, n)$ be the usual residuals. Let M_r be the median of the residuals and let MAD_r be the median of the values $|r_1 - M_r|, \ldots, |r_n - M_r|$. Then the ith point (x_i, y_i) is declared a regression outlier if

$$|r_i - M_r| > \frac{2(\text{MAD}_r)}{0.6745}. \tag{10.10}$$

The final estimate of the slope and intercept is obtained by applying the Theil–Sen estimator to those points not declared regression outliers. When there are p predictors, again compute the residuals based on STS and use Eq. (10.10) to eliminate any points with large residuals. This will be called method *TSTS*.

The STS estimator might have high execution time, in which case it might be preferable to modify the TSTS estimator by replacing the STS estimator with the SNM estimator in Section 10.13.1. This will be called the TSSNM estimator.

Another variation, called an *adjusted M-estimator*, is to proceed as in Table 10.1, but in step 2, set $w_{i,k} = 0$ if $(y_i, \mathbf{x}_i)$ is a regression outlier based, for example, on the regression outlier detection method in Rousseeuw and van Zomeren (1990); see Section 10.15. This estimator will be labeled $\hat{\beta}_{\text{ad}}$.

Gervini and Yohai (2002) use an alternative approach for determining whether any residuals are outliers and they derived some general theoretical results for this class of estimators. In particular, they describe conditions under which the asymptotic breakdown point is not less than the initial estimator, and they find that good efficiency is obtained under normality and homoscedasticity.

The method begins by obtaining an initial fit to the data; Gervini and Yohai focus on LMS and S-estimators to obtain this initial fit. Then the absolute value of the residuals are checked for outliers, any such points are eliminated, and the least squares estimator is applied to the remaining data. But unlike other estimators in this section, an adaptive method for detecting outliers, which is based on the empirical distribution of the residuals, is used to detect outliers. An interesting result is that under general conditions, if the errors are normally distributed, the estimator has full asymptotic efficiency.

To outline the details, let R_i be the residuals based on an initial estimator, and let

$$r_i = \frac{R_i}{S}$$

be the standardized residuals, where S is some measure of scale applied to the R_i values. Following Gervini and Yohai, S is taken to be MADN (MAD divided by 0.6745). Let $|r|_{(1)} \leq \cdots \leq |r|_{(n)}$ be the absolute values of the standardized residuals written in ascending order and let $i_0 = \max\{|r|_{(i)} < \eta\}$ for some constant η; Gervini and Yohai use $\eta = 2.5$. Let

$$d_n = \max \left\{ \Phi(|r|_{(i)}) - \frac{i-1}{n} \right\},$$

where the maximum is taken over all $i > i_0$ and Φ is the cumulative standard normal distribution. In the event $d_n < 0$, set $d_n = 0$, and let $i_n = n - [d_n]$, where $[d_n]$ is the greatest integer less than or equal to d_n. The point corresponding to $|r|_{(i)}$ is eliminated (is given zero weight) if $i > i_n$, and the least squares estimator is applied to the data that remain.

For the situations in Table 10.2, the Gervini–Yohai estimator does not compete well in terms of efficiency when using LMS regression as the initial estimate when checking for regression outliers. Switching to the LTS estimator as the initial estimator does not improve efficiency for the situations considered. (Also see Section 10.14.)

10.13.4 R Functions mbmreg, tstsreg, tssnmreg, and gyreg

The function

$$\text{tstsreg(x,y,sc=pbvar,xout=FALSE,outfun=outpro,plotit=FALSE,...)}$$

computes the E-type estimator described in Section 10.13.3. The argument sc indicates which measure of scale will be used when method STS is employed to detect regression outliers, and the default measure of scale is the percentage bend midvariance. Setting the argument xout=TRUE, leverage points are removed, which are identified using the outlier detection method indicated by the argument outfun, and then the E-type estimator is applied using the remaining data. This function can have relatively high execution time, particularly with more than one covariate. In this case, the function

$$\text{tssnmreg(x,y,sc=pbvar,xout=FALSE,outfun=out,plotit=FALSE,...)}$$

might be preferred; it applies the TSSNM estimator, which uses the SNM estimator to detect outliers among the residuals.

The function

$$mbmreg(x,y,iter = 20, bend = (2 *sqrt(ncol(x)+1))/nrow(x), xout=FALSE, outfun=outpro,...)$$

computes the so-called adjusted M-estimator. Finally, the function

$$gyreg(x,y,rinit = lmsreg, K = 2.5)$$

computes the Gervini–Yohai estimator where the argument rinit indicates which initial estimator will be used to detect regression outliers. By default, the LMS estimator is used. The argument K corresponds to η.

10.13.5 Methods Based on Robust Covariances

A general approach to regression, briefly discussed by Huber (1981), is based on estimating a robust measure of covariance, which in turn can be used to estimate the parameters of the model. For the one-predictor case, the slope of the OLS regression line is

$$\beta_1 = \frac{\sigma_{xy}}{\sigma_x^2},$$

where σ_{xy} is the usual covariance between x and y. This suggests estimating the slope with

$$\hat{\beta}_1 = \frac{\hat{\tau}_{xy}}{\hat{\tau}_x^2},$$

where $\hat{\tau}_{xy}$ estimates τ_{xy}, some measure of covariance chosen to have good robustness properties, and $\hat{\tau}_x^2$ is an estimate of some measure of scale. The intercept can be estimated with

$$\hat{\beta}_0 = \hat{\theta}_y - \hat{\beta}_1\hat{\theta}_x$$

for some appropriate estimate of location, θ. As noted in Chapter 9, there are many measures of covariance. Here, the biweight midcovariance is employed, which is described in Section 9.3.8 and is motivated in part by results in Lax (1985), who found that the biweight midvariance is relatively efficient.

For the more general case where $p \geq 1$, let

$$\mathbf{A} = (s_{byx_1}, \ldots, s_{byx_p})'$$

be the vector of sample biweight midcovariances between y and the p predictors. The quantity s_{byx_j} estimates the biweight midcovariance between y and the jth predictor, x_j. Let

$$\mathbf{C} = (s_{bx_j x_k})$$

be the p-by-p matrix of estimated biweight midcovariances among the p predictors. By analogy with OLS, the regression parameters $(\beta_1, \ldots, \beta_p)'$ are estimated with

$$(\hat{\beta}_1, \ldots, \hat{\beta}_p)' = \mathbf{C}^{-1}\mathbf{A}, \tag{10.11}$$

and an estimate of the intercept is

$$\hat{\beta}_0 = \hat{\mu}_{my} - \hat{\beta}_1 \hat{\mu}_{m1} - \cdots - \hat{\beta}_p \hat{\mu}_{mp}, \tag{10.12}$$

where $\hat{\mu}_{my}$ is taken to be the one-step M-estimator based on the y values, and $\hat{\mu}_{mj}$ is the one-step M-estimator based on the n values corresponding to the jth predictor.

The estimation procedure just described performs reasonably well when there is a homoscedastic error term, but it can give poor results when the error term is heteroscedastic. For example, Wilcox (1996e) reports a situation where the error term is heteroscedastic, $\beta_1 = 1$, yet with $n = 100,000$, the estimated slope is approximately equal to 0.5. Apparently the simple estimation procedure described in the previous paragraph is not even consistent when the error term is heteroscedastic. However, a simple iterative procedure corrects this problem.

Set $k = 0$, let $\hat{\boldsymbol{\beta}}_k$ be the $p + 1$ vector of estimated slopes and intercepts using Eqs. (10.10) and (10.11), and let r_{ki} be the resulting residuals. Let $\hat{\boldsymbol{\delta}}_k$ be the $p + 1$ vector of estimated slopes and intercepts when using $\mathbf{x}$ to predict the residuals. That is, replace y_i with r_{ki} and compute the regression slopes and intercepts using Eqs. (10.10) and (10.11). Then an updated estimate of $\boldsymbol{\beta}$ is

$$\hat{\boldsymbol{\beta}}_{k+1} = \hat{\boldsymbol{\beta}}_k + \hat{\boldsymbol{\delta}}_k, \tag{10.13}$$

and this process can be repeated until convergence. That is, compute a new set of residuals using the estimates just computed, increase k by 1, and use Eqs. (10.10) and (10.11) to get a new adjustment, $\hat{\boldsymbol{\delta}}_k$. The iterations stop when all of the $p + 1$ values in $\hat{\boldsymbol{\delta}}_k$ are close to zero, say within 0.001. The final estimate of the regression parameters is denoted by $\hat{\boldsymbol{\beta}}_{\text{mid}}$.

It is noted that for certain measures of scatter, this iteration scheme does not appear to be necessary. For example, Zu and Yuan (2010) use the multivariate measure of scatter derived by Maronna (1976), and checks on this approach indicate that when there is heteroscedasticity, using Eqs. (10.10) and (10.11) suffices.

A concern about some robust covariances is that they do not take into account the overall structure of the data and so might be influenced by properly placed outliers. One could use the

Table 10.2:   Estimates of R using covariance methods, $n = 20$.

VP	x and ε normal			x normal, ε heavy-tailed			x heavy-tailed, ε normal		
	b_{mid}	$\gamma = 0.1$	$\gamma = 0.2$	b_{mid}	$\gamma = 0.1$	$\gamma = 0.2$	b_{mid}	$\gamma = 0.1$	$\gamma = 0.2$
1	0.94	0.92	0.81	1.10	2.55	2.64	0.61	0.78	0.57
2	1.80	1.69	2.17	2.28	4.49	6.46	24.41	9.08	19.58
3	18.25	13.82	10.26	13.79	9.62	9.64	13.20	2.83	3.57

skipped correlations described in Chapter 6 for the situation at hand, but there are no results regarding the small-sample properties of this approach.

Another approach to estimating regression parameters is to replace the biweight midcovariance with the Winsorized covariance in the biweight midregression method (cf. Yale and Forsythe, 1976). This will be called *Winsorized regression.* An argument for this approach is that the goal is to estimate the Winsorized mean of y, given x, and the Winsorized mean satisfies the Bickel–Lehmann condition described in Chapter 2. In terms of probability coverage, it seems that there is little or no reason to prefer Winsorized regression over the biweight midregression procedure.

It should be noted that Srivastava et al. (2010) report results on what they call Winsorized regression where all variables are Winsorized as described in Section 9.3.6, after which they apply the usual least squares estimator using the Winsorized values. They demonstrate via simulations that the resulting estimator can be more efficient than least squares. Evidently there are no results on how the efficiency of this estimator compares to that of other robust estimators in this chapter. And no results were reported regarding the effects of heteroscedasticity.

To provide some indication of how the biweight midregression and Winsorized regression methods compare to OLS regression with respect to efficiency, Table 10.2 shows estimates of R, the standard error of OLS regression divided by the standard error of the competing method. The columns headed by $\gamma = 0.1$ are the values when 10% Winsorization is used, and $\gamma = 0.2$ is 20%. These estimates correspond to three types of error terms: $\lambda(x) = 1$, $\lambda(x) = x^2$, and $\lambda(x) = 1/|x|$. For convenience, these three choices are labeled VP 1, VP 2, and VP 3. In general, the biweight and Winsorized regression methods compare well to OLS, and in some cases they offer a substantial advantage. Note, however, that when x has a heavy-tailed distribution and ϵ is normal, OLS offers better efficiency when the error term is homoscedastic. In some cases, $\hat{\beta}_{\text{mid}}$ has better efficiency versus $\hat{\beta}_{\text{ad}}$, but in other situations the reverse is true.

10.13.6 R Functions bireg, winreg, and COVreg

The R function

$$\text{bireg}(x,y,\text{iter}=20,\text{bend}=1.28)$$

is supplied for performing the biweight midregression method just described. As usual, x can be a vector when dealing with simple regression ($p = 1$), or it is an n-by-p matrix for the more general case where there are $p \geq 1$ predictors. The argument iter indicates the maximum number of iterations allowed. It defaults to 20, which is more than sufficient for most practical situations. If convergence is not achieved, the function prints a warning message. The argument bend is the bending constant, K, used in Huber's Ψ when computing the one-step M-estimator. If unspecified, $K = 1.28$ is used. The function returns estimates of the coefficients in bireg$coef, and the residuals are returned in bireg$resid.

The R function

$$\text{winreg}(x,y,\text{iter}=20,\ \text{tr}=0.2,\ \text{iter}=20,\text{tr}=0.2,\text{xout}=\text{FALSE},\ \text{outfun}=\text{outpro},...)$$

performs Winsorized regression, where tr indicates the amount of Winsorizing, which defaults to 20%. Again, iter is the maximum number of iterations allowed, which defaults to 20.

The R function

$$\text{COVreg}(x,y,\text{cov.fun}=\text{MARest},\text{loc.fun}=\text{MARest},\text{xout}=\text{FALSE},\text{outfun}=\text{out},...)$$

estimates the slopes and intercept via Eqs. (10.11) and (10.12) without iterations and defaults to using Maronna's M-estimator in Section 6.3.13.

■ **Example**

For the star data shown in Fig. 6.3 of Chapter 6, bireg estimates the slope and intercept to be 2.66 and -6.7, respectively. The OLS estimates are -0.41 and 6.79. The function winreg estimates the slope to be 0.31 using 10% Winsorization (tr=0.1), and this is considerably smaller than the estimate of 2.66 returned by bireg. Also, winreg reports a warning message that convergence was not obtained in 20 iterations. This problem seems to be very rare. Increasing iter to 50, convergence is obtained, and again the slope is estimated to be 0.31, but the estimate of the intercept drops from 3.62 to 3.61. Using the default 20% Winsorization (tr=0.2), the slope is now estimated to be 2.1 and convergence problems are eliminated.

■

10.13.7 L-Estimators

A reasonable approach to regression is to compute some initial estimate of the parameters, compute the residuals, and then re-estimate the parameters based in part on the trimmed residuals. This strategy was employed by Welsh (1987a, 1987b) and expanded upon by De Jongh et al. (1988). The small-sample efficiency of this approach does not compare well with other estimators such as $\hat{\beta}_m$ or M regression with Schweppe weights (Wilcox, 1996d). In terms of achieving high efficiency when there is heteroscedasticity, comparisons with the better estimators in this chapter have not been made. So even though the details of the method are not described here, it is not suggested that Welsh's estimator be abandoned.

10.13.8 L_1 and Quantile Regression

Yet another approach is to estimate the regression parameters by minimizing $\sum |r_i|$, the so-called L_1 norm, which is just the sum of the absolute values of the residuals. This approach predates OLS by about 50 years. This is, of course, a special case of the LTA estimator in Section 10.5. The potential advantage of L_1 (or least absolute value) regression over OLS, in terms of efficiency, was known by Laplace (1818). The L_1 approach reduces the influence of outliers, but the breakdown point is still $1/n$. More precisely, L_1 regression protects against unusual y values, but not leverage points, which can have a large influence on the fit to data. Another concern is that a relatively large weight is being given to observations with the smallest residuals (Mosteller and Tukey, 1977, p. 366). For these reasons, further details are omitted. (For a review of L_1 regression, see Narula, 1987, as well as Dielman and Pfaffenberger, 1982.) Hypothesis testing procedures are described by Birkes and Dodge (1993), but no results are given on how the method performs when the error term is heteroscedastic.

An interesting generalization of the L_1 estimator was proposed by Koenker and Bassett (1978), which is aimed at estimating the qth quantile of y given x. Let

$$\psi_q(u) = u(q - I_{u<0}), \tag{10.14}$$

where I is the indicator function. Then the slope and intercept of the regression line are determined by minimizing

$$\sum \psi_q(r_i).$$

So $q = 0.5$ corresponds to the least absolute value (or L_1) estimator and yields an estimate of the median of y, given x (cf. Koenker and Portnoy, 1987; Gutenbrunner and Jurečková, 1992). For a quantile regression estimator that takes into account auxiliary information, see Tang and Leng (2012).

A (rank inversion) method for testing hypotheses about the individual parameters was studied by Koenker (1994) and can be applied via the R package quantreg. (Also see Koenker and Xiao, 2002.) The method is limited to testing at the $\alpha = 0.05$ level. The R function rqfit, described in the next section, applies the method. When the goal is to test at some other level, or if an omnibus test is to be performed, see Section 11.1.6.

Neykov et al. (2012) consider a generalization of the quantile regression estimator where subsets of the data are considered, with each subset created by removing k points, with k specified by the user. The final estimate of the parameters is based on the subset that minimizes $\sum \psi_q(r_i)$.

10.13.9 R Functions qreg, rqfit, and qplotreg

The R function

qreg(x,y,qval=0.5, q=NULL, pr=FALSE, xout=FALSE, outfun=out, plotit=FALSE, xlab='X', ylab='Y',...)

computes the Koenker–Bassett quantile regression estimator. The argument qval (or the argument q) determines the quantile to be used. As usual, if the argument xout=TRUE, leverage points are removed. (The R function trq computes a trimmed generalization of the quantile regression estimator derived by Koenker and Portnoy, 1987.) On rare occasions, computational issues arise when using qreg, which can be avoided by using the R function Qreg at the expense of slightly higher execution time.

The R function

rqfit(x,y,qval=0.5,alpha=0.05,xout=FALSE, outfun=out, res=TRUE,...)

tests hypotheses about the individual parameters. As usual, setting the argument xout=TRUE eliminates leverage points. If the argument alpha is not equal to 0.05, an error message is printed indicating that the R function qregci (described in Section 11.1.6) should be used.

For convenience, the function

qplotreg(x, y, qval = c(0.2, 0.8), q = NULL, plotit = TRUE, xlab = 'X', ylab = 'Y', xout = FALSE, outfun = out, ...)

plots the quantile regression lines indicated by the argument qval. (If the argument q is specified, qval is taken to have the values stored in q.) By default, the 0.2 and 0.8 quantile regression lines are plotted. (The R function qregplots can be used as well.)

10.13.10 Methods Based on Estimates of the Optimal Weights

It is well known that when using weighted least squares, the optimal (most efficient) method of estimating regression parameters, when there is heteroscedasticity, is to use weights $w_i = 1/\sigma_i^2$, where σ_i^2 is the variance of ϵ_i. Several estimation procedures have been designed to handle heteroscedastic error terms based on this result. One such procedure was proposed by Cohen et al. (1993). The general strategy is to start with some robust estimator and then use the residuals to estimate appropriate weights based on an estimate of how the variance of ϵ_i varies with the predictor. Only a single predictor has been considered so far. The method can have high efficiency compared to OLS, and efficiency is very close to OLS when both x and ϵ have standard normal distributions. It remains unclear, in terms of efficiency, whether the method ever offers a practical advantage over various alternative estimators covered here. The method might be particularly effective when the predictor values are fixed and evenly spaced. Also, unlike many robust estimators, the percentile bootstrap method does not provide reasonably accurate confidence intervals for the parameters when n is small (Wilcox, 1996d). Because the practical value of the method needs more research, the lengthy computational details are not given here.

Wilcox (1996d) suggests a method of estimating σ_i^2 using a running interval smoother. (Smoothers are described in Chapter 11.) The efficiency of the resulting estimator compares well to OLS, and in various situations it offers a substantial advantage. At the moment, there seems to be little practical advantage to using this approach over other robust estimators that have good efficiency under heteroscedasticity.

Robinson (1987) suggests yet another estimator that uses a smoother to estimate σ_i^2. All indications are that it offers little advantage over OLS (Wilcox, 1996d), so no details are given. For a method based on the assumption that σ_i is given by some *known* function depending of x, β_1, and perhaps some additional unknown parameters, see Carroll and Ruppert (1982). For results on how the method performs when the function is incorrectly specified, see Mak (1992).

10.13.11 Projection Estimators

Maronna and Yohai (1993) derive yet another regression estimator called *projection regression*. Let $T(\mathbf{x}, y)$ be any estimating functional through the origin that is scale and affine equivariant. Let $s(\lambda'\mathbf{x})$ be a measure of scale based on the projection $\lambda'\mathbf{x}$. For any vectors β and λ having length p, let

$$A(\beta, \lambda) = |T(\lambda'\mathbf{x}, y - \beta\lambda')|s(\lambda'\mathbf{x}),$$

$$C(\beta) = \sup A(\beta, \lambda),$$

where the supremum is taken over all λ satisfying $\|\lambda\| = 1$. The projection estimate is the vector β that minimizes $C(\beta)$. Several variations of this estimator are considered by Maronna and Yohai. The variation given by their Eq. (3.11) was considered here but found to have relatively unsatisfactory efficiency under heteroscedasticity. Other variations have not been considered.

10.13.12 Methods Based on Ranks

Naranjo and Hettmansperger (1994) suggest estimating regression coefficients by minimizing

$$\sum_{i<j} c_{ij}|r_i - r_j|,$$

where c_{ij} are (Mallows) weights given by $c_{ij} = h(\mathbf{x}_i)h(\mathbf{x}_j)$,

$$h(\mathbf{x}_i) = \min\{1, (c/(\mathbf{x_i} - \mathbf{m_x})'\mathbf{C}^{-1}(\mathbf{x_i} - \mathbf{m_x}))^{\mathbf{a}/2}\}.$$

Letting $d_i = (\mathbf{x_i} - \mathbf{m_x})'\mathbf{C}^{-1}(\mathbf{x_i} - \mathbf{m_x})$, they suggest using $c = \text{med}\{d_i\} + 3\text{MAD}\{d_i\}$, where $\text{MAD}\{d_i\}$ means that MAD is computed using the d_i values and med is the median. They report that $a = 2$ is effective in uncovering outliers. The quantities $\mathbf{m_x}$ and $\mathbf{C}$ are the minimum volume ellipsoid estimators of location and scale described in Chapter 6. When $c_{ij} \equiv 1$, their method reduces to Jaeckel's (1972) method, which minimizes a sum that is a function of the ranks of the residuals. For the one-predictor case, they replace the minimum volume ellipsoid estimators $\mathbf{m}_x$ and $\mathbf{C}$ with $M_x = \text{med}\{x_i\}$ and $C = (1.483\text{MAD}\{x_i\})^2$, where $\text{MAD}\{x_i\}$ is the median of $|x_1 - M_x|, \ldots |x_n - M_x|$, and M_x is the median of the x_i values. The resulting breakdown point appears to be at least 0.15. Evidently, this method can be used to get good control over the probability of a Type I error when testing hypotheses about the regression parameters, even when the error term is heteroscedastic, but its power can be poor (Wilcox, 1995c). For rank-based diagnostic tools, see McKean et al. (1990). For other results and methods based on ranks, see Cliff (1994), Tableman (1990), Hettmansperger (1984), Gutenbrunner et al. (1993), Hettmansperger and McKean (1977), and Dixon and McKean (1996). (For results on a multivariate linear model, see Davis and McKean, 1993.)

The method derived by Hettmansperger and McKean (1977) does not protect against leverage points, but their method can be of interest when trying to detect curvature (McKean et al., 1990, 1993). Consequently, a brief discussion of their method seems warranted. Let $R(y_i - \mathbf{x}_i'\beta)$ be the rank associated with the ith residual. They determine β by minimizing Jaeckel's (1972) dispersion function given by

$$D(\beta) = \sum a(R(y_i - \mathbf{x}_i'\boldsymbol{\beta}))R(y_i - \mathbf{x}_i'\boldsymbol{\beta}),$$

where

$$a(i) = \phi(i/(n+1))$$

for a non-decreasing function ϕ defined on (0, 1) such that $\int \phi(u)du = 0$ and $\int \phi^2(u)du = 1$. Two common choices for ϕ are $\phi(u) = \sqrt{12}(u - 0.5)$ and $\phi(u) = \text{sign}(u - 0.5)$. The slope parameters can be estimated by minimizing $D(\beta)$, but the intercept cannot. One approach to estimating the intercept, which can be used when the error term has a skewed distribution, is to use the median of the residuals after the slope parameters have been estimated. Estimation and hypothesis testing can be done via the R function Rfit, described in Section 10.13.13, but as illustrated, there might be concerns when tied values occur among the dependent variable.

Another approach is to estimate β to be the vector of values that minimizes

$$\frac{1}{n} \sum a(R(y_i - \mathbf{x}_i'\boldsymbol{\beta}))|r_i|$$

(Hössjer, 1994). Hössjer shows that this estimator can be chosen with a breakdown point of 0.5, and he establishes asymptotic normality. However, Hössjer notes that poor efficiency can result with a breakdown point of 0.5 and suggests designing the method so that its breakdown point is between 0.2 and 0.3, but under normality the asymptotic relative efficiency is only 0.56. Despite this, the method might have a practical advantage when the error term is heteroscedastic, but this has not been determined. Yet another rank-based method was derived by Chang et al. (1999). It can have a breakdown point of 0.5, but direct comparisons with some of the better estimators in this chapter have not been made. For some related results, see Hettmansperger and McKean (2011) and Gong and Abebe (2012), as well as Kloke et al. (2009). (For some results on R estimators, see McKean and Sheather, 1991.) For results dealing with simultaneous confidence intervals, see Mansouri and Li (2019).

10.13.13 R Functions Rfit and Rfit.est

The R function

$$\text{Rfit(x,y)}$$

estimates regression parameters via Jaeckel's dispersion function given in the previous section, assuming that the R package Rfit (Kloke and McKean, 2012) has been installed. The function also tests hypotheses about the parameters assuming homoscedasticity. Heteroscedasticity can result in Type I error probabilities well above or below the nominal level. For example, if $y = \epsilon(|x| + 1)$, where both x and ϵ have standard normal distributions, the actual Type I error probability when testing H_0: $\beta_1 = 0$ at the 0.05 level, $n = 100$, is approximately 0.16. If $y = \epsilon/(|x| + 1)$, the actual level is approximately 0.06. Another concern is situations where tied values occur among the dependent variable: Power might be relatively poor. The R function

$$\text{Rfit.est(x,y)}$$

returns estimates of the slopes and intercepts in $coef, which is convenient when using a bootstrap method to test hypotheses. The function

$$\text{rfitv2(formula, data, subset, yhat0 = NULL, scores = Rfit::wscores, symmetric = FALSE,}$$
$$\text{TAU = 'F0', ...)}$$

avoids a computational problem that sometimes occurs with the R function Rfit.

10.13.14 Empirical Likelihood Type and Distance-Constrained Maximum Likelihood Estimators

Bondell and Stefanski (2013) derive an empirical likelihood type estimator. The derivation of the method assumes homoscedasticity. The efficiency of their estimator compares well to the MM-estimator when the sample size is small and the error term has a symmetric, heavy-tailed distribution. It is unknown how well their estimator performs when the error term has a skewed distribution or when there is heteroscedasticity. And currently there is no R function for applying it.

Yet another approach was derived by Maronna and Yohai (2015) using what they call a distance-constrained maximum likelihood estimator. Unlike the typical robust estimator, their approach assumes that a parametric family of distributions can be specified and they use some distance or discrepancy measure between densities. Their model assumes a homoscedastic error term.

10.13.15 Ridge Estimators: Dealing With Multicollinearity

Collinearity refers to a situation where two or more predictor variables are closely related to one another. For two variables, some measure of association might be used to detect collinearity, but it is possible for collinearity to exist between three or more variables, even if no pair of variables has a particularly high correlation. This is called multicollinearity. Generally, multicollinearity is a practical concern because it can result in relatively high standard errors when estimating the slope parameters of a linear regression model.

There are diagnostic tools for detecting multicollinearity, but encountering relatively high standard errors also depends on the nature of the matrix $\mathbf{C} = \mathbf{X}'\mathbf{X}$, where $\mathbf{X}$ is the $n \times p$ matrix containing the values of the independent variables. Spanos (2019) provides a detailed discussion of this issue. In particular, high correlations are neither necessary nor sufficient for there to be numerical concerns related to $\mathbf{C}$.

Hoerl and Kennard (1970) suggest using a generalization of the least squares regression estimator to deal with multicollinearity. The strategy is to find values for $\beta_0, \ldots, \beta_p$ that minimize

$$\sum \left(y_i - \beta_0 - \sum_{j=1}^{p} \beta_j x_{ij} \right)^2 + k \sum_{j=1}^{p} \beta_j^2, \tag{10.15}$$

where the non-negative bias parameter k is to be determined and the second term, $k \sum_{j=1}^{p} \beta_j^2$, is called a shrinkage penalty term. Increasing k has the effect of shrinking the estimates of the slope parameters toward zero. A so-called ridge estimator is used:

$$\hat{\boldsymbol{\beta}}(k) = (\mathbf{C} + k\mathbf{I}_p)^{-1}\mathbf{X}'\mathbf{y}, \tag{10.16}$$

which has been studied extensively. When $k > 0$, $\hat{\boldsymbol{\beta}}(k)$ is an unbiased estimator when $\beta_1 = \cdots = \beta_p = 0$, but otherwise it is biased. In essence, when $k > 0$, a ridge estimator downweights the least squares estimate toward zero (e.g., Montgomery et al., 2012).

Here, the bias parameter, k, is estimated based on results in Kibria (2003), which was also used by Lukman et al. (2014) when dealing with a robust analog of the ridge estimator. Based on some regression estimator, let

$$\hat{\sigma}^2 = \frac{1}{n - p - 1} \sum r_i^2,$$

where $r_1, \ldots, r_p$ are the usual residuals. Determine the orthogonal matrix $\mathbf{D}$ such that $\mathbf{D}'\mathbf{C}\mathbf{D} = \boldsymbol{\Lambda}$, where $\boldsymbol{\Lambda}$ is a diagonal matrix containing the eigenvalues of the matrix $\mathbf{C}$. Let $\boldsymbol{\gamma} = \boldsymbol{\Lambda}^{-1}(\mathbf{XD})'\mathbf{Y}$. Then the estimate of k is taken to be

$$\hat{k} = \frac{\hat{\sigma}^2}{(\prod_{j=1}^{p} \gamma_j)^{1/p}} \tag{10.17}$$

(cf. Suhail et al., 2020).

Liu (1993) proposes a generalization of the ridge estimator:

$$\hat{\boldsymbol{\beta}}(k, d) = (\mathbf{C} + k\mathbf{I}_p)^{-1}(\mathbf{C} + d\mathbf{I}_p)\hat{\boldsymbol{\beta}}_{\text{ols}}. \tag{10.18}$$

For results on estimating d, see Qasim et al. (2019). Aslam and Ahmad (2020) derive a modification of this estimator but currently robust versions have not been studied.

As is commonly done (e.g., Kan et al., 2013), a robust ridge estimator is taken to be

$$\tilde{\beta} = (\mathbf{C} + k)\mathbf{I})^{-1}\mathbf{C}\hat{\beta}_R, \tag{10.19}$$

where $\hat{\beta}_R$ is some robust regression estimator. For results when dealing with high-dimensional data, see Maronna (2011). Ertaş et al. (2017) derive a method for dealing with multicollinearity based on the M-estimator in Section 10.6 assuming homoscedasticity.

Two versions of a robust Liu estimator were derived by Ertaş et al. (2017). Let

$$W_1(k, d) = (\mathbf{C} + k\mathbf{I})^{-1}(\mathbf{C} - d\mathbf{I})$$

and

$$W_2(k, d) = (\mathbf{C} + k)\mathbf{I})^{-1}(\mathbf{C} - d(\mathbf{C} + k\mathbf{I})^{-1}\mathbf{C},$$

where $k > 0$ and $-\infty < d < \infty$. Then robust Liu estimators are given by

$$\hat{\beta}_L = W_1(k, d)\hat{\beta}_R \tag{10.20}$$

and

$$\tilde{\beta}_L = W_2(k, d)\hat{\beta}_R. \tag{10.21}$$

Ertaş et al. focus on the situation where $\hat{\beta}_R$ is an M-estimator. As usual, no single estimator dominates. But when there is sufficiently severe multicollinearity plus regression outliers, a robust Liu estimator can make a practical difference.

10.13.16 R Functions ols.ridge, rob.ridge, and rob.ridge.liu

The R function

$$\text{ols.ridge(x, y, k = NULL, xout = FALSE, outfun = outpro)}$$

computes the basic ridge estimator given by Eq. (10.15).

The R function

$$\text{rob.ridge(x, y, Regfun = tsreg, k = NULL, xout = FALSE, outfun = outpro, plotit = FALSE,}$$
$$\text{STAND = TRUE, INT = FALSE, locfun = median, ...)}$$

computes a robust ridge estimator where the choice for $\hat{\beta}_R$ is controlled via the argument Regfun. STAND=TRUE means the variables are standardized. The argument locfun determines how the intercept is estimated. The R function

$$\text{rob.ridge.liu(x, y, k = NULL, d = NULL, OP = TRUE, Regfun = tsreg, locfun = median, xout}$$
$$\text{= FALSE, outfun = outpro, STAND = FALSE)}$$

computes the robust Liu estimators. The argument OP=TRUE means that $\tilde{\beta}_L$ given by Eq. (10.21) is computed and OP=FALSE returns $\hat{\beta}_L$.

10.13.17 Robust Elastic Net and Lasso Estimators: Reducing the Number of Independent Variables

A close competitor of the ridge estimator is the lasso (e.g., Efron and Hastie, 2016). An appeal of the lasso is that it is better suited for variable selection (e.g., James et al., 2017, p. 219). Typically the lasso estimates a number of the slope parameters to be equal to zero, suggesting that the corresponding independent variables can be eliminated, which in turn results in a more interpretable linear regression model.

However, Zou and Hastie (2005) state that if there are high correlations between predictors, it has been empirically observed that the prediction performance of the lasso is dominated by ridge regression. James et al. (2017, pp. 234–235) illustrate the advantage of the ridge estimator over the lasso in terms of mean squared error. Frank and Friedman (1993) compare the performances of several estimators in terms of prediction error, including partial least squares (PLS) and principal components regression (PCR). They conclude that "in all of the situations studied, the ridge estimator dominated the other methods, closely followed by PLS and PCR, in that order."

The basic lasso estimator is based on finding values for $\beta_0, \ldots, \beta_p$ that minimize

$$\sum \left(y_i - \beta_0 - \sum_{j=1}^{p} \beta_j x_{ij} \right)^2 + k \sum_{j=1}^{p} |\beta_j|. \tag{10.22}$$

The only difference from the ridge estimator is the second term. A generalization is the adaptive lasso derived by Zou (2006). Now the goal is to determine values for $\beta_0, \ldots, \beta_p$ that minimize

$$\sum \left(y_i - \beta_0 - \sum_{j=1}^{p} \beta_j x_{ij} \right)^2 + k \sum_{j=1}^{p} w_j |\beta_j|, \tag{10.23}$$

where the weight w_j is based in part on the basic lasso estimate of β_j given by Eq. (10.22).

Various generalizations of the basic lasso estimator have been derived in an effort to achieve robustness. One approach is to replace the left term of Eq. (10.22) with a Huber-type loss function (Lambert-Lacroix and Zwald, 2011). Another approach is to replace Eq. (10.23) with

$$\alpha \sum \left(y_i - \beta_0 - \sum_{j=1}^{p} \beta_j x_{ij} \right)^2 + (1-\alpha) \sum \left| Y_i - \beta_0 - \sum_{j=1}^{p} \beta_j X_{ij} \right| + k \sum_{j=1}^{p} w_j |\beta_j| \quad (10.24)$$

(Zheng et al., 2016). This is called the robust adaptive (RA) lasso. Setting $\alpha = 0$ and $k = n$ yields the least absolute deviation (LAD) lasso estimator proposed by Wang et al. (2007). Estimation of α assumes that the error term has a symmetric distribution. Assuming the error term has a symmetric distribution, results reported by Zheng et al. indicate that in any given situation, something other than the Huber-type estimator performs better in terms of prediction error as well as the correct number of coefficients estimated to be zero. Some additional methods are a so-called outlier shift method derived by Jung et al. (2016) and MM-lasso methods derived by Smucler and Yohai (2017). The relative merits of these methods need further study.

Zou and Hastie (2005) derive an improvement of the basic lasso method, called an elastic net, that encourages a grouping effect where strongly correlated predictors tend to be in or out of the model together. In addition, it is better able to handle situations where p is larger than n. What they term the naive elastic net determines $\beta_0, \ldots, \beta_p$ that minimize

$$\sum \left(y_i - \beta_0 - \sum_{j=1}^{p} \beta_j x_{ij} \right)^2 + \lambda_1 \sum_{j=1}^{p} |\beta_j| + \lambda_2 \sum_{j-1}^{p} \beta_j^2, \quad (10.25)$$

where λ_1 and λ_2 are fixed non-negative constants. The final estimate is a rescaled estimate of the naive estimate. Robust versions have been derived by Kurnaz et al. (2017), Yi and Huang (2016), and Liu et al. (2015). Let r_i^s denote the standardized residuals and let $w_i = 1$ if $|r_i^s| \le 2.2414$; otherwise $w_i = 0$ $(i = 1, \ldots, n)$. The reweighted estimator derived by Kurnaz et al. are the values of $\beta_0, \ldots, \beta_p$ that minimize

$$\sum w_i \left(y_i - \beta_0 - \sum_{j=1}^{p} \beta_j x_{ij} \right)^2 + \lambda n_w \left(\alpha \sum_{j=1}^{p} |\beta_j| + 0.5(1-\alpha) \sum_{j-1}^{p} \beta_j^2 \right), \quad (10.26)$$

where $n_w = \sum w_i$ and α and λ are tuning parameters determined via cross-validation. The Yi and Huang estimator replaces (10.26) with

$$\frac{1}{n} \sum \ell \left(y_i - \beta_0 - \sum_{j=1}^{p} \beta_j x_{ij} \right) + \lambda \left(\alpha \sum_{j=1}^{p} |\beta_j| + 0.5(1-\alpha) \sum_{j-1}^{p} \beta_j^2 \right), \quad (10.27)$$

where $\ell\left(y_i - \beta_0 - \sum_{j=1}^{p}\beta_j x_{ij}\right)$ is the Huber loss function in Table 2.1. A quantile regression estimator is obtained by replacing the Huber loss function with the loss function given by Eq. (10.14).

As previously noted, when a slope is estimated to be zero via the methods in this section, this suggests that it is not relevant. However, these methods do not reflect the strength of the empirical evidence that some subset of the independent variables is indeed superior to the other independent variables that might be used. For example, with $p = 4$ independent variables, the RA lasso might suggest that independent variables 1 and 3 are relatively important and that independent variables 2 and 4 are not. Method IBS, described in Section 11.10.5, provides a test of the hypothesis that variables 1 and 3 are indeed more important. The method can be applied via the R function regIVcom, described in Section 11.10.6. The resulting p-value reflects the strength of the empirical evidence that a decision can be made that variables 1 and 3 are indeed more important. However, results on how well the R function regIVcom performs are limited to situations where the number of explanatory variables, p, is relatively small. The methods in this section were developed with the goal of dealing with p relatively large, assuming that a linear model is reasonable.

10.13.18 R Functions lasso.est, lasso.rep, RA.lasso, LAD.lasso, H.lasso, HQreg, and LTS.EN

The R function

   lasso.est(x, y, xout = FALSE, STAND = TRUE, outfun = outpro, regout = FALSE,
                    lam = NULL, ...)

applies the basic lasso method given by Eq. (10.22). Setting the argument regout=TRUE, regression outliers are removed. Due to the algorithm used, running this function twice on the same data can alter which slopes are estimated to be zero. This occurs because the R function uses a different seed for the random number generator when used again. Consequently, the R function

   lasso.rep(x, y, lasso.fun = lasso.est, xout = FALSE, outfun = outpro, iter = 20, EST = FALSE,
                    SEED = TRUE)

is supplied. It calls lasso.est several times. If any of the slope estimates are equal to zero, the final estimate is set equal zero. The R function

   RA.lasso(x, y, STAND = TRUE, grid = seq(log(0.01), log(1400), length.out = 100),
                    xout = FALSE, outfun = outpro, ...)

performs the robust adaptive lasso derived by Zheng et al. (2016). The LAD lasso derived by Wang et al. (2007) can be applied via the R function

LAD.lasso(x, y, lam = NULL, WARN = FALSE, xout = FALSE, outfun = outpro, STAND = TRUE).

The R function

H.lasso(x, y, lambda.lasso.try = NULL, k = 1.5, STAND = TRUE, xout = FALSE, outfun = outpro, ...)

performs the Huber-type lasso.

The robust elastic net method derived by Yi and Huang (2016) can be computed via the R function

HQreg(x,y,alpha=1,xout=FALSE,method='huber',tau=.5,outfun=outpro,..).

The argument alpha=1, which corresponds to α in Eq. (10.27), means that the robust lasso method is used. Using alpha=0 results in the ridge estimator, and alpha=0.5 is a balance between the two methods. Setting the argument method='quantile', quantile regression is performed where the argument tau indicates the quantile to be used. The R function

LTS.EN(x,y,xout=FALSE,family='gaussian', alphas=NULL, lambdas=NULL, outfun=outpro,...)

performs the robust elastic net derived by Kurnaz et al. (2017). By default, it uses cross-validation to determine the values of the tuning parameters α and λ in Eq. (10.26) that minimize the root mean squared prediction error. (Root mean squared prediction error is the square root of the average of the squared residuals.) In contrast, Yi and Huang use the mean absolute prediction error and they also consider error as measured by Eq. (10.14). These last two R functions are based on the R packages hqreg and enetLTS, respectively, which have more options than indicated here.

10.14 Comments About Various Estimators

A few additional comments about the various regression estimators in this chapter might help. We have seen illustrations that certain estimators can have high efficiency compared to OLS when the error term is heteroscedastic. Here, some additional results relevant to this issue are summarized. Let R be the standard error of the OLS estimator divided by the standard error of some competing estimator. So if R is less than 1, least squares tends to be more accurate, while $R > 1$ indicates the opposite. Suppose observations are generated according to the model $y = x + \lambda(x)\epsilon$, where the function $\lambda(x)$ reflects heteroscedasticity. Setting $\lambda(x) = 1$ corresponds to the homoscedastic case. Table 10.3 shows estimates of R (based on simulations with 5,000 replications) for the Theil–Sen (TS) estimator, the MGV estimator, the deepest regression line estimator (T^*), TSTS (described in Section 10.13.3), and the MM-estimator, where VP 1 corresponds to $\lambda(X) = 1$, VP 2 is where $\lambda(x) = x^2$, and VP 3 means $\lambda(x) = 1/|x|$. So for VP 2, the error term has more variance corresponding to extreme x values and VP 3 is a situation where the opposite is true. The results are limited to situations where the distribution for x is symmetric, but very similar results are obtained when x has an asymmetric distribution instead. In Table 10.3, the distributions for ϵ are taken to be normal (N), a g-and-h distribution with $(g, h) = (0, 0.5)$, which is symmetric and heavy-tailed (SH), a g-and-h distribution with $(g, h) = (0.5, 0)$, which is asymmetric and relatively light-tailed (AL), and a g-and-h distribution with $(g, h) = (0.5, 0.5)$, which is asymmetric and relatively heavy-tailed (AH). Note that all five estimators in Table 10.3 generally compete well with the OLS estimator, the main exception being a situation in which x has a symmetric, heavy-tailed distribution and ϵ has a normal distribution. The MMreg estimator is a popular choice among some statisticians and it performs relatively well for VP 3 for the situations considered in Table 10.3. But for VP 1 and VP 2, the MM-estimator is less satisfactory. Also some caution is needed when using the MM-estimator for reasons described in Section 10.14.1. The OP-estimator, described in Section 10.10, is not included in Table 10.3, but it is noted that it performs in a manner very similar to the MGV estimator.

It is not being suggested that if there is heteroscedasticity, it necessarily follows that the robust regression estimators considered here will have a substantially smaller standard error than the least squares estimator. In some situations these robust estimators offer little or no advantage in terms of efficiency. Also, there is a connection between the types of heteroscedasticity considered here and regression outliers. Variance pattern VP 3, for example, has a tendency to generate regression outliers that can have a relatively large effect on the standard error of the least squares estimator.

10.14.1 Contamination Bias

Clearly, one goal underlying robust regression is to avoid situations where a small number of points can completely dominate an estimator. In particular, a goal is to avoid getting a poor fit to the bulk of the points. An approach toward achieving this goal is to require that an estimator have a reasonably high finite sample breakdown point. But there are some regression

Table 10.3: Estimated ratios of standard errors, x distribution symmetric, $n = 20$.

x	ϵ	VP	TS	MGV	T^*	TSTS	MM
N	N	1	0.91	0.91	0.76	0.88	0.98
		2	2.64	2.62	3.11	2.36	1.95
		3	96.53	77.70	67.72	100.86	116.27
N	SH	1	4.28	4.27	4.42	3.51	1.13
		2	10.67	10.94	11.03	8.66	1.97
		3	70.96	65.20	57.84	82.019	83.61
N	AL	1	1.13	1.13	0.92	1.05	0.95
		2	3.21	3.21	3.69	2.84	1.96
		3	96.52	77.69	67.72	100.86	116.26
N	AH	1	8.89	8.85	16.41	7.05	1.18
		2	26.66	27.07	25.81	20.89	1.94
		3	70.96	65.20	57.84	82.01	83.61
SH	N	1	0.81	0.80	0.61	0.76	0.95
		2	40.57	42.30	55.47	27.91	7.28
		3	106.50	49.79	58.82	106.28	151.61
SH	SH	1	3.09	2.78	2.88	2.41	1.12
		2	78.43	83.56	90.84	47.64	5.94
		3	106.50	49.79	58.82	106.28	151.61
SH	AL	1	0.99	0.87	0.73	0.90	1.20
		2	46.77	49.18	63.60	31.46	2.54
		3	106.50	49.79	58.82	106.28	151.61
SH	AH	1	6.34	5.64	6.75	4.62	1.19
		2	138.53	146.76	108.86	78.35	2.47
		3	81.95	42.27	53.54	92.65	107.26

N=Normal; SH=Symmetric, heavy-tailed;
AL=Asymmetric, light-tailed; AH=Asymmetric, heavy-tailed.

estimators that, despite having a breakdown point reasonably close to 0.5, can be greatly influenced by a few outliers.

■ **Example**

Twenty points were generated where both x and ϵ have a standard normal distribution, and $y = x + \epsilon$ was computed, so the true slope is 1. Then two aberrant points were added to the data at $(x, y) = (2.1, -2.4)$. Fig. 10.2 shows a scatterplot of the points plus the LTS regression line, which has an estimated slope of -0.94. So in this case, LTS is a complete disaster in terms of detecting how the majority of the points were

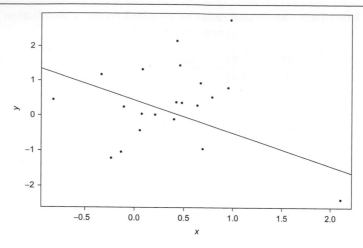

Figure 10.2: Scatterplot of 20 points where $y = x + \epsilon$, with x and ϵ having independent, standard normal distributions, plus two outliers at $(2.1, -2.4)$. The straight line is the LTS regression line, which poorly estimates the slope used to generate the majority of the points.

generated. The least squares estimate is -0.63. The Coakley–Hettmansperger estimator relies on LTS as an initial estimate of the slope, and despite its high breakdown point, the estimate of the slope is -0.65. The MM-estimator, described in Section 10.9.1, estimates the slope to be -0.12. In contrast, the MGV estimate of the slope is 0.97 and the OP-estimate is 0.89. The deepest regression line estimate is 0.66, and the STS estimator (described in Section 10.13.1) performs poorly in this particular case, the estimate being -0.98. The LMS estimate of the slope is 1.7, so it performs poorly as well in this instance. The TSTS estimator, which is an E-type estimator described in Section 10.13.3, yields an estimate of -0.05, and the Gervini–Yohai estimator, described in the same section, estimates the slope to be 1.49. The main point is that the choice of robust estimator is not an academic issue, but this one example is not intended as an argument that the estimators that perform poorly in this particular case should be excluded from consideration. Rather, the point is that despite the robust properties they enjoy, they can perform poorly in some situations where other methods do well. Also, although both the OP and MGV estimators do very well here, this is not to suggest that they be used to the exclusion of all other methods. ∎

To add perspective, the process used to generate the data in Fig. 10.2 was repeated 500 times, and estimates of the slope were computed using least squares, the M-estimator with Schweppe weights (using the R function bmreg), the Coakley–Hettmansperger estimator (chreg), the Theil–Sen estimator (tsreg), and the deepest regression line (depreg). Boxplots of the results

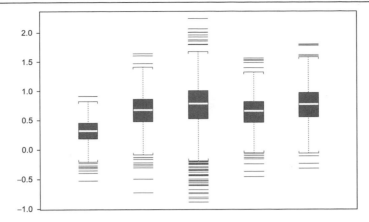

Figure 10.3: Boxplots of 500 estimates of the slope when data are generated as in Fig. 10.2. From left to right, the boxplots are based on the OLS estimator, an M-estimator with Schweppe weights, the Coakley–Hettmansperger estimator, Theil–Sen, and the deepest regression line. All five estimators suffer from contamination bias.

are shown in Fig. 10.3. Notice that the median of all these estimators differs from 1, the value being estimated. This illustrates that these estimators can be sensitive to a type of *contamination bias*. That is, despite having a reasonably high finite sample breakdown point, it is possible for a few unusual points to result in a poor fit to the bulk of the observations. So these estimators, plus many other robust estimators, can provide substantial advantages versus least squares, but they do not eliminate all practical concerns.

Fig. 10.4 shows the results when using the LTS, LTA, MGV, and OP-estimators (with default settings). The LTA estimator gives results similar to LTS, and the OP-estimator gives results similar to MGV. In contrast to the estimators in Fig. 10.3, the median of all the estimators is approximately 1. So in this particular situation, these estimators do a better job of avoiding contamination bias. Note that there is considerably more variation among the LTS estimates based on a breakdown point of 0.5.

There is some evidence that generally, the STS estimator gives a better fit to the majority of points than LTS and LMS. In particular, it seems common to encounter situations where STS is less affected by a few aberrant points. However, exceptions occur, as is illustrated next, so again it seems that multiple methods should be considered.

■ Example

Fig. 10.5 shows 20 points that were generated in the same manner as in Fig. 10.2. So the two aberrant points located at $(x, y) = (2.1, -2.4)$ are positioned relatively far from

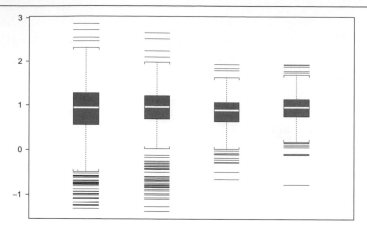

Figure 10.4: Boxplots of 500 estimates of the slope. From left to right, the boxplots are based on the LTS estimator, the LTA estimator, the OP-estimator, and the MGV estimator. In contrast to the estimators used in Fig. 10.3, all four estimators avoid the contamination bias problem; each is an approximately median unbiased estimator of the slope.

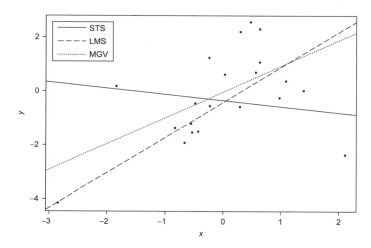

Figure 10.5: Scatterplot of 20 points where $y = x + \epsilon$, with x and ϵ having independent, standard normal distributions, plus two outliers at $(2.1, -2.4)$. This illustrates that only two outliers can have a substantial impact on the STS estimator. The LMS estimate of the slope is 1.3 and the MGV estimate is 0.96.

the true regression line, which again has a slope of 1 and an intercept of 0. Also shown in Fig. 10.5 are the STS, LMS, and MGV estimates of the regression line. In this particular case, STS performs poorly; the estimated slope is -0.23. The LMS estimate of the slope is 1.3 and the estimated slope and intercept based on the MGV estimator are 0.96 and -0.03, respectively, which are closer to the true values compared with the other estimates considered here. The OP-estimates are 1.26 and -0.34. The estimated slope based on the TSTS estimator is 0.58 and least squares returns an estimate of 0.4. So once again we see that the choice of which robust estimator to use can make a substantial difference in how the association between x and y is summarized.

It is stressed, however, that while several estimators compete well with least squares, it is fairly easy to find faults with any estimator that has been proposed. For example, both the MGV and OP-estimators have several practical advantages over many other estimators that might be used. But using software written exclusively in R, the execution time required for both of these estimators can be relatively high when using a bootstrap method to test hypotheses. But access to a multicore processor can substantially reduce execution time.

It should be noted that in the theoretical literature, the term contamination bias is used in a different manner. To begin with a simple case, first focus on location estimators. Let T be any functional as described in Chapter 2 and let $T(F) = \theta$. The *contamination bias* associated with T is

$$\sup|T((1 - \epsilon)F + \epsilon G) - \theta|,$$

where $0 \leq \epsilon \leq 1$ and the supremum is taken overall all distributions G. Huber (1964) establishes results on the contamination bias of the median and various extensions have appeared in the literature (e.g., He and Simpson, 1993).

The idea can be extended to regression estimators. Following, for example, Maronna and Yohai (1993), let **V** be any affine equivariant scatter matrix associated with F, let $H = (1 - \epsilon)F + \epsilon G$, and let $T(F) = \beta$ be the vector of regression parameters. The bias at G is

$$[(T(H) - \beta)'\mathbf{V}^{-1}(T(H) - \beta)]^{1/2}.$$

Maronna and Yohai (1993) derive results related to the maximum bias of the projection estimator. Variations on this approach are described by He and Simpson (1993).

10.15 Outlier Detection Based on a Robust Fit

Several outlier detection methods have been proposed that are based in part on first fitting a robust regression model assuming that Eq. (10.1) is true. Typically these methods assume a homoscedastic error term. Given the issue of contamination bias, it would seem that they should be used with caution. And the issue of how to deal with a heteroscedastic error term seems to warrant consideration. Billor and Kiral (2008) compare several techniques. No single method dominates and the most effective method depends to some extent on where the outliers occur. Nurunnabi et al. (2016) suggest using a six-step process to identify influential points.

10.15.1 Detecting Regression Outliers

Rousseeuw and van Zomeren (1990) suggest using the LMS estimator to detect what are called *regression outliers*. Roughly, these are points that deviate substantially from the linear pattern for the bulk of the points under study. Their method begins by computing the residuals associated with LMS regression, $r_1, \ldots, r_n$. Next, let M_r be the median of $r_1^2, \ldots, r_n^2$, the squared residuals, and let

$$\hat{\tau} = 1.4826 \left(1 + \frac{5}{n - p - 1} \right) \sqrt{M_r}.$$

The point $(y_i, x_{i1}, \ldots, x_{ip})$ is labeled a regression outlier if the corresponding standardized residual is large. In particular, Rousseeuw and van Zomeren label the ith vector of observations a *regression outlier* if $|r_i| / \hat{\tau} > 2.5$.

10.15.2 R Functions reglev and rmblo

The R function

$$\text{reglev(x,y,plotit=TRUE)}$$

is provided for detecting regression outliers and leverage points using the method described in the previous section. If the ith vector of observations is a regression outlier, the function stores the value of i in the R variable reglev$regout. If $\mathbf{x}_i$ is an outlier based on the method in Section 6.4.3, it is declared a leverage point and the function stores the value of i in reglev$levpoints. The plot created by this function can be suppressed by setting plotit=FALSE. The R function

$$\text{rmblo(x,y)}$$

removes bad leverage points and returns the remaining data.

■ **Example**

If the reading data, described in Section 10.8.1, are stored in the R variables x and y, the command reglev(x,y) returns

```
$levpoints:
[1] 8

$regout:
```

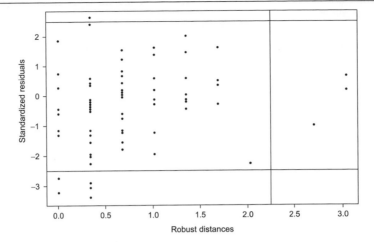

Figure 10.6: The plot created by the function reglev based on the reading data. Points to the right of the vertical line located at 2.24 on the x-axis are declared leverage points. Points beyond the two horizontal lines are declared regression outliers.

```
[1]   12 44 46 48 59 80
```

This says that x_8 is flagged as a leverage point (it is an outlier among the x values), and the points (y_{12}, x_{12}), (y_{44}, x_{44}), (y_{46}, x_{46}), (y_{48}, x_{48}), (y_{59}, x_{59}), and (y_{80}, x_{80}) are regression outliers. Note that even though x_8 is an outlier, the point (y_8, x_8) is not a regression outlier. For this reason, x_8 is called a *good leverage point*. (Recall that extreme x values can lower the standard error of an estimator.) If (y_8, x_8) had been a regression outlier, x_8 would be called a *bad leverage point*. Regression outliers, for which x is not a leverage point, are called *vertical outliers*. In the illustration all of the regression outliers are vertical outliers as well.

■

■ Example

The reading data used in the last example are considered again, only the predictor is now taken to be the data in column 3 of the file read.dat, which is another measure of phonological awareness called sound blending. The plot created by reglev is shown in Fig. 10.6. Points below the horizontal line that intersects the y-axis at -2.24 are declared regression outliers, as are points above the horizontal line that intersects the y-axis at 2.24. There are three points that lie to the right of the vertical line that intersects the x-axis at $\sqrt{\chi^2_{0.975,p}} = 2.24$; these points are flagged as leverage points. These three points are not flagged as regression outliers, so they are deemed to be good leverage points.

■

10.16 Logistic Regression and the General Linear Model

A common situation is where the outcome variable y is binary. In the context of regression, a general approach is to assume that

$$P(y = 1|\mathbf{X} = \mathbf{x}) = F(\mathbf{x}'\boldsymbol{\beta}), \tag{10.28}$$

where F is some strictly increasing cumulative distribution function and $\boldsymbol{\beta}$ is a vector of unknown parameters. A common choice for F is

$$F(t) = \frac{\exp(t)}{1 + \exp(t)},$$

which yields what is generally known as the logistic regression model. That is, assume that

$$P(y = 1|\mathbf{X} = \mathbf{x}) = \frac{\exp(\beta_0 + \beta_1 x_1 + \cdots + \beta_p x_p)}{1 + \exp(\beta_0 + \beta_1 x_1 + \cdots + \beta_p x_p)}. \tag{10.29}$$

The maximum likelihood estimator of $\boldsymbol{\beta} = (\beta_0, \ldots, \beta_p)$ is the vector $\mathbf{b} = (b_0, \ldots, b_p)$ that minimizes

$$\sum D(y_i, \mathbf{x}'_i \mathbf{b}),$$

where

$$D(y_i, \mathbf{x}'_i \mathbf{b}) = -y_i \ln(F(\mathbf{x}'_i \mathbf{b})) - (1 - y_i) \ln(1 - F(\mathbf{x}'_i \mathbf{b})).$$

For results on dealing with multicollinearity, see Asar and Wu (2019), as well as Etran and Akay (2020). This maximum likelihood estimator is routinely used, but it is not robust. Roughly, leverage points can have an inordinate influence on the estimates. Croux et al. (2002) discuss its breakdown point. Here the focus is on a variation of a robust estimator derived by Bianco and Yohai (1996), which is motivated by results in Croux and Haesbroeck (2003). Also see Bianco and Martínez (2009). The robust estimate is the value of $\mathbf{b}$ that minimizes

$$\sum w_i \phi(y_i, \mathbf{x}_i \mathbf{b}), \tag{10.30}$$

where

$$\phi(y,t) = y\rho(-\ln(F(t))) + (1-y)\rho(-\ln(1-F(t))) + G(F(t)) + G(1-F(t)) - G(1),$$

$$G(t) = \int_0^t \psi(-\ln u)du,$$

$\psi(t) = \rho'(t)$, and

$$\rho(t) = \begin{cases} te^{-\sqrt{c}}, & \text{if } t \leq c, \\ -2e^{-\sqrt{c}}(1+\sqrt{t}) + e^{-\sqrt{c}}(2(1+\sqrt{c})+c), & \text{if } t > c, \end{cases}$$

and c is a constant. Following the suggestion by Croux and Haesbroeck, $c = 0.5$ is used here. (Croux and Haesbroeck also provide an analytic form for $G(t)$.) For additional results related to robust estimators for the logistic regression model, some of which are derived in the more general framework of the general linear model outlined in Section 10.16.2, see Pregibon (1987), Carroll and Pedersen (1993), Christmann (1994), Rousseeuw and Christmann (2003), Stefanski et al. (1986), Künsch et al. (1989), Morgenthaler (1992), and Bondell (2005, 2008). When computing confidence intervals for the parameters in this model, the percentile bootstrap method in Section 11.1.3 is recommended. The R function wlogregci, described in Section 11.1.4, performs the calculations (cf. Victoria-Feser, 2002). For methods aimed at assessing the fit of the model, see Shu and Wenqing (2017). For results on a ridge estimator, see for example Varathan and Wijekoon (2019).

For a single independent variable, there is a well-known method for computing a confidence interval for $P(y = 1|x)$ (e.g., Brand et al., 1973; Khorasani and Milliken, 1982). The method is reasonable if there is no association. If there is an association, even when the model is slightly inaccurate, the actual probability coverage can differ substantially from the nominal level (Wilcox, 2019e). A method aimed at dealing with this issue is described in Section 11.5.4.

10.16.1 R Functions glm, logreg, logreg.pred, wlogreg, logreg.plot, logreg.P.ci, and logistic.lasso

The built-in R function glm can be used to compute the maximum likelihood estimate of the parameters in the logistic regression model. And the R function summary tests hypotheses. If, for example, the data are stored in the R variables x and y, use the commands

fit=glm(formula=y~x,family=binomial),

summary(fit).

For convenience, the R function

logreg(x, y, xout = FALSE, outfun = outpro, plotit = FALSE)

is provided, which performs both of the R commands glm and summary. The function also removes any leverage points if the argument xout=TRUE; it will use the outlier detection method indicated by the argument outfun. By default, the projection method in Section 6.4.9 is used. For a single predictor, if the argument plotit=TRUE, the regression line will be plotted. To compute the probability of success for one or more values of the independent variable, the R function

logreg.pred(x, y, pts, xout = FALSE, outfun = outpro)

can be used.

A weighted version of the Bianco–Yohai estimator suggested by Croux and Haesbroeck (2003), and supported by results reported by Ahmad et al. (2010), can be computed with the R function

wlogreg(x, y).

The function wlogreg returns estimates of the standard errors, but using them to compute confidence intervals and test hypotheses is not recommended. (Use the R function wlogregci, which is described in Chapter 11.) Finally, the R function

logreg.plot(x, y, MLE = FALSE, ROB = TRUE, xlab = 'X', ylab = 'P(X)')

plots the robust estimate of the regression line, assuming there is a single predictor. To plot the usual (maximum likelihood) estimate simultaneously, set the argument MLE=TRUE. If ROB=FALSE, the robust regression line is not plotted.

Note that there are two ways of dealing with leverage points. Use the R function logreg with xout=TRUE, or use the Bianco–Yohai estimator via the R function wlogreg. In terms of achieving a relatively small standard error, all indications are that the Bianco–Yohai estimator is preferable to using the R function logreg with xout=TRUE. However, each method reacts differently to outliers and it is not completely clear which is preferable for general use in terms of achieving relatively high power and short confidence intervals.

The R function

logreg.P.ci(x,y, alpha = 0.05, plotit = TRUE, xlab = 'X', ylab = 'P(Y=1|X)', xout = FALSE, outfun = outpro, ...)

computes a confidence interval for $P(y = 1|x)$ for each value stored in the argument x. The output includes a plot of the confidence intervals. When the logistic regression model is true, such as when there is no association, the resulting confidence intervals are relatively accurate. But when there is an association, also consider the R function rplot.binCI, described in Section 11.5.9.

A lasso version of the logistic regression estimator can be computed via the R package glmnet. To simplify the removal of leverage points when it is desired to do so, the R function

$$\text{logistic.lasso(x,y,xout=FALSE,outfun=outpro)}$$

can be used.

10.16.2 The General Linear Model

Briefly, the *general linear model* model consists of three components. The first is the assumption that an outcome variable y has a distribution that belongs to the exponential family. This family of distributions includes the normal, binomial, Poisson, and gamma distributions as special cases. (In practice, one specifies which of these distributions will be assumed.) It is further assumed that the independent random variables $y_1, \ldots, y_n$ have the same distribution. In the context of regression, typically homoscedasticity is assumed. (But some generalized linear models are designed to allow heteroscedasticity.) The second component is a set of p predictors $\mathbf{x}$ and associated parameters β. And the third component is a monotone link function g that satisfies

$$g(\mu) = \mathbf{x}\beta.$$

If the link function is taken to be the identity function, we get the usual linear model given by Eq. (10.1). A class of M-estimators for the generalized linear model has been derived, a summary of which can be found in Heritier et al. (2009, Section 5.3). For results on testing hypotheses, see Cantoni and Ronchetti (2001). Here it is noted that the generalized linear model provides yet another approach to logistic regression, and it has the advantage of being able to handle discrete data assuming that y has a Poisson distribution.

10.16.3 R Function glmrob

Robust estimation and hypothesis testing can be performed via the generalized linear model just described using the R function

$$\text{glmrob(formula, family, data),}$$

which belongs to the R package robustbase. (Hypothesis testing is accomplished with the R function summary.) Mallows- or Huber-type robust estimators, as described in Cantoni and Ronchetti (2001), are used. In principle, this class of M-estimators can handle continuous outcomes, but currently the R function glmrob only allows discrete outcomes where y has a binomial or Poisson distribution. That is, the argument family can be equal to "binomial" or "poisson." (When dealing with continuous outcomes, methods for testing hypotheses that perform well when there is heteroscedasticity are described in Chapter 11.)

10.17 Multivariate Regression

Consider a regression problem where there are p predictors $\mathbf{x}' = (x_1, \ldots, x_p)$ and q responses $\mathbf{y} = (y_1, \ldots, y_q)$. The usual multivariate regression model is

$$\mathbf{y} = \mathbf{B}'\mathbf{x} + \mathbf{a} + \boldsymbol{\epsilon}, \tag{10.31}$$

where $\mathbf{B}$ is a $(p \times q)$ slope matrix, $\mathbf{a}$ is a q-dimensional intercept vector, and the errors $\boldsymbol{\epsilon} = (\epsilon_1, \ldots, \epsilon_q)$ are independent and identically distributed with mean $\mathbf{0}$ and covariance matrix Σ_ϵ, a positive definite matrix of size q. Let $\boldsymbol{\mu}$ be some measure of location associated with the joint distribution of $(\mathbf{x}, \mathbf{y})$ and let Σ be some measure of scatter. Partitioning $(\mathbf{x}, \mathbf{y})$ and Σ in an obvious way yields

$$\boldsymbol{\mu} = \begin{pmatrix} \mu_x \\ \mu_y \end{pmatrix} \quad \text{and} \quad \Sigma = \begin{pmatrix} \Sigma_{xx} & \Sigma_{xy} \\ \Sigma_{yx} & \Sigma_{yy} \end{pmatrix}.$$

In practice, Eq. (10.31) is typically assumed and the most common choice for $\boldsymbol{\mu}$ is the population mean, which is estimated with the usual sample mean, say $\hat{\boldsymbol{\mu}}$, and the estimate of Σ is typically taken to be the usual covariance matrix, say $\hat{\Sigma}$. The resulting estimates of $\mathbf{B}$ and $\mathbf{a}$ are

$$\hat{\mathbf{B}} = \hat{\Sigma}_{xx}^{-1} \hat{\Sigma}_{xy} \tag{10.32}$$

and

$$\hat{\mathbf{a}} = \hat{\boldsymbol{\mu}}_y - \hat{\mathbf{B}}' \hat{\boldsymbol{\mu}}_x, \tag{10.33}$$

respectively. The estimate of the covariance matrix associated with the error term, ϵ, is

$$\hat{\Sigma}_\epsilon = \hat{\Sigma}_{yy} - \hat{\mathbf{B}}' \hat{\Sigma}_{xx} \hat{\mathbf{B}}. \tag{10.34}$$

It is well known, however, that this classic estimator is extremely sensitive to outliers. Another concern is that when $q = 1$, it is known that the efficiency of the least squares estimator can be poor relative to other estimators that might be used.

Another point worth mentioning is that the estimator just described does not take into account the overall structure of the y values. Indeed, it is tantamount to simply computing the least squares regression line for each of the q response variables $y_1, \ldots, y_q$ (e.g., Jhun and Choi, 2009). From a robustness point of view, a simple strategy is to mimic this approach with some robust estimator. The remainder of this section summarizes some alternative estimators that have been proposed. (Also see She and Chen, 2017.)

10.17.1 The RADA Estimator

Let $\mathbf{z} = (\mathbf{x}, \mathbf{y})$ and let $\mathbf{z}_i$ $(i = 1, \ldots, n)$ be a random sample of size n. Rousseeuw et al. (2004) propose three robust alternatives to Eqs. (10.32) and (10.33), and they recommend one for general use based on simulation estimates of its efficiency. They begin by computing the MCD estimate based on $\mathbf{z}_i$ $(i = 1, \ldots, n)$. Recall that the MCD estimator searches for a subset $\{\mathbf{z}_{i_1}, \ldots, \mathbf{z}_{i_h}\}$ of size h whose covariance matrix has the smallest determinant, where $\lceil n/2 \rceil \leq h \leq n$. Let $\gamma = (n - h)/n$, so $0 \leq \gamma \leq 0.5$. The estimated center is

$$\hat{\boldsymbol{\theta}} = \sum_{j=1}^{h} \mathbf{z}_{i_j} / h,$$

and the estimated scatter is

$$\hat{\boldsymbol{\Xi}} = c_n c_\gamma \frac{1}{h} \sum_{j=1}^{h} (\mathbf{z}_{i_j} - \hat{\boldsymbol{\theta}})(\mathbf{z}_{i_j} - \hat{\boldsymbol{\theta}})',$$

where c_n is a small-sample correction factor and c_γ is a consistency factor (Pison et al., 2002). For the problem at hand, Rousseeuw et al. found that $h \approx 3n/4$ provides relatively good efficiency and this choice is used here unless stated otherwise. Once the MCD estimates of location and scatter, based on $\mathbf{z}$, are available, their values are used in Eqs. (10.32) and (10.33) to get estimates of the slopes and intercepts. But efficiency can be relatively low.

Rousseeuw et al. consider two strategies for improving efficiency. Briefly, their first strategy uses weighted measures of location and scatter, with the weights computed as follows. Let $d(\mathbf{z}) = ((\mathbf{z}_i - \hat{\boldsymbol{\theta}})' \hat{\boldsymbol{\Xi}}^{-1} (\mathbf{z}_i - \hat{\boldsymbol{\theta}}))^{1/2}$ and $w_i = I(d^2(\mathbf{z}) \leq q)$, where q is the 0.975 quantile of a chi-squared distribution with $p + q$ degrees of freedom. Then the weighted measures of location and scatter (omitting a consistency factor) are

$$\hat{\boldsymbol{\theta}}_1 = \frac{\sum w_i \mathbf{z}_i}{\sum w_i} \tag{10.35}$$

and

$$\hat{\mathbf{\Xi}}_1 = \frac{\sum w_i (\mathbf{z}_i - \hat{\boldsymbol{\theta}}_1)(\mathbf{z}_i - \hat{\boldsymbol{\theta}}_1)'}{\sum w_i}, \tag{10.36}$$

respectively. Their second and recommended method uses updated weights based on the residuals associated with Eqs. (10.35) and (10.36). Let $\mathbf{r}_i$ be the residuals. Then the weights are taken to be $w_i = (\mathbf{r}_i' \hat{\mathbf{\Sigma}}_\epsilon \mathbf{r}_i)^{1/2}$. One appealing feature of this reweighting scheme is that good leverage points (outliers among the $\mathbf{x}$ values for which the corresponding residual is not an outlier) are not downweighted. This will be called the RADA estimator henceforth.

Wilcox (2009a) compares the RADA estimator to several other estimators, including situations where the error term is heteroscedastic. The RADA estimator does not dominate, but it performs reasonably well, particularly when there is dependence among the $\mathbf{y}$ values.

10.17.2 The Least Distance Estimator

Bai et al. (1990) propose another estimator that takes into account the dependence among the outcome variables, $\mathbf{y}$, called the *least distance estimator*. The regression parameters are estimated with the matrix $\mathbf{B}$ that minimizes

$$\sum_{i=1}^{n} \|\mathbf{y}_i - \mathbf{B}'\mathbf{x}_i\|, \tag{10.37}$$

where now the design matrix $\mathbf{x}$ is assumed to have a column of 1s when the model includes an intercept term. The least distance estimator generalizes the spatial median estimator of multivariate location. Jhun and Choi (2009) confirm that the efficiency of the least distance estimator compares well to the least absolute estimator, meaning that the univariate least absolute regression estimator is applied for each of the q outcome variables. In particular, the efficiency of the least distance estimator, relative to the least absolute regression estimator, improves under normality as the correlation among the outcome variables, $\mathbf{y}$, increases.

There is some indication that the least distance estimator competes well with the RADA estimator, in terms of mean squared error, when the error term is homoscedastic. When the error term is heteroscedastic, the reverse might be true. A negative feature of the least distance estimator is that it can be a bit biased with $n = 40$, while bias is negligible when using RADA. It is stressed, however, that a systematic comparison of these two estimators has not been made.

10.17.3 R Functions MULMreg, mlrreg, and Mreglde

The R function

$$MULMreg(x,y, regfun=MMreg, xout = FALSE, outfun = outpro, ...)$$

computes the regression parameters for each column of y using the regression estimator indicated by the argument regfun. The R function

$$mlrreg(x,y,cov.fun=cov.mcd)$$

computes the RADA multivariate regression estimator. By default, it uses the MCD estimator, but some other covariance matrix can be used via the argument cov.fun. The function assumes the argument y is a matrix with two or more columns. The R function

$$Mreglde(x,y,xout=FALSE, eout=FALSE, outfun=outpro)$$

computes the least distance estimator. If the argument eout=TRUE, the function combines the columns of data in the arguments x and y into a single matrix and then removes all outliers detected by the method indicated by the argument outfun. If xout=TRUE, the function removes any row of data from both x and y for which the row in x is declared an outlier. By default the projection-type outlier detection method is used. The function

$$MULR.yhat(x,y,pts=x,regfun=MULMreg, xout=FALSE,outfun=outpro,...)$$

computes the predicted value of the dependent variables for the points stored in the argument pts using the regression estimator indicated by the argument regfun.

■ **Example**

A practical issue is whether situations are encountered where the choice of a robust multivariate regression estimator can result in estimates that appear to differ substantially. This can indeed occur as illustrated here using the reading data mentioned in Section 10.8.1. Consider the first two predictors (stored in columns 2 and 3) and the first two outcome variables of interest (stored in columns 8 and 9). The estimates returned by the R function mlrreg (the RADA estimator) are

```
                Y1          Y2
Intercept 66.000739  68.2289879
V2         1.027754   0.6587633
V3         2.587086   2.1111645
```

The estimates returned by Mreglde (the least distance estimator) are

```
              Y            Y
INTER 95.444444  89.987654
SLOPE  7.344828   3.444856
SLOPE  7.045528   6.303742
```

∎

10.17.4 Multivariate Least Trimmed Squares Estimator

Agulló et al. (2008) suggest another approach to multivariate regression based on what they call the LTS estimators. For the usual least squares estimate of $\mathbf{B}$, say $\hat{\mathbf{B}}_{LS}$, let

$$\hat{\boldsymbol{\Sigma}}_{LS} = \frac{1}{n-p}(\mathbf{Y} - \mathbf{X}\hat{\mathbf{B}}_{LS})'(\mathbf{Y} - \mathbf{X}\hat{\mathbf{B}}_{LS}).$$

Consider any subset of the $\mathbf{z}_i$ vectors (defined as in Section 10.17.1) having cardinality h. For this subset of the data and some choice for $\mathbf{B}$, let $\mathbf{r}_i = \mathbf{y_i} - \mathbf{B}'\mathbf{x}_i$ be the matrix of residuals and

$$\text{cov}(\mathbf{B}) = \frac{1}{h}\sum(\mathbf{r}_i - \bar{\mathbf{r}})(\mathbf{r}_i - \bar{\mathbf{r}})',$$

where $\bar{\mathbf{r}} = \sum \mathbf{r}_i/h$. Their strategy is to first search for the subset of the data that minimizes $|\hat{\boldsymbol{\Sigma}}_{LS}|$. Their multivariate LTS (MLTS) estimator, $\hat{\mathbf{B}}_{MLTS}$, is the least squares estimate based on this subset of the data. They establish that this is tantamount to choosing $\mathbf{B}$ so as to minimize the determinant of the MCD scatter matrix estimate based on the residuals from $\mathbf{B}$.

A criticism is that the efficiency of this estimator can be relatively low. Agulló et al. deal with this issue by using a one-step reweighted estimator. Let

$$\hat{\boldsymbol{\Sigma}}_{MLTS} = \frac{1}{n-p}(\mathbf{Y} - \mathbf{X}\hat{\mathbf{B}}_{MLTS})'(\mathbf{Y} - \mathbf{X}\hat{\mathbf{B}}_{MLTS}).$$

Let $J = \{j : d_j^2(\hat{\mathbf{B}}_{MLTS}, \hat{\boldsymbol{\Sigma}}_{MLTS}) \le q_\delta\}$, where

$$d_j^2(\mathbf{B}, \boldsymbol{\Sigma}) = \mathbf{r}_i'\boldsymbol{\Sigma}^{-1}\mathbf{r}_i.$$

Agulló et al. take $\delta = 0.01$ and c_δ equal to the $1 - \delta$ quantile of a chi-squared distribution with q degrees of freedom. The reweighted estimate is taken to be $\hat{\mathbf{B}}_{RMLTS}$, the least squares estimate based on the vectors of observations corresponding to the set J.

10.17.5 R Function MULtsreg

The R function

$$\text{MULtsreg}(x, y, tr = 0.2, \text{RMLTS} = \text{TRUE})$$

computes the MLTS estimator. The argument RMLTS = TRUE means the reweighted estimate is returned; otherwise the MLTS estimate is returned.

10.17.6 Other Robust Estimators

Not all multivariate regression estimators, which have been proposed, are listed here. But in case it helps, two others are mentioned. The first uses Eqs. (10.32) and (10.33) to estimate the slopes and intercept but with the usual mean and covariance matrices replaced by some robust analog. Zhou (2009) studies this approach when using the projection estimate of location and scatter in Section 6.3.7. (The form of the Stahel–Donoho W-estimator suggested by Zuo et al., 2004a, was used.) Currently, execution time can be an issue and little is known about how this approach compares to the estimators in Sections 10.17.1 and 10.17.2. Yet another approach was derived by Ben et al. (2006). The regression coefficients and the covariance matrix of the errors are estimated simultaneously by minimizing the determinant of the covariance matrix, subject to a constraint on a robust scale of the Mahalanobis norms of the residuals. They use a τ-estimate of scale. Ben et al. report simulation results indicating that their estimator compares favorably to S-estimates. Finally, for results on an estimator that has a high breakdown point and simultaneously deals with missing values, see Statti et al. (2018).

10.18 Exercises

1. The average LSAT scores (x) for the 1973 entering classes of 15 American law schools and the corresponding grade point averages (y) are as follows:

   x: 576 635 558 578 666 580 555 661 651 605 653 575 545 572 594
   y: 3.39 3.30 2.81 3.03 3.44 3.07 3.00 3.43 3.36 3.13 3.12 2.74 2.76 2.88 2.96

   Using the R function lsfitci, verify that the 0.95 confidence interval for the slope, based on the least squares regression line, is $(0.0022, 0.062)$.
2. Discuss the relative merits of $\hat{\beta}_{ch}$.
3. Using the data in Exercise 1, show that the estimate of the slope given by $\hat{\beta}_{ch}$ is 0.057. In contrast, the OLS estimate is 0.0045, and $\hat{\beta}_m = 0.0042$. Comment on the difference among the three estimates.

4. Let T be any regression estimator that is affine equivariant. Let $\mathbf{A}$ be any non-singular square matrix. Argue that the predicted y values, $\hat{y}_i$, remain unchanged when $\mathbf{x}_i$ is replaced by $\mathbf{x}_i \mathbf{A}$.

5. For the data in Exercise 1, use the R function reglev to comment on the advisability of using M regression with Schweppe weights.

6. Compute the hat matrix for the data in Exercise 1. Which x values are identified as leverage points? Relate the result to the previous exercise.

7. The example in Section 6.6.1 reports the results of drinking alcohol for two groups of subjects measured at three different times. Using the group 1 data, compute an OLS estimate of the regression parameters for predicting the time 1 data using the data based on times 2 and 3. Compare the results to the estimates given by $\hat{\beta}_m$ and $\hat{\beta}_{ch}$.

8. For the data used in the previous exercise, compute 0.95 confidence intervals for the parameters using OLS as well as M regression with Schweppe weights.

9. Referring to Exercise 6, how do the results compare to the results obtained with the R function reglev?

10. For the data in Exercise 6, verify that the 0.95 confidence interval for the regression parameters, using the R function regci with M regression and Schweppe weights, are $(-0.2357, 0.3761)$ and $(-0.0231, 1.2454)$. Also verify that if regci is used with OLS, the confidence intervals are $(-0.4041, 0.6378)$ and $(0.2966, 1.7367)$. How do the results compare to the confidence intervals returned by lsfitci? What might be wrong with confidence intervals based on regci when the OLS estimator is used?

11. The file read.dat contains reading data collected by Doi. Of interest is predicting WWISST2, a word identification score (stored in column 8), using TAAST1, a measure of phonological awareness stored in column 2, and SBT1 (stored in column 3), another measure of phonological awareness. Compare the OLS estimates to the estimates given by $\hat{\beta}_m$, $\hat{\beta}_{ad}$, and $\hat{\beta}_{mid}$.

12. For the data used in Exercise 11, compute the hat matrix and identify any leverage points. Also check for leverage points with the R function reglev. How do the results compare?

13. For the data used in Exercise 11, RAN1T1 and RAN2T1 (stored in columns 4 and 5) are measures of digit naming speed and letter naming speed. Use M regression with Schweppe weights to estimate the regression parameters when predicting WWISST2. Use the function elimna, described in Chapter 1, to remove missing values. Compare the results with the OLS estimates and $\hat{\beta}_{ch}$.

14. For the data in Exercise 13, identify any leverage points using the hat matrix. Next, identify leverage points with the function reglev. How do the results compare?

15. Graphically illustrate the difference between a regression outlier and a good leverage point. That is, plot some points for which $y = \beta_1 x + \beta_0$ and then add some points that represent regression outliers and good leverage points.

16. Describe the relative merits of the OP and MGV estimators in Section 10.10.

17. For the star data in Fig. 6.3, which are stored in the file star.dat, eliminate the four outliers in the upper left corner of the plot by restricting the range of the x values. Then using the remaining data, estimate the standard error of the least squares estimator, the M-estimator with Schweppe weights, and the OP and MGV estimators. Comment on the results.

## CHAPTER 11

More Regression Methods

This chapter describes some additional robust regression methods that have been found to have practical value, including some inferential techniques that perform well in simulation studies even when the error term is heteroscedastic. Also covered are methods for testing the hypothesis that two or more of the regression parameters are equal to zero, a method for comparing the slope parameters of independent groups, measures of association based on a given fit to the data, methods for comparing dependent correlations, and methods for dealing with curvilinear relationships. Chapter 10 describes some methods for determining the relative importance of explanatory variables, but they do not provide an indication of the strength of the empirical evidence that certain explanatory variables are more important than others. Section 11.10.5 describes techniques for dealing with this issue.

11.1 Inferences About Robust Regression Parameters

This section deals with testing hypotheses about the parameters in the regression model

$$y_i = \beta_0 + \beta_1 x_{i1} + \cdots + \beta_p x_{ip} + \epsilon_i,$$

where the error term might be heteroscedastic and some robust regression estimator is used. (Section 10.1 describes methods designed specifically for the situation where the least squares estimator is used.) A common goal is testing

$$H_0: \beta_1 = \cdots = \beta_p = 0, \tag{11.1}$$

the hypothesis that all of the slope parameters are equal to zero, but the methods described here can also be used to test

$$H_0: \beta_1 = \cdots = \beta_q = 0,$$

the hypothesis that $q < p$ of the parameters are equal to zero. A more general goal is to test the hypothesis that q parameters are equal to some specified value, and the method described here accomplishes this goal as well. And there is the goal of computing confidence intervals for the individual parameters. It is noted that Barber and Candès (2015) derive an approach aimed at determining which slope parameters are significant when p is large. They focus on controlling the false discovery rate when testing which parameters should be retained. Their

Introduction to Robust Estimation and Hypothesis Testing
https://doi.org/10.1016/B978-0-12-820098-8.00017-8

method assumes normality and homoscedasticity. There are some indications that it continues to perform well under non-normality, but there appear to be no results on the impact of heteroscedasticity.

Before continuing, a word of caution is in order. The linear model used in this section is routinely adopted, and situations are encountered where it appears to provide a reasonably accurate characterization of the association. It is suggested, however, that when dealing with regression, it is prudent to also consider more flexible methods for modeling an association, which can be done via the methods in Section 11.5. This is particularly important when dealing with more than one independent variable. Even with one independent variable, assuming a straight regression line can be misleading and miss important features of an association. A simple strategy is to include a quadratic term in the linear model used here. But often, a more flexible approach is needed, as will be illustrated.

11.1.1 Omnibus Tests for Regression Parameters

This section begins with testing Eq. (11.1). Generally, a type of bootstrap method performs well when using a robust regression estimator. An exception is when using a robust ridge estimator as in Section 10.13.15, with the goal of dealing with multicollinearity. The end of this section describes how to deal with this special case.

When working with robust regression, three strategies for testing hypotheses have received attention in the statistical literature and should be mentioned. The first is based on the so-called *Wald scores*, the second is a *likelihood ratio test*, and the third is based on a measure of *drop in dispersion*. Details about these methods can be found in Markatou et al. (1991), as well as Heritier and Ronchetti (1994). Coakley and Hettmansperger (1993) suggest using a Wald scores test in conjunction with their estimation procedure, assuming that the error term is homoscedastic. The method estimates the standard error of their estimator, which can be used to get an appropriate test statistic for which the null distribution is chi-squared. When both x and ϵ are normal, and the error term is homoscedastic, the method provides reasonably good control over the probability of a Type I error when $n = 50$. However, if ϵ is non-normal, the actual probability of a Type I error can exceed 0.1 when testing at the $\alpha = 0.05$ level, even with $n = 100$ (Wilcox, 1994e). Consequently, details about the method are not described. Birkes and Dodge (1993) describe drop-in-dispersion methods when working with M regression methods that do not protect against leverage points. Little is known about how this approach performs under heteroscedasticity, so it is not discussed either. Instead, attention is focused on a method that has been found to perform well when there is a heteroscedastic error term. It is not suggested that the method described here outperforms all other methods that might be used, only that it gives good results over a relatively wide range of situations, and based on extant simulation studies, it is the best method available.

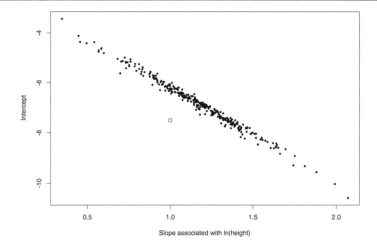

Figure 11.1: Scatterplot of bootstrap estimates using the tree data. The square marks the null values.

The basic strategy is to generate B bootstrap estimates of the parameters, and then determine whether the vector of values specified by the null hypothesis is far enough away from the bootstrap samples to warrant rejecting H_0. This strategy is illustrated with the tree data in the Minitab handbook (Ryan et al., 1985, p. 329). The data consist of tree volume (V), tree diameter (d), and tree height (h). If the trees are cylindrical or cone-shaped, then a reasonable model for the data is $y = \beta_1 x_1 + \beta_2 x_2 + \beta_0$, where $y = \ln(V)$, $x_1 = \ln(d)$, $x_2 = \ln(h)$, with $\beta_1 = 2$ and $\beta_2 = 1$ (Fairley, 1986). The ordinary least squares (OLS) estimate of the intercept is $\hat{\beta}_0 = -6.632$, $\hat{\beta}_1 = 1.98$, and the estimates of the slopes are $\hat{\beta}_2 = 1.12$. Using M regression with Schweppe weights (the R function bmreg), the estimates are -6.59, 1.97, and 1.11, respectively.

Suppose bootstrap samples are generated as described in Section 10.1.1 (cf. Salibian-Barrera and Zamar, 2002). That is, rows of data are sampled with replacement. Fig. 11.1 shows a scatterplot of 300 bootstrap estimates, using M regression with Schweppe weights, of the intercept, β_0, and β_2, the slope associated with log height. The square marks the hypothesized values, $(\beta_0, \beta_2) = (-7.5, 1)$. These bootstrap values provide an estimate of a confidence region for (β_0, β_2) that is centered at the estimated values $\hat{\beta}_0 = -6.59$ and $\hat{\beta}_2 = 1.11$. Fig. 11.1 suggests that the hypothesized values might not be reasonable. That is, the point $(-7.5, 1)$ might be far enough away from the bootstrap values to suggest that it is unlikely that β_0 and β_2 simultaneously have the values -7.5 and 1, respectively. The problem is measuring the distance between the hypothesized values and the estimated values, and then finding a decision rule that rejects the null hypothesis with probability α when H_0 is true.

For convenience, temporarily assume the goal is to test the hypothesis given by Eq. (11.1). A simple modification of the method in Section 8.2.5 can be used where bootstrap samples are obtained by resampling with replacement n rows from

$$
\begin{pmatrix}
y_1, x_{11}, \ldots, x_{1J} \\
\vdots \\
y_n, x_{n1}, \ldots, x_{nJ}
\end{pmatrix},
$$

yielding

$$
\begin{pmatrix}
y_1^*, x_{11}^*, \ldots, x_{1J}^* \\
\vdots \\
y_n^*, x_{n1}^*, \ldots, x_{nJ}^*
\end{pmatrix}.
$$

Let $\hat{\beta}_{jb}^*$, $j = 1, \ldots, p$, $b = 1, \ldots, B$, be an estimate of the jth parameter based on the bth bootstrap sample and any robust estimator described in Chapter 10. Then an estimate of the covariance between $\hat{\beta}_j$ and $\hat{\beta}_k$ is

$$
v_{jk} = \frac{1}{B-1} \sum_{b=1}^{B} (\hat{\beta}_{jb}^* - \bar{\beta}_j^*)(\hat{\beta}_{kb}^* - \bar{\beta}_k^*),
$$

where $\bar{\beta}_j^* = \sum \hat{\beta}_{jb}^* / B$. Here, $\hat{\beta}_j$ can be any estimator of interest. Now, the distance between the bth bootstrap estimate of the parameters and the estimate based on the original observations can be measured with

$$
d_b^2 = (\hat{\beta}_{1b}^* - \hat{\beta}_1, \ldots, \hat{\beta}_{pb}^* - \hat{\beta}_p) \mathbf{V}^{-1} (\hat{\beta}_{1b}^* - \hat{\beta}_1, \ldots, \hat{\beta}_{pb}^* - \hat{\beta}_p)',
$$

where $\mathbf{V}$ is the p-by-p covariance matrix with the element in the jth row and kth column equal to v_{jk}. That is, $\mathbf{V} = (v_{jk})$ is the sample covariance matrix based on the bootstrap estimates of the parameters. The square root of d_b^2, d_b, represents a simple generalization of the Mahalanobis distance. If the point corresponding to the vector of hypothesized values is sufficiently far from the estimated values, relative to the distances d_b, reject H_0. This strategy is implemented by putting the d_b values in ascending order, yielding $d_{(1)} \leq \cdots \leq d_{(B)}$, setting $M = [(1 - \alpha)B]$, and letting m be the value of M rounded to the nearest integer. The null hypothesis is rejected if

$$
D > d_{(m)}, \tag{11.2}
$$

where

$$
D = \sqrt{(\hat{\beta}_1, \ldots, \hat{\beta}_p) \mathbf{V}^{-1} (\hat{\beta}_1, \ldots, \hat{\beta}_p)'}.
$$

The method just described is easily generalized to testing

$$H_0: \beta_1 = \beta_{10}, \beta_2 = \beta_{20}, \dots, \beta_q = \beta_{q0},$$

the hypothesis that q of the $p + 1$ parameters are equal to specified constants, $\beta_{10}, \dots, \beta_{q0}$. Proceed as before, only now

$$d_b = \sqrt{(\hat{\beta}_{1b}^* - \hat{\beta}_1, \dots, \hat{\beta}_{qb}^* - \hat{\beta}_q)\mathbf{V}^{-1}(\hat{\beta}_{1b}^* - \hat{\beta}_1, \dots, \hat{\beta}_{qb}^* - \hat{\beta}_q)'},$$

and $\mathbf{V}$ is a q-by-q matrix of estimated covariances based on the B bootstrap estimates of the q parameters being tested. The critical value, $d_{(m)}$, is computed as before, and the test statistic is

$$D = \sqrt{(\hat{\beta}_1 - \beta_{10}, \dots, \hat{\beta}_q - \beta_{q0})\mathbf{V}^{-1}(\hat{\beta}_1 - \beta_{10}, \dots, \hat{\beta}_q - \beta_{q0})'}.$$

The (generalized) p-value is

$$\hat{p}^* = \frac{1}{B}\sum I(D \le d_b),$$

where $I(D \le d_b) = 1$ if $D \le d_b$ and $I(D \le d_b) = 0$ if $D > d_b$.

The hypothesis testing method just described can be used with any regression estimator. When using the OLS estimator, it has advantages over the conventional F test, but problems remain. This is illustrated by Table 11.1, which shows the estimated probability of a Type I error for various situations when testing $H_0: \beta_1 = \beta_2 = 0$, $\alpha = 0.05$, and where

$$y = \beta_1 x_1 + \beta_2 x_2 + \lambda(x_1, x_2)\epsilon. \tag{11.3}$$

(For results on testing hypotheses when the function λ is known, see Zhao and Wang, 2009.) In Table 11.1, VP 1 corresponds to $\lambda(x_1, x_2) = 1$ (a homoscedastic error term), VP 2 is $\lambda(x_1, x_2) = |x_1|$, and VP 3 is $\lambda(x_1, x_2) = 1/(|x_1| + 1)$. Both x_1 and x_2 have identical g-and-h distributions, with the g and h values specified by the first two columns. In some cases, the conventional F test performs well, but it performs poorly for VP 2. The bootstrap method improves matters considerably, but the probability of a Type I error exceeds 0.075 in various situations. In practical terms, when testing hypotheses using OLS, use the methods in Section 10.1.1 rather than the bootstrap method described here.

Table 11.2 shows $\hat{\alpha}$, the estimated probability of a Type I error when using $\hat{\beta}_{\text{mid}}$, the biweight midregression estimator with $n = 20$, and the goal is to test $H_0: \beta_1 = \beta_2 = 0$, with $\alpha = 0.05$. Now, the probability of a Type I error is less than or equal to the nominal level, but in some cases, it is too low, particularly for VP 3, where it drops as low as 0.002. (For more details about the simulations used to create Tables 11.1 and 11.2, see Wilcox, 1996f.) Simulations indicate that switching to the Theil–Sen estimator improves the control over the Type I error probability when dealing with heavy-tailed distributions (Wilcox, 2004c).

Table 11.1: Estimated Type I error probabilities using OLS, $\alpha = 0.05$, $n = 20$.

x		ϵ		VP 1		VP 2		VP 3	
g	h	g	h	Boot	F	Boot	F	Boot	F
0.0	0.0	0.0	0.0	0.072	0.050	0.097	0.181	0.009	0.015
0.0	0.0	0.0	0.5	0.028	0.047	0.046	0.135	0.004	0.018
0.0	0.0	0.5	0.0	0.052	0.049	0.084	0.174	0.009	0.018
0.0	0.0	0.5	0.5	0.028	0.043	0.042	0.129	0.005	0.019
0.0	0.5	0.0	0.0	0.022	0.055	0.078	0.464	0.003	0.033
0.0	0.5	0.0	0.5	0.014	0.074	0.042	0.371	0.002	0.038
0.0	0.5	0.5	0.0	0.017	0.048	0.072	0.456	0.005	0.032
0.0	0.5	0.5	0.5	0.011	0.070	0.039	0.372	0.005	0.040
0.5	0.0	0.0	0.0	0.054	0.044	0.100	0.300	0.013	0.032
0.5	0.0	0.0	0.5	0.024	0.057	0.049	0.236	0.010	0.038
0.5	0.0	0.5	0.0	0.039	0.048	0.080	0.286	0.010	0.033
0.5	0.0	0.5	0.5	0.017	0.058	0.046	0.217	0.010	0.040
0.5	0.5	0.0	0.0	0.013	0.054	0.083	0.513	0.006	0.040
0.5	0.5	0.0	0.5	0.009	0.073	0.043	0.416	0.002	0.048
0.5	0.5	0.5	0.0	0.013	0.053	0.079	0.505	0.005	0.043
0.5	0.5	0.5	0.5	0.006	0.067	0.036	0.414	0.005	0.050

Table 11.2: Values of $\hat{\alpha}$ using biweight midregression, $\alpha = 0.05$, $n = 20$.

x		ϵ				
g	h	g	h	VP 1	VP 2	VP 3
0.0	0.0	0.0	0.0	0.047	0.039	0.015
0.0	0.0	0.0	0.5	0.018	0.024	0.008
0.0	0.0	0.5	0.0	0.038	0.037	0.011
0.0	0.0	0.5	0.5	0.021	0.025	0.003
0.0	0.5	0.0	0.0	0.016	0.018	0.002
0.0	0.5	0.0	0.5	0.009	0.018	0.002
0.0	0.5	0.5	0.0	0.015	0.016	0.002
0.0	0.5	0.5	0.5	0.009	0.012	0.003
0.5	0.0	0.0	0.0	0.033	0.037	0.012
0.5	0.0	0.0	0.5	0.020	0.020	0.006
0.5	0.0	0.5	0.0	0.024	0.031	0.009
0.5	0.0	0.5	0.5	0.015	0.021	0.005
0.5	0.5	0.0	0.0	0.015	0.021	0.002
0.5	0.5	0.0	0.5	0.008	0.011	0.002
0.5	0.5	0.5	0.0	0.014	0.017	0.002
0.5	0.5	0.5	0.5	0.006	0.007	0.002

Methods Based on a Ridge Estimator

As noted in Section 10.13.15, when there is multicollinearity, a regression estimator can have relatively high standard errors, which in turn impacts power. Ridge estimators introduced in Section 10.13.15 are designed to deal with this issue, but an alternative to Eq. (11.2) is needed when testing Eq. (11.1).

Method RT1

First consider the non-robust ridge estimator. Let the values of the independent variables be denoted by the $n \times p$ matrix $\mathbf{X}$. Let $\mathbf{C} = \mathbf{X}'\mathbf{X}$ and $\mathbf{V} = \text{diag}(r_i^2/(1 - h_{ii}))$ $(i = 1, \ldots, n)$, where $h_{ii} = \mathbf{x}_i(\mathbf{X}'\mathbf{X})^{-1}\mathbf{x}_i'$, $\mathbf{x}_i$ is the ith row of $\mathbf{X}$ and $r_i = Y_i - \hat{Y}_i$ $(i = 1, \ldots, n)$ are the residuals. As done in Section 10.13.15, let k denote the bias parameter used by the ridge estimator and let

$$\mathbf{S}(k) = (\mathbf{C} + k\mathbf{I}_p)^{-1}\mathbf{X}'\mathbf{V}\mathbf{X}(\mathbf{C} + k\mathbf{I}_p)^{-1}. \tag{11.4}$$

Wilcox (2019a) notes that $s_{jj}(k)$, the jth diagonal element of $\mathbf{S}(k)$, estimates the squared standard error of $\hat{\beta}_j(k)$ and suggests testing H_0: $\beta_j = 0$ $(j = 1, \ldots, p)$ using the test statistic

$$T_j = \frac{\hat{\beta}_j}{\sqrt{s_{jj}}}, \tag{11.5}$$

where the null distribution is approximated with a Student's t distribution, with $n - p - 1$ degrees of freedom. The familywise error rate is controlled via Hochberg's method. If any of these p hypotheses is rejected, reject the global hypothesis given by Eq. (11.1). This is reasonable because when Eq. (11.1) is true, the ridge estimator is unbiased. But otherwise it is biased, in which case, any confidence interval based on Eq. (11.5) can be highly inaccurate. It remains unknown how to compute confidence intervals for the individual slopes based on the ridge estimator when one or more differ from zero. In practical terms, if H_0: $\beta_j = 0$ is rejected for any j, reject (11.1) but make no inferences about which slopes differ from zero.

Method RT2

Now consider a robust ridge estimator $\tilde{\beta}$ given by Eq. (10.19). Again the goal is to test Eq. (11.1). Consider the test statistic

$$T_R^2 = \frac{p(n-1)}{n-p}\tilde{\beta}\mathbf{S}^{-1}\tilde{\beta}', \tag{11.6}$$

which is an analog of Hotelling's T^2 statistic. An approach to controlling the Type I error probability is to momentarily assume homoscedasticity, the error term has a normal distribution, and the independent variables have a multivariate normal distribution, with all correlations equal to zero. Then, for a given sample size, a simulation can be used to determine the null distribution, which yields a critical value. The issue is whether this critical value controls the Type I error probability reasonably well when dealing with an error term that has a non-normal distribution and when there is heteroscedasticity. Simulations reported by Wilcox (2019b) indicate that when using a robust ridge estimator, with $\hat{\beta}_R$ in Eq. (10.19) taken to be the Theil–Sen estimator, reasonably good control of the Type I error probability is achieved. The MM-estimator performs nearly as well, but in some situations, the actual level was found to be well above the nominal level when leverage points are retained. When leverage points are removed, the actual level was found to be less than or equal to the nominal level, but in some cases, it is less than 0.025 when testing at the 0.05 level. In terms of power, T_R^2 does not dominate Eq. (11.2), but generally T_R^2 competes well with Eq. (11.2). Extant simulations indicate that Eq. (11.2) never offers a striking advantage, but T_R^2 can offer a distinct advantage even when there is little or no indication of multicollinearity (cf. Spanos, 2019).

11.1.2 R Functions regtest, ridge.test and ridge.Gtest

The R function

> ridge.test(x, y, k = NULL, alpha = 0.05, pr = TRUE, xout = FALSE, outfun = outpro,
> STAND = TRUE, method = 'hoch', locfun = mean, scat = var, ...)

applies method RT1, described in the previous section. If any hypothesis is rejected, reject the global hypothesis given by (11.1), but make no decision about which of the individual slopes differ from zero.

The R function

> ridge.Gtest(x, y, k = NULL, regfun = tsreg, xout = FALSE, outfun = outpro,
> STAND = FALSE, PV = FALSE, iter = 5000, locfun = mean, scat = var, MC = FALSE, ...)

tests the hypothesis given by Eq. (11.1) using a robust ridge estimator via method RT2. The Theil–Sen version is used by default. When testing at the $\alpha = 0.05$ level, an approximation of the critical value is used to reduce execution time. Critical values were determined for sample sizes 20, 30, 40, 50, 75, 100, 200, and 500. Interpolation is used for samples between any two of these values. To get a p-value, or to test at some level other than 0.05, set the argument PV=TRUE. The number of replications in the simulation used to determine the null distribution is controlled by the argument iter. If a multicore processor is available, setting MC=TRUE will reduce execution time.

The R function

> regtest(x,y,regfun=tsreg,nboot=600, alpha=0.05,plotit=TRUE,grp=c(1:ncol(x)), nullvec =
> c(rep(0, length(grp))))

tests the hypothesis that all of the slopes are zero using the bootstrap method described in the previous section.

■ **Example**

For the tree data used to create Fig. 11.1, suppose there is reason to believe that $\beta_0 = -7.5$ and $\beta_2 = 1$. If the logarithms of the predictor values are stored in mtree and the logarithms of the volume (the y values) are stored in ytree, then the command

```
regtest(mtree,ytree,regfun=bmreg,grp=c(0,2),nullvec=c(-7.5,1))
```

will test the hypothesis that H_0: $\beta_0 = -7.5$ and $\beta_2 = 1$ using the R function bmreg to estimate the parameters. The function regtest reports a test statistic of 97.24, with a 0.05 critical value of 7.98, so H_0 is rejected. Using the Theil–Sen estimator, the test statistic is 93.8, the 0.05 critical value is 7.06, and the (generalized) p-value is 0.

■

■ **Example**

This next example is based on data stemming from the Well Elderly study described in Section 11.6.1. One specific goal was to understand the association between a measure of depressive symptoms (the dependent variable) and two explanatory variables, age and the cortisol awakening response (CAR), which is the change in cortisol levels between the moment of awakening and 30–60 minutes later. Pearson's correlation between the two independent variables is 0.06, and the skipped correlation in Section 9.4.3 is -0.03. As noted, for example, by James et al. (2017, p. 101), multicollinearity can be an issue even when the independent variables have a low correlation. In terms of detecting multicollinearity, they suggest using instead the variance inflation factor (VIF), which is just the ratio of the variance of $\hat{\beta}_j$ when fitting the full model divided by the variance of $\hat{\beta}_j$ if fit on its own. They further suggest that a ratio that exceeds 5 or 10 indicates a problematic amount of collinearity. Taking $\hat{\beta}_j$ to be the OLS estimator, the largest ratio for the two independent variables used here is 1.57. Using the Theil–Sen estimator gives similar results. (A bootstrap estimate of the squared standard errors was used via the R function regse.) Testing the hypothesis that both slopes are zero, using the R function ridge.Gtest, the p-value is less than 0.001. Using the Theil–Sen estimator, in conjunction with the first method in Section 11.1.1, the p-value is 0.805. This demonstrates that even when the VIF is estimated to be relatively low, using a robust ridge estimator can make a practical difference.

■

11.1.3 Inferences About Individual Parameters

When the goal is to compute a confidence interval for the individual parameters in a regression model, a simple percentile bootstrap method appears to perform well, in terms of probability coverage, when used with some robust regression estimators. But as previously noted, when using a robust ridge estimator, currently it is unknown how to do this in a satisfactory manner because when any of the slopes differ from zero, the ridge estimator is biased.

Briefly, generate a bootstrap sample as described in Section 11.1.1, and compute an estimate of the slopes. Repeat this process B times, yielding $\hat{\beta}_{j1}^*, \ldots, \hat{\beta}_{jB}^*$. Then for fixed j, the $1 - \alpha$ confidence interval for β_j is

$$(\hat{\beta}_{j(\ell+1)}^*, \hat{\beta}_{j(u)}^*), \tag{11.7}$$

where $\ell = \alpha B / 2$, rounded to the nearest integer, $u = B - \ell$, and $\hat{\beta}_{j(1)}^* \leq \cdots \leq \hat{\beta}_{j(B)}^*$ are the B bootstrap estimates of β_j written in ascending order. In other words, use the standard percentile bootstrap method, as opposed to the modified method used when working with OLS. A (generalized) p-value can be computed in the usual way. Let $\hat{p}^*$ be the proportion of bootstrap estimates greater than zero. Then the p-value is

$$\hat{p}_m^* = 2\min(\hat{p}^*, 1 - \hat{p}^*).$$

(Under certain circumstances, when there is one predictor only, an alternative approach to computing a confidence interval for the slope that might have practical value is described by Adrover and Salibian-Barrera, 2010.)

To provide some indication of how well the method performs when using the M-estimator $\hat{\beta}_m$, Table 11.3 shows values of $\hat{\alpha}$, simulation estimates of one minus the actual probability coverage, when $n = 20$ and $\alpha = 0.05$. The notation VP 1 indicates a homoscedastic error term ($\lambda(x) = 1$ in Eq. (10.2), VP 2 is a heteroscedastic error term, where the variance of the error term increases as x moves away from its median value ($\lambda(x) = x^2$), and VP 3 is where the variance decreases as x moves away from its median ($\lambda(x) = 1 + 2/(|x| + 1)$). The $\hat{\alpha}$ values never exceed 0.075, but for VP 2, they can exceed 0.070. (For results when using the Theil–Sen estimator, see Wilcox, 1998a, 1998b. For results related to the OP-estimator in Section 10.10, see Wilcox, 2004c.)

To provide a bit more perspective, Table 11.4 shows simulation estimates of the probability of a Type I error when using the Theil–Sen estimator, with $p = 2$. Under VP 1, the first entry is the estimated probability of a Type I error using Eq. (11.2), and the second entry is the probability of at least one Type I error when using Eq. (11.7). The same is true for the columns headed by VP 2 and VP 3. For brevity, only results where x has a symmetric distribution are shown. So for n small, using Eq. (11.2) can result in Type I error probabilities considerably smaller than the nominal level.

Table 11.3:   Values of $\hat{\alpha}$ using $\hat{\beta}_m$, $\alpha = 0.05$, $n = 20$.

x		ϵ				
g	*h*	*g*	*h*	VP 1	VP 2	VP 3
0.0	0.0	0.0	0.0	0.054	0.065	0.050
0.0	0.0	0.0	0.5	0.051	0.064	0.051
0.0	0.0	0.5	0.0	0.057	0.066	0.066
0.0	0.0	0.5	0.5	0.055	0.065	0.049
0.0	0.5	0.0	0.0	0.058	0.070	0.034
0.0	0.5	0.0	0.5	0.057	0.067	0.035
0.0	0.5	0.5	0.0	0.058	0.069	0.036
0.0	0.5	0.5	0.5	0.059	0.069	0.037
0.5	0.0	0.0	0.0	0.049	0.071	0.051
0.5	0.0	0.0	0.5	0.049	0.064	0.047
0.5	0.0	0.5	0.0	0.051	0.068	0.053
0.5	0.0	0.5	0.5	0.050	0.065	0.050
0.5	0.5	0.0	0.0	0.054	0.072	0.043
0.5	0.5	0.0	0.5	0.054	0.071	0.047
0.5	0.5	0.5	0.0	0.056	0.071	0.044
0.5	0.5	0.5	0.5	0.056	0.071	0.044

Table 11.4:   Values of $\hat{\alpha}$ using Eqs. (11.2) and (11.7) and the Theil–Sen estimator, $\alpha = 0.05$, $n = 20$.

x	ϵ		VP 1		VP 2		VP 3	
h	*g*	*h*	(11.2)	(11.7)	(11.2)	(11.7)	(11.2)	(11.7)
0.0	0.0	0.0	0.036	0.033	0.037	0.043	0.017	0.030
0.0	0.0	0.0	0.010	0.031	0.013	0.038	0.007	0.029
0.0	0.0	0.5	0.020	0.033	0.034	0.041	0.013	0.030
0.0	0.0	0.5	0.008	0.032	0.010	0.037	0.001	0.031
0.5	0.0	0.0	0.015	0.033	0.036	0.039	0.007	0.026
0.5	0.0	0.0	0.008	0.032	0.029	0.036	0.004	0.032
0.5	0.0	0.5	0.008	0.035	0.029	0.036	0.004	0.032
0.5	0.0	0.5	0.004	0.031	0.013	0.039	0.002	0.032

11.1.4 R Functions regci, regciMC and wlogregci

The R function

regci(x,y,regfun=tsreg,nboot=599,alpha=0.05, SEED=TRUE, pr=TRUE, null.val=NULL, xout=FALSE, outfun=outpro, plotit=FALSE, xlab='Predictor 1',ylab='Predictor 2',...)

is supplied for computing confidence intervals for regression parameters with the percentile bootstrap method just described. The R function

regciMC(x,y,regfun=tsreg,nboot=599,alpha=0.05, SEED=TRUE, pr=TRUE, null.val=NULL,
    xout=FALSE, outfun=outpro, plotit=FALSE, xlab='Predictor 1', ylab='Predictor 2',...)

is the same as regci, only it takes advantage of a multicore processor, assuming one is available. (The R function regcits is the same as regciMC, only it defaults to using the Harrell–Davis estimator, which might increase power when there are tied values for the dependent variable.) Here, x can be a vector or a matrix having n rows and p columns. The optional argument regfun can be any R function that estimates regression parameters and returns the results in regfun$coef. The first element of regfun$coef is assumed to be the estimated intercept, the second element is the estimate of β_1, and so on. Regression methods that come with R follow this convention, as do all of the R regression functions written for this book. For example, bmreg returns the estimated values in bmreg$coef. If unspecified, regfun is tsreg, which is the Theil–Sen estimator. The default value for nboot, which is the number of bootstrap samples to be used, B, is 599.

For the special case where y is binary and the robust estimator given by Eq. (10.26) is used, the R function

wlogregci(x,y,nboot=400, alpha=0.05, SEED=TRUE, MC=FALSE, xout=FALSE,
    outfun=out,...)

can be used.

■ **Example**

As a simple illustration, consider the data

x: −80, 79, −90, 11, 137, 141, 116, −54, 92, −58, −9, −96, −27, −135, 76, −56, 19, −93, −19, −158,

y: 7, 56, −84, −69, 88, 103, −102, −82, 25, 84, −69, −78, −127, 50, 210, −51, 120, −212, 174, −72.

The 0.95 confidence interval for β_1, returned by the command regci(x,y), is (−0.95, 1.14). (This result is based on an older version of the random number generator used by R.) If the function lsfitci is used instead, the 0.95 confidence interval is (−0.024, 1.13). These intervals are similar, as expected, because both x and ϵ were generated from normal distributions with $\beta_1 = 1$.

■

The command regci(x,y,regfun=lsfit) would return a confidence interval based on the OLS estimator, using the standard percentile bootstrap method, but this is not recommended for reasons already explained. However, regci appears to give good results when working with nearly all of the robust regression methods described in this chapter. (Exceptions are noted in Sections 11.1.5 and 11.1.7.)

■ Example

For the star data in Fig. 6.3, the 0.95 confidence interval for the slope returned by the R function regci, using the default estimator, is $(-0.78, 5.03)$. Eliminating leverage points by setting the argument xout=TRUE, now the 0.95 confidence interval is $(2.1, 4.0)$. If the bounded influence M regression method is used instead, $\hat{\beta}_m$, the 0.95 confidence interval for the slope is $(-1.076, 2.436)$, with xout=FALSE. So even among robust estimators, the estimator used, as well as eliminating leverage points, can make a practical difference when computing confidence intervals.

■

■ Example

For the tree data used in the last example of Section 11.1.2, the hypothesis H_0: $(\beta_0, \beta_2) = (-7.5, 1)$ was rejected using M regression with Schweppe weights. The 0.95 confidence intervals for these two parameters returned by regci, again using M regression with Schweppe weights (i.e., setting regfun=bmreg when using regci) are $(-9.1, -4.9)$ and $(0.65, 1.76)$, respectively, suggesting that the hypothesized values for β_0 and β_2 are reasonable. This illustrates the well-known result that confidence intervals can fail to reject when an omnibus test rejects. (The reason is that the confidence region used by the omnibus test is an ellipse, versus a rectangular confidence region when computing confidence intervals for the individual parameters. See Fairley, 1986, for more details.)

■

11.1.5 Methods Based on the Quantile Regression Estimator

Section 10.13.8 describes a non-bootstrap R function for making inferences about the parameters associated with a quantile regression estimator. Two limitations of the method are that it can be used only when testing at the $\alpha = 0.05$ level, and it does not provide a way of testing the omnibus hypothesis that two or more parameters are equal to zero. Switching to a percentile bootstrap method, simulations indicate that Type I error probabilities greater than the nominal level are avoided. But the actual level can drop well below the nominal level when the sample size is small. A slightly better approach appears to be one based in part on a bootstrap estimate of the standard errors, but again the actual level can be lower than intended. Here, an adjustment is made for dealing with this problem that was suggested by Wilcox and Costa (2009).

Generate B bootstrap estimates of the slope, yielding $b_1^*, \ldots, b_B^*$. Then an estimate of the squared standard error of b_1 is

$$S^2 = \frac{1}{B-1} \sum_{b=1}^{B} (b_b^* - \bar{b})^2,$$

where $\bar{b} = \sum b_b^* / B$. So an approximate $1 - \alpha$ confidence interval for β_1 is

$$b_1 \pm z_{1-\alpha/2} S,$$

where $z_{1-\alpha/2}$ is the $1 - \alpha/2$ quantile of a standard normal distribution.

To avoid Type I error probabilities well below the nominal level when the sample sizes are small, Wilcox and Costa found that the following adjusted critical values perform reasonably well in simulations:

1. If $\alpha = 0.1$, $z_a = 1.645 - 1.19/\sqrt{n}$,
2. If $\alpha = 0.05$, $z_a = 1.96 - 1.37/\sqrt{n}$,
3. If $\alpha = 0.025$, $z_a = 2.24 - 1.18/\sqrt{n}$,
4. If $\alpha = 0.01$, $z_a = 2.58 - 1.69/\sqrt{n}$.

That is, an approximate $1 - \alpha$ confidence interval for β_1 is taken to be

$$b_1 \pm z_a S.$$

This approximation appears to work well when estimating the γth quantile regression line when $0.2 \leq \gamma \leq 0.8$.

As for testing the global hypothesis given by Eq. (11.1), that all slope parameters are equal to zero, take a bootstrap sample in the usual manner, and label the resulting estimate of the slopes b_k^*, $k = 1, \ldots, p$. Repeat this process B times, yielding $b_{1k}^*, \ldots b_{Bk}^*$. An estimate of the variances and covariances associated with $b_1, \ldots, b_p$ is

$$\mathbf{S} = \frac{1}{B-1} \sum_{c=1}^{B} (\mathbf{b_c^*} - \bar{\mathbf{b}})^2,$$

where $\mathbf{b}_c^* = (b_{c1}^*, \ldots, b_{cp}^*)$, $\bar{\mathbf{b}} = (\bar{b}_1^*, \ldots, \bar{b}_p^*)$, and $\bar{b}_k^* = \sum b_{bk}^*/B$. A reasonable test statistic is

$$T^2 = n\bar{\mathbf{b}}'\mathbf{S}^{-1}\bar{\mathbf{b}}. \tag{11.8}$$

And from basic principles, a natural strategy is to reject if

$$T^2 \geq \frac{n-1}{n-p} f_{p,n-p},$$

where $f_{p,n-p}$ is the $1 - \alpha$ quantile of an F distribution, with p and $n - p$ degrees of freedom. All indications are that the actual probability of a Type I error is less than the nominal level when the sample size is small, particularly as the number of predictors increases. For example, when $\gamma = 0.5$, $p = 2$, $n = 20$, $\alpha = 0.05$, and x_1 and x_2 have a bivariate normal distribution with correlation $\rho = 0$, the actual Type I error probability is approximately 0.026. Increasing p to 6, the estimate is now 0.001. But with $n = 60$, the actual probability of a Type I error has been found to be reasonably close to 0.05 (Wilcox, 2007). Adjusted critical values when $n < 60$ and $\alpha = 0.1, 0.05, 0.025$, and 0.01 are reported by the R function rqtest, described in the next section.

It is briefly noted that He and Zhu (2003) derive a method for testing the hypothesis that a specified family of quantile regression models fits the data. In particular, one can test the hypothesis that for some choice for $\beta_0, \ldots, \beta_p$, the γ quantile of y, given $x_1, \ldots, x_p$, is given by

$$y = \beta_0 + \beta_1 x_1 + \cdots + \beta_p x_p.$$

A simple variation of their method has been found to reduce execution time considerably (Wilcox, 2008b). The details are omitted, but an R function (qrchk) is supplied for performing the analysis.

11.1.6 R Functions rqtest, qregci, and qrchk

The R function

```
rqtest(x,y,qval=0.5, nboot=200, alpha=0.05, SEED=TRUE, xout=FALSE, outfun=out,...)
```

tests the hypothesis that p slope parameters are equal to zero, assuming the parameters are estimated via the quantile regression method. Reject if the reported p-value is less than or equal to the value stored in adjusted.alpha. For situations where an adjusted critical value cannot be computed, or when $n > 60$, the function sets adjusted.alpha equal to alpha.

The R function

   qregci(x,y,qval=0.5, nboot=200, alpha=0.05, SEED=TRUE,xout=FALSE, outfun=out,...)

computes confidence intervals for the slope parameters. If there is a single independent variable, the function will compute a confidence interval for all of the quantiles indicated by the argument qval. If there is more than one independent variable, only the first quantile stored in qval is used. For computing confidence intervals for both the slopes and the intercepts, for a single quantile, use the R function regci with the argument regfun=qval.

The R function

   qrchk(x, y, qval = 0.5, q=NULL,nboot = 1000, com.pval = FALSE, SEED = TRUE, tr=0.2,
                pr = TRUE, xout = FALSE, outfun = out, MC=FALSE,...)

tests the hypothesis that for some $\beta_0, \ldots, \beta_p$, the γ quantile of y, given $x_1, \ldots, x_p$, is given by

$$y = \beta_0 + \beta_1 x_1 + \cdots + \beta_p x_p.$$

The quantile to be used is specified by the argument qval or the argument q. (If q is specified, this quantile is used regardless of the value indicated by qval.) The function contains approximate critical values when dealing with the 0.5 quantile and testing at the 0.1, 0.05, 0.025, and 0.01 levels. A p-value can be computed by setting the argument com.pval=TRUE, which will increase execution time considerably. However, a substantial reduction in execution time can be achieved by setting MC=TRUE, assuming that a multicore processor is available. Reject the null hypothesis, if the test statistic exceeds the critical value.

■ **Example**

Data from the Well Elderly 2 study (Clark et al., 2011; Jackson et al., 2009) are used to illustrate the R function qrchk. A general goal was to assess the efficacy of an intervention strategy aimed at improving the physical and emotional health of older adults. A portion of the study was aimed at understanding the association between a measure of depressive symptoms (CESD) and the CAR. Cortisol is measured upon awakening and about 30–60 minutes later. The CAR refers to the change in cortisol (the level upon awakening minus the level taken 30–60 minutes later). Using the data obtained after six months of intervention, the p-value returned by qrchk is 0.052, suggesting that using a straight regression line might be misleading.

■

11.1.7 Inferences Based on the OP-Estimator

When using the skipped estimators in Section 10.10, and when $p > 1$, the bootstrap methods in Sections 11.1.1 and 11.1.3 tend to be too conservative in terms of Type I errors when the sample size is small. That is, when testing at the 0.05 level, the actual probability of a Type I error tends to be considerably less than 0.05 when the sample size is less than 60 (Wilcox, 2004c). Accordingly, the following modifications are suggested when using the OP-estimator. When testing Eq. (11.1), the hypothesis that all slope parameters are zero, let $\hat{p}^*$ be the bootstrap estimate of the p-value given in Section 11.1.1. Let $n_a = n$, if $20 \leq n \leq 60$. If $n < 20$, $n_a = 20$, and if $n > 60$, $n_a = 60$. Then the adjusted p-value used here is

$$\hat{p}_a^* = \frac{\hat{p}^*}{2} + \left(\frac{n_a - 20}{40}\right)\frac{\hat{p}^*}{2},$$

and the null hypothesis is rejected if $\hat{p}_a^* \leq \alpha$.

As for testing hypotheses about the individual slope parameters, let

$$C = 1 - \frac{60 - n_a}{80},$$

and let $\hat{p}_m^*$ be computed as in Section 11.1.3. Then the adjusted p-value is

$$\hat{p}_a^* = C\hat{p}_m^*.$$

To control FWE (the familywise error, meaning the probability of at least one Type I error), Hochberg's (1988) method is used. For convenience, let Q_j be the adjusted p-value associated with the bootstrap test of H_0: $\beta_j = 0$. Put the Q_j values in descending order, yielding $Q_{[1]} \geq Q_{[2]} \geq \cdots \geq Q_{[p]}$. Beginning with $k = 1$, reject all hypotheses if

$$Q_{[k]} \leq \frac{\alpha}{k}.$$

That is, reject all hypotheses if the largest p-value is less than or equal to α. If $Q_{[1]} > \alpha$, proceed as follows:

1. Increment k by 1. If

$$Q_{[k]} \leq \frac{\alpha}{k},$$

   stop and reject all hypotheses having a p-value less than or equal to $Q_{[k]}$.
2. If $Q_{[k]} > \alpha/k$, repeat step 1.
3. Repeat steps 1 and 2 until a significant result is obtained or all p hypotheses have been tested.

■ Example

Suppose x_1, x_2, and ϵ are independent and have standard normal distributions, and that the goal is to test H_0: $\beta_2 = 0$ at the 0.05 level, with $n = 20$, assuming that $y = x_1 + x_2 + \epsilon$. Further imagine that unknown to us, $y = x_1 + x_1x_2 + \epsilon$. Using the conventional Student's t test of H_0: $\beta_2 = 0$, the actual probability of rejecting is approximately 0.16 (based on a simulation with 1,000 replications using the built-in R functions lm and summary). Increasing n to 100, the actual probability of rejecting is again approximately 0.16. Using instead the R function opregpb (described in the next section), with $n = 20$, the probability of rejecting is approximately 0.049.

■

11.1.8 R Functions opregpb and opregpbMC

The R function

> opregpb(x,y,nboot=1000, tr=0.2,om=TRUE, ADJ=TRUE, nullvec=rep(0, ncol(x) + 1),
>   plotit=TRUE, gval = sqrt(qchisq(0.95,ncol(x) + 1)))

tests hypotheses based on the OP-estimator and the modified bootstrap method just described. Both an omnibus test and confidence intervals for the individual parameters are reported. To avoid the omnibus test, set om=FALSE. The argument gval is the critical value used by the projection-type outlier detection method. Setting ADJ=FALSE, the adjustments of the p-values, described in Section 11.1.7, are not made. The function

> opregpbMC(x,y,nboot=1000,tr=0.2,om=TRUE,ADJ= TRUE,nullvec=rep(0, ncol(x) + 1),
>   plotit=TRUE, gval = sqrt(qchisq(0.95,ncol(x) + 1)))

is the same as opregpb, only it uses a multicore processor, assuming that one is available, and that the R package parallel has been installed.

11.1.9 Hypothesis Testing When Using a Multivariate Regression Estimator RADA

Consider again the multivariate regression model in Section 10.17.1, where

$$\mathbf{y} = \mathbf{B}'\mathbf{x} + \mathbf{a} + \epsilon, \tag{11.9}$$

$\mathbf{B}$ is a $(p \times q)$ slope matrix, $\mathbf{a}$ is a q-dimensional intercept vector, and the errors $\epsilon = (\epsilon_1, \ldots, \epsilon_q)$ are independent and identically distributed with mean $\mathbf{0}$ and covariance matrix

Σ_ϵ, a positive definite matrix of size q. (That is, for any non-zero vector $\mathbf{x}$, $\mathbf{x}'\Sigma_\epsilon\mathbf{x} > 0$.) When using the multivariate regression estimator RADA, in Section 10.17.1, consider the issue of testing

$$H_0: \mathbf{B} = \mathbf{0}.$$

A natural guess is to proceed along the lines in Section 11.1.1 and use a simple modification of the R function regtest. But this method has been found to be unsatisfactory in simulations. A percentile bootstrap method appears to avoid Type I error probabilities above the nominal level. However, the actual level can be substantially smaller than the nominal level, suggesting that power might be relatively poor. Imagine, for example, that with $n = 40$, $p = 2$, and $q = 3$, a p-value is computed in the usual way. Under normality, an actual Type I error probability of 0.05 is achieved, if the null hypothesis is rejected when the estimated p-value is less than or equal to 0.16.

Currently, only one method has been found that performs reasonably well in terms of controlling the Type I error probability, including situations where there is heteroscedasticity. Briefly, let $\mathbf{C}$ be a row vector of length pq containing the pq slope estimates. Let $\hat{\Sigma}$, a pq-by-pq matrix, be a bootstrap estimate of the variances and covariances associated with $\mathbf{C}$. The test statistic is

$$\frac{1}{pq}\mathbf{C}\hat{\Sigma}^{-1}\mathbf{C}',$$

with the null distribution taken to be an F distribution, with $\nu_1 = pq - 1$ and $\nu_2 = n - pq$ degrees of freedom.

11.1.10 R Function mlrGtest

The R function

$$\text{mlrGtest(x,y,regfun=mlrreg,nboot=300,SEED=T)}$$

tests the hypothesis

$$H_0: \mathbf{B} = \mathbf{0},$$

using the method just described. By default, the RADA estimator is used. Any multivariate regression estimator could be used via the argument regfun. Currently, however, simulation results regarding the ability of the method to control the probability of a Type I error are limited to the RADA estimator.

11.1.11 Robust ANOVA via Dummy Coding

As noted in Section 10.1.4, a well-known approach to comparing the means of multiple groups is via least squares regression coupled with dummy coding, also known as indicator variables (e.g., Montgomery et al., 2012). An issue of interest is whether this approach might be generalized by replacing the least squares regression estimator with one of the robust estimators described in Chapter 10. When using the Theil–Sen estimator, simulations do not support this strategy: Control over the Type I error probability can be poor (Ng, 2009b). Whether a similar problem occurs when using some other robust regression estimator has not been investigated. Talib and Midi (2009) study a robust regression estimator aimed at situations where both continuous and categorical regressors are present. But it is unknown how well it performs for the situation at hand.

11.1.12 Confidence Bands for the Typical Value of y, Given x

This section deals with computing an approximation of a confidence band, sometimes called prediction bands, for $m(x) = \beta_0 + \beta_1 x$, the typical value of y, given x, that allows heteroscedasticity. More precisely, if the parameters β_0 and β_1 are estimated based on the random sample $(x_1, y_1), \ldots, (x_n, y_n)$, the goal is to compute a confidence interval for $m(x_i)$ $(i = 1, \ldots, n)$ such that the simultaneous probability coverage is approximately $1 - \alpha$. And there is the related goal of testing the n hypotheses

$$H_0: m(x_i) = \theta_0, \tag{11.10}$$

where θ_0 is some specified constant. (For a review of methods based on the least squares estimator that assume normality and homoscedasticity, see Liu et al., 2008.)

The basic strategy mimics the approach used by the two-sample version of Student's t test. Begin by assuming normality and homoscedasticity, determine an appropriate critical value based on the sample size and the regression estimator that is used in conjunction with an obvious test statistic, and then study the impact of non-normality and heteroscedasticity via simulations.

First consider a single value for the covariate, x. Let τ^2 denote the squared standard error of $\hat{y} = b_0 + b_1 x$, an estimate of $m(x)$, where b_0 and b_1 are estimates of β_0 and β_1, respectively, based on some regression estimator to be determined. A basic percentile bootstrap method is used to estimate τ^2 (e.g., Efron and Tibshirani, 1993). More precisely, generate a bootstrap sample by randomly sampling with replacement n pairs of points from $(x_1, y_1), \ldots, (x_n, y_n)$, yielding $(x_1^*, y_1^*), \ldots, (x_n^*, y_n^*)$. Based on this bootstrap sample, estimate the intercept and slope, and label the results b_0^* and b_1^*, which yields $\hat{y}^* = b_0^* + b_1^* x$. Repeat this B times, yielding $\hat{y}_1^*, \ldots, \hat{y}_B^*$, in which case an estimate of τ^2 is

$$\hat{\tau}^2 = \frac{1}{B-1} \sum (\hat{y}_b^* - \bar{y}^*)^2,$$

where $\bar{y}^* = \sum \hat{y}_b^* / B$. (In terms of controlling the probability of a Type I error, $B = 100$ appears to suffice.) Then the hypothesis given by Eq. (11.10) can be tested with

$$W = \frac{\hat{y} - \theta_0}{\hat{\tau}}$$

once an appropriate critical value has been determined.

Momentarily assume that W has a standard normal distribution, in which case a p-value can be determined for each x_i, $i = 1, \ldots, n$. Denote the resulting p-values by $p_1, \ldots, p_n$, and let $p_m = \min(p_1, \ldots, p_n)$. As is evident, if p_α, the α quantile of p_m, can be determined, the probability of one or more Type I errors can be controlled simply by rejecting the ith hypothesis, if and only if $p_i \le p_\alpha$. And in addition, confidence intervals for each $m(x_i)$ can be computed that have simultaneous probability coverage $1 - \alpha$.

The distribution of p_m is approximated in the following manner. Momentarily assume that both the error term ϵ and x have a standard normal distribution, and consider the case $\beta_0 = \beta_1 = 0$. Then a simulation can be performed, yielding an estimate of the α quantile of the distribution of p_m. In effect, generate n pairs of observations from a bivariate normal distribution having correlation zero, yielding $(x_1, y_1), \ldots, (x_n, y_n)$. Compute p_m, and repeat this process A times, yielding $p_{m1}, \ldots, p_{mA}$. Put these A values in ascending order, yielding $p_{m(1)} \le \cdots \le p_{m(A)}$, and let $k = \alpha A$ rounded to the nearest integer. Then the α quantile of p_m, p_α, is estimated with $p_{m(k)}$. Moreover, the simultaneous probability coverage among the n confidence intervals

$$\hat{y}_i \pm z \hat{\tau}_i \ (i = 1, \ldots, n) \tag{11.11}$$

is approximately $1 - \alpha$, where z is the $1 - p_\alpha/2$ quantile of a standard normal distribution, $\hat{y}_i = b_0 + b_1 x_i$, and $\hat{\tau}_i$ is the corresponding estimate of the standard error. Here are some estimates of p_α when $1 - \alpha = 0.95$ and when using the Theil–Sen (TS) estimator, the modification of the Theil–Sen estimator based on the Harrell–Davis estimator (TSHD), OLS, and the quantile regression estimator (QREG):

n	TS	OLS	TSHD	QREG
10	0.011	0.001	0.009	0.011
20	0.010	0.004	0.008	0.009
50	0.010	0.008	0.009	0.009
100	0.010	0.008	0.009	0.008
400	0.011	0.011	0.012	0.009
600	0.010	0.011	0.010	0.010

As can be seen, the value depends on the sample size when using least squares regression, as expected. In contrast, when using the robust regression estimators, the estimated values suggest that there is little or no variation in the value of p_α as a function of the sample size, at least when $10 \leq n \leq 600$.

Of course, a crucial issue is how well the method performs when dealing with non-normality and heteroscedasticity. Simulations indicate that it performs well when testing at the 0.05 level and $n = 20$ (Wilcox, 2016c). Even OLS performed tolerably well, but generally using the Theil–Sen estimator or the quantile regression estimator provides better control over the Type I error probability. (When using least squares regression, Faraway and Sun, 1995, derived an alternative method that allows heteroscedasticity.)

11.1.13 R Functions regYhat, regYci, and regYband

The R function

regYhat(x,y, xr=x, regfun=tsreg, xout=FALSE, outfun=outpro,...)

computes $\hat{y}$ for every x indicated by the argument xr. By default, the Theil–Sen estimator is used.

The R function

regYci(x, y, regfun = tsreg, pts = x, nboot = 100, ADJ = FALSE, xout = FALSE, outfun = out,
    SEED = TRUE, tr=0.2, crit = NULL, null.value = 0, plotPV = FALSE, scale = FALSE,
  span = 0.75, xlab = 'X', xlab1 = 'X1', xlab2 = 'X2', ylab = 'p-values', theta = 50, phi = 25,
                    MC = FALSE, nreps = 1000, pch = '*', ...)

computes a $1 - \alpha$ confidence interval for $m(x)$ for every x indicated by the argument pts. So by default, a confidence interval is computed for each value stored in the argument x. Setting plotPV=TRUE, the p-values are plotted for each value stored in x. Setting the argument ADJ=TRUE, the confidence intervals are adjusted so that the simultaneous probability coverage is approximately equal to $1 - \alpha$, where α is controlled via the argument alpha. The adjusted critical values, when ADJ=TRUE, have been determined when the argument reg-fun=tsreg (Theil–Sen), ols, tshdreg (the modification of the Theil–Sen estimator based on the Harrell–Davis estimator), and qreg (quantile regression). When using some other regression estimator, an adjusted critical value is computed with the R function

regYciCV(n, alpha=0.05, nboot=1000, regfun=tsreg, SEED=TRUE, MC=FALSE, null.value=0, xout=FALSE,...)

again assuming that there is a single independent variable. (That is, the R function regYci calls the R function regYciCV.) If, for example, alpha=0.05, and this function returns 0.01, then, in effect, regYci is used with the argument alpha=0.01, in which case, the probability of one or more Type I errors is approximately 0.05. If the argument alpha differs from 0.05, again the R function regYci uses the R function regYciCV to determine the critical value that is used. However, execution time can be quite high. Execution time can be reduced by setting the argument MC=TRUE, assuming that a multicore processor is available.

The R function

regYband(x, y, regfun = tsreg, npts = NULL, nboot = 100, xout = FALSE, outfun = outpro, SEED = TRUE, tr=0.2, crit = NULL, xlab = 'X', ylab = 'Y', SCAT = TRUE, ADJ = TRUE, pr = TRUE, nreps = 1000, MC = FALSE, ...)

plots the regression line as well as a confidence band for $m(x)$. If the argument ADJ=FALSE, regYband computes a confidence interval for x values evenly spaced between $\min(x)$ and $\max(x)$. The number of points used is controlled by the argument npts and defaults to 20. The ends of the resulting confidence intervals, which are computed with the R function regYci, are used to plot the confidence band. The probability coverage for each confidence interval is 1-alpha, where the argument alpha defaults to 0.05. To get a confidence interval for each value stored in x that has simultaneous probability coverage 0.95, set the argument ADJ=TRUE. So if ADJ=TRUE, when testing n hypotheses about $m(x)$ for each value stored in the argument x, the probability of one or more Type I errors is approximately 0.05. This adjustment assumes that there is a single independent variable. There are indications that with more than one independent variable a different adjustment is required, but this issue is in need of more study before a recommendation can be made.

■ **Example**

Consider again the example at the end of Section 11.1.6, which deals with the CAR and its association with a measure of depressive symptoms (CESD). Focusing on positive CAR values (cortisol decreases after awakening), a positive association is found between the CAR and CESD based on the Theil–Sen estimator (with leverage points removed). The p-value is 0.021. So a simple interpretation is that the more cortisol decreases, the higher typical CESD scores tend to be. But to what extent is this a serious health concern? CESD scores greater than 15 are often taken to indicate mild depression. A score greater than 21 indicates the possibility of major depression. Is it the case that CESD scores greater than 15 occur for the typical participant over the range of CAR values that are available? Based on a confidence band having approximately simultaneous probability coverage 0.95, the results indicate that for CAR between 0 and 0.10, the typical CESD score is significantly less than 15. Using OLS, the range is 0 to 0.07. For CAR greater than 0.23, the typical CESD score is estimated to be greater than 15. But over the entire range of available CAR values (0–0.265), the typical CESD scores are not significantly greater than 15. Perhaps for CAR values greater than 0.265, this would no longer be the case, but this remains to be determined.

■

11.1.14 R Function regse

When estimating the slope and intercept based on some robust regression estimator, situations might arise where there is an explicit interest in estimates of the standard errors. The R function

regse(x,y,xout=FALSE,regfun=tsreg,outfun=outpro,nboot=200,SEED=TRUE,...)

is supplied for accomplishing this goal using a bootstrap method. As can be seen, the Theil–Sen estimator is used by default, but any regression estimator can be used via the argument regfun.

11.2 Comparing the Regression Parameters of $J \geq 2$ Groups

This section describes methods for comparing the regression parameters associated with two or more groups. First the focus is on comparing independent groups, and then some methods for comparing dependent groups are described.

11.2.1 Methods for Comparing Independent Groups

For $J \geq 2$ independent groups, let $(y_{ij}, \mathbf{x}_{ij})$ be the ith vector of observations in the jth group, $i = 1, \ldots, n_j$, $j = 1, \ldots, J$. Suppose

$$y_{ij} = \beta_{0j} + \mathbf{x}'_{ij}\boldsymbol{\beta}_j + \lambda_j(\mathbf{x}_j)\epsilon_{ij},$$

where $\boldsymbol{\beta}_j = (\beta_{1j}, \ldots, \beta_{pj})'$ is a vector of slope parameters for the jth group, $\lambda_j(\mathbf{x}_j)$ is some unknown function of $\mathbf{x}_j$, and ϵ_{ij} has variance σ_j^2. This section deals with the problem of testing

$$H_0: \beta_{k1} = \cdots = \beta_{kJ}, \forall k, 0 \le k \le p. \tag{11.12}$$

That is, the hypothesis is that all J regression lines are identical. The method described here can also be used to test the hypothesis that slope parameters are equal. That is, the goal is to test Eq. (11.12) ignoring the intercepts.

Classic methods assume $\sigma_j^2 = \sigma_k^2$ for any $j \ne k$ (between group homoscedasticity) and $\lambda_j(\mathbf{x}_j) \equiv 1$ (within group homoscedasticity). Methods that assume both types of homoscedasticity (e.g., Huitema, 2011) can perform poorly, in terms of controlling the probability of a Type I error, when the homoscedasticity assumptions are violated. For example, Ng and Wilcox (2012) illustrate that in terms of Type I errors, classic methods can perform very poorly when the within group homoscedasticity assumption is violated. Conerly and Mansfield (1988) provide references to other solutions. Included is Chow's (1960) likelihood ratio test, which also is known to fail. The methods in this section allow both within group and between group heteroscedasticity. First, attention is focused on methods that are based on least squares regression, and then robust regression estimators are considered.

Methods Based on the Least Squares Regression Estimator

As previously noted, the least squares regression estimator is not robust. One concern is that outliers among the dependent variable y can result in a poor fit for the bulk of the points and poor power. Bad leverage points are also a concern, but they can be addressed simply by removing them. Also, in some situations, there might be a particular interest in the conditional mean of y, given $\mathbf{x}$ even though the population mean is not robust.

The method for testing Eq. (11.12), based on the least squares estimator, is based in part on a simple adaptation of a multivariate analysis of variance (MANOVA) test statistic that was derived by Johansen (1980), with the goal of developing a heteroscedastic method for comparing means, which was studied by Wilcox and Clark (2015).

For the jth group, let

$$\hat{\mathbf{B}}_j = (\hat{\beta}_{0j}, \ldots, \hat{\beta}_{pj})$$

be the OLS estimate of $(\beta_{0j}, \ldots, \beta_{pj})$, and let $\mathbf{S}_j$ denote the HC4 estimate of the covariance matrix associated with $\hat{\mathbf{B}}_j$, the least squares estimate of $(\beta_{0j}, \ldots, \beta_{pj})$. (The HC4 estimator is described in Section 10.1.1.) Let

$$\mathbf{W}_j = \mathbf{S}_j^{-1}$$

and

$$\mathbf{W} = \sum \mathbf{W}_j.$$

The estimate of the regression parameters, assuming H_0 is true, is

$$(\tilde{\beta}_0, \ldots, \tilde{\beta}_p)' = \mathbf{W}^{-1} \sum \mathbf{W}_j \hat{\mathbf{B}}_j'.$$

The test statistic is

$$F = \sum_{j=1}^{J} \sum_{k=0}^{p} \sum_{m=0}^{p} w_{mkj} (\hat{\beta}_{mj} - \tilde{\beta}_m)(\hat{\beta}_{kj} - \tilde{\beta}_k), \qquad (11.13)$$

where w_{mkj} is the $(m+1)(k+1)$th element of $\mathbf{W}_j$. Based on results in Johansen (1980), when the null hypothesis is true, F has, approximately, a chi-squared distribution, with degrees of freedom $v = (p+1)(J-1)$. However, an adjusted critical value derived by Johansen (1980) has been found to perform better in simulations (Wilcox and Clark, 2015), which is computed as follows:

$$A = \frac{1}{2} \sum_{j=1}^{J} [\{tr(\mathbf{I} - \mathbf{W}^{-1}\mathbf{W}_j)\}^2 + tr\{(\mathbf{I} - \mathbf{W}^{-1}\mathbf{W}_j)^2\}]/f_j, \qquad (11.14)$$

where $f_j = n_j - 1$ and $\mathbf{I}$ is the identity matrix. The null hypothesis is rejected if

$$F \geq c + \frac{c}{2p(J-1)} \left\{ A + \frac{3cA}{p(J-1)+2} \right\}, \qquad (11.15)$$

where c is the $1 - \alpha$ quantile of a chi-squared distribution, with degrees of freedom $v = (p+1)(J-1)$. Note that a p-value is readily computed using a pinch process. That is, determine the smallest level α that yields a significant result based on Eq. (11.13). Wilcox and Clark (2015) found that this method controls the probability of a Type I error reasonably well when the groups have a common sample size $n = 20$ and $p = 1$ or 2. Removing leverage points, the actual Type I error probability can exceed 0.075 when testing at the 0.05 level. Increasing n to 40 corrected this problem among the situations considered. Consequently, because checking on the impact of leverage points can be crucial, it is recommended that the method just described be used with $n \geq 40$.

For completeness, another approach to comparing the slopes is to use a wild bootstrap method in conjunction with the HC4 estimate of the standard errors (Ng and Wilcox, 2010, 2012). Simulations indicate that, again, good control over the Type I error probability is achieved with a sample size of $n = 20$. However, the impact of removing leverage points on the probability of a Type I error is unknown, and the methods studied were limited to a single covariate. So for the moment, the method based on the test statistic F, given by Eq. (11.13), is recommended.

Multiple Comparisons

A related goal is performing all pairwise comparisons of the slopes in a manner that controls FWE, the probability of one or more Type I errors. And another goal is, for each $j < k$, to test the hypothesis that regression lines associated with groups j and k are identical. Regarding the goal of comparing the slopes, when there is a single independent variable, use the test statistic

$$T_w = \frac{\hat{\beta}_{1j} - \hat{\beta}_{1k}}{\sqrt{S_j^2 + S_k^2}},$$

where again S_j^2 and S_k^2 are the HC4 estimates of the squared standard errors. FWE is controlled well via Hochberg's methods (Wilcox and Ma, 2015). In principle, the method can be used with more than one independent variable, but there are no simulation results on how it performs. As for testing the hypothesis that the regression lines are identical, the test statistic F used in conjunction with the adjusted critical value, given by Eq. (11.15), performed well where again Hochberg's method is used to control FWE.

Methods Based on Robust Estimators

First consider the goal of computing a confidence interval for the difference between the slopes. When using any robust estimator with a reasonably high breakdown point, the percentile bootstrap technique appears to give reasonably accurate confidence intervals for a fairly broad range of non-normal distributions and heteroscedastic error terms. This suggests a method for addressing the goals considered here, and simulations support their use. Briefly, the procedure begins by generating a bootstrap sample from the jth group as described, for example, in Section 11.1.1. That is, for the jth group, randomly sample n_j vectors of observations, with replacement, from $(y_{1j}, \mathbf{x}_{1j}), \ldots, (y_{n_j j}, \mathbf{x}_{n_j j})$. Let $d_k^* = \hat{\beta}_{k1}^* - \hat{\beta}_{k2}^*$ be the difference between the resulting estimates of the kth independent variable, $k = 1, \ldots, p$. Repeat this process B times, yielding $d_{k1}^*, \ldots, d_{kB}^*$. For each k, put these B values in ascending order, yielding $d_{k(1)}^* \le \cdots \le d_{k(B)}^*$. Let $\ell = \alpha B/2$, $u = (1 - \alpha/2)B$, rounded to the nearest integer, in which case an approximate $1 - \alpha$ confidence interval for $\beta_{k1} - \beta_{k2}$ is $(d_{k(\ell+1)}^*, d_{k(u)}^*)$.

To provide some indication of how well the method performs when computing a 0.95 confidence interval, Tables 11.5 and 11.6 show $\hat{\alpha}$, an estimate of one minus the probability coverage, when $n = 20$, $p = 1$, and M regression with Schweppe weights is used. In these tables, VP refers to three types of error terms: $\lambda(x) = 1$, $\lambda(x) = x^2$, and $\lambda(x) = 1 + 2/(|x| + 1)$. For convenience, these three variance patterns are labeled VP 1, VP 2, and VP 3, respectively. The situation VP 2 corresponds to large error variances when the value of x is in the tails of its distribution, and VP 3 is the reverse. Three additional conditions are considered as well. The first, called C1, is where x_{i1} and x_{i2}, as well as ϵ_{i1} and ϵ_{i2}, have identical distributions. The second condition, C2, is the same as the first condition, only $\epsilon_{i2} = 4\epsilon_{i1}$. The third condition,

Table 11.5:   Values of $\hat{\alpha}$, using the method in Section 11.2.1 when x has a symmetric distribution, $n = 20$.

x		ϵ		VP	Condition		
g	h	g	h		C1	C2	C3
0.0	0.0	0.0	0.0	1	0.029	0.040	0.040
				2	0.042	0.045	0.045
				3	0.028	0.039	0.039
0.0	0.0	0.0	0.5	1	0.029	0.036	0.036
				2	0.045	0.039	0.039
				3	0.025	0.037	0.037
0.0	0.0	0.5	0.0	1	0.026	0.040	0.041
				2	0.043	0.043	0.043
				3	0.029	0.040	0.040
0.0	0.0	0.5	0.5	1	0.028	0.036	0.036
				2	0.042	0.040	0.040
				3	0.023	0.037	0.037
0.0	0.5	0.0	0.0	1	0.024	0.035	0.040
				2	0.051	0.058	0.046
				3	0.014	0.023	0.039
0.0	0.5	0.0	0.5	1	0.023	0.035	0.036
				2	0.049	0.054	0.039
				3	0.013	0.020	0.037
0.0	0.5	0.5	0.0	1	0.022	0.039	0.040
				2	0.050	0.039	0.043
				3	0.014	0.022	0.040
0.0	0.5	0.5	0.5	1	0.024	0.037	0.036
				2	0.052	0.058	0.040
				3	0.013	0.020	0.037

C3, is where for the first group, both x_{i1} and ϵ_{i1} have standard normal distributions, but for the second group, both x_{i2} and ϵ_{i2} have a g-and-h distribution.

Notice that $\hat{\alpha}$ never exceeds 0.06, and, in general, it is less than 0.05. There is room for improvement, however, because in some situations, $\hat{\alpha}$ drops below 0.020. This happens when ϵ has a heavy-tailed distribution, as would be expected based on results in Chapters 4 and 5. Also, VP 3, which corresponds to large error variances when x is near the center of its distribution, plays a role. Despite this, all indications are that, in terms of probability coverage, the bootstrap method in conjunction with $\hat{\beta}_m$ (M regression with Schweppe weights) performs reasonably well over a broader range of situations than any other method that has

Table 11.6: Values of $\hat{\alpha}$, x has a skewed distribution, $n = 20$.

x		ϵ			Condition		
g	h	g	h	VP	C1	C2	C3
0.5	0.0	0.0	0.0	1	0.026	0.040	0.040
				2	0.044	0.048	0.046
				3	0.032	0.037	0.039
0.5	0.0	0.0	0.5	1	0.028	0.039	0.036
				2	0.041	0.047	0.039
				3	0.030	0.031	0.037
0.5	0.0	0.5	0.0	1	0.025	0.040	0.040
				2	0.046	0.052	0.040
				3	0.032	0.038	0.043
0.5	0.0	0.5	0.5	1	0.024	0.038	0.036
				2	0.041	0.045	0.040
				3	0.031	0.034	0.037
0.5	0.5	0.0	0.0	1	0.018	0.032	0.040
				2	0.049	0.050	0.046
				3	0.019	0.020	0.039
0.5	0.5	0.0	0.5	1	0.019	0.031	0.036
				2	0.045	0.049	0.039
				3	0.015	0.018	0.037
0.5	0.5	0.5	0.0	1	0.019	0.031	0.040
				2	0.050	0.050	0.043
				3	0.014	0.020	0.040
0.5	0.5	0.5	0.5	1	0.022	0.027	0.036
				2	0.046	0.051	0.040
				3	0.016	0.019	0.037

been proposed, and using M regression offers the additional advantage of a relatively efficient estimator for the situations considered. It appears that when using other robust estimators, such as Theil–Sen, probability coverage greater than or equal to the nominal level is again obtained. With $n = 30$, there are situations where the actual probability coverage can be as high as 0.975 when computing a 0.95 confidence interval. That is, when testing the hypothesis of equal slopes, the actual level can be as low as 0.025 when testing at the 0.05 level (cf. Luh and Guo, 2000).

As for testing the global hypothesis given by Eq. (11.12), a modification of the test statistic F, given by Eq. (11.13) can be used. Simply replace the HC4 estimator with a bootstrap estimate of the variances and covariances among the estimators that are being used. More precisely, for the jth group, generate a bootstrap sample by randomly sampling with replacement n_j vectors of observations from $(y_{ij}, x_{ij1}, \ldots, x_{ijp})$, yielding $(y_{ij}^*, x_{ij1}^*, \ldots, x_{ijp}^*)$ $(i = 1, \ldots, n_j)$. Based on this bootstrap sample, denote the estimated intercept and slopes by $(b_{0j}^*, \ldots, b_{pj}^*)$. Repeat this process B times, yielding $(b_{0jb}^*, \ldots, b_{pjb}^*)$ $(b = 1, \ldots, B)$. Using $B = 100$ appears to suffice. Then the covariance between b_{kj}^* and $b_{\ell j}^*$ is estimated with

$$s_{k\ell} = \frac{1}{B-1} \sum_{b=1}^{B} (b_{kjb}^* - \bar{b}_{kj})(b_{\ell jb}^* - \bar{b}_{\ell j}),$$

where $\bar{b}_{kj} = \sum b_{kjb}^*/B$. The corresponding covariance matrix $\mathbf{S}_j$ replaces the HC4 estimate when computing F, and of course the least squares estimate of the slopes and intercept are replaced by some robust estimate. Unlike the method based on OLS, an adjusted critical value does not appear to be necessary, and in terms of controlling the Type I error probability, removing leverage points does not appear to be an issue. But for reasons noted in Section 10.14.1, it can be important to check on the impact of removing leverage points.

For the special case where $J = 2$ groups are compared, Moses and Klockars (2012) compare several techniques that include the percentile bootstrap method described here, but no results based on the bootstrap estimate of the covariances are reported. They found that a variation of the Theil–Sen estimator (Luh and Guo, 2000) performed well in simulations. The breakdown point of this variation of the Theil–Sen estimator is unknown. The method uses an explicit expression of the standard errors rather than a bootstrap estimator, and it appears to perform well even when there is heteroscedasticity. They considered three non-normal distributions, only one of which was skewed; the skewness was 1.6. The largest kurtosis value was 2.86. The simulation results in Tables 11.5 and 11.6 are based on much more extreme departures from normality. Because situations are encountered where the estimated skewness and kurtosis are substantially larger than those considered by Moses and Klockars, it would be of interest to know whether their approach continues to perform well for the situations considered in Tables 11.5 and 11.6.

As for performing multiple comparisons based on the intercept, simply use Hochberg's method to control the probability of one or more Type I errors. The same can be done when dealing with any of the slopes.

11.2.2 R Functions reg2ci, reg1way, reg1wayISO, ancGpar, ols1way, ols1wayISO, olsJmcp, olsJ2, reg1mcp, and olsWmcp

The R function

```
reg2ci(x, y, x1, y1, regfun = tsreg, nboot = 599, alpha = 0.05, plotit = TRUE, SEED = TRUE,
       xout = FALSE, outfun = outpro, xlab = 'X', ylab = 'Y', pr = FALSE, ...)
```

computes a $1 - \alpha$ confidence interval for the difference between the regression parameters corresponding to two independent groups using a basic percentile bootstrap method in conjunction with a robust estimator. That is, it tests Eq. (11.16) for each $k = 0, \ldots, p$, where p is the number of independent variables. (The R function reg2ciMC takes advantage of a multi-core processor if one is available.) The function assumes that a robust regression estimator is used. The first two arguments contain the data for the first group, and the data for group 2 are contained in x2 and y2. As usual, x1 and x2 can be vectors or matrices, or they can be a data frame. The optional argument regfun indicates the regression estimator to be used. If not specified, regfun=tsreg, meaning that the Theil–Sen estimator is used. Setting regfun=MMreg, for example, results in using the MM-estimator. The default number of bootstrap samples indicated by the argument nboot is $B = 599$, and alpha, which is α, defaults to 0.05 if unspecified. When the argument plotit equals TRUE, the function also creates a scatterplot that includes the regression lines for both groups.

Here is an example of what the output looks like when using reg2ci:

```
$output
      Parameter   ci.lower   ci.upper      p.value Group 1    Group 2
[1,]          0  -4.047503   4.700254   0.93989983       8   8.125679
[2,]          1 -62.149548  -7.052186   0.01836394       0  28.672477
```

The first row gives the results for the intercepts, and the second row contains the results for the slopes. For example, the estimate of the slope for the first group is zero, and for the second it is 28.67. This result is based on the Well Elderly data used in Section 11.5.5, with data split into two groups based on whether the CAR values are less than or greater than zero. As noted in Section 11.5.5, Fig. 11.6 suggests that the nature of the association differs substantially depending on whether the CAR is positive or negative. The result reported here supports that conclusion.

The R function

$$\text{reg1way(x,y,regfun=tsreg, nboot=100, SEED=TRUE, xout=FALSE, outfun=outpro,}$$
$$\text{alpha=0.05, pr=TRUE,...)}$$

tests the hypothesis that $J \geq 2$ independent groups have identical regression lines. A robust regression estimator is used in conjunction with the test statistic given by Eq. (11.17). The arguments x and y are assumed to have list mode. For example, x[[1]] is assumed to contain the values for the independent variables associated with group 1. If, for example, there are two independent variables, x[[1]] contains a matrix with n_1 rows and two columns and y[[1]]

contains the corresponding values of the dependent variable. The R function

$$\text{reg1wayISO(x,y,regfun=tsreg, nboot=100, SEED=TRUE, xout=FALSE,}$$
$$\text{outfun=outpro,alpha=0.05,pr=TRUE,...)}$$

tests the hypothesis that all J groups have identical slopes. So for a single covariate, the hypothesis is that the regression lines are parallel. The R function

$$\text{regGmcp(x,y,regfun=tsreg, SEED=TRUE, nboot=100, xout=FALSE, outfun=outpro, tr=0.2,}$$
$$\text{pr=TRUE, MC=FALSE, ISO=TRUE,...)}$$

performs all pairwise global comparisons of the regression parameters. By default, ISO=TRUE meaning that for every pair of groups, the hypothesis given by Eq. (11.12) is tested excluding the intercepts. If ISO=FALSE, Eq. (11.12) is tested including the intercepts. That is, the hypothesis is that the intercepts as well as the slopes are identical.

For convenience, the R function

$$\text{ancGpar(x1,y1,x2,y2, regfun=tsreg, nboot=100, SEED=TRUE, xout=FALSE, outfun=outpro,}$$
$$\text{plotit=TRUE, xlab='X, ylab='Y', ISO=FALSE,...)}$$

is supplied for the case where only two groups are to be compared. So unlike reg1way, the data are stored in vectors rather than in list mode. If there are $p \geq 2$ independent variables, the arguments x1 and x2 can be matrices with p columns, or they can be a data frame. By default, the argument ISO=FALSE and has the same meaning as when using regGmcp. The R function ancGparMC is the same as ancGpar, only it uses a multicore processor if one is available.

The R function

$$\text{ols2ci(x1,y1,x2,y2, xout=FALSE, outfun=outpro, alpha=0.05, HC3=FALSE, plotit=TRUE,}$$
$$\text{xlab='X', ylab='Y',...)}$$

is like reg2ci, only it uses the least squares estimator in conjunction with the HC4 estimate of the standard errors. (Setting the argument HC3=TRUE results in using the HC3 estimator instead.) So this function performs $p + 1$ tests. It compares the intercepts, the slopes associated with the first independent variable, the slopes associated with the second independent variable, and so forth.

The R function

$$\text{ols1way(x,y, xout=FALSE, outfun=outpro, alpha=0.05,pr=TRUE, BLO = FALSE,}$$
$$\text{HC3 = FALSE,...)}$$

is exactly like reg1way, only it uses the least squares estimator and the adjusted critical value indicated by Eq. (11.14). So, again, x and y are assumed to have list mode. BLO=TRUE means that only bad leverage points, based on the method in Section 10.15.1, are removed. Using xout=TRUE, all leverage points are removed based on the R function indicated by the argument outfun. If BLO=TRUE, xout=FALSE is used. To compare the slopes, ignoring the intercepts, use the function

ols1wayISO(x,y, xout=FALSE, outfun=outpro, alpha=0.05, pr=TRUE,BLO=FALSE,...).

All pairwise comparisons among $J > 2$ groups are performed by

olsJmcp(x,y, xout=FALSE, outfun=outpro, alpha=0.05, pr=TRUE,...).

That is, for each pair of groups, the global hypothesis of identical distributions is tested. For convenience, the R function

olsJ2(x1, y1, x2, y2, xout = FALSE, outfun = outpro, plotit = TRUE, xlab = 'X', ylab = 'Y', ISO = FALSE, ...)

can be used when comparing two groups only. The argument ISO=FALSE means that the hypothesis of identical regression lines is tested. Setting ISO=TRUE, the hypothesis of identical slopes is tested.

It is noted that the 3rd edition of this book describes R functions that use bootstrap methods to compare the slopes and intercepts based on the least squares estimator: tworegwb and hc4wmc. These functions are still available in the R packages WRS and Rallfun, but at the moment, there is no known advantage to using these functions rather than the non-bootstrap functions just described.

The R function

reg1mcp(x,y,regfun=tsreg, SEED=TRUE, nboot=100, xout=FALSE, outfun=outpro, alpha=0.05, pr=TRUE, MC=FALSE,...)

performs all pairwise comparisons for each of the regression parameters. Here is an example of what the output looks like when there are two independent variables and three groups, and leverage points are removed by setting the argument xout=TRUE:

```
$n
[1] 40 40 50

$n.keep
[1] 37 40 48

$output
          Group Group    p.value      p.crit       ci.low        ci.hi Sig
Intercept    1     2 0.84808013 0.05000000 -0.9885736  0.77725213   0
Intercept    1     3 0.63439065 0.01666667 -1.0380991  0.71060887   0
Intercept    2     3 0.78797997 0.02500000 -0.5668638  0.46445251   0
slope 1      1     2 0.17362270 0.02500000 -0.2492054  1.00707139   0
slope 1      1     3 0.79465776 0.05000000 -0.6273464  0.45520566   0
slope 1      2     3 0.07345576 0.01666667 -1.0448311  0.05611244   0
slope 2      1     2 0.79465776 0.05000000 -0.8326233  0.63402653   0
slope 2      1     3 0.54424040 0.02500000 -0.4555235  0.86602003   0
slope 2      2     3 0.19031720 0.01666667 -0.1753694  0.73816752   0
```

So the initial sample sizes are 40, 40, and 50, and after leverage points are removed, the sample sizes are 37, 40, and 48. For each parameter, the column headed by p.crit indicates the critical p-value based on Hochberg's method. So for each parameter, $(J^2 - J)/2$ tests are performed, where J indicates the number of groups. Here, three tests are performed for each parameter, and the goal is to have the probability of one or more Type I errors equal to the value indicated by the argument alpha. That is, make a decision about which group has the larger intercept if the p.value is less than or equal to the p.crit value. For convenience, the final column headed by Sig indicates whether a significant result occurred. A zero indicates that the null hypothesis is not rejected, and a one indicates that the p.value is less than or equal to the p.crit value.

The R function

$$\text{olsWmcp(x,y, xout=TRUE, outfun=outpro, alpha=0.05, pr=TRUE, BLO=FALSE,}$$
$$\text{HC3=FALSE,...)}$$

is the same as the function reg1mcp, only it uses the least squares regression estimator.

■ Example

Consider again the Well Elderly 2 study (Clark et al., 2011; Jackson et al., 2009) described in the example at the end of Section 11.1.6. Another goal was to understand the impact of intervention on a measure of meaningful activities (MAPA), again taking into account the CAR. Based on the Theil–Sen estimator, and with leverage points removed, the estimated slope before intervention (MAPA is the dependent variable, and the CAR is the independent variable) is 4.49 and -10.23. The R function reg2ci returns a p-value equal to 0.012. The regression lines cross approximately where the CAR is equal to zero, suggesting that when cortisol increases after awakening (the CAR is negative), MAPA scores tend to be higher, and when cortisol decreases, the reverse is true. A similar result was obtained using the OLS estimator via the R function ols2ci again with leverage points removed. A crossover design was used, resulting in an opportunity to replicate the result just described. MAPA scores were found to be significantly higher after intervention, regardless of the value of the CAR, but the slopes were not significantly different. Indeed, now both slopes are virtually zero. There are, however, some consistent features of how the groups compare when using a non-parametric regression estimator, which are illustrated in Chapter 12.

11.2.3 Methods for Comparing Two Dependent Groups

Next, methods designed for comparing two dependent groups are described, which were studied by Wilcox and Clark (2014) as well as Wilcox (2015c). Evidently, robust methods for testing Eq. (11.12) when there are more than two dependent groups have not been investigated. For simplicity, the method is described when dealing with a single independent variable, but the method is readily generalized to situations where there is more than one independent variable. Let $(x_{i1}, y_{i1}, x_{i2}, y_{i2})$, $i = 1, \ldots, n$, be a random sample where, for example, (x_{i1}, y_{i1}) might indicate pairs of observations taken at time 1 and (x_{i2}, y_{i2}) are pairs of observations taken at time 2. The resulting estimates of the slopes at times 1 and 2 are denoted by b_{11} and b_{12}, respectively, and the corresponding estimates of the intercepts are b_{01} and b_{02}.

Methods Based on a Robust Estimator

A version of the percentile bootstrap method is used to test Eq. (11.12), the hypothesis that the regression lines are identical. Begin by resampling with replacement n vectors of observations from $(x_{i1}, y_{i1}, x_{i2}, y_{i2})$, yielding $(x_{i1}^*, y_{i1}^*, x_{i2}^*, y_{i2}^*)$. Let b_{11}^*, b_{12}^*, b_{01}^*, and b_{02}^* be the resulting estimates of the slopes and intercepts, respectively, based on this bootstrap sample. Let $d_k^* = b_{k1}^* - b_{k2}^*$ ($k = 0, 1$). Repeat this process B times, yielding d_{kb}^* ($b = 1, \ldots, B$). Let D_b denote some measure reflecting the distance of $\mathbf{d}^* = (d_0^*, d_1^*)$ from the center of the bootstrap data cloud, where the center of the data cloud is estimated with some robust estimator. One possibility is to use the Mahalanobis distance based on an estimate of the variances and covariances associated with the bootstrap values d_{kb}^*. But when using a robust regression estimator, situations are encountered where the sample covariance matrix based on a bootstrap sample is singular, which rules out using the Mahalanobis distance. Here, to avoid this problem, projection distances are used as described in Section 6.2.5 and computed via the R function pdis. Let D_0 denote the distance of the null vector from the center of the bootstrap cloud. Then from general theoretical results in Liu and Singh (1997), a p-value is given by

$$\frac{1}{B} \sum I_{D_0 < D_b},$$

where the indicator function $I_{D_0 < D_b} = 1$, if $D_0 < D_b$; otherwise, $I_{D_0 < D_b} = 0$. This method is readily extended to more than one independent variable, but simulation results on how well the method performs are limited to a single independent variable.

Next, consider the goal of testing the hypothesis

$$H_0: \beta_{k1} = \beta_{k2}. \tag{11.16}$$

So for $k = 0$, the null hypothesis is that the intercepts are equal, and for $k = 1$, the hypothesis is that the slopes are equal. This can be accomplished using a basic percentile bootstrap method. For each k, put the d_k^* values in ascending order, yielding $d_{k(1)}^* \leq \cdots \leq d_{k(B)}^*$. Let $\ell = \alpha B / 2$, $u = (1 - \alpha/2)B$, rounded to the nearest integer, in which case, an approximate $1 - \alpha$ confidence interval for $\beta_{k1} - \beta_{k2}$ is $(d_{k(\ell+1)}^*, d_{k(u)}^*)$.

Methods Based on the Least Squares Estimator

Now the goal is to test the hypothesis that the regression lines are identical when using a least squares regression estimator. Let $\mathbf{V}$ be the sample covariance matrix based on the d_{kb}^* values, where b_{k1}^* and b_{k2}^* are the least squares estimates based on a bootstrap sample. That is,

$$v_{k\ell} = \frac{1}{B-1} \sum (d_{kb}^* - \bar{d}_k^*)(d_{\ell b}^* - \bar{d}_\ell^*),$$

where $\bar{d}_k^* = \sum d_{kb}/B$. The test statistic is based on a simple modification of Hotelling's T^2 statistic for testing the hypothesis that a multivariate normal distribution has a mean of zero:

$$H = \frac{n(n-2)}{2(n-1)} (d_0, d_1) \mathbf{V}^{-1} (d_0, d_1)'. \tag{11.17}$$

Reject the null hypothesis given by Eq. (11.12), at the α level, if H is greater than or equal to the $1 - \alpha$ quantile of an F distribution, with $\nu_1 = 2$ and $\nu_2 = n - 2$ degrees of freedom. In the event there are p independent variables, proceed in the manner just described, only now $\nu_1 = p + 1$.

Now, consider the goal of testing the hypotheses given by Eq. (11.16). The estimated squared standard error of d_k is $v_{k+1,k+1}$, and the test statistic is

$$T_k = \frac{d_k}{\sqrt{v_{k+1,k+1}}}.$$

The null distribution is approximated by a Student's t distribution, with $n-1$ degrees of freedom. When the sample size is small, say $n = 20$, removing the data associated with leverage points can improve the control over the Type I error probability in some situations when there is heteroscedasticity (Wilcox, 2015d). The method is readily generalized to situations where there are $p > 1$ independent variables.

11.2.4 R Functions DregG, difreg, and DregGOLS

The R function

DregG(x1,y1,x2,y2, regfun=tshdreg, nboot=500, xout=FALSE, outfun=outpro, SEED=TRUE, plotit=FALSE, pr=TRUE,...)

tests the hypothesis given by Eq. (11.12) when dealing with two dependent groups and when using a robust regression estimator. To test the hypothesis given by Eq. (11.16) for each $k = 0, \ldots, p$, where p is the number of independent variables, use the R function

difreg(x1,y1,x2,y2, regfun=tshdreg, nboot=500, xout=FALSE, outfun=outpro, SEED=TRUE, plotit=FALSE, pr=TRUE,...).

The function returns confidence intervals as well as p-values.

The R functions

DregGOLS(x1,y1,x2,y2, xout=FALSE, outfun=outpro, SEED=TRUE, nboot=200,...)

and

difregOLS(x1,y1,x2,y2,regfun=lsfit,xout=FALSE,outfun=outpro,nboot=200, alpha=0.05,SEED=TRUE,plotit=FALSE,xlab='X',ylab='Y',...)

test the same hypotheses as DregG and difreg, respectively, only they use the least squares estimator.

11.3 Detecting Heteroscedasticity

Yet another way of establishing dependence between some outcome variable y and some predictor x is to test the hypothesis that the (conditional) variation of y, given x, does not vary with x. Among the many methods that have been proposed, most do not perform well in simulations (Lyon and Tsai, 1996). Two that do perform well are described in this section, and a third method, which performs well when there is curvature, is described in Section 11.6.7.

It is *not* suggested that the methods in this section be used to justify homoscedastic techniques. That is, if a test of the assumption of homoscedasticity fails to reject, it is not recommended that a homoscedastic regression model would then be used. The reason is that it is unclear when the power of the methods in this section will be high enough to detect a departure from the usual homoscedastic regression model that is important. Rather, the methods in this section are intended as a method for establishing that a certain type of dependence is present. Situations are encountered where the methods in this section reject, yet methods aimed at testing the hypothesis that the slope parameters are equal to zero fail to reject. And methods for testing the hypothesis of a zero correlation can fail to reject as well.

11.3.1 A Quantile Regression Approach

Let

$$y_\gamma = \alpha_\gamma + \beta_\gamma x$$

be the regression line for predicting the γth quantile of y, given x. The goal is to test

$$H_0: \beta_{0.2} = \beta_{0.8},$$

which represents a type of homoscedasticity. Failing to reject does not mean it is safe to assume there is homoscedasticity. Rejecting this hypothesis means that it is reasonable to make an inference about how the variation in Y depends on X, assuming that the regression lines are straight. The R function qhdsm in Section 11.5.7 provides a partial check on this assumption. Of course, other quantiles might be used. The strategy is to choose quantiles different enough to help achieve relatively high power. But if the quantiles are too close to 0 and 1, controlling the Type I error probability can be difficult. Wilcox and Keselman (2006) consider several methods and find the following to be best among those that are considered.

Compute a bootstrap estimate of the standard error of $d = b_{0.2} - b_{0.8}$, and label the result s_d^*, where b_γ is an estimate of β_γ. Here, $B = 100$ is used. Then an appropriate test statistic is

$$T = \frac{d}{s_d^*}.$$

Assuming that this test statistic has, approximately, a standard normal distribution, the actual Type I error probability was found to be less than the nominal level among the situations considered by Wilcox and Keselman (2006) when testing at the 0.05 level. Even for $n = 100$, the actual level can drop well below the nominal level. (The same problem occurs when using a percentile bootstrap method.) When testing at the 0.05 level, Wilcox and Keselman suggest using an approximate critical value given by

$$q = \Phi^{-1}\left(\frac{-0.104}{\sqrt{n}} + 0.975\right),$$

where Φ^{-1} is the inverse of the cumulative standard normal distribution. That is, reject if $|T| \geq q$. Nothing is known about how to adjust the critical value when dealing with quantiles other than 0.2 and 0.8 or when testing at some level other than 0.05.

11.3.2 The Koenker–Bassett Method

In terms of controlling the probability of a Type I error when testing the hypothesis of homoscedasticity, the Koenker and Bassett (1981) method has been found to perform well in simulations by Lyon and Tsai (1996) as well as Wilcox and Keselman (2006). The method begins by fitting an OLS regression line. Let r_i be the usual residuals ($i = 1, \ldots, n$). If the null hypothesis is true, then

$$\hat{\sigma}^2 = \frac{1}{n}\sum r_i^2$$

provides an estimate of the common variance. Let $A = \sum(r_i^2 - \hat{\sigma}^2)^2/n$ and $\tilde{y} = \sum \hat{y}_i/n$. The test statistic is

$$V = \frac{\{\sum r_i^2(\hat{y}_i - \tilde{y})\}^2}{A\sum(\hat{y}_i - \tilde{y})^2},$$

which has, approximately, a chi-squared distribution, with 1 degree of freedom when the null hypothesis is true. The method can be applied with the R function khomreg, described in the next section.

11.3.3 R Functions qhomt and khomreg

The R function

```
qhomt(x,y,nboot=100, alpha=0.05, qval=c(0.2,0.8), plotit=TRUE, SEED=TRUE,
xlab='X',ylab='Y', xout=FALSE, outfun=outpro, pr=TRUE, WARN=FALSE,...)
```

tests the hypothesis

$$H_0: \beta_{0.2} = \beta_{0.8}.$$

The quantiles that are used can be altered via the argument qval. For example, qval=c(0.25, 0.75) would test $H_0: \beta_{0.25} = \beta_{0.75}$. For more than one independent variable, use the R function

qhomtv2(x, y, nboot = 100, alpha = 0.05, qval = c(0.2, 0.8), SEED = TRUE).

The R function

khomreg(x, y)

performs the Koenker–Bassett test.

11.4 Curvature and Half-Slope Ratios

This section describes an approach to dealing with curvature where the strategy is to attempt to straighten a regression line by replacing the x values with x^a for some a to be determined. By curvature is meant that the regression line is not straight, or more generally, some specified linear parametric model poorly captures the nature of the association. Here, the so-called half-slope ratio is used to help suggest an appropriate choice for a. Another general approach when dealing with curvature is to use some type of non-parametric regression method described in Section 11.5.

Temporarily consider the situation where there is only one predictor ($p = 1$), and let $m = n/2$, rounded down to the nearest integer. Suppose the x values are divided into two groups: x_L, the m smallest x values, and x_R, the $n - m$ largest. Let y_L and y_R be the corresponding y values. For example, if the (x, y) points are $(1, 6)$, $(8, 4)$, $(12, 9)$, $(2, 23)$, $(11, 33)$, and $(10, 24)$, then the (x_L, y_L) values are $(1, 6)$, $(2, 23)$, and $(8, 4)$, and the (x_R, y_R) values are $(10, 24)$, $(11, 33)$, and $(12, 9)$. That is, the x values are sorted into two groups containing the lowest half and the highest half of the values, and the y values are carried along.

Suppose some regression method is used to estimate the slope using the (x_L, y_L) values, yielding, say $\hat{\beta}_L$. Similarly, let $\hat{\beta}_R$ be the slope corresponding to the other half of the data. The *half-slope ratio* is $H = \hat{\beta}_R/\hat{\beta}_L$. If the regression line is straight, then H should have a value reasonably close to 1. In principle, the method can be extended to $p \geq 1$ predictors. But the practical utility of the method is unclear. Simply choose a particular predictor, say the kth, and then divide the vectors of observations into two groups, the first containing the lowest m values of the kth predictor and the second containing the $n - m$ highest. Simultaneously, the y values and all of the remaining predictor values are carried along. That is, rows of data are sorted according to the values of the predictor being considered. Next, estimate the p slope parameters for the first group, yielding $\hat{\beta}_{Lk1}, \ldots, \hat{\beta}_{Lkp}$, where the second subscript, k, indicates that the data are divided into two groups based on the kth predictor. Do the same for the second group of observations, yielding $\hat{\beta}_{Rk1}, \ldots, \hat{\beta}_{Rkp}$, and the half-slope ratios are

$$H_{k\ell} = \hat{\beta}_{Rk\ell}/\hat{\beta}_{Lk\ell},$$

$k = 1, \ldots, p$, $\ell = 1, \ldots, p$. That is, $H_{k\ell}$ is the ratio of the estimated regression slopes for the ℓth predictor when the data are split using the kth predictor.

It is stressed that the half-slope ratio is an exploratory tool that should be used in conjunction with other techniques such as the smoothers described in Section 11.5. A practical advantage is that it might suggest a method of straightening the regression line. In simple regression ($p = 1$), it can suggest a choice for a such that the regression line $y = \beta_1 x^a + \beta_0$ gives a better fit to data, as will be illustrated. However, even when the half-slope ratio appears to be substantially different from 1, replacing x with x^a might have little practical advantage, as will be seen. Also, the half-slope ratio can be highly misleading when, for example, the usual linear model is correct but the slope parameters are close to zero.

11.4.1 R Function hratio

The R function

$$\text{hratio(x,y,regfun=bmreg)}$$

computes the half-slope ratios as just described. As usual, x is an n-by-p matrix containing the predictors. The optional argument regfun can be any R regression function that returns the estimated coefficients in regfun$coef. If regfun is not specified, bmreg (the bounded influence M regression estimator with Schweppe weights) is used.

The function returns a p-by-p matrix. The first row reports the half-slope ratios when the data are divided into two groups using the first predictor. The first column is the half-slope ratio for the first predictor, the second column is the half-slope ratio for the second predictor, and so forth. The second row contains the half-slope ratios when the data are divided into two groups using the second predictor, and so on.

Table 11.7: Breast cancer rate versus solar radiation.

City	Rate	Daily calories	City	Rate	Daily calories
New York	32.75	300	Chicago	30.75	275
Pittsburgh	28.00	280	Seattle	27.25	270
Boston	30.75	305	Cleveland	31.00	335
Columbus	29.00	340	Indianapolis	26.50	342
New Orleans	27.00	348	Nashville	23.50	354
Washington, DC	31.20	357	Salt Lake City	22.70	394
Omaha	27.00	380	San Diego	25.80	383
Atlanta	27.00	397	Los Angeles	27.80	450
Miami	23.50	453	Fort Worth	21.50	446
Tampa	21.00	456	Albuquerque	22.50	513
Las Vegas	21.50	510	Honolulu	20.60	520
El Paso	22.80	535	Phoenix	21.00	520

■ **Example**

Table 11.7 shows some data on the rate of breast cancer per 100,000 women and the amount of solar radiation received (in calories per square centimeter) within the indicated city. (These data are also stored in the file cancer.dat. See Section 1.8.) The half-slope ratio is estimated to be 0.64 using the default regression function, bmreg. The estimates of the slope and intercept, based on all of the data, are -0.030 and 35.3, respectively.

■

When the half-slope ratio is between 0 and 1, it might be possible to straighten the regression line by replacing x with x^a, where $0 < a < 1$, but there is no explicit equation for determining what a should be. However, it is a simple matter to try a few values and see what effect they have on the half-slope ratio. When the half-slope ratio is greater than 1, try $a > 1$. If $a < 0$, often it is impossible to find an a that gives a better fit to the data (e.g., Velleman and Hoaglin, 1981). In the illustration, the half-slope ratio is less than 1, so try $a = 0.5$. Replacing x with x^a, the half-slope ratio increases to 0.8, and $a = 0.2$ increases it to 0.9. In this latter case, the slope and intercept are now estimated to be -1.142 and 46.3. However, a check of the residuals indicates that using $a = 0.2$, rather than $a = 1$, offers little advantage for the data at hand. This is not surprising, based on a cursory examination of a scatterplot of the data, and the smoothers in Section 11.5 also suggest that $a = 1$ will provide a reasonable fit to the data. However, in some situations, this process proves to be valuable.

■ **Example**

Doi conducted a study on variables that predict reading ability. Two predictors of interest were TAAST1, a measure of phonological awareness (auditory analysis), and SBT1, another measure of phonological awareness (sound blending). (The data are stored on the author's web site in the file read.dat in columns 2 and 3.) One goal was to predict an individual's score on a word identification test, WWISST2 (column 8 in the file read.dat), and, more generally, to understand the association among the three variables. An issue is whether the model $y = \beta_0 + \beta_1 x_1 + \beta_2 x_2 + \epsilon$ provides a reasonable approximation of the regression surface. The function hratio returns

```
            [,1]          [,2]
[1,] -0.03739762   0.8340422
[2,]  0.34679647  -0.9352855
```

That is, dividing the data into two groups using the first predictor, the half-slope ratio for the first predictor is estimated to be -0.037, and the second predictor has an estimated half-slope ratio of 0.83. Dividing the data into two groups using the second predictor, the estimates are 0.35 and -0.93. This suggests that a regression plane might be an unsatisfactory representation of how the variables are related. (A method for testing the hypothesis that the regression surface is a plane is described in Section 11.6.1.)

■

11.5 Curvature and Non-Parametric Regression

Roughly, non-parametric regression deals with the problem of estimating a conditional measure of location associated with y, given the p predictors $x_1, \ldots, x_p$, assuming only that this conditional measure of location is given by some unknown function $m(x_1, \ldots, x_p)$. The problem, then, is estimating the function m. This is in contrast to specifying m in terms of unknown parameters, where the best-known approach assumes

$$y_i = \beta_0 + \beta_1 x_{i1} + \cdots + \beta_p x_{ip} + \epsilon_i.$$

There is a vast literature on estimating m non-parametrically, with most methods assuming that the measure of location of interest is the mean (e.g., Efromovich, 1999; Eubank, 1999; Fan and Gijbels, 1996; Fox, 2001; Green and Silverman, 1993; Györfi et al., 2002; Härdle, 1990; Hastie and Tibshirani, 1990). For $p = 1$ and 2, these techniques provide useful graphical methods for studying curvature. Complete details about all methods cannot be covered here. Instead, the focus is on a few methods that appear to have considerable practical value when the focus is on robust measures of location.

11.5.1 Smoothers

Methods for estimating the unknown function m are generally based on what are called smoothing techniques. The basic idea is that if $m(x_1, \ldots, x_p)$ is a smooth function, then among n observations, those points near $(x_1, \ldots, x_p)$ should contain information about the value of m at $(x_1, \ldots, x_p)$. So a crude description of smoothing techniques is that they identify which points, among n vectors of observations, are close to $(x_1, \ldots, x_p)$, and then some measure of location is computed based on the corresponding y values. The result is $\hat{m}(x_1, \ldots, x_p)$, an estimate of the measure of location associated with y at the point $(x_1, \ldots, x_p)$. The estimator $\hat{m}$ is called a *smoother*, and the outcome of a smoothing procedure is called a *smooth* (Tukey, 1977). A slightly more precise description of smoothers is that they are weighted averages of the y values, with the weights a function of how close the vector of predictor values is to the point of interest. (For results on estimating a robust measure of scale that allows the usual error term to be heteroscedastic, see Boente et al., 2010. For general results on computing a confidence band, when $p = 1$ and $m(x_1)$ is taken to be the conditional mean of y given x_1, see Hall and Horowitz, 2013.)

11.5.2 Kernel Estimators and Cleveland's LOWESS

To elaborate, first consider the case of a single predictor ($p = 1$), and suppose it is desired to estimate some measure of location for y, given x. Let w_i be some measure of how close x_i is to x. Then, generally, the estimate of $m(x)$ is taken to be

$$\hat{m}(x) = \sum w_i y_i, \tag{11.18}$$

and the goal is to choose the w_i in some reasonable manner.

Kernel Smoothing

One approach is based on what is called *kernel smoothing*. Let the kernel $K(u)$ be a continuous, bounded, and symmetric real function such that

$$\int K(u)du = 1.$$

An example is the Epanechnikov kernel in Section 3.2.4. Then an estimate of $m(x)$ is given by Eq. (11.18), where

$$w_i = \frac{1}{W_s} K\left(\frac{x - x_i}{h}\right),$$

$$W_s = \sum K \left(\frac{x - x_i}{h} \right),$$

and h is the span described in Section 3.2.4. Even within the class of kernel smoothers, many variations are possible. One of these variations is outlined here; it represents a very slight modification of the kernel regression estimator in Fan (1993). (In essence, the description given by Bjerve and Doksum, 1993, is used, but with the span taken to be min$\{s, \text{IQR}/1.34\}$.)

Again, let $K(u)$ be the Epanechnikov kernel given in Section 3.2.4, and let

$$h = \min(s, \text{IQR}/1.34).$$

Then given x, $m(x)$ is estimated with $\hat{m}(x) = b_0 + b_1 x$, where b_0 and b_1 are estimated via weighted least squares, with weights $w_i = K((x_i - x)/h)$. A smooth can be created by taking x to be a grid of points and plotting the results. (The method can be extended to multiple predictors using the multivariate extension of the Epanechnikov kernel described at the end of Section 6.9, but often the smooth seems to be a bit lumpy. Altering the span might improve matters, but this has not been investigated.)

Cleveland's LOWESS

Another approach to smoothing was developed by Cleveland (1979) and is generally known as locally weighted scatterplot smoothing (LOWESS). Briefly, let

$$\delta_i = |x_i - x|.$$

Next, sort the δ_i values, and retain the fn pairs of points that have the smallest δ_i values, where f is a number between 0 and 1 and plays the role of a span. Let δ_m be the largest δ_i value among the retained points. Let

$$Q_i = \frac{|x - x_i|}{\delta_m},$$

and if $0 \le Q_i < 1$, set

$$w_i = (1 - Q_i^3)^3;$$

otherwise, set

$$w_i = 0.$$

Next, use weighted least squares to predict y using w_i as weights (cf. Fan, 1993). That is, determine the values b_1 and b_0 that minimize

$$\sum w_i (y_i - b_0 - b_1 x_i)^2$$

and estimate the mean of y corresponding to x to be $\hat{y} = b_0 + b_1 x$. Because the weights (the w_i values) change with x, generally, a different regression estimate of y is used when x is altered. Finally, let $\hat{y}_i$ be the estimated mean of y, given that $x = x_i$ based on the method just described. Then an estimate of the regression line is obtained by the line connecting the points $(x_i, \hat{y}_i)$ $(i = 1, \ldots, n)$. (For some interesting comments relevant to LOWESS versus kernel regression methods, see Hastie and Loader, 1993.)

Cleveland (1979) also discusses a robust version of this method. In effect, extreme y values get little or no weight, the result being that multiple outliers among the y values have little or no impact on the smooth. R provides access to a function, called lowess, that performs the computations. (An outline of the computations can be found in Härdle, 1990, p. 192.) For a smoothing method that is based in part on L_1 regression, see Wang and Scott (1994). (Related results are given by Fan and Hall, 1994.) For a method based in part on M-estimators, see Verboon (1993).

11.5.3 R Functions lplot, lplot.pred, and kerreg

It is a fairly simple matter to create a plot of the smooth returned by the built-in R function lowess, which applies Cleveland's method described in the previous section. But to facilitate the use of lowess, a function is supplied that creates the plot automatically. It has the form

lplot(x, y, low.span=2/3, span=0.75, pyhat=FALSE, eout=FALSE, xout=FALSE, outfun=out, plotit=TRUE, expand=0.5, varfun=pbvar, cor.op=FALSE, cor.fun=pbcor, pr=TRUE, scale=FALSE, xlab='X', ylab='Y', zlab='', theta=50, phi=25, family='gaussian', duplicate='error', pc='*',ticktype='simple'),

where the argument low.span is the span, f, when there is a single independent variable. More than one predictor can be handled using a method outlined in Section 11.5.12. Now the span is indicated by the argument span. With two predictors, setting the argument ticktype='detailed' will result in values indicated on the axes when a plot is created. If the argument pyhat=TRUE and the number of predictors is less than or equal to 4, the function returns the $\hat{m}(x_i)$ values, $i = 1, \ldots, n$. If eout=TRUE, the function first eliminates any outliers among the (x_i, y_i) values using the outlier detection method specified by the argument outfun. If xout=TRUE instead, the function removes outliers (leverage points) among the x_i values only. To suppress the plot, set plotit=FALSE. (The argument family is relevant only when $p = 2$; see Section 11.5.12.) The arguments theta and phi can be used to rotate or tilt a three-dimensional plot. The arguments xlab, ylab, and zlab indicate labels for the x-axis, y-axis, and z-axis, respectively. The argument varfun is explained in Section 11.9. When dealing with a single predictor, the argument pc controls how points will be represented in the scatterplot. By default, an * is used. But with n large, this can make it difficult seeing the regression line, in which case, pc='.' might be more satisfactory.

The R function

$$\text{lplot.pred(x, y, pts = x, xout = FALSE, outfun = outpro, span = 2/3, ...)}$$

can be used to compute predicted y values based on one or more x vectors. (This function can be accessed from the R function smpred as well.) The argument pts is assumed to be a matrix with the number of columns equal to the number of independent variables. By default, all of the points stored in the argument x are used, as is done by the R function lplot. An advantage of the R function lplot.pred, over the R function lplot, is that it can be used when there is an interest in predicting the typical value of y based on points not included in the observed data stored in the R variable x. For example, if there are two independent variables and the goal is to compute $\hat{y}$ for the point (0,0), the R commands

$$\text{pts=matrix(c(0,0),ncol=2),}$$
$$\text{lplot.pred(x,y,pts=pts)}$$

accomplish this goal. This assumes, of course, that there are a sufficient number of points among the observed data, stored in x, that are close to (0, 0). If this is not the case, the value for $\hat{y}$ is returned as NA, meaning not available.

The function

$$\text{kerreg(x,y,pyhat=FALSE, pts=NA, plotit=TRUE, theta=50, phi=25, expand=0.5,}$$
$$\text{scale=FALSE, zscale=FALSE, eout=FALSE, xout=FALSE,}$$
$$\text{outfun=out,np=100,xlab='X',ylab='Y',zlab='Z',}$$
$$\text{varfun=pbvar,e.pow=TRUE,pr=TRUE,ticktype='simple')}$$

creates a smooth using a slight modification of the method derived by Fan (1993). The arguments are the same as those used by the function lplot, except the argument np, which determines how many x values are used when creating the smooth. (Details about the argument np can be found in the R function locreg.) If the argument pyhat=TRUE, the function returns the predicted value of y for each value stored in the argument pts.

11.5.4 The Running-Interval Smoother

Now, we consider how smoothers might be generalized to robust measures of location. To help fix ideas, momentarily focus on the single-predictor case ($p = 1$). One approach to estimating m and exploring curvilinearity is with the so-called running-interval smoother. To be concrete, suppose the goal is to use the data in Table 11.7 to estimate the 20% trimmed mean of the breast cancer rate, given that solar radiation is 390. The strategy behind the running-interval smoother is to compute the 20% trimmed mean using all of the y_i values for which the corresponding x_i values are close to the x value of interest, 390. The immediate problem is finding a rule for determining which y values satisfy this criterion.

Let f be some constant that is chosen in a manner to be described and illustrated. Then the point x is said to be close to x_i if

$$|x_i - x| \leq f \times \text{MADN},$$

where MADN is computed using $x_1, \ldots, x_n$. So for normal distributions, x is close to x_i if x is within f standard deviations of x_i. Let

$$N(x_i) = \{j : |x_j - x_i| \leq f \times \text{MADN}\}.$$

That is, $N(x_i)$ indexes the set of all x_j values that are close to x_i. Let $\hat{\theta}_i$ be an estimate of some parameter of interest based on the y_j values such that $j \in N(x_i)$. That is, use all of the y_j values for which x_j is close to x_i. For example, if x_3, x_8, x_{12}, x_{19}, and x_{21} are the only values close to $x = 390$, then the 20% trimmed mean of y, given that $x = 390$, is estimated by computing the 20% sample trimmed mean using the corresponding y values y_3, y_8, y_{12}, y_{19}, and y_{21}. To get a graphical representation of the regression line, compute $\hat{\theta}_i$, the estimate of y, given that $x = x_i$, $i = 1, \ldots, n$, and then plot the points $(x_1, \hat{\theta}_1), \ldots, (x_n, \hat{\theta}_n)$. This process will be called a *running-interval smoother*. (For an alternative approach for creating smooths using M-estimators and trimmed means based on generalizations of kernel smoothers for means, see Härdle, 1990, Chapter 6. Also see Hall and Jones, 1990.)

A practical problem is choosing f. If there are no ties among the x values and f is chosen small enough, the running-interval smoother produces a scatterplot of the points. If f is too large, the horizontal line $\hat{y} = \hat{\theta}$ is obtained, where $\hat{\theta}$ is the estimate of θ using all n of the y values. The problem, then, is to choose f large enough so that the resulting plot is reasonably smooth, but not too large so as to mask any non-linear relationship between x and y. Often the choice $f = 1$ gives good results, but both larger and smaller values might be better, particularly when n is small. As with all smoothers, a good method is to try some values within an interactive-graphics environment, the general strategy being to find the smallest f so that the plot of points is reasonably smooth.

Note that when θ_i is taken to be the trimmed mean of y, a confidence interval for θ_i can be computed via the Tukey–McLaughlin method. Wilcox (2017c) suggests a method for computing a collection of confidence intervals that have, approximately, simultaneous probability coverage $1 - \alpha$. Two techniques were studied. The first, M1, chooses K points such that $\#N(x_k)$ is greater than 12, where $\#N(x_k)$ is the cardinality of $N(x_k)$. The focus was on $K = 25$ and $\alpha = 0.05$, where points are evenly spaced between x_1 and x_{25}, inclusive, where x_1 is taken to be the smallest x_i value such that $\#N(x_i) \geq 12$, and x_{25} is taken to be the largest X_i value such that $\#N(X_i) \geq 12$. Let p_i be the p-value when testing $H_0: \theta_i = 0$. The strategy is to determine the distribution of $p_{\min} = \min(p_1, \ldots, p_K)$ when $y = \epsilon$, where ϵ has a standard normal distribution, and then compute q_α, the α quantile of $p_{\min}$. This was done via a simulation for sample sizes 50, 60, 70, 80, 100, 150, 200, 300, 400, 500, 600, 800, and 1000. Then $1 - q_\alpha$ confidence intervals were computed for each of the K points. Simulations indicate that this approach works reasonably well when $n \geq 50$, $y = x^a + \epsilon$, $a = 0$, 1, and 2, and when the span is taken to be 0.5. This includes situations where ϵ is non-normal.

The second method, M2, is like M1, only now all points for which $\#N(x_k) \geq 12$ are used. Now $n \geq 100$ is required, and the span was taken to be 0.2. When the span is taken to be 0.5, situations were found where the actual simultaneous probability coverage is less than 0.9.

The smoother is first illustrated with some data generated from a known model, to demonstrate how well it performs, and then some additional illustrations are given using data from actual studies. For convenience, it is again assumed that the goal is to predict the 20% trimmed mean of y given x. First consider the situation where both x and ϵ are standard normal and $\beta_1 = \beta_0 = 0$. Then the correct regression line is $y = 0$. Fig. 11.2 shows the running interval smoother, plus several other smoothers for $n = 20$ points, where both x and ϵ were generated from standard normal distributions. The solid line (labeled rl trim) is the running interval smoother. Note that the running interval smoother does a relatively good job of capturing the true regression line. The additional smoothers that come with R include a kernel smoother, a super smoother (labeled supsmu), Cleveland's method described in Section 11.5.2, and a smoothing spline. (Super smoothers and smoothing splines are discussed in manuals, but the details go beyond the scope of this book.) Of course, this one example is not convincing evidence that the running interval smoother has practical value.

A challenge for any smoother is correctly identifying a straight line when in fact the regression line is straight and n is small. Fig. 11.3 illustrates some of the problems that can arise using both the running interval smoother and lowess, described in Section 11.5.2. The upper left panel of Fig. 11.3 is based on the same data used in Fig. 11.2, but only lowess and the running interval smoother are shown. The upper right panel of Fig. 11.3 is the same as the upper left panel, but two of the y values were altered so that they are now outliers. Note that lowess suggests a curved regression line, although the curvature is small enough that it

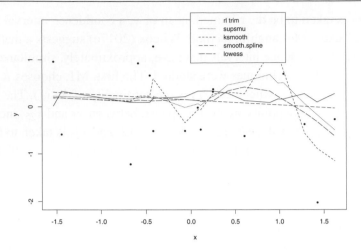

Figure 11.2: Various smoothers, $n = 20$, $f = 1$. The straight line is based on a spline method and gives the best results in this instance. But in other situations, alternative smoothers give superior results.

might be discounted. The lower left panel of Fig. 11.3 shows the same data as in the upper right panel, only the furthest point to the right is moved to $(x, y) = (-2.5, -2.5)$. That is, both x and y are outliers. The curvature in lowess is more pronounced, but even the running interval smoother suggests that there is curvature. This is because the largest x value is so far removed from the other x values, the corresponding trimmed mean of y is based on only one value, $y = -2.5$. One obvious way of dealing with this problem is to check for any outlying or isolated x values, remove them, and see what effect this has on the smoother. The lower right panel of Fig. 11.3 shows what happens when lowess is replaced with a kernel smoother used by R.

The left panel of Fig. 11.4 shows a running interval smooth, with $f = 0.75$, based on $n = 20$ points generated from the model $y = x^2 + \epsilon$, with both x and ϵ having standard normal distributions. (Using the default $f = 1$ is a bit less satisfactory.) The right panel is based on $n = 40$ and $f = 1$. The dashed line in both panels is the true regression line, $y = x^2$. The solid, ragged line is the estimate of the regression line using the running interval smoother. (For more about the running interval smoother, see Wilcox, 1995d.)

Fig. 11.5 shows the results of applying the smoother to various data sets. The first scatterplot is based on data from a study of diabetes in children (Sockett et al., 1987). The upper left panel of Fig. 11.5 shows the age in months versus the logarithm of serum C-peptide. Also shown is the smoother resulting from the lowess command in R. As is evident, they give similar results. One interesting feature of the data is that the half-slope ratio is approximately zero, so a regression model of the form $\hat{y} = \beta_1 x^a + \beta_0$, for some appropriately chosen a, is not very satisfactory.

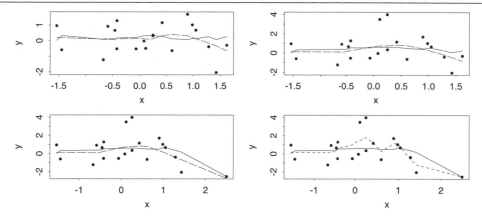

Figure 11.3: A comparison of a running interval smooth, indicated by the solid line, versus some smoothers for means, $n = 20$. The upper left panel shows the running interval smooth versus lowess. The upper right panel is the same as the upper left panel but with two of the y values increased so that they are outliers. The lower left panel is the same as the upper right panel but now with three outliers and one leverage point. The lower right panel shows the same points as the lower left panel, with lowess replaced by a kernel smooth.

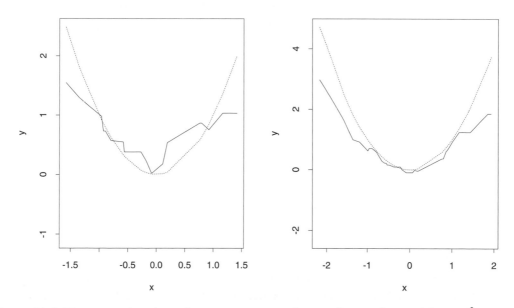

Figure 11.4: Two smooths where data were generated according to the model $y = x^2 + \epsilon$. The left panel is with a span of $f = 0.75$ and the right is with $f = 1$.

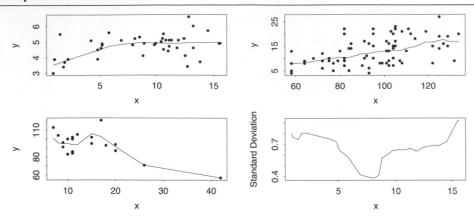

Figure 11.5: The upper left panel is a smooth for predicting the association of the logarithm of C-peptide levels with age. The upper right panel is a smooth for predicting WWISST2 with SBT1. The lower left panel is a smooth of the Gesell score with age. And the lower right panel is a smooth of the standard deviation of log(C-peptide) versus age.

The upper right panel of Fig. 11.5 shows data from a reading study, where one of the goals is to consider how well a measure of phonological awareness (sound blending as measured by the variable SBT1 in the data file read.dat) predicts a word identification score (WWISST2). It appears that a straight line does a reasonably good job of capturing the relationship between the two random variables being investigated.

The lower left panel of Fig. 11.5 shows a scatterplot of data reported by Mickey et al. (1967) for $n = 21$ children, where the goal is to predict a child's Gesell adaptive score based on age in months when a child utters its first word. There is the suggestion that the regression line decreases sharply for older children, but there are too few observations to be sure. Clearly the two largest x values are outliers. If the outliers are eliminated, the running interval smoother returns a nearly flat line for children aged 12 months or younger, but for older children a decreasing regression line appears again. (The details are left as an exercise.) Of course, with only 19 observations left, most of which correspond to children under the age of 12 months, more data are needed to resolve this issue.

An appeal of the running interval smoother is its versatility. Consider the diabetes data in the upper left panel of Fig. 11.5. The lower right panel of Fig. 11.5 shows a running interval smoother, where the goal is to predict the standard deviation of the log C-peptide values based on the child's age. Note that the standard deviation drops dramatically, until about the age of 7, and then increases rapidly. Based on results in Chapter 10, this suggests that even if the error term has a normal distribution, OLS regression might be relatively inefficient compared to various robust estimators.

In some situations, particularly when the sample size is small, the running interval smooth can be somewhat ragged compared to other smoothers. A method that might help correct this problem is bootstrap *bagging* (e.g., Davison et al., 2003; Bühlmann and Yu, 2002; Breiman, 1996a, 1996b). In the present context, the method begins by applying the running interval smoother, yielding, say, $m(x|d_n)$, where $d_n = (x_i, y_i), i = 1, \ldots, n$. That is, $m(x|d_n)$ is some measure of location for y, given x, that is based on the n pairs of observations that are available. Generate a bootstrap sample by randomly sampling, with replacement, n pairs of points from d_n. Label the results d^*. Repeat this B times, yielding $d_1^*, \ldots, d_B^*$. Then the bagged estimate of $m(x)$ is

$$\hat{m}(x|d) = \frac{1}{B} \sum_{b=1}^{B} m(x|d_b^*).$$

That is, use the average of the bootstrap estimates of $m(x)$.

Another strategy for dealing with a running interval smoother that is somewhat ragged, which appears to be more effective than bootstrap bagging, is to simply smooth it again using LOWESS, described in Section 11.5.2. That is, create a smooth based on $(x_i, \hat{\theta}_i)$ $(i = 1, \ldots, n)$, where $\hat{\theta}_i$ is the estimated value of y, given that $x = x_i$ based on the running interval smoother.

Another situation where bagging can be useful is when there is considerable curvature. Suppose $y = x^2 + \epsilon$, where both x and ϵ have standard normal distributions. Using the default span in the R function rplot, described in Section 11.5.5, can result in a poor fit. Reducing the span can help, but there are two other approaches that can be used, both of which involve bagging. See the R functions rplotsm and RFreg in Section 11.5.11.

11.5.5 R Functions rplot, runYhat, rplotCI, and rplotCIv2

The R function

```
rplot(x, y, est = tmean, scat = TRUE, fr = NA, plotit = TRUE, pyhat = FALSE, efr = 0.5, theta
= 50, phi = 25, scale = TRUE, expand = 0.5, SEED = TRUE, varfun = pbvar, outfun = outpro,
  nmin = 0, xout = FALSE, eout = FALSE, xlab = 'X', ylab = 'Y', zscale = FALSE, zlab = ' ',
pr = TRUE, duplicate = 'error', ticktype = 'simple', LP = TRUE, OLD = FALSE, pch = '.', ...)
```

applies the running interval smoother. It consolidates several variations of the running interval smoother that were described in the 3rd edition of this book. (They are the R functions runmean, rungen, and runhat.) The function rplot defaults to using a 20% trimmed mean. That is, the goal is to estimate the trimmed mean of y corresponding to x_i $(i = 1, \ldots, n)$, but any estimator can be used via the argument est. For example, to use the Harrell–Davis estimate of the median, use the command rplot(x,y,est=hd). The argument scat=TRUE means that a scatterplot is created when plotting the regression line, and the argument pch indicates the symbol used by the scatterplot. Setting scat=FALSE eliminates the scatterplot. This might be done, for example, when the goal is to see how a measure of scale associated with y varies with x. Care must be taken because scat=FALSE means that a plot of the smoothed values versus x is created, and this might affect one's perspective on the degree of curvature. (See Exercise 14 at the end of this chapter.)

The argument LP=TRUE means that the initial fit is smoothed again using LOWESS. Setting the argument pyhat=TRUE, the function returns the estimate of the typical value of y for each value in the argument x. The argument fr corresponds to the span, f, which is used to determine which of the x_i values is close to a given point. If not specified, fr=1 is used when there is a single independent variable. (For two independent variables, fr=0.8 is used by default.) For example, the R command rplot(x,y,fr=0.75) would use $f = 0.75$ when creating the running interval smoother. If unsure, first try fr=1, and if the line seems smooth and straight, try fr=0.75 to see what happens. Similarly, the command rplot(x,y,tr=0.1) would cause the running interval smoother to use the 10% trimmed mean. The command rplot(x,y,tr=0.1,plotit=FALSE) would suppress the plot of the smooth. Most of the smooths in Figs. 11.1–11.4 were created in this manner.

As usual, the argument xout eliminates leverage points using the function indicated by the argument outfun, which defaults to outpro. So for a single independent variable, the MAD-median rule is used. Setting eout=TRUE eliminates outliers among both the independent and dependent variables. The arguments varfun and OLD control how the explanatory strength of the association is estimated, which is covered in Section 11.9. (The arguments efr, theta, phi, and scale have to do with situations where there are two independent variables, which is covered in Section 11.5.11.)

The last argument, ..., can be any additional arguments required by the function associated with the argument est. For example, the command rplot(x,y,est=hd,q=0.25) would result in a running interval smoother that predicts the 0.25 quantile of y given x.

A smoother does not provide an explicit equation for predicting y, given x. So the R function

   runYhat(x, y, pts = x, est = tmean, fr = 1, nmin = 1, xout = FALSE, outfun = outpro, ...),

which is the same as the function rplot.pred, is supplied. It computes $\hat{m}(x)$ for each of the values stored in the vector pts. This can also be done via the R function smpred. The argument est defaults to the function tmean, which computes a 20% trimmed mean. If, for example, it is desired to compute $\hat{m}(x)$ for $x = 1$ and 3 using a 10% trimmed mean, the command

runYhat(x,y,pts=c(1,3),mean, tr=0.1, xout = FALSE, outfun = outpro, ...)

accomplishes this goal. (The R function runhat is the same as runYhat when there is one independent variable. Unlike the function runhat, runYhat can be used with more than one independent variable.)

The R function

rplotCI(x, y, tr = 0.2, fr = 0.5, p.crit = NA, plotit = TRUE, scat = TRUE, SEED = TRUE, pyhat = FALSE, npts = 25, xout = FALSE, xlab = "X", ylab = "Y", low.span = 2/3, nmin = 12, outfun = outpro, LP = FALSE, LPCI = FALSE, MID = TRUE, alpha = 0.05, pch = '.', ...)

computes confidence intervals for the trimmed mean of y, given x, using method M1 in the previous section. The R function

rplotCIv2(x, y, tr = 0.2, fr = 0.5, p.crit = NA, MC = FALSE, plotit = TRUE, scat = TRUE, pyhat = FALSE, SEED = TRUE, nmin = 12, nreps = 4000, pts = NULL, xout = FALSE, xlab = 'x', ylab = 'y', outfun = out, LP = TRUE, LPCI = FALSE, alpha = 0.05, nullval = 0, pch = '.', pr = TRUE, ...)

applies method M2, which defaults to using $K = 25$ covariate values, with the goal of computing confidence intervals having simultaneous probability coverage 0.95. Using $K = 10$, execution time is again very low. But otherwise, the function must compute adjusted critical levels, which can increase execution time considerably. An alternative approach is to adjust the confidence intervals via the Studentized maximum modulus distribution. This is the approach used by the R function

rplotCIsmm(x, y, tr = 0.2, fr = 0.5, plotit = TRUE, scat = TRUE, pyhat = FALSE, SEED = TRUE, dfmin = 2, pts = NULL, npts = 25, nmin = 12, eout = FALSE, xout = FALSE, xlab = 'x', ylab = 'y', outfun = out, LP = TRUE, MID = TRUE, alpha = 0.05, pch = '.', ...),

but the function rplotCIv2 provides shorter confidence intervals and is generally more satisfactory. An advantage of rplotCIsmm is that it reduces execution time when $\alpha \neq 0.05$ or when the number of covariate points differs from both 10 and 25.

■ **Example**

The Well Elderly study, described in Section 11.1.6 and analyzed again in Section 11.1.13, is used to illustrate the R function rplotCIv2. Again, the goal is to understand the association between the CAR and a measure of depressive symptoms (CESD). Fig. 11.6 shows the confidence intervals for the regression line, which have, approximately, 0.95

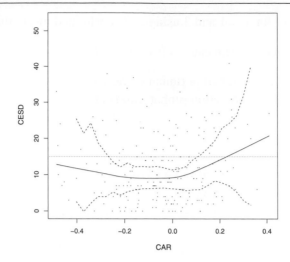

Figure 11.6: Confidence intervals based on 20% trimmed means and the Well Elderly 2 data. The simultaneous probability coverage is approximately 0.95. The horizontal dotted line corresponds to CESD=15.

simultaneous probability coverage. The horizontal dashed line corresponds to a CESD score of 15. CESD scores greater than 15 are taken to be an indication of mild depression or worse. Note that for CAR sufficiently high, typical CESD scores are estimated to be greater than 15. But the confidence intervals make it clear that there is no strong indication that this is the case. The example in Section 11.1.13 shows that there is a significant positive association, based on the Theil–Sen estimator, when the CAR is positive. Fig. 11.6 indicates that the nature of the association is distinctly different when the CAR is negative: It appears to be much weaker and possibly negative. The example based on the R function reg2ci, described in Section 11.2.2, adds support to this conclusion.

Bootstrap bagging can be applied via the R function rplotsm, described in Section 11.5.11. This might provide a more accurate estimate of the regression surface when the sample size is small. The relative merits of using bootstrap bagging versus smoothing the initial using rplot with LP=TRUE are unknown.

11.5.6 Smoothers for Estimating Quantiles

This section deals with smoothers designed for estimating conditional quantiles associated with some dependent variable y, given x. An approach that appears to perform relatively well is to use the running interval smoother in Section 11.5.4 in conjunction with the Harrell–Davis estimator.

Another approach is based on what are called splines (e.g., Kim, 2007). They are a compromise between polynomial regression, which has been criticized due to the global nature of its fit, and other smoothers that have an explicit local nature. Regression splines compromise by employing a piecewise polynomial. The region that defines the pieces are separated by a sequence of knots or breakpoints. (For a summary of data-driven methods for choosing the knots, see, for example, Hastie and Tibshirani, 1990, Chapter 9. For general results on B-splines, see, for example, de Boor, 1978.) A common goal is to force the piecewise polynomials to join smoothly at the knots. One popular choice consists of piecewise cubic polynomials constrained to be continuous and to have continuous first and second derivatives at the knots. Informal comparisons with other smoothers suggest that certain variations of methods based on splines are not quite as satisfactory as other smoothers that might be used (Härdle, 1990). A variation designed specifically for quantiles (e.g., He and Ng, 1999; Koenker and Ng, 2005; cf. Doksum and Koo, 2000) is called constrained B-spline smoothing (COBS). The Koenker–Ng method improves on a computational method studied by He and Ng (1999) and builds upon results in Koenker et al. (1994). The method can be applied via the R package COBS, but using default settings, it can poorly approximate the true regression line. More specifically, it might indicate substantially more curvature than is actually present (Wilcox, 2016b). Wilcox found that generally the running interval smoother gives more satisfactory results. Another advantage of the running interval smoother is that it can be used with two covariates based on the strategy described in Section 11.5.10. COBS can only be applied with a single covariate. For results based on a K nearest neighbor approach, see Ma et al. (2016). An alternative approach was derived by Koenker and Park (1996). Ahmed et al. (2020) derive an estimator based on kernel density estimation, but R software for applying their method is not yet available. For yet another approach, see Kai et al. (2010).

11.5.7 R Function qhdsm

The R function

    qhdsm(x, y, qval = 0.5, q = NULL, pr = FALSE, xout = FALSE, outfun = outpro, plotit =
    TRUE, xlab = 'X', ylab = 'Y', zlab = 'Z', pyhat = FALSE, fr = NULL, LP = TRUE, theta =
    50, phi = 25, ticktype = "simple", nmin = 0, scale = TRUE, pr.qhd = TRUE, pch = '.', ...)

creates a quantile smooth based on the running interval smoother in conjunction with the Harrell–Davis estimator. The argument qval (or q) can be used to specify the quantiles of interest. By default, the 0.5 quantile regression line is plotted. If there is one covariate, multiple quantile regression lines can be plotted. For example,

$$qhdsm(x,y,q=c(0.25,0.5,0.75))$$

will plot the regression lines for predicting the 0.25, 0.5, and 0.75 quartiles. (When the argument pr.qhd = TRUE, the function prints a message about the argument scale. To avoid the message, set pr.qhd = FALSE.) The R function qsmcobs creates a smooth using COBS, but it should be used with caution for reasons explained in the previous section.

11.5.8 Special Methods for Binary Outcomes

When y is binary, now $m(x)$ is taken to be the (conditional) probability that $y = 1$, given x. Smoothers based on means can, again, be used, but some smoothers cannot be recommended. Examples are Cleveland's LOWESS estimator and the kernel estimator in Section 11.5.2. Both of these estimators can yield an estimate of $m(x)$ that is substantially smaller than 0 or larger than 1. However, there are estimators that deal explicitly with binary outcomes that guarantee that $0 \leq m(x) \leq 1$. One relevant study is by Copas (1983). Hosmer and Lemeshow (1989, p. 85) suggest using an estimator that is motivated in part by general results in Kay and Little (1987). Here, a slight modification of the Hosmer–Lemeshow estimator is used. The estimate of $m(x)$ is taken to be

$$\hat{m}(x) = \frac{\sum w_i y_i}{\sum w_i},\qquad(11.19)$$

where

$$w_i = I_h e^{-(x_i - x)^2},$$

and $I_h = 1$ if $|x_i - x| < h$; otherwise, $I_h = 0$. Also, when computing the weights, it is assumed that the x values have been standardized by subtracting the median and dividing by MADN. That is, if the observed predictors are $X_1, \ldots, X_n$, use $x_i = (X_i - M)/\text{MADN}$. If the predictors are not standardized, a change in scale can have a major impact on $\hat{m}$, yielding highly inaccurate and misleading results. Using $h = 1.2$ appears to perform relatively well. Yet another approach is to use the running interval smoother in Section 11.5.4 with the amount of trimming set equal to zero.

Note that Eq. (11.19) is readily extended to situations where y is discrete with a relatively small sample space. That is, y has a multinomial distribution. Suppose, for example, the possible values of y are the integers 1–5. Then Eq. (11.19) can be used to estimate $P(y = j|x)$ for each j, $j = 1, \ldots, 5$. The R function multsm, described in Section 11.5.9, deals with this situation.

A limitation of Eq. (11.19) is that it can handle only a single predictor. A slight variation of this estimator, which can handle more than one predictor, is to take

$$w_i = I_h e^{-d_i},\qquad(11.20)$$

where d_i is the squared Mahalanobis distance between $\mathbf{x}_i$ and $\mathbf{x}$ but with the usual covariance matrix replaced by the MVE estimator. That is,

$$d_i = (\mathbf{x}_i - \mathbf{x})'\mathbf{S}^{-1}(\mathbf{x}_i - \mathbf{x}),$$

where $\mathbf{S}$ is the MVE measure of scatter. When using Eq. (11.20), now $h = 2$ appears to be a good choice for general use with $I_h = 1$, if $\sqrt{d_i} < h$; otherwise, $I_h = 0$. Of course, the MVE estimator could be replaced by some other robust measure of scatter, but the practical advantages of doing so are unknown.

Other variations have been studied by Signorini and Jones (2004) that are based in part on kernel density estimators. Let $f(x)$ be the probability density function of x, given that $y = 1$, and let $g(x)$ be the density, given that $y = 0$. One of the estimators they studied has the form

$$\hat{m}(x) = \frac{n_1 f(x)}{n_1 \hat{f}(x) + n_0 \hat{g}(x)}, \qquad (11.21)$$

where n_j is the number of times $y = j$, $j = 0, 1$. (So, for example, n_1 is the observed number of successes.) Here, $\hat{f}(x)$ and $\hat{g}(x)$ are taken to be adaptive kernel estimators, described in Section 3.2.4. But limited results suggest that this offers little advantage over other estimators in terms of mean squared error and bias, and situations arise where the reverse is true.

None of the estimators listed in this section dominate in terms of mean squared error and bias, but the estimator given by Eq. (11.20) appears to perform relatively well, with the running interval smoother as another good choice (Wilcox, 2012).

Section 10.16 noted that when dealing with a single predictor, computing a confidence interval for $P(y = 1|x)$ based on the logistic regression model can be unsatisfactory, even if the model is slightly inaccurate. Wilcox (2019e) found that a confidence interval based on the running interval smoother can provide more accurate results. Basically, determine the x_i values that are close to x, and then use the Clopper and Pearson (1934) confidence interval on the corresponding y values, if $n \geq 80$. Otherwise, use the Agresti–Coull method. (The Clopper–Pearson and Agresti–Coull methods are described in Section 4.8.)

11.5.9 R Functions logSM, logSM2g, logSMpred, rplot.bin, runbin.CI, rplot.binCI, and multsm

The R functions in this section are designed with the explicit goal of creating a smooth when the outcome variable y is binary. The function

```
logSM(x,y,pyhat=FALSE, plotit=TRUE,xlab='X',ylab='Y', zlab='Z', xout=FALSE,
outfun=outpro, pr=TRUE, theta=50, phi=25, duplicate='error', expand=0.5, scale=FALSE,
                               fr=2,...)
```

computes the smooth given by Eq. (11.19), where the argument fr is the span, h. (This function replaces the R function logrsm, which is limited to a single independent variable.) When there is more than one independent variable, the function applies the method based on Eq. (11.20) and appears to be a relatively good choice when y is binary. The R function

logSMpred(x, y, pts, fr = 2, LP = TRUE, xout = FALSE, outfun = outpro, SEED = TRUE, ...)

can be used to estimate the probability of y, given that the independent variable has the values stored in the argument pts. The R function

logSM2g(x1, y1, x2, y2, fr = 2, xout = FALSE, outfun = outpro, xlab = 'X', ylab = 'Y')

plots a smooth for two groups. A single explanatory variable is assumed.

An alternative approach is to use an appropriate version of the *running interval smoother*, which includes the ability of dealing with more than one predictor. This can be accomplished with the R function

rplot.bin(x,y,est=mean,scat=TRUE,fr=NULL,plotit=TRUE,pyhat=FALSE,pts=x,
theta=50,phi=25,scale=TRUE,expand=0.5,SEED=TRUE,
nmin=0,xout=FALSE,outfun=outpro,xlab=NULL,ylab=NULL,
zlab='P(Y=1)',pr=TRUE,duplicate='error',...).

By default, a plot is created. Setting the argument pyhat=TRUE, the function returns estimates of the probability of success for the explanatory values indicated by the argument pts. The R function

runbin.CI(x,y,pts=NULL,fr=1.2,xout=FALSE,outfun=outpro)

computes confidence intervals for each point indicated by the argument pts, which defaults to using all of the unique points in x. That is, based on the running interval, the y values corresponding to the nearest neighbors of a specified point are used. More than one explanatory variable is allowed.

Finally, the R function

multsm(x, y, pts = x, fr = 0.5, xout = FALSE, outfun = outpro, plotit = TRUE, xlab = 'X',
ylab = 'Prob', ylab2 = 'Y', zlab = 'Prob', ticktype = 'det', vplot = NULL, scale = TRUE,
L = TRUE, ...)

deals with the situation where y has a discrete distribution, with a relatively small sample space. That is, y has a multinomial distribution. It returns a smooth for each unique value stored in y using Eq. (11.19). In addition, for each possible value for y, the function estimates the probability of getting this value given x, where the values for x are indicated by the argument pts. The R function

multireg.pro(x, y, pts = x, xout = FALSE, outfun = outpro, plotit = TRUE, xlab = 'X', ylab = 'Prob', zlab = 'Prob', ticktype = 'det', vplot = NULL, L = TRUE, scale = TRUE, ...)

computes $P(Y = k|x)$ using a multinomial logit model via the R package nnet. The argument vplot indicates the value for k that will be used. By default, the value of vplot is taken to be the maximum value stored in y.

11.5.10 Smoothing With More Than One Predictor

The running interval smoother can be generalized to more than one predictor by replacing MADN with the minimum volume ellipsoid estimate of scatter, $\mathbf{M}$, introduced in Chapter 6, and by measuring the distance between $\mathbf{x}_i$ and $\mathbf{x}_j$ with

$$D_{ij} = \sqrt{(\mathbf{x}_i - \mathbf{x}_j)'\mathbf{M}^{-1}(\mathbf{x}_i - \mathbf{x}_j)}.$$

When trying to predict y, given $\mathbf{x}_i$, simply compute the trimmed mean of all y_j values such that $\mathbf{x}_j$ is close to $\mathbf{x}_i$. More formally, compute the trimmed mean of all the y_j values for which the subscript j satisfies $D_{ij} \leq f$. The choice $f = 1$ or 0.8 often gives good results. When there are only two predictors, adjustments can be made as in the previous subsection. That is, start with $f = 1$, generate a graph of the three-dimensional smooth, and try other choices for f to see how the graph is affected. (For $p = 2$ and when estimating quantiles, also see He et al., 1998.)

The running interval smoother described in this section performs reasonably well with two predictors. As the number of predictors increases, there are concerns due to the so-called *curse of dimensionality*: Neighborhoods with a fixed number of points become less local as the dimensions increase (Bellman, 1961). Currently, little is known about how well any smoother performs when dealing with $p = 3$ predictors. Note that the linear model in Section 11.1, which is commonly used, simply ignores any concerns about the curse of dimensionality.

To provide some indication of how well the method performs with two predictors, first suppose $y = x_1 + x_2 + \epsilon$. The left panel of Fig. 11.7 shows a smooth based on $f = 1$ and $n = 20$ observations, where x_1, x_2, and ϵ all have a standard normal distribution. As can be seen, the

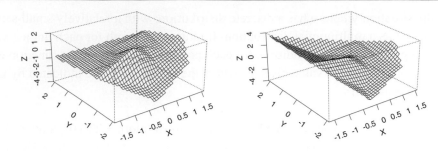

Figure 11.7: Illustrations of how runm3d performs under normality with a small sample size.

shape of the regression plane is captured reasonably well. The right panel of Fig. 11.7 shows a smooth when $y = x_1^2 + x_2 + \epsilon$; otherwise, the situation is the same as before. Again the shape of the regression surface is captured. Of course, it is not suggested that the correct surface is always reflected with only 20 points. Even with only one predictor, a smooth might suggest there is some curvature when data are generated from a straight line. Also, any smooth might be unreliable for extreme x_1 and x_2 values simply because there might be few points available for estimating the trimmed mean of y.

A possible concern with using D_{ij}, a robust analog of the Mahalanobis distance, is that an ellipsoid is being used to identify the points close to $\mathbf{x}_i$. This might suffice, but a more flexible approach is to use projection distances instead. That is, use approximation A1 in Section 6.2.3.

A criticism of the running interval smoother is that with a small sample size, the regression surface can be relatively ragged when it should be smooth. One way of improving the method is to apply the bootstrap bagging method as described at the end of Section 11.5.4. A seemingly more effective strategy is to simply smooth the initial smooth using LOESS as mentioned at the end of Section 11.5.4. But the relative merits of these two approaches are not well understood.

Bagging can also be useful when there is a fair amount of curvature. For example, when there is a single independent variable, the regression line differs substantially from a straight line; or when there are two independent variables, the regression surface is poorly approximated by a plane. One option is to use the bootstrap bagging method as described at the end of Section 11.5.4 again.

Another option is to use an extension of the random forest classification method in Section 6.16. Meinshausen (2006) describes how this can be done when estimating the conditional quantiles of y, given $\mathbf{x}$. The R code for doing this is readily extended to any location estimator. How this approach compares to bagging the running interval smoother has not been determined. The relative merits of this approach, versus LOESS and the running interval smoother, are not well understood. A few results when there is no association suggest that the random forest approach is less satisfactory.

11.5.11 R Functions rplot, runYhat, rplotsm, runpd, and RFreg

The R function

> rplot(x,y,est=tmean, scat=TRUE, fr=NA, plotit=TRUE, pyhat=FALSE,efr=0.5, theta=50,
> phi=25, scale=TRUE, expand=0.5, SEED=TRUE, varfun=pbvar,outfun=outpro, nmin=0,
> xout=FALSE, out=FALSE, eout=FALSE, xlab='X',ylab='Y', zscale=FALSE, zlab=' ',
> pr=TRUE,duplicate='error', ticktype='simple', LP=TRUE, OLD=FALSE, pch='.',...),

introduced in Section 11.5.5, can be used to create a smooth when there are two independent variables. (This function replaces the R functions runm3d, run3hat, and rung3d, which were used in the 3rd edition of this book.) When x is an n-by-2 matrix, the function automatically plots the estimated regression surface via the R package akima. (This R package is easy to use and is made even easier via the R function plot3D.) To avoid the plot, set the argument plotit=FALSE. For a three-dimensional plot, setting the argument ticktype='detailed' will create ticks as done when creating a two-dimensional plot. If the argument pyhat=TRUE and the argument est=tmean, the function returns the estimated 20% trimmed mean of y for each of the n vectors of predictors stored in the n-by-p matrix, x. Setting est=hd, for example, the median of y would be returned based on the Harrell–Davis estimator. If the data are not stored in an R variable having matrix mode, the function prints an error message and terminates.

The argument nmin can be used to modify how the regression surface is estimated. By default, nmin is 0, meaning that the regression surface is estimated using all n rows of x. If, for example, nmin=2, the regression surface is estimated using only those points $\mathbf{x}_i$ for which the number of points close to $\mathbf{x}_i$ is greater than 2. Put another way, the regression surface is estimated using only those points for which the sample trimmed mean of y is based on more than nmin values. Setting the argument xout=TRUE eliminates outliers among the **x** values before creating the plot, and eout=TRUE removes outliers among the (**x**, y) vectors.

When there is no association, and the regression surface is a flat, horizontal plane, using scale=FALSE typically gives the best visual representation. But when there is an association, often scale=TRUE provides a better perspective. The arguments theta and phi control the orientation of the plot. The argument theta controls the azimuthal direction and phi the co-latitude. The left graph in Fig. 11.8 shows a plot of $y = x_1 + x_2$ using the default values for theta and phi. The right panel is the same plot but with theta=20. (Changing the argument phi tilts the plot forward or backwards.)

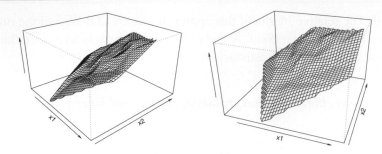

Figure 11.8: An illustration of what happens when the argument theta is altered in the R function runm3d.

The R function

runYhat(x, y, pts = x, est = tmean, fr = 1, nmin = 1, xout = FALSE, outfun = outpro, ...),

introduced in Section 11.5.5, can be used to compute $\hat{m}(x)$ for each vector stored in the R argument pts, where now pts is assumed to be a matrix with two or more columns. (This function replaces the R function rung3hat.) As usual, the measure of location is indicated by the argument est, which defaults to a 20% trimmed mean.

■ **Example**

If tp is a 2-by-3 matrix with the elements of the first row equal to zero and the second equal to 1, the command runYhat(x,y,est=onstep,pts=tp) returns two values in $Y.hat: the predicted one-step M-estimate of y, given that x is equal to $(0, 0, 0)$ and the predicted value when x is equal to $(1, 1, 1)$. The function also returns, in the R variable $nval, the number of y values used to compute the measure of location. Here, the first value in $nval is the number of predictors that are close to $(0, 0, 0)$, and the second value is the number of predictors close to $(1, 1, 1)$. For example, if the first value in nval is 8, there are eight points close to $(0, 0, 0)$, which in turn means that the predicted value of y is based on eight values as well.

■

The R function

rplotsm(x, y, est = tmean, fr = 1, plotit = TRUE, pyhat = FALSE, nboot = 40, atr = 0, nmin =
0, outfun = out, eout = FALSE, xlab = 'X', ylab = 'Y', scat = TRUE, SEED = TRUE,RUE
expand = 0.5, scale = FALSE, varfun = pbvar, pr = TRUE, ticktype='simple',...)

can be used to create a bagged version of the running interval smoother; see the end of Section 11.5.4. This function can give substantially better results, compared to rplot, when the sample size is relatively small. The relative merits of using rplotsm, versus rplot with LP=TRUE, are not well understood. The R function

RFreg(x, y, pts = x, newdata = pts, loc.fun = tmean, xout = FALSE, plotit = TRUE, outfun = outpro, span = 0.75, ZLIM = FALSE, scale = TRUE, xlab = 'X', ylab = 'Y', ticktype = 'simple', frame = TRUE, pyhat = FALSE, eout = FALSE, zlab = "", theta = 50, phi = 25, ...)

plots a smooth when there are one or two independent variables. (This can also be done via the R function smpred.) Setting the argument pyhat=TRUE returns estimates of a measure of location associated with the data in y, given values of the independent variables stored in the argument pts. The method uses bagging via a random forest method (Meinshausen, 2006). By default, pts is taken to be the values in the argument x. The measure of location is indicated by the argument loc.fun and defaults to a 20% trimmed mean.

■ Example

This example illustrates that a smooth, based on bagging, can make a practical difference. Using R, 50 values were generated from a standard normal distribution for both x and ϵ, and $y = x + \epsilon$ was computed. Then a smooth of the conditional variance of y, given x, was created with the command rplot(x,y,est=var,scat=FALSE). The result is shown in the left panel of Fig. 11.9. Then a bagged version of the smooth was created with the command rplotsm(x,y,est=var,scat=FALSE), and the result is shown in the right panel of Fig. 11.9. As is evident, the bagged version gives a much more accurate indication of the conditional variance of y, given x.

■

When dealing with $p > 1$ predictors, the R functions previously described in this section determine which points are close to some specified **x** using a robust analog of the Mahalanobis distance based on the MVE covariance matrix. So the closest points to **x** are based on ellipsoids. A more flexible approach to identifying the closest points is to use projection distances instead. This is done by the R function

runpd(x, y, pts = x, est = tmean, fr = 0.8, plotit = TRUE, pyhat = FALSE, nmin = 0, scale = FALSE, expand = 0.5, xout = FALSE, outfun = out, pr = TRUE, xlab = 'X1', ylab = 'X2', zlab = ' ', LP=TRUE, theta = 50, phi = 25, duplicate = 'error', MC = FALSE, ...).

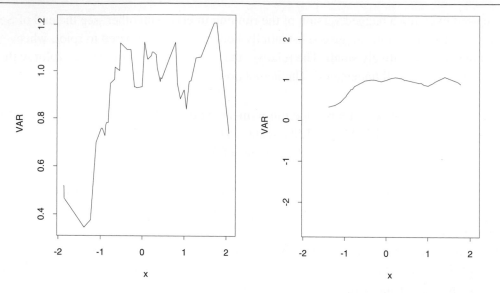

Figure 11.9: The left panel shows a smooth of y, given x, where the conditional variance of y, given any x, is one. The right panel shows a bagged version of the smooth created by the function rplotsm.

The function runpd uses the R function

$$pdclose(x, pts = x, fr = 1, MM = FALSE, MC = FALSE)$$

to determine which points stored in x are close to the points stored in the argument pts. The argument MM controls how the projection distances are scaled. See the description of the R function pdis in Section 6.2.6 for further details. At least in some situations, using a robust analog of the Mahalanobis distance, via the R function rplot, gives better results. For example, if $n = 100$, $y = x_1 + x_2^2 + \epsilon$, where both x and ϵ have standard normal distributions, the R function rplot tends to capture the shape of the regression surface better than the R function runpd. However, extensive comparisons of these two methods have not been made.

The R function

$$rung3hatCI(x,y, pts=x, tr=0.2, alpha=0.05, fr=1, nmin=12, ADJ=FALSE, iter=1000,..)$$

computes confidence intervals for the trimmed mean of the dependent variable, given x. This is done for the points indicated by the argument pts for which the number of nearest neighbors is greater than the value of the argument nmin. By default, the probability coverage for each confidence interval is 0.95. To get confidence intervals having approximately simultaneous probability coverage $1 - \alpha$, set the argument ADJ=TRUE, which will increase the execution time. The method for adjusting the confidence intervals is similar to methods M1 and M2, described in Section 11.5.4. When this adjustment is made, or when dealing with the curse of dimensionality, the safest way of interpreting the results is that the function returns confidence intervals for the trimmed mean of y based on points close to x. This interpretation can be more accurate compared to saying that the confidence intervals are for the conditional trimmed mean of y, given x.

11.5.12 LOESS

There is an extension of the smoother lowess (described in Section 11.5.2) to multiple predictors that was derived by Cleveland and Devlin (1988); it can be applied with the function loess, which comes with R. (In recent years, the terms lowess and loess have been used interchangeably.) The R function lplot, described in Section 11.5.3, uses loess to create a plot when there are $p = 2$ independent variables. Like lowess, the goal is to estimate the conditional mean of y, but unlike lowess (which handles $p = 1$ only), when using the default settings of the function, a single outlier can grossly distort the estimate of the regression surface and non-normality can greatly influence the plot. One way of addressing this problem is to set the argument family="symmetric" when using the function lplot. Another possibility is to eliminate all outliers by setting the argument eout=TRUE, and use the default value for the argument family.

■ **Example**

As an illustration, $n = 100$ points were generated from the model $y = x_1 + x_2 + \epsilon$, where x_1 and x_2 are independent standard normal random variables and ϵ has a g-and-h distribution with $g = h = 0.5$. The left panel of Fig. 11.10 shows the plot created by lplot using the default settings, and the right panel is the plot with eout=TRUE, which eliminates all outliers before creating the plot.

■

■ **Example**

Using the reading data described in Section 11.4.1, with the two independent variables taken to be TAAST1 and SBT1 (which are measures of phonological awareness and stored in columns 2 and 3 of the file read.dat), and the dependent variable taken to be OCT2 (a measure of orthographic ability and stored in column 10), Fig. 11.11 shows an estimate of the regression surface using four different smoothers. The upper left graph was created by lplot using the default values for the arguments. The upper right graph was created by rplot, again using the default values. The lower left graph was created

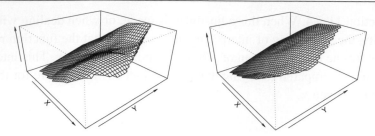

Figure 11.10: An illustration of how non-normality might affect smooths created by lowess. The left panel shows a smooth using the default settings of the function lplot. The right panel is a plot of the same data, but with outliers removed by setting the argument eout=TRUE.

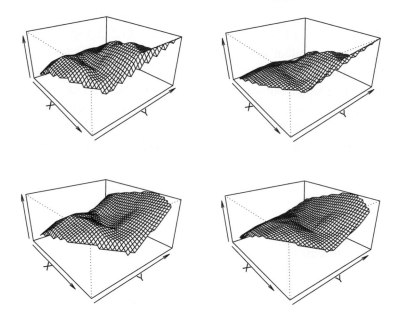

Figure 11.11: Four different smooths based on the reading data.

by lplot but with xout=TRUE so that leverage points are eliminated before creating the smooth. The lower right graph was created by rplot but with xout=TRUE.

∎

■ **Example**

To illustrate runYhat, again using the reading data, suppose it is desired to estimate WWISST2 (a word identification score stored in column 8) when TASST1 (stored in column 2) is 15 and SBT1 (stored in column 3) is 8. Then there is only one point of interest, so store the values 15 and 8 in any 1-by-2 matrix. For example, the R command val=matrix(c(15,8),1,2) could be used. Assuming the values of the predictors are stored in the R variable x and the WWISST2 values are stored in y, the command runYhat(x,y,val) returns the value 106.2 in the R variable $Y.hat. That is, the estimated 20% trimmed mean of WWISST2, given that TASST1 is 15 and SBT1 is 8, is equal to 107.5. If instead, it is desired to compute $\hat{y}$ for the points $(15, 8)$ and $(15, 9)$, now use the R command val=matrix(c(15,8,15,9),2,2,byrow=TRUE). Then the first row of the matrix val contains $(15, 8)$, the second row contains $(15, 9)$, and the command runYhat(x,y,val) returns the values 107.5 and 108.6.

■

■ **Example**

Kyphosis is a postoperative spinal deformity. R has built-in data, stored in the R variable kypho, reporting the presence or absence of kyphosis versus the age of the patient, in months, the number of vertebrae involved in the spinal operation, and a variable called start, which is the beginning of the range of vertebrae involved. Suppose it is desired to estimate the probability of kyphosis based on age and the number of vertebrae involved. The function rplots accomplishes this goal by setting the argument tr equal to zero, or the function rplot.bin could be used. (Of course, another option is to use the smoother in Section 11.5.8 via the R function logSM in Section 11.5.9.) The top two graphs in Fig. 11.12 show the resulting estimate of the regression surface using rplot (with tr=0 and LP=FALSE) and lplot (shown on the right). The bottom two graphs were again created by rplot and lplot, but both functions used xout=TRUE to eliminate any outliers among the independent variables. (Three outliers were found using the MVE method.) Also, rplot used fr=1.1 to smooth the plot. (Standard logistic regression is typically used when y is binary. See Section 10.16 for some robust alternatives.)

■

11.5.13 Other Approaches

Yet another approach when dealing with two or more predictors is to use what is called a *generalized additive model*. That is, assume that

$$y = \beta_0 + \sum_{j=1}^{p} g_j(x_j) + \epsilon, \tag{11.22}$$

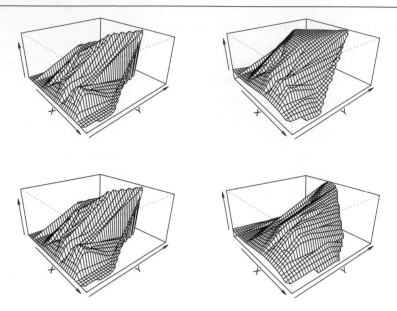

Figure 11.12: Four smooths based on the kyphosis data.

where $g_1(x_1), \ldots, g_p(x_p)$ are unknown functions to be estimated based on the available data. This is in contrast to assuming

$$y = g(x_1, \ldots, x_p) + \epsilon. \tag{11.23}$$

A concern about the more general model given by Eq. (11.23) is that the *curse of dimensionality*, which was described in Section 11.5.10, comes into play. Regardless of the extent to which Eq. (11.23) improves upon Eq. (11.22), the additive model provides an interesting generalization of the usual linear model $y_i = \beta_0 + \beta_1 x_{i1} + \cdots + \beta_p x_{ip} + \epsilon_i$ when testing hypotheses and trying to gain insight into any associations that might exist. (Illustrations are given in Section 11.6.) A general concern with the generalized additive model is that it might not be flexible enough so as to capture the nature of the regression surface in a reasonably accurate manner. (It assumes that there is no interaction as described in Section 11.7.)

When dealing with robust measures of location, the generalized additive model given by Eq. (11.22) can be fit to data using the running interval smoother in conjunction with the so-called *backfitting algorithm* (e.g., Friedman and Stuetzle, 1981). More generally, virtually any smoother can be used, including the many smoothers designed specifically for means. The backfitting algorithm is applied as follows. Set $k = 0$, and let g_j^0 be some initial estimate of g_j. Here, $g_j^0 = S_j(y|x_j)$, where $S_j(y|x_j)$ is the running interval smooth based on the jth predictor, ignoring the other $p - 1$ predictors that are available. Next, iterate as follows.

1. Increment k by 1.
2. For each j, $j = 1, \ldots, p$, let

$$g_j^k = S_j(y - \sum_{\ell \neq j} g_\ell^{k-1} | x_j).$$

3. Repeat steps 1 and 2 until convergence.

(For general theoretical results on the backfitting algorithm, see Buja et al., 1989.)

Finally, estimate β_0 with

$$b_0 = m(y - \sum g_j^k),$$

where m indicates the measure of location used when computing the smooths. R contains functions that estimate the generalized additive model given by Eq. (11.22) when the goal is to estimate the mean of y. Again, when the goal is to get a more robust version of these methods, a simple approach is to remove any outliers before using these R functions.

Methods have been derived that are a blend of both a parametric model and a non-parametric smoother (e.g., Ruppert et al., 2003). For results on how this approach might be implemented in a robust manner, see Boente and Rodriguez (2010).

11.5.14 R Functions adrun, adrunl, gamplot, and gamplotINT

The R function

adrun(x, y, est = tmean, iter = 10, pyhat = FALSE, plotit = TRUE, fr = 1, xlab = 'X', ylab = 'Y', zlab = ' ', theta = 50, phi = 25, expand = 0.5, scale = FALSE, zscale = TRUE, xout = FALSE, eout = xout, outfun = out, ticktype = 'simple', ...)

fits the additive model given by Eq. (11.22) in conjunction with the running interval smoother. As usual, the arguments theta and phi control the orientation of the plot; see Section 11.5.14. The measure of location is specified by the argument est and defaults to a 20% trimmed mean. The command adrun(x,y,est=mean,tr=0.1), for example, would result in a smooth based on a 10% trimmed mean instead. Setting the argument pyhat=T causes the function to return the estimates of y for each design point, and fr specifies the span. For bivariate data, the function plots the smooth, if plotit=TRUE. To avoid the plot, set plotit=FALSE. As p, the number of predictors, gets large, caution must be exercised. Situations can arise where the fit to data is wildly inaccurate due to the span being too small. So, at a minimum, it is suggested to check the output with pyhat=T to make sure the function is returning reasonable results. The function

adrunl(x, y, est = tmean, iter = 10, pyhat = FALSE, plotit = TRUE, fr = 0.8, xlab = 'x1',
ylab = 'x2', zlab =' ', theta = 50, phi = 25, expand = 0.5, scale = FALSE, zscale = TRUE,
xout = FALSE, outfun = out, ticktype = 'simple', ...)

is like the function adrun, only the running interval smoother is replaced by lowess.

The R function

gamplot(x, y, sop = TRUE, pyhat = FALSE, eout = FALSE, xout = FALSE, outfun = out,
plotit = TRUE, xlab = 'X', ylab = ' ', zlab = ' ', theta = 50, phi = 25, expand = 0.5,
scale = TRUE, ticktype = "simple")

creates a plot based on an additive fit for means that is computed via a call to the built-in R
function gam. (Splines are used to create the smooth. With the argument sop=FALSE, the
usual linear model is used.) The R function gam has many more options for modeling the
regression surface than are used by the function gamplot. For two predictors, the function
gamplot is intended as a way of graphing the regression surface assuming that Eq. (11.22)
holds. The current version is limited to $p = 4$. The other arguments are the same as those described in Section 11.5.11 in conjunction with rplot. (When y is binary, the function logadr
fits a generalized additive model in conjunction with Copas's method previously mentioned in
Section 11.5.8.) The R function

gamplotINT((x, y, pyhat = FALSE, plotit = TRUE, theta = 50, phi = 25, expand = 0.5,
xout = FALSE, SCALE = FALSE, zscale = TRUE, eout = FALSE, outfun = out,
ticktype = 'simple', xlab = 'X', ylab = 'Y', zlab = ' ', ...)

is like gamplot, only it is limited to $p = 2$ predictors and is based on the model $y = g_1(x_1) +
g_2(x_2) + g_3(x_1, x_2) + \epsilon$ rather than $y = g_1(x_1) + g_2(x_2) + \epsilon$. This is useful when checking for
interactions as described in Section 11.7.

11.5.15 Detecting and Describing Associations via Quantile Grids

There is the issue of detecting an association, and if one exists, describing it in a useful manner. Experience with smoothers clearly suggests that the regression surface can be complex
to the point that achieving these goals can be difficult. Section 9.5 describes methods for detecting an association without specifying a particular parametric model. But these methods
provide no details about the nature of the association. When there are one or two independent

variables, one possibility is to examine a smooth, with the goal of determining whether there are subsets of the data where a linear model is reasonable, but it is not always possible to do this. Another possibility is to use quantile grids based on the independent variables and essentially turn the problem into an ANOVA design. For example, when there are two independent variables, one could split the data into four groups based on their medians. Then multiple comparisons could be performed using methods in Chapter 7. Of course, other quantiles, and more than one quantile could be used. For example, the data could be split into groups based on the quartiles of the first independent variable. An issue is whether this approach ever makes a practical difference in terms of detecting and describing an association. The example in the next section demonstrates that the answer is yes.

11.5.16 R Functions smgridAB, smgridLC, smgrid, smtest, and smbinAB

The R function

smgridAB(x, y, IV = c(1, 2), Qsplit1 = 0.5, Qsplit2 = 0.5, tr = 0.2, PB = FALSE, est = tmean, nboot = 1000, pr = TRUE, xout = FALSE, outfun = outpro, SEED = TRUE, ...)

splits the data into groups based on quantiles specified by the arguments Qsplit1 and Qsplit2, and then compares the resulting groups based on trimmed means. By default, the splits are based on the medians of two of the independent variables. The argument IV indicates which of the two independent variables will be used. The first two columns of the argument x, a matrix or a data frame, are used by default. For each row of the first factor (the splits based on the first independent variable), all pairwise comparisons are made among the levels of the second factor (the splits based on the second independent variable). In similar manner, for each level of the second factor (the splits based on the second independent variable), all pairwise comparisons among the levels of the first factor are performed. Setting PB=TRUE, a percentile bootstrap method is used, which makes it possible to use a robust measure of location other than a trimmed mean via the argument est. Measures of effect size are returned as well.

The R function

smgridLC(x, y, IV = c(1, 2), Qsplit1 = 0.5, Qsplit2 = 0.5, PB = FALSE, est = tmean, tr = 0.2, nboot = 1000, pr = TRUE, con = NULL, xout = FALSE, outfun = outpro, SEED = TRUE, ...)

can be used to test hypotheses about linear contrasts. Linear contrast coefficients can be specified via the argument con. See, for example, Section 7.4.4. By default, con=NULL, meaning that all relevant interactions are tested. If it is desired to split the data based on a single independent variable, this can be done with the R function

> smtest(x, y, IV = 1, Qsplit = 0.5, nboot = 1000, est = tmean, tr = 0.2, PB = FALSE, xout = FALSE, outfun = outpro, SEED = TRUE, ...).

To perform all pairwise comparisons, use the R function

> smgrid(x, y, IV = c(1, 2), Qsplit1 = 0.5, Qsplit2 = 0.5, tr = 0.2, PB = FALSE, est = tmean, nboot = 1000, pr = TRUE, xout = FALSE, outfun = outpro, SEED = TRUE, ...).

If the dependent variable is binary, use the function

> smbinAB(x, y, IV = c(1, 2), Qsplit1 = 0.5, Qsplit2 = 0.5, tr = 0.2, method = 'KMS', xout = FALSE, outfun = outpro, ...),

which is like the function smgridAB, only the KMS method for comparing two binomial distributions, described in Section 5.2, is used by default. To use method SK, also described in Section 5.2, set the argument method='SK'. The R function

> smbin.inter(x, y, IV = c(1, 2), Qsplit1 = 0.5, Qsplit2 = 0.5, alpha = 0.05, con = NULL, xout = FALSE, outfun = outpro, SEED = TRUE, ...)

also deals with a binary dependent variable. By default, all interactions are tested, but other linear contrasts can be tested via the argument con.

■ Example

This example is based on the Well Elderly data mentioned in Sections 11.1.6 and 11.2.2. The goal here is to understand the association between a measure of meaningful activities (MAPA) and two covariates: a measure of life satisfaction (LSIZ) and the CAR, which was described in Section 11.1.6. All analyses were done with leverage points removed. The sample size is n=246. Testing the hypothesis that the linear model is correct, the p-value is 0.45. Least squares regression, with leverage points removed, yields a significant association for LSIZ (p-value < 0.001) but not for the CAR (p-value = 0.68). The HC4 method was used for dealing with heteroscedasticity. Switching to the Theil–Sen estimator in conjunction with a percentile bootstrap method, again, LSIZ is significant at the 0.05 level (p-value < 0.001) and CAR is not (p-value = 0.51). The basic lasso estimator given by Eq. (10.21), as well as the Huber-type lasso, also indicates that CAR is not relevant. That is, both of these methods estimate the slope for CAR to be zero.

■

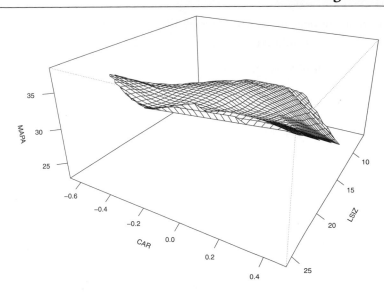

Figure 11.13: A smooth illustrating that in some situations, simply using grids can detect associations that are missed by a basic linear model.

However, the smooth in Fig. 11.13 suggests that the usual linear regression model might poorly reflect the true nature of the association. (The plot was rotated by setting the argument theta=120, which provides a better view of the association.) In particular, it suggests that the nature of the association depends in part on whether cortisol increases after awakening and whether LSIZ is relatively high or low. Here is a portion of the output from smgridAB, based on default settings and leverage points removed, with LSIZ the first independent variable and CAR the second:

```
$A[[1]]
     Group Group  psihat  ci.lower ci.upper   p.value     Est.1    Est.2
[1,]     1     2 2.38529 0.4991086 4.271472 0.01393357 33.02632 30.64103


$A[[2]]
     Group Group   psihat   ci.lower  ci.upper   p.value     Est.1    Est.2
[1,]     1     2 -1.435165 -3.579639 0.7093088 0.1863744 34.30769 35.74286
```

Here, A[[1]] refers to the first level of the first factor, which corresponds to LSIZ scores less than the median. The reported results are for the two levels of the second factor, which correspond to low and high CAR values. The median CAR is −0.032. The results indicate that for low LSIZ scores, typical MAPA scores differ significantly for CAR less than the median versus CAR greater than the median. Roughly, for low LSIZ scores, MAPA scores tend to be higher when the CAR is negative (cortisol increases upon awakening). When LSIZ is high, which corresponds to A[[2]], now no significant difference is found. That is, contrary to the results based on linear models, the CAR has a significant association with MAPA when taking into account whether LSIZ is relatively high or low.

Another portion of the output looks like this:

```
$B[[1]]
     Group Group    psihat  ci.lower  ci.upper   p.value    Est.1    Est.2
[1,]     1     2 -1.281377 -3.220148 0.6573951 0.1917055 33.02632 34.30769


$B[[2]]
     Group Group    psihat  ci.lower  ci.upper      p.value    Est.1    Est.2
[1,]     1     2 -5.101832 -7.199571 -3.004092 6.971174e-06 30.64103 35.74286
```

This indicates that for high CAR values, B[[2]], it is reasonable to decide that typical MAPA scores are higher when LSIZ scores are relatively high, but no decision is made for low CAR values based on the results labeled B[[1]].

It is noted that smgridLC indicates that there is an interaction. The p-value is 0.009. As indicated in Section 11.7, a common strategy for modeling an interaction is to simply include a product term involving the two independent variables. But often this approach is not flexible enough to adequately detect any interaction that might exist. For the situation at hand, this approach does not yield a significant result. Using least squares regression via the R function olsci, the p-value is 0.57. Using the Theil–Sen estimator via the R function regci, the p-value is 0.91.

11.6 Checking the Specification of a Regression Model

Typically, when testing hypotheses, a particular parametric form for a regression model is specified and inferences are made about the parameters assuming that the model is correct. A practical concern is that the assumed parametric form might represent a poor approximation of the true regression surface, which in turn can lead to erroneous conclusions. As a simple example, values for x were generated from a bivariate normal distribution, with $\rho = 0$, the marginal distributions as well as ϵ had a standard normal distribution, $n = 20$, and the error term was homoscedastic. Now imagine we assume that $y = \beta_0 + \beta_1 x_1 + \beta_2 x_2 + \epsilon$ and the goal is to test H_0: $\beta_2 = 0$. Furthermore, based on how the data were generated, power is approximately 0.26 when testing at the 0.05 level. Is it reasonable to conclude that the model is a good approximation of how the data were generated, and that indeed, $\beta_2 \neq 0$? Here, such a conclusion would be erroneous; the data were generated using the model $y = \beta_1 x_1 + \beta_2 x_2^2 + \epsilon$. So an issue is whether it is reasonable to assume that for some β_0, β_1, and β_2, $y = \beta_0 + \beta_1 x_1 + \beta_2 x_2 + \epsilon$. Of course, exploratory graphical methods, covered in

Section 11.5, help address this issue. Here, the goal is to describe some additional tools for dealing with this problem.

There are, in fact, many methods for testing the hypothesis that a regression equation has a particular parametric form. Typically, these methods are based on estimates of the conditional mean of y, given $\mathbf{x}$. Included are methods that begin with a kernel-type smooth, and then compare the fitted y values to those obtained by an assumed parametric model. Miles and Mora (2003) summarize and compare a variety of these methods assuming normality. More generally, there is the problem of testing the hypothesis that a regression surface belongs to some particular family of models. For example, can we rule out the possibility that a generalized additive model generated the data? Samarov (1993) provides an interesting overview of various models and how they might be investigated. It seems that few results are available on how extensions of these methods to robust estimators perform.

11.6.1 Testing the Hypothesis of a Linear Association

Given p predictors, $x_1, \ldots, x_p$, let $\mathcal{M}$ be the family of all regression equations having the form $y = \beta_0 + \beta_1 x_1 + \cdots + \beta_p x_p + \epsilon$, where the error term may be heteroscedastic. This section describes a test of the hypothesis

$$H_0\colon m(\mathbf{x}) \in \mathcal{M}, \tag{11.24}$$

where, as usual, $m(\mathbf{x})$ represents some conditional measure of location given $\mathbf{x}$. That is, the null hypothesis is that the data are generated from the model $y = \beta_0 + \beta_1 x_1 + \cdots + \beta_p x_p + \epsilon$. If, for example, $y = \beta_0 + \beta_1 x_1^2 + \epsilon$, the null hypothesis is false. The method described here stems from Stute et al. (1998).

Let $\hat{y}$ be some regression estimate of y. Least squares could be used, but it has been shown that this can lead to problems in terms of controlling the probability of a Type I error (Wilcox, 1999), so, it is suggested that some robust estimator be used instead. For fixed j ($1 \le j \le n$), set $I_i = 1$ if $\mathbf{x}_i \le \mathbf{x}_j$ (all p elements of $\mathbf{x}_i$ are less than or equal to the corresponding elements of $\mathbf{x}_j$), otherwise, $I_i = 0$, and let

$$\begin{aligned} R_j &= \tfrac{1}{\sqrt{n}} \sum I_i (y_i - \hat{y}_i) \\ &= \tfrac{1}{\sqrt{n}} \sum I_i r_i, \end{aligned} \tag{11.25}$$

where $r_i = y_i - \hat{y}_i$ are the usual residuals. The (Kolmogorov) test statistic is the maximum absolute value of all the R_j values. That is, the test statistic is

$$D = \max|R_j|, \tag{11.26}$$

where max means that D is equal to the largest of the $|R_j|$ values. As in Section 9.5, a Cramér–von Mises test statistic can be used instead where now

$$D = \frac{1}{n} \sum R_j^2. \tag{11.27}$$

A critical value is determined using the wild bootstrap method. Generate n observations from a uniform distribution, and label the results $U_1, \ldots, U_n$. Next, for $i = 1, \ldots, n$, set

$$V_i = \sqrt{12}(U_i - .5),$$

$$r_i^* = r_i V_i,$$

and

$$y_i^* = \hat{y}_i + r_i^*.$$

Then based on $(\mathbf{x}_1, y_1^*), \ldots, (\mathbf{x}_n, y_n^*)$, compute the test statistic, and label it D^*. Repeat this process B times, and label the resulting test statistics $D_1^*, \ldots, D_B^*$. Finally, put these B values in ascending order, yielding $D_{(1)}^* \leq \cdots \leq D_{(B)}^*$. The critical value is $D_{(u)}^*$, where $u = (1 - \alpha)B$, rounded to the nearest integer. That is, reject if

$$D \geq D_{(u)}^*.$$

A limitation of this method is that if it fails to reject, this does not necessarily mean that the null hypothesis should be accepted. (Wang and Qu, 2007, propose another approach, but it is unknown how it compares to the method covered here. For a method aimed specifically at L_1 regression, see Horowitz and Spokoiny, 2002. For yet another method dealing with quantile regression, see He and Zhu, 2003.)

11.6.2 R Function lintest

The R function

$$\text{lintest(x,y,regfun=tsreg,nboot=500,alpha=0.05)}$$

tests the hypothesis that a regression surface is a plane (more generally that the regression surface corresponds to a linear model) using the method just described. (Execution is fairly fast with one predictor, but it might be slow when there are multiple predictors. This problem can be greatly reduced by using regfun=chreg, which uses the Coakley–Hettmansperger M-estimator.) When reading the output, the Kolmogorov test statistic is labeled dstat, and its critical value is labeled critd. The Cramér–von Mises test statistic is labeled wstat. The default regression method (indicated by the argument regfun) is Theil–Sen.

■ **Example**

For the diabetes data shown in Fig. 11.5, suppose the goal is to test the hypothesis that there is a linear association between the logarithm of the C-peptide values and age. That is, the hypothesis is that, for some β_0 and β_1, $y = \beta_0 + \beta_1 x + \epsilon$, where x is age. The Kolmogorov test statistic returned by the R version of lintest is $D = 0.179$, and it reports a 0.05 critical value of 0.269, so it fails to reject. If both predictors (age and base deficit) are used, again we fail to reject at the 0.05 level.

■

11.6.3 Testing the Hypothesis of a Generalized Additive Model

This section describes a variation and extension of the test of linearity given in Section 11.6.1. Here, rather than test the hypothesis of a linear association, the goal is to test the hypothesis that the data were generated from a generalized additive model. More formally, given p predictors, $x_1, \ldots, x_p$, now let $\mathcal{M}$ be the family of all regression equations having the form given by Eq. (11.22). The goal is to test the hypothesis

$$H_0: m(\mathbf{x}) \in \mathcal{M}, \tag{11.28}$$

where, as usual, $m(\mathbf{x})$ represents some conditional measure of location, given $\mathbf{x}$. When there are $p = 2$ independent variables, rejecting this hypothesis is one way of detecting an interaction. That is, approximating the regression surface appears to require some function of both independent variables rather than the simple additive model given by Eq. (11.22). There are various ways this problem might be addressed. For example, some obvious extension of the method in Dette (1999) might be used, but so far no such variation has been found that performs well in simulations. Another approach is suggested by results in Samarov (1993), but again there are no simulation results supporting this strategy. Another possibility is to fit the additive model, and test the hypothesis that the regression surface for the residuals, versus $\mathbf{x}$, is a horizontal plane, which can be done along the lines noted in Section 9.5, or one might compare the fit of the additive model to the fit obtained by the method in Section 11.5.10. Wild bootstrap methods based on these last two strategies have, so far, proven to be rather unsatisfactory in simulations.

Currently, the only method that performs well in simulations, when the sample size is small, is applied exactly as in Section 11.6.1, only rather than compute $\hat{y}$ based on some robust regression estimator, use $\hat{y} = \hat{m}(\mathbf{x})$ based on the additive fit described in Section 11.5.13 (Wilcox, 2003e). Here, it is assumed that the additive fit is obtained using the 20% trimmed mean. The method does not perform well when using means, and nothing is known about how it performs when using an M-estimator.

There is, however, a practical concern about the choice of the span when applying the running interval smoother to get the additive fit. If $p = 2$, and the span is too large, the actual Type I error probability can drop well below the nominal level. For this special case, and when testing at the 0.05 level, approximations of a good choice for the span corresponding to the sample sizes 20, 30, 50, 80, and 150 are 0.4, 0.36, 0.18, 0.15, and 0.09, respectively. It is suggested that when $20 \leq n \leq 150$, interpolation based on these values be used, and for $n > 150$, simply use a span equal to 0.09. So for n sufficiently large, perhaps the actual Type I error probability might be well below the nominal level, but exactly how the span should be modified when $n > 150$ is an issue that is in need of further investigation. For $p = 3$, the choice of the span seems less sensitive to the sample size, with a span of $f = 0.8$ being a reasonable choice for $n < 100$. What happens when $p > 3$ has not been investigated.

11.6.4 R Function adtest

The R function

$$adtest(x,y,est=tmean,nboot=100, alpha=0.05, fr=NA,xout=FALSE, outfun=out,$$
$$SEED=TRUE,...)$$

tests the hypothesis given by Eq. (11.28). This is one way of testing the hypothesis of no interaction. If xout=TRUE, outliers among the $\mathbf{x}$ values are first identified, and $(y_i, \mathbf{x}_i)$ is eliminated, if $\mathbf{x}_i$ is flagged an outlier.

11.6.5 Inferences About the Components of a Generalized Additive Model

Inferences about the components of a generalized additive model, based on the running interval smoother, can be made as follows. For convenience, assume the goal is to test

$$H_0: g_1(x_1) = 0.$$

Fit the generalized additive model, yielding

$$\hat{y}_i = b_0 + \hat{g}_2(x_{i2}) + \cdots + \hat{g}_p(x_{ip}).$$

Let $r_i = y_i - \hat{y}_i$, $i = 1, \ldots, n$. The strategy is to test the hypothesis that the association between the residuals and x_1 is a straight horizontal line, and this can be done with the wild bootstrap method in Section 9.5 (cf. Härdle and Korostelev, 1996).

When using the running interval smoother, the choice of the span can be crucial in terms of controlling the probability of a Type I error (Wilcox, 2006a). Let f be the span used in Section 11.5.4. The choice for f when using means or a 20% trimmed mean is as follows:

n	20% trimming	mean
20	1.20	0.80
40	1.00	0.70
60	0.85	0.55
80	0.75	0.50
120	0.65	0.50
160	0.65	0.50

So, for example, if $n = 60$, and a generalized additive model based on the running interval smoother and a 20% trimmed mean is to be used to test H_0, choose the span to be $f = 0.85$.

In principle, the method is readily extended to situations where something other than the running interval smoother is used to fit the generalized additive model, but currently, there are no results on the resulting probability of a Type I error.

11.6.6 R Function adcom

The R function

$$\text{adcom(x, y, est = mean, tr = 0, nboot = 600, tr=0.2, fr = NA, jv = NA,...)}$$

tests hypotheses about the components of a generalized additive model using the method just described. With the argument fr=NA, the function chooses the appropriate span, as a function of the sample size, using linear interpolation where necessary. By default, all components are tested. The argument jv can be used to limit which components are tested. For example, jv=2 would test only H_0: $g_2(x_2) = 0$.

11.6.7 Detecting Heteroscedasticity Based on Residuals

This section describes a method for testing the homoscedasticity assumption based on the residuals associated with some fit to the data. (This approach has an obvious connection with what is known as the Tukey–Anscombe plot.) Let $m(x)$ denote some conditional measure of location associated with y, given x, and let $r_i = y_i - \hat{m}(x_i)$ $(i = 1, \ldots, n)$ denote the usual residuals based on some estimate of $m(x)$. Here, $m(x)$ is estimated using the running interval smoother in Section 11.5.4 or the Theil–Sen estimator. Homoscedasticity implies that a regression line used to predict $|r|$, given x, will be a straight horizontal line, and there are several ways of testing the hypothesis that this regression line is indeed straight and horizontal. One way is to assume the regression line is straight with an unknown slope β_r, and test the hypothesis H_0: $\beta_r = 0$. Here, this hypothesis is tested using the percentile bootstrap method in Section 11.1.3 in conjunction with the Theil–Sen estimator or the running interval smoother. Given the residuals, a second strategy is to test the hypothesis that the Winsorized correlation between $|r|$ and x is zero. A third possibility is to test the hypothesis that the regression line for $|r|$ and x is both straight and horizontal using method INDT in Section 9.5. (For relevant simulation results, see Wilcox, 2006g.) There are, of course, many other variations that might be used, but there are no results on the extent to which they provide a practical advantage over the variations just described.

11.6.8 R Function rhom

The R function

$$\text{rhom(x,y,op=1, op2=FALSE,tr=0.2, plotit=TRUE, xlab='NA',ylab='NA', zlab='ABS(res)',}$$
$$\text{est=median, sm=FALSE,SEED=TRUE, xout=FALSE, outfun=outpro,...)}$$

tests the hypothesis that there is homoscedasticity using the method described in the previous section. There are three choices for the argument op:

- op=1: Test H_0: $\beta_r = 0$ using the Theil–Sen estimator in conjunction with the percentile bootstrap method in Section 11.1.3.
- op=2: Test the hypothesis that the 20% Winsorized correlation between $|r|$ and x is zero using the method in Section 9.3.6.
- op=3: Test the hypothesis that the regression line for predicting $|r|$ with x is both straight and horizontal using method INDT in Section 9.5.

If the argument op2=FALSE, the Theil–Sen estimator is used when computing the residuals; otherwise, the running interval smoother is used.

11.7 Regression Interactions and Moderator Analysis

Regarding the method in Section 11.6.3, note that it provides a flexible approach to the so-called regression interaction problem. Consider the two-predictor case, and let c_1 and c_2 be two distinct values for the second predictor, x_2. Roughly, no interaction refers to a situation where the regression line between y and x_1, given that $x_2 = c_1$, is parallel to the regression line between y and x_1, given that $x_2 = c_2$. An early approach to modeling interactions assumes that

$$y = \beta_0 + \beta_1 x_1 + \beta_2 x_2 + \beta_3 x_1 x_2 + \epsilon, \tag{11.29}$$

where an interaction is said to exist, if $\beta_3 \neq 0$ (e.g., Saunders, 1956). This model often plays a role in what is called a *moderator analysis*, roughly meaning that the goal is to determine the extent to which knowing the value of one variable, x_2 here, alters the association between y and x_1. Note that Eq. (11.29) can be written as

$$y = (\beta_0 + \beta_2 x_2) + (\beta_1 + \beta_3 x_2)x_1 + \epsilon,$$

so the slope for x_1 changes as a linear function of x_2. (An R function, called ols.plot.inter, described in Section 11.7.1, plots the regression surface when using the least squares estimate of the parameters.) Currently, a commonly used method for testing the hypothesis of no interaction is to test $H_0\!: \beta_3 = 0$, meaning that the slope for x_1 does not depend on x_2.

A more general approach to testing the hypothesis of no interaction is to use a variation of the method in Section 11.6.1 to test the hypothesis that for some functions g_1 and g_2, $y = g_1(x_1) + g_2(x_2) + \epsilon$. This can be done with the function adtest in Section 11.6.4. Another way of stating the problem is described, for example, by Barry (1993), who uses an ANOVA-type decomposition. Essentially, write

$$m(x_1, x_2) = \beta_0 + g_1(x_1) + g_2(x_2) + g_3(x_1, x_2) + \epsilon,$$

in which case, the hypothesis of no interaction is

$$H_0\!: g_3(x_1, x_2) \equiv 0.$$

Barry (1993) derives a Bayesian-type test of this hypothesis, assuming the mean of y is to be estimated, and that prior distributions for g_1, g_2, and g_3 can be specified. Another approach is outlined by Samarov (1993), but when dealing with robust measures of location, the details have not been investigated. Note that this last hypothesis can be tested with the function adcom in Section 11.6.6. How this approach compares to using the function adtest is unknown.

Next, graphical methods are described that might be useful when studying interactions. The first simply plots a smooth of y versus x_1, given a particular value for x_2. So if there is no interaction, and this plot is created at, say $x_2 = c_1$ and $x_2 = c_2$, $c_1 \neq c_2$, the regression lines should be parallel. Here, creating this plot is tackled using a simple extension of the kernel estimator (the modification of Fan's method) described in Section 11.5.2. (Many alternative versions are possible and might have practical value.)

Momentarily, consider a single predictor x. In Section 11.5.2, an estimate of the conditional mean of y at x is obtained using weighted least squares, with weights $K\{(x - x_i)/h\}$. One possibility for extending this method to estimating $m(x_1)$, given that $x_2 = c$, which is used here, begins with a bivariate Epanechnikov kernel, where, if $1 - x_1^2 - x_2^2 < 1$,

$$K(x_1, x_2) = \frac{2}{\pi}(1 - x_1^2 - x_2^2);$$

otherwise, $K(x_1, x_2) = 0$. An estimate of the bivariate density $f(\mathbf{x})$, based on (x_{i1}, x_{i2}), $i = 1, \ldots, n$, is

$$\hat{f}(\mathbf{x}) = \frac{1}{nh^2} \sum_{i=1}^{n} K\left\{ \frac{1}{h}(\mathbf{x} - \mathbf{x}_i) \right\},$$

where, as usual, h is the span. For the jth predictor, let $u_j = \min(s_j, \text{IQR}_j/1.34)$, where s_j and IQR_j are the sample standard deviation and interquartile range (estimated with the ideal fourths) based on $x_{1j}, \ldots, x_{nj}$. Here, the span is taken to be

$$h = 1.77n^{-1/6}\sqrt{u_1^2 + u_2^2}.$$

(See Silverman, 1986, pp. 86–87.) Then an estimate of $m(x_{i1})$, given that $x_{i2} = c$, is obtained via weighted least squares applied to (y_i, x_{i1}), $i = 1, \ldots, n$, with weights

$$w_i = \frac{K(x_{i1}, x_{i2} = c)}{K_2(x_{i2} = c)},$$

where K_2 is the Epanechnikov kernel used to estimate the probability density function of x_2.

Let $\hat{y}_i$ be the estimate of y based on (x_{i1}, x_{i2}) and the generalized additive model given by Eq. (11.22). Another approach to gaining insight regarding any interaction is to plot (x_{i1}, x_{i2}) versus the residuals $y_i - \hat{y}_i$, $i = 1, \ldots, n$.

Yet one more possibility is to split the data into two groups according to whether x_{i2} is less than some constant. For example, one might let M_2 be the median of the x_{i2} values, and then take the first group to be the (x_{i1}, y_i) values, for which $x_{i2} < M_2$, and the second group would be the (x_{i1}, y_i) values, for which $x_{i2} \geq M_2$, and then a smooth for both groups could be created. If there is no interaction, the two smooths should be reasonably parallel. Using quantile grids as described in Section 11.5.15 is another approach that has practical value as illustrated in Section 11.5.16.

11.7.1 R Functions kercon, riplot, runsm2g, ols.plot.inter, olshc4.inter, reg.plot.inter, and regci.inter

The R functions in Section 11.5.14 can be used to get some graphical information about how regression surfaces compare when no interaction is assumed versus situations where an interaction term is included. This section summarizes some additional R functions that might be useful.

The R function

$$ols.plot.inter(x, y, pyhat = FALSE, eout = FALSE, xout = FALSE, outfun = out,$$
$$plotit = TRUE, expand = 0.5, scale = FALSE, xlab = `X', ylab = `Y', zlab = ` `, theta = 50,$$
$$phi = 25, family = `gaussian', duplicate = `error', ticktype = `simple')$$

plots the regression surface assuming that Eq. (11.29) is true, and that the least squares estimates of the parameters are used. Because this model is often used, an issue of interest is how the estimated regression surface compares to other plots that are based on a more flexible nonparametric estimator.

The R function

$$reg.plot.inter(x, y, regfun=tsreg, pyhat = FALSE, eout = FALSE, xout = FALSE, outfun = out,$$
$$plotit = TRUE, expand = 0.5, scale = FALSE, xlab = `X', ylab = `Y', zlab = ` `, theta = 50,$$
$$phi = 25, family = `gaussian', duplicate = `error', ticktype = `simple',)$$

is exactly like the function ols.plot.inter, only it can be used with any regression estimator that returns the residuals in $residuals. By default, the Theil–Sen estimator is used.

■ **Example**

A portion of a study conducted by Shelley Tom and David Schwartz deals with the association between a Totagg score and two predictors: grade point average (GPA) and a measure of academic engagement. The Totagg score is a sum of peer nomination items that are based on an inventory that includes descriptors focusing on adolescents' behaviors and social standing. (The peer nomination items were obtained by giving children a roster sheet and asking them to nominate a certain amount of peers who fit particular behavioral descriptors.) The sample size is $n = 336$. The left panel of Fig. 11.14 shows the plot of the regression surface created with the R function ols.plot.inter. Compare this to the right panel, which is an estimate of the regression surface using LOESS and created by the R function lplot. This suggests that using the usual interaction model is unsatisfactory for the situation at hand. Testing $H_0: \beta_3 = 0$, assuming Eq. (11.29) is true, and using OLS, the resulting p-value returned by the R function olshc4 is 0.64. The R function adtest returns a p-value less than 0.01, indicating that an interaction exists.

■

The R function

$$kercon(x,y,cval=NA, eout=FALSE, xout=FALSE, outfun=out,xlab=`X',ylab=`Y',pch`.')$$

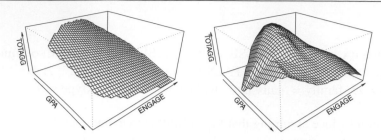

Figure 11.14: Plots of the estimated regression surface based on the peer nomination data. The left panel shows the plot created by ols.plot.inter, which assumes that an interaction can be modeled with $Y = \beta_0 + \beta_1 X_1 + \beta_2 X_2 + \beta_3 X_1 X_2 + e$, and where the least squares estimate of the parameters is used. The right panel shows an approximation of the regression surface based on the R function lplot.

creates a plot using the first of the two methods described in the previous section. It assumes there are two predictors and terminates with an error message if this is not the case. For convenience, let x1 and x2 represent the data in columns 1 and 2 of the R variable x. The function estimates the quartiles of the data stored in x2 using the ideal fourths, and then creates three smooths between y and x1. By default, the smooths correspond to the regression lines between y and $x1$, given that x2 is equal to the estimated lower quartile, the median, and the upper quartile. If it is desired to use other values for $x2$, this can be done via the argument cval. The arguments are used in the same manner as described, for example, in Section 11.5.5.

The R function

$$riplot(x,y,adfun=adrun,plotfun=lplot,eout=FALSE, xout=FALSE)$$

fits a model to data using the function specified by the argument adfun, which defaults to the generalized additive model given by Eq. (11.22). It then computes the residuals and plots them versus the data in x. Again, x must be a matrix with two columns of data. Note that the function automatically removes leverage points.

The R function

$$runsm2g(x1,y1,x2, val=median(x2), est=tmean, sm=FALSE,...)$$

splits the x1 and y1 values into two groups according to whether x2 is less than the value stored in the argument val. By default, val is the median of the values stored in x2. It then creates a smooth for both groups. Setting the argument sm=T results in a bagged version of the smooths. With small sample sizes, setting sm=T can be beneficial.

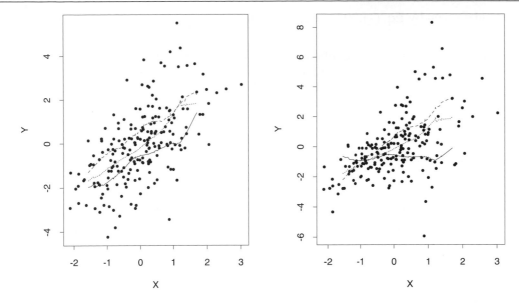

Figure 11.15: An illustration of the plot created by the function kercon.

■ **Example**

Two hundred values were generated for x_1, x_2, and ϵ, where x_1, x_2, and ϵ are independent and have standard normal distributions. The left panel of Fig. 11.15 shows the output from kercon, when $y = x_1 + x_2 + \epsilon$. The solid line is the smooth for y and x_1, given that x_2 is equal to the estimate of its lower quartile. The middle line is the smooth, given that x_2 is equal to its estimated median, and the upper line is the smooth for the upper quartile. The right panel shows the output, where now $y = x_1 + x_2 + x_1 x_2 + \epsilon$. ■

The R function

$$\text{olshc4.inter(x,y, tr=0.2, xout = FALSE, outfun = out, ...)}$$

tests hypotheses about the regression parameters in the interaction model given by Eq. (11.29) using the least squares regression estimator and the HC4 method in Section 10.1.1. Because this method of modeling an interaction is often inadequate, and because OLS is not robust, this function should be used with caution. It is supplied primarily for convenience in situations where it is desired to compare the results based on this standard model to the results based on some other method. To test hypotheses based on some robust estimator, again, based on the model given by Eq. (11.29), the R function

> regci.inter(x, y, regfun = tsreg, nboot = 599, tr=0.2, SEED = TRUE, pr = TRUE, xout = FALSE, outfun = out, ...)

can be used.

11.7.2 Mediation Analysis

This section provides some very brief comments about what is generally known as *mediation analysis*. (For books dedicated to this topic, see MacKinnon, 2008, as well as Vanderweele, 2015.) Mediation analysis is similar to a moderator analysis in the sense that the goal is to understand how the association between two variables is related to a third (mediating) variable. (Blends of the two methods, yielding what are called moderated-mediation analyses, have been proposed as well. See, for example, Preacher et al., 2007.) In the parlance of researchers working on this problem, an *indirect effect*, also known as a *mediation effect*, refers to a situation where two variables of interest are associated via a third variable. For example, stress and obesity are believed to be associated through cortisol secretion (Rosmond et al., 1998). The strategy behind a mediation analysis is to assume that the three variables of interest satisfy three linear models. The first is that two primary variables of interest x and y (e.g., stress and obesity) are related via the usual linear model

$$y = \beta_{01} + \beta_{11}x + \epsilon_1. \tag{11.30}$$

The second assumption is that the mediating variable (cortisol in the example), which here is labeled x_m, is related to x via

$$x_m = \beta_{02} + \beta_{12}x + \epsilon_2. \tag{11.31}$$

And finally, it is assumed that

$$y = \beta_{03} + \beta_{13}x + \beta_{23}x_m + \epsilon_3. \tag{11.32}$$

Briefly, there are four steps associated with a mediation analysis:

1. Establish that there is an association between y and x. This step establishes that there is an effect that might be mediated.
2. Establish that there is an association between x and x_m.
3. Establish that there is an association between y and x_m.
4. To establish that x_m completely mediates the association between x and y relationship, the association between x and y controlling for x_m should be zero.

There are many issues associated with a mediation analysis that go beyond the scope of this book. Roughly, if $\beta_{13} = 0$, this is said to constitute full mediation (Judd and Kenny, 1981a, 1981b). If $\beta_{13} < \beta_{11}$, there is said to be partial mediation. (A possible way of assessing whether the strength of the association between x and y is reduced, when the mediator is included, is to use explanatory power in conjunction with method IBS in Section 11.10.6.)

Various strategies have been proposed for assessing whether x_m mediates the association between y and x (e.g., Zhao et al., 2010). One is to focus on testing H_0: $\beta_{11} = \beta_{13}$. Another is to focus on the product $\beta_{12}\beta_{23}$, which has been called the *mediated effect* or *indirect effect*. This latter approach arises by noting that if Eq. (11.31) is substituted into Eq. (11.32), the total effect represented by the slope in Eq. (11.30) satisfies $\beta_{11} = \beta_{12}\beta_{23} + \beta_{13}$. (See MacKinnon et al., 1995, for more details.) Consequently, a common goal is testing

$$H_0: \beta_{12}\beta_{23} = 0. \tag{11.33}$$

Under normality and homoscedasticity, a bootstrap method for testing this hypothesis, using the least squares estimator, has been found to perform reasonably well in simulations. But under non-normality, or when there is heteroscedasticity, this is no longer the case (Ng, 2009a; Ng and Lin, 2016). Replacing the least squares estimator with the Theil–Sen estimator, Ng (2009a) found that a percentile bootstrap method performs well in simulations when $\beta_{12} = \beta_{23} = 0$. But otherwise, control over the probability of a Type I error can be unsatisfactory in some situations. Biesanz et al. (2010) compare several alternative methods. But the results relevant to non-normality are limited to a single non-normal distribution that is skewed with a relatively light tail. No results on the effects of heteroscedasticity were reported.

There is an issue that should be mentioned. Given the goal of establishing that there is no association between x and y, given x_m, Shah and Peters (2020) establish that numerous techniques are unsatisfactory. For the special case where a linear model is assumed, they establish that a method based on partial correlation is unsatisfactory. That is, these methods are unsatisfactory for establishing that x_m completely mediates the association between x and y.

Another approach when performing a mediation analysis is to compute a confidence interval for $\beta_{11} - \beta_{13}$ using some robust regression estimator and a percentile bootstrap method. Briefly, take a bootstrap sample in the usual way assuming Eq. (11.32) is true, which yields a bootstrap estimate of β_{13}, say b_{13}^*. Using this same bootstrap sample, compute a bootstrap estimate of β_{11} assuming that Eq. (11.30) is true, yielding b_{11}^*. Let $d^* = b_{11}^* - b_{13}^*$. Repeating this process B times yields a confidence interval for $\beta_{11} - \beta_{13}$, and a p-value when testing H_0: $\beta_{11} = \beta_{13}$, by proceeding along the lines noted in Section 11.2. Limited simulation studies suggest that when testing at the 0.05 level, the actual level can drop well below 0.05 when the sample size is less than or equal to 40. With $n = 80$, this does not seem to be an issue.

Zu and Yuan (2010) derive an approach to testing Eq. (11.33) based on a Huber-type M-estimator that is used in conjunction with a percentile bootstrap method. Briefly, their method begins by computing the multivariate measure of location and scatter derived by Maronna (1976) based on (x_i, x_{mi}, y_i), $i = 1, \ldots, n$, yielding, say $\hat{\mu}$ and $\hat{\Sigma}$. They then estimate the regression parameters via the method in Section 10.13.5. Finally, Eq. (11.33) is tested via a percentile bootstrap method. (Zu and Yuan also consider hypothesis testing techniques based on an estimate of the standard errors. For results on a method that assumes homoscedasticity, see Yang and Yuan, 2016.) The percentile bootstrap method appears to perform relatively well in terms of controlling the probability of a Type I error, but situations are encountered where it can be unsatisfactory. For example, under normality, with $n = 40$, $\beta_{23} = 0.5$, and $\beta_{12} = 0$, if there is heteroscedasticity in the form where the error term is $\epsilon_3/(|x| + 1)$, the actual level of the test is approximately 0.09 when testing at the 0.05 level. Increasing n to 60, the actual level drops to about 0.056. But with $n = 60$ and a homoscedastic error term, if two additional points are added at $(x, x_m, y) = (3, -2, -3)$, the actual level is again approximately 0.09. Using instead the Theil–Sen estimator in conjunction with a percentile bootstrap method for testing H_0: $\beta_{11} = \beta_{13}$, the actual level is approximately 0.025. But a criticism of this latter approach is that, in various situations, the actual level can drop well below the nominal level. Currently, the best method for dealing with these problems is to modify slightly the Zu and Yuan method. In particular, use their method after excluding any (x_i, x_{mi}, y_i), for which x_i is an outlier among the values $x_1, \ldots, x_n$. Another seemingly natural strategy is to instead eliminate any (x_i, x_{mi}, y_i), for which (x_i, x_{mi}) is an outlier. But this can result in poor control over the probability of a Type I error.

Yuan and MacKinnon (2014) report simulation results when using the least absolute value regression estimator to test Eq. (11.33). They found that a bias corrected bootstrap method performed well in their simulations, which were based on three symmetric distributions: normal, Student's t with two degrees of freedom, and the mixed normal given by Eq. (1.2). Good control over the Type I error probability is maintained when there is heteroscedasticity.

Green et al. (2010) have raised some concerns about mediation analyses in the context of establishing causality. The stated goal in the abstract of their paper is "to puncture the widely held view that it is a relatively simple matter to establish the mechanism by which causality is transmitted. This means puncturing the faith that has been placed in commonly used statistical methods of establishing mediation." Other concerns and how they might be addressed are summarized by Cole and Maxwell (2003).

11.7.3 R Functions ZYmediate, regmed2, and regmediate

The R function

ZYmediate(x, y, nboot = 2000, tr=0.2, kappa = 0.05, SEED = TRUE, xout = FALSE, outfun = out)

tests the hypothesis given by Eq. (11.33) using the method derived by Zu and Yuan (2010), which was outlined in the previous section. By default, the function eliminates any point for which x_i is an outlier. This improves control over the probability of a Type I error when there is heteroscedasticity. Currently, it seems to be one of the better methods when the sample size is small.

In case it helps, the R function

regmed2(x, y, regfun = tsreg, nboot = 400, tr=0.2, xout = FALSE, outfun = out, MC = FALSE, SEED = TRUE, pr = TRUE, ...)

tests the two hypotheses H_0: $\beta_{12} = 0$ and H_0: $\beta_{23} = 0$, which are relevant to a mediation analysis as explained in the previous section. By default, the Theil–Sen estimator is used, but other regression estimators can be used via the argument regfun. As usual, setting the argument xout=TRUE results in leverage points being removed.

The R function

regmediate(x,y,regfun=tsreg,nboot=400,alpha=0.05, xout=FALSE, outfun=out, MC=FALSE, SEED=TRUE,...)

computes a confidence interval for $\beta_{11} - \beta_{13}$, and a p-value when testing H_0: $\beta_{11} = \beta_{13}$ is returned as well. Again by default, the Theil–Sen estimator is used. For relevant software, beyond what is covered here, see MacKinnon et al. (2004).

11.8 Comparing Parametric, Additive, and Non-Parametric Fits

One way of comparing two different fits to data is to simply compute $m(\mathbf{x}_i)$, $i = 1, \ldots, n$, using both methods, and then plot the results. That is, if $\hat{y}_{i1}$ is $m(\mathbf{x}_i)$ based on the fit using the first method and $\hat{y}_{i2}$ is $m(\mathbf{x}_i)$ based on the second fit, plot $\hat{y}_{i1}$ versus $\hat{y}_{i2}$. So, for example, if data are generated according to a generalized additive model, then a plot of $\hat{y}_{i1}$ obtained by a method that assumes a generalized additive model generated the data versus $\hat{y}_{i2}$ obtained by the running interval smooth in Section 11.5.10 should consist of points that are reasonably close to a line having slope 1 and intercept 0. For one or two independent variables, another approach to comparing two fits is to simply inspect a plot of the regression lines or surfaces. This can be done with the R functions regplot, lplot, and rplot. The approach described in this section can be used when dealing with more than two independent variables, but there is the issue of the curse of dimensionality described in Section 11.5.10. With more than two independent variables, perhaps both parametric and non-parametric fits are unsatisfactory.

11.8.1 R Functions reg.vs.rplot, reg.vs.lplot, and logrchk

The R function

reg.vs.rplot(x, y, xout = FALSE, outfun = outpro, fr = 1, est = median, regfun = tsreg, xlab = 'Reg.Est', ylab = 'Rplot.Est', pch = '.', pr = TRUE, nmin = 1, ...)

is designed to compare predicted values based on the running interval smoother versus the parametric regression method indicated by the argument regfun, which defaults to the Theil–Sen estimator. The R function

reg.vs.lplot(x, y, xout = FALSE, outfun = outpro, fr = 1, est = median, regfun = tsreg, xlab = 'Reg.Est', ylab = 'Rplot.Est', pch = '.', pr = TRUE, nmin = 1, ...)

is like reg.vs.rplot, only now, the smoother LOESS is used. (These functions replace the R function pmodchk.)

The R function

logrchk(x, y, FUN=lplot, xout = FALSE, outfun = outpro, xlab = 'X', ylab = 'Y', ...)

is like the functions reg.vs.rplot and reg.vs.lplot, only it is designed specifically for the logistic regression model. The function creates where the x-axis shows the estimated probabilities based on a smooth and the y-axis shows the corresponding estimates based on the logistic regression model. The argument FUN indicates the method used to create a smooth of the two estimates.

11.9 Measuring the Strength of an Association Given a Fit to the Data

The measures of association covered in Chapter 9 are not based on any particular regression model or fit to the data. Pearson's correlation has a well-known connection to the least squares regression line, but for the bulk of the robust correlations, there is no explicit connection to any of the robust regression methods covered in Chapter 10. This section is aimed at filling this gap. There are, in fact, various ways one might proceed. The immediate goal is to describe how this might be done based on simple generalizations of the notion of explanatory power, which was studied in a general context by Doksum and Samarov (1995).

Let $\hat{y}$ be some predicted value of y, given the values of p predictors $x_1, \ldots, x_p$. Explanatory power is

$$\frac{\sigma^2(\hat{y})}{\sigma^2(y)},$$

the usual variance of the predicted values divided by the variance of the observed y values. If $\hat{y}$ is based on the usual least squares regression line, and there is $p = 1$ predictor, explanatory power reduces to ρ^2, the coefficient of determination. To see this, note that from basic principles, the least squares regression line can be written as

$$\hat{y} = \beta_0 + \rho \frac{\sigma_y}{\sigma_x} x.$$

So $\sigma^2(\hat{y}) = \rho^2(\sigma_y^2/\sigma_x^2)\sigma_x^2 = \rho^2\sigma_y^2$. Dividing this last quantity by σ_y^2 yields ρ^2.

A robust generalization of explanatory power consists of simply replacing the usual variance with some robust analog and taking $\hat{y}$ to be the predicted value of y based on any regression estimator or smoother. In symbols, let $\tau^2(y)$ be any measure of variation. Then a robust analog of explanatory power is

$$\eta^2 = \frac{\tau^2(\hat{y})}{\tau^2(y)}. \tag{11.34}$$

The explanatory strength of association is the (positive) square root of explanatory power, η. From Chapter 3, there are several reasonable choices for τ^2. Here, unless stated otherwise, τ^2 is taken to be the percentage bend midvariance, which is computed as described in Table 3.8. Perhaps other robust measures of variation offer a practical advantage when measuring the strength of association, but this has not been explored. R functions previously described that report the explanatory strength of association include lplot (LOWESS) and tsreg (the Theil–Sen estimator).

In principle, explanatory power can be estimated when using any regression method or smoother. First, compute the percentage bend midvariance based on predicted y values, say

$\hat{\tau}^2(\hat{y})$, and compute the percentage bend midvariance based on the observed y values, $\hat{\tau}^2(y)$, in which case, the estimate of η^2 is

$$\hat{\eta}^2 = \frac{\hat{\tau}^2(\hat{y})}{\hat{\tau}^2(y)}. \tag{11.35}$$

But a fundamental issue is whether the choice of method for obtaining the predicted y values makes a practical difference when estimating η^2. For small to moderate sample sizes, it has been found that it does (e.g., Wilcox, 2010b). Two regression estimators that seem to perform relatively well, given the goal of estimating η^2, are the Theil–Sen estimator when the regression surface is a plane and Cleveland's smoother (LOWESS), described in Section 11.5.2, when there is curvature.

Section 11.5.3 described an R function, lplot, for plotting Cleveland's non-parametric regression line (LOESS). One of the arguments is varfun, which can now be explained. It indicates the measure of variation used when estimating explanatory power and defaults to the percentage bend midvariance. The R function tsreg, which computes the Theil–Sen estimator, also contains the argument varfun, which again indicates how explanatory power is computed. When using the running interval smoother, what appears to be a relatively good estimate of explanatory power is to first compute $\hat{y}_i$ using a leave-one-out cross-validation method ($i = 1, \ldots, n$), and then use these values to compute $\hat{\tau}^2(\hat{y})$.

Renaud and Victoria-Feser (2010) compare several other robust analogs of R^2, the coefficient of determination, which are based in part on a fit to the data obtained via the MM-estimator in Section 10.9.1. Their approach represents a generalization of a measure of association suggested by Maronna et al. (2006, p. 171). For yet another approach to getting a robust version of R^2, see Croux and Dehon (2003).

Let $\Psi(r_i; c)$ be defined as in Section 10.9.1. In principle, some other choice for Ψ, associated with some M-estimator, could be used, but the focus here is on the choice used by the MM-estimator. The measure of association proposed by Maronna et al. (2006) is

$$R_{MM}^2 = 1 - \frac{\sum \Psi\left(\frac{r_i}{\hat{\tau}}\right)}{\sum \Psi\left(\frac{y_i - \hat{\mu}}{\hat{\tau}}\right)},$$

where $\hat{\mu}$ is some robust measure of location, taken here to be the M-measure of location associated with Ψ.

For convenience, write $w_i = \Psi(r_i; c)$. The generalization of R_{MM}^2, suggested by Renaud and Victoria-Feser, is

$$R_w^2 = \frac{\sum w_i(\hat{y} - \tilde{y})^2}{\sum w_i(\hat{y}_i - \tilde{y})^2 + a\sum w_i(y_i - \hat{y}_i)^2},$$

where a is a correction factor for achieving consistency and $\tilde{y} = (1/\sum w_i)\sum w_i \hat{y}_i$ and $\hat{y}_i$ are the predicted y values produced by the MM-estimator. The motivation for this generalization is that it reduces the bias associated with R^2_{MM}. Following Renaud and Victoria-Feser, $a = 1.2067$ is used.

Note that when dealing with a quantile regression estimator, as described in Section 10.13.8, a measure of association similar to R^2_{MM} could be used for any quantile q. Let $\rho_q(u)$ be defined as indicated in Section 10.13.8. Let $r_1, \ldots, r_n$ be the residuals based on the quantile regression estimator, and let $m_i = y_i - \hat{\theta}_q$ $(i = 1, \ldots, n)$, where $\hat{\theta}_q$ is the Harrell–Davis estimate of the qth quantile based on $y_1, \ldots, y_n$. (Of course, some other quantile estimator could be used, the relative merits of which have not been investigated.) The strength of an association is estimated with

$$r^2_q = 1 - \frac{\sum \rho_q(r_i)}{\sum \rho_q(m_i)}. \tag{11.36}$$

Another approach to measuring the strength of the association, based on a quantile regression estimator, was suggested by Li et al. (2015). Let $Q_{\tau,Y}$ be the τth (unconditional) quantile of Y, and assume that the τth quantile of Y, given X, is $Q_{\tau,Y}(X) = \beta_0 + \beta_1 X$. A method for estimating the unknown parameters β_0 and β_1, for any q, was derived by Koenker and Bassett (1978), which is discussed in more detail in Section 10.13.8. Briefly, the slope and intercept are estimated by the values b_0 and b_1 that minimize

$$\sum \psi_q(r_i),$$

where $r_1, \ldots, r_n$ are the usual residuals,

$$\psi_q(u) = u(q - I_{u<0}),$$

and I is the indicator function. That is, $I_{u<0} = 1$, if $u < 0$; otherwise, $I_{u<0} = 0$. A correlation based on this regression estimator was derived by Li et al. (2015). They begin by defining a quantile covariance:

$$\text{cov}_q(X, Y) = E\{\psi_q(Y - Q_{\tau,Y})(X - E(X))\}.$$

The quantile correlation is defined to be

$$\rho_q(X, Y) = \frac{\text{cov}_q(X, Y)}{\sqrt{(\tau - \tau^2)\sigma^2_X}}, \tag{11.37}$$

where $\sigma_X^2 = \text{VAR}(X)$. It follows that $-1 \le \rho_q(X, Y) \le 1$. The quantile correlation is estimated with

$$r_q(X, Y) = \frac{1}{\sqrt{(\tau - \tau^2)s_X^2}} \frac{1}{n} \sum \psi_q(Y_i - \hat{Q}_{\tau,Y})(X_i - \bar{X}).$$

Here, $\hat{Q}_{\tau,Y}$ is based on a single order statistic, $Y_{(k)}$, where k is $qn + 0.5$ rounded down to the nearest integer and $Y_{(1)} \le \cdots \le Y_{(n)}$. Any practical advantages of using some alternative estimate of $Q_{\tau,Y}$ have not been established.

It should be noted that $r_q(X, Y)$ is not necessarily equal to $r_q(Y, X)$, in contrast to the correlation coefficients described in Chapter 9. Indeed, situations are encountered where $r_q(Y, X)$ is positive but $r_q(X, Y)$ is negative.

Li et al. establish that $\rho_q(X, Y) = 0$ corresponds to $\beta_1 = 0$. So the hypothesis

$$H_0: \rho_q(X, Y) = 0$$

can be tested with techniques described in Section 10.13.9. Another approach is to use a percentile bootstrap method, but simulation results regarding the small-sample properties of this approach have not been studied. An alternative approach to measuring the strength of the association when dealing with quantiles is to use Eq. (11.34).

11.9.1 R Functions RobRsq, qcorp1, and qcor

The R function

$$\text{RobRsq(x,y)}$$

computes R_w^2, the measure of association derived by Renaud and Victoria-Feser (2010). The R function

$$\text{qcorp1(x,y,q=0.5, xout=FALSE, outfun=outpro, plotit=FALSE)}$$

computes r_q, the square root of the measure of association based on the quantile regression estimator given by Eq. (11.36), where the argument x can be a vector or a matrix. Quantiles associated with the marginal distributions are estimated with the Harrell–Davis estimator. The R function

$$\text{qcor(x,y, q=0.5, qfun=qest, xout=FALSE, outfun=outpro)}$$

computes the quantile correlation given by Eq. (11.37). The quantiles of the marginal distribution are estimated using the function denoted by argument qfun. By default, a single order statistic is used as described by Eq. (3.10). To use the Harrell–Davis estimator, set qfun=hd.

11.9.2 Comparing Two Independent Groups via the LOWESS Version of Explanatory Power

For two independent groups, let η_j^2 be the explanatory power associated with the jth group ($j = 1, 2$) based on the LOWESS smoother described in Section 11.5.2. Here, it is assumed that for each group, there is a single covariate. This section describes a modified percentile bootstrap method for testing

$$H_0: \eta_1^2 = \eta_2^2. \tag{11.38}$$

A simple strategy is to use a percentile bootstrap method. That is, generate bootstrap samples from the jth group, estimate η_j^2, yielding, say $\tilde{\eta}_j^2$, repeat this B times, yielding $\tilde{\eta}_{jb}^2$ ($b = 1, \ldots, B$), in which case, a p-value is $p = 2\min(P, 1 - P)$, where P is the proportion of times $\tilde{\eta}_1^2 > \tilde{\eta}_2^2$. Imagine that the goal is to test at the $\alpha = 0.05$ level, in which case, H_0 is rejected, if $p \leq 0.05$. Then the actual level of the percentile bootstrap method just described is very close to 0.05, with sample sizes $n_1 = n_2 = 200$ (Wilcox, 2009b). But for smaller sample sizes, the actual level is substantially smaller than 0.05, particularly when both sample sizes are less than 100. However, Wilcox (2009b) found that the actual level of the test was fairly stable among the non-normal distributions that were considered, which suggests a simple modification: Determine an adjusted level α_a, with the goal of achieving a 0.05 Type I error probability if the null hypothesis is rejected when $p \leq \alpha_a$. First, consider $n_1 = n_2 = n$. For standard normal distributions, it was found that for $n = 30, 50$, and 100, $\alpha_a = 0.3, 0.21$, and 0.08, respectively. For other sample sizes, simple linear interpolation is suggested. More precisely, if $30 < n < 50$, use linear interpolation based on n and the α_a values 0.3 and 0.21. For $50 < n < 100$, interpolate using the α_a values 0.21 and 0.08, and for $100 < n < 200$, use the values 0.08 and 0.05. As for $n_1 \neq n_2$, let α_1 and α_2 be the values of α_a corresponding to n_1 and n_2, respectively. Then the adjusted level is taken to be $(n_2\alpha_1 + n_1\alpha_2)/(n_1 + n_2)$. For example, with $n_1 = 30$ and $n_2 = 100$, this yields, $\alpha_a = 0.249$, which is nearly equal to the estimate of α_a based on simulations, namely 0.24. And the simulation estimate of the actual Type I error probability remains 0.05. For $n_1 = 30$ and $n_2 = 50$, this yields 0.266, the simulation estimate is 0.26, and the level of the test using 0.266 is again 0.05.

A related goal is testing

$$H_0: \eta_1 = \eta_2, \tag{11.39}$$

which generalizes methods for testing the hypothesis that two independent groups have equal Pearson correlations. When there is curvature, an obvious way of attaching a sign to the square root of $\hat{\eta}^2$ is to use the positive square root, if the association is monotonic increasing; otherwise, use the negative square root. If the regression line is not monotonic, a possibility is to attach a sign indicating whether, in general, the regression line is increasing. For convenience, assume $x_1 \leq \cdots \leq x_n$, let

$$S = \sum_{i=2}^{n} \text{sign}(\hat{y}_i - \hat{y}_{i-1}),$$

and let $I = 1$, if $S \geq 0$; otherwise, $I = -1$. Then use $I\hat{\eta}$ as the measure of association. (Choosing the sign in this manner has similarities to Kendall's tau.)

As for testing Eq. (11.39), a slight modification of the method for testing Eq. (11.38) is needed to avoid Type I error probabilities well above the nominal level when the sample size is small. For $n \geq 50$, determine α_a exactly as done when testing Eq. (11.38). But for $n < 50$, extrapolate using the α_a values 0.21 and 0.08, which correspond to the sample sizes 50 and 100, respectively.

11.9.3 R Functions smcorcom and smstrcom

The R function

$$\text{smcorcom(x1, y1, x2, y2, nboot = 200, pts = NA, plotit = TRUE, SEED = TRUE, varfun = pbvar)}$$

tests Eq. (11.38). If the argument plotit=TRUE, the two regression lines are plotted (by calling the R function lplot2g in Section 12.2.6). The R function

$$\text{smstrcom(x1, y1, x2, y2, nboot = 200, plotit = TRUE, SEED = TRUE, varfun = pbvar, xout=FALSE, outfun=out,...)}$$

tests the hypothesis given by Eq. (11.39).

11.10 Comparing Predictors

When dealing with two or more predictors, an issue that has received considerable attention is determining which predictors are best. Numerous methods have been proposed, many of which are known to be unsatisfactory. Relatively well-known methods that have proven to be unsatisfactory include stepwise regression (e.g., Montgomery et al., 2012, Section 7.2.3; Derksen and Keselman, 1992), a related (forward selection) method (Kuo and Mallick, 1998;

Huberty, 1989; Chatterjee and Hadi, 1988), methods based on R^2 (the squared multiple correlation), and the classic F statistic that tests the hypothesis that all slopes are zero. A homoscedastic approach based on

$$C_p = \frac{1}{\hat{\sigma}^2} \sum (y_i - \hat{y}_i)^2 - n + 2p,$$

called Mallows (1973) C_p criterion, cannot be recommended either (Miller, 1990). Another approach is based on ridge regression introduced in Section 10.13.15, but it suffers from problems listed by Breiman (1995). Briefly, ridge regression is not scale-invariant. If the scale of the predictors is changed, the ridge coefficients do not change inversely proportional to the changes in the variable scale. An approach to this criticism is to standardize each predictor so that they each have mean 0 and variance 1. Breiman notes, for example, that if the interquartile range were used instead of the usual variance to normalize the predictors, this would give different results. Breiman (1995) derived the non-negative garrote technique for dealing with these concerns. For results on a robust version of the non-negative garrote, see Gijbels and Vrinssen (2015). Yet another method for identifying the best predictors are lasso estimators, introduced in Section 10.13.17. And a related approach is least angle regression; see Efron et al. (2004). For a review of the literature dealing with least angle regression, see Zhang and Zamar (2014). Also see Wang and Leng (2007), as well as Radchenko and James (2011). Two alternative approaches, versions of which are described later in this section, are cross-validation and bootstrap methods known as the 0.632 estimator. Efron and Tibshirani (1993) provide additional details regarding the 0.632 estimator.

This section begins by describing methods for comparing Pearson correlations as well as robust analogs. A criticism of these methods is that they do not take into account which independent variables are included in the model. That is, the nature of the association between y and x_1, for example, can depend on whether x_2 is included in the model, a fact that is not addressed when comparing correlations. Methods for dealing with this issue are described as well.

11.10.1 Comparing Correlations

This section describes methods for comparing measures of association. The first portion of this section considers methods for comparing two measures of association. This is followed by a method for determining whether it is reasonable to make decision about which of p independent variables has the highest correlation with some dependent variable. Pearson's correlation is considered first, followed by methods based on robust measures of association. As previously stressed, Pearson's correlation is not robust, but perhaps there are situations where comparing Pearson correlations has practical value. Many methods have been derived, comparisons of which are reported in Wilcox (2009c).

The first method based on Pearson's correlation deals with what is known as the overlapping case. It is based on a slight modification of a method derived by Zou (2007). Rather than use Fisher's r-to-z transformation to compute confidence intervals for the individual correlations, as done by Zou, the HC4 method or the bootstrap version of the HC4 method, described in Section 9.3.13, is used.

For notational convenience, let ρ_{jk} be the correlation between x_j and x_k ($j = 1, 2, 3, k = 1, 2, 3$), where all three variables are possibly dependent. The goal is to test

$$H_0: \rho_{12} = \rho_{13}. \tag{11.40}$$

Typically, x_1 is some outcome variable y, while x_2 and x_3 are two predictor variables. This situation is often called the overlapping case because both r_{12} and r_{13} involve the first variable x_1. Let (l_1, u_1) and (l_2, u_2) be $1 - \alpha$ confidence intervals for ρ_{12} and ρ_{13}, respectively, which are based on the HC4 method in Section 9.3. Then a $1 - \alpha$ confidence interval for $\rho_{12} - \rho_{13}$ is

$$(L, U),$$

where

$$L = r_{12} - r_{13} - \sqrt{(r_{12} - l_1)^2 + (u_2 - r_{13})^2 - 2\widehat{corr}(r_{12}, r_{13})(r_{12} - l_1)(u_2 - r_{13})},$$

$$U = r_{12} - r_{23} + \sqrt{(u_1 - r_{12})^2 + (r_{23} - l_2)^2 - 2\widehat{corr}(r_{12}, r_{13})(u_1 - r_{12})(r_{23} - l_2)},$$

and

$$\widehat{corr}(r_{12}, r_{13}) = \frac{(r_{23} - 0.5r_{12}r_{23})(1 - r_{12}^2 - r_{13}^2 - r_{23}^2) + r_{23}^2}{(1 - r_{12}^2)(1 - r_{13})^2}.$$

In terms of avoiding a Type I error well above the nominal level, this method performs well. But if all three variables have a reasonably strong association, and the HC4 method is used to compute a confidence interval for individual correlations, the actual Type I error probability can be well below the nominal level, even under normality and homoscedasticity (Wilcox, 2009c). For example, with $n = 30$ and $\rho_{jk} = 0.5$ for all $j \neq k$, the actual level is 0.002 when testing at the 0.05 level, raising the concern that this results in relatively low power. This problem persists with $n = 100$. The situation can be improved by replacing the HC4 method with the bootstrap analog of the HC4 method described in Section 9.3.13. But outliers can have a negative impact on this approach. In terms of computing a confidence interval for which the actual probability coverage is reasonably close to the nominal level, a completely satisfactory method has not been found when using Pearson's correlation.

Now consider four variables x_1, x_1, x_3, and x_4, and again let ρ_{jk} be the correlation between x_j and x_k ($j = 1, 2, 3, 4$, $k = 1, 2, 3, 4$). The goal is to compute a confidence interval for

$$\rho_{12} - \rho_{34},$$

which is sometimes called comparing non-overlapping dependent correlations. Now (l_1, u_1) and (l_2, u_2) are $1 - \alpha$ confidence intervals for ρ_{12} and ρ_{34}, respectively, which are based on the modified bootstrap method in Section 9.3.13. Using, instead, the HC4 estimator in Section 10.1.1 can result in poor control over the probability of a Type I error (the actual level can be substantially less than the nominal level) and relatively low power. The $1 - \alpha$ confidence interval for $\rho_{12} - \rho_{34}$ is (L, U), where

$$L = r_{12} - r_{34} - \sqrt{(r_{12} - l_1)^2 + (u_2 - r_{34})^2 - 2\widehat{corr}(r_{12}, r_{34})(r_{12} - l_1)(u_2 - r_{34})},$$

$$U = r_{12} - r_{34} + \sqrt{(u_1 - r_{12})^2 + (r_{34} - l_2)^2 - 2\widehat{corr}(r_{12}, r_{34})(u_1 - r_{12})(r_{34} - l_2)},$$

$$\widehat{corr}(r_{12}, r_{13}) = \frac{T1 - T2}{T3},$$

$$T1 = 0.5 r_{12} r_{34}(r_{13}^2 - r_{14}^2 + r_{23}^2 + r_{24}^2) + r_{13} r_{24} + r_{14} r_{23},$$

$$T2 = r_{12} r_{13} r_{14} + r_{12} r_{23} r_{24} + r_{13} r_{23} r_{34} + r_{14} r_{24} r_{34},$$

and

$$T3 = (1 - r_{12}^2)(1 - r_{34})^2.$$

When using a robust measure of association, a basic percentile bootstrap method has been found to perform well in simulations (Wilcox, 2016d). That is, proceed as follows:

1. Generate a bootstrap sample in the usual manner by resampling with replacement n points from (x_{i1}, x_{i2}, x_{i3}) ($i = 1, \ldots, n$).
2. Compute robust correlation coefficients based on this bootstrap sample, yielding $\hat{\xi}_{12}^*$ and $\hat{\xi}_{13}^*$, the correlation between x_1 and x_2 and the correlation between x_1 and x_3, respectively, and let $d^* = \hat{\xi}_{12}^* - \hat{\xi}_{13}^*$.
3. Repeat steps 1 and 2 B times, and let d_b^* ($b = 1, \ldots, B$) denote the resulting d^* values.
4. Put the $d_1^*, \ldots, d_B^*$ values in ascending order, and label the results $d_{(1)}^* \leq \cdots \leq d_{(B)}^*$.
5. Let $\ell = \alpha B / 2$, rounded to the nearest integer, and $u = B - \ell$. Then the $1 - \alpha$ confidence interval for $\xi_1 - \xi_2$ is

$$(d_{(\ell+1)}^*, d_{(u)}^*). \tag{11.41}$$

Simulations indicate that when using a 20% Winsorized correlation or Spearman's correlation, good control over the Type I error probability is achieved, even when all three variables have a reasonably strong association, in contrast to the method based on Pearson's correlation.

A concern, however, is that these (type M) measures of association do not take into account the overall structure of the data cloud. The skipped correlation in Section 9.4.3 is one way of dealing with this issue. That is, remove points flagged as outliers using the projection method, and apply some measure of association to the remaining data. When using Pearson's correlation, the actual level can drop well below 0.025 when testing at the 0.05 level, with $n = 40$. Switching to a bias corrected accelerated bootstrap (BCa) method (Efron and Tibshirani, 1993), control over the Type I error probability is very poor when still using Pearson's correlation. However, when using the 20% Winsorized correlation or Kendall tau after outliers are removed, the BCa bootstrap method has been found to perform well in simulations. When using Spearman's rho, $n = 60$ can be required.

The methods just described can be used for the situation where the goal is to compare the association between y and x_j to the association between y and x_k. That is, there are p independent variables, and the goal is to test $H_0: \xi_{yj} = \xi_{yk}$ for each $j < k$, where ξ_{yj} is a measure of the association between y and x_j. Simply apply the method just described, and control the probability of one or more Type I errors using Hochberg's method or Hommel's method.

When dealing with type M measures of association, the non-overlapping case can be addressed with a percentile bootstrap method. But when using the Pearson version of the skipped correlation in Section 9.4.3 in conjunction with a percentile bootstrap method, the actual level is generally less than or equal to 0.02 when testing at the 0.05 level, with sample size $n = 20$ or 40. Results for the overlapping case suggest that this problem can be corrected by using the Winsorized version of the skipped correlation in conjunction with the BCa method, but this has not be been investigated.

11.10.2 R Functions TWOpov, TWOpNOV, corCOMmcp, twoDcorR, and twoDNOV

The R function

$$\text{TWOpov(x, y,alpha=0.05, BOOT=TRUE,nboot=499,SEED=TRUE)}$$

computes a confidence interval for the difference between Pearson correlations for the overlapping case using the method in Section 11.10.1. The argument x is assumed to be a matrix with two columns corresponding to two predictors. In the notation of the previous section, r_{12} is Pearson's correlation based on the data in x[,1] and y. And r_{13} is Pearson's correlation based on the data in x[,2] and y. The argument BOOT=TRUE means that the bootstrap version of the HC4 method is used to compute confidence intervals for the individual correlations. The R function

$$\text{TWOpovPV}(x, y, \text{BOOT=TRUE}, \text{alpha=0.05})$$

is exactly like the R function TWOpov, only a p-value is returned as well. The R function

$$\text{corCOMmcp}(x,y, \text{corfun=wincor}, \text{alpha=0.05}, \text{nboot=500}, \text{SEED=TRUE}, \text{MC=FALSE},$$
$$\text{xout=FALSE}, \text{outfun=outpro}, \text{method='hommel'},...)$$

tests H_0: $\xi_{yj} = \xi_{yk}$ for each $j < k$, and it reports adjusted p-values using the technique indicated by the argument method. The R function

$$\text{TWOpNOV}(x,y,\text{HC4=FALSE},\text{alpha=0.05})$$

deals with the non-overlapping case, again using Pearson's correlation. By default, it uses the modified bootstrap method, which is limited to testing at the 0.05 level. Setting HC4=TRUE, the function uses instead the HC4 method, which can be used when testing at some level other than 0.05. As previously indicated, using HC4=TRUE when testing at the 0.05 level can result in a loss of power. The R function

$$\text{TWOpNOVPV}(x, y, \text{tr=0.2})$$

is exactly like the R function TWOpNOV, only a p-value is returned as well. By necessity, TWOpNOVPV uses HC4=TRUE, so the results might differ from those obtained with TWOpNOV when HC4=TRUE rather than the default value HC4=FALSE.

The R function

$$\text{twoDcorR}(x,y,\text{corfun=wincor},\text{alpha=0.05},\text{nboot=500},\text{SEED=TRUE},\text{MC=FALSE}),$$

deals with the overlapping case when using a robust correlation. By default, a 20% Winsorized correlation is used, but any other robust estimator can be used via the argument corfun. Setting MC=TRUE, the function takes advantage of a multicore processor, if one is available, which can be useful in terms of reducing execution time when using a skipped correlation.

When dealing with the skipped correlation in Section 9.4.3, use the R function

$$\text{cor.skip.com}(x,y,\text{corfun=wincor},\text{outfun=outpro}, \text{alpha=0.05}, \text{nboot=1000}, \text{SEED=TRUE},....).$$

It is based on the BCa bootstrap method. Note that by default, a Winsorized correlation is used after removing outliers. This avoids the concern that the Winsorized correlation does not take into account the overall structure of the data cloud when dealing with outliers. Using Pearson's correlation instead, control over the Type I error probability can be very poor. The R function

$$\text{corskip.comPV(x,y,corfun=wincor,outfun=outpro, alpha=0.05, nboot=1000,}$$
$$\text{SEED=TRUE,....)}$$

computes a p-value at the expense of higher execution time.

The R function

$$\text{twoDNOV(x,y,corfun=wincor,alpha=0.05,nboot=500,SEED=TRUE,MC=FALSE)}$$

deals with the non-overlapping case. The arguments x and y are assumed to be matrices with two columns. The function compares the correlation associated with the data stored in x with the correlation associated with the data in y.

11.10.3 Methods for Characterizing the Relative Importance of Independent Variables

Section 10.13.17 describes approaches to determine which independent variables are relatively unimportant. This section describes some additional methods for comparing predictors. The first is based on the notion of prediction error, which is estimated in one of two ways. The first is called the 0.632 bootstrap method, which allows heteroscedasticity. The other uses a leave-one-out cross-validation method. A limitation of this method is that it does not yield any information about the strength of the empirical evidence that the optimal set of predictors has been correctly identified. The same criticism applies to the methods in Section 10.13.17. Another approach is to determine whether it is reasonable to make a decision about which independent variable has the strongest association with the independent variable. A method for dealing with this issue is described in this section. But a criticism is that it does not take into account the fact that the importance of, say, the first predictor can depend on which other independent variables are included. Section 11.10.5 deals with this issue. See, in particular, method IBS.

Prediction Error

Imagine that the n pairs of values $(x_1, y_1), \ldots, (x_n, y_n)$ are used to determine the regression line $\hat{y} = b_0 + b_1 x$. Now imagine that a new x value is observed, which is labeled x_0, in which case, the predicted value of y, based on the original n pairs of points, is $\hat{y}_0 = b_0 + b_1 x_0$. *Prediction error* refers to the discrepancy between the predicted value of y, $\hat{y}_0$, and the actual value of y, y_0, if it could be observed. One way of measuring the typical amount of prediction error is with

$$E[(y_0 - \hat{y}_0)^2],$$

the expected squared difference between the observed and predicted value of Y. And another possibility is

$$E[|y_0 - \hat{y}_0|],$$

the expected absolute error. As is evident, the notion of a prediction error is easily generalized to multiple predictors. The basic idea is that, via some method, we get a predicted value for y, which we label $\hat{y}$, and the goal is to measure the discrepancy between $\hat{y}_0$ (the predicted value of y based on a future collection of x values) and the actual value of y, y_0, if it could be observed.

A simple estimate of prediction error is the apparent error rate, which is just the average error when predicting the observed y values with $\hat{y}$. More formally, let $Q(y, \hat{y})$ be some measure of the discrepancy between an observation, y, and its predicted value, $\hat{y}$. So the squared error corresponds to

$$Q(y, \hat{y}) = (y - \hat{y})^2.$$

The goal is to estimate the typical amount of error for future observations. In symbols, the goal is to estimate

$$\eta = E[Q(y_0, \hat{y}_0)],$$

the expected error between a predicted value for y, based on a future value of x, and the actual value of y, y_0, if it could be observed. A simple estimate of η is the *apparent error*:

$$\hat{\eta}_{ap} = \frac{1}{n} \sum Q(y_i, \hat{y}_i).$$

So for the squared error, the apparent error is

$$\hat{\eta}_{ap} = \frac{1}{n} \sum (y_i - \hat{y}_i)^2,$$

the average of the squared residuals. (For results on estimating prediction error when using the MM-estimator, see Khan et al., 2010.)

A practical concern is that the apparent error is biased downward because the data used to come up with a prediction rule ($\hat{y}$) are also used to estimate error (Efron and Tibshirani, 1993). That is, it tends to underestimate the true error rate, η. The so-called 0.632 bootstrap estimator is designed to address this problem.

The 0.632 Estimator

The 0.632 estimator is applied as follows. Generate a bootstrap sample, only rather than sample n vectors of observations with replacement, as is typically done, sample $m < n$ vectors of observations instead. (Setting $m = n$, Shao, 1996, shows that the probability of selecting the correct model may not converge to 1 as n gets large.) Here, $m = 5\log(n)$ is used, which was derived from results reported by Shao (1996). Let $\hat{y}_i^*$ be the estimate of y_i based on the bootstrap sample, $i = 1, \ldots, n$. Repeat this process B times, yielding $\hat{y}_{ib}^*$, $b = 1, \ldots, B$. Then an estimate of η is

$$\hat{\eta}_{\text{Boot}} = \frac{1}{nB} \sum_{b=1}^{B} \sum_{i=1}^{n} Q(y_i, \hat{y}_{ib}^*).$$

A refinement of $\hat{\eta}_{\text{Boot}}$ is to take into account whether a y_i value is contained in the bootstrap sample used to compute $\hat{y}_{ib}^*$. Let

$$\hat{\epsilon}_0 = \frac{1}{n} \sum_{i=1}^{n} \frac{1}{B_i} \sum_{b \in C_i} Q(y_i, \hat{y}_{ib}^*),$$

where C_i is the set of indices of the bth bootstrap sample not containing y_i and B_i is the number of such bootstrap samples. Then the 0.632 estimate of the prediction error is

$$\hat{\eta}_{0.632} = 0.368\hat{\eta}_{\text{ap}} + 0.632\hat{\epsilon}_0. \tag{11.42}$$

This estimator arises in part from a theoretical argument showing that 0.632 is approximately the probability that a given observation appears in a bootstrap sample of size n.

The Leave-One-Out Cross-Validation Method

Prediction error using the leave-one-out cross-validation method is applied as follows. Momentarily, omit the ith point (x_i, y_i), and fit a regression model to the data. Based on this fit, let $\hat{y}_{-i}$ be the estimate of y using x_i. Let $e_i = y_i - \hat{y}_{-i}$. Then a prediction error is measured via some measure of variation applied to the e_i values $(i = 1, \ldots, n)$. Here, the percentage bend midvariance is used unless stated otherwise, but this is not to suggest that alternative measures of variation should be ruled out.

Which Independent Variable Has the Strongest Association With the Independent Variable?

Let τ_j $(j = 1, \ldots, p)$ be some measure of the strength of the association between x_j, the jth independent variable, and the dependent variable y. Let $\tau_{(1)} \leq \cdots \leq \tau_{(p)}$ denote these measures of association written in ascending order. Let $\hat{\tau}_{(j)}$ be some estimate of τ_j and denote the ordered estimates by $\hat{\tau}_{(1)} \leq \cdots \leq \hat{\tau}_{(p)}$. The goal is to determine the extent to which it is reasonable to conclude that the independent variable corresponding to $\hat{\tau}_{(p)}$ is indeed the variable corresponding to $\tau_{(p)}$, the variable with the highest population measure of association. The focus here is on the 20% Winsorized correlation. The basic strategy can be extended to other measures of association, but extant simulations are limited to the 20% Winsorized correlation.

Simple variations of methods RS1 and RS2, described in Sections 8.1.8 and 7.4.11, can be used for the problem at hand. Currently, it appears that RS2 works relatively well, it avoids certain limitations of RS1, so only RS2 is considered here. Basically, use a percentile bootstrap method to compute p-values when testing

$$H_0: \tau_{\pi(j)} = \tau_{\pi(p)}.$$

That is, test the hypothesis that the variable with the largest estimate has the same correlation as the variable associated with the jth largest estimate, $j = 1, \ldots, p - 1$. If the largest p-value is less than or equal to α, make a decision.

Note that if the first independent variable has $\tau = 0.25$ and the second has $\tau = -0.5$, a correct decision is that the first independent variable has a larger correlation with the independent variable. That is, there are situations where the method does not reflect the strongest association, regardless of whether it is positive or negative. If there is compelling evidence that an independent variable has a negative association with the dependent variable, one possibility is to multiply that variable by -1, making the correlation positive, and then apply method RS2 to determine whether this alters the ability to make a decision about which independent variable has the largest association.

When dealing with Pearson's correlation, an analog of method RS2 can be used, which is limited to the 0.05 level. Now, for any two independent variables, compute confidence intervals as described in Section 6.14. The R function TWOpov would be used. If none of the relevant confidence intervals contain zero, make a decision about which independent variable has the largest Pearson correlation with the dependent variable.

11.10.4 R Functions regpre, regpreCV, corREG.best.DO, and PcorREG.best.DO

The R function

```
regpre(x, y, regfun=lsfit, error=absfun, nboot=100, adz=TRUE,
mval=round(5*log(length(y))), model=NULL, locfun=mean, pr=TRUE, xout=FALSE,
outfun=out, plotit=TRUE, xlab='Model Number', ylab='Prediction Error', SEED=TRUE, ...)
```

estimates the prediction error using the 0.632 bootstrap method. By default, least squares regression is used, but results in Wilcox (2008d) indicate that the Theil–Sen estimator is better for general use. This can be done by setting the argument regfun=tsreg. With adz=TRUE, the function includes an estimate of prediction error based on using only the measure of location indicated by the argument locfun. That is, no predictors are used. The argument mval is m, the number of observations sampled when generating bootstrap samples. The argument error=absfun means that the absolute error is used by default. Setting error=sqfun, the squared error is used. Other robust measures of variation might be used as well. For example, setting error=winvar, the 20% Winsorized variance is used.

The R function

   regpreCV(x, y, regfun=tsreg, varfun=pbvar, adz=TRUE, model=NULL, locfun=mean, xout=FALSE, outfun=out, plotit=TRUE, xlab='Model Number', ylab = 'Prediction Error', ...)

performs the leave-one-out cross-validation estimate of prediction error based on the regression estimator indicated by the argument regfun.

■ **Example**

The R function regpre is illustrated with the reading data described in the first example of Section 10.8.1. Here, we consider how well the first three predictors, stored in columns 2–4, compare when predicting a word identification score (stored in column 8 of the file read.dat). Assuming the data are stored in the R variable read, the command regpre(read[,2:4],read[,8],regfun=tsreg,locfun=median) returns

```
$estimates
      apparent.error  boot.est   err.632  var.used  rank
[1,]         12.48052  13.36920  13.25599         1     4
[2,]         11.99610  12.74614  12.70693         2     2
[3,]         14.87278  16.12612  16.08528         3     8
[4,]         11.47575  12.51257  12.55914        12     1
[5,]         12.76048  14.33530  14.27367        13     6
[6,]         11.90024  13.24456  13.28122        23     5
[7,]         11.30447  13.30425  13.21919       123     3
[8,]               NA        NA  14.30223         0     7
```

The column headed by var.used indicates the predictors used in the model. The entry 12 means that both predictors 1 and 2 were used, ignoring predictor 3. The entry 123 indicates that all three predictors are used. The last column provides an easy way of identifying which combination of predictors produced the lowest prediction error. Here, using both predictors 1 and 2 performed best. The worst model was obtained using predictor 3 and ignoring the other predictors. The last row is for the case where all predictors are ignored. So here, using predictor 3 is worse than using no predictors at all, meaning that one simply uses the median of the y values to predict future observations. Using instead the R function regpreCV, the results are:

```
       est.error var.used rank
[1,]   302.4256         1    5
[2,]   248.9536         2    4
[3,]   459.2516         3    8
[4,]   231.9329        12    2
[5,]   308.1499        13    6
[6,]   245.4202        23    3
[7,]   225.8654       123    1
[8,]   403.2488         0    7
```

The R regpre function automatically generates all possible combinations of predictors, assuming the number of predictors is at most five. The argument model, which is assumed to have list mode, can be used to analyze only models that are of specific interest. For example, setting model[[1]]=1 and model[[2]]=c(1,2,3), and then setting the argument model=model, the prediction error would be estimated when using predictor 1, as well as using predictors 1, 2, and 3 simultaneously, but no other models would be considered.

The R function

$$\text{corREG.DO(x,y,corfun=wincor,alpha=0.05,nboot=500,SEED=TRUE,neg.col=NULL,}$$
$$\text{LARGEST=TRUE,xout=FALSE,outfun=outpro,...)}$$

applies the percentile bootstrap version of method RS2 to determine whether a decision can be made about which independent variable has the largest robust correlation with the dependent variable. The argument neg.col can be used to change the sign of a correlation. For example, if the independent variables 2 and 4 have a negative correlation, neg.col=c(2,4) would change the correlations so that they are positive.

The R function

PcorREG.best.DO(x,y,neg.col=NULL, LARGEST=TRUE,xout=FALSE,outfun=outpro,...)

is like the function corREG.best.DO, only it is designed specifically for Pearson's correlation. It is limited to $\alpha = 0.05$.

11.10.5 Inferences About Which Predictors Are Best

The methods in Sections 11.10.3 and 10.13.17 suggest which independent variables are most important. But these methods do not address a fundamental issue: How strong is the empirical evidence that the first independent variable is indeed more or less important than the second independent variable when both independent variables are included in the model?

To elaborate on how these issues might be addressed, first, focus on the model

$$y = \beta_0 + \beta_1 x_1 + \beta_2 x_2 + (\lambda_1(x_1) + \lambda_2(x_2))\epsilon,$$

where ϵ has some unknown variance σ^2, $E(\epsilon) = 0$, and ϵ is independent of x_1 and x_2. The functions λ_1 and λ_2 are used to model heteroscedasticity. For convenience, let

$$m(x_1, x_2) = \beta_1 x_1 + \beta_2 x_2$$

denote some conditional measure of location, given x_1 and x_2. Recall that the numerator of explanatory power (introduced in Section 11.9) is based on some measure of dispersion associated with $m(x_1, x_2)$. Let $m_j(x_j) = \beta_j x_j$, where it is stressed that β_j is the slope associated with x_j when the other independent variable is included in the model. Let η_j^2 be some measure of variation associated with $m_j(x_j)$ over the range of possible x_j values. That is, η_j^2 is the numerator of explanatory power given by Eq. (11.34) based on the slope associated with the jth independent variable. The goal is to test

$$H_0: \eta_1^2 = \eta_2^2. \tag{11.43}$$

Note that testing this hypothesis can play a role in a mediation analysis as described in Section 11.7.2.

Now consider the linear model

$$y = \beta_0 + \beta_1 x_1 + \beta_2 x_2 + \beta_3 x_3 + (\lambda_1(x_1) + \lambda_2(x_2) + \lambda_2(x_3))\epsilon.$$

Let $m_{12}(x_1, x_2) = \beta_1 x_1 + \beta_2 x_2$, and let $m_3(x_3) = \beta_3 x_3$, where β_1, β_2, and β_3 are the slopes when all three independent variables are included in the model. Let η_{12}^2 be some measure of dispersion associated with $m_{12}(x_1, x_2)$, and let η_3^2 be some measure of dispersion associated with $m_3(x_3)$. To determine whether the first two independent variables are more important than the third independent variable, one can test

$$H_0: \eta_{12}^2 = \eta_3^2. \tag{11.44}$$

Or in terms of Tukey's three-decision rule, is it reasonable to make a decision about whether η_{12}^2 is less than or greater than η_3^2 based on the available data? As is evident, this approach is readily generalized to more than three independent variables. For a related technique based on the logistic regression model, see Wilcox (2018e).

There is, however, a limitation that should be stressed. While the method in this section provides a valid way of making inferences about the relative importance of x_1 and x_2 taken together, versus x_3, the method is meaningless regarding the relative importance of x_1 and x_2 taken together, versus x_2. That is, it is pointless testing $H_0: \eta_{12}^2 = \eta_2^2$ because it is generally the case that $\eta_{12}^2 > \eta_2^2$.

Method IBS

Based on results previously summarized, a natural way of dealing with non-normality is to use some robust estimate of the slopes and use some robust measure of dispersion, and then use a basic percentile bootstrap method. Here, the measure of dispersion is taken to be the 20% Winsorized variance. When using the quantile regression estimator, this approach performs poorly in terms of controlling the Type I error probability. When testing at the 0.05 level, when the slopes differ from zero, the actual level can exceed 0.10.

Consider the goal of testing Eq. (11.43). Wilcox (2018b) considers a slight modification of the standard percentile bootstrap method that consists of using independent bootstrap samples. That is, generate a bootstrap sample and estimate η_1^2, yielding $\hat{\eta}_2^2$. Then generate a new bootstrap sample and estimate η_2^2, yielding $\hat{\eta}_2^2$. Repeat this B times, and use the results to estimate $P(\hat{\eta}_1^2 < \hat{\eta}_2^2)$. Also, rather than use the standard estimate of this probability, the method described in Section 5.7.1 is used, which is denoted by $\hat{p}$. So in effect, B^2 pairs of values are used when computing $\hat{p}$. Then a p-value is $2 \min(\hat{p}, 1 - \hat{p})$ (cf. Racine and MacKinnon, 2007b). Extant simulations indicate that now, even when the slopes differ from zero, the actual Type I error probability is substantially smaller than the nominal level, still using the quantile regression estimator. Wilcox found, however, that using the Theil–Sen estimator instead, reasonably good control over the Type I error probability is achieved. Limited simulations suggest that when using the MM-estimator, again, control over the Type I error probability is good. As is evident, a similar approach can be used when testing Eq. (11.44). This method is substantially better than a related technique derived by Wilcox (2011b).

Table 11.8: Estimated Type I error probabilities, $p = 2$, $n = 50$, $\alpha = 0.05$.

g	h	VP	$\beta_1 = \beta_2 = 0$		$\beta_1 = \beta_2 = 1$	
			$\rho = 0$	$\rho = 0.6$	$\rho = 0$	$\rho = 0.6$
0.0	0.0	1	0.001	0.002	0.022	0.066
		2	0.002	0.001	0.050	0.065
		3	0.001	0.001	0.029	0.046
0.0	0.2	1	0.001	0.002	0.030	0.067
		2	0.001	0.001	0.035	0.060
		3	0.002	0.001	0.031	0.052
0.2	0.0	1	0.001	0.001	0.032	0.056
		2	0.002	0.001	0.038	0.065
		3	0.001	0.001	0.030	0.045
0.2	0.2	1	0.001	0.001	0.035	0.067
		2	0.001	0.001	0.036	0.058
		3	0.002	0.001	0.031	0.051

Table 11.8 shows some estimated Type I error probabilities when testing Eq. (11.43) at the 0.05 level and when using the Theil–Sen estimator in conjunction with the 20% Winsorized variance. The sample size is $n = 50$. The independent variables were generated from a bivariate normal distribution, having correlation $\rho = 0$ or 0.6. The error term was generated from a g-and-h distribution. Heteroscedasticity was modeled with $\lambda_j(x_j) = |x_j| + 1$ ($j = 1$, 2), labeled VP 2 in Table 11.8, or $\lambda_j(x_j) = 1/(|x_j| + 1)$ (VP 3). VP 1 corresponds to homoscedasticity. As can be seen, the Type I error probability is estimated to be well below the nominal level when $\beta_1 = \beta_2 = 0$. This was expected because for this special case, there is no distinction between the strength of the association when the slope estimates have the same absolute value. The main point is that when the slope parameters differ from zero, control over the Type I error probability is reasonably good, the estimates ranging between 0.022 and 0.066. Similar results were obtained when the independent variables have a chi-squared distribution, with four degrees of freedom.

Simulations also indicate that method IBS continues to perform reasonably well when testing Eq. (11.44), again using the Theil–Sen estimator and the 20% Winsorized variance. Table 11.9 shows some estimates of the actual Type I error probability, again testing at the 0.05 level, where now VP 2 corresponds to an error term having the form

$$(\lambda_1(x_1) + \lambda_2(x_2) + \varrho\lambda_3(x_3))\epsilon,$$

Table 11.9: Estimated Type I error probabilities, $p = 3$, $n = 50$, $\alpha = 0.05$.

g	h	VP	$\beta_1 = \beta_2 = 0$		$\beta_1 = \beta_2 = 1, \beta_3 = \sqrt{2}$	
			$\rho = 0$	$\rho = 0.6$	$\rho = 0$	$\rho = 0.6$
0.0	0.0	1	0.003	0.001	0.028	0.039
		2	0.002	0.001	0.034	0.041
		3	0.000	0.001	0.026	0.044
0.0	0.2	1	0.003	0.000	0.031	0.035
		2	0.002	0.001	0.018	0.025
		3	0.004	0.001	0.025	0.045
0.2	0.0	1	0.003	0.000	0.032	0.029
		2	0.001	0.001	0.034	0.042
		3	0.003	0.002	0.028	0.048
0.2	0.2	1	0.002	0.000	0.029	0.036
		2	0.002	0.000	0.019	0.026
		3	0.003	0.001	0.028	0.049

VP 3 corresponds to

$$\left(\frac{1}{\lambda_1(x_1)} + \frac{1}{\lambda_2(x_2)} + \frac{\varrho}{\lambda_3(x_3)} \right) \epsilon,$$

$\lambda_j(x_j)$ is defined as in the previous paragraph, and ϱ is a constant chosen so that the variance of $\hat{m}_{12}(x_1, x_2)$ is equal to the variance of $\hat{m}_3(x_3)$. (Using $\beta_3 = \sqrt{2 + 2\rho_{12}}$ follows from the fact that the marginal distributions have variance 1.) Again, the estimates are well below the nominal level when all three slope parameters are equal to zero. Otherwise, the estimates never exceed the nominal level, and there are only two instances where the estimates are less than 0.025. Simulations with $\varrho = 1$ (a type of heteroscedasticity) result in estimates similar to those in Table 11.9. Limited simulations indicate that the method continues to perform well when using the MM-estimator or when dealing with four independent variables.

Method BTS

Yet another approach to comparing two explanatory variables is to estimate the strength of the association based on the Theil–Sen estimator, and then use a percentile bootstrap method to test

$$H_0: \eta_1^2 = \eta_2^2,$$

where now η_j^2 is explanatory power when using predictor x_j. The approach used here differs from method IBS in a crucial way: The slope associated with x_1 is the slope obtained when the other explanatory variable, x_2, is ignored. That is, x_2 is not used when estimating β_1. In a similar manner, the slope associated with x_2 is the slope when x_1 is ignored. If the correlations have the same sign, one could instead compare correlations as described in Section 11.10.1. The method described here might be of interest when the correlations have opposite signs.

Like method IBS, the standard percentile bootstrap method performs poorly. As the correlation between x_1 and x_2 increases, the actual level of this method can drop well below the nominal level, even with a sample size of $n = 100$. Taking independent bootstrap samples, the first from (x_{i1}, y_i) and the second from (x_{i2}, y_i), has been found to improve matters (Wilcox, 2011b). Imagine that this is done, yielding $D^* = \tilde{\eta}_1^2 - \tilde{\eta}_2^2$, the difference between the two estimates of η^2. Repeating this process B times yields $D_1^*, \ldots, D_B^*$, which can be used to estimate $P = P(D < 0)$ in the manner already described, which in turn yields the generalized p value. But again, control over the probability of a Type I error has been found to be not quite satisfactory: When testing at the 0.05 level, the actual level can be substantially smaller than 0.05. Some improvement is obtained if, rather than estimate $P = P(D < 0)$ with the bootstrap samples in the usual way, a kernel density estimate is used instead, a strategy motivated by results in Racine and MacKinnon (2007b). When computing a confidence interval, adjustments are also made based on an estimate of Kendall's tau for the two independent variables. The method is limited to computing a 0.95 confidence interval; it is unknown how to adjust the confidence interval when computing a $1 - \alpha$ confidence interval, $\alpha \neq 0.05$. In contrast to method IBS, there are no results regarding how method BTS might be extended to more than two independent variables.

Here, the adaptive kernel estimator in Section 3.2.4 is used, in which case the distribution of D is estimated with

$$\hat{f}(d) = \frac{1}{nh} \sum_{i=1}^{B} K\left(\frac{d - D_i}{h}\right),$$

where K is taken to be the Epanechnikov kernel and h is the span. An estimate of $P(D < 0)$ is

$$\hat{P}(D < 0) = \frac{1}{nh} \sum_{i=1}^{n} \int_{\ell}^{0} K\left(\frac{t - D_i}{h}\right) dt.$$

The method just described performs well in simulations when $\rho_{12} = 0$, but, again, the actual level can drop well below the nominal level when $\rho_{12} \neq 0$. Let ρ_{k12} be Kendall's tau for x_1 and x_2. Compute a 0.95 confidence interval for ρ_{k12} using the method in Section 9.3.10. If this interval contains 0, let $\tilde{p} = 0$; otherwise

$$\tilde{p} = 0.352|r_{k12}| + 0.049. \tag{11.45}$$

For $n \leq 100$, reject at the 0.05 level, if the p-value is less than or equal to $\tilde{p}$. For $n > 100$, use the p-value in the usual manner.

Method SM

Method BTS can be extended to the situation where explanatory power is estimated via LOWESS, described in Section 11.5.2, when testing at the 0.05 level. Again there are no results on how to deal with more than two independent variables or $\alpha \neq 0.05$. Let

$$\check{p} = 0.25|r_{k12}| + 0.05 + (100 - n)/10000, \tag{11.46}$$

$\check{p} = \max(0.05, \check{p})$, and reject if $p \leq \check{p}$. For $n > 200$, $\check{p}$ is taken to be 0.05. Also see Wilcox (2018f).

11.10.6 R Functions reglVcom, reglVcommcp, loglVcom, ts2str, and sm2strv7

The R function

> regIVcom(x,y, IV1=1, IV2=2, regfun=qreg, nboot=200, xout=FALSE, outfun=outpro, SEED=TRUE, MC=FALSE, tr=0.2,...)

performs method IBS. The arguments IV1 and IV2 indicate which independent variables are to be compared. A key feature of this method is that it compares the importance of independent variables when all p independent variables are included in the model. For example, the R commands IV1=c(1,2) and IV2=3 would compare (i) the strength of the association between independent variables 1 and 2 and the dependent variable to (ii) the strength of the association between the third independent variable and the dependent variable. That is, the function would test Eq. (11.44). The argument tr indicates the amount of Winsorizing. The function returns a p-value, estimates of η^2 labeled est.1 and est.2, their ratio labeled ratio, estimates of explanatory power labeled e.pow1 and e.pow2, the square root of e.pow1 and e.pow2 labeled strength.assoc.1 and strength.assoc.2 (analogs of Pearson's correlation), and the ratio of these values, which is labeled strength.ratio. For convenience, the function

> regIVcommcp((x, y, regfun = tsreg, nboot = 200, xout = FALSE, outfun = outpro, SEED = TRUE, MC = FALSE, tr = 0.2, ...)

is provided. For each pair of independent variables, it tests the hypothesis that they are equally important when all p independent variables are included in the model. The function

> logIVcom(x, y, IV1 = 1, IV2 = 2, nboot = 500, xout = FALSE, outfun = outpro, SEED = TRUE, val = NULL, ...)

is like the function regIVcom, only it is designed for the logistic regression model. Caution must be taken, however, because experience with smoothers covered in Section 11.5.8 suggests that the logistic regression model can be unsatisfactory.

The R function

$$ts2str(x, y, nboot = 400, SEED = TRUE)$$

performs method BTS, where the argument x is assumed to be a matrix, with two columns containing the data for two predictors, and the function

$$sm2strv7(x, y, nboot = 100, SEED = TRUE, xout = FALSE, outfun = outpro, ...)$$

performs method SM.

11.11 Marginal Longitudinal Data Analysis: Comments on Comparing Groups

There is a vast literature dealing with *longitudinal data* analyses (e.g., Diggle et al., 2002; Molenberghs and Verbeke, 2005). Roughly, the goal is to deal with situations where measures are taken over time. As a concrete example, consider again the data in Section 1.9 dealing with an orthodontic growth study. The measure of interest is the distance between the pituitary and pterygomaxillary fissure, which was measured at ages 8, 10, 12, and 14 years. As noted in Section 1.9, the first 10 rows of the data are:

```
     distance  age  Subject   Sex
1       26.0    8     M01   Male
2       25.0   10     M01   Male
3       29.0   12     M01   Male
4       31.0   14     M01   Male
5       21.5    8     M02   Male
6       22.5   10     M02   Male
7       23.0   12     M02   Male
8       26.5   14     M02   Male
9       23.0    8     M03   Male
10      22.5   10     M03   Male
```

There are 16 males and 11 females.

Many goals arise when dealing with longitudinal data. And often, one of three models is used. The first is a marginal regression model where the goal is to understand the typical outcome (distance in the example), given the value of some explanatory variable, which here is age. The other two models are a random effects model and a transitional model, where the covariate effects and the within subject association are modeled through a single equation. Transitional models go beyond the scope of this book. Indeed, only a few comments are made about a narrow range of problems that are relevant to longitudinal data. From a robustness point of view, progress has been made, but more research in this area is needed for reasons outlined at the end of this section.

A common approach to longitudinal data is to fit some type of linear regression model that takes into account in some manner the time at which measures were taken. A simple approach is to assume

$$y_{ij} = \beta_0 + \beta_1 t_j + \epsilon_{ij},$$

where t_j is some measure taken at the jth time point. In the orthodontic example, $t_1, \ldots, t_4$ correspond to ages 8, 10, 12, and 14, respectively. So a single slope and intercept are used to characterize the association between y and t for the population of individuals under study. A semiparametric regression model for longitudinal data was studied by Chen and Zhong (2010), where an empirical likelihood method is used to test hypotheses. Their simulations indicate that the method performs well under normality, but it seems that a robust version of this approach has not been derived. Recall from Section 4.7 that in the one-sample case, the empirical likelihood method can be relatively unsatisfactory when dealing with heavy-tailed distributions.

A variation of this approach fits a regression line for each individual. Diggle et al. (2002) describe a number of situations where this approach appears to be reasonable. For example, for each participant in the orthodontic growth data, the strategy would be to fit a regression model that relates distance to the age of the child. In more formal terms, assume that for the ith participant

$$y_{ij} = \beta_{0i} + \beta_{1i} t_{ij} + \epsilon_{ij}.$$

So each individual is characterized by a slope and intercept, with the slopes and intercepts possibly varying among the population of participants.

Imagine that the goal is to compare males and females. One approach is to use the between-by-within ANOVA method described in Section 8.6. Another approach is to compare the groups using a multivariate measure of location as described, for example, in Section 6.8, 6.9, or 6.11. In terms of the orthodontic data, we have four measures for each individual, and so the groups could be compared, for example, based on the multivariate OP measure of location. For the kth group, let $(\theta_{0k}, \theta_{1k})$ represent some measure of location associated with $(\beta_{01k}, \beta_{11k}), \ldots, (\beta_{0n_k k}, \beta_{1n_k k})$, where β_{0ik} and β_{1ik} are the intercept and slope, respectively, associated with the kth group ($k = 1, 2$). For example, θ_{1k} might be the median of the slopes associated with group k. Yet another approach is to test

$$H_0: (\theta_{01}, \theta_{11}) = (\theta_{02}, \theta_{12}). \tag{11.47}$$

So the p-variate data have been reduced to two variables, and these two variables could be compared, for example, using the methods in Section 6.8 or 6.9.

A broader, more involved approach toward longitudinal data, based on a marginal model and the MM-estimator in Section 10.9.1, is summarized by Heritier et al. (2009, Section 6.2). Included is an inferential technique that is based in part on appropriate estimates of the standard errors; the test statistic is assumed to be approximately standard normal. Evidently, there are no results on the ability of this approach to control Type I errors when dealing with skewed distributions or heteroscedasticity. In simpler situations, skewness is a serious concern when using M-estimators, and a hypothesis testing method is based on a (non-bootstrap) technique that is a function of estimated standard errors. So caution seems warranted for the situation at hand. For robust methods based on a random effects model, see Mills et al. (2002), Sinha (2004), and Noh and Lee (2007). The basic strategy is to estimate parameters assuming observations are randomly sampled from a class of distributions that includes normal distributions as a special case. For example, Mills et al. assume that sampling is from a mixture of normal and t distributions, which results in a bounded influence function. Certainly these methods are an improvement on methods that assume normality. Again, what is unclear is the extent to which skewness and heteroscedasticity affect efficiency and Type I error probabilities. How well do these methods handle contamination bias as described in Section 10.14.1? If practical problems are found, perhaps some bootstrap method can provide more satisfactory results, but this remains to be determined.

11.11.1 R Functions long2g, longreg, longreg.plot, and xyplot

The R function

long2g(x, x.col, y.col, s.id, grp.id, regfun = tsreg, MAR = TRUE, tr = 0.2)

compares two groups based on estimates of the slope and intercept for each participant. The data are assumed to be stored in a matrix or data frame as illustrated by the orthodontic data in the previous section. The arguments x.col and y.col indicate the columns of x where the covariate and outcome variables are stored, respectively. The argument s.id is the column containing the subject's identification, and grp.id is the column indicating group membership, which is assumed to have two possible values only. The regression line for each participant is fitted with the regression estimator indicated by the argument regfun, which defaults to tsreg, the Theil–Sen estimator. If MAR=TRUE, the slopes and intercepts are compared using Yuen's method for trimmed means, which is described in Section 5.3. If MAR=FALSE, the hypothesis given by Eq. (11.47) is tested using the OP-estimator in conjunction with the method in Section 6.8.

The R function

$$\text{longreg}(x, x.col, y.col, s.id, regfun = tsreg, est = tmean)$$

computes the slope and intercept for each participant, using the regression estimator indicated by the argument regfun, and returns the results in a matrix labeled S.est. The typical slope and intercept are returned as well, which are based on the estimator indicated by the argument est.

The R function

$$\text{longreg.plot}(x,x.col,y.col,s.id,regfun=tsreg,scat=TRUE,xlab='X', ylab='Y')$$

plots the regression lines based on the R function longreg. A scatterplot of the points can be created with the R function xyplot, which is in the R package lattice. If the orthodontic data are stored in the R variable x, the R command library(lattice) followed by

$$\text{xyplot}(x[,1]\tilde{}x[,2],group=x[,3])$$

accomplishes this goal. To add line segments connecting the responses for each participant, include the argument type='b'. (Spaghetti plots can also be created using the R function spag.plot in Section 6.4.14.) The command

$$\text{long2g}(x,2,1,3,4)$$

would compare the slopes and intercepts using Yuen's method.

11.12 Exercises

1. For the data in Exercise 1 of Chapter 10, the 0.95 confidence interval for the slope, based on the least squares regression line, is $(0.0022, 0.0062)$. Using R, the 0.95 confidence interval for the slope returned by lsfitci is $(0.003, 0.006)$. The 0.95 confidence interval returned by the R function regci (using the Theil–Sen estimator) is $(0.003, 0.006)$. Verify these results.

2. Section 8.6.2 reports data on the effects of consuming alcohol on three different occasions. Using the data for group 1, suppose it is desired to predict the response at time 1 using the responses at times 2 and 3. Test H_0: $\beta_1 = \beta_2 = 0$ using the R function regtest and $\hat{\beta}_m$.

3. For the data in Exercise 1, test H_0: $\beta_1 = 0$ with the functions regci and regtest. Comment on the results.

4. Use the function winreg to estimate the slope and intercept of the star data using 20% Winsorization. (The data are stored in the file star.dat. See Section 1.8 on how to obtain the data.)

5. The example at the end of Section 11.5.5 are based on the data stored in the file A3B3C_dat.txt, which can be downloaded as described in Section 1.10. Verify the results in Section 11.5.5 when leverage points are removed.

6. For the reading data in file read.dat, let x be the data in column 2 (TAAST1), and suppose it is desired to predict y, the data in column 8 (WWISST2). Speculate on whether there are situations where it would be beneficial to use x^2 to predict y taking into account the value stored in column 3 (SBT1). Use the functions in this chapter to address this issue.

7. For the reading data in the file read.dat, use the R function rplot to investigate the shape of the regression surface when predicting the 20% trimmed mean of WWISST2 (the data in column 8) with RAN1T1 and RAN2T1 (the data in columns 4 and 5).

8. The data in the lower left panel of Fig. 11.5 are stored in the file agegesell.dat. Remove the two pairs of points having the largest x value, and create a running interval smoother using the data that remain.

9. For the reading data in the upper right panel of Fig. 11.5, recreate the smooth. If you wanted to find a parametric regression equation, what might be tried? Examine how well your suggestions perform.

10. Generate 25 observations from a standard normal distribution, and store the results in the R variable x. Generate 25 more observations, and store them in y. Use rungen to plot a smooth based on the Harrell–Davis estimator of the median. Also create a smooth with the argument scat=F. Comment on how the two smooths differ.

11. Generate 25 pairs of observations from a bivariate normal distribution having correlation 0, and store them in x. (The R function rmul, written for this book, can be used.) Generate 25 more observations, and store them in y. Create a smooth using rplot using scale=T, and compare it to the smooth when scale=FALSE.

12. Generate data from a bivariate normal distribution with the R command x=rmul(200). Then enter the R command y=x[,1]+x[,2]+x[,1]*x[,2]+rnorm(200), and examine the plot returned by the R command gamplot(x,y,scale=TRUE). Compare this to the plot returned by the R command gamplotINT(x,y,scale=TRUE).

## CHAPTER 12

ANCOVA

This chapter deals with what is commonly known as analysis of covariance (ANCOVA), where the goal is to compare groups in terms of some measure of location while taking into account some covariate. Consider, for example, two independent groups. For the jth group, let $m_j(x)$ be some conditional population measure of location associated with y given x. Given x, a basic problem is determining how the typical value of y in the first group compares to the typical value in the second.

There is a vast body of literature on ANCOVA. For an entire book devoted to the subject, see Huitema (2011). Also see Rutherford (1992) and Harwell (2003), as well as Ceyhan and Goad (2009). Obviously all relevant methods cannot be described here. Rather, attention is focused on the methods that currently seem best in terms of dealing with heteroscedasticity, non-parallel regression lines, non-normality, outliers, and curvature. As done in Section 11.4, curvature refers to a regression line that is not straight, or, more generally, some specified linear parametric model poorly captures the nature of the association.

The classic and best-known ANCOVA method arises as follows. Assume that for the jth group,

$$y_{ij} = \beta_{0j} + \beta_{1j}x + \lambda(x_j)\epsilon_j, \tag{12.1}$$

where ϵ_j has mean zero and variance σ_j^2 and $\lambda(x_j)$ is some unknown function that models how the conditional variance of y, given x, varies with x. So it is assumed that the conditional mean of y_{ij}, given x, is

$$m_j(x) = \beta_{0j} + \beta_{1j}x.$$

That is, it is assumed that a straight regression line provides a reasonably good approximation of the true regression line. The standard ANCOVA method also assumes that $\sigma_1^2 = \sigma_2^2$, called between group homoscedasticity, and that $\lambda(x_j) \equiv 1$, which is called within group homoscedasticity, $\beta_{11} = \beta_{12}$, parallel regression lines, and that the error term, ϵ_j, has a normal distribution. If these assumptions are true, then the groups can be compared simply by comparing the intercepts using in part the least squares regression estimator. But violating even one of these assumptions can render this classic ANCOVA method unsatisfactory in terms of both Type I errors and power, and violating more than one of these assumptions only makes matters worse. Cao et al. (2019) derive a method based on means that deals with between group heteroscedasticity. A positive feature is that it can be used with any number of covariates. But it does not deal with violations of the other assumptions made by the standard ANCOVA method.

The assumption of parallel regression lines can be tested using the method in Section 11.2, but how much power should this test have in order to be reasonably certain that the regression lines are, for all practical purposes, sufficiently parallel? Also, if the slopes are not parallel, what should be done instead? There is a solution based on means and the assumption that the error term within each group is homoscedastic (Wilcox, 1987b), but one of the goals here is to allow the error term to be heteroscedastic. Also, testing the homoscedasticity assumptions can be unsatisfactory: The ability to detect situations where heteroscedasticity is a practical issue can be poor.

Based on results covered in Chapter 11, as well as Chapter 5, a reasonable speculation is that a percentile bootstrap procedure will provide fairly accurate probability coverage when working with some robust regression estimator. However, simulations do not support this approach. In fact, probability coverage can be poor, at least when $n \leq 50$. An alternative approach, which has been found to perform reasonably well in simulations, is to use a bootstrap estimate of the standard error of $\hat{m}_j(x)$, where $\hat{m}_j(x)$ is some estimate of $m_j(x)$ based on some regression estimator, and then use a pivotal test statistic in conjunction with some appropriate critical value with the goal of making inferences about $m_1(x) - m_2(x)$. Details are given in Section 12.1.

Yet another problem is determining what to do if there is curvature. In some cases it might help to replace x with x^a, for some constant a, but as noted in Section 11.5, this approach is often ineffective. Several non-parametric methods have been proposed for testing H_0: $m_1(x) = m_2(x)$ for all x (e.g., Bowman and Young, 1996; Delgado, 1993; Dette and Neumeyer, 2001; Ferreira and Stute, 2004; Härdle and Marron, 1990; Hall and Hart, 1990; Hall et al., 1997; King et al., 1991; Kulasekera, 1995; Kulasekera and Wang, 1997; Munk and Dette, 1998; Neumeyer and Dette, 2003; Young and Bowman, 1995; Srihera and Stute, 2010; Zou et al., 2010). Typically they are based on kernel-type regression estimators where $m(x)$ is the conditional mean of y given x. Many of these methods make rather restrictive assumptions, such as homoscedasticity or equal design points, but recent efforts have yielded methods that remove these restrictions (e.g., Dette and Neumeyer, 2001; Neumeyer and Dette, 2003). The method derived by Srihera and Stute (2010) allows heteroscedasticity, but it is unknown how well it performs under non-normality. There are various ways these methods might be extended to robust measures of location. So far simulations do not support their use, but many variations have yet to be investigated. Feng et al. (2015) derive a robust method based on a generalized likelihood ratio test that uses a Wilcoxon-type artificial likelihood function. The method assumes that the usual error term has a symmetric distribution. Evidently there are no results on how well the method performs when this assumption is violated.

Boente and Pardo-Fernández (2016) derive a method that deals effectively with curvature that is aimed at testing the global hypothesis that H_0: $m_1(x) = m_2(x)$, against one-sided alternatives, for all x in some subset of the sample space. Their method uses a robust smoother. Their hypothesis testing method assumes that the error term has a symmetric distribution and is homoscedastic. For a method based on splines that allows heteroscedasticity via a wild bootstrap method, see Zhao et al. (2020).

The immediate goal is to describe robust methods based on a linear model. This is followed by a description of methods that are based on more flexible methods for modeling the association via smoothers. Smoothers also provide convenient ways of estimating measures of effect size.

12.1 Methods Based on Specific Design Points and a Linear Model

This section describes methods for testing

$$H_0: m_1(x) = m_2(x) \tag{12.2}$$

for one or more values of the covariate x assuming that

$$m_j(x) = \beta_{0j} + \beta_{1j}x.$$

Unlike the classic ANCOVA method, both between group and within group heteroscedasticity are allowed and the assumption of parallel regression lines is not required. Currently, the only known methods that perform well in simulations when the sample sizes are relatively small use a bootstrap estimate of the standard error of $\hat{m}_j(x)$ followed by a pivotal test statistic (Wilcox, 2013).

For fixed j, generate a bootstrap sample from the random sample

$$(x_{1j}, y_{1j}), \ldots, (x_{n_j j}, y_{n_j j}),$$

yielding

$$(x^*_{1j}, y^*_{1j}), \ldots, (x^*_{n_j j}, y^*_{n_j j}).$$

Estimate the slope and intercept based on this bootstrap sample and some appropriate regression estimator, yielding b^*_{1j} and b^*_{0j}, respectively. For x specified, let $\hat{m}^*_j(x) = b^*_{0j} + b^*_{1j}x$. Repeat this process B times, yielding $\hat{m}^*_{jb}(x)$ ($b = 1, \ldots, B$). Then, from basic principles (e.g., Efron and Tibshirani, 1997), an estimate of the squared standard error of $\hat{m}_j(x) = b_{0j} + b_{1j}x$ is

$$\hat{\tau}^2_j = \frac{1}{B-1} \sum (\hat{m}^*_{jb}(x) - \bar{m}^*_j(x))^2, \tag{12.3}$$

where $\bar{m}_j^*(x) = \sum \hat{m}_{jb}^*(x)/B$. In terms of controlling the probability of a Type I error, $B = 100$ appears to suffice. Letting z be the $1 - \alpha/2$ quantile of a standard normal distribution, an approximate $1 - \alpha$ confidence interval for $m_1(x) - m_2(x)$ is

$$\hat{m}_1(x) - \hat{m}_2(x) \pm z\sqrt{\hat{\tau}_1^2 + \hat{\tau}_2^2}. \tag{12.4}$$

(This is a robust generalization of a technique derived by Johnson and Neyman, 1936.)

There remains the issue of choosing the covariate values for which Eq. (12.2) will be tested and then controlling the probability of one or more Type I errors. Of course, there might be interest in specific values based on substantive reasons. Here, two strategies for choosing the covariate values are considered, which are called methods S1 and S2.

12.1.1 Method S1

The first strategy is to pick $K = 5$ covariate values and then control the probability of one or more Type I errors using the K-variate Studentized maximum modulus distribution with infinite degrees of freedom. One strategy would be to pick the smallest and largest covariate values that were observed as well as three points evenly spaced between these two extremes. (Another strategy is to pick the covariate values as described in Section 12.2.1.) Simulation results indicate that fairly good control over the Type I error probability is obtained with sample sizes of 30 when using the Theil–Sen estimator, the MM-estimator, the rank-based estimator in Section 10.13.12 that is based on Jaeckel's dispersion function, and somewhat surprisingly, even the least squares estimator (Ma, 2015). Ma found that the least trimmed squares estimator in Section 10.4 does not perform well. The actual Type I error probability was estimated to be considerably less than the nominal level and it had relatively low power. Another strategy is to control FWE (the probability of one or more Type I errors) using Hochberg's method, which was described in Section 7.4.7. With $n_1 = n_2 = 20$, the first approach seems to have a slight advantage over Hochberg's method (Wilcox, 2013). For $n_1 = n_2 = 30$, Hochberg's method performs reasonably well, even when using the least squares estimator (Ma, 2015). In general, all indications are that there is little difference between using the Studentized maximum distribution versus Hochberg's method.

12.1.2 Method S2

A concern about using only $K = 5$ covariate values is that this might not provide sufficient detail about where and how the regression lines differ. In particular, some other choice for the covariate values might have an important impact on power. One could of course increase K, but when using the Studentized maximum distribution or Hochberg's method for controlling the FWE rate, all indications are that an alternative approach, called method S2, generally offers a distinct power advantage (Wilcox, 2017a). By default, method S2 uses $K = 25$ covariate values evenly spaced between $\min(x_{ij})$ and $\max(x_{ij})$, the minimum and maximum values among the observed covariate values.

Imagine that the goal is to test Eq. (12.2) for each of the covariate values $x_1, \ldots, x_K$. Unlike method S1, FWE is controlled in the following manner. Let $\hat{p}_k$ be the p-value when testing Eq. (12.2) based on x_k $(k = 1, \ldots, K)$ and the confidence interval given by Eq. (12.4). Let

$$p_{\min} = \min\{\hat{p}_1, \ldots, \hat{p}_K\}. \tag{12.5}$$

Let p_c be the α quantile associated with the distribution of $p_{\min}$. When all of the K hypotheses are true, the probability of one or more Type I errors is α if Eq. (12.2) is rejected for any covariate value x_k for which $p_k \le p_c$.

The strategy used to determine p_c is to proceed in a manner similar in spirit to Student's t test: determine p_c when sampling from normal distributions where there is no association and there is homoscedasticity. This is done via simulations. An interesting feature of this approach is that as K increases, p_c does not appear to converge to zero. Estimates of p_c, based on a simulation with 4,000 replications, are reported in Table 12.1 when using the Theil–Sen estimator and testing at the $\alpha = 0.05$ level. Very similar results are obtained with $K = 10$ or when using the quantile regression estimator in Section 10.13.8. These results suggest that for $\min(n_1, n_2) \ge 30$, $\alpha = 0.05$, and $K \ge 10$, the estimate of p_c is approximately 0.015 when using the Theil–Sen estimator, but there is no proof that this is the case. It is stressed, however, that different results are obtained when using ordinary least squares (OLS). The estimates are smaller.

There is, of course, the issue of how well the method performs when there is heteroscedasticity and non-normality. Table 12.2 reports some estimates of the actual Type I error probability when testing at the 0.05 level based on the Theil–Sen estimator and the quantile regression estimator. Three choices for λ were used in Eq. (12.1): $\lambda(x) = 1$, $\lambda(x) = |x| + 1$, and $\lambda(x) = 1/(|x| + 1)$. For convenience, these three choices are denoted by variance patterns (VPs) 1, 2, and 3. As is evident, VP 1 corresponds to the usual homoscedasticity assumption. Data were generated where both x and ϵ have one of four g-and-h distributions: standard normal ($g = h = 0$), a symmetric heavy-tailed distribution ($h = 0.2$, $g = 0.0$), an asymmetric distribution with relatively light tails ($h = 0$, $g = 0.2$), and an asymmetric distribution with heavy tails ($g = h = 0.2$). As can be seen, control over the Type I error probability is generally good. The main difficulty is that with a heavy-tailed distribution, the estimates drop below 0.025 for VP 3. The lowest estimates are 0.020 and 0.014 using the Theil–Sen and quantile estimator, respectively.

Table 12.1: Estimates of p_c when testing at the $\alpha = 0.05$ level, $n_1 = n_2 = n$.

n	K	p_c
20	25	0.018
20	100	0.019
30	25	0.015
30	100	0.015
50	25	0.014
50	100	0.015
100	25	0.014
100	100	0.015
200	25	0.016
200	100	0.015
300	25	0.015
300	100	0.014
500	25	0.013
500	100	0.016

Table 12.2: Estimated probability of one or more Type I errors, $\alpha = 0.05$, $n = 30$, $K = 25$ covariate values.

g	h	VP	TS	QREG
0.0	0.0	1	0.050	0.050
0.0	0.0	2	0.041	0.053
0.0	0.0	3	0.032	0.032
0.0	0.2	1	0.032	0.035
0.0	0.2	2	0.046	0.043
0.0	0.2	3	0.023	0.017
0.2	0.0	1	0.041	0.043
0.2	0.0	2	0.056	0.056
0.2	0.0	3	0.032	0.033
0.2	0.2	1	0.032	0.032
0.2	0.2	2	0.041	0.041
0.2	0.2	3	0.020	0.014

QREG, Quantile regression; *TS*, Theil–Sen.

To provide at least some indication of how the power of S1 compares to S2, consider the situation where both the covariate and the error term have one of the four g-and-h distributions used in Table 12.2. For the first group, the slope is $\beta_{11} = 0$ and for the second group the slope is $\beta_{12} = 0.8$. Both intercepts are taken to be zero. Estimated power when using S1 with

Table 12.3: Estimated power when using methods S1 and S2, $\alpha = 0.05$, $n_1 = n_2 = 50$.

g	h	S1 ($K = 5$)	S1 ($K = 25$)	S2
0.0	0.0	0.490	0.673	0.825
0.0	0.2	0.396	0.769	0.894
0.2	0.0	0.455	0.679	0.852
0.2	0.2	0.349	0.742	0.894

$K = 5$, S1 with $K = 25$, and S2 are reported in Table 12.3. If instead both slopes are 0 and the regression lines differ only in terms of the intercepts, again method S2 has a distinct power advantage.

12.1.3 Linear Contrasts for a One-Way or Higher Design

The basic idea behind method S2 is readily extended to testing hypotheses about linear contrasts when dealing with $J \geq 2$ independent groups. That is, for the kth value of the covariate, the goal is to test

$$H_0: \Psi_k = 0 \tag{12.6}$$

($k = 1, \ldots, K$), where $\Psi_k = \sum c_j m_j(x_k)$ and $c_1, \ldots, c_J$ are linear contrast coefficients such that $\sum c_j = 0$. Here it is convenient to choose the covariate values in the following manner. Let $L_j = \min\{x_{1j}, \ldots, x_{n_j}\}$ and $U_j = \max\{x_{1j}, \ldots, x_{n_j}\}$, $j = 1, \ldots, J$. Suppose, for example, $(L_1, L_2, L_3) = (2, 4, 6)$. This indicates that for the third group, there is no information about the nature of the association when the covariate has a value less than 6. That is, for the third group, predicting the value of the dependent variable, when the covariate variable is equal to 4, or 2, is open to the criticism that it might be highly inaccurate due to extrapolation. A similar concern applies to the U_j values. To avoid this criticism, let $L = \max\{L_1, \ldots, L_J\}$ and $U = \min\{U_1, \ldots, U_J\}$. Then K points are used. The first and last are L and U, the idea being to use only points within the range of the covariate values of all J covariate variables. That is, the goal is to avoid extrapolation when predicting the typical value of the dependent variable based on a covariate value outside the range of values that were observed. The other values are evenly spaced between L and U. Compared to method S2, this has the effect of increasing the critical p-value needed to control FWE. Simulation estimates of a critical p-value are done in a manner similar to method S2. Now the test statistic is

$$Z_k = \frac{\hat{\Psi}_k}{\sqrt{\tilde{\tau}_k^2}},$$

where $\hat{\Psi}_k = \sum c_j \hat{m}_j(x_k)$, $\hat{\tau}_k^2 = \sum \hat{\tau}_j^2$, and again $\hat{\tau}_j^2$ is a bootstrap estimate of the squared standard error of $\hat{m}_j(x_k)$. The resulting p-value, p_k, is based on assuming that Z has a standard normal distribution. As done with method S2, a simulation is performed to determine p_c so that when all K hypotheses are true and the kth hypothesis is rejected when $p_k \leq p_c$, the probability of one or more Type I errors among the K hypotheses will be approximately equal to some specified value. But in practice, the execution time needed to estimate p_c can be extremely high. To deal with this, an approximation of p_c is used that performs reasonably based on extant simulations.

The approximation of the null distribution of $p_{\min}$, given by Eq. (12.5), was derived as follows. Based on simulations, the main impact on the value of p_c is the sample size with the value of p_c decreasing very slowly as the sample size increases. Consistent with the results in Table 12.1, the value of p_c does not appear to decrease further when $n \geq 100$. Consequently, the strategy was to estimate the distribution of $p_{\min}$ given by Eq. (12.5) when all K hypotheses are true, $J = 5$, the error term has a standard normal distribution, and there is homoscedasticity. This was done for $n = 50$ and 100. If the minimum sample size is less than or equal to 75, the estimated distribution of $p_{\min}$ based on $n = 50$ is used to estimate p_c. Otherwise the distribution of $p_{\min}$ based on $n = 100$ is used. Note that an adjusted p-value can be computed as well. The R function ancJN.LN in Section 12.1.5 does this using estimates of the quantiles of the null distribution of $p_{\min}$. The 0.001(0.001)0.1 and 0.01(0.01)0.99 quantiles are used. Given a p-value when testing Eq. (12.6), these quantiles can be used to determine an adjusted p-value aimed at controlling FWE. For simulation results regarding how well the method performs under non-normality and heteroscedasticity, see Wilcox (2019c). With sufficient computing power, perhaps this method can be improved, but this remains to be determined.

12.1.4 Dealing With Two or More Covariates

First focus on the situation where there are two covariates. Once covariate points are chosen, the strategy used by method S2 when dealing with one covariate can be extended to two covariates in an obvious way. However, choosing covariate points is not as straightforward and the number of covariate points impacts the critical p-value, p_c. A simple strategy is to use the deepest point as well as those points on the 0.5 depth contour associated with the covariate for group 1. However, a concern is that typically this results in using a fairly small number of covariate points, which in turn might result in an unsatisfactory sense of where and how the regression surfaces differ. Another approach is to use the deepest half of the covariate values associated with one of the groups or even both groups. Here, the deepest half is determined by projection distances, which are computed as described in connection with approximation A1 of halfspace depth in Section 6.2.3. Yet another strategy is to use all of the covariate points associated with one of the groups, which results in a decrease in the critical p-value, p_c. If the regression planes are parallel, this can result in a slight decrease in power compared to using the deepest half of the covariate points. But otherwise there are situations where using all of covariate points can have substantially higher power (Wilcox, 2018a).

Consider the situation where the quantile regression estimator is used, the deepest half of the covariate values are used, and the desired FWE is 0.05. An R function for computing p_c is described in the next section. But execution time might be an issue. To make the method more practical, it is noted that using $\hat{p}_c = 0.012$, when testing at the 0.05 level, controls the probability of a Type I error reasonably well, at least for sample sizes as large as 500. For larger sample sizes it might be better to use an estimate of p_c obtained via the R function ancJNmpcp, described in the next section.

When the number of covariates is between three and six, the sample sizes are at least 100, and at most 25 tests are performed. Results in Wilcox (2018d) indicate that good control over the Type I error probability can be achieved by using critical values based on the Studentized maximum modulus distribution with infinite degrees of freedom. With sample sizes less than or equal to 50, the actual level can be well below the nominal level. For more than 25 tests, adjusted critical values can be used when the goal is to have an FWE rate of 0.05. Letting p denote the number of covariates, for $p = 2, 3, 4, 5,$ and 6, the critical p-values are 0.00615847, 0.002856423, 0.00196, 0.001960793, and 0.001120947, respectively.

A concern about the methods in this section, particularly when dealing with more than one covariate, is the assumption that a linear model provides an adequate fit. Non-parametric methods can make a substantial difference, as illustrated by the next to last example in Section 12.3.4. If the linear model is correct, the methods described in this section can have substantially more power, but when the linear model is incorrect, the methods in Section 12.3.4 can have more power.

12.1.5 R Functions ancJN, ancJNmp, ancJNmpcp, anclin, reg2plot, reg2g.p2plot, and ancJN.LC

The R function

ancJN(x1,y1,x2,y2, pts=NULL, Dpts=FALSE,regfun=tsreg, fr1=1, fr2=1, SCAT = TRUE, pch1 = '+', pch2 = 'o', alpha=0.05, plotit=TRUE, xout=FALSE, outfun=out, nboot=100, SEED=TRUE, xlab='X', ylab='Y', ...)

performs the robust generalization of the Johnson–Neyman method based on method S1 described in Section 12.1.2. By default, the Theil–Sen estimator is used but other estimators can be used via the argument regfun. Least squares regression can be used by setting the argument regfun=ols. The covariate values, for which the hypothesis given by Eq. (12.2) is to be tested, can be specified by the argument pts. By default, the function picks five covariate values evenly spaced between the smallest and largest covariate values stored in the arguments x1 and x2. Setting Dpts=TRUE, covariate points are chosen as described in Section 12.2.1. (Earlier versions of this function used the method in Section 12.2.1 by default.) The probability of one or more Type I errors is controlled by using a critical value based on a Studentized maximum modulus distribution with infinite degrees of freedom. Adjusted p-values, based on Hochberg's method and related techniques described in Section 7.4.7, can be computed via the R function p.adjust in Section 7.4.8. The arguments pch1 and pch2 determine the symbols used when creating a scatterplot. Setting SCAT=FALSE, no scatterplot is created, only the regression lines are plotted.

The R function

> anclin(x1,y1,x2,y2,regfun=tsreg, pts=NULL, ALL=FALSE, npts=25, plotit=TRUE, SCAT=TRUE, pch1='*', pch2='+', nboot=100, ADJ=TRUE, xout=FALSE, outfun=out, SEED=TRUE, p.crit=0.015, alpha=0.05, crit=NULL, null.value=0, plotPV=FALSE, scale=TRUE, span=0.75, xlab='X', ylab='p-values',ylab2='Y', MC=FALSE, nreps=1000, pch='*', ...)

applies method S2, described in Section 12.1.2, when there is one covariate only. The argument ALL=FALSE means that the covariate values are chosen to be values evenly spaced between the minimum value and maximum values observed. The number of covariate values is controlled by the argument npts. If ALL=TRUE, the hypothesis given by Eq. (12.2) is tested for each of the unique values among all of the covariate values. Again, FWE is controlled using the method described in Section 12.1.2. If the desired probability of one or more Type I errors differs from 0.05, the function computes an estimate of p_c with the number of replications used in the simulation controlled by the argument nreps. If a multicore processor is available, setting MC=TRUE can reduce execution time considerably. By default, the regression lines are plotted with the labels for the x-axis and y-axis controlled by the arguments xlab and ylab2, respectively. If the argument plotPV=TRUE and plot=FALSE, the p-values are plotted and now the argument ylab controls the label for the y-axis. The arguments pch1 and pch2 control the symbol used when creating a scatterplot for group 1 and group 2, respectively. If the argument SCAT=FALSE, no scatterplot is created.

The R function

> ancJNPVAL(x1, y1, x2, y2, regfun = MMreg, p.crit = NULL, DEEP = TRUE, plotit = TRUE, xlab = 'X', ylab = 'X2', null.value = 0, WARNS = FALSE, alpha = 0.05, pts = NULL, SEED = TRUE, nboot = 100, xout = FALSE, outfun = outpro, ...)

is like the R functions ancJN and anclin, only it is designed for two or more covariates. (It replaces the function ancJNmp.) By default, the covariate points are taken to be the deepest half of the combined points in x1 and x2. Setting DEEP=FALSE, the deepest covariate point and the covariate points that lie on the 0.5 depth contour (as described in Section 12.1.3) are used. The covariate points can be specified via the argument pts.

Some of the functions previously described in this section have an option for plotting the regression lines being compared. In case it helps, the R function

reg2plot(x1,y1,x2,y2, regfun=tsreg, xlab='X', ylab='Y', xout=FALSE,outfun=out, STAND=TRUE, ...)

can be used to plot two regression lines based on the regression estimator indicated by the argument regfun. The R function

reg2g.p2plot(x1,y1,x2,y2, xout=FALSE, outfun=out, xlab='Var 1', ylab='Var 2', zlab='Var 3', regfun=tsreg, COLOR=TRUE, STAND=TRUE, tick.marks=TRUE, type='p', pr=TRUE, ticktype='simple', ...)

can be used to plot the regression planes when there are two independent variables. The R function

reg2difplot(x1, y1, x2, y2, regfun = tsreg, pts = x1, xlab = 'VAR 1', ylab = 'VAR 2', zlab = 'Group 2 minus Group 1', xout = FALSE, outfun = out, ALL = TRUE, pts.out = FALSE, SCAT = FALSE, theta = 50, phi = 25, ...)

plots the regression plane where $\hat{y}_1 - \hat{y}_2$ is the dependent variable, assuming that there are two independent variables. By default, the data for the independent variables, which are used when creating the plot, are the vectors stored in both x1 and x2. If it is desired to use only the covariate points in say x1, set the argument ALL=FALSE. (This function uses the R function rplot, which returns a measure of association equal to NULL. The strength of the association, based on rplot, is meaningless for the situation at hand.)

The R function

ancJN.LC(x, y, pts = NULL, con = NULL, regfun = tsreg, fr = rep(1, 4), nmin = 12, npts = 5, alpha = 0.05, xout = FALSE, outfun = out, nboot = 100, SEED = TRUE, pr = TRUE, ...)

tests hypotheses about linear contrasts using the method in Section 12.1.3. The arguments x and y are assumed to be matrices with *J* columns, or they can have list mode.

12.2 Methods When There Is Curvature and a Single Covariate

This section describes several methods for making inferences about how $m_1(x)$ compares to $m_2(x)$, where for the jth group, again $m_j(x)$ is some conditional measure of location associated with y, given x. But unlike the methods in Section 12.1, no parametric assumption is made about the shape of the regression line. In particular, when there is a single covariate, it is not assumed that the regression lines are straight. Furthermore, complete heteroscedasticity is allowed, meaning that the error term for each group can be heteroscedastic, and nothing is assumed about how the variance of the error term in the first group is related to the variance of the error term associated with the second group. The first general goal is to test

$$H_0: m_1(x) = m_2(x), \text{ for each } x \in \{x_1, \ldots, x_K\}, \tag{12.7}$$

where $x_1, \ldots, x_K$ are K specified values for the covariate. Methods for testing the global hypothesis

$$H_0: m_1(x) = m_2(x), \forall x \in \{x_1, \ldots, x_K\}, \tag{12.8}$$

are described as well. The general strategy is to approximate the regression lines with a running interval smoother and then use the components of the smoother to make comparisons at appropriate design points. There are many variations of these methods that might prove to be useful, as will become evident. In principle, $m_j(x)$ can be any measure of location. The primary focus is on the 20% trimmed mean, but bootstrap versions of the basic strategy can be used with other measures of location.

12.2.1 Method Y

This section describes the most basic version, called method Y, which is based in part on the running interval smoother in Section 11.5.4. Subsequent sections describe some alternative methods that might increase power or provide a more detailed understanding of where and how the regression lines differ.

First consider the situation where a single covariate value x has been chosen with the goal of computing a confidence interval for $m_1(x) - m_2(x)$. For the jth group, let x_{ij}, $i = 1, \ldots, n_j$, be values of the covariate variables that are available. For the moment, assume that $m_j(x)$ is estimated with the trimmed mean of the y_{ij} values such that i is an element of the set

$$N_j(x) = \{i : |x_{ij} - x| \leq f_j \times \text{MADN}_j\}.$$

That is, for fixed j, estimate $m_j(x)$ using the y_{ij} values corresponding to the x_{ij} values that are close to x. As noted in Chapter 11, the choice $f_j = 0.8$ or $f_j = 1$ generally gives good results, but some other value might be desirable. Let $M_j(x)$ be the cardinality of the set $N_j(x)$.

That is, $M_j(x)$ is the number of points in the jth group that are close to x, which in turn is the number of y_{ij} values used to estimate $m_j(x)$. When $m_j(x)$ is the 20% trimmed mean of y, given x, the two regression lines are defined to be *comparable* at x if $M_1(x) \geq 12$ and $M_2(x) \geq 12$. The idea is that if the sample sizes used to estimate $m_1(x)$ and $m_2(x)$ are sufficiently large, then a reasonably accurate confidence interval for $m_1(x) - m_2(x)$ can be computed using the methods in Chapter 5. Yuen's method often gives satisfactory results. As is evident, a bootstrap-t or percentile bootstrap could be used as well, including situations where some robust estimator other than a 20% trimmed mean is of interest.

When comparing the regression lines at more than one design point, confidence intervals for $m_1(x) - m_2(x)$, having simultaneous probability coverage approximately equal to $1 - \alpha$, can be computed as described in Chapter 7. When this is done for the situation at hand, the value for the span, f, that is used is related to how close the actual simultaneous probability coverage is to the nominal level.

Suppose it is desired to compare the regression lines at five x values: z_1, z_2, z_3, z_4, and z_5. Of course, in practice, an investigator might have some substantive reason for picking certain design points, but this process is difficult to study via simulations. For illustrative purposes, suppose the design points are chosen using the following process. First, for notational convenience, assume that for fixed j, the x_{ij} values are in ascending order. That is, $x_{1j} \leq \cdots \leq x_{n_j j}$. Suppose z_1 is taken to be the smallest x_{i1} value for which the regression lines are comparable. That is, search the first group for the smallest x_{i1} such that $M_1(x_{i1}) \geq 12$. If $M_2(x_{i1}) \geq 12$, in which case the two regression lines are comparable at x_{i1}, set $z_1 = x_{i1}$. If $M_2(x_{i1}) < 12$, consider the next largest x_{i1} value and continue until it is simultaneously true that $M_1(x_{i1}) \geq 12$ and $M_2(x_{i1}) \geq 12$. Let i_1 be the value of i. That is, i_1 is the smallest integer such that $M_1(x_{i_1 1}) \geq 12$ and $M_2(x_{i_1 1}) \geq 12$. Similarly, let z_5 be the largest x value in the first group for which the regression lines are comparable. That is, z_5 is the largest x_{i1} value such that $M_1(x_{i1}) \geq 12$ and $M_2(x_{i1}) \geq 12$. Let i_5 be the corresponding value of i. Let $i_3 = (i_1 + i_5)/2$, $i_2 = (i_1 + i_3)/2$, and $i_4 = (i_3 + i_5)/2$. Round i_2, i_3, and i_4 down to the nearest integer and set $z_2 = x_{i_2 1}$, $z_3 = x_{i_3 1}$, and $z_4 = x_{i_4 1}$. Finally, consider computing confidence intervals for $m_1(z_q) - m_2(z_q)$, $q = 1, \ldots, 5$, by applying the methods for trimmed means described in Chapter 5. One possibility is to perform Yuen's test using the y values for which the corresponding x values are close to the design point z_q and control the probability of at least one Type I error among the five tests by using the critical value given by the five-variate Studentized maximum modulus distribution. Another possibility is to replace Yuen's method with the percentile bootstrap method and control the probability of at least one Type I error via Hochberg's method or Hommel's method. An advantage of this last approach is that robust measures of location other than a trimmed mean can be used. A bootstrap-t method can be used as well.

Table 12.4:   Estimated Type I error probabilities, $\alpha = 0.05$.

X		ϵ		$Y = X$ $n = 30$ $f = 1$	$Y = X^2$ $n = 40$ $f = 1$	$Y = X^2$ $n = 30$ $f = 0.75$	$Y = X^2$ $n = 40$ $f = 0.75$	$Y = X^2$ $n = 40$ $f = 0.5$
g	h	g	h					
0	0	0	0	0.046	0.049	0.045	0.045	0.039
0	0	0	0.5	0.024	0.034	0.030	0.027	0.023
0	0	0.5	0	0.034	0.055	0.045	0.043	0.037
0	0	0.5	0.5	0.022	0.038	0.032	0.031	0.024
0	0.5	0	0	0.042	0.065	0.045	0.059	0.041
0	0.5	0	0.5	0.027	0.049	0.030	0.037	0.031
0	0.5	0.5	0	0.041	0.073	0.043	0.052	0.031
0	0.5	0.5	0.5	0.027	0.041	0.035	0.026	0.021
0.5	0	0	0	0.042	0.071	0.029	0.059	0.039
0.5	0	0	0.5	0.027	0.046	0.021	0.037	0.031
0.5	0	0.5	0	0.041	0.066	0.026	0.052	0.035
0.5	0	0.5	0.5	0.026	0.043	0.018	0.026	0.021
0.5	0.5	0	0	0.044	0.082	0.039	0.060	0.044
0.5	0.5	0	0.5	0.033	0.056	0.024	0.036	0.033
0.5	0.5	0.5	0	0.045	0.078	0.041	0.057	0.035
0.5	0.5	0.5	0.5	0.032	0.053	0.028	0.031	0.023

Table 12.4 shows some simulation results when x and ϵ are generated from various g-and-h distributions and Yuen's method is used. Column 5 shows a simulation estimate of the actual probability of at least one Type I error, $\hat{\alpha}$, when $\alpha = 0.05$, $y = x + \epsilon$, $n = 30$, and $f = 1$. The control over the probability of a Type I error is reasonably good; the main problem being that the actual probability of a Type I error can drop slightly below 0.025 when ϵ has a heavy-tailed distribution. Simulations are not reported for $n = 20$ because situations arise where five design points cannot always be found for which the regression lines are comparable.

Column 6 shows the results when $y = x^2 + \epsilon$, $n = 40$, and $f = 1$. When x is highly skewed and has a very heavy-tailed distribution ($g = h = 0.5$), $\hat{\alpha}$ can exceed 0.075, the highest estimate being equal to 0.082. With $n = 30$, not shown in Table 12.3, the estimate goes as high as 0.089. There is the additional problem that $f = 1$ might not be sufficiently small, as previously illustrated. If $f = 0.75$ is used, $\hat{\alpha}$ never exceeds 0.05, but in a few cases it drops below 0.025, the lowest estimate being equal to 0.018. Increasing n to 40, the lowest estimate is 0.026. Results in Chapter 5 suggest that even better probability coverage can be obtained using a bootstrap method, but the extent to which the probability coverage is improved for the problem at hand has not been determined. (For more details about the simulations, see Wilcox, 1997b.)

Another positive feature of the method described here is that its power compares well with the conventional approach to ANCOVA when the standard assumptions of normality, homogeneity of variance, and parallel regression lines are true. For example, if both groups have sample sizes of 40, $y_{i1} = x_{i1} + \epsilon_{i1}$, but $y_{i2} = x_{i2} + \epsilon_{i2} + 1$, the conventional approach to ANCOVA has power approximately equal to 0.867 when testing at the 0.05 level. The method described here has power 0.828, where power is the probability that the null hypothesis is rejected for at least one of the empirically chosen design points, $z_1, \ldots, z_5$. If the error terms have heavy-tailed distributions, the traditional approach has relatively low power, as is evident from results described in previous chapters. Even under normality and homoscedasticity, the classic ANCOVA method, described in the introduction to this chapter, can have relatively low power. For example, if $y_{i1} = x_{i1} + \epsilon_{i1}$, but $y_{i2} = 0.5x_{i2} + \epsilon_{i2}$, standard ANCOVA has power 0.039 versus 0.225 for the method based on trimmed means, again with sample sizes of 40. Even when the classic ANCOVA method has relatively good power, an advantage of using a trimmed mean with a running interval smoother is that the goal is to determine where the regression lines differ and by how much, and this is done without assuming that the regression lines are straight. A negative feature is that if indeed the regression line is straight, method S1 in Section 12.1.1 can have more power (e.g., Ma, 2015). There are, however, several ways of improving the power of method Y. How well these improvements compare to methods that assume a straight line, when indeed the regression line is approximately straight, has not been determined.

12.2.2 Method BB: Bootstrap Bagging

An early attempt at improving the power of method Y combines the running-interval smoother with bootstrap bagging. In effect, a nested bootstrap method is used where for each bootstrap sample, bootstrap bagging is applied. Let $D = \hat{m}_1^*(x) - \hat{m}_2^*(x)$, where $\hat{m}_1^*(x)$ and $\hat{m}_2^*(x)$ are estimates of $m_1(x)$ and $m_2(x)$, respectively, based on bootstrap bagging. Repeat this process B times, yielding $D_1, \ldots, D_B$. Then a (generalized) p-value when testing H_0: $m_1(x) = m_2(x)$ is

$$P = \frac{1}{B} \sum_{b=1}^{B} (I_{D_b < 0} + 0.5 I_{D_b = 0}),$$

(12.9)

where the indicator function $I_{D_b < 0} = 1$ if $D_b < 0$; otherwise $I_{D_b < 0} = 0$. This will be called method BB henceforth.

12.2.3 Method UB

Currently, it appears that there are two alternative methods that perform better than method BB in terms of power, at the expense of relatively high execution time. (Execution time can be reduce substantially when a multicore processor is available.) The first, called method UB, is based on what might be termed an unconditional bootstrap method, which is based on a simple modification of a method studied by Wilcox (2014). The second is method TAP, described in the next section.

Rather than take bootstrap samples based on the y_{ij} values such that i is an element of the set

$$N_j(x) = \{i : |x_{ij} - x| \leq f_j \times \text{MADN}_j\},$$

as done by method Y, method UB uses bootstrap samples based on all of the data. That is, begin by generating a bootstrap sample by resampling with replacement n_1 rows of data from (x_{i1}, y_{i1}) and n_2 rows of data from (x_{i2}, y_{i2}), yielding (x_{ij}^*, x_{ij}^*), $i = 1, \ldots, n_j$, $j = 1, 2$. Let $m_j^*(x)$ be the estimate of $m_j(x)$ based on this bootstrap sample and let $D^* = m_1^*(x) - m_2^*(x)$. Repeat this process B times, yielding $D_1^*, \ldots, D_B^*$. Let

$$\tilde{p} = \frac{1}{B} \sum I(D_b^* < 0) + 0.5I(D_b^* = 0),$$

where the indicator function $I(D_b^* < 0) = 1$ if $D_b^* < 0$; otherwise $I(D_b^* < 0) = 0$. Then a (generalized) p-value is $\hat{p} = 2\min(\tilde{p}, 1 - \tilde{p})$.

Method UB chooses covariate values in one of two ways. The first is to proceed in the same manner as method Y, in which case $K = 5$ covariate values are chosen. The second approach is to choose the covariate values based on estimates of the quantiles associated with the first group. The default approach is to use the median as well as the lower and upper quartiles, but this can be altered when using the R function ancovaUB in Section 12.2.6. Let $\hat{p}_k$ be the p-value when testing Eq. (12.7) and $x = z_k$ ($k = 1, \ldots, K$), let

$$p_{\min} = \min\{\hat{p}_1, \ldots, \hat{p}_K\},$$

and let p_c be the α quantile associated with the distribution of $p_{\min}$. Then a test of H_0: $m_1(z_k) = m_2(z_k)$, with the goal that the probability of one or more Type I errors is less than or equal to α, would be to reject for any k such that $p_k \leq p_c$ (cf. Martínez-Camblor, 2014). The strategy used to determine p_c is to proceed in a manner similar in spirit to the derivation of Student's t test: Determine p_c when sampling from normal distributions. This is done via simulations for the situation where there is no association. More precisely, generate data x_{ij} and y_{ij} ($i = 1, \ldots, n_j$, $j = 1, 2$) from a standard normal distribution, compute $p_{\min}$ as previously described, and label the result $\tilde{p}$. Repeat this process G times, yielding $\tilde{p}_1, \ldots, \tilde{p}_G$. Here $G = 1000$ is used by default and p_c is taken to be the α quantile of the distribution of $p_{\min}$, which is estimated based on $\tilde{p}_1, \ldots, \tilde{p}_G$ and the Harrell–Davis estimator.

12.2.4 Method TAP

By default, both methods Y and UB are based on five covariate values that are chosen as described in Section 12.2.1. A possible concern is that a relatively small number of covariate values is used, which might result in missing important details about where and how the regression lines differ. A simple way of dealing with this concern is to use a larger number of covariate values, but this might result in relatively low power when using the K-variate Studentized maximum modulus distribution as done by method Y. When $K = 5$ covariate values are used, in which case five hypotheses are tested, FWE (the probability of one or more Type I errors) tends to be reasonably close to the nominal level. But as the number of covariate values increases, this is no longer the case: The actual level can be substantially smaller than the nominal level, which raises the concern that power will be relatively low as well. In principle, method UB can be used when $K > 5$ covariate values are used, but here a slight variation of method UB is used instead, called method TAP, which has been found to perform well in simulations even when dealing with two covariates.

Method TAP (Wilcox, 2017a) uses an adjusted p-value in a manner similar to method UB. The default number of covariate values used by method TAP is $K = 25$, which are taken to be the points evenly spaced between z_1 and z_5 as defined in connection with method Y. Interestingly, the adjusted p-value appears to be virtually the same when for example $K = 100$ is used. More broadly, as K gets large, the adjusted p-value used here does not appear to go to zero, but rather a value that can have a practical advantage in terms of power due to the more detailed comparison of the groups. However, there is no proof that the adjusted p-value does not converge to zero. Like method UB, a critical p-value, p_c, is determined via simulations assuming normality and that there is no association. Two main differences are that TAP uses Yuen's test for each of the K tests to be performed, rather than a bootstrap method as done by method UB, and a larger number of covariate values is used. Of course, an issue is whether good control over the probability of a Type I error can be achieved under non-normality as well as curvature and all indications are that it performs very well. Another issue is how TAP compares to other methods in terms of power. Simulations indicate that at least in some situations it has a distinct advantage compared to method Y.

Recall that z_1 and z_5 are the smallest and largest covariate values used by method Y, which are computed as described in Section 12.2.1. Method TAP uses K covariate values evenly spaced between z_1 and z_5. For notational convenience, relabel z_1 and z_5 as d_1 and d_K, in which case the goal is to test Eq. (12.7) for each of the K covariate values $d_1 \leq \cdots \leq d_K$. As previously noted, $K = 25$ is assumed but larger values for K can be accommodated. Proceeding in a manner similar to method UB, let $p_1, \ldots, p_K$ be the K p-values based on method Y, let

$$p_{\min} = \min\{\hat{p}_1, \ldots, \hat{p}_K\},$$

and let p_c be the α quantile associated with the distribution of p_{min}. An estimate of p_c is obtained in exactly the same manner as method UB, except that the p-values $p_1, \ldots, p_K$ are based on Yuen's method rather than a percentile bootstrap method. Then a test of H_0: $m_1(d_k) = m_2(d_k)$ for each $k = 1, \ldots, K$, with the goal that the probability of one or more Type I errors is less than or equal to α, can be performed: Reject for any k such that $p_k \leq p_c$. Measures of effect size can be computed via the R function ancM.COV.ES in Section 12.3.4.

To provide at least some perspective, imagine that method Y is applied with $K = 25$ covariate values. Even with infinite degrees of freedom, in effect each test is performed at the 0.002 level. In contrast, with sample sizes $n_1 = n_2 = 30$, method TAP performs each test at the $p_c = 0.008$ level. (The simulation estimate of p_c was based on 5,000 replications.) For $K = 100$, again using TAP, each test is performed at the $p_c = 0.007$ level. Here are some estimates of p_c when $n_1 = n_2 = n$:

n	p_c	n	p_c	n	p_c	n	p_c	n	p_c
30	0.00824	50	0.00581	60	0.00544	70	0.00550	80	0.00476
100	0.00417	150	0.00441	200	0.00441	300	0.00381	400	0.00365
500	0.00345	600	0.00363	700	0.00337	800	0.00335		

12.2.5 Method G

The method in this section is like method Y, only it tests the global hypothesis given by Eq. (12.8). By default, it is based on the same $K = 5$ covariate values used by method Y. The method can have more power than method Y, but based on limited results, it does not appear to have a striking advantage over methods BB, UB, and TAP. (For relevant simulation results, see Wilcox, 2016a.)

Let $\hat{\theta}_{jk}$ be some location estimator based on the y_{ij} values for which $i \in N_j(x_k)$. Let $\hat{\delta}_k = \hat{\theta}_{1k} - \hat{\theta}_{2k}$ and let δ_k denote the population analog of $\hat{\delta}_k$ ($k = 1, \ldots, K$). Then the global hypothesis given by Eq. (12.8) corresponds to

$$H_0: \delta_1 = \delta_2 = \cdots = \delta_K = 0. \tag{12.10}$$

The basic strategy for testing Eq. (12.10) is to generate bootstrap samples from each group, compute $\hat{\delta}_k$ based on these bootstrap samples, repeat this B times, and then measure how deeply the null vector $\mathbf{0}$ is nested in the bootstrap cloud of points via Mahalanobis distance.

To elaborate, let (x_{ij}^*, y_{ij}^*) be a bootstrap sample from the jth group, which is obtained by resampling with replacement n_j pairs of points from (x_{ij}, y_{ij}) ($i = 1, \ldots, n_j$, $j = 1, 2$). Let $\hat{\delta}_k^*$ be the estimate of δ_k based on the bootstrap samples from the two groups. Repeat this process B times, yielding $\hat{\Delta}_b^* = (\hat{\delta}_{1b}^*, \ldots, \hat{\delta}_{pb}^*)$, $b = 1, \ldots, B$. Let $\mathbf{S}$ be the covariance matrix based on the B vectors $\hat{\Delta}_1^*, \ldots, \hat{\Delta}_B^*$. Note that the center of the bootstrap cloud being estimated by these B bootstrap samples is known. It is $\hat{\Delta} = (\hat{\delta}_1, \ldots, \hat{\delta}_K)$, the estimate of $\Delta = (\delta_1, \ldots, \delta_p)$ based on the (x_{ij}, y_{ij}) values. Let

$$d_b^2 = (\hat{\Delta}_b^* - \hat{\Delta})\mathbf{S}^{-1}(\hat{\Delta}_b^* - \hat{\Delta})',$$

where for $b = 0$, $\hat{\Delta}_0^*$ is taken to be the null vector $\mathbf{0}$. Then a (generalized) p-value is

$$\frac{1}{B}\sum_{b=1}^{B} I(d_0^2 \leq d_b^2), \tag{12.11}$$

where the indicator function $I(d_0^2 \leq d_b^2) = 1$ if $d_0^2 \leq d_b^2$; otherwise $I(d_0^2 \leq d_b^2) = 0$.

The obvious decision rule is to reject the null hypothesis if the p-value is less than or equal to the nominal level. When testing at the $\alpha = 0.05$ level, simulations indicate that this approach performs well, in term of controlling the Type I error probability, when $K = 3$ and the x_k values are taken to be the quartiles corresponding to the x_{i1} values. But when $K = 5$ and the x_k values are chosen as described in connection with method Y, the actual level can exceed 0.075 when testing at the $\alpha = 0.05$ level with $n_1 = n_2 = 30$. This problem persists with $n_1 = n_2 = 50$. However, Wilcox found that the actual level is relatively stable among the situations considered in simulations. This suggests using a strategy similar to Gosset's (Student's) approach to comparing means: Assume normality, determine an appropriate critical value using a reasonable test statistic and continue using this critical value when dealing with non-normal distributions.

Given n_1 and n_2, this strategy is implemented first by generating, for each j, n_j pairs of observations from a bivariate normal distribution having a correlation $\rho = 0$. Based on these generated data, determine $K = 5$ values of the covariate as done in connection with method Y and then compute the p-value given by Eq. (12.11). Denote this p-value by $\hat{p}$. Repeat this process A times, yielding $\hat{p}_1, \ldots, \hat{p}_A$. Then an α level critical p-value, say $\hat{p}_c$, is taken to be the α quantile of the $\hat{p}_1, \ldots, \hat{p}_A$ values, which here is estimated via the Harrell–Davis estimator. That is, letting p_o denote the p-value based on Eq. (12.11), reject Eq. (12.10) if $p_o \leq \hat{p}_c$.

Note that once p_c has been determined, a $1 - \alpha$ confidence region for $\Delta = (\delta_1, \ldots, \delta_K)$ can be computed. A confidence region consists of the convex hull containing the $(1 - \hat{p}_c)B$ $\hat{\Delta}_b$ vectors that have the smallest d_b^2 values. This confidence region provides a perspective on why the global test considered here can have more power than method Y. Situations are encountered where the null vector is not contained in the confidence region, yet the confidence intervals for each of the K differences contain zero.

A possible appeal of method G is that it can have more power than method Y, sometimes substantially so, but it does not dominate in terms of maximizing power (Wilcox, 2016c).

Consider, for example, the situation where $n_1 = n_2 = 50$ and both x and the error term have standard normal distributions. If for the first group $\beta_1 = \beta_0 = 0$, while for the second group $y = 0.5 + \epsilon$, method G has power approximately equal to 0.51 when using a 20% trimmed mean, while for method Y the probability of detecting one or more true differences is 0.38. Using method BB instead, the probability of detecting one or more true differences is 0.46. However, at least in some situations, method UB can have more power than methods G and BB. For the situation at hand, but with $y = x + \epsilon$ for the second group, the power of methods G, BB, and UB is approximately 0.65, 0.69, and 0.84, respectively. If method UB is used in conjunction with adjusted p-values via Hochberg's method, rather than using the critical p-value $\hat{p}_c$, the power is 0.69. So based on very limited information, a speculation is that method UB might tend to have higher power than methods Y and BB. But the extent to which this is the case needs further research.

When the hypothesis given by Eq. (12.8) is rejected, a reasonable decision rule is to reject H_0: $m_1(x_k) = m_2(x_k)$ corresponding to the smallest p-value. However, this leaves open the issue of whether the null hypothesis can be rejected for other covariate values of interest. So an argument might be that methods BB, UB, and TAP are better for general use.

12.2.6 A Method Based on Grids

Section 11.5.16 illustrates that in terms of detecting and describing an association when there are two independent variables, the simple strategy of using quantile grids can have practical value. In particular, associations can be detected that other methods completely miss. Using quantile grids when dealing with ANCOVA might be useful as well. One possibility is to divide the data into subgroups based on the median of the first covariate in group 1. For example, if the median is 12, the data for both groups 1 and 2 could be divided into subgroups based on whether the first covariate is less than or greater than 12. These subgroups could be divided again based on whether the second covariate is less than or greater than its median. Then, for each region, the resulting groups could be compared with methods in Chapter 5. Smoothers, described in Chapter 11, might provide a better sense of where the data should be divided.

12.2.7 R Functions ancova, anc.ES.sum, anc.grid, anc.grid.bin, anc.grid.cat, ancovaWMW, ancpb, rplot2g, runmean2g, lplot2g, ancdifplot, ancboot, ancbbpb, qhdsm2g, ancovaUB, ancovaUB.pv, ancdet, ancmg1, and ancGLOB

Several R functions are supplied with the hope that one of them matches the needs of the reader when dealing with ANCOVA. The first is

ancova(x1, y1, x2, y2, fr1 = 1, fr2 = 1, tr = 0.2, alpha = 0.05, plotit = TRUE, pts = NA, sm =
FALSE, method = 'EP', SEED = TRUE, pr = TRUE, xout = FALSE, outfun = out, LP =
TRUE, SCAT = TRUE, xlab = 'X', ylab = 'Y', pch1 = '*', pch2 = '+', skip.crit = FALSE,
nmin = 12, crit.val = 1.09, ...),

which compares trimmed means using Method Y in Section 12.2.1. The data for group 1 are
stored in x1 and y1, and for group 2 they are stored in x2 and y2. The arguments fr1 and fr2
are the values of f (the span) for the first and the second group, respectively, that are used by
the running interval smoother, which default to 1 if unspecified. If the degree of curvature is
sufficiently high, setting the spans to 0.8 might improve the control over the Type I error probability, as illustrated in Table 12.4. The argument method indicates which measure of effect
size will be computed. See Section 5.3.5 for the choices that are available or use the R function anc.ES.sum, described below, to compute several measures of effect size simultaneously.
The argument tr is the amount of trimming, which defaults to 0.2. The default value for the argument alpha (α), the probability of at least one Type I error, is 0.05. If the argument pts=NA,
five design points are chosen, as described in connection with method Y in the previous section. The results are returned in the matrix ancova$output, as illustrated in the next example.
If values are stored in pts, the function compares groups at the values specified. So if pts contains the values 5, 8, and 12, the function will test H_0: $m_1(x) = m_2(x)$ at $x = 5$, 8, and 12, and
it controls the probability of a Type I error by determining a critical value based on the Studentized maximum modulus distribution (as described in Chapter 7). When plotit=TRUE, the
function creates a scatterplot and smooth for both groups by calling the function

runmean2g(x1, y1, x2, y2, fr = 0.8, est = tmean, xlab = 'X', ylab = 'Y', SCAT = TRUE, sm =
FALSE, nboot = 40, SEED = TRUE, eout = FALSE, xout = FALSE, outfun = out, LP =
TRUE, pch1 = '*', pch2 = '+', ...).

(The R function rplot2g can be used in place of runmean2g.) The smooth for the first group
is indicated by a solid line and a dashed line is used for the other. Setting the argument
sm=TRUE results in using a bagged version of the smooth, which can be useful when the
sample size is small. More precisely, a running interval smoother can be a bit ragged looking.
Setting LP=TRUE means that the initial smooth is smoothed again using LOESS. Currently,
this seems to be more satisfactory than using bootstrap bagging. If the argument xout=TRUE,
leverage points are removed when plotting the regression lines. The arguments pch1 and pch2
control the symbol used when creating a scatterplot. For example, pch1='o' indicates that
points associated with the data in x1 and y1 will be indicated by an o.

The R function ancova returns a measure of effect size, which can be chosen via the argument
method. The methods that are available are summarized in Section 5.3.5 in conjunction with

the R function ESfun. To get several measures of effect size simultaneously, including confidence intervals, the R function

anc.ES.sum(x1, y1, x2, y2, fr1 = 1, fr2 = 1, tr = 0.2, alpha = 0.05, pts = NA, SEED = TRUE,
nboot = 1000, pr = TRUE, xout = FALSE, outfun = out, nmin = 12, NULL.V = c(0, 0, 0.5,
0.5, 0.5, 0), REL.M = NULL, n.est = 1e+06, ...)

can be used. It computes measures of effect size via the R function ES.summary.CI, described in Section 5.7.3. This is done for each value of the covariate, values indicated by the argument pts, where the regression lines are compared. Again, if pts=NA, five design points are chosen as described in connection with method Y in the previous section.

Assuming there are two covariates, the R function

anc.grid(x1,y1,x2,y2, alpha=0.05, Qsplit1=0.5, Qsplit2=0.5, SV1=NULL,SV2=NULL, tr=0.2,
PB=FALSE,est=tmean,nboot=1000, xout=FALSE,outfun=outpro,SEED=TRUE,...)

divides the data into groups as described in the previous section. By default, the medians of the covariates associated with the first group are used to partition the data. The arguments Qsplit1 and Qsplit2 can be used to split the data based on some other quantile. Then for each region, the trimmed means of the resulting groups are compared using Yuen's method. Measures of effect size are returned based on the R function ES.summary. If PB=TRUE, a percentile bootstrap method is used instead. Rather than use quantiles to split the data into groups, values can be specified instead via the arguments SV1 and SV2. For example, SC1=0 would partition both groups according to whether the first covariate is less than or greater than zero. When the dependent variable is binary, the R function

anc.grid.bin(x1,y1,x2,y2, alpha=0.05, method='KMS', Qsplit1=0.5, Qsplit2=0.5,
SV1=NULL, SV2=NULL, xout=FALSE, outfun=outpro, SEED=TRUE)

can be used. Choices for the argument method are KMS, SK, and ECP. See Section 5.8.1 for details. When the dependent variable is discrete with a relatively small sample space, the R function

anc.grid.cat(x1,y1,x2,y2, alpha=0.05, KMS=FALSE, Qsplit1=0.5, Qsplit2=0.5, SV1=NULL,
SV2=NULL, pr=TRUE, xout=FALSE, outfun=outpro)

can be used. It uses the method in Section 5.8.2 for comparing two multinomial distributions.

The function

$$\text{ancovaWMW}(x1,y1,x2,y2, fr1=1, fr2=1, alpha=0.05, plotit=TRUE, pts=NA,xout=FALSE,}$$
$$\text{outfun=out,LP=TRUE, sm=FALSE, est=hd, ...)}$$

is like the function ancova, only it compares groups in terms of the likelihood that a randomly sampled observation from the first group is less than a random sampled observation from the second group. This is done via Cliff's method in Section 5.7.1, which represents an improvement on the Wilcoxon–Mann–Whitney test (cf. De Schryver and De Neve, 2019). The argument est indicates which measure of location will be used when plotting the regression lines.

The R function

$$\text{ancovaUB}(x1=NULL,y1=NULL,x2=NULL,y2=NULL, fr1=1, fr2=1, p.crit=NULL,}$$
$$\text{padj=FALSE, pr=TRUE, method='hochberg', FAST=TRUE, est=tmean,}$$
$$\text{alpha=0.05,plotit=TRUE, xlab='X', ylab='Y', pts=NULL, qpts=FALSE, qvals=c(0.25, 0.5,}$$
$$\text{0.75), sm=FALSE, xout=FALSE, eout=FALSE, outfun=out, LP=TRUE, nboot=500,}$$
$$\text{SEED=TRUE, nreps=2000, MC=FALSE, nmin=12, q=0.5, SCAT=TRUE,}$$
$$\text{pch1='*',pch2='+', ...)}$$

applies method UB. By default, FAST=TRUE, meaning that the critical p-value, $\hat{p}_c$, will be computed quickly if in addition to the argument alpha=0.05. Otherwise $\hat{p}_c$ must be estimated via a simulation, which can require a relatively high execution time. The argument nreps controls how many replications are used when computing $\hat{p}_c$. Setting the argument MC=TRUE can reduce execution time considerably if a multicore processor is available. The function ancovaUB determines the critical p-value, p_c, by calling the function

$$\text{ancovaUB.pv}(n1,n2, nreps=2000, MC=FALSE, qpts=FALSE, qvals = c(0.25, 0.5, 0.75),}$$
$$\text{nboot=500, SEED=TRUE, est=tmean, alpha=0.05).}$$

The value of $\hat{p}_c$ depends only on the sample sizes, so once it is known, it can be specified via the argument p.crit. If the argument qpts=TRUE, the covariate points are chosen to be the quantiles associated with the data in x1; the quantiles used are controlled by the argument qvals. If qpts=FALSE, the covariate values are chosen as done by method Y in Section 12.2.1.

If the argument padj=TRUE, the function reports adjusted p-values using the method indicated by the argument method. Hochberg's method is used by default. Hommel's method can be used by setting the argument method='hommel.' A possible appeal of using padj=TRUE is that execution time is very low when the goal is to test hypotheses at some level other than

0.05, but power might be reduced. If qpts=TRUE, covariate values are chosen based on the quantiles indicated by the argument qvals in conjunction with the data in the argument x1.

The R function

ancdet(x1,y1,x2,y2, fr1=1, fr2=1, tr=0.2, alpha=0.05, plotit=TRUE, plot.dif=FALSE, pts=NA,
     sm=FALSE, pr=TRUE, xout=FALSE, outfun=out, MC=FALSE, npts=25, p.crit=NULL,
     nreps=5000, SEED=TRUE, EST=FALSE, SCAT=TRUE, xlab='X', ylab='Y', pch1='*',
                                   pch2='+',...)

applies method TAP. The argument npts indicates how many covariate values will be used, which defaults to 25. With plotit=TRUE, the function plots a smooth for both regression lines. Setting plot.dif=TRUE, the function plots the difference between the regression lines, $\hat{m}_1(x) - \hat{m}_2(x)$, based on the covariate values that were used. A confidence band is also plotted based on the adjusted p-value, $\hat{p}_c$. That is, the simultaneous probability coverage among the K confidence intervals is approximately $1 - \alpha$, where α is specified via the argument alpha, which defaults to 0.05. The value of $\hat{p}_c$ can be indicated via the argument p.crit. If not specified, the function determines $\hat{p}_c$ in the following manner: When the argument EST=FALSE, and if the argument alpha=0.05, $\hat{p}_c$ is approximated using a smooth (LOESS) in conjunction with values reported at the end of Section 12.2.4. Two estimates of $\hat{p}_c$ are computed based on the smooth. The first estimate uses $1/n_1$ to estimate $\hat{p}_c$, the second is based on $1/n_2$, and the results are averaged. If EST=TRUE or alpha differs from 0.05, a simulation estimate of $\hat{p}_c$ is used where the number of replications is controlled by the argument nreps. Setting MC=TRUE can reduce execution if a multicore processor is available. The confidence intervals that are returned are adjusted so that the simultaneous probability coverage is approximately equal to alpha. The last column of the output indicates whether a significant result is obtained based on the corresponding confidence interval.

■ Example

Data from the Well Elderly 2 study, described in Section 11.1.5, are used to illustrate the R functions described in this section. Of particular interest is how the results compare to a method that assumes the usual linear model where the least squares regression estimator is used. After six months of intervention, one of the goals was to compare males and females in terms of a life satisfaction measure (SF36) using the cortisol awakening response (CAR) as a covariate. First, suppose the groups are compared using method S1 in Section 12.1.1 using the least squares regression estimator. Assuming that the data for males are stored in x1 and y1 and the data for females are stored in x2 and y2, this was accomplished with the command

```
ancJN(x1,y1,x2,y2, xout=TRUE, outfun=outbox, xlab='CAR',ylab='SF36', plotit=TRUE,
                              regfun=ols),
```

which performs the heteroscedastic method in Section 12.1 using the least squares regression estimator with leverage points removed. Here is a portion of the output:

```
$output
                 X      Est1      Est2       DIF       TEST        se      ci.low
[1,] -0.32160000 47.47286 40.78677 6.6860850 2.4272872 2.754550 -0.3931095
[2,] -0.14490072 45.75017 40.79015 4.9600156 2.8930128 1.714481  0.5537994
[3,] -0.01307176 44.46493 40.79267 3.6722575 2.4353167 1.507918 -0.2030914
[4,]  0.08277492 43.53049 40.79451 2.7359887 1.5076466 1.814741 -1.9278967
[5,]  0.26549698 41.74909 40.79800 0.9510864 0.3216437 2.956956 -6.6482914
           ci.hi      p.value
[1,] 13.765279 0.015212206
[2,]  9.366232 0.003815657
[3,]  7.547606 0.014878761
[4,]  7.399874 0.131644995
[5,]  8.550464 0.747722624
```

So based on the confidence intervals, assuming a linear model, a significant result is obtained at the 0.05 level for one covariate value only, namely the second value, -0.14490072. As can be seen, the corresponding p-value is 0.0038. The adjusted p-value (via the R function p.adjust using Hochberg's method) is 0.019. All of the other four adjusted p-values are greater than 0.059. The left panel of Fig. 12.1 shows the plot created by the function ancJN, where the solid line is the regression line for males. Using instead the Theil–Sen estimator, a significant result is obtained for the first four covariate values.

Now the groups are compared using method UB in Section 12.2.3 by setting the argument padj=TRUE when using the R function ancovaUB. Here are the results:

```
$output
                 X    p.values  p.adjusted
[1,] -0.32160000 0.009029345 0.02708804
[2,] -0.14490072 0.000000000 0.00000000
[3,] -0.01307176 0.004000000 0.01600000
[4,]  0.08277492 0.088000000 0.17600000
[5,]  0.26549698 0.952380952 0.95238095
```

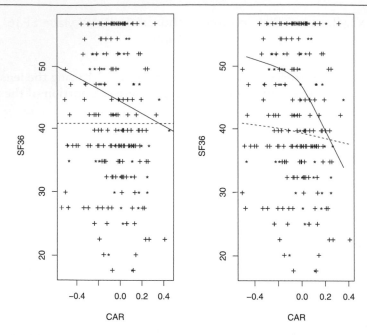

Figure 12.1: The left panel shows the plot returned by ancJN when using the least squares regression estimator. The solid regression line corresponds to males. The right panel shows the plot returned by ancovaV2 based on the Harrell–Davis estimator. In contrast to the left panel, the nature of the association for males appears to change rather abruptly approximately where CAR is zero.

So for the first three covariate values, a significant result is obtained at the 0.05 level, in contrast to the results based on the least squares estimator. From the point of view of Tukey's three-decision rule, decide that for the negative CAR values males have higher SF36 scores compared to females. For the positive CAR values, no decision is made. The right panel of Fig. 12.1 shows the plot created by ancovaUB. Note that compared to the left panel, this plot gives a decidedly different impression about the nature of the association for males. For CAR negative, the two regression lines in the right panel appear to be nearly parallel. But when the CAR is greater than zero, differences between the two regression lines diminish as the CAR increases. Method TAP, applied via the R function ancdet, indicates a significant difference for CAR values ranging between -0.32 and -0.077.

■ **Example**

The last example in Section 11.2.2 dealt with the Well Elderly 2 study where the goal was to compare a control group to an intervention group based on a measure of mean-

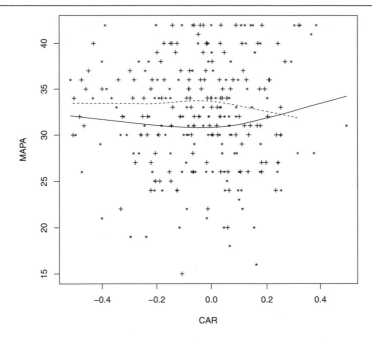

Figure 12.2: Shown are the smooths for predicting MAPA scores based on the CAR. The solid line is the smooth prior to intervention.

ingful activities (MAPA) and the CAR. Two analyses were reported, each using two independent groups of participants. The first analysis indicated crossing regression lines, assuming that the regression lines are straight. The second analysis, essentially aimed at determining the extent to which the first analysis can be replicated, found that the regression lines do not cross. Indeed, the regression lines appear to be virtually parallel. Fig. 12.2 shows a smooth using the data from this latter situation. The output from the R function ancdet indicates that MAPA scores differ significantly for CAR values between -0.13 and 0.048. That is, using a method that deals with curvature in a relatively flexible manner, significant differences are found, in contrast to an analysis that assumes the regression lines are straight.

The function

$$\text{ancpb(x1,y1,x2,y2, est=hd, pts=NA, nboot=599, plotit=TRUE, LP=TRUE,...)}$$

is like the function ancova, only any measure of location can be used, which is specified by the argument est, and a percentile bootstrap method is used to compute confidence intervals.

By default, the Harrell–Davis estimate of the median is used. The arguments are the same as those associated with the R function ancova except the argument nboot, which indicates B, how many bootstrap samples are used. When LP=TRUE, the initial smooth is smoothed again using LOESS.

The function

ancboot(x1,y1,x2,y2, fr1=1, fr2=1, tr=0.2, nboot=599, pts=NA, plotit=TRUE)

is exactly like the function ancova, only a bootstrap-t method is used to compute confidence intervals and test hypotheses based on trimmed means.

The R function

ancbbpb(x1, y1, x2, y2, fr1 = 1, fr2 = 1, est=tmean, nboot = 200, pts = NA, plotit = TRUE,
SEED = TRUE, tr=0.2, RNA = TRUE,...)

is like the R function ancova, only bootstrap bagging is used. So it has the potential of more power at the cost of higher execution time. There are some indications that method UB, applied via the R function ancovaUB, is better for general use, but a more detailed study is needed to better understand their relative merits.

The R function

lplot2g(x1,y1,x2,y2,fr=0.8, est=tmean, xlab='X', ylab='Y', xout=FALSE, eout=FALSE,
outfun=out,...)

is like runmean2g, only it plots the regression lines corresponding to two groups using Cleveland's smoother instead.

The R function

ancdifplot(x1,y1,x2,y2,fr1=1, fr2=1, tr=0.2, alpha=0.05, pr=TRUE, xout=FALSE, outfun=out,
LP=TRUE, nmin=8, scat=TRUE, xlab='X', ylab='Y', report=FALSE,...)

plots the estimated difference, $m_1(x) - m_2(x)$, for each of the covariate values used by the R function ancova. Confidence intervals, having simultaneous probability coverage approximately equal to the value of the argument alpha, are plotted as well. The R function

qhdsm2g(x1, y1, x2, y2, q = 0.5, qval = NULL, LP = TRUE, fr = 0.8, xlab = 'X', ylab = 'Y',
xout = FALSE, outfun = outpro, ...)

plots two quantile regression lines using the running interval smoother in conjunction with the Harrell–Davis estimator. (The function cobs2g uses COBS to plot the regression lines, but it should be used with caution for reasons summarized in Section 11.5.6.) The argument q indicates which quantile will be used. Setting q=0.75, for example, the plots are based on estimates of the upper quartiles.

The R function

```
ancmg1(x, y, pool = TRUE, jcen = 1, fr = 1, depfun = fdepth, nmin = 8, op = 3, tr=0.2,
    SEED = TRUE, pr = TRUE, pts = NA, con = 0, nboot = NA, tr=0.2, bhop = FALSE)
```

can be used to compare multiple groups when there is a single covariate. The arguments x and y can be matrices with J columns, where J is the number of groups, or they can have list mode with length J. The argument op determines how the groups are compared. There are four options:

- op=1: Omnibus test for trimmed means, based on the R function t1way, with the amount of trimming controlled via the argument tr.
- op=2: Omnibus test for medians based on the R function med1way. (Not recommended when there are tied values; use op=4.)
- op=3: Multiple comparisons using trimmed means and a percentile bootstrap via the R function linconpb.
- op=4: Multiple comparisons using medians and percentile bootstrap via the R function medpb.

(When there are two covariates, the function ancgm in Section 12.3.4 can be used.) The results for the first covariate point are returned in the R variable $points[[1]], the results for the second covariate point are in $points[[2]], and so on. So by default, $points[[k]] contains the results for all pairwise comparisons among the J groups based on the kth covariate point, $k = 1, \ldots, p$. The argument con can be used to specify the linear contrasts of interest. For example, in a 2-by-2 design, the hypothesis of no interaction can be tested by setting con=con2way(2,2)$conAB.

The R function

```
ancGLOB(x1,y1,x2,y2, xout=FALSE, outfun=outpro, est=tmean, p.crit=NULL, nreps=500,
    alpha=0.05, pr=TRUE, nboot=500, SEED=TRUE, MC=FALSE, CR=FALSE, nmin=12,
    pts=NULL, fr1=1, fr2=1,plotit=TRUE, SCAT=TRUE, pch1='', pch2='o', xlab='X',
    ylab='Y', LP=TRUE,...)
```

applies method G in Section 12.2.5. The argument nreps determines how many replications are used to determine the critical p-value, $\hat{p}_c$. The critical p-value depends only on the sample sizes. Once it is known, it can be specified via the argument p.crit, which reduces execution time to a very low value. When the critical p-value is not known, setting MC=TRUE can reduce execution time considerably assuming a multicore processor is available. As usual, the argument nboot controls the number of bootstrap samples and xout determines whether leverage points are removed.

12.3 Dealing With Two or More Covariates When There Is Curvature

Experience with the robust smoothers in Section 11.5 suggest that when dealing with more than one covariate, standard linear models are more likely to provide an unsatisfactory approximation of the regression surface. When there is curvature (the regression surface is not well approximated by a plane or more generally by some parametric linear model), there are various ways the methods described in Section 12.2 might be extended to the case of multiple covariates. Due to the curse of dimensionality, mentioned in Section 11.5.3, there are limits to what can be done as the number of covariates increases.

This section describes three methods for dealing with two covariates (methods MC1, MC2, and MC3) followed by a fourth method that deals with three or four covariates. The first, MC1, is a simple generalization of method Y in Section 12.2.1. Some possible concerns with method MC1 are described, followed by methods MC2 and MC3, which are designed to address these concerns. Method MC4 is designed for three or four covariates.

12.3.1 Method MC1

Momentarily focus on the ith value of the covariate in the first group, $\mathbf{x}_{i1}$. Then proceeding along the lines in Section 11.5.10, it is a simple matter to determine the set

$$N_1(\mathbf{x}_{i1}) = \{j : D_{1ij} \leq f\}, \tag{12.12}$$

where D_{1ij} is some measure of the distance between $\mathbf{x}_{j1}$ and $\mathbf{x}_{i1}$ and f is some constant to be determined, in which case the set N_1 identifies all $\mathbf{x}_{j1}$ values that are close to $\mathbf{x}_{i1}$. One possible choice for D_{1ij} is

$$D_{1ij} = \sqrt{(\mathbf{x}_{i1} - \mathbf{x}_{j1})'\mathbf{M}^{-1}(\mathbf{x}_{i1} - \mathbf{x}_{j1})},$$

where $\mathbf{M}$ is some measure of covariance. The default choice used here is the MVE covariance matrix in Section 6.3.1. All of the points in the second group that are close to $\mathbf{x}_{i1}$ can be identified in a similar manner.

Let

$$N_2(\mathbf{x}_{i1}) = \{j : D_{2ij} \leq f\}, \qquad (12.13)$$

where D_{2ij} measures the distance between $\mathbf{x}_{i1}$ and $\mathbf{x}_{j2}$. Then the y_{j1} values, such that $j \in N_1$, can be compared to the y_{j2} values, $j \in N_2$, using some measure of location. If there is interest in some particular $\mathbf{x}_{i1}$, this approach is readily implemented, but otherwise how should $\mathbf{x}_{i1}$ be chosen?

The approach used by MC1 is to use the covariate point having the largest halfspace depth plus the points on the 0.5 depth contour. Another possibility is to pool $\mathbf{x}_{i1}$ and $\mathbf{x}_{i2}$ and compare the groups based on the covariate point having the largest halfspace depth among the pooled data. Of course, covariate points might be chosen because they are particularly important for the situation at hand.

When testing H_0: $m_1(\mathbf{x}) = m_2(\mathbf{x})$ using Yuen's method for trimmed means, critical values based on the Studentized maximum modulus distribution can be used to control the probability of one or more Type I errors. Another approach is to use the Hochberg or Hommel improvements on the Bonferroni method, which were described in Section 7.4.7. An advantage of this latter approach is that it can be used for a broader range of situations, such as when a percentile bootstrap method is used.

Notice that the method described in this section is readily extended to comparing more than two groups. Again pick design points for each of the J groups to be compared, determine observed covariate values that are close to the chosen design points, and then use methods in Chapter 7 to test hypotheses about the corresponding y values.

12.3.2 Method MC2

A fundamental issue associated with method MC1 is whether the design points have been chosen so as to detect any true differences, and perhaps more importantly whether they reveal where and by how much the groups differ. When choosing the deepest point as well those points that lie on the 0.5 depth contour, for example, typically very few points are chosen, which raises the concern that important details might be missed. A way of dealing with these concerns is to simply use more design points, but as the number points increases, this can negatively impact power when controlling the probability of one or more Type I errors with Hochberg's method or some related technique. The approach in this section is to test the global hypothesis given by Eq. (12.8) based on a function of the p-values stemming from the K hypotheses given by Eq. (12.7). Method MC3, described in the next section, provides a way of testing the K hypotheses indicated by Eq. (12.7) that improves on Hochberg's method for controlling the FWE rate.

Roughly, method MC2 begins by testing the K hypotheses given by Eq. (12.7) for all of the points $\mathbf{x}$ that are deeply nested among the cloud of covariate points associated with the first group, which are determined using approximation A1 of halfspace depth described in Section 6.2.3. (The R function fdepth is used.) Another possibility is to use the deepest covariate points among all $n_1 + n_2$ covariate points. The extent to which this latter approach offers a practical advantage is unknown. Deeply nested points generally correspond to situations where the regression surfaces can be estimated in a relatively accurate manner. If a point $\mathbf{x}$ is not deeply nested in the cloud of covariate values, finding a sufficiently large number of other points that are close to $\mathbf{x}$ might be impossible. The R functions based on MC2, described in the next section, are not limited to comparing trimmed means. Included are R functions where the dependent variable is binary.

Note that from an exploratory point of view, plotting the p-values might help provide an overall sense of where it is reasonable to make a decision about which group has the larger measure of location. And of course a plot of the estimated differences among all of the design points can be used as well.

Consider the situation where H_0: $m_1(\mathbf{x}) = m_2(\mathbf{x})$ is tested for each $\mathbf{x} \in \{\mathbf{x}_1, \ldots, \mathbf{x}_K\}$, where K/n_1 might be relatively large and n_1 is the sample size associated with the first group. Here, $K/n_1 = 1/2$ is used by default, but this fraction can be altered when using the R functions in the next section. Let $p_1, \ldots, p_K$ be the resulting p-values and consider how these p-values might be used to test the global hypothesis given by Eq. (12.8). Perhaps the best-known method based on these p-values is a technique derived by Fisher (1932). But there are two concerns. First, the method assumes the p-values are independent, which is not necessarily the case here because the intersection of $N_j(x_k)$ and $N_j(x_\ell)$, $k < \ell$, is not necessarily empty. Second, Zaykin et al. (2002) note that the ordinary Fisher product test loses power in cases where there are a few large p-values. They suggest using instead a truncated product method (TPM), which is based on the test statistic

$$W = \prod_{k=1}^{K} p_k^{I(p_k \leq \tau)}, \tag{12.14}$$

where I is the indicator function. Setting $\tau = 1$ yields Fisher's method, but Zaykin et al. suggest using $\tau = 0.05$ (cf. Li and Siegmund, 2015). Zaykin et al. derive the null distribution of W when all K tests are independent. But again the K tests performed here are not necessarily independent. Yet one more possible test statistic is

$$\bar{Q} = \frac{1}{K} \sum_{k=1}^{K} p_k. \tag{12.15}$$

To deal with the dependence among the p-values, method MC2 proceeds as follows, still assuming that there are two covariates. Momentarily assume that for each j, $(y_{ij}, \mathbf{x}_{ij})$ ($i = 1, \ldots n_j$, $j = 1, 2$) has a trivariate normal distribution where all correlations are zero. Generate a random sample from both groups and compute the test statistic W. Repeat this process B times, yielding $W_1, \ldots, W_B$. Put these B values in ascending order, yielding $W_{(1)} \leq \cdots \leq W_{(B)}$. Then the α level critical value is estimated to be $W_{(c)}$, where c is αB rounded to the nearest integer. That is, Eq. (12.8) is rejected at the α level if $W \leq w$. Of course the same process can be used in conjunction with the test statistic $\bar{Q}$. Simulations reported by Wilcox (2016a) indicate that both versions of this method perform well in terms of controlling the Type I error probability when dealing with non-normal distributions, including situations where there is curvature. The simulations also suggest that generally, $\bar{Q}$ tends to have more power than W, but the only certainty is that the reverse can occur. Moreover, the p-values associated with W and $\bar{Q}$ can differ substantially, as will be illustrated. In particular, situations are encountered where W rejects at the 0.01 level but $\bar{Q}$ fails to reject at the 0.10 level.

12.3.3 Methods MC3 and MC4

A limitation of method MC2 is that it might suggest where significant results appear to occur, but in essence little can be said about how reasonable it is to make a decision about whether $m_1(\mathbf{x})$ is larger than $m_2(\mathbf{x})$ among the covariate points that were used. A way of dealing with this issue, called method MC3, is to use an analog of method TAP in Section 12.2.4 (Wilcox, 2018a). First, choose covariate points as done by method MC2. Based on this process for choosing covariate points, determine p_c, the α quantile of the distribution of the minimum p-value returned by method MC2. This is done via a simulation under normality and when there is no association among six variables. Then, from the point of view of Tukey's three-decision rule, make a decision about whether $m_1(\mathbf{x})$ is larger than $m_2(\mathbf{x})$ for any covariate point for which the corresponding p-value is less than or equal to p_c. Otherwise, no decision is made. So, compared to method MC2, method MC3 has the potential of providing more detail about where the regression surfaces differ. At least in some situations method MC2 can have substantially more power than MC3, where power is taken to be the probability of at least one significant difference. But an example in the next section demonstrates that situations are encountered where MC3 rejects when MC2 does not.

Simulation estimates of p_c can result in relatively high execution time. To help deal with this issue, Table 12.5 reports some estimates of p_c when $n_1 = n_2 = n$. The estimates are based on 4,000 replications. For unequal sample sizes, p_c is estimated as indicated in Section 12.3.4 in connection with the R function ancdet2C.

Method MC4 is a variation of MC3 that deals with three or four covariates. The sets N_1 and N_2 are determined as done by method MC1. Again let $M_j(\mathbf{x}_{i1})$ denote the number of points

Table 12.5:
Estimates of p_c based on 4,000 replications and $\alpha = 0.05$.

n	p_c
50	0.004585
55	0.003120
60	0.002820
70	0.002594
80	0.002481
100	0.001861
200	0.001420
300	0.001423
400	0.001314
500	0.001352
600	0.001075
800	0.000959

in the set $N_j(\mathbf{x}_{i1})$ ($j = 1, 2$). MC4 has the ability to compare the groups for each $\mathbf{x}_{i1}$ such that both $M_1(\mathbf{x}_{i1})$ and $M_2(\mathbf{x}_{i1})$ are greater than or equal to 12. When $f = 1$ (the span) in Eqs. (12.12) and (12.13), $p = 3$, and $n = 50$, the probability of finding any point satisfying these two conditions can be quite low due to the curse of dimensionality described in Section 11.5.10. With $n = 80$ there is a good chance of finding at least one point and with $n = 150$ the probability is quite high. Again let K denote the number of points for which both M_1 and M_2 are greater than or equal to 12. Increasing the span f even slightly can increase K substantially, but extant simulations for how well method MC4 controls the FWE rate are limited to $f = 1$. Situations are encountered where $f \leq 1$ is required to get a reasonably accurate approximation of the regression surface. So using $f > 1$, caution is needed when interpreting the results. Rather than making an inference about the typical response at some point x_{i1}, the results reflect the typical difference for points in $N_1(\mathbf{x}_{i1})$. For sufficiently large sample sizes, K can be quite large with $f = 1$. If $K \leq 25$, Hochberg's method (in Section 7.4.7) is used to control the FWE rate. For $25 < K \leq 100$, when the goal is to have an FWE rate equal to 0.05, results in Wilcox (2018d) indicate using a critical p-value given by

$$p_c = \frac{0.0806452604}{K} - 0.0002461736. \tag{12.16}$$

For $K > 100$, use

$$p_c = \frac{0.06586286}{K} - 0.00004137143. \tag{12.17}$$

These results are limited to situations where each test is performed with Yuen's method based on a 20% trimmed mean.

12.3.4 R Functions ancovamp, ancovampG, ancmppb, ancmg, ancov2COV, ancdes, ancdet2C, ancdetM4, and ancM.COV.ES

The R function

$$\text{ancovamp(x1,y1,x2,y2, fr1=1, fr2=1, tr=0.2, alpha=0.05, pts=NA)}$$

compares two groups based on trimmed means and takes into account multiple covariates using method MC1 in the previous section. The arguments are the same as those associated with the R function ancova in Section 12.2.5, only now x1 and x2 are assumed to be matrices with two or more columns. By default, pts=NA, meaning that the points among the covariates at which the groups will be compared are determined by the function ancdes; it finds a point among the x1 values that has the deepest halfspace depth, plus the points on the 0.5 depth contour, and the groups are compared at these points provided that the corresponding sample sizes are at least 10. Should one want to pool the data and then find the deepest point, plus the points on the 0.5 depth contour, this can be done as indicated by some of the R functions to be described. The function controls the FWE rate by determining a critical value via the Studentized maximum modulus distribution (as described in Chapter 7). The MVE covariance matrix is used when determining the points that are close to $\mathbf{x}_k$.

The R function

$$\text{ancovampG(x1,y1,x2,y2, fr1=1, fr2=1, tr=0.2, alpha=0.05, pts=NULL, SEED=TRUE,}$$
$$\text{test=yuen, DH=FALSE, FRAC=0.5, cov.fun=skip.cov, ZLIM=TRUE, pr=FALSE, q=0.5,}$$
$$\text{plotit=FALSE, LP=FALSE, theta=50, xlab='X1', ylab='X2', SCAT=FALSE,zlab='p.value',}$$
$$\text{ticktype='detail',...)}$$

also applies method MC1, but unlike the R function ancovamp, it is not limited to comparing trimmed means and it is more flexible in other ways as well. The argument test can be used to specify the method for comparing measures of location that will be used, which defaults to Yuen's method for trimmed means. Two other choices are qcomhd and qcomhdMC, which can be used to compare quantiles, in which case the argument q determines the quantile that will be used. When the dependent variable is binary, for example, test=qcomhd and q=0.25 would compare the lower quartiles using the Harrell–Davis estimator. Setting plotit=TRUE, the function plots the p-values as a function of the covariate points. If the number of points is reasonably large, a smooth of the p-values can be created by setting LP=TRUE. The plot can be rotated via the argument theta. The argument cov.fun controls the covariance matrix that is used when determining the points that are close to $\mathbf{x}_k$ using a robust analog of Mahalanobis distance. By default a skipped covariance matrix is used. Consequently, although both ancovampG and ancovamp compare 20% trimmed means by default, the results typically differ due to different strategies for determining the covariate points that are close to $\mathbf{x}_k$. If DH=TRUE, the covariate points are chosen as described in Section 12.3.2 in connection with method MC2. (The covariate points are chosen via the R function ancdes.) As a result, the hypothesis given by Eq. (12.7) is tested for every $\mathbf{x}$ that is in the deepest half of the covariate points. If, for example, FRAC=0.3, the deepest 70% of the covariate points would be used. But the probability of one or more Type I errors is controlled via Hochberg's method, which can result in relatively low power compared to using method MC3, which can be applied with the R function ancdet2C described later in this section.

The R function

> ancmppb(x1,y1,x2,y2,fr1 = 1, fr2 = 1, tr=0.2, pts = NA, est = tmean, nboot = NA, bhop = FALSE, SEED = TRUE, cov.fun = skip, cop = NULL, COV.both=FALSE,...)

is like ancovamp, only a percentile bootstrap method is used and any measure of location can be employed. The argument cov.fun indicates which measure of location will be used to determine the center of the covariate data, assuming the estimate is returned in $location. By default, a 20% trimmed mean is used to compare the two groups. In essence, the function determines groups as described in connection with method MC2, and then it compares the corresponding y values by calling the function pbmcp in Section 7.6.3, where the argument bhop is explained. (It determines the approach used to control the probability of at least one Type I error among the tests performed.) The argument pts can be used to specify the covariate points where the groups will be compared. If it is desired to pool the data in x1 and x2 when determining the covariate points where the groups are to be compared, set the argument COV.both=TRUE.

The R function

> ancmg(x, y, pool = TRUE, jcen = 1, fr = 1, depfun = fdepth, nmin = 8, op = 3, tr=0.2, pts = NA, SEED = TRUE, pr = TRUE, cop = 3, con = 0, nboot = NA, tr=0.2, bhop = FALSE)

can be used to compare multiple groups when there are multiple covariates. The arguments are basically the same as the arguments for the function ancmg1 in Section 12.2.6. The main difference is that now the argument x can be a matrix with Jp columns where J is the number of groups and p is the number of covariates. The first p columns correspond to group 1, the next p columns correspond to group 2, and so on. Or x can have list mode, where x[[j]]

contains the covariate values for the jth group, which can be a matrix with p columns. For the moment, this function should be used only when $p = 2$. How it performs with $p > 2$ covariates is unknown. Presumably, as the number of covariates increases, at some point it will perform poorly due to the curse of dimensionality. The argument op determines how the groups are compared. The four options are the same as those described in Section 12.2.6. As was the case with the R function ancmg1, the results for the first covariate point are returned in the R variable $points[[1]], the results for the second covariate point are in $points[[2]], and so on. As can be seen, the default is op=3, meaning that multiple comparisons are performed based on trimmed means via the R function linconpb. Like the R function linconpb, the argument con can be used to specify the linear contrasts of interest. For example, in a 2-by-2 design, the hypothesis of no interaction can be tested by setting con=con2way(2,2)$conAB.

Here is an example of some of the output from ancmg when dealing with three groups:

```
$points.chosen
             [,1]         [,2]
[1,] -0.1168167  -0.1021651
[2,] -0.7254685   0.7561923
[3,] -0.3523684  -1.1417145


$sample.sizes
       [,1] [,2] [,3]
[1,]    21   15   19
[2,]    11    8   10
[3,]     8    9   10


$point
$point[[1]]
       con.num       psihat p.value       p.crit    ci.lower   ci.upper
[1,]         1   0.16028098   0.721 0.01666667  -0.9078780  1.0857720
[2,]         2   0.07353133   0.831 0.02500000  -0.7167843  0.8219238
[3,]         3  -0.08674966   0.871 0.05000000  -1.0805465  0.9926640


$point[[2]]
       con.num       psihat p.value       p.crit    ci.lower  ci.upper
[1,]         1 0.12370772   0.895 0.05000000  -1.3665729  1.435669
[2,]         2 0.17065563   0.735 0.01666667  -0.9817933  1.438490
[3,]         3 0.04694791   0.857 0.02500000  -1.0503887  1.458355
```

```
$point[[3]]
     con.num     psihat p.value     p.crit  ci.lower   ci.upper
[1,]       1  0.09949583   0.820 0.05000000 -1.128347  1.2756405
[2,]       2 -0.62844012   0.171 0.01666667 -1.565212  0.4625049
[3,]       3 -0.72793596   0.192 0.02500000 -1.847244  0.5851383

$contrast.coef
     [,1] [,2] [,3]
[1,]    1    1    0
[2,]   -1    0    1
[3,]    0   -1   -1
```

For example, for the first covariate point, indicated by row 1 of $points.chosen, the results for the multiple comparisons are returned in $point[[1]]. More generally, the results for the *k*th point are returned in $point[[k]]. When dealing with multiple comparisons, the corresponding linear contrast coefficients are returned in $contrast.coef.

The R function

ancov2COV(x1, y1, x2, y2, tr=0.2, test = yuen, cr = NULL, pr = TRUE, DETAILS = FALSE, cp.value = FALSE, plotit = FALSE, xlab = 'X', ylab = 'Y', zlab = NULL, span = 0.75, PV = TRUE, FRAC = 0.5, MC = FALSE, q = 0.5, iter = 1000, tr=0.2, TPM = FALSE, tau = 0.05, est = tmean, fr = 1, ...)

compares groups based on points chosen via method MC2. The function can be used to test the global hypothesis given by Eq. (12.8) via Eq. (12.14) or (12.15). By default, trimmed means are compared via Yuen's method, but quantiles can be compared by setting the argument test=qcomhd or qcomhdMC, in which case the argument q controls the quantile that will be used. (Quantiles are estimated via the Harrell–Davis estimator.) The default is q=0.5, the median. The function does not have an option for eliminating leverage points; this is not necessary because the method performs the analysis on the deepest covariate values. The default test statistic is $\bar{Q}$ given by Eq. (12.15) in Section 12.3.2. The critical value can be determined quickly when testing at the 0.05 level. When testing at the $\alpha \neq 0.05$ level or when using the test statistic given by Eq. (12.14) (method TPM), a critical value must be computed, which can increase execution time considerably. Execution time can be reduced by setting MC=TRUE, assuming that a multicore processor is available and that the R package parallel has been installed.

If plotit=TRUE and PV=FALSE, the function plots $m_1(\mathbf{x}) - m_2(\mathbf{x})$ as a function of the two covariates using LOESS. If PV=TRUE, the function creates a plot of the p-values as a function of the two covariates. If the argument DETAILS=TRUE, all p-values are returned, in which case they can be adjusted using Hochberg's or Hommel's method (via the R function p.adjust) with the goal of controlling the probability of one or more Type I errors. Setting the argument cp.value=TRUE, the function returns a p-value based on the test statistic that was used to test the global hypothesis given by Eq. (12.8). This can increase execution time considerably. Again, execution time can be reduced by setting MC=TRUE, assuming that a multicore processor is available.

The covariate points, where the regression surfaces are compared, are determined via the function

$$\text{ancdes(x,depfun=fdepth, DH=FALSE, FRAC=0.5,...).}$$

This function determines the halfspace depth of the points in x, where the depth of points is determined using the function indicated by the argument depfun, which defaults to fdepth, described in Section 6.2.4. The argument FRAC controls the proportion of the least deep points that are ignored. For example, setting FRAC=0.3, the deepest 70% of the covariate points would be used. The critical value is known when using FRAC=0.5, the default value. But otherwise the critical value must be computed, which increases execution time considerably. Again, setting MC=TRUE can reduce execution time considerably.

■ **Example**

Consider again the Well Elderly 2 study, described in Section 11.1.5, where the general goal was to assess the efficacy of an intervention program aimed at improving the health and wellbeing of older adults. One of the goals was to compare a measure of meaningful activities (MAPA) associated with a control group and a group receiving intervention. Here, these two groups are compared using method MC2 where a measure of depressive symptoms (CESD) and the CAR are the covariates. Here is a portion of output returned by the R function ancov2COV when cp.value=TRUE:

```
$num.points.used
[1] 74

$test.stat
[1] 0.1193853

$crit.value
[1] 0.2459744
```

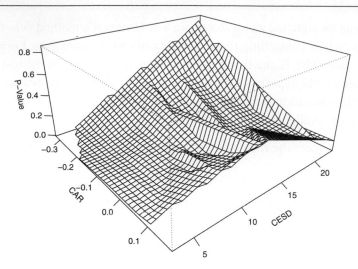

Figure 12.3: Shown are the p-values when comparing MAPA scores before and after intervention using two covariates: CESD and CAR.

```
$GLOBAL.p.value
[1] 0.008

$min.p.val.point
[1] -0.2176042   4.0000000

$min.p.value
     p.value
0.002299862
```

So 74 covariate points were chosen. The test statistic is $\bar{Q} = 0.1193853$, the 0.05 critical value is 0.2459744 (the approximate critical value when cp.value=FALSE is 0.27), and the p-value is 0.008. The minimum p-value among the 74 hypotheses that were tested is 0.002299862, which occurred at CAR=-0.2176042 and MAPA=4. Using TPM, the global p-value is now 0.021. Fig. 12.3 shows a plot of the p-values. So the data indicate that the groups differ significantly and the strongest evidence that the control group and the experimental group differ significantly occurs when CESD is low. Both ancovampG and ancovamp use only three covariate points when using default settings. The function ancovamp returns a significant result for one of these points when testing at the 0.05 level; ancovampG fails to reject.

■ **Example**

This next example illustrates that when using method MC2 in Section 12.3.2, the choice between TPM and $\bar{Q}$ can yield a fairly different result. Again the Well Elderly 2 data are used, only now the dependent variable is taken to be a measure of life satisfaction. Using $\bar{Q}$ via the R function ancov2COV, the global p-value is 0.126. In contrast, using TPM, it is 0.009. So despite simulation results suggesting that TPM tends to have lower power, situations are encountered where TPM is highly significant when $\bar{Q}$ fails to reject at the 0.05 level. Both ancovampG and ancovamp fail to reject using default settings. The power of these two methods depends on the pattern of the p-values. Also, assuming the regression surfaces are a plane and using the function ancJNmp in Section 12.1.4 does not give significant results. That is, using a method that allows curvature can make a practical difference.

■

The R function

ancdet2C(x1, y1, x2, y2, fr1 = 1, fr2 = 1, tr=0.2, test = yuen, q = 0.5, tr=0.2, plotit = TRUE,
  op = FALSE, pts = NA, sm = FALSE, FRAC = 0.5, pr = TRUE, xout = FALSE,
  outfun = outpro, MC = FALSE, p.crit = NULL, nreps = 2000, SEED = TRUE,
  FAST = TRUE, ticktype = 'detail', xlab = 'X1', ylab = 'X2', zlab = 'Y', pch1 = '*',
  pch2 = '+', ...)

applies method MC3. By default, trimmed means are compared using Yuen's method. Setting the argument test=qcomhd or qcomhdMC, quantiles will be compared instead based on the Harrell–Davis estimator. The argument q controls the quantile that will be used, which defaults to 0.5, the median. If the argument plotit=TRUE, the function creates a scatterplot of the covariate values. If the argument op=FALSE, covariate points where a significant result was obtained are indicated by the symbol given by the argument pch2. Points where a non-significant result was obtained are indicated by the symbol given by the argument pch1. If op=TRUE, a smooth is created where the z-axis indicates an estimate of $m_1(\mathbf{x}) - m_2(\mathbf{x})$ given values for the two covariates. The argument p.crit corresponds to p_c, which depends only on the sample sizes n_1 and n_2. An approximation of p_c is used with the argument FAST=TRUE and alpha=0.05. The approximation is based on a smooth (LOESS) applied to the sample sizes and estimates of p_c in Table 12.5. (If $n_1 = n_2 = n$, the smooth uses $1/n$ as the independent variable. If $n_1 \neq n_2$, p_c is estimated using $1/n_1$ as well as $1/n_2$, and the results are averaged. If FAST=FALSE or alpha is not equal to 0.05, a simulation estimate of p_c is used which is computed via the R function

ancdet2C.pv(n1, n2, nreps = 2000, tr=0.2, FRAC =0.5, tr=0.2, MC = FALSE, SEED = TRUE).

The number of replications is controlled via the argument nreps. To provide some sense of how well FAST=TRUE performs when using unequal sample sizes, suppose $n_1 = 50$ and $n_2 = 500$. Then ancdet2C.pv estimates p_c to be 0.0021. If $n_1 = 500$ and $n_2 = 50$, the estimate is 0.002. Using FAST=TRUE, p_c is estimated to be 0.0027.

The R function

ancdetM4(x1, y1, x2, y2, fr1 = 1, fr2 = 1, tr = 0.2, alpha = 0.05, pts = NA, pr = TRUE, xout = FALSE, outfun = outpro, MC = FALSE, p.crit = 0.05, BOTH = FALSE, ...)

applies method MC4. By default, all covariate points are used for which both $M_1(\mathbf{x}_{i1})$ and $M_2(\mathbf{x}_{i1})$ are greater than or equal to 12. Individual tests are performed via Yuen's method. The function reports which points satisfy these two conditions. A subset of these points can be used via the argument pts. For example,

a=ancdetM4(x1,y1,x2,y2)

would use all of the points. To rerun the analysis using just the first five points, use the command

ancdetM4(x1,y1,x2,y2,pts=a$selected.points[1:5,]).

The R function

ancM.COV.ES(x1,y1,x2,y2, fr1=1,fr2=1, tr=0.2, pts=NULL, xout=FALSE, outfun=outpro,...)

computes measures of effect size. By default, it uses the covariate points that are significant, as indicated by ancdetM4. Any set of points can be used via the argument pts. As long as the number of nearest neighbors is at least 10, effect sizes are estimated. Note that the three functions just described assume there is a single covariate that is used for both groups.

■ Example

The previous example is repeated, only now method MC3 is applied via the R function ancdet2C. No significant results are found when testing at the 0.05 level. However, if the dependent variable (life satisfaction) is replaced by a measure of perceived physical health, now MC2 fails to reject even though the minimum observed p-value is 3.591197e−07. In contrast, method MC3 finds nine significant results among the 74 covariate points that were used. They occur where the CAR is negative and CESD is relatively low.

∎

12.4 Some Global Tests

The methods in the previous section are aimed at comparing groups at specific design points. This section describes methods where the goal is to test the hypothesis that two independent groups do not differ for any design point.

12.4.1 Method TG

Method TG is a global test based on trimmed means and is limited to a single covariate. It is assumed that x has been standardized based on some robust measure of location and scale. Unless stated otherwise, the median and median absolute deviation are used. That is, if z indicates the covariate, work with

$$x = \frac{z - M}{\text{MADN}}.$$

The goal is to test

$$H_0: m_1(x) = m_2(x) \ \forall x. \tag{12.18}$$

The method is based in part on a simple generalization of the notion of regression depth. Recall from Section 10.11 that when estimating the slope and intercept of the usual linear model, a candidate fit, (b_0, b_1), is called a non-fit if a partition of the x values can be found such that all of the residuals for the lower x values are negative (positive), but for all of the higher x values the residuals are positive (negative). For the random sample $(x_1, y_1), \ldots, (x_n, y_n)$, letting $r_i = y_i - b_0 - b_1 x_i$, a candidate fit is called a non-fit if and only if a value for v can be found such that

$$r_i < 0 \quad \text{for all } x_i < v$$

and

$$r_i > 0 \quad \text{for all } x_i > v$$

or

$$r_i > 0 \quad \text{for all } x_i < v$$

and

$$r_i < 0 \quad \text{for all } x_i > v.$$

Rousseeuw and Hubert define the regression depth of a fit (b_1, b_0), relative to $(x_1, y_1), \ldots,$ (x_n, y_n), as the smallest number of observations that need to be removed to make (b_1, b_0) a non-fit. Their deepest regression line estimator corresponds to the values of b_1 and b_0 that maximize regression depth.

Now consider any fit $\hat{y}_i = m(x_i)$, which might be obtained via any of the non-parametric regression methods previously described. Given a fit, the depth of the fit can be measured using a simple extension of the Rousseeuw and Hubert approach because their notion of depth is based entirely on the residuals and the values of the covariate. In particular, it does not require that the regression line be straight. That is, now $r_i = y_i - \hat{y}_i$, and given $x_1, \ldots, x_n$, depth is defined as before.

Using a simple modification of the computational algorithm in Rousseeuw and Hubert (1999), the depth of a non-parametric regression line can be computed as follows. First, reorder the covariate values so that $x_1 \leq \cdots \leq x_n$. Then the regression depth of $m(x)$ is

$$D = \min_{1 \leq i \leq n} (\min\{L^+(x_i) + R^-(x_i), R^+(x_i) + L^-(x_i)\}),$$

where

$$L^+(v) = \#\{j; x_j \leq v \text{ and } r_j \geq 0\},$$
$$R^-(v) = \#\{j; x_j > v \text{ and } r_j \leq 0\},$$

where # denotes the cardinality of the set and L^- and R^+ are defined accordingly. Note that regression depth is scale-invariant. From Theorem 1 in Rousseeuw and Hubert, it follows that the maximum possible value for D is greater than or equal to $\lceil n/3 \rceil$ and less than or equal to n, where the ceiling $\lceil z \rceil$ is the smallest integer $\geq z$.

Let (y_{ij}, x_{ij}) be a random sample from the jth group $(i = 1, \ldots, n_j)$, and imagine that a non-parametric regression line is fitted to the data in the first group. Then in general, this fit can be used to estimate y given any value for x simply by computing the trimmed mean of the y_{i1} values for which the corresponding x_{i1} values are close to x. In particular, an estimate can be computed for the covariate values corresponding to the second group: $x_{i2}, i = 1, \ldots, n_2$. This assumes, of course, that given x_{i2}, there are one or more x_{i1} values that are close to x_{i2} that can be used to compute $m_1(x_{i2})$.

Let

$$\hat{y}_{ijk} = m_j(x_{ik}; x_{1j}, \ldots, x_{n_j j})$$

be the predicted value of y corresponding to the ith observation in the kth group using the fit obtained from the jth group. That is, $\hat{y}_{ijk}$ is the 20% trimmed mean of the y_{ij} values for which x_{ij} is close to x_{ik}. Let $r_{ijk} = y_{ijk} - \hat{y}_{ijk}$, and let D_{jk} be the resulting regression depth. So D_{11}, for example, is the depth of the first smooth relative to the points in the first group, and D_{12} is the depth of the first smooth relative to the second group. If H_0 is true, it should be the case that $D_{11} - D_{21}$, as well as $D_{12} - D_{22}$, is relatively small, suggesting the test statistic

$$T = D_{11} - D_{21} + D_{12} - D_{22}. \tag{12.19}$$

Because regression depth is scale-invariant, T is scale-invariant as well. If H_0 is rejected when $T \geq t$, the problem is determining t so as to control the probability of a Type I error. Note that T has a discrete distribution, so in general choosing t so that the Type I error probability is exactly α cannot be accomplished in most cases.

Momentarily assume the running-interval smoother is used. The only known method that has been found to perform well in simulations begins by pooling the data from both groups and using bootstrap samples to estimate the null distribution of T (Wilcox, 2010a). Note that if the null hypothesis is true, the individual smooths estimate the same regression line estimated by the pooled estimate. Let $N = n_1 + n_2$ and generate a bootstrap sample by sampling with replacement N pairs of points from the pooled data. Based on this bootstrap sample, use the first n_1 pairs of points to compute the depths D_{11} and D_{21}, and label the results D_{11}^* and D_{21}^*. In a similar manner, the remaining n_2 points are used to compute D_{12} and D_{22} and the results are labeled D_{12}^* and D_{22}^*. Let

$$T^* = D_{11}^* - D_{21}^* + D_{12}^* - D_{22}^*.$$

Repeat this process B times, yielding $T_1^*, \ldots, T_B^*$, and let

$$P = \frac{1}{B} \sum I_{T > T^*},$$

where $I_{T > T^*} = 1$ if $T > T^*$; otherwise $I_{T > T^*} = 0$. Then a (generalized) p-value is

$$p = 1 - P.$$

If H_0 is rejected when $p \leq 0.05$, simulations indicate that, generally, the actual level is reasonably close to 0.05 when the span is $f = 1$ and both sample sizes are between 40 and 150. However, for smaller or larger sample sizes, an alternative choice for f is required. If the smallest sample size is greater than 150, $f = 0.2$ was found to give good results with sample sizes as large as 800. If $\max(n_1, n_2) < 35$, use f=0.5. For sample sizes between 150 and 180, both $f = 0.2$ and $f = 1$ perform well. But for $\min(n_1, n_2) > 200$, f=0.2 should be used.

The choice of smoother is important. It is unknown, for example, how to control Type I errors reasonably well in simulations when the running interval smoother is replaced by LOWESS. Simulations do indicate that using the quantile smoother COBS, the probability of a Type I error will not exceed the nominal level when using sample sizes of at least 30. However, when testing at the 0.05 level, the actual level can be as low as 0.01 in some situations. Also, for reasons summarized in Section 11.5.6, COBS should be used with caution.

It might seem that regression depth could be used to determine a non-parametric regression line, but this strategy is unsatisfactory. Note that if $x_i \neq x_j$ for all $i \neq j$, then the maximum possible depth, $D = n$, is achieved by taking $m(x_i) = y_i$. The point here is that given a non-parametric fit, based on some appropriate choice for the span, its regression depth can be computed, which in turn can be used to test the hypothesis given by Eq. (12.18).

There is a feature of the global tests in this section that should be stressed, which is relevant to the classic ANCOVA method as well. Imagine that for the first group, the range of the covariate values, x, is 0–10, and for the second group the range is 30–40. Classic ANCOVA would assume that the regression lines are parallel and compare the intercepts. If the usual linear model holds, the methods in this section are aimed at testing the hypothesis that the slopes, as well as the intercepts, are equal. More generally, the methods in this section ignore the fact that the two groups do not have any covariate values in common. But based on the ANCOVA methods in Section 12.1, comparisons would not be made. For instance, it might be of interest to determine whether the groups differ when the covariate $x = 5$. Because data are not available for the second group when $x = 5$, comparisons cannot be made based on the methods in Section 11.11.1. Perhaps comparisons should not be made by imposing assumptions such as those made by the classic ANCOVA model.

12.4.2 R Functions ancsm and Qancsm

The R function

ancsm(x1, y1, x2, y2, nboot = 200, SEED = TRUE, est = tmean, fr = NULL, plotit = TRUE,
sm = FALSE, tr=0.2, xout=FALSE, outfun=out,...)

applies method TG using the running interval smoother. By default, a 20% trimmed mean is used. Setting sm=TRUE, the plots of the regression lines will be based on bootstrap bagging. (But the test of the null hypothesis of identical regression lines is based on the running interval smoother without bootstrap bagging.) The function

Qancsm(x1,y1,x2,y2, crit.mat=NULL, nboot=200, SEED=TRUE, REP.CRIT=FALSE,
qval=0.5, xlab='X',ylab='Y',plotit=TRUE, pr=TRUE, xout=FALSE, outfun=out,...)

is like ancsm, only COBS is used to estimate the quantile regression lines. The argument qval determines the quantile that is used and defaults to 0.5, the median. However, this function should be used with caution for reasons outlined in Section 11.5.6.

12.5 Methods for Dependent Groups

This section considers the situation where there are two outcome measures that are possibly dependent, and there is also a covariate. What is observed is $(x_{i1}, y_{i1}, x_{i2}, y_{i2})$, $i = 1, \ldots, n$. For example, y_{i1} and y_{i2} might be measures taken at two different times and the goal is to take into account covariates x_{i1} and x_{i2}. Let θ_1 and θ_2 be some population measure of location associated with y_{i1} and y_{i2}, respectively, and let θ_d be the corresponding measure of location associated with $y_{i1} - y_{i2}$. As noted in Section 5.9, under general conditions $\theta_1 - \theta_2 \neq \theta_d$. So when dealing with a covariate, some thought is required about which perspective is most appropriate. And there is a third perspective that might be of interest, as noted in Section 5.9.9. For methods similar in spirit to rank-based methods in Section 8.6.12, see, for example, Tsangari and Akritas (2004), as well as Fan and Zhang (2017).

12.5.1 Methods Based on a Linear Model

Let $y_{di} = y_{i1} - y_{i2}$ $(i = 1, \ldots, n)$ and assume that the typical value of y_d, given x, is given by $m(x) = \beta_0 + \beta_1 x$. (Of course, x could be a difference score as well.) Then the method in Section 11.1.12 and the R functions in Section 11.1.13 can be used to make inferences about $m(y_d)$. In particular, $H_0: m(x) = 0$ can be tested for a range of x values via the R function in Section 11.1.13.

It is briefly noted that in the context of a pretest-posttest study, there are concerns about using pretest scores as the covariate. For a discussion of this issue, see for example Erikkson and Häggström (2014).

As for testing $H_0: m_1(x) = m_2(x)$ when y_{i1} and y_{i2} are dependent, the ANCOVA methods described in Section 12.1 are readily modified to handle this situation. Let $\hat{m}_j(x)$ be some estimate of $m_j(x)$ based on (x_{ij}, y_{ij}) $(j = 1, 2)$ and let $\delta = m_1(x) - m_2(x)$. The first step is to estimate the squared standard error of $\hat{\delta} = \hat{m}_1(x) - \hat{m}_2(x)$ in a manner that takes into account that the corresponding estimates are dependent.

Let $(x_{i1}^*, y_{i1}^*, x_{i2}^*, y_{i2}^*)$, $i = 1, \ldots, n$, be a bootstrap sample. Based on this bootstrap sample, and for a single value of the covariate, x, let $\hat{m}_j^*(x) = b_0^* + b_1^* x$, where b_0^* and b_1^* are estimates of the intercept and slope, respectively, based on (x_{ij}^*, y_{ij}^*), $j = 1, 2$. Let

$$D^* = \hat{m}_1^*(x) - \hat{m}_2^*(x).$$

Repeat this process B times, yielding D_b^* ($b = 1, \ldots, B$). Then an estimate of the squared standard error of $\hat{\delta}$ is

$$\hat{\tau}^2 = \frac{1}{B-1} \sum (D_b^* - \bar{D}^*)^2,$$

where $\bar{D}^* = \sum D_b / B$. An appropriate test statistic for testing $H_0: \delta = \theta_0$, where θ_0 is some specified constant, is

$$W = \frac{D - \theta_0}{\hat{\tau}}.$$

When using the Theil–Sen estimator, simulations indicate that W has, approximately, a standard normal distribution. When using least squares regression, assume W has, approximately, a Student's t distribution with $n - 1$ degrees of freedom (Wilcox and Clark, 2014).

One strategy for choosing the covariate values when testing $H_0: m_1(x) = m_2(x)$ is to proceed along the lines in Section 12.2.1 using the data associated with the first covariate. This approach is assumed unless stated otherwise. As for controlling the probability of one or more Type I errors when testing K hypotheses based on K values of the covariate, using the Studentized maximum modulus distribution with infinite degrees of freedom appears to perform well when $K = 5$. (As for least squares regression, again the degrees of freedom are taken to be $n - 1$.)

12.5.2 R Functions Dancts and Dancols

The R function

```
Dancts(x1,y1,x2,y2,pts=NULL,regfun=tsreg, fr1=1, fr2=1, alpha=0.05, plotit=TRUE,
        xout=FALSE, outfun=out, BLO=FALSE, nboot=100, SEED=TRUE, xlab='X',
                           ylab='Y',pr=TRUE,...)
```

tests $H_0: m_1(x) = m_2(x)$. When the argument pts=NULL, it picks five covariate values where the regression lines are compared. These covariate values are chosen based on the values stored in x1 in conjunction with the method described in Section 12.2.1. The R function

```
Dancols(x1, y1, x2, y2, pts = NULL, fr1 = 1, fr2 = 1, plotit = TRUE, xout = FALSE, outfun =
        out, nboot = 100, SEED = TRUE, xlab = 'X', ylab = 'Y', CR = FALSE, ...)
```

is like the function Dancts, only it is designed specifically for the least squares regression estimator.

12.5.3 Dealing With Curvature: Methods DY, DUB, and DTAP and a Method Based on Grids

The ANCOVA methods designed to deal with curvature when comparing independent groups are readily extended to situations where dependent groups are compared.

Consider, for example, method Y in Section 12.2.1. Covariate values can be chosen in the exact same manner for the situation at hand and the K hypotheses indicated by Eq. (12.7) can be tested based on a trimmed mean by replacing Yuen's test with the method in Section 5.9.5. This will be called method DY. Other robust measures of location can be used via a percentile bootstrap method as described in Section 5.9.11.

Method UB, described in Section 12.2.3, which is based on the running interval smoother, can be extended to the case where there is a single covariate measured at two different times and there are two outcome measures that are dependent. Now bootstrap samples are based on re-sampling with replacement from $(x_{i1}, y_{i1}, x_{i2}, y_{i2})$, yielding $(x_{i1}^*, y_{i1}^*, x_{i2}^*, y_{i2}^*)$ $(i = 1, \ldots, n)$. For some specified value for the covariate, x, let $m_j^*(x)$ be the estimate of $m_j(x)$ $(j = 1, 2)$ based on this bootstrap sample and let $D^* = m_1^*(x) - m_2^*(x)$. Repeat this process B times, yielding $D_1^*, \ldots, D_B^*$. Let

$$\tilde{p} = \frac{1}{B} \sum I(D_b^* < 0) + 0.5 I(D_b^* = 0),$$

where the indicator function $I(D_b^* < 0) = 1$ if $D_b^* < 0$; otherwise $I(D_b^* < 0) = 0$. Then a (generalized) p-value is $\hat{p} = 2\min(\tilde{p}, 1 - \tilde{p})$. This will be called method DUB. The probability of one or more Type I errors is controlled via Hochberg's method unless stated otherwise.

A positive feature of method DUB is that it can have higher power than method DY. However, both methods, by default, use relatively few covariate points. Method DTAP uses a relatively large number of covariate points at the expense of higher execution time. Execution time can be reduced substantially if a multicore processor is available. Method DTAP is exactly like method TAP, described in Section 12.2.4, only it compares trimmed means using the method in Section 5.9.5 (Wilcox, 2014). That is, it uses a non-bootstrap method for comparing the trimmed means of two dependent variables. DTAP estimates p_c (the critical p-value as defined in Section 12.2.3) assuming that the four variables are independent, each having a standard normal distribution with a common correlation $\rho = 0.0$. There are indications that power might be improved if the correlation among the variables under study could be taken into account when estimating p_c, but this issue is in need of more study.

One remaining issue is how to deal with two covariates when there is curvature. Simple modifications of the methods in Section 12.3, aimed at dealing with dependent groups, might be used. An alternative approach is to use grids similar to the approach used in Section 12.2.6. Basically, split the data into groups based on the two covariates and compare the resulting dependent groups using methods in Section 5.9.

12.5.4 R Functions Dancova, Dancova.ES.sum, Dancovapb, DancovaUB, Dancdet, Dancovamp, Danc.grid, and ancDEP.MULC.ES

The first five functions in this section deal with a single covariate. The final three deal with situations where there are two or more covariates.

The R function

$$\text{Dancova(x1, y1, x2=x1, y2, fr1 = 1, fr2 = 1, tr=0.2, tr=0.2, plotit = TRUE, pts = NA, sm =}$$
$$\text{FALSE, xout = FALSE, outfun = out, DIF = FALSE, LP = TRUE, xlab = 'X', ylab = 'Y',}$$
$$\text{pch1 = '*', pch2 = '+', ...)}$$

performs method DY, where the arguments y1 and y2 are the dependent variables. By default, the marginal trimmed means are compared. Setting the argument DIF=TRUE, difference scores are used instead. The default is x2=x1, meaning that only the covariate values in x1 are used. Consider a study where participants are measured before an intervention and after an intervention. If the covariate measures are taken only before the intervention, x2=x1 would be used. If the covariate measures are measured after intervention, they can be passed to Dancova via the argument x2.

The R function

$$\text{Dancova.ES.sum(x1, y1, x2=x1, y2, fr1 = 1, fr2 = 1, tr = 0.2, alpha = 0.05, pts = NA, xout =}$$
$$\text{FALSE, outfun = out, REL.MAG = NULL, SEED = TRUE, nboot = 1000, ...)}$$

computes measures of effect size described in Section 5.9.19.

The R function

$$\text{Dancovapb(x1, y1, x2=x1, y2, fr1 = 1, fr2 = 1, est = hd, tr=0.2, nboot = 500, pr = TRUE,}$$
$$\text{SEED = TRUE, plotit = TRUE, pts = NA, sm = FALSE, xout = FALSE, outfun = out, DIF =}$$
$$\text{FALSE, na.rm = TRUE, ...)}$$

is the same as Dancova, only a bootstrap method is used when comparing the marginal measures of location or when testing hypotheses based on the difference scores. If there are missing values, setting na.rm=FALSE and DIF=FALSE, the function uses all of the available data rather than performing casewise deletion.

The R function

DancovaUB(x1 = NULL, y1 = NULL, x2 = NULL, y2 = NULL, xy = NULL, fr1 = 1, fr2 = 1, est = tmean, tr=0.2, plotit = TRUE, xlab = 'X', ylab = 'Y', qvals = c(0.25, 0.5, 0.75), sm = FALSE, xout = FALSE, eout = FALSE, outfun = out, DIF = FALSE, LP = TRUE, method='hochberg',nboot = 500, SEED = TRUE, nreps = 2000, MC = TRUE, SCAT = TRUE, pch1 = '*', pch2 = '+', nmin = 12, q = 0.5, ...)

applies method DUB. If the data are stored in a matrix with four columns, the argument xy can be used. This assumes that the columns correspond to x1, y1, x2, and y2, respectively. Adjusted p-values are reported based on the method indicated by the argument method, which defaults to Hochberg's method. By default, MC=TRUE, meaning that the function uses parallel processing via the R package parallel, assuming that a multicore processor is available.

■ **Example**

Consider again the Well Elderly 2 study. Another goal was to assess the impact of intervention on a measure of perceived health (SF36) before and after intervention using the CAR as a covariate. Fig. 12.4 shows a plot of the smooths. Prior to intervention, the regression line appears to be reasonably straight, but after intervention this is no longer the case. Assuming the data prior to intervention are stored in xx1 (CAR) and yy1 (SF36) and the data after intervention are stored in xx2 and yy2, the command

DancovaUB(xx1,yy1,xx2,yy2,cpp=TRUE)

performs method DUB and creates the plot shown in Fig. 12.4. The function returns a significant result for CAR=−0.099 and 0.317. So the results suggest that when the CAR is negative, perceived health is higher after intervention. For CAR positive, at some point the reverse appears to be true.

■

The R function

Dancdet(x1, y1, x2=x1, y2, fr1 = 1, fr2 = 1, tr=0.2, DIF = TRUE, tr=0.2, plotit = TRUE, plot.dif = FALSE, pts = NA, sm = FALSE, pr = TRUE, xout = FALSE, outfun = out, MC = FALSE, npts = 25, p.crit = NULL, nreps = 2000, SEED = TRUE, SCAT = TRUE, xlab = 'X', ylab = 'Y', pch1 = '*', pch2 = '+', ...)

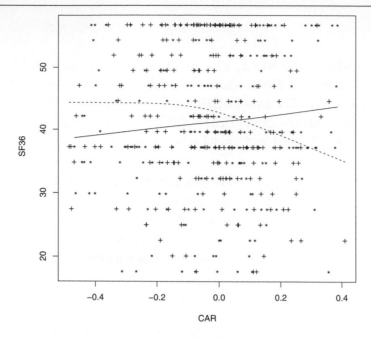

Figure 12.4: The regression lines for predicting the typical SF36 score as a function of the CAR. The solid line is the regression line prior to intervention and the dashed line is the regression line after intervention. Points prior to intervention are indicated by an "o."

applies method DTAP. The argument npts controls how many covariate values will be used. The argument p.crit corresponds to p_c, the adjusted level so that the probability of one or more Type I errors is approximately equal to the value given by the argument alpha. By default p.crit is NULL, which means that it will be computed based on a simulation where the number of replications is controlled via the argument nreps. Execution time can be high, but it can be reduced substantially if a multicore processor is available by setting the argument MC=TRUE. The estimate of p_c depends only on the sample size, so once it has been estimated, specifying its value via the argument p.crit results in very low execution time.

Functions for Two or More Covariates

The R function

$$Dancovamp(x1,y1,x2=x1,y2, fr1=1,fr2=1, tr=0.2, alpha=0.05, pts=NULL,$$
$$SEED=TRUE,DIF=TRUE,cov.fun=skipcov,...)$$

is designed to deal with two or more covariates. It is basically an extension of the R function Dancova. It determines the nearest neighbors via the R function near3d, based on the robust covariance matrix indicated by the argument cov.fun. (In R, type Dancovamp and hit return. At the top are possible choices for cov.fun.)

The R function

$$\text{Danc.grid(x,y1,y2, alpha=0.05, DIF=TRUE, METHOD='TR', AUTO=TRUE,}$$
$$\text{PVSD=FALSE, Qsplit1=0.5, Qsplit2=0.5, SV1=NULL, SV2=NULL, tr=0.2, PB=FALSE,}$$
$$\text{est=tmean, nboot=1000, xout=FALSE, outfun=outpro, SEED=TRUE,...)}$$

splits the data into groups in the same manner as described in Section 12.2.6. Note that this function assumes there is a single covariate that is used for both groups. The argument method indicates the method used to compare the two dependent groups based. When DIF=TRUE, the choices are:

- TR (compare trimmed means using the Tukey–McLaughlin method),
- TRPB (compare trimmed means using a percentile bootstrap),
- MED (inference based on the median of the difference scores),
- AD (inference based on the median of the distribution of the typical difference, see Section 5.9), and
- SIGN (sign test based on an estimate of probability that for a random pair, the first is less than the second).

When DIF=FALSE, only trimmed means are used. When using the sign test, see Section 5.9.18 for an explanation of the arguments AUTO and PVSD. Effect sizes are estimated via the R function

$$\text{ancDEP.MULC.ES(x1,y1,x2,y2, fr1=1.5, fr2=1.5, tr=0.2, pts=NULL, xout=FALSE,}$$
$$\text{outfun=outpro, cov.fun=skip.cov,...).}$$

If pts=NULL, effect sizes are reported only for the covariate points where a significant result is obtained via the function ancdetM4. If no significant results are found, the function returns NA.

12.6 Exercises

1. Comment on the relative merits of using a linear model versus a smoother in the context of ANCOVA.
2. Repeat the first example in Section 12.2.6, only use the measure of meaningful activities stored in column 214 of the Well Elderly data. (The variable label is MAPAFREQ_SUM and gender is indicated by the variable BK_SEX in column 268. The data are stored on the author's web page in the files B1 and A3.)

3. Using the same variables as in the last exercise, perform the analysis using the R function ancova.

4. Comment generally on why method UB and TAP can give different results. Why will they tend to differ in terms of power?

5. For the variables used in Exercise 2, use method UB and TAP. Explain any differences in the results.

6. Repeat the first example in Section 12.2.6, only use the measure of meaningful activities stored in column 253 of the Well Elderly data. (The variable label is MAPAGLOB.)

7. Repeat the previous exercise, but now use method TAP. Compare the plot returned by the R function ancdet to the plot created by ancJN and comment on the results. (Gender is indicated by the variable BK_SEX in column 268.)

References

Acion, L., Peterson, J.J., Temple, S., Arndt, S., 2006. Probabilistic index: an intuitive non-parametric approach to measuring the size of treatment effects. Statistics in Medicine 25, 591–602.

Adam, L., Bejda, P., 2018. Robust estimators based on generalization of trimmed mean. Communications in Statistics—Simulation and Computation 47, 2139–2151.

Adrover, J., Salibian-Barrera, M., 2010. Globally robust confidence intervals for simple linear regression. Computational Statistics & Data Analysis 54, 2899–2913.

Adrover, J., Yohai, V., 2002. Projection estimates of multivariate location. Annals of Statistics 30, 1760–1781.

Agostinelli, C., Leung, A., Yohai, V., Zimmerman, R., 2015. Robust estimation of multivariate location and scatter in the presence of cellwise and casewise contamination. Test 24, 441–461.

Agresti, A., Caffo, B., 2000. Simple and effective confidence intervals for proportions and differences of proportions result from adding two successes and two failures. American Statistician 54, 280–288.

Agresti, A., Coull, B.A., 1998. Approximate is better than "exact" for interval estimation of binomial proportions. American Statistician 52, 119–126.

Agresti, A., Pendergast, J., 1986. Comparing mean ranks for repeated measures data. Communications in Statistics—Theory and Methods 15, 1417–1433.

Agulló, J., Croux, C., Van Aelst, S., 2008. The multivariate least trimmed squares estimator. Journal of Multivariate Analysis 99, 311–338.

Ahmad, S., Ramli, N.M., Midi, H., 2010. Robust estimators in logistic regression: a comparative simulation study. Journal of Modern Applied Statistical Methods 9 (2). https://doi.org/10.22237/jmasm/1288585020.

Ahmed, H., Salha, R., EL-Sayed, H., 2020. Adaptive weighted Nadaraya–Watson estimation of the conditional quantiles by varying bandwidth. Communications in Statistics—Simulation and Computation 49, 1105–1117.

Akbani, R., Kwek, S., Japkowicz, N., 2004. Applying support vector machines to imbalanced datasets. In: ECML, pp. 39–50.

Akritas, M.G., Arnold, S.F., 1994. Fully nonparametric hypotheses for factorial designs I: multivariate repeated measures designs. Journal of the American Statistical Association 89, 336–343.

Akritas, M.G., Arnold, S.F., Brunner, E., 1997. Nonparametric hypotheses and rank statistics for unbalanced factorial designs. Journal of the American Statistical Association 92, 258–265.

Algina, J., Olejnik, S.F., 1984. Implementing the Welch-James procedure with factorial designs. Educational and Psychological Measurement 44, 39–48.

Algina, J., Oshima, T.C., Lin, W.-Y., 1994. Type I error rates for Welch's test and James's second-order test under nonnormality and inequality of variance when there are two groups. Journal of Educational and Behavioral Statistics 19, 275–291.

Algina, J., Keselman, H.J., Penfield, R.D., 2005. An alternative to Cohen's standardized mean difference effect size: a robust parameter and confidence interval in the two independent groups case. Psychological Methods 10, 317–328.

Ammann, L.P., 1993. Robust singular value decompositions: a new approach to projection pursuit. Journal of the American Statistical Association 88, 505–514.

Anaya-Izquierdo, K., Critchley, F., Vines, K., 2011. Orthogonal simple component analysis: a new, exploratory approach. Annals of Applied Statistics 5, 486–522.

Andrews, D.F., Bickel, P.J., Hampel, F.R., Huber, P.J., Rogers, W.H., Tukey, J.W., 1972. Robust Estimates of Location. Princeton University Press, Princeton.

Archimbaud, A., Nordhausen, K., Ruiz-Gazen, A., 2018. ICS for multivariate outlier detection with application to quality control. Computational Statistics & Data Analysis 128, 184–199.

Arcones, M.A., Chen, Z., Gine, E., 1994. Estimators related to U-processes with applications to multivariate medians: asymptotic normality. Annals of Statistics 44, 587–601.

Arnold, B.C., Balakrishnan, N., Nagaraja, H.N., 1992. A First Course in Order Statistics. Wiley, New York.

Asar, Y., Wu, J., 2019. An improved and efficient biased estimation technique in logistic regression model. Communications in Statistics—Theory and Methods 49 (9). https://doi.org/10.1080/03610926.2019.1568494.

Aslam, M., Ahmad, S., 2020. The modified Liu-ridge-type estimator: a new class of biased estimators to address multicollinearity. Communications in Statistics—Simulation and Computation. https://doi.org/10.1080/03610918.2020.1806324.

Atkinson, A.C., 1994. Fast very robust methods for the detection of multiple outliers. Journal of the American Statistical Association 89, 1329–1339.

Babu, G.J., 1986. A note on bootstrapping the variance of sample quantile. Annals of the Institute of Statistical Mathematics 38, 439–443. https://doi.org/10.1007/BF02482530.

Bai, Z.-D., He, X., 1999. Asymptotic distributions of the maximal depth estimators for regression and multivariate location. Annals of Statistics 27, 1616–1637.

Bai, Z.-D., Chen, X.R., Miao, B.Q., Rao, C.R., 1990. Asymptotic theory of least distance estimate in multivariate linear model. Statistics 21, 503–519.

Bailey, B.J.R., 1980. Accurate normalizing transformations of Student's t variate. Applied Statistics 29, 304–306.

Bakker, M., Wicherts, J.M., 2014. Outlier removal, sum scores, and the inflation of the type I error rate in t tests. Psychological Methods 19, 409–427.

Banik, S., Kibria, B.M.G., 2010. Comparison of some parametric and nonparametric type one sample confidence intervals for estimating the mean of a positively skewed distribution. Communications in Statistics—Simulation and Computation 39, 361–380.

Bansal, N.K., Bhandry, M., 1994. Robust M-estimation of the intraclass correlation coefficient. Australian Journal of Statistics 36, 287–301.

Barber, R.F., Candès, E.J., 2015. Controlling the false discovery rate via knockoffs. Annals of Statistics 43, 2055–2085.

Barrett, J.P., 1974. The coefficient of determination—some limitations. Annals of Statistics 28, 19–20.

Barry, D., 1993. Testing for additivity of a regression function. Annals of Statistics 21, 235–254.

Basu, S., DasGupta, A., 1995. Robustness of standard confidence intervals for location parameters under departures from normality. Annals of Statistics 23, 1433–1442.

Bathke, A.C., Solomon, W.H., Madden, L.V., 2008. How to compare small multivariate samples using nonparametric tests. Computational Statistics & Data Analysis 52, 4951–4965.

Baumgartner, W., Weiss, P., Schindler, H., 1998. A nonparametric test for the general two-sample problem. Biometrics 54, 1129–1135.

Beal, S.L., 1987. Asymptotic confidence intervals for the difference between two binomial parameters for use with small samples. Biometrics 43, 941–950. https://doi.org/10.2307/2531547.

Beasley, T.M., 2000. Nonparametric tests for analyzing interactions among intra-block ranks in multiple group repeated measures designs. Journal of Educational and Behavioral Statistics 25, 20–59.

Beasley, T.M., Zumbo, B.D., 2003. Comparison of aligned Friedman rank and parametric methods for testing interactions in split-plot designs. Computational Statistics & Data Analysis 42, 569–593.

Bechhofer, R.E., 1954. A single-sample multiple decision procedure for ranking means of normal populations with known variances. Annals of Mathematical Statistics 25, 16–39.

Bechhofer, R.E., Dunnett, C.W., 1982. Multiple comparisons for orthogonal contrasts. Technometrics 24, 213–222.

Becker, R.A., Chambers, J.M., Wilks, A.R., 1988. The New S Language. Wadsworth & Brooks/Cole, Pacific Grove, CA.

Bedall, P.J., Zimmermann, H., 1979. AS 143: the median centre. Applied Statistics 28, 325–328.

Bellman, R.E., 1961. Adaptive Control Processes. Princeton University Press, Princeton, NJ.

Belsley, D.A., Kuh, E., Welsch, R.E., 1980. Regression Diagnostics: Identifying Influential Data and Sources of Collinearity. Wiley, New York.

Ben, M.G., Martínez, E., Yohai, V.J., 2006. Robust estimation for the multivariate linear model based on a τ-scale. Journal of Multivariate Analysis 90, 1600–1622.

Benjamini, Y., 1983. Is the t test really conservative when the parent distribution is long-tailed? Journal of the American Statistical Association 78, 645–654.

Benjamini, Y., Hochberg, Y., 1995. Controlling the false discovery rate: a practical and powerful approach to multiple testing. Journal of the Royal Statistical Society, B 57, 289–300.

Berger, R.L., 1996. More powerful tests from confidence interval p values. American Statistician 50, 314–318.

Bernhardson, C., 1975. Type I error rates when multiple comparison procedures follow a significant F test of ANOVA. Biometrics 31, 719–724.

Bernholdt, T., Fischer, P., 2004. The complexity of computing the MCD-estimator. Theoretical Computer Science 326, 383–393.

Bertsimas, D., Mazumder, R., 2014. Least quantile regression via modern optimization. Annals of Statistics 42, 2494–2525.

Bessel, F.W., 1818. Fundamenta Astronomiae pro anno MDCCLV deducta ex observationibus viri incomparabilis James Bradley in specula astronomica Grenovicensi per annos 1750–1762 institutis. Friedrich Nicolovius, Königsberg.

Bianco, A.M., Martínez, E., 2009. Robust testing in the logistic regression model. Computational Statistics & Data Analysis 53, 4095–4105.

Bianco, A.M., Yohai, V.J., 1996. Robust estimation in the logistic regression model. In: Reider, H. (Ed.), Robust Statistics, Data Analysis, and Computer Intensive Methods. In: Lecture Notes in Statistics, vol. 109. Springer, New York, pp. 17–34.

Biau, D.J., Brigitte, M., Jolles, M.J., Porcher, R., 2010. P value and the theory of hypothesis testing: an explanation for new researchers. Clinical Orthopaedics and Related Research 468, 885–892.

Bickel, P.J., Lehmann, E.L., 1975. Descriptive statistics for nonparametric models II. Location. Annals of Statistics 3, 1045–1069.

Bickel, P.J., Lehmann, E.L., 1976. Descriptive statistics for nonparametric models III. Dispersion. Annals of Statistics 4, 1139–1158.

Biesanz, J.C., Falk, C.F., Savalei, V., 2010. Assessing mediational models: testing and interval estimation for indirect effects. Multivariate Behavioral Research 45, 661–701.

Billor, N., Kiral, G., 2008. A comparison of multiple outlier detection methods for regression data. Communications in Statistics—Simulation and Computation 37, 521–545.

Birkes, D., Dodge, Y., 1993. Alternative Methods of Regression. Wiley, New York.

Bischl, B., Lang, M., Kotthoff, L., Schiffner, J., Richter, J., Studerus, E., Casalicchio, G., Jones, Z.M., 2016. mlr: machine learning in R. Journal of Machine Learning Research 17, 1–5. http://jmlr.org/papers/v17/15-066.html.

Bishara, A., Hittner, J.A., 2012. Testing the significance of a correlation with nonnormal data: comparison of Pearson, Spearman, transformation, and resampling approaches. Psychological Methods 17, 399–417.

Bishara, A., Li, J., Nash, T., 2017. Asymptotic confidence intervals for the Pearson correlation via skewness and kurtosis. British Journal of Mathematical and Statistical Psychology 71, 167–185.

Bjerve, S., Doksum, K., 1993. Correlation curves: measures of association as functions of covariate values. Annals of Statistics 21, 890–902.

Blair, R.C., Lawson, S.B., 1982. Another look at the robustness of the product-moment correlation coefficient to population non-normality. Florida Journal of Educational Research 24, 11–15.

Blyth, C.R., 1986. Approximate binomial confidence limits. Journal of the American Statistical Association 81, 843–855.

Boente, G., Pardo-Fernández, J.C., 2016. Robust testing for superiority between two regression curves. Computational Statistics & Data Analysis 97, 151–168.

Boente, G., Rodriguez, D., 2010. Robust inference in generalized partially linear models. Computational Statistics & Data Analysis 54, 2942–2966.

Boente, G., Ruiz, M., Zamar, R.H., 2010. On a robust local estimator for the scale function in heteroscedastic nonparametric regression. Statistics & Probability Letters 80, 1185–1195.

Boik, R.J., 1987. The Fisher-Pitman permutation test: a non-robust alternative to the normal theory F test when variances are heterogeneous. British Journal of Mathematical and Statistical Psychology 40, 26–42.

Bondell, H.D., 2005. Minimum distance estimation for the logistic regression model. Biometrika 92, 724–731.

Bondell, H.D., 2008. A characteristic function approach to the biased sampling model, with application to robust logistic regression. Journal of Statistical Planning and Inference 138, 742–755.

Bondell, H.D., Stefanski, L.A., 2013. Efficient robust regression via two-stage generalized empirical likelihood. Journal of the American Statistical Association 108, 644–655.

Booth, J.G., Sarkar, S., 1998. Monte Carlo approximation of bootstrap variances. American Statistician 52, 354–357.

Bowman, A., Young, S., 1996. Graphical comparison of nonparametric curves. Applied Statistics 45, 83–98.

Box, G.E.P., 1954. Some theorems on quadratic forms applied in the study of analysis of variance problems, I. Effect of inequality of variance in the one-way model. Annals of Mathematical Statistics 25, 290–302.

References

Bradley, J.V., 1978. Robustness? British Journal of Mathematical and Statistical Psychology 31, 144–152. https://doi.org/10.1111/j.2044-8317.1978.tb00581.x.

Brand, R.J., Pinnock, D.E., Jackson, K.L., 1973. Large sample confidence bands for the logistic response curve and its inverse. American Statistician 27, 157–160.

Breiman, L., 1995. Better subset regression using the nonnegative garrote. Technometrics 37, 373–384.

Breiman, L., 1996a. Heuristics of instability and stabilization in model selection. Annals of Statistics 24, 2350–2383.

Breiman, L., 1996b. Bagging predictors. Machine Learning 24, 123–140.

Breiman, L., 2001a. Random forests. Machine Learning 45, 5–32.

Breiman, L., 2001b. Statistical modeling: the two cultures. Statistical Science 16, 199–215.

Brown, B.M., 1983. Statistical uses of the spatial median. Journal of the Royal Statistical Society, B 45, 25–30.

Brown, L., Li, X., 2005. Confidence intervals for two sample binomial distribution. Journal of Statistical Planning and Inference 130, 359–375.

Brown, L.D., Cai, T.T., DasGupta, A., 2002. Confidence intervals for a binomial proportion and asymptotic expansions. Annals of Statistics 30, 160–201.

Brown, M.B., Forsythe, A., 1974. The small sample behavior of some statistics which test the equality of several means. Technometrics 16, 129–132.

Bruffaerts, C., Verardi, V., Vermandele, C., 2014. A generalized boxplot for skewed and heavy-tailed distributions. Statistics & Probability Letters 95, 110–117.

Brunner, E., Munzel, U., 2000. The nonparametric Behrens-Fisher problem: asymptotic theory and small-sample approximation. Biometrical Journal 42, 17–25.

Brunner, E., Dette, H., Munk, A., 1997. Box-type approximations in non-parametric factorial designs. Journal of the American Statistical Association 92, 1494–1502.

Brunner, E., Domhof, S., Langer, F., 2002a. Nonparametric Analysis of Longitudinal Data in Factorial Experiments. Wiley, New York.

Brunner, E., Munzel, U., Puri, M.L., 2002b. The multivariate nonparametric Behrens–Fisher problem. Journal of Statistical Planning and Inference 108, 37–53.

Brunner, E., Bathke, A.C., Konietschke, F., 2019. Rank and Pseudo Rank Procedures for Independent Observations in Factorial Designs. Springer, Cham Switzerland.

Brys, G., Hubert, M., Struyf, A., 2004. A robust measure of skewness. Journal of Computational and Graphical Statistics 13, 996–1017.

Bühlmann, P., Yu, B., 2002. Analyzing bagging. Annals of Statistics 30, 927–961.

Buja, A., Hastie, T., Tibshirani, R., 1989. Linear smoothers and additive models (with discussion). Annals of Statistics 17, 453–555.

Bulut, H., 2020. Mahalanobis distance based on minimum regularized covariance determinant estimators for high dimensional data. Communications in Statistics—Theory and Methods. https://doi.org/10.1080/03610926.2020.1719420.

Büning, H., 2001. Kolmogorov-Smirnov and Cramer von Mises type two-sample tests with various weights. Communications in Statistics—Theory and Methods 30, 847–866.

Butler, R.W., Davies, P.L., Jhun, M., 1993. Asymptotics for the minimum covariance determinant estimator. Annals of Statistics 21, 1385–1400.

Calhoun, P., Hallett, M.J., Su, X., et al., 2020. Random forest with acceptance-rejection trees. Computational Statistics 35, 983–999. https://doi.org/10.1007/s00180-019-00929-4.

Cantoni, E., Ronchetti, E., 2001. Robust inference for generalized linear models. Journal of the American Statistical Association 96, 1022–1030.

Cao, C., Pauly, M., Konietschke, F., 2019. The Behrens-Fisher problem with covariates and baseline adjustments. https://doi.org/10.1007/s00184-019-00729-2.

Cao, Y., Xu, Y., Ming, T., Tan, M.T., Chen, P., Chongyang Duan, C., 2020. A simple and improved score confidence interval for a single proportion. Communications in Statistics—Theory and Methods. https://doi.org/10.1080/03610926.2020.1779747.

Carling, K., 2000. Resistant outlier rules and the non-Gaussian case. Computational Statistics & Data Analysis 33, 249–258.

Carroll, R.J., Pedersen, S., 1993. On robustness in the logistic regression model. Journal of the Royal Statistical Society, B 55, 693–706.

Carroll, R.J., Ruppert, D., 1982. Robust estimation in heteroscedastic linear models. Annals of Statistics 10, 429–441.

Carroll, R.J., Ruppert, D., 1988. Transformation and Weighting in Regression. Chapman and Hall, New York.

Carroll, R.J., Welsh, A.H., 1988. A note on asymmetry and robustness in linear regression. American Statistician 42, 285–287.

Celik, N., 2020. Welch's ANOVA: heteroskedastic skew-t error terms. Communications in Statistics—Theory and Methods. https://doi.org/10.1080/03610926.2020.1788084.

Cerioli, A., 2010. Multivariate outlier detection with high-breakdown estimators. Journal of the American Statistical Association 105, 147–156.

Cerioli, A., Farcomeni, A., 2011. Error rates for multivariate outlier detection. Computational Statistics & Data Analysis 55, 544–553.

Cevallos-Valdiviezo, H., Van Aelst, S., 2019. Fast computation of robust subspace estimators. Computational Statistics & Data Analysis 134, 171–185. https://doi.org/10.1016/j.csda.2018.12.013.

Ceyhan, E., Goad, C.L., 2009. A comparison of analysis of covariate-adjusted residuals and analysis of covariance. Communications in Statistics—Simulation and Computation 38, 2019–2038.

Chambers, J.M., 1998. Programming with Data. A Guide to the S Language. Springer-Verlag, New York.

Chambers, J.M., Hastie, T.J., 1992. Statistical Models in S. Chapman & Hall, New York.

Chang, W.H., McKean, J.W., Naranjo, J.D., Sheather, S.J., 1999. High-breakdown rank regression. Journal of the American Statistical Association 94, 205–219.

Chaouch, M., Goga, C., 2010. Design-based estimation for geometric quantiles with applications to outlier detection. Computational Statistics & Data Analysis 54, 2214–2229.

Chatterjee, S., Hadi, A.S., 1988. Sensitivity Analysis in Linear Regression Analysis. Wiley, New York.

Chaudhuri, A., Hu, W., 2019. A fast algorithm for computing distance correlation. Computational Statistics & Data Analysis 135, 15–24.

Chaudhuri, P., 1996. On a geometric notion of quantiles for multivariate data. Journal of the American Statistical Association 91, 862–872.

Chawla, N., Bowyer, K., Hall, L., Kegelmeyer, W., 2002. SMOTE: synthetic minority over-sampling technique. Journal of Artificial Intelligence Research 16, 341–378.

Chen, L., 1995. Testing the mean of skewed distributions. Journal of the American Statistical Association 90, 767–772.

Chen, S.X., Van Keilegom, I., 2009. A review on empirical likelihood methods for regression. Test 18, 415–447.

Chen, S.X., Zhong, P.-S., 2010. ANOVA for longitudinal data with missing values. Annals of Statistics 38, 3630–3659.

Chen, T., Martin, E., Montague, G., 2009. Robust probabilistic PCA with missing data and contribution analysis for outlier detection. Computational Statistics & Data Analysis 53, 3706–3716.

Chen, T.-C., Victoria-Feser, M.-P., 2002. High-breakdown estimation of multivariate mean and covariance with missing observations. British Journal of Mathematical and Statistical Psychology 55, 317–336.

Chen, Z., Tyler, D.E., 2002. The influence function and maximum bias of Tukey's median. Annals of Statistics 30, 1737–1760.

Chernick, M.R., 1999. Bootstrap Methods: A Practitioner's Guide. Wiley, New York.

Choi, K., Marden, J., 1997. An approach to multivariate rank tests in multivariate analysis of variance. Journal of the American Statistical Association 92, 1581–1590.

Chow, G.C., 1960. Tests of equality between sets of coefficients in two linear regressions. Econometrika 28, 591–606.

Chowdhury, N., Cook, D., Hofmann, H., Majumder, M., Lee, E-K., Toth, A., 2015. Using visual statistical inference to better understand random class separations in high dimension, low sample size data. Computational Statistics 30, 293–316.

Christmann, A., 1994. Least median of weighted squares in logistic regression with large strata. Biometrika 81, 413–417.

Chung, E., Romano, J.P., 2013. Exact and asymptotically robust permutation tests. Annals of Statistics 41, 484–507.

Claeskens, G., Hubert, M., Slaets, L., Vakili, K., 2014. Multivariate functional halfspace depth. Journal of the American Statistical Association 109, 411–423.

Clark, F., Jackson, J., Carlson, M., Chou, C.-P., Cherry, B.J., Jordan-Marsh, M., Knight, B.G., Mandel, D., Blanchard, J., Granger, D.A., Wilcox, R.R., Lai, M.Y., White, B., Hay, J., Lam, C., Marterella, A., Azen, S.P., 2011. Effectiveness of a lifestyle intervention in promoting the well-being of independently living older people: results of the Well Elderly 2 Randomise Controlled Trial. Journal of Epidemiology and Community Health 66, 782–790. https://doi.org/10.1136/jech.2009.099754.

Clements, A., Hurn, S., Lindsay, K., 2003. Mobius-like mappings and their use in kernel density estimation. Journal of the American Statistical Association 98, 993–1000.

Cleveland, W.S., 1979. Robust locally weighted regression and smoothing scatterplots. Journal of the American Statistical Association 74, 829–836. https://doi.org/10.2307/2286407.

Cleveland, W.S., 1985. The Elements of Graphing Data. Chapman & Hall, New York.

Cleveland, W.S., Devlin, S.J., 1988. Locally-weighted regression: an approach to regression analysis by local fitting. Journal of the American Statistical Association 83, 596–610.

Cliff, N., 1993. Dominance statistics: ordinal analyses to answer ordinal questions. Psychological Bulletin 114, 494–509.

Cliff, N., 1994. Predicting ordinal relations. British Journal of Mathematical and Statistical Psychology 47, 127–150.

Cliff, N., 1996. Ordinal Methods for Behavioral Data Analysis. Erlbaum, Mahwah, NJ.

Clopper, C., Pearson, E.S., 1934. The use of confidence or fiducial limits illustrated in the case of the binomial. Biometrika 26, 404–413. https://doi.org/10.1093/biomet/26.4.404.

Coakley, C.W., Hettmansperger, T.P., 1993. A bounded influence, high breakdown, efficient regression estimator. Journal of the American Statistical Association 88, 872–880. https://doi.org/10.2307/2290776.

Coe, P.R., Tamhane, A.C., 1993. Small sample confidence intervals for the difference, ratio, and odds ratio of two success probabilities. Communications in Statistics—Simulation and Computation 22, 925–938.

Cohen, J., 1988. Statistical Power Analysis for the Behavioral Sciences, 2nd ed. Academic Press, New York.

Cohen, M., Dalal, S.R., Tukey, J.W., 1993. Robust, smoothly heterogeneous variance regression. Applied Statistics 42, 339–354.

Cole, D.A., Maxwell, S.E., 2003. Testing meditational models with longitudinal data: questions and tips in the use of structural equation modeling. Journal of Abnormal Psychology 112, 558–577.

Conerly, M.D., Mansfield, E.R., 1988. An approximate test for comparing heteroscedastic regression models. Journal of the American Statistical Association 83, 811–817.

Cook, R.D., Hawkins, D.M., 1990. Discussion of Unmasking multivariate outliers and leverage points by P. Rousseuw and B. van Zomeren. Journal of the American Statistical Association 85, 640–644.

Cook, R.D., Weisberg, S., 1992. Residuals and Influence in Regression. Chapman and Hall, New York.

Cook, R.D., Hawkins, D.M., Weisberg, S., 1992. Comparison of model misspecification diagnostics using residuals from least mean of squares and least median of squares fit. Journal of the American Statistical Association 87, 419–424.

Copas, J.B., 1983. Plotting p against x. Applied Statistics 32, 25–31.

Copt, S., Heritier, S., 2007. Robust alternatives to the F-Test in mixed linear models based on MM-estimates. Biometrics 63, 1045–1052.

Cramér, H., 1946. Mathematical Methods of Statistics. Princeton University Press, Princeton.

Crawley, M.J., 2007. The R Book. Wiley, New York.

Cressie, N.A.C., Whitford, H.J., 1986. How to use the two sample t-test. Biometrical Journal 28, 131–148.

Cribari-Neto, F., 2004. Asymptotic inference under heteroscedasticity of unknown form. Computational Statistics & Data Analysis 45, 215–233.

Cribari-Neto, F., Lima, M., 2014. New heteroskedasticity-robust standard errors for the linear regression model. Brazilian Journal of Probability and Statistics 28, 83–95.

Cribari-Neto, F., Souza, T.C., Vasconcellos, K.L.P., 2007. Inference under heteroskedasticity and leveraged data. Communications in Statistics—Theory and Methods 36, 1977–1988 Erratum: Communications in Statistics—Theory and Methods 37, 3329–3330 (2008).

Cribbie, R.A., Fiksenbaum, L., Keselman, H.J., Wilcox, R.R., 2012. Effects of nonnormality on test statistics for one-way independent groups designs. British Journal of Mathematical and Statistical Psychology 65, 56–73.

Croux, C., 1994. Efficient high-breakdown M-estimators of scale. Statistics & Probability Letters 19, 371–379.

Croux, C., Dehon, C., 2003. Estimators of the multiple correlation coefficient: local robustness and confidence intervals. Statistical Papers 44, 315–334.

Croux, C., Dehon, C., 2010. Influence functions of the Spearman and Kendall correlation measures. Statistical Methods & Applications 19, 497–515.

Croux, C., Haesbroeck, G., 2000. Principal component analysis based on robust estimators of the covariance or correlation matrix: influence functions and efficiencies. Biometrika 87, 603–618.

Croux, C., Haesbroeck, G., 2003. Implementing the Bianco and Yohai estimator for logistic regression. Computational Statistics & Data Analysis 44, 273–295.

Croux, C., Ruiz-Gazen, A., 2005. High breakdown estimators for principal components: the projection-pursuit approach revisited. Journal of Multivariate Analysis 95, 206–226.

Croux, C., Rousseeuw, P.J., Hössjer, O., 1994. Generalized S-estimators. Journal of the American Statistical Association 89, 1271–1281.

Croux, C., Flandre, C., Haesbroeck, G., 2002. The breakdown behavior of the maximum likelihood estimator in the logistic regression model. Statistics & Probability Letters 60, 377–386.

Cuesta-Albertos, J.A., Nieto-Reyes, A., 2008. The random Tukey depth. Computational Statistics & Data Analysis 52, 4979–4988.

Cuesta-Albertos, J.A., Gordaliza, A., Matran, C., 1997. Trimmed k-means: an attempt to robustify quantizers. Annals of Statistics 25, 553–576.

Cuevas, A., Febrero, M., Fraiman, R., 2004. An anova test for functional data. Computational Statistics & Data Analysis 47, 111–122.

Cui, X., Lin, L., Yang, G.R., 2008. An extended projection data depth and its applications to discrimination. Communications in Statistics—Theory and Methods 37, 2276–2290.

Cushny, A.R., Peebles, A.R., 1904. The action of optical isomers II. Hyoscines. Journal of Physiology 32, 501–510.

Dahlquist, G., Björck, A., 1974. Numerical Methods. Prentice Hall, Englewood Cliffs, NJ.

Dana, E., 1990. Salience of the self and salience of standards: attempts to match self to standard. Unpublished Ph.D. dissertation. University of Southern California.

Danilov, M., Yohai, V.J., Zamar, R.H., 2012. Robust estimation of multivariate location and scatter in the presence of missing data. Journal of the American Statistical Association 107, 1178–1186.

Davidson, R., MacKinnon, J.G., 2000. Bootstrap tests: how many bootstraps? Econometric Reviews 19, 55–68.

Davies, L., Gather, U., 1993. The identification of multiple outliers (with discussion). Journal of the American Statistical Association 88, 782–792.

Davies, P.L., 1987. Asymptotic behavior of S-estimates of multivariate location parameters and dispersion matrices. Annals of Statistics 15, 1269–1292.

Davies, P.L., 1990. The asymptotics of S-estimators in the linear regression model. Annals of Statistics 18, 1651–1675.

Davies, P.L., 1993. Aspects of robust linear regression. Annals of Statistics 21, 1843–1899.

Davis, J.B., McKean, J.W., 1993. Rank-based method for multivariate linear models. Journal of the American Statistical Association 88, 245–251.

Davison, A.C., Hinkley, D.V., 1997. Bootstrap Methods and Their Application. Cambridge University Press, Cambridge, UK.

Davison, A.C., Hinkley, D.V., Young, G.A., 2003. Recent developments in bootstrap methodology. Statistical Science 18, 141–157.

de Boor, C., 1978. A Practical Guide to Splines. Springer-Verlag, New York.

De Jongh, P.J., De Wet, T., Welsh, A.H., 1988. Mallows-type bounded-influence-regression trimmed means. Journal of the American Statistical Association 83, 805–810.

De Neve, J., Thas, O., 2017. A Mann–Whitney type effect measure of interaction for factorial designs. Communications in Statistics—Theory and Methods 46 (22), 11243–11260. https://doi.org/10.1080/03610926.2016.1263739.

De Schryver, M., De Neve, J., 2019. A tutorial on probabilistic index models: regression models for the effect size $P(Y1 < Y2)$. Psychological Methods 24, 403–418. https://doi.org/10.1037/met0000194.

Debruyne, M., Hubert, M., van Horebeek, J.V., 2010. Detecting influential observations in kernel PCA. Computational Statistics & Data Analysis 54, 3007–3019.

Debruyne, M., Höppner, S., Serneels, S., Verdonck, T., 2019. Outlyingness: which variables contribute most? Statistical Computation 29, 707–723.

Delattre, S., Roquain, E., 2015. New procedures controlling the false discovery proportion via Romano–Wolf's heuristic. Annals of Statistics 43, 1141–1177.

Delgado, M.A., 1993. Testing the equality of nonparametric regression curves. Statistics & Probability Letters 17, 199–204.

Derksen, S., Keselman, H.J., 1992. Backward, forward and stepwise automated subset selection algorithms: frequency of obtaining authentic and noise variables. British Journal of Mathematical and Statistical Psychology 45, 265–282.

Dette, H., 1999. A consistent test for the functional form of a regression based on a difference of variances estimator. Annals of Statistics 27, 1012–1040.

Dette, H., Neumeyer, N., 2001. Nonparametric analysis of covariance. Annals of Statistics 29, 1361–1400.

Devlin, S.J., Gnanadesikan, R., Kettenring, J.R., 1981. Robust estimation of dispersion matrices and principal components. Journal of the American Statistical Association 76, 354–362.

Devroye, L., Lugosi, G., 2001. Combinatorial Methods in Density Estimation. Springer-Verlag, New York.

DiCiccio, C.J., Romano, J.P., 2017. Robust permutation tests for correlation and regression coefficients. Journal of the American Statistical Association 112, 1211–1220.

DiCiccio, T., Hall, P., Romano, J., 1991. Empirical likelihood is Bartlett-correctable. Annals of Statistics 19, 1053–1061.

Dielman, T., Pfaffenberger, R., 1982. LAV (least absolute value) estimation in linear regression: a review. In: Zanakis, S.H., Rustagi, J. (Eds.), Optimization on Statistics. North-Holland, New York.

Dielman, T., Lowry, C., Pfaffenberger, R., 1994. A comparison of quantile estimators. Communications in Statistics—Simulation and Computation 23, 355–371.

Dietz, E.J., 1987. A comparison of robust estimators in simple linear regression. Communications in Statistics—Simulation and Computation 16, 1209–1227.

Dietz, E.J., 1989. Teaching regression in a nonparametric statistics course. American Statistician 43, 35–40.

Diggle, P.J., Heagerty, P.J., Liang, K.-Y., Zeger, S.L., 2002. Analysis of Longitudinal Data, 2nd ed. Oxford University Press, Oxford.

Dixon, S.L., McKean, J.W., 1996. Rank-based analysis of the heteroscedastic linear model. Journal of the American Statistical Association 91, 699–712.

Dixon, W.J., Tukey, J.W., 1968. Approximate behavior of the distribution of Winsorized t (Trimming/Winsorization 2). Technometrics 10, 83–98.

Doksum, K.A., 1974. Empirical probability plots and statistical inference for nonlinear models in the two-sample case. Annals of Statistics 2, 267–277.

Doksum, K.A., 1977. Some graphical methods in statistics. A review and some extensions. Statistica Neerlandica 31, 53–68.

Doksum, K.A., Koo, J.-Y., 2000. On spline estimators and prediction intervals in nonparametric regression. Computational Statistics & Data Analysis 35, 67–82.

Doksum, K.A., Samarov, A., 1995. Nonparametric estimation of global functionals and a measure of the explanatory power of covariates in regression. Annals of Statistics 23, 1443–1473.

Doksum, K.A., Sievers, G.L., 1976. Plotting with confidence: graphical comparisons of two populations. Biometrika 63, 421–434.

Doksum, K.A., Wong, C.-W., 1983. Statistical tests based on transformed data. Journal of the American Statistical Association 78, 411–417.

Don, H., Olive, D., 2019. Bootstrapping analogs of the one way MANOVA test. Communications in Statistics—Theory and Methods 48, 5546–5558. https://doi.org/10.1080/03610926.2018.1515363.

Donoho, D.L., 1982. Breakdown properties of multivariate location estimators. PhD qualifying paper. Harvard University.

Donoho, D.L., Gasko, M., 1992. Breakdown properties of the location estimates based on halfspace depth and projected outlyingness. Annals of Statistics 20, 1803–1827.

Duncan, G.T., Layard, M.W., 1973. A Monte-Carlo study of asymptotically robust tests for correlation. Biometrika 60, 551–558.

Dunnett, C.W., 1980. Pairwise multiple comparisons in the unequal variance case. Journal of the American Statistical Association 75, 796–800.

Dutta, S., Ghosh, A.K., 2012. On robust classification using projection depth. Annals of the Institute of Statistical Mathematics 64, 657–676.

Dyckerhoff, R., Mozharovskyi, P., 2016. Exact computation of the half space depth. Computational Statistics & Data Analysis 98, 19–30.

Edgell, S.E., Noon, S.M., 1984. Effect of violation of normality on the t test of the correlation coefficient. Psychological Bulletin 95, 576–583.

Efromovich, S., 1999. Nonparametric Curve Estimation: Methods, Theory and Applications. Springer-Verlag, New York.

Efron, B., 1987. Better bootstrap confidence intervals. Journal of the American Statistical Association 82, 171–185.

Efron, B., 2020. Prediction, estimation, and attribution. Journal of the American Statistical Association 115, 636–655. https://doi.org/10.1080/01621459.2020.1762613.

Efron, B., Hastie, T., 2016. Computer Age Statistical Inference. Cambridge University Press, New York.

Efron, B., Tibshirani, R.J., 1993. An Introduction to the Bootstrap. Chapman & Hall, New York.

Efron, B., Tibshirani, R.J., 1997. Improvements on cross-validation: the 632+ bootstrap method. Journal of the American Statistical Association 92, 548–560.

Efron, B., Hastie, T., Johnstone, I., Tibshirani, R., 2004. Least angle regression (with discussion and rejoinder). Annals of Statistics 32, 407–499.

Elashoff, J.D., Snow, R.E., 1970. A case study in statistical inference: reconsideration of the Rosenthal-Jacobson data on teacher expectancy. Technical report no. 15. School of Education, Stanford University.

Ellis, R.L., 1844. On the method of least squares. Transactions of the Cambridge Philosophical Society 8, 204–219l.

Emerson, J.D., Stoto, M.A., 1983. Transforming data. In: Hoaglin, D.C., Mosteller, F., Tukey, J.W. (Eds.), Understanding Robust and Exploratory Data Analysis. Wiley, New York.

Engelen, S., Hubert, M., Vanden Branden, K., 2005. A comparison of three procedures for robust PCA in high dimensions. Australian Journal of Statistics 2, 117–126.

Erceg-Hurn, D.M., Steed, L.G., 2011. Does exposure to cigarette health warnings elicit psychological reactance in smokers? Journal of Applied Social Psychology 41, 219–237. https://doi.org/10.1111/j.1559-1816.2010.00710.x.

Erikkson, K., Häggström, O., 2014. Lord's paradox in a continuous setting and a regression artifact in numerical cognition research. PLoS ONE 9, e95949. https://doi.org/10.1371/journal.pone.0095949.

Ertaş, H., Kaçiranlar, S., Güer, H., 2017. Robust Liu-type estimator for regression based on M-estimator. Communications in Statistics—Simulation and Computation 46, 3907–3932. https://doi.org/10.1080/03610918.2015.1045077.

Etran, E., Akay, K.U., 2020. A new Liu-type estimator in binary logistic regression models. Communications in Statistics—Theory and Methods. https://doi.org/10.1080/03610926.2020.1813777.

Eubank, R.L., 1999. Nonparametric Regression and Spline Smoothing. Marcel Dekker, New York.

Everitt, B.S., Landau, S., Leese, M., Stahl, D., 2011. Cluster Analysis, 5th ed. Wiley, New York.

Fairley, D., 1986. Cherry trees with cones? American Statistician 40, 138–139.

Fan, C., Zhang, D., 2017. Rank repeated measures analysis of covariance. Communications in Statistics—Theory and Methods 46, 1158–1183.

Fan, J., 1993. Local linear smoothers and their minimax efficiencies. Annals of Statistics 21, 196–216.

Fan, J., 1996. Test of significance based on wavelet thresholding and Neyman's truncation. Journal of the American Statistical Association 91, 674–688.

Fan, J., Gijbels, I., 1996. Local Polynomial Modeling and Its Applications. CRC Press, Boca Raton, FL.

Fan, J., Hall, P., 1994. On curve estimation by minimizing mean absolute deviation and its implications. Annals of Statistics 22, 867–885.

Faraway, J.J., Sun, J., 1995. Simultaneous confidence bands for linear regression with heteroscedastic error terms. Journal of the American Statistical Association 90, 1094–1098.

Febrero, M., Galeano, P., González-Manteiga, W., 2008. Outlier detection in functional data by depth measures, with application to identify abnormal nox levels. Environmetrics 19, 331–345.

Febrero-Bande, M., de la Fuente, M.O., 2012. Statistical computing in functional data analysis: the R package fda.usc. Journal of Statistical Software 51. http://www.jstatsoft.org/.

Feng, L., Zou, C., Wang, Z., Zhu, L., 2015. Robust comparison of regression curves. Test 24, 185–204.

Fenstad, G.U., 1983. A comparison between U and V tests in the Behrens-Fisher problem. Biometrika 70, 300–302.

Ferraty, F., Vieu, P., 2006. Nonparametric Functional Data Analysis: Theory and Practice. Springer, New York.

Ferreira, E., Stute, W., 2004. Testing for differences between conditional means in a time series context. Journal of the American Statistical Association 99, 169–174.

Ferretti, N., Kelmansky, D., Yohai, V.J., Zamar, R.H., 1999. A class of locally and globally robust regression estimates. Journal of the American Statistical Association 94, 174–188.

Filzmoser, P., Maronna, R., Werner, M., 2008. Outlier identification in high dimensions. Computational Statistics & Data Analysis 52, 1694–1711.

Filzmoser, P., Höppner, S., Ortner, I., Serneels, S., Verdonck, T., 2020. Cellwise robust M regression. Computational Statistics & Data Analysis 147, 106944. https://doi.org/10.1016/j.csda.2020.106944.

Fisher, R.A., 1922. On the mathematical foundations of theoretical statistics. Philosophical Transactions of the Royal Astronomical Society of London, Series A 222, 309–368.

Fisher, R.A., 1932. Statistical Methods for Research Workers. Oliver and Boyd, London.

Fix, E., Hodges, J.L., 1951. Discriminatory analysis–nonparametric discrimination: consistency properties. Technical report 4. USAF School of Aviation Medicine, Randolph Field, Texas.

Fligner, M.A., Policello II, G.E., 1981. Robust rank procedures for the Behrens-Fisher problem. Journal of the American Statistical Association 76, 162–168.

Flores, P., Ocaña, J., 2019. Pretesting strategies for homoscedasticity when comparing means. Their robustness facing non-normality. Communications in Statistics—Simulation and Computation. https://doi.org/10.1080/03610918.2019.1649698.

Flores, S., 2010. On the efficient computation of robust regression estimators. Computational Statistics & Data Analysis 54, 3044–3056.

Fox, J., 1999. Applied Regression Analysis, Linear Models, and Related Methods. Sage, Thousands Oaks, CA.

Fox, J., 2001. Multiple and Generalized Nonparametric Regression. Sage, Thousands Oaks, CA.

Fox, J., 2002. An R and S-PLUS Companion to Applied Regression. Sage, Thousands Oaks, CA.

Frahm, G., Jaekel, U., 2010. A generalization of Tyler's M-estimators to the case of incomplete data. Computational Statistics & Data Analysis 54, 374–393.

Frank, I.E., Friedman, J.H., 1993. A statistical view of some chemometrics regression tools. Technometrics 2, 109–135.

Freedman, D., Diaconis, P., 1981. On the histogram as density estimator: L_2 theory. Zeitschrift für Wahrscheinlichkeitstheorie und Verwandte Gebiete 57, 453–476.

Freedman, D., Diaconis, P., 1982. On inconsistent M-estimators. Annals of Statistics 10, 454–461.

Freidlin, B., Gastwirth, J.L., 2000. Should the median test be retired from general use? American Statistician 54, 161–164.

Frey, J., Zhang, Y., 2017. What do interpolated nonparametric confidence intervals for population quantiles guarantee? American Statistician 71, 305–309.

Friedman, J.H., Stuetzle, W., 1981. Projection pursuit regression. Journal of the American Statistical Association 76, 817–823.

Friedrich, S. Frank, Konietschke, F., Pauly, M., 2017. A wild bootstrap approach for nonparametric repeated measurements. Computational Statistics & Data Analysis 113, 38–52.

Frigge, M., Hoaglin, D.C., Iglewicz, B., 1989. Some implementations of the Boxplot. American Statistician 43, 50–54.

Fritz, H., Filzmoser, P., Croux, C., 2012. A comparison of algorithms for the multivariate L_1 median. Computational Statistics 27, 393–410.

Fung, K.Y., 1980. Small sample behaviour of some nonparametric multi-sample location tests in the presence of dispersion differences. Statistica Neerlandica 34, 189–196.

Fung, W.-K., 1993. Unmasking outliers and leverage points: a confirmation. Journal of the American Statistical Association 88, 515–519.

Gail, M.H., Santner, T.J., Brown, C.C., 1980. An analysis of comparative carcinogenesis experiments with multiple times to tumor. Biometrics 36, 255–266.

Galeano, P., Joseph, E., Lillo, R.E., 2015. The Mahalanobis distance for functional data with applications to classifications. Technometrics 57, 281–291.

Gao, X., Alvo, M., 2005. A nonparametric test for interaction in two-way layouts. Canadian Journal of Statistics 33, 529–543.

Gather, U., Hilker, T., 1997. A note on Tyler's modification of the MAD for the Stahel-Donoho estimator. Annals of Statistics 25, 2024–2026.

Gatto, R., Ronchetti, E., 1996. General saddlepoint approximations of marginal densities and tail probabilities. Journal of the American Statistical Association 91, 666–673.

Genton, M.G., Lucas, A., 2003. Comprehensive definitions of breakdown points for independent and dependent observations. Journal of the Royal Statistical Society, B 65, 81–94.

Gervini, D., 2002. The influence function of the Stahel-Donoho estimator of multivariate location and scatter. Statistics & Probability Letters 60, 425–435.

Gervini, D., 2012. Outlier detection and trimmed estimation for general functional data. Statistica Sinica 22, 1639–1660.

Gervini, D., Yohai, V.J., 2002. A class of robust and fully efficient regression estimators. Annals of Statistics 30, 583–616.

Ghosh, A.K., Chaudhuri, P., 2005. On maximum depth and related classifiers. Scandinavian Journal of Statistics 32, 327–350.

Ghosh, M., Parr, W.C., Singh, K., Babu, G.J., 1985. A note on bootstrapping the sample median. Annals of Statistics 12, 1130–1135. https://doi.org/10.1214/aos/1176346731.

Gibbons, J., Olkin, I., Sobel, M., 1987. Selecting and Ordering Populations: A New Statistical Methodology. Society for Industrial and Applied Mathematics.

Gignac, G., Szodorai, E., 2016. Effect size guidelines for individual differences researchers. Personality and Individual Differences 102, 74–78.

Gijbels, I., Vrinssen, I., 2015. Robust nonnegative garrote variable selection in linear regression. Computational Statistics & Data Analysis 85, 1–22.

Gleason, J.R., 1993. Understanding elongation: the scale contaminated normal family. Journal of the American Statistical Association 88, 327–337.

Glen, N.L., Zhao, Y., 2007. Weighted empirical likelihood estimates and their robustness properties. Computational Statistics & Data Analysis 51, 5130–5141.

Gnanadesikan, R., Kettenring, J.R., 1972. Robust estimates, residuals and outlier detection with multiresponse data. Biometrics 28, 81–124.

Godfrey, L.G., 2006. Tests for regression models with heteroskedasticity of unknown form. Computational Statistics & Data Analysis 50, 2715–2733.

Goldberg, K.M., Iglewicz, B., 1992. Bivariate extensions of the boxplot. Technometrics 34, 307–320.

Golub, G.H., van Loan, C.F., 1983. Matrix Computations. Johns Hopkins University Press, Baltimore, MD.

Gong, Y., Abebe, A., 2012. On the iteratively reweighted rank regression estimator. Communications in Statistics—Simulation and Computation 41, 155–166.

Good, P., 2000. Permutation Tests. Springer-Verlag, New York.

Górecki, T., Smaga, L., 2015. A comparison of tests for the one-way ANOVA problem for functional data. Computational Statistics 30, 987–1010. https://doi.org/10.1007/s00180-015-0555-0.

Graybill, F.A., 1976. Theory and Application of the Linear Model. Wadsworth, Belmont, CA.

Graybill, F.A., 1983. Matrices with Applications in Statistics. Wadsworth, Belmont, CA.

Green, D.P., Ha, S.E., Bullock, J.G., 2010. Enough already about 'black box' experiments: studying mediation is more difficult than most scholars suppose. Annals of the American Academy of Political and Social Science 628, 200–208.

Green, P.J., Silverman, B.W., 1993. Nonparametric Regression and Generalized Linear Models: A Roughness Penalty Approach. CRC Press, Boca Raton, FL.

Gribkova, N., 2017. Cramér-type moderate deviations for intermediate trimmed means. Communications in Statistics—Theory and Methods 46, 11918–11932. https://doi.org/10.1080/03610926.2017.1285930.

Grissom, R.J., 2000. Heterogeneity of variance in clinical data. Journal of Consulting and Clinical Psychology 68, 155–165.

Gul, A., Perperoglou, A., Khan, Z., Mahmoud, O., Miftahuddin, M., Adler, W., Lausen, B., 2018. Ensemble of a subset of kNN classifiers. Advances in Data Analysis and Classification 12, 827–840. https://doi.org/10.1007/s11634-015-0227-5.

Guo, J.H., Luh, W.M., 2000. An invertible transformation two-sample trimmed t-statistic under heterogeneity and nonnormality. Statistics & Probability Letters 49, 1–7.

Guo, J.H., Billard, L., Luh, W.-M., 2011. New heterogeneous test statistics for the unbalanced fixed-effect nested design. British Journal of Mathematical and Statistical Psychology 64, 259–276.

Guo, W., He, L., Sarkar, S.K., 2014. Further results on controlling the false discovery proportion. Annals of Statistics 42, 1070–1101.

Gupta, A.K., Rathie, P.N., 1983. On the distribution of the determinant of sample correlation matrix from multivariate Gaussian population. Metron 61, 43–56.

Gupta, S.S., Panchapakesan, S., 1987. Multiple Decision Procedures: Theory and Methodology of Selecting and Ranking Populations. SIAM.

Gutenbrunner, C., Jurečková, J., 1992. Regression rank scores and regression quantiles. Annals of Statistics 20, 305–330.

Gutenbrunner, C., Jurečková, J., Koenker, R., Portnoy, S., 1993. Tests of linear hypotheses based on regression rank scores. Journal of Nonparametric Statistics 2, 307–331.

Györfi, L., Kohler, M., Krzyzk, A., Walk, H., 2002. A Distribution-Free Theory of Nonparametric Regression. Springer Verlag, New York.

Haldane, J.B.S., 1948. Note on the median multivariate distribution. Biometrika 35, 414–415.

Hall, P., 1986. On the number of bootstrap simulations required to construct a confidence interval. Annals of Statistics 14, 1431–1452.

Hall, P., 1988a. On symmetric bootstrap confidence intervals. Journal of the Royal Statistical Society, B 50, 35–45.

Hall, P., 1988b. Theoretical comparison of bootstrap confidence intervals. Annals of Statistics 16, 927–953.

Hall, P., 1992. On the removal of skewness by transformation. Journal of the Royal Statistical Society, B 54, 221–228.

Hall, P., Hart, J.D., 1990. Bootstrap test for difference between means in nonparametric regression. Journal of the American Statistical Association 85, 1039–1049.

Hall, P., Horowitz, J., 2013. A simple bootstrap method for constructing nonparametric confidence bands for functions. Annals of Statistics 41, 1892–1921.

Hall, P., Jones, M.C., 1990. Adaptive M-estimation in nonparametric regression. Annals of Statistics 18, 1712–1728.

Hall, P., Presnell, B., 1999. Biased bootstrap methods for reducing the effects of contamination. Journal of the Royal Statistical Society, B 61, 661–680.

Hall, P., Sheather, S.J., 1988. On the distribution of a Studentized quantile. Journal of the Royal Statistical Society, B 50, 380–391.

Hall, P., Welsh, A.H., 1985. Limit theorems for the median deviation. Annals of the Institute of Statistical Mathematics 37 (A), 27–36.

Hall, P., Huber, C., Speckman, P.L., 1997. Covariate-matched one-sided tests for the difference between functional means. Journal of the American Statistical Association 92, 1074–1083.

Hall, P.G., Hall, D., 1995. The Bootstrap and Edgeworth Expansion. Springer Verlag, New York.

Hallin, M., Paindaveine, D., Verdebout, T., 2014. Efficient R-estimation of principal and common principal components. Journal of the American Statistical Association 109, 1071–1083.

Hamilton, L.C., 1992. Regression with Graphics: A Second Course in Applied Statistics. Brooks/Cole, Pacific Grove, CA.

Hampel, F.R., 1968. Contributions to the theory of robust estimation. Unpublished Ph.D. dissertation. University of California, Berkeley.

Hampel, F.R., 1973. Robust estimation: a condensed partial survey. Zeitschrift für Wahrscheinlichkeitstheorie und Verwandte Gebiete 27, 87–104.

Hampel, F.R., 1974. The influence curve and its role in robust estimation. Journal of the American Statistical Association 62, 1179–1186.

Hampel, F.R., 1975. Beyond location parameters: robust concepts and methods (with discussion). Bulletin of the ISI 46, 375–391.

Hampel, F.R., Ronchetti, E.M., Rousseeuw, P.J., Stahel, W.A., 1986. Robust Statistics. Wiley, New York.

Hand, A., 1998. A History of Mathematical Statistics from 1750 to 1930. Wiley, New York.

Handschin, E., Schweppe, F.C., Kohlas, J., Fiechter, A., 1975. Bad data analysis for power system state estimation. IEEE Transactions of Power Apparatus and Systems PAS-94, 329–337.

Härdle, W., 1990. Applied Nonparametric Regression. Econometric Society Monographs, vol. 19. Cambridge University Press, Cambridge, UK.

Härdle, W., Korostelev, A., 1996. Search for significant variables in nonparametric additive regression. Biometrika 83, 541–549.

Härdle, W., Marron, J.S., 1990. Semiparametric comparison of regression curves. Annals of Statistics 18, 63–89.

Harpole, J.K., Woods, C.M., Rodebaugh, T.L., Levinson, C.A., Lenze, E.J., 2014. How bandwidth selection algorithms impact exploratory data analysis using kernel density estimation. Psychological Methods 19, 428–443.

Harrar, S.W., Ronchi, F., Salmaso, L., 2018. A comparison of recent nonparametric methods for testing effects in two-by-two factorial designs. Journal of Applied Statistics DOI. https://doi.org/10.1080/02664763.2018.1555575.

Harrell, F.E., Davis, C.E., 1982. A new distribution-free quantile estimator. Biometrika 69, 635–640.

Harwell, M., 2003. Summarizing Monte Carlo results in methodological research: the single-factor, fixed effects ANCOVA case. Journal of Educational and Behavioral Statistics 28, 45–70.

Hastie, T., Tibshirani, R., Friedman, J., 2001. The Elements of Statistical Learning. Springer, New York.

Hastie, T.J., Loader, C., 1993. Local regression: automatic kernel carpentry. Statistical Science 8, 120–143.

Hastie, T.J., Tibshirani, R.J., 1990. Generalized Additive Models. Chapman and Hall, New York.

Hawkins, D.G., Simonoff, J.S., 1993. Algorithm AS 282: high breakdown regression and multivariate estimation. Applied Statistics 42, 423–431.

Hawkins, D.M., Olive, D., 1999. Applications and algorithms for least trimmed sum of absolute deviations regression. Computational Statistics & Data Analysis 32, 119–134.

Hawkins, D.M., Olive, D., 2002. Inconsistency of resampling algorithms for high-breakdown regression estimators and a new algorithm. Journal of the American Statistical Association 97, 136–147.

Hayes, A.F., Cai, L., 2007. Further evaluating the conditional decision rule for comparing two independent means. British Journal of Mathematical and Statistical Psychology 60, 217–244.

He, X., Ng, P., 1999. Quantile splines with several covariates. Journal of Statistical Planning and Inference 75, 343–352.

He, X., Portnoy, S., 1992. Reweighted LS estimators converge at the same rate as the initial estimator. Annals of Statistics 20, 2161–2167.

He, X., Simpson, D.G., 1993. Lower bounds for contamination bias: global minimax versus locally linear estimation. Annals of Statistics 21, 314–337.

He, X., Wang, G., 1997. Convergence of depth contours for multivariate data sets. Annals of Statistics 25, 495–504.

He, X., Zhu, L.-X., 2003. A lack-of-fit test for quantile regression. Journal of the American Statistical Association 98, 1013–1022.

He, X., Simpson, D.G., Portnoy, S.L., 1990. Breakdown robustness of tests. Journal of the American Statistical Association 85, 446–452.

He, X., Ng, P., Portnoy, S., 1998. Bivariate quantile smoothing splines. Journal of the Royal Statistical Society, B 60, 537–550.

Headrick, T.C., Kowalchuk, R.K., Sheng, Y., 2008. Parametric probability densities and distribution functions for Tukey g-and-h transformations and their use for fitting data. Applied Mathematical Sciences 2, 449–462.

Hedges, L.V., Olkin, I., 1985. Statistical Methods for Meta-Analysis. Academic Press, New York, NY.

Hennig, C., Viroli, C., 2016. quantileDA: quantile classifier. https://CRAN.R-project.org/package=quantileDA. R package version 1.1.

Herbert, R.D., Hayen, A., Macaskill, P., Walter, S.D., 2011. Interval estimation for the difference of two independent variances. Communications in Statistics—Simulation and Computation 40, 744–758.

Heritier, S., Ronchetti, E., 1994. Robust bounded-influence tests in general linear models. Journal of the American Statistical Association 89, 897–904.

Heritier, S., Cantoni, E., Copt, S., Victoria-Feser, M.-P., 2009. Robust Methods in Biostatistics. Wiley, New York.

Herwindiati, D.E., Djauhari, M.A., Mashuri, M., 2007. Robust multivariate outlier labeling. Communications in Statistics—Simulation and Computation 36, 1287–1294.

Hettmansperger, T.P., 1984. Statistical Inference Based on Ranks. Wiley, New York.

Hettmansperger, T.P., McKean, J.W., 1977. A robust alternative based on ranks to least squares in analyzing linear models. Technometrics 19, 275–284.

Hettmansperger, T.P., McKean, J.W., 1998. Robust Nonparametric Statistical Methods. Arnold, London.

Hettmansperger, T.P., McKean, J.W., 2011. Robust Nonparametric Statistical Methods, 2nd ed. Arnold, London.

Hettmansperger, T.P., Sheather, S.J., 1986. Confidence interval based on interpolated order statistics. Statistical Probability Letters 4, 75–79.

Hill, M., Dixon, W.J., 1982. Robustness in real life: a study of clinical laboratory data. Biometrics 38, 377–396.

Hill, R.W., 1977. Robust regression when there are outliers in the carriers. Unpublished PhD dissertation. Harvard University.

Hill, R.W., Holland, P.W., 1977. Two robust alternatives to robust regression. Journal of the American Statistical Association 72, 828–833.

Hilton, J.F., Mehta, C.R., Patel, N.R., 1994. An algorithm for conducting exact Smirnov tests. Computational Statistics & Data Analysis 19, 351–361.

Hoaglin, D.C., 1985. Summarizing shape numerically: the g-and-h distribution. In: Hoaglin, D., Mosteller, F., Tukey, J. (Eds.), Exploring Data Tables Trends and Shapes. Wiley, New York, pp. 461–515.

Hoaglin, D.C., Iglewicz, B., 1987. Fine-tuning some resistant rules for outlier labeling. Journal of the American Statistical Association 82, 1147–1149.

Hoaglin, D.C., Welsch, R., 1978. The hat matrix in regression and ANOVA. American Statistician 32, 17–22.

Hochberg, Y., 1975. Simultaneous inference under Behrens-Fisher conditions: a two sample approach. Communications in Statistics 4, 1109–1119.

Hochberg, Y., 1988. A sharper Bonferroni procedure for multiple tests of significance. Biometrika 75, 800–802.

Hochberg, Y., Tamhane, A.C., 1987. Multiple Comparison Procedures. Wiley, New York.

Hodges, J.L., Lehmann, E.L., 1963. Estimates of location based on rank tests. Annals of Mathematical Statistics 34, 598–611.

Hoerl, A.E., Kennard, R.W., 1970. Ridge regression: biased estimation for nonorthogonal problems. Technometrics 12, 55–67. https://doi.org/10.2307/1271436.

Hogg, R.V., 1974. Adaptive robust procedures: a partial review and some suggestions for future applications and theory. Journal of the American Statistical Association 69, 909–922.

Holladay, J.T., Wilcox, R.R., Koch, D.D., Wang, L., 2020. Review and recommendations for univariate statistical analysis of spherical equivalent prediction error for intraocular lens power calculation. Journal of Cataract and Refractive Surgery 47, 65–77. https://doi.org/10.1097/j.jcrs.0000000000000370.

Hollander, M., Wolfe, D.A., 1973. Nonparametric Statistical Methods. Wiley, New York.

Holm, S., 1979. A simple sequentially rejective multiple test procedure. Scandinavian Journal of Statistics 6, 65–70.

Hommel, G., 1988. A stagewise rejective multiple test procedure based on a modified Bonferroni test. Biometrika 75, 383–386.

Horowitz, J.L., Spokoiny, V.G., 2002. An adaptive, rate-optimal test of linearity for median regression models. Journal of the American Statistical Association 97, 822–835.

Hosmer, D.W., Lemeshow, S., 1989. Applied Logistic Regression. Wiley, New York.

Hössjer, O., 1992. On the optimality of S-estimators. Statistics & Probability Letters 14, 413–419.

Hössjer, O., 1994. Rank-based estimates in the linear model with high breakdown point. Journal of the American Statistical Association 89, 149–158.

Hössjer, O., Croux, C., 1995. Generalizing univariate signed rank statistics for testing and estimating a multivariate location parameter. Journal of Nonparametric Statistics 4, 293–308.

Hsu, J.C., 1981. Simultaneous confidence intervals for all distances from the 'best'. Annals of Statistics 9, 1026–3104. https://doi.org/10.1214/aos/1176345582.

Hu, X., Jung, A., Qin, G., 2020. Interval estimation of the correlation coefficient. American Statistician 74, 29–36. https://doi.org/10.1080/00031305.2018.1437077.

Huber, P.J., 1964. Robust estimation of location parameters. Annals of Mathematical Statistics 35, 73–101.

Huber, P.J., 1981. Robust Statistics. Wiley, New York.

Huber, P.J., Ronchetti, E., 2009. Robust Statistics, 2nd ed. Wiley, New York.

Hubert, M., Rousseeuw, P.J., 1998. The catline for deep regression. Journal of Multivariate Analysis 66, 270–296.

Hubert, M., Vandervieren, E., 2008. An adjusted boxplot for skewed distributions. Computational Statistics & Data Analysis 52, 5186–5201.

Hubert, M., Rousseeuw, P.J., Verboven, S., 2002. A fast method for robust principal components with applications to chemometrics. Chemometrics and Intelligent Laboratory Systems 60, 101–111.

Hubert, M., Rousseeuw, P.J., Vanden Branden, K., 2005. ROBPCA: a new approach to robust principal component analysis. Technometrics 47, 64–79.

Hubert, M., Rousseeuw, P.J., Verdonck, T., 2012. A deterministic algorithm for robust location and scatter. Journal of Computational and Graphical Statistics 21, 618–637.

Hubert, M., Rousseeuw, P.J., Segaert, P., 2015a. Multivariate functional outlier detection. Statistical Methods & Applications 24, 177–202.

Hubert, M., Rousseeuw, P.J., Vanpaemel, D., Verdonck, T., 2015b. The DetS and DetMM estimators for multivariate location and scatter. Computational Statistics & Data Analysis 81, 64–75.

Hubert, M., Rousseeuw, P.J., Segaert, P., 2017. Multivariate and functional classification using depth and distance. Advances in Data Analysis and Classification 11, 445–466. https://doi.org/10.1007/s11634-016-0269-3.

Hubert, M., Rousseeuw, P.J., Van den Bossche, W., 2019. MacroPCA: an all-in-one PCA method allowing for missing values as well as cellwise and rowwise outliers. Technometrics 61, 459–473. https://doi.org/10.1080/00401706.2018.1562989.

Huberty, C.J., 1989. Problems with stepwise methods—better alternatives. Advances in Social Science Methodology 1, 43–70.

Huberty, C.J., 1994. Applied Discriminate Analysis. Wiley, New York.

Huitema, B.E., 2011. The Analysis of Covariance and Alternatives, 2nd ed. Wiley, New York.

Hussain, S.S., Sprent, P., 1983. Non-parametric regression. Journal of the Royal Statistical Society 146, 182–191.

Hwang, J., Jorn, H., Kim, J., 2004. On the performance of bivariate robust location estimators under contamination. Computational Statistics & Data Analysis 44, 587–601.

Hyndman, R.J., Fan, Y., 1996. Sample quantiles in statistical packages. American Statistician 50, 361–365.

Hyndman, R.J., Shang, H.L., 2010. Rainbow plots, bagplots, and boxplots for functional data. Journal of Computational and Graphical Statistics 19, 29–45.

Iglewicz, B., 1983. Robust scale estimators and confidence intervals for location. In: Hoaglin, D., Mosteller, F., Tukey, J. (Eds.), Understanding Robust and Exploratory Data Analysis. Wiley, New York, pp. 404–431.

Ivokić, I., Rajić, V., Stanojević, J., 2020. Coverage probabilities of confidence intervals for the slope parameter of linear regression model when the error term is not normally distributed. Communications in Statistics—Theory and Methods 49, 147–158. https://doi.org/10.1080/03610918.2018.1476702.

Jackson, J., Mandel, D., Blanchard, J., Carlson, M., Cherry, B., Azen, S., Chou, C.-P., Jordan-Marsh, M., Forman, T., White, B., Granger, D., Knight, B., Clark, F., 2009. Confronting challenges in intervention research with ethnically diverse older adults: the USC Well Elderly II trial. Clinical Trials 6, 90–101.

Jaeckel, L.A., 1972. Estimating regression coefficients by minimizing the dispersion of residuals. Annals of Mathematical Statistics 43, 1449–1458.

James, G., Witten, D., Hastie, T., Tibshirani, R., 2017. An Introduction to Statistical Learning: With Applications in R (7th printing). Springer, New York.

Janssen, A., Pauls, T., 2005. How do bootstrap and permutation tests work? Annals of Statistics 31, 786–806.

Jeyaratnam, S., Othman, A.R., 1985. Test of hypothesis in one-way random effects model with unequal error variances. Journal of Statistical Computation and Simulation 21, 51–57.

Jhun, Myoungshic, Choi, Inkyung, 2009. Bootstrapping least distance estimator in the multivariate regression model. Computational Statistics & Data Analysis 53, 4221–4227.

Jiang, Y., Wang, Y.-G., Fu, L., Wang, X., 2019. Robust estimation using modified Huber's functions with new tails. Technometrics 61, 111–123.

Jöckel, K.-H., 1986. Finite sample properties and asymptotic efficiency of Monte Carlo tests. Annals of Statistics 14, 336–347.

Johansen, S., 1980. The Welch-James approximation of the distribution of the residual sum of squares in weighted linear regression. Biometrika 67, 85–92.

Johnson, N.J., 1978. Modified t tests and confidence intervals for asymmetrical populations. Journal of the American Statistical Association 73, 536–544.

Johnson, N.L., Kotz, S., 1970. Distributions in Statistics: Continuous Univariate Distributions-2. Wiley, New York.

Johnson, P., Neyman, J., 1936. Tests of certain linear hypotheses and their application to some educational problems. Statistical Research Memoirs 1, 57–93.

Johnson, W.D., Romer, J.E., 2016. Hypothesis testing of population percentiles via the Wald test with bootstrap variance estimates. Open Journal of Statistics 6, 14–24. https://doi.org/10.4236/ojs.2016.61003.

Jones, L.V., Tukey, J.W., 2000. A sensible formulation of the significance test. Psychological Methods 5, 411–414.

Jorgensen, J.O., Gilles, R.B., Hunt, D.R., Caplehorn, J.R.M., Lumley, T., 1995. A simple and effective way to reduce postoperative pain after laparoscopic cholecystectomy. Australian and New Zealand Journal of Surgery 65, 466–469.

Judd, C.M., Kenny, D.A., 1981a. Estimating the Effects of Social Interventions. Cambridge University Press, New York.

Judd, C.M., Kenny, D.A., 1981b. Process analysis: estimating mediation in treatment evaluations. Evaluation Review 5, 602–619.

Jung, Y., Lee, S., Hu, J., 2016. Robust regression for highly corrupted response by shifting outliers. Statistical Modeling 16, 1–23.

Jurečková, J., Portnoy, S., 1987. Asymptotics for one-step M-estimators with application to combining efficiency and high breakdown point. Communications in Statistics—Theory and Methods 16, 2187–2199.

Kai, B., Runze, L., Zou, H., 2010. Local composite quantile regression smoothing: an efficient and safe alternative to local polynomial regression. Journal of the Royal Statistical Society, B 72, 49–69.

Kaizar, E.E., Li, Y., Hsu, J.C., 2011. Permutation multiple tests of binary features do not uniformly control error rates. Journal of the American Statistical Association 106, 1067–1074.

Kallenberg, W.C., Ledwina, T., 1999. Data-driven rank tests for independence. Journal of the American Statistical Association 94, 285–310.

Kan, B., Alpu, Ö., Yazici, B., 2013. Robust ridge and robust Liu estimator for regression based on the LTS estimator. Journal of Applied Statistics 40, 644–655. https://doi.org/10.1080/02664763.2012.750285.

Kay, R., Little, S., 1987. Transformation of the explanatory variables in the logistic regression model for binary data. Biometrika 74, 495–501.

Kent, J.T., Tyler, D.E., 1996. Constrained M-estimation for multivariate location and scatter. Annals of Statistics 24, 1346–1370.

Keselman, H.C., Keselman, J.C., Lix, L.M., 1995. The analysis of repeated measurements: univariate tests, multivariate tests, or both? British Journal of Mathematical and Statistical Psychology 48, 319–338.

Keselman, H.C., Huberty, C.J., Lix, L.M., Olejnik, S., Cribbie, R.A., Donahue, B., Kowalchuk, R.K., Lowman, L.L., Petoskey, M.D., Keselman, J.C., Levin, J.R., 1998. Statistical practices of educational researchers: an analysis of their ANOVA, MANOVA and ANCOVA analyses. Review of Educational Research 68, 350–386.

Keselman, H.C., Wilcox, R.R., Othman, A.R., Fradette, K., 2002. Trimming, transforming statistics, and bootstrapping: circumventing the biasing effects of heteroscedasticity and nonnormality. Journal of Modern Applied Statistical Methods 1, 288–309.

Keselman, H.C., Wilcox, R.R., Lix, L.M., Algina, J., Fradette, K., 2007. Adaptive robust estimation and testing. British Journal of Mathematical and Statistical Psychology 60, 267–294.

Keselman, H.J., Carriere, K.C., Lix, L.M., 1993. Testing repeated measures hypotheses when covariance matrices are heterogeneous. Journal of Educational Statistics 18, 305–319.

Keselman, H.J., Algina, J., Kowalchuk, R.K., Wolfinger, R.D., 1999. A comparison of recent approaches to the analysis of repeated measurements. British Journal of Mathematical and Statistical Psychology 52, 62–78.

Keselman, H.J., Algina, J., Wilcox, R.R., Kowalchuk, R.K., 2000. Testing repeated measures hypotheses when covariance matrices are heterogeneous: revisiting the robustness of the Welch-James test again. Educational and Psychological Measurement 60, 925–938.

Keselman, H.J., Othman, A.R., Wilcox, R.R., Fradette, K., 2004. The new and improved two-sample t test. Psychological Science 15, 47–51.

Keselman, H.J., Miller, C.W., Holland, B., 2011. Many tests of significance: new methods for controlling type I errors. Psychological Methods 16, 420–431. https://doi.org/10.1037/a0025810.

Keselman, H.J., Othman, A.R., Wilcox, R.R., 2016. Generalized linear model analyses for treatment group equality when data are non-normal. Journal of Modern Applied Statistical Methods 15, 32–61.

Khan, J.A., van Aelst, S., Zamar, R., 2010. Fast robust estimation of prediction error based on resampling. Computational Statistics & Data Analysis 54, 3121–3130.

Khorasani, F., Milliken, G.A., 1982. Simultaneous confidence bands for nonlinear regression models. Communications in Statistics—Theory and Methods 11, 1241–1253.

Khuri, A.I., 1992. Tests concerning a nested mixed model with heteroscedastic random effects. Journal of Statistical Planning and Inference 30, 33–44.

Kibria, B.M.G., 2003. Performance of some new ridge regression estimators. Communications in Statistics—Simulation and Computation 32, 419–435. https://doi.org/10.1081/SAC-120017499.

Kim, M., 2007. Quantile regression with varying coefficients. Annals of Statistics 35, 92–108.

Kim, P.J., Jennrich, R.I., 1973. Tables of the exact sampling distribution of the two-sample Kolmogorov-Smirnov criterion, D_{mn}, $m \leq n$. In: Harter, H.L., Owen, D.B. (Eds.), Selected Tables in Mathematical Statistics, Vol. I. American Mathematical Society, Providence, Rhode Island.

Kim, S.-J., 1992a. A practical solution to the multivariate Behrens-Fisher problem. Biometrika 79, 171–176.

Kim, S.-J., 1992b. The metrically trimmed mean as a robust estimator of location. Annals of Statistics 20, 1534–1547.

Kim, Y.J., Cribbie, R.A., 2017. ANOVA and the variance homogeneity assumption: exploring a better gatekeeper. British Journal of Mathematical and Statistical Psychology 71, 1–12.

King, E.C., Hart, J.D., Wherly, T.E., 1991. Testing the equality of two regression curves using linear smoothers. Statistics & Probability Letters 12, 239–247.

Kirk, R.E., 1995. Experimental Design. Brooks/Cole, Pacific Grove, CA.

Kloke, J., McKean, J., Rashid, M., 2009. Rank-based estimation and associated inferences for linear models with cluster correlated errors. Journal of the American Statistical Association 104, 384–390.

Kloke, J.D., McKean, J.W., 2012. Rfit: rank-based estimation for linear models. The R Journal 4, 57–64.

Kmetz, J.L., 2019. Correcting corrupt research: recommendations for the profession to stop misuse of p-values. American Statistician 73 (sup1), 36–45. https://doi.org/10.1080/00031305.2018.1518271.

Knight, K., 1998. Limiting distributions for L_1 regression estimators under general conditions. Annals of Statistics 26, 755–770.

Koenker, R., 1994. Confidence intervals for regression quantiles. In: Mandl, P., Huskova, M. (Eds.), Asymptotic Statistics. In: Proceedings of the Fifth Prague Symposium, pp. 349–359.

Koenker, R., 2005. Quantile Regression. Cambridge University Press, New York.

Koenker, R., Bassett, G., 1978. Regression quantiles. Econometrika 46, 33–50.

Koenker, R., Bassett, G., 1981. Robust tests for heteroscedasticity based on regression quantiles. Econometrika 50, 43–61.

Koenker, R., Ng, P., 2005. Inequality constrained quantile regression. Sankhya: The Indian Journal of Statistics 67, 418–440.

Koenker, R., Park, B.J., 1996. An interior point algorithm for nonlinear quantile regression. Journal of Econometrics 71, 265–283.

Koenker, R., Portnoy, S., 1987. L-estimation for linear models. Journal of the American Statistical Association 82, 851–857.

Koenker, R., Xiao, Z.J., 2002. Inference on the quantile regression process. Econometrica 70, 1583–1612.

Koenker, R., Ng, P., Portnoy, S., 1994. Quantile smoothing splines. Biometrika 81, 673–680.

Koller, M., 2013. Robust Estimation of Linear Mixed Models. Unpublished PhD dissertation. Dept. of Mathematics, ETH, Zurich.

Koller, M., 2016. robustlmm: an R package for robust estimation of linear mixed-effects models. Journal of Statistical Software 75 (6). https://doi.org/10.18637/jss.v075.i06.

Koller, M., Stahel, W.A., 2011. Sharpening Wald-type inference in robust regression for small samples. Computational Statistics & Data Analysis 55, 2504–2515.

Koshevoy, G., Mosler, K., 1997. Zonoid trimming for multivariate distributions. Annals of Statistics 25, 1998–2017.

Kosinski, A., 1999. A procedure for the detection of multivariate outliers. Computational Statistics & Data Analysis 29, 145–161.

Kowalchuk, R.K., Headrick, T.C., 2010. Simulating multivariate g-and-h distributions. British Journal of Mathematical and Statistical Psychology 63, 63–74.

Kowalski, C.J., 1972. On the effects of non-normality on the distribution of the sample product-moment correlation coefficient. Applied Statistics 21, 1–12.

Kraemer, H.C., Kupfer, D.J., 2006. Size of treatment effects and their importance to clinical research and practice. Biological Psychiatry 59, 990–996.

Krasker, W.S., 1980. Estimation in linear regression models with disparate data points. Econometrika 48, 1333–1346.

Krasker, W.S., Welsch, R.E., 1982. Efficient bounded influence regression estimation. Journal of the American Statistical Association 77, 595–604.

Krause, A., Olson, M., 2002. The Basics of S-PLUS. Springer-Verlag, New York.

Krishnamoorthy, K., Lu, F., Mathew, T., 2007. A parametric bootstrap approach for ANOVA with unequal variances: fixed and random models. Computational Statistics & Data Analysis 51, 5731–5742.

Krzyśko, M., Lukasz, S., 2019. Robust multivariate functional discriminant coordinates. Communications in Statistics—Simulation and Computation. https://doi.org/10.1080/03610918.2019.1580731.

Kulasekera, K.B., 1995. Comparison of regression curves using quasi-residuals. Journal of the American Statistical Association 90, 1085–1093.

Kulasekera, K.B., Wang, J., 1997. Smoothing parameter selection for power optimality in testing of regression curves. Journal of the American Statistical Association 92, 500–511.

Kulinskaya, E., Dollinger, M.B., 2007. Robust weighted one-way ANOVA: improved approximation and efficiency. Journal of Statistical Planning and Inference 137, 462–472. https://doi.org/10.1016/j.jspi.2006.01.008.

Kulinskaya, E., Staudte, R.G., 2006. Interval estimates of weighted effect sizes in the one-way heteroscedastic ANOVA. British Journal of Mathematical and Statistical Psychology 59, 97–111.

Kulinskaya, E., Morgenthaler, S., Staudte, R., 2008. Meta Analysis: A Guide to Calibrating and Combining Statistical Evidence. Wiley, New York.

Kulinskaya, E., Morgenthaler, S., Staudte, R., 2010. Variance stabilizing the difference of two binomial proportions. American Statistician 64, 350–356.

Künsch, H., Stefanski, L., Carroll, R., 1989. Conditionally unbiased bounded influence estimation in general regression models, with applications to generalized linear models. Journal of the American Statistical Association 84, 460–466.

Kuo, L., Mallick, B., 1998. Variable selection for regression models. Sankhya, Series B 60, 65–81.

Kuonen, D., 2005. Studentized bootstrap confidence intervals based on M-estimates. Journal of Applied Statistics 32, 443–460.

Kurnaz, F.S., Hoffmann, I., Filzmoser, P., 2017. Robust and sparse estimation methods for high dimensional linear and logistic regression. Chemometrics and Intelligent Laboratory Systems.

Lai, Y., McLeod, I., 2020. Ensemble quantile classifier. Computational Statistics & Data Analysis 144, 106849. https://doi.org/10.1016/j.csda.2019.106849.

Laird, N.M., Ware, J.H., 1982. Random-effects models for longitudinal data. Biometrics 38, 963–974.

Lambert, D., 1985. Robust two-sample permutation test. Annals of Statistics 13, 606–625.

Lambert-Lacroix, S., Zwald, L., 2011. Robust regression through the Huber's criterion and adaptive lasso penalty. Electronic Journal of Statistics 5, 1015–1053.

Laplace, P.S. de, 1818. Deuxieme Supplement a la Theorie Analytique des Probabilites. Courcier, Paris.

Lax, D.A., 1985. Robust estimators of scale: finite-sample performance in long-tailed symmetric distributions. Journal of the American Statistical Association 80, 736–741.

Lee, H., Fung, K.F., 1985. Behavior of trimmed F and sine-wave F statistics in one-way ANOVA. Sankhya: The Indian Journal of Statistics 47, 186–201.

Léger, C., Romano, J.P., 1990a. Bootstrap adaptive estimation: the trimmed mean example. Canadian Journal of Statistics 18, 297–314.

Léger, C., Romano, J.P., 1990b. Bootstrap choice of tuning parameters. Annals of the Institute of Mathematical Statistics 42, 709–735.

Léger, C., Politis, D.N., Romano, J.P., 1992. Bootstrap technology and applications. Technometrics 34, 378–398.

Lehmann, E.L., Romano, J.P., 2005. Generalizations of the familywise error rate. Annals of Statistics 33, 1138–1154.

Levy, P., 1967. Substantive significance of significant differences between two groups. Psychological Bulletin 67, 37–40.

Li, G., 1985. Robust regression. In: Hoaglin, D., Mosteller, F., Tukey, J. (Eds.), Exploring Data Tables, Trends and Shapes. Wiley, New York.

Li, G., Chen, Z., 1985. Projection-pursuit approach to robust dispersion and principal components: primary theory and Monte Carlo. Journal of the American Statistical Association 80, 759–766.

Li, G., Tiwari, R.C., Wells, M.T., 1996. Quantile comparison functions in two-sample problems, with application to comparisons of diagonal markers. Journal of the American Statistical Association 91, 689–698.

Li, G., Li, Y., Tsai, C.-L., 2015. Quantile correlations and quantile autoregressive modeling. Journal of the American Statistical Association 110, 246–261.

Li, H., Bradic, J., 2018. Boosting in the presence of outliers: adaptive classification with nonconvex loss functions. Journal of the American Statistical Association 113, 660–674. https://doi.org/10.1080/01621459.2016.1273116.

Li, J., Siegmund, D., 2015. Higher criticism: p-values and criticism. Annals of Statistics 43, 1323–1350.

Li, J., Cuesta-Albertos, J., Liu, R., 2012. DD-classifier: nonparametric classification procedure based on DD-plot. Journal of the American Statistical Association 107, 737–753.

Liebscher, S., Kirschstein, T., Becker, C., 2012. The flood algorithm—a multivariate, self-organizing-map-based, robust location and covariance estimator. Statistical Computation 22, 325–336. https://doi.org/10.1007/s11222-011-9250-3.

Lin, P., Stivers, L., 1974. On the difference of means with missing values. Journal of the American Statistical Association 61, 634–636.

Little, R.J.A., Rubin, D.B., 2002. Statistical Analysis with Missing Data. Wiley, New York.

Liu, K., 1993. A new class of biased estimate in linear regression. Communications in Statistics—Theory and Methods 22, 393–402. https://doi.org/10.1080/03610929308831027.

Liu, R.C., Brown, L.D., 1993. Nonexistence of informative unbiased estimators in singular problems. Annals of Statistics 21, 1–14.

Liu, R.G., Singh, K., 1997. Notions of limiting P values based on data depth and bootstrap. Journal of the American Statistical Association 92, 266–277. https://doi.org/10.2307/2291471.

Liu, R.Y., 1990. On a notion of data depth based on random simplices. Annals of Statistics 18, 405–414.

Liu, R.Y., Singh, K., 1993. A quality index based on data depth and multivariate rank tests. Journal of the American Statistical Association 88, 252–260.

Liu, R.Y., Parelius, J.M., Singh, K., 1999. Multivariate analysis by data depth: descriptive statistics, graphics and inference. Annals of Statistics 27, 783–858.

Liu, W., Lin, S., Piegorsch, W.W., 2008. Construction of exact simultaneous confidence bands for a simple linear regression model. International Statistical Review 76 (1), 39–57. https://doi.org/10.1111/j.1751-5823.2007.00027.x.

Liu, W., Lin, R., Yang, M., 2015. Robust elastic net regression. https://arxiv.org/abs/1511.04690.

Liu, X., 2017. Approximating the projection depth median of dimensions $p \geq 3$. Communications in Statistics—Simulation and Computation 46, 3756–3768.

Livacic-Rojas, P., Vallejo, G., Fernández, P., 2010. Analysis of type I error rates of univariate and multivariate procedures in repeated measures designs. Communications in Statistics—Simulation and Computation 39, 624–640.

Lix, L.M., Keselman, H.J., 1998. To trim or not to trim: tests of mean equality under heteroscedasticity and nonnormality. Educational and Psychological Measurement 58, 409–429.

Lix, L.M., Keselman, H.J., Hinds, A., 2005. Robust tests for the multivariate Behrens–Fisher problem. Computer Methods and Programs in Biomedicine 77, 129–139.

Lloyd, C.J., 1999. Statistical Analysis of Categorical Data. Wiley, New York.

Locantore, N., Marron, J.S., Simpson, D.G., Tripoli, N., Zhang, J.T., 1999. Robust principal components for functional data. Test 8, 1–28.

Loh, W.-Y., 1987a. Calibrating confidence coefficients. Journal of the American Statistical Association 82, 155–162.

Loh, W.-Y., 1987b. Does the correlation coefficient really measure the degree of clustering around a line? Journal of Educational Statistics 12, 235–239.

Lombard, F., 2005. Nonparametric confidence bands for a quantile comparison function. Technometrics 47, 364–369.

Long, J.S., Ervin, L.H., 2000. Using heteroscedasticity consistent standard errors in the linear regression model. American Statistician 54, 217–224. https://doi.org/10.2307/2685594.

López-Pintado, S., Romo, J., 2009. On the concept of depth for functional data. Journal of the American Statistical Association 104, 718–734.

Lopuhaä, H.P., 1989. On the relation between S-estimators and M-estimators of multivariate location and covariance. Annals of Statistics 17, 1662–1683.

Lopuhaä, H.P., 1991. τ-estimators for location and scatter. Canadian Journal of Statistics 19, 307–321.

Lopuhaä, H.P., 1999. Asymptotics of reweighted estimators of multivariate location and scatter. Annals of Statistics 27, 1638–1665.

Lopuhaä, H.P., Rousseeuw, P.J., 1991. Breakdown points of affine equivariant estimators of multivariate location and covariance matrices. Annals of Statistics 19, 229–248.

Luepsen, H., 2018. Comparison of nonparametric analysis of variance methods: a vote for van der Waerden. Communications in Statistics—Simulation and Computation 47, 2547–2576.

Luh, W., Guo, J., 2000. Approximate transformation trimmed mean methods to the test of simple linear regression slope equality. Journal of Applied Statistics 27, 843–857.

Luh, W.M., Guo, J.H., 1999. A powerful transformation trimmed mean method for one-way fixed effects ANOVA model under non-normality and inequality of variance. British Journal of Mathematical and Statistical Psychology 52, 303–320.

Luh, W.M., Guo, J.H., 2010. Approximate sample size formulas for the two-sample trimmed mean test with unequal variances. British Journal of Mathematical and Statistical Psychology 60, 137–146.

Lukman, A., Arowolo, O., Ayinde, K., 2014. Some robust ridge regression for handling multicollinearity and outlier. International Journal of Sciences: Basic and Applied Research 16, 192–202.

Lumley, T., 1996. Generalized estimating equations for ordinal data: a note on working correlation structures. Biometrics 52, 354–361.

Lunneborg, C.E., 2000. Data Analysis by Resampling: Concepts and Applications. Duxbury, Pacific Grove, CA.

Lyon, J.D., Tsai, C.-L., 1996. A comparison of tests for homogeneity. The Statistician 45, 337–350.

Ma, J., 2015. Robustness of rank-based and other robust estimators when comparing groups. Unpublished dissertation. Department of Psychology, University of Southern California.

Ma, J., Wilcox, R.R., 2013. Robust within Groups ANOVA: Dealing with Missing Values. Mathematics and Statistics, vol. 1. Horizon Research Publishing, pp. 1–4.

Ma, X., He, X., Shia, X., 2016. A variant of K nearest neighbor quantile regression. Journal of Applied Statistics 43, 526–537. https://doi.org/10.1080/02664763.2015.1070807.

MacKinnon, D.P., 2008. Introduction to Statistical Mediation Analysis. Psychology Press, Clifton, New Jersey.

MacKinnon, D.P., Warsi, G., Dwyer, J.H., 1995. A simulation study of mediated effect measures. Multivariate Behavioral Research 30, 41–62.

MacKinnon, D.P., Lockwood, C.M., Williams, J., 2004. Confidence limits for the indirect effect: distribution of the product and resampling methods. Multivariate Behavioral Research 39, 99–128. https://doi.org/10.1207/s15327906mbr3901_4.

MacKinnon, J.G., White, H., 1985. Some heteroskedasticity consistent covariance matrix estimators with improved finite sample properties. Journal of Econometrics 29, 53–57.

Mair, P., Wilcox, R., 2019. Robust statistical methods in R using the WRS2 package. Behavior Research Methods. https://doi.org/10.3758/s13428-019-01246-w.

Mak, T.K., 1992. Estimation of parameters in heteroscedastic linear models. Journal of the Royal Statistical Society, B 54, 649–655.

Makinde, O., Chakraborty, B., 2018. On some classifiers based on multivariate ranks. Communications in Statistics—Theory and Methods 47, 3955–3969.

Makinde, O.S., Fasoranbaku, O.A., 2018. On maximum depth classifiers: depth distribution approach. Journal of Applied Statistics 45, 1106–1117. https://doi.org/10.1080/02664763.2017.1342783.

Malec, P., Schienle, M., 2014. Nonparametric kernel density estimation near the boundary. Computational Statistics & Data Analysis 72, 57–76.

Mallows, C.L., 1973. Some comments on C_p. Technometrics 15, 661–675.

Mallows, C.L., 1975. On some topics in robustness. Technical memorandum. Bell Telephone Laboratories.

Mammen, E., 1993. Bootstrap and wild bootstrap for high dimensional linear models. Annals of Statistics 21, 255–285.

Mansouri, H., Li, B., 2019. On simultaneous confidence intervals based on rank-estimates with application to analysis of gene expression data. Communications in Statistics—Theory and Methods 48, 4339–4349.

Marazzi, A., 1993. Algorithms, Routines, and S Functions for Robust Statistics. Chapman and Hall, New York.

Mardia, K.V., Kent, J.T., Bibby, J.M., 1979. Multivariate Analysis. Academic Press, San Diego, CA.

Maritz, J.S., Jarrett, R.G., 1978. A note on estimating the variance of the sample median. Journal of the American Statistical Association 73, 194–196.

Markatou, M., Hettmansperger, T.P., 1990. Robust bounded-influence tests in linear models. Journal of the American Statistical Association 85, 187–190.

Markatou, M., Stahel, W.A., Ronchetti, E., 1991. Robust M-type testing procedures for linear models. In: Stahel, W., Weisberg, S. (Eds.), Directions in Robust Statistics Diagnostics (Part I). Springer-Verlag, New York, pp. 201–220.

Markowski, C.A., Markowski, E.P., 1990. Conditions for the effectiveness of a preliminary test of variance. American Statistician 44, 322–326.

Marmolejo-Ramos, F., Tian, T.S., 2010. The shifting boxplot. A boxplot based on essential summary statistics around the mean. International Journal of Psychological Research 3, 37–45.

Maronna, R.A., 1976. Robust M-estimators of multivariate location and scatter. Annals of Statistics 4, 51–67.

Maronna, R.A., 2005. Principal components and orthogonal regression based on robust scales. Technometrics 47, 264–273.

Maronna, R.A., 2011. Robust ridge regression for high-dimensional data. Technometrics 53, 44–53. https://doi.org/10.2307/40997291.

Maronna, R.A., Yohai, V.J., 1993. Bias-robust estimates of regression based on projections. Annals of Statistics 21, 965–990.

Maronna, R.A., Yohai, V.J., 1995. The behavior of the Stahel-Donoho robust estimator. Journal of the American Statistical Association 90, 330–341.

Maronna, R.A., Yohai, V.J., 2015. High finite-sample efficiency and robustness based on distance-constrained maximum likelihood. Computational Statistics & Data Analysis 83, 262–272.

Maronna, R.A., Yohai, V.J., 2010. Correcting MM estimates for "fat" data sets. Computational Statistics & Data Analysis 54, 3168–3173.

Maronna, R.A., Zamar, R.H., 2002. Robust estimates of location and dispersion for high-dimensional datasets. Technometrics 44, 307–317.

Maronna, R.A., Martin, D.R., Yohai, V.J., 2006. Robust Statistics: Theory and Methods. Wiley, New York.

Marozzi, M., Reiczigel, J., 2018. A progressive shift alternative to evaluate nonparametric tests for skewed data. Communications in Statistics—Simulation and Computation 47 (10), 3083–3094. https://doi.org/10.1080/03610918.2017.1371745.

Martin, R.A., Zamar, R.H., 1993. Efficiency-constrained bias-robust estimation of location. Annals of Statistics 21, 338–354.

Martin, R.A., Yohai, V.J., Zamar, R.H., 1989. Asymptotically min-max bias robust regression. Annals of Statistics 17, 1608–1630.

Martinez, W., Gray, J.B., 2016. Noise peeling methods to improve boosting algorithms. Computational Statistics & Data Analysis 93, 483–497.

Martínez-Camblor, P., 2014. On correlated z-values distributions in hypothesis testing. Computational Statistics & Data Analysis 79, 30–43.

Massé, J.-C., Plante, J.-F., 2003. A Monte Carlo study of the accuracy and robustness of ten bivariate location estimators. Computational Statistics & Data Analysis 42, 1–26.

McCulloch, C.E., 1987. Tests for equality of variance for paired data. Communications in Statistics—Theory and Methods 16, 1377–1391.

McKean, J.W., Schrader, R.M., 1984. A comparison of methods for studentizing the sample median. Communications in Statistics—Simulation and Computation 13, 751–773.

McKean, J.W., Sheather, S.J., 1991. Small sample properties of robust analyses of linear models based on R-estimates: a survey. In: Stahel, W., Weisberg, S. (Eds.), Directions in Robust Statistics and Diagnostics, part II. Springer-Verlag, New York.

McKean, J.W., Sheather, S.J., Hettmansperger, T.P., 1990. Regression diagnostics for rank-based methods. Journal of the American Statistical Association 85, 1018–1029.

McKean, J.W., Sheather, S.J., Hettmansperger, T.P., 1993. The use and interpretation of residuals based on robust estimation. Journal of the American Statistical Association 88, 1254–1263.

Mee, R.W., 1990. Confidence intervals for probabilities and tolerance regions based on a generalization of the Mann-Whitney statistic. Journal of the American Statistical Association 85, 793–800.

Meinshausen, N., 2006. Quantile regression forests. Journal of Machine Learning Research 7, 983–999.

Messer, K., Goldstein, L., 1993. A new class of kernels for nonparametric curve estimation. Annals of Statistics 21, 179–195.

Micceri, T., 1989. The unicorn, the normal curve, and other improbable creatures. Psychological Bulletin 105, 156–166.

Mickey, M.R., Dunn, O.J., Clark, V., 1967. Note on the use of stepwise regression in detecting outliers. Computational Biomedical Research 1, 105–111.

Miles, D., Mora, J., 2003. On the performance of nonparametric specification tests in regression models. Computational Statistics & Data Analysis 42, 477–490.

Miller, J., 1988. A warning about median reaction time. Journal of Experimental Psychology: Human Perception and Performance 14 (3), 539.

Miller, A.J., 1990. Subset Selection in Regression. Chapman and Hall, London.

Mills, J.E., Field, C.A., Dupuis, D.J., 2002. Marginally specified generalized mixed models: a robust approach. Biometrics 58, 727–734.

Mizera, I., 2002. On depth and deep points: a calculus. Annals of Statistics 30, 1681–1736.

Molenberghs, G., Verbeke, G., 2005. Models for Discrete Longitudinal Data. Springer, New York.

Montgomery, D.C., Peck, E.A., Vining, G.G., 2012. Introduction to Linear Regression Analysis, 5th ed. Wiley, New York.

Mooney, C.Z., Duval, R.D., 1993. Bootstrapping: A Nonparametric Approach to Statistical Inference. Sage, Newbury Park, CA.

Morgenthaler, S., 1992. Least-absolute deviations fit for generalized linear models. Biometrika 79, 747–754.

Morgenthaler, S., Tukey, J.W., 1991. Configural Polysampling. Wiley, New York.

Moser, B.K., Stevens, G.R., Watts, C.L., 1989. The two-sample t-test versus Satterthwaite's approximate F test. Communications in Statistics—Theory and Methods 18, 3963–3975.

Moses, T., Klockars, A.J., 2012. Traditional and proposed tests of slope homogeneity for non-normal and heteroscedastic data. British Journal of Mathematical and Statistical Psychology 65, 402–426.

Mosteller, F., Tukey, J.W., 1977. Data Analysis and Regression. Addison-Wesley, Reading, MA.

Möttönen, J., Oja, H., 1995. Multivariate spatial sign and rank methods. Journal of Nonparametric Statistics 5, 201–213.

Mount, D.M., Netanyahu, N.S., Piatko, A.Y., Wu, A.R., Silverman, R., 2016. A practical approximation algorithm for the LTS estimator. Computational Statistics & Data Analysis 99, 148–170.

Mudholkar, G.S., Wilding, G.E., Mietlowski, W.L., 2003. Robustness properties of the Pitman–Morgan test. Communications in Statistics—Theory and Methods 32, 1801–1816.

Muirhead, R.J., 1982. Aspects of Multivariate Statistical Theory. Wiley, New York.

Mukhopadhyay, N., Solanky, T., 1994. Multistage Selection and Ranking Procedures: Second Order Asymptotics. Marcel Dekker, New York.

Multach, M., Wilcox, R.R., 2017. Some results on a Wilcoxon–Mann–Whitney type measure of interaction. Advances in Social Sciences Research Journal 4 (20). https://doi.org/10.14738/assrj.420.2017.

Munk, A., Dette, H., 1998. Nonparametric comparison of several regression functions: exact and asymptotic theory. Annals of Statistics 26, 2339–2386.

Munzel, U., Brunner, E., 2000. Nonparametric test in the unbalanced multivariate one-way design. Biometrical Journal 42, 837–854.

Nanayakkara, N., Cressie, N., 1991. Robustness to unequal scale and other departures from the classical linear model. In: Stahel, W., Weisberg, S. (Eds.), Directions in Robust Statistics and Diagnostics, Part II. Springer-Verlag, New York, pp. 65–113.

Naranjo, J.D., Hettmansperger, T.P., 1994. Bounded influence rank regression. Journal of the Royal Statistical Society, B 56, 209–220.

Narula, S.C., 1987. The minimum sum of absolute errors regression. Journal of Quality Technology 19, 37–45.

Navruz, G., Özdemir, F., 2018. Quantile estimation and comparing two independent groups with an approach based on percentile bootstrap. Communications in Statistics—Simulation and Computation 47, 2119–2138. https://doi.org/10.1080/03610918.2017.1335410.

Nelder, J.A., Mead, R., 1965. A simplex method for function minimization. Computer Journal 7, 308–313.

Neubert, K., Brunner, E., 2007. A studentized permutation test for the non-parametric Behrens–Fisher problem. Computational Statistics & Data Analysis 51, 5192–5204.

Neuhäuser, M., 2003. A note on the exact test based on the Baumgartner-Weiss-Schindler statistic in the presence of ties. Computational Statistics & Data Analysis 42, 561–568.

Neuhäuser, M., Lösch, C., Jöckel, K-H., 2007. The Chen-Luo test in case of heteroscedasticity. Computational Statistics & Data Analysis 51, 5055–5060.

Neumeyer, N., Dette, H., 2003. Nonparametric comparison of regression curves: an empirical process approach. Annals of Statistics 31, 880–920.

Newcomb, S., 1882. Discussion and results of observations on transits of Mercury from 1677 to 1881. Astronomical Papers 1, 363–487.

Newcomb, S., 1886. A generalized theory of the combination of observations so as to obtain the best result. American Journal of Mathematics 8, 343–366.

Newcombe, R.G., 1998. Improved confidence intervals for the difference between binomial proportions based on paired data. Statistics in Medicine 17, 2635–2650.

Newcombe, R.G., 2006a. Confidence intervals for an effect size measure based on the Mann-Whitney statistic. Part 1: general issues and tail-area-based methods. Statistics in Medicine 25, 543–557.

Newcombe, R.G., 2006b. Confidence intervals for an effect size measure based on the Mann-Whitney statistic. Part 2: asymptotic methods and evaluation. Statistics in Medicine 25, 559–573.

Neykov, N.M., Čížek, P., Filzmoser, P., Neytchev, P.N., 2012. The least trimmed quantile regression. Computational Statistics & Data Analysis 56, 1757–1770.

Ng, M., 2009a. A comparison of bootstrap techniques for evaluating indirect effects. Unpublished technical report. Dept. of Psychology, University of Southern California.

Ng, M., 2009b. Significance Testing in Regression Analyses. Unpublished doctoral dissertation. Dept. of Psychology, University of Southern California.

Ng, M., Lin, J., 2016. Testing for mediation effects under non-normality and heteroscedasticity: a comparison of classic and modern methods. International Journal of Quantitative Research in Education 3, 24–40.

Ng, M., Wilcox, R.R., 2009. Level robust methods based on the least squares regression estimator. Journal of Modern Applied Statistical Methods 8, 384–395.

Ng, M., Wilcox, R.R., 2010. Comparing the slopes of regression lines. British Journal of Mathematical and Statistical Psychology 63, 319–340.

Ng, M., Wilcox, R.R., 2011. A comparison of two-stage procedures for testing least-squares coefficients under heteroscedasticity. British Journal of Mathematical and Statistical Psychology 64, 244–258.

Ng, M., Wilcox, R.R., 2012. Bootstrap methods for comparing independent regression slopes. British Journal of Mathematical and Statistical Psychology 65, 282–301.

Noh, M., Lee, Y., 2007. Robust modeling for inference from generalized linear model classes. Journal of the American Statistical Association 102, 1059–1072.

Nurunnabi, A.A.M., Nasser, M., Imon, A.H.M.R., 2016. Identification and classification of multiple outliers, high leverage points and influential observations in linear regression. Journal of Applied Statistics 43, 509–525. https://doi.org/10.1080/02664763.2015.1070806.

Oja, H., Randles, R.H., 2006. Multivariate nonparametric tests. Statistical Science 19, 598–605.

Olive, D.J., 2004. A resistant estimator of multivariate location and dispersion. Computational Statistics & Data Analysis 46, 93–102.

Olive, D.J., 2010. The number of samples for resampling algorithms. Preprint. www.math.siu.edu/olive/preprints.htm.

Olive, D.J., Hawkins, D.M., 2010. Robust multivariate location and dispersion. Preprint. www.math.siu.edu/olive/preprints.htm.

Olsson, D.M., 1974. A sequential simplex program for solving minimization problems. Journal of Quality Technology 6, 53–57.

Olsson, D.M., Nelson, L.S., 1975. The Nelder-Mead simplex procedure for function minimization. Technometrics 17, 45–51.

O'Reilly-Shah, V., Gentry, K.R., Walters, A.M., Zovot, J., Anderson, C.T., Tighe, P.J., 2020. Bias and ethical considerations in machine learning and the automation of perioperative risk assessment. British Journal of Anaesthesia 125, 843–846. https://doi.org/10.1016/j.bja.2020.07.040.

Othman, A.R., Keselman, H.J., Wilcox, R.R., Fradette, K., Padmanabhan, A.R., 2002. A test of symmetry. Journal of Modern Applied Statistical Methods 1, 310–315.

Owen, A., 2001. Empirical likelihood for linear models. Annals of Statistics 19, 1725–1747.

Özdemir, A.F., 2013. Comparing two independent groups: a test based on a one-step M-estimator and bootstrap-t. British Journal of Mathematical and Statistical Psychology 66, 322–337.

Özdemir, A.F., Wilcox, R.R., 2012. New results on the small-sample properties of some robust univariate estimators. Communications in Statistics—Simulation and Computation 41, 1544–1556.

Özdemir, A.F., Wilcox, R.R., Yildiztepe, E., 2013. Comparing measures of location: some small-sample results when distributions differ in skewness and kurtosis under heterogeneity of variances. Communications in Statistics—Simulation and Computation 42, 407–424.

Özdemir, A.F., Wilcox, R.R., Yildiztepe, E., 2018. Comparing J independent groups with a method based on trimmed means. Communications in Statistics—Simulation and Computation 47, 852–863. https://doi.org/10.1080/03610918.2017.1295152.

Özdemir, A.F., Yildiztepe, E., Wilcox, R.R., 2020. A new test for comparing J independent groups by using one-step M-estimator and bootstrap-t. Technical report.

Paindaveine, D., Verdebout, T., 2016. On high-dimensional sign tests. Bernoulli 22, 1745–1769.

Pajari, M., Tikanmäki, M., Makkonen, L., 2019. Probabilistic evaluation of quantile estimators. Communications in Statistics—Theory and Methods. https://doi.org/10.1080/03610926.2019.1696975.

Park, C., Kim, H., Wang, M., 2020. Investigation of finite-sample properties of robust location and scale estimators. Communications in Statistics—Simulation and Computation. https://doi.org/10.1080/03610918.2019.1699114.

Parra-Frutos, I., Molera, L., 2019. Removing skewness and kurtosis by transformation when testing for mean equality. Communications in Statistics—Simulation and Computation. https://doi.org/10.1080/03610918.2019.1610439.

Parrish, R.S., 1990. Comparison of quantile estimators in normal sampling. Biometrics 46, 247–257.

Patel, K.M., Hoel, D.G., 1973. A nonparametric test for interaction in factorial experiments. Journal of the American Statistical Association 68, 615–620.

Patel, K.R., Mudholkar, G.S., Fernando, J.L.I., 1988. Student's t approximations for three simple robust estimators. Journal of the American Statistical Association 83, 1203–1210.

Pawar, S.D., Shirke, S.T., 2019. Nonparametric tests for multivariate locations based on data depth. Communications in Statistics—Simulation and Computation 48, 753–776. https://doi.org/10.1080/03610918.2017.1397165.

Pearson, E.S., Please, N.W., 1975. Relation between the shape of the population distribution and the robustness of four simple statistics. Biometrika 62, 223–241.

Pedersen, W.C., Miller, L.C., Putcha-Bhagavatula, A.D., Yang, Y., 2002. Evolved sex differences in sexual strategies: the long and the short of it. Psychological Science 13, 157–161.

Peña, D., Prieto, F.J., 2001. Multivariate outlier detection and robust covariance matrix estimation. Technometrics 43, 286–299.

Peng, C.-Y., Chen, L.-T., 2014. Beyond Cohen's d: alternative effect size measures for between-subject designs. Journal of Experimental Education 82, 22–50.

Peng, H., Wang, S., Wang, X., 2008. Consistency and asymptotic distribution of the Theil–Sen estimator. Journal of Statistical Planning and Inference 138, 1836–1850.

Pernet, C.R., Wilcox, R., Rousselet, G.A., 2013. Robust correlation analyses: a Matlab toolbox for psychology research. Frontiers in Quantitative Psychology and Measurement. https://doi.org/10.3389/fpsyg.2012.00606.

Pesarin, F., 2001. Multivariate Permutation Tests. Wiley, New York.

Pison, G., Van Aelst, S., Willems, G., 2002. Small sample corrections for LTS and MCD. Metrika 55, 111–123.

Politis, D.N., Romano, J.P., 1997. Multivariate density estimation with general flat-top kernels of infinite order. Journal of Multivariate Analysis 68, 1–25.

Poon, W.-Y., Lew, S-F., Poon, Y.S., 2000. A local influence approach to identifying multiple outliers. British Journal of Mathematical and Statistical Psychology 53, 255–273.

Potthoff, R.F., Roy, S.N., 1964. A generalized multivariate analysis of variance model useful especially for growth curve problem. Biometrika 51, 313–326.

Pratt, J.W., 1964. Robustness of some procedures for the two-sample location problem. Journal of the American Statistical Association 59, 665–680.

Pratt, J.W., 1968. A normal approximation for binomial, F, beta, and other common, related tail probabilities, I. Journal of the American Statistical Association 63, 1457–1483.

Preacher, K.J., Rucker, D.D., Hayes, A.F., 2007. Addressing moderated mediation hypotheses: theory, methods, and prescriptions. Multivariate Behavioral Research 42, 185–227.

Pregibon, D., 1987. Resistant fits for some commonly used logistic models with medical applications. Biometrics 38, 485–498.

Price, R.M., Bonett, D.G., 2001. Estimating the variance of the median. Journal of Statistical Computation and Simulation 68, 295–305.

Qasim, M., Amin, M., Omer, T., 2019. Performance of some new Liu parameters for the linear regression model. Communications in Statistics—Theory and Methods. https://doi.org/10.1080/03610926.2019.1595654.

R Development Core Team, 2010. R: A Language and Environment for Statistical Computing. R Foundation for Statistical Computing, Vienna, Austria. ISBN 3-900051-07-0. http://www.R-project.org.

Racine, J., MacKinnon, J.G., 2007a. Simulation-based tests than can use any number of simulations. Communications in Statistics—Simulation and Computation 36, 357–365.

Racine, J., MacKinnon, J.G., 2007b. Inference via kernel smoothing of bootstrap P values. Computational Statistics & Data Analysis 51, 5949–5957.

Radchenko, P., James, G., 2011. Improved variable selection with forward-lasso adaptive shrinkage. Annals of Applied Statistics 5, 427–448.

Raine, A., Buchsbaum, M., LaCasse, L., 1997. Brain abnormalities in murderers indicated by positron emission tomography. Biological Psychiatry 42, 495–508.

Ramsay, J.O., Silverman, B.W., 2005. Functional Data Analysis, 2nd ed. Springer Verlag, New York.

Ramsey, P.H., 1980. Exact type I error rates for robustness of Student's t test with unequal variances. Journal of Educational Statistics 5, 337–349.

Randal, J.A., 2008. A reinvestigation of robust scale estimation in finite samples. Computational Statistics & Data Analysis 52, 5014–5021.

Randles, R.H., Wolfe, D.A., 1979. Introduction to the Theory of Nonparametric Statistics. Wiley, New York.

Rao, C.R., 1948. Tests of significance in multivariate analysis. Biometrika 35, 58–79.

Rao, P.S.R.S., Kaplan, J., Cochran, W.G., 1981. Estimators for one-way random effects model with unequal error variances. Journal of the American Statistical Association 76, 89–97.

Raper, S., 2020. Two cultures. Significance 17, 34–37.

Rasch, D., Teuscher, F., Guiard, V., 2007. How robust are tests for two independent samples? Journal of Statistical Planning and Inference 137, 2706–2720.

Rasmussen, J.L., 1989. Data transformation, type I error rate and power. British Journal of Mathematical and Statistical Psychology 42, 203–211.

Reed, J.F., 1998. Contributions to adaptive estimation. Journal of Applied Statistics 25, 651–669.

Reed, J.F., Stark, D.B., 1996. Hinge estimators of location: robust to asymmetry. Computer Methods and Programs in Biomedicine 49, 11–17.

Reiczigel, J., Zakarás, I., Rózsa, L., 2005. A bootstrap test of stochastic equality of two populations. American Statistician 59, 156–161.

Reiczigel, J., Abonyi-Tóth, Z., Singer, J., 2008. An exact confidence set for two binomial proportions and exact unconditional confidence intervals for the difference and ratio of proportions. Computational Statistics & Data Analysis 52, 5046–5053. https://doi.org/10.1016/j.csda.2008.04.032.

Reider, H., 1994. Robust Asymptotic Statistics. Springer-Verlag, New York.

Ren, H., Chen, N., Zou, C., 2017. Projection-based outlier detection in functional data. Biometrika 104, 111–423. https://doi.org/10.1093/biomet/asx012.

Renaud, O., Victoria-Feser, M-P., 2010. A robust coefficient of determination for regression. Journal of Statistical Planning and Inference 140, 1852–1862.

Rivest, L.-P., 1994. Statistical properties of Winsorized means for skewed distributions. Biometrika 81, 373–383.

Rizzo, M.L., Székely, G., 2010. DISCO analysis: a nonparametric extension of analysis of variance. Annals of Applied Statistics 4, 1034–1055.

Roberts, S., Martin, M.A., Zheng, L., 2015. An adaptive, automatic multiple-case deletion technique for detecting influence in regression. Technometrics 57, 408–417.

Robinson, J., Ronchetti, E., Young, G.A., 2003. Saddlepoint approximations and tests based on multivariate M-estimates. Annals of Statistics 31, 1154–1169.

Robinson, P.M., 1987. Asymptotically efficient estimation in the presence of heteroskedasticity of unknown form. Econometrica 55, 875–891.

Rocke, D.M., 1996. Robustness properties of S-estimators of multivariate location and shape in high dimension. Annals of Statistics 24, 1327–1345.

Rocke, D.M., Woodruff, D.L., 1996. Identification of outliers in multivariate data. Journal of the American Statistical Association 91, 1047–1061.

Rom, D.M., 1990. A sequentially rejective test procedure based on a modified Bonferroni inequality. Biometrika 77, 663–666.

Romano, J.P., 1990. On the behavior of randomization tests without a group invariance assumption. Journal of the American Statistical Association 85, 686–692.

Romano, J.P., Wolfe, M., 2005. Exact and approximate stepdown methods for multiple hypothesis testing. Journal of the American Statistical Association 100, 94–108. https://doi.org/10.1198/016214504000000539.

Rosenbusch, H., Hilbert, L., Evans, A., Zeelenberg, M., 2020. StatBreak: identifying 'lucky' data points through genetic algorithms. Advances in Methods and Practices in Psychological Science 3 (2). https://doi.org/10.1177/2515245920917950.

Rosmond, R., Dallman, M.F., Björntorp, P., 1998. Stress-related cortisol secretion in men: relationships with abdominal obesity and endocrine, metabolic and hemodynamic abnormalities. Journal of Clinical Endocrinology & Metabolism 83, 1853–1859.

Rousseeuw, P.J., 1984. Least median of squares regression. Journal of the American Statistical Association 79, 871–880. https://doi.org/10.1080/01621459.1984.10477105.

Rousseeuw, P.J., Christmann, A., 2003. Robustness against separation and outliers in logistic regression. Computational Statistics & Data Analysis 43, 315–332.

Rousseeuw, P.J., Croux, C., 1993. Alternative to the median absolute deviation. Journal of the American Statistical Association 88, 1273–1283.

Rousseeuw, P.J., Hubert, M., 1999. Regression depth. Journal of the American Statistical Association 94, 388–402.

Rousseeuw, P.J., Leroy, A.M., 1987. Robust Regression & Outlier Detection. Wiley, New York.

Rousseeuw, P.J., Ruts, I., 1996. AS 307: bivariate location depth. Applied Statistics 45, 516–526.

Rousseeuw, P.J., Struyf, A., 1998. Computing location depth and regression depth in higher dimensions. Statistics and Computing 8, 193–203.

Rousseeuw, P.J., van Driessen, K., 1999. A fast algorithm for the minimum covariance determinant estimator. Technometrics 41, 212–223.

Rousseeuw, P.J., van Zomeren, B.C., 1990. Unmasking multivariate outliers and leverage points. Journal of the American Statistical Association 85, 633–639. https://doi.org/10.2307/2289995.

Rousseeuw, P.J., Verboven, S., 2002. Robust estimation in very small samples. Computational Statistics & Data Analysis 40, 741–758.

Rousseeuw, P.J., Yohai, V., 1984. Robust Regression by Means of S-Estimators. Nonlinear Time Series Analysis. Lecture Notes in Statistics, vol. 26. Springer, New York, pp. 256–272.

Rousseeuw, P.J., Ruts, I., Tukey, J.W., 1999. The bagplot: a bivariate boxplot. American Statistician 53, 382–387.

Rousseeuw, P.J., Van Aelst, S., Van Driessen, K., Gulló, J.A., 2004. Robust multivariate regression. Technometrics 46, 293–305.

Rousselet, G.A., Wilcox, R.R., 2020. Reaction times and other skewed distributions: problems with the mean and the median. Meta-Psychology 4, MP.2019.1630. https://doi.org/10.15626/MP.2019.1630.

Rousselet, G.A., Pernet, C.R., Wilcox, R.R., 2017. Beyond differences in means: robust graphical methods to compare two groups in neuroscience. European Journal of Neuroscience. https://doi.org/10.1111/ejn.13610.

Rubin, A.S., 1983. The use of weighted contrasts in analysis of models with heterogeneity of variance. Proceedings of the Business and Economics Statistics Section, American Statistical Association, 347–352.

Ruppert, D., Wand, M.P., Carroll, R.J., 2003. Semiparametric Regression. Cambridge University Press, Cambridge.

Ruscio, J., Mullen, T., 2012. Confidence intervals for the probability of superiority effect size measure and the area under a receiver operating characteristic curve. Multivariate Behavioral Research 47, 201–223.

Rust, S.W., Fligner, M.A., 1984. A modification of the Kruskal-Wallis statistic for the generalized Behrens-Fisher problem. Communications in Statistics—Theory and Methods 13, 2013–2027.

Rutherford, A., 1992. Alternatives to traditional analysis of covariance. British Journal of Mathematical and Statistical Psychology 45, 197–224.

Ryan, B.F., Joiner, B.L., Ryan, T.A., 1985. MINITAB Handbook. PWS-KENT, Boston.

Ryu, E., Agresti, A., 2008. Modeling and inference for an ordinal effect size measure. Statistics in Medicine 27, 1703–1717.

Sakaori, F., 2002. Permutation test for equality of correlation coefficients in two populations. Communications in Statistics—Simulation and Computation 31, 641–652.

Salibian-Barrera, M., Yohai, V., 2006. A fast algorithm for S-regression estimates. Journal of Computational and Graphical Statistics 15, 414–427.

Salibian-Barrera, M., Zamar, R.H., 2002. Bootstrapping robust estimates of regression. Annals of Statistics 30, 556–582.

Salibián-Barrera, M., Van Aelst, S., Willems, G., 2006. PCA based on multivariate MM-estimators with fast and robust bootstrap. Journal of the American Statistical Association 101, 1198–1211.

Salk, L., 1973. The role of the heartbeat in the relations between mother and infant. Scientific American 235, 26–29.

Samarov, A.M., 1993. Exploring regression structure using nonparametric functional estimation. Journal of the American Statistical Association 88, 836–847.

Santner, T.J., Pradhan, V., Senchaudhur, P., Mehta, C., Tamhane, A., 2007. Small-sample comparisons of confidence intervals for the difference of two independent binomial proportions. Computational Statistics & Data Analysis 51, 5791–5799.

Sarkar, S.K., 2008. Generalizing Simes' test and Hochberg's stepup procedure. Annals of Statistics 36, 337–363.

Saunders, D.R., 1956. Moderator variables in prediction. Educational and Psychological Measurement 16, 209–222.

Sawilowsky, S.S., 2002. The probable difference between two means when $\sigma_1 \neq \sigma_2$: the Behrens-Fisher problem. Journal of Modern Applied Statistical Methods 1, 461–472.

Sawilowsky, S.S., Blair, R.C., 1992. A more realistic look at the robustness and type II error properties of the t test to departures from normality. Psychological Bulletin 111, 352–360.

Schapire, R.E., Freund, Y., 2012. Boosting: Foundations and Algorithms. MIT Press, Cambridge, MA.

Schilling, M., Doi, J., 2014. A coverage probability approach to finding an optimal binomial confidence procedure. American Statistician 68, 133–145.

Schlölkopf, S., Smola, A., Müller, K.R., 1998. Nonlinear component analysis as a kernel eigenvalue problem. Neural Computation 10, 1299–1319.

Schnys, M., Haesbroeck, G., Critchley, F., 2010. RelaxMCD: smooth optimisation for the minimum covariance determinant estimator. Computational Statistics & Data Analysis 54, 843–857.

Scholz, F.W., 1978. Weighted median regression estimates. Annals of Statistics 6, 603–609.

Schrader, R.M., Hettmansperger, T.P., 1980. Robust analysis of variance. Biometrika 67, 93–101.

Schroër, G., Trenkler, D., 1995. Exact and randomization distributions of Kolmogorov-Smirnov tests two or three samples. Computational Statistics & Data Analysis 20, 185–202.

Schwertman, N.C., de Silva, R., 2007. Identifying outliers with sequential fences. Computational Statistics & Data Analysis 51, 3800–3810.

Scott, D.W., 1979. On optimal and data-based histograms. Biometrika 66, 605–610.

Scott, D.W., 1992. Multivariate Density Estimation. Theory, Practice, and Visualization. Wiley, New York.

Sen, P.K., 1968. Estimate of the regression coefficient based on Kendall's tau. Journal of the American Statistical Association 63, 1379–1389. https://doi.org/10.2307/2285891.

Serfling, R., Wang, Y., 2016. On Liu's simplicial depth and Randles' interdirection. Computational Statistics & Data Analysis 99, 235–247.

Serfling, R.J., 1980. Approximation Theorems of Mathematical Statistics. Wiley, New York.

Serneels, S., Verdonck, T., 2008. Principal components analysis for data containing outliers and missing elements. Computational Statistics & Data Analysis 52, 1712–1727.

Sfakianakis, M.E., Verginis, D.G., 2008. A new family of nonparametric quantile estimators. Communications in Statistics—Simulation and Computation 37, 337–345.

Shah, R., Peters, J., 2020. The hardness of conditional independence testing and the generalised covariance measure. Annals of Statistics 48, 1514–1538.

Shao, J., 1996. Bootstrap model selection. Journal of the American Statistical Association 91, 655–665.

Shao, J., Tu, D., 1995. The Jackknife and the Bootstrap. Springer-Verlag, New York.

Shao, W., Zuo, Y., 2019. Computing the halfspace depth with multiple try algorithm and simulated annealing algorithm. Computational Statistics. https://doi.org/10.1007/s00180-019-00906-x.

She, Y., Chen, K., 2017. Robust reduced-rank regression. Biometrika 104, 633–647. https://doi.org/10.1093/biomet/asx032.

Sheather, S.J., Marron, J.S., 1990. Kernel quantile estimators. Journal of the American Statistical Association 85, 410–416.

Sheather, S.J., McKean, J.W., 1987. A comparison of testing and confidence intervals for the median. Statistical Probability Letters 6, 31–36.

Shoemaker, L.H., 2003. Fixing the F test for equal variances. American Statistician 57, 105–114.

Shoemaker, L.H., Hettmansperger, T.P., 1982. Robust estimates and tests for the one- and two-sample scale models. Biometrika 69, 47–54.

Shu, D., Wenqing, H., 2017. A new method for logistic model assessment. International Journal of Statistics and Probability 6 (6). ISSN 1927-7032, E-ISSN 1927-7040.

Sievers, G.L., 1978. Weighted rank statistics for simple linear regression. Journal of the American Statistical Association 73, 628–631.

Signorini, D.F., Jones, M.C., 2004. Kernel estimators for univariate binary regression. Journal of the American Statistical Association 99, 119–126.

Silverman, B.W., 1986. Density Estimation for Statistics and Data Analysis. Chapman and Hall, New York.

Simonoff, J.S., 1996. Smoothing Methods in Statistics. Springer, New York.

Sing, T., Sander, O., Beerenwinkel, N., Lengauer, T., 2005. ROCR: visualizing classifier performance in R. Bioinformatics 21, 3940–3941.

Singh, K., 1998. Breakdown theory for bootstrap quantiles. Annals of Statistics 26, 1719–1732.

Sinha, S.K., 2004. Robust analysis of generalized linear mixed models. Journal of the American Statistical Association 99, 451–460.

Smucler, E., Yohai, V., 2017. Robust and sparse estimators for linear regression models. Computational Statistics & Data Analysis 111, 116–130.

Snedecor, G.W., Cochran, W., 1967. Statistical Methods, 6th ed. University Press, Ames, IA.

Sockett, E.B., Daneman, D., Carlson, C., Ehrich, R.M., 1987. Factors affecting and patterns of residual insulin secretion during the first year of type I (insulin dependent) diabetes mellitus in children. Diabetes 30, 453–459.

Spanos, A., 2019. Near-collinearity in linear regression revisited: the numerical vs. the statistical perspective. Communications in Statistics—Theory and Methods 48, 5492–5516.

Srihera, R., Stute, W., 2010. Nonparametric comparison of regression functions. Journal of Multivariate Analysis 101, 2039–2059.

Srivastava, D.K., Pan, J., Sarkar, I., Mudholkar, G.S., 2010. Robust Winsorized regression using bootstrap approach. Communications in Statistics—Simulation and Computation 39, 45–67.

Srivastava, M.S., Awan, H.M., 1984. On the robustness of the correlation coefficient in sampling from a mixture of two bivariate normals. Communications in Statistics—Theory and Methods 13, 371–382.

Stahel, W.A., 1981. Breakdown of covariance estimators. Research report 31. Fachgruppe für Statistik, E.T.H Zürich.

Statti, F., Sued, M., Yohai, V.J., 2018. High breakdown point robust estimators with missing data. Communications in Statistics—Theory and Methods 47, 5145–5162.

Staudte, R.G., Sheather, S.J., 1990. Robust Estimation and Testing. Wiley, New York.

Steele, C.M., Aronson, J., 1995. Stereotype threat and the intellectual test performance of African Americans. Journal of Personality and Social Psychology 69, 797–811.

Stefanski, L., Carroll, D., Ruppert, D., 1986. Optimally bounded score functions for generalized linear models with applications to logistic regression. Biometrika 73, 413–424.

Stein, C., 1945. A two-sample test for a linear hypothesis whose power is independent of the variance. Annals of Statistics 16, 243–258.

Sterne, T.E., 1954. Some remarks on confidence or fiducial limits. Biometrika 41, 275–278.

Stigler, S.M., 1973. Simon Newcomb, Percy Daniel, and the history of robust estimation 1885–1920. Journal of the American Statistical Association 68, 872–879.

Storer, B.E., Kim, C., 1990. Exact properties of some exact test statistics for comparing two binomial proportions. Journal of the American Statistical Association 85, 146–155.

Struyf, A., Rousseeuw, P.J., 2000. High-dimensional computation of the deepest location. Computational Statistics & Data Analysis 34, 415–426.

Stute, W., Gonzalez Manteiga, W.G., Presedo Quindimil, M.P., 1998. Bootstrap approximations in model checks for regression. Journal of the American Statistical Association 93, 141–149.

Suhail, M., Chand, S., Kibria, B.M.G., 2020. Quantile based estimation of biasing parameters in ridge regression model. Communications in Statistics—Simulation and Computation 49 (10). https://doi.org/10.1080/03610918.2018.1530782.

Sun, Y., Genton, M.G., 2011. Functional boxplots. Journal of Computational and Graphical Statistics 20, 316–334.

Tableman, M., 1990. Bounded-influence rank regression: a one-step estimator based on Wilcoxon scores. Journal of the American Statistical Association 85, 508–513.

Tableman, M., 1994. The asymptotics of the least trimmed absolute deviation (LTAD) estimators. Statistics & Probability Letters 19, 387–398.

Talib, B.A., Midi, H., 2009. Robust estimator to deal with regression models having both continuous and categorical regressors: a simulation study. Malaysian Journal of Mathematical Sciences 3, 161–181.

Talwar, P.P., 1991. A simulation study of some non-parametric regression estimators. Computational Statistics & Data Analysis 15, 309–327.

Tamura, R., Boos, D., 1986. Minimum Hellinger distance estimation for multivariate location and covariance. Journal of the American Statistical Association 81, 223–229.

Tan, W.Y., 1982. Sampling distributions and robustness of t, F, and variance-ratio of two samples and ANOVA models with respect to departure from normality. Communications in Statistics—Theory and Methods 11, 2485–2511.

Tang, C.Y., Leng, C., 2012. An empirical likelihood approach to quantile regression with auxiliary information. Statistics & Probability Letters 82, 29–36.

Theil, H., 1950. A rank-invariant method of linear and polynomial regression analysis. Indagationes Mathematicae 12, 85–91.

Thompson, G.L., Amman, L.P., 1990. Efficiencies of interblock rank statistics for repeated measures designs. Journal of the American Statistical Association 85, 519–528.

Thomson, A., Randall-Maciver, R., 1905. Ancient Races of the Thebaid. Oxford University Press, Oxford, UK.

Tian, T., Wilcox, R.R., 2009. A comparison of the Brunner-Puri and Agresti-Pendergast methods. Unpublished technical report. Dept. of Psychology, University of Southern California.

Tingley, M., Field, C., 1990. Small-sample confidence intervals. Journal of the American Statistical Association 85, 427–434.

Todorov, V., Filzmoser, P., 2010. Robust statistic for the one-way MANOVA. Computational Statistics & Data Analysis 54, 37–48.

Todorov, V., Templ, M., Filzmoser, P., 2011. Detection of multivariate outliers in business survey data with incomplete information. Advances in Data Analysis and Classification 5, 37–56.

Tomarken, A., Serlin, R., 1986. Comparison of ANOVA alternatives under variance heterogeneity and specific noncentrality structures. Psychological Bulletin 99, 90–99.

Tsangari, H., Akritas, M., 2004. Nonparametric models and methods for ANCOVA with dependent data. Journal of Nonparametric Statistics 16, 403–420.

Tukey, J.W., 1960. A survey of sampling from contaminated normal distributions. In: Olkin, I., Ghurye, S., Hoeffding, W., Madow, W., Mann, H. (Eds.), Contributions to Probability and Statistics. Stanford University Press, Stanford, CA, pp. 448–485.

Tukey, J.W., 1975. Mathematics and the picturing of data. In: Proceedings of the International Congress of Mathematicians, 2, pp. 523–531.

Tukey, J.W., 1977. Exploratory Data Analysis. Addison-Wesley, Reading, MA.

Tukey, J.W., 1991. The philosophy of multiple comparisons. Statistical Science 6, 100–116.

Tukey, J.W., McLaughlin, D.H., 1963. Less vulnerable confidence and significance procedures for location based on a single sample: trimming/Winsorization 1. Sankhya, Series A 25, 331–352.

Tyler, D.E., 1994. Finite sample breakdown points of projection based multivariate location and scatter statistics. Annals of Statistics 22, 1024–1044.

Vakili, K., Schmitt, E., 2014. Finding multivariate outliers with FastPCS. Computational Statistics & Data Analysis 69, 54–66. https://doi.org/10.1016/j.csda.2013.07.021.

Vallejo, G., Ato, M., 2012. Robust tests for multivariate factorial designs under heteroscedasticity. Behavior Research Methods 44, 471–489.

Van Aelst, S., Vandervieren, E., Willems, G., 2012. A Stahel-Donoho estimator based on Huberized outlyingness. Computational Statistics & Data Analysis 56, 531–542.

Vanden Branden, K., Verboven, S., 2009. Robust data imputation. Computational Biology and Chemistry 33, 7–13.

Vanderweele, T.J., 2015. Explanation in Causal Inference: Method for Mediation and Interaction. Oxford University Press, New York, Oxford.

Varathan, N., Wijekoon, P., 2019. Modified almost unbiased Liu estimator in logistic regression. Communications in Statistics—Simulation and Computation. https://doi.org/10.1080/03610918.2019.1626888.

Vargha, A., Delaney, H.D., 2000. A critique and improvement of the CL common language effect size statistics of McGraw and Wong. Journal of Educational and Behavioral Statistics 25, 101–132.

Velina, M., Valeinis, J., Luta, G., 2019. Empirical likelihood-based inference for the difference of two location parameters using smoothed M-estimators. Journal of Statistical Theory and Practice 13 (34). https://doi.org/10.1007/s42519-019-0037-8.

Velleman, P.F., Hoaglin, D.C., 1981. Applications, Basics, and Computing of Exploratory Data Analysis. Duxbury Press, Boston, MA.

Venables, W.N., Ripley, B.D., 2000. S Programming. Springer, New York.

Venables, W.N., Ripley, B.D., 2002. Modern Applied Statistics with S. Springer, New York.

Venables, W.N., Smith, D.M., 2002. An Introduction to R. Network Theory Ltd., Bristol, UK.

Verboon, P., 1993. Robust nonlinear regression analysis. British Journal of Mathematical and Statistical Psychology 46, 77–94.

Verzani, J., 2004. Using R for Introductory Statistics. CRC Press, Boca Raton, FL.

Vexler, A., Liu, S., Kang, L., Hutson, A.D., 2009. Modifications of the empirical likelihood interval estimation with improved coverage probabilities. Communications in Statistics—Simulation and Computation 38, 2171–2183.

Victoria-Feser, M.-P., 2002. Robust inference with binary data. Psychometrika 67, 21–32.

Villacorta, P.J., 2017. The welchADF package for robust hypothesis testing in unbalanced multivariate mixed models with heteroscedastic and non-normal data. The R Journal 9, 309–328.

Walker, M.L., Dovoedo, Y.H., Chakraborti, S., Hilton, C.W., 2018. An improved boxplot for univariate data. American Statistician 72, 348–353.

Wand, M.P., Jones, M.C., 1995. Kernel Smoothing. Chapman & Hall, London.

Wang, F.T., Scott, D.W., 1994. The L_1 method for robust nonparametric regression. Journal of the American Statistical Association 89, 65–76.

Wang, H., Leng, C., 2007. Unified LASSO estimation by least squares approximation. Journal of the American Statistical Association 102, 1039–1048.

Wang, H., Li, G., Jiang, G., 2007. Robust regression shrinkage and consistent variable selection through the LAD-lasso. Journal of Business & Economic Statistics 25, 347–355. https://doi.org/10.1198/073500106000000251.

Wang, L., 2016. Bartlett-corrected two-sample adjusted empirical likelihood via resampling. Communications in Statistics—Theory and Methods. https://doi.org/10.1080/03610926.2016.1257720.

Wang, L., Qu, A., 2007. Robust tests in regression models with omnibus alternatives and bounded influence. Journal of the American Statistical Association 102, 347–358.

Wang, N., Raftery, A.E., 2002. Nearest-neighbor variance estimation (NNVE): robust covariance estimation via nearest-neighbor cleaning. Journal of the American Statistical Association 97, 994–1006.

Wang, Y., Pham, T., Nguyen, D., Kim, E.S., Chen, Y.-H., Kromrey, J., Yin, Y., 2018. Evaluating the efficacy of conditional analysis of variance under heterogeneity and non-normality. Journal of Modern Applied Statistical Methods 17, eP2701. https://doi.org/10.22237/jmasm/1555340224.

Wasserstein, R.L., Schirm, A.L., Lazar, N.A., 2019. Moving to a world beyond '$p < 0.05$'. American Statistician 73 (sup1), 1–19. https://doi.org/10.1080/00031305.2019.1583913.

Welch, B.L., 1938. The significance of the difference between two means when the population variances are unequal. Biometrika 29, 350–362.

Welch, B.L., 1951. On the comparison of several mean values: an alternative approach. Biometrika 38, 330–336.

Welsch, R.E., 1980. Regression sensitivity analysis and bounded-influence estimation. In: Kmenta, J., Ramsey, J.B. (Eds.), Evaluation of Econometric Models. Academic Press, New York, pp. 153–167.

Welsh, A.H., 1987a. One-step L-estimators for the linear model. Annals of Statistics 15, 626–641.

Welsh, A.H., 1987b. The trimmed mean in the linear model (with discussion). Annals of Statistics 15, 20–45.

Welsh, A.H., Morrison, H.L., 1990. Robust L estimation of scale with an application in astronomy. Journal of the American Statistical Association 85, 729–743.

Westfall, P.H., Young, S.S., 1993. Resampling Based Multiple Testing. Wiley, New York.

Wilcox, R.R., 1986. Improved simultaneous confidence intervals for linear contrasts and regression parameters. Communications in Statistics—Simulation and Computation 15, 917–932.

Wilcox, R.R., 1987a. New designs in analysis of variance. Annual Review of Psychology 38, 29–60.

Wilcox, R.R., 1987b. Pairwise comparisons of J independent regression lines over a finite interval, simultaneous comparison of their parameters, and the Johnson-Neyman technique. British Journal of Mathematical and Statistical Psychology 40, 80–93.

Wilcox, R.R., 1989. Percentage points of a weighted Kolmogorov-Smirnov statistics. Communications in Statistics—Simulation and Computation 18, 237–244.

Wilcox, R.R., 1990a. Comparing the means of two independent groups. Biometrical Journal 32, 771–780.

Wilcox, R.R., 1990b. Determining whether an experimental group is stochastically larger than a control. British Journal of Mathematical and Statistical Psychology 43, 327–333.

Wilcox, R.R., 1991a. Bootstrap inferences about the correlation and variance of paired data. British Journal of Mathematical and Statistical Psychology 44, 379–382.

Wilcox, R.R., 1991b. Testing whether independent treatment groups have equal medians. Psychometrika 56, 381–396.

Wilcox, R.R., 1991c. A step-down heteroscedastic multiple comparison procedure. Communications in Statistics—Theory and Methods 20, 1087–1097.

Wilcox, R.R., 1992. Comparing one-step M-estimators of location corresponding to two independent groups. Psychometrika 57, 141–154.

Wilcox, R.R., 1993a. Comparing the biweight midvariances of two independent groups. The Statistician 42, 29–35.

Wilcox, R.R., 1993b. Some results on a Winsorized correlation coefficient. British Journal of Mathematical and Statistical Psychology 46, 339–349.

Wilcox, R.R., 1993c. Analyzing repeated measures or randomized block designs using trimmed means. British Journal of Mathematical and Statistical Psychology 46, 63–76.

Wilcox, R.R., 1993d. Comparing one-step M-estimators of location when there are more than two groups. Psychometrika 58, 71–78.

Wilcox, R.R., 1994a. Some results on the Tukey-McLaughlin and Yuen methods for trimmed means when distributions are skewed. Biometrical Journal 36, 259–306.

Wilcox, R.R., 1994b. A one-way random effects model for trimmed means. Psychometrika 59, 289–306.

Wilcox, R.R., 1994c. Estimating Winsorized correlations in a univariate or bivariate random effects model. British Journal of Mathematical and Statistical Psychology 47, 167–183.

Wilcox, R.R., 1994d. The percentage bend correlation coefficient. Psychometrika 59, 601–616.

Wilcox, R.R., 1994e. Computing confidence intervals for the slope of the biweight midregression and Winsorized regression lines. British Journal of Mathematical and Statistical Psychology 47, 355–372.

Wilcox, R.R., 1995a. Comparing two independent groups via multiple quantiles. The Statistician 44, 91–99.

Wilcox, R.R., 1995b. Comparing the deciles of two dependent groups. Unpublished technical report. Dept. of Psychology, University of Southern California.

Wilcox, R.R., 1995c. A regression smoother for resistant measures of location. British Journal of Mathematical and Statistical Psychology 48, 189–204.

Wilcox, R.R., 1995d. Simulation results on solutions to the multivariate Behrens-Fisher problem via trimmed means. The Statistician 44, 213–225.

Wilcox, R.R., 1995e. ANOVA: the practical importance of heteroscedastic methods, using trimmed means versus means, and designing simulation studies. British Journal of Mathematical and Statistical Psychology 48, 99–114.

Wilcox, R.R., 1996a. Statistics for the Social Sciences. Academic Press, San Diego, CA.

Wilcox, R.R., 1996b. A note on testing hypotheses about trimmed means. Biometrical Journal 38, 173–180.

Wilcox, R.R., 1996c. Confidence intervals for the slope of a regression line when the error term has non-constant variance. Computational Statistics & Data Analysis 22, 89–98.

Wilcox, R.R., 1996d. Estimation in the simple linear regression model when there is heteroscedasticity of unknown form. Communications in Statistics—Theory and Methods 25, 1305–1324.

Wilcox, R.R., 1996e. Confidence intervals for two robust regression lines with a heteroscedastic error term. British Journal of Mathematical and Statistical Psychology 49, 163–170.

Wilcox, R.R., 1996f. Comparing the variances of dependent groups. Unpublished technical report. Dept. of Psychology, University of Southern California.

Wilcox, R.R., 1996g. Simulation results on performing pairwise comparisons of trimmed means. Unpublished technical report. Dept. of Psychology, University of Southern California.

Wilcox, R.R., 1997a. Pairwise comparisons using trimmed means or M-estimators when working with dependent groups. Biometrical Journal 39, 677–688.

Wilcox, R.R., 1997b. ANCOVA based on comparing a robust measure of location at empirically determined design points. British Journal of Mathematical and Statistical Psychology 50, 93–103.

Wilcox, R.R., 1998a. A note on the Theil-Sen regression estimator when the regressor is random and the error term is heteroscedastic. Biometrical Journal 40, 261–268.

Wilcox, R.R., 1998b. Simulation results on extensions of the Theil-Sen regression estimator. Communications in Statistics—Simulation and Computation 27, 1117–1126.

Wilcox, R.R., 1999. Comments on Stute, Manteiga, and Quindimil. Journal of the American Statistical Association 94, 659–660.

Wilcox, R.R., 2000. Rank-based tests for interactions in a two-way design when there are ties. British Journal of Mathematical and Statistical Psychology 53, 145–153.

Wilcox, R.R., 2001a. Pairwise comparisons of trimmed means for two or more groups. Psychometrika 66, 343–356.

Wilcox, R.R., 2001b. Comments on Long and Ervin. American Statistician 55, 374–375.

Wilcox, R.R., 2002. Comparing the variances of independent groups. British Journal of Mathematical and Statistical Psychology 55, 169–176.

Wilcox, R.R., 2003a. Approximating Tukey's depth. Communications in Statistics—Simulation and Computation 32, 977–985.

Wilcox, R.R., 2003b. Two-sample, bivariate hypothesis testing methods based on Tukey's depth. Multivariate Behavioral Research 38, 225–246.

Wilcox, R.R., 2003c. Applying Contemporary Statistical Techniques. Academic Press, San Diego, CA.

Wilcox, R.R., 2003d. Inferences based on multiple skipped correlations. Computational Statistics & Data Analysis 44, 223–236.

Wilcox, R.R., 2003e. Testing the hypothesis that a regression model is additive. Unpublished technical report. Dept. of Psychology, University of Southern California.

Wilcox, R.R., 2003f. Multiple hypothesis testing based on the ordinary least squares estimator when there is heteroscedasticity. Educational and Psychological Measurement 63, 758–764.

Wilcox, R.R., 2004a. Extension of Hochberg's two-stage multiple comparison method. In: Mukhopadhyay, N., Datta, S., Chattopoadhyay, S. (Eds.), Applied Sequential Methodologies: Real World Examples with Data Analysis. Dekker, New York.

Wilcox, R.R., 2004b. An extension of Stein's two-stage method to pairwise comparisons among dependent groups based on trimmed means. Sequential Analysis 23, 63–74.

Wilcox, R.R., 2004c. Some results on extensions and modifications of the Theil-Sen regression estimator. British Journal of Mathematical and Statistical Psychology 57, 265–280.

Wilcox, R.R., 2005. Depth and a multivariate generalization of the Wilcoxon-Mann-Whitney test. American Journal of Mathematical and Management Sciences 25, 343–364.

Wilcox, R.R., 2006a. Inference about the components of a generalized additive model. Journal of Modern Applied Statistical Methods 5, 309–316.

Wilcox, R.R., 2006b. Pairwise comparisons of dependent groups based on medians. Computational Statistics & Data Analysis 50, 2933–2941.

Wilcox, R.R., 2006c. Comparing medians. Computational Statistics & Data Analysis 51, 1934–1943.

Wilcox, R.R., 2006d. A note on inferences about the median of difference scores. Educational and Psychological Measurement 66, 624–630.

Wilcox, R.R., 2006e. Comparing robust generalized variances and comments on efficiency. Statistical Methodology 3, 211–223.

Wilcox, R.R., 2006f. Some results on comparing the quantiles of dependent groups. Communications in Statistics—Simulation and Computation 35, 893–900.

Wilcox, R.R., 2006g. Testing the hypothesis of a homoscedastic error term in simple, non-parametric regression. Educational and Psychological Measurement 66, 85–92.

Wilcox, R.R., 2007. An omnibus test when using a quantile regression estimator with multiple predictors. Journal of Modern Applied Statistical Methods 6, 361–366.

Wilcox, R.R., 2008a. Some small-sample properties of some recently proposed multivariate outlier detection techniques. Journal of Statistical Computation and Simulation 78, 701–712.

Wilcox, R.R., 2008b. Quantile regression: a simplified approach to a lack-of-fit test. Journal of Data Science 6, 547–556.

Wilcox, R.R., 2008c. Robust principal components: a generalized variance perspective. Behavior Research Methods 40, 102–108.

Wilcox, R.R., 2008d. Post-hoc analyses in multiple regression based on prediction error. Journal of Applied Statistics 35, 9–17.

Wilcox, R.R., 2008e. On a test of independence via quantiles that is sensitive to curvature. Journal of Modern Applied Statistical Methods 7, 11–20.

Wilcox, R.R., 2009a. Robust multivariate regression when there is heteroscedasticity. Communications in Statistics—Simulation and Computation 38, 1–13.

Wilcox, R.R., 2009b. Comparing robust measures of association estimated via a smoother. Communications in Statistics—Simulation and Computation 38, 1969–1979.

Wilcox, R.R., 2009c. Comparing Pearson correlations: dealing with heteroscedasticity and non-normality. Communications in Statistics—Simulation and Computation 38, 2220–2234.

Wilcox, R.R., 2010a. Comparing robust nonparametric regression lines via regression depth. Journal of Statistical Computation and Simulation 80, 379–387.

Wilcox, R.R., 2010b. Measuring and detecting associations: methods based on robust regression estimators or smoothers that allow curvature. British Journal of Mathematical and Statistical Psychology 63, 379–393.

Wilcox, R.R., 2010c. A note on principal components via a robust generalized variance. Unpublished technical report. Dept. of Psychology, University of Southern California.

Wilcox, R.R., 2010d. Regression: comparing predictors and groups of predictors based on robust measures of association. Journal of Data Science 8, 429–441.

Wilcox, R.R., 2010e. Inferences about the population mean: empirical likelihood versus bootstrap-t. Journal of Modern Applied Statistical Methods 9, 9–14.

Wilcox, R.R., 2011a. Nested ANOVA design: Methods that are robust and allow heteroscedasticity. Unpublished technical report. Dept. of Psychology, University of Southern California.

Wilcox, R.R., 2011b. Comparing the strength of association of two predictors via smoothers or robust regression estimators. Journal of Modern Applied Statistical Methods 10, 8–18.

Wilcox, R.R., 2011c. Inferences about a probabilistic measure of effect size when dealing with more than two groups. Journal of Data Science 9, 471–486.

Wilcox, R.R., 2012. Nonparametric regression when estimating the probability of success. Journal of Statistical Theory and Practice 6, 1–9.

Wilcox, R.R., 2013. A heteroscedastic method for comparing regression lines at specified design points when using a robust regression estimator. Journal of Data Science 11, 281–291.

Wilcox, R., 2014. Within groups ANCOVA: multiple comparisons at specified design points using a robust measure of location when there is curvature. Journal of Statistical Computation and Simulation. https://doi.org/10.1080/00949655.2014.962536.

Wilcox, R.R., 2015a. Comparing the variances of two dependent variables. Journal of Statistical Distributions and Applications 2 (7). https://doi.org/10.1186/s40488-015-0030-z. http://www.jsdajournal.com/content/2/1/7.

Wilcox, R.R., 2015b. Global comparisons of medians and other quantiles in a one-way design when there are tied values. Communications in Statistics—Simulation and Computation. https://doi.org/10.1080/03610918.2015.1071388.

Wilcox, R.R., 2015c. Within groups comparisons of least squares regression lines when there is heteroscedasticity. Unpublished technical report. Dept. of Psychology, University of Southern California.

Wilcox, R.R., 2015d. Inferences about the skipped correlation coefficient: dealing with heteroscedasticity and non-normality. Journal of Modern Applied Statistical Methods 14, 2–8.

Wilcox, R.R., 2016a. ANCOVA: a heteroscedastic global test when there is curvature and two covariates. Computational Statistics, 1–14. http://link.springer.com/article/10.1007/s00180-015-0640-4.

Wilcox, R.R., 2016b. Comparisons of two quantile regression smoothers. Journal of Modern Applied Statistical Methods 15, 62–77.

Wilcox, R.R., 2016c. ANCOVA: a global test based on a robust measure of location or quantiles when there is curvature. Journal of Modern Applied Statistical Methods 15. http://digitalcommons.wayne.edu/jmasm/vol15/iss1/3.

Wilcox, R.R., 2016d. Comparing dependent robust correlations. British Journal of Mathematical and Statistical Psychology 69, 215–224. https://doi.org/10.1111/bmsp.12069.

Wilcox, R.R., 2017a. Robust ANCOVA: heteroscedastic confidence bands that have some specified simultaneous probability coverage. Journal of Data Science 15, 313–328.

Wilcox, R.R., 2017b. Understanding and Applying Basic Statistical Methods Using R. Wiley, New York.

Wilcox, R.R., 2017c. The running interval smoother: a confidence band having some specified simultaneous probability coverage. International Journal of Statistics: Advances in Theory and Applications 1, 21–43.

Wilcox, R.R., 2017d. Modern Statistics for the Social and Behavioral Sciences: A Practical Introduction, 2nd ed. Chapman & Hall/CRC Press, Boca Raton, FL.

Wilcox, R.R., 2018a. Robust ANCOVA: confidence intervals that have some specified simultaneous probability coverage when there is curvature and two covariates. Journal of Modern Applied Statistical Methods. http://digitalcommons.wayne.edu/jmasm.

Wilcox, R.R., 2018b. Robust regression: an inferential method for determining which independent variables are most important. Journal of Applied Statistics 45, 100–111. https://doi.org/10.1080/02664763.2016.1268105.

Wilcox, R.R., 2018c. A robust nonparametric measure of effect size based on an analog of Cohen's d, plus inferences about the median of the typical difference. Journal of Modern Applied Statistical Methods 17 (2), eP2726. https://doi.org/10.22237/jmasm/1551905677.

Wilcox, R.R., 2018d. Robust ANCOVA, curvature and the curse of dimensionality. Journal of Modern Applied Statistical Methods 17 (2), eP2682. https://doi.org/10.22237/jmasm/1551906370.

Wilcox, R., 2018e. Logistic regression: an inferential method for identifying the best predictors. Journal of Modern Applied Statistical Methods 17 (2), eP3061. https://doi.org/10.22237/jmasm/155190690.

Wilcox, R.R., 2018f. An inferential method for determining which of two independent variables is most important when there is curvature. Journal of Modern Applied Statistical Methods 17, eP2588. https://doi.org/10.22237/jmasm/1525132920.

Wilcox, R.R., 2019a. Multicollinearity and ridge regression: results on type I errors, power and heteroscedasticity. Journal of Applied Statistics 46, 946–957. https://doi.org/10.1080/02664763.2018.1526891.

Wilcox, R.R., 2019b. Robust regression: testing global hypotheses about the slopes when there is multicollinearity or heteroscedasticity. British Journal of Mathematical and Statistical Psychology 72, 355–369. https://doi.org/10.1111/bmsp.12152.

Wilcox, R.R., 2019c. Robust ANCOVA: an approach to one-way and higher designs based on linear contrasts. Unpublished technical report. Dept. of Psychology, University of Southern California.

Wilcox, R.R., 2019d. Inferences about which of J dependent groups has the largest robust measure of location. British Journal of Mathematical and Statistical Psychology. https://doi.org/10.1111/bmsp.12205.

Wilcox, R.R., 2019e. Inferences about the probability of success, given the value of a covariate, using a nonparametric smoother. Journal of Modern Applied Statistical Methods.

Wilcox, R.R., 2020a. Comparing the variances or robust measures of scale of two dependent variables. Communications in Statistics—Simulation and Computation. https://doi.org/10.1080/03610918.2020.1807568.

Wilcox, R.R., 2020b. Some new results on computing a confidence interval for $P(X<Y)$. Unpublished technical report. Dept. of Psychology, University of Southern California.

Wilcox, R.R., in press. A note on determining which of J parameters has the largest or smallest value. Journal of Modern Applied Statistical Methods.

Wilcox, R.R., Clark, F., 2013. Robust regression estimators when there are tied values. Journal of Modern Applied Statistical Methods 12, 20–34.

Wilcox, R.R., Clark, F., 2014. Comparing robust regression lines associated with two dependent groups when there is heteroscedasticity. Computational Statistics 29, 1175–1186. http://link.springer.com/article/10.1007/s00180-014-0485-2.

Wilcox, R.R., Clark, F., 2015. Heteroscedastic global tests that the regression parameters for two or more independent groups are identical. Communications in Statistics—Simulation and Computation 44, 773–786. https://doi.org/10.1080/03610918.2013.784986.

Wilcox, R.R., Costa, K., 2009. Quantile regression: on inferences about the slopes corresponding to one, two or three quantiles. Journal of Modern Applied Statistical Methods 8, 368–375.

Wilcox, R.R., Erceg-Hurn, D., 2012. Comparing two dependent groups via quantiles. Journal of Applied Statistics 39, 2655–2664.

Wilcox, R.R., Keselman, H.J., 2002. Within groups multiple comparisons based on robust measures of location. Journal of Modern Applied Statistical Methods 1, 281–287.

Wilcox, R.R., Keselman, H.J., 2006. Detecting heteroscedasticity in a simple regression model via quantile regression slopes. Journal of Statistical Computation and Simulation 76, 705–712.

Wilcox, R.R., Ma, J., 2015. Heteroscedastic methods for performing all pairwise comparisons of regression lines associated with J independent groups. Methodology 11, 110–115. https://doi.org/10.1027/1614-2241/a000097.

Wilcox, R.R., Muska, J., 1999. Measuring effect size: a nonparametric analog of ω^2. British Journal of Mathematical and Statistical Psychology 52, 93–110.

Wilcox, R.R., Muska, J., 2001. Inferences about correlations when there is heteroscedasticity. British Journal of Mathematical and Statistical Psychology 54, 39–47.

Wilcox, R.R., Rousselet, G.A., 2018. A guide to robust statistical methods in neuroscience. Current Protocols in Neuroscience 82 (1), 8.42.1–8.42.30. https://doi.org/10.1002/cpns.41.

Wilcox, R.R., Tian, T., 2011. Measuring effect size: a robust heteroscedastic approach for two or more groups. Journal of Applied Statistics 38, 1359–1368.

Wilcox, R.R., Charlin, V.L., Thompson, K.L., 1986. New Monte Carlo results on the robustness of the ANOVA F, W, and F^* statistics. Communications in Statistics—Simulation and Computation 15, 933–944.

Wilcox, R.R., Rousselet, G.A., Pernet, C.R., 2018. Improved methods for making inferences about multiple skipped correlations. Journal of Statistical Computation and Simulation 88, 3116–3131. https://doi.org/10.1080/00949655.2018.1501051.

Willems, G., Pison, G., Rousseeuw, P., van Aelst, S., 2002. A robust Hotelling test. Metrika 55, 125–138.

Williams, V.S.L., Jones, L.V., Tukey, J.W., 1999. Controlling error in multiple comparisons, with examples from state-to-state differences in educational achievement. Journal of Educational and Behavioral Statistics 24, 42–69.

Wilson, E.B., 1927. Probable inference, the law of succession, and statistical inference. Journal of the American Statistical Association 22, 209–212.

Witten, I.H., Frank, E., Hall, M.A., Pal, C.J., 2017. Data Mining: Practical Machine Learning Tools and Techniques. Morgan Kaufmann, Cambridge, MA.

Woodruff, D.L., Rocke, D.M., 1994. Computable robust estimation of multivariate location and shape in high dimension using compound estimators. Journal of the American Statistical Association 89, 888–896.

Wright, S.P., 1992. Adjusted P-values for simultaneous inference. Biometrics 48, 1005–1013.

Wu, C.F.J., 1986. Jackknife, bootstrap, and other resampling methods in regression analysis. Annals of Statistics 14, 1261–1295.

Wu, M., Zuo, Y., 2009. Trimmed and Winsorized means based on a scaled deviation. Journal of Statistical Planning and Inference 139, 350–365.

Wu, P.-C., 2002. Central limit theorem and comparing means, trimmed means, one-step M–estimators and modified one-step M–estimators under non-normality. Unpublished doctoral dissertation. Dept. of Education, University of Southern California.

Xu, G., Genton, M.G., 2015. Efficient maximum approximated likelihood inference for Tukey's g-and-h distribution. Computational Statistics & Data Analysis 91, 78–91.

Xu, Y., Iglewicz, B., Chervoneva, I., 2014. Robust estimation of the parameters of g-and-h distributions, with applications to outlier detection. Computational Statistics & Data Analysis 75, 66–80.

Yale, C., Forsythe, A.B., 1976. Winsorized regression. Technometrics 18, 291–300.

Yanagihara, H., Yuan, K.H., 2005. Three approximate solutions to the multivariate Behrens-Fisher problem. Communications in Statistics—Simulation and Computation 34, 975–988.

Yang, L., Marron, J.S., 1999. Iterated transformation-kernel density estimation. Journal of the American Statistical Association 94, 580–589.

Yang, M., Yuan, K.-H., 2016. Robust methods for moderation analysis with a two-level regression model. Multivariate Behavioral Research 51, 757–771.

Yi, C., Huang, J., 2016. Semismooth Newton coordinate descent algorithm for elastic-net penalized Huber loss regression and quantile regression. Journal of Computational and Graphical Statistics. http://www.tandfonline.com/doi/full/10.1080/10618600.2016.1256816.

Yohai, V.J., 1987. High breakdown point and high efficiency robust estimates for regression. Annals of Statistics 15, 642–656.

Yohai, V.J., Zamar, R.H., 1988. High breakdown point estimates of regression by means of the minimization of an efficient scale. Journal of the American Statistical Association 83, 406–414.

Yohai, V.J., Zamar, R.H., 2004. Robust non-parametric inference for the median. Annals of Statistics 32, 1841–1857.

Yoshizawa, C.N., Sen, P.K., Davis, C.E., 1985. Asymptotic equivalence of the Harrell-Davis median estimator and the sample median. Communications in Statistics—Theory and Methods 14, 2129–2136.

Young, S.G., Bowman, A.W., 1995. Nonparametric analysis of covariance. Biometrics 51, 920–931.

Yuan, Y., MacKinnon, D.P., 2014. Robust mediation analysis based on median regression. Psychological Methods 19, 1–20.

Yuen, K.K., 1974. The two sample trimmed t for unequal population variances. Biometrika 61, 165–170. https://doi.org/10.2307/2334299.

Zaykin, D.V., Zhivotovsky, L.A., Westfall, P.H., Weir, B.S., 2002. Truncated product method for combining p-values. Genetic Epidemiology 22, 170–185.

Zhang, H., Zamar, R.H., 2014. Least angle regression for model selection. WIREs Computational Statistics 6, 116–123. https://doi.org/10.1002/wics.1288.

Zhang, J., Olive, D.J., Ye, P., 2012. Robust covariance matrix estimation with canonical correlation analysis. International Journal of Statistics and Probability 1, 119–136.

Zhang, J.-T., 2011. Two-way MANOVA with unequal cell sizes and unequal cell covariance matrices. Technometrics 53, 426–439.

Zhang, J.-T., 2013. Analysis of Variance for Functional Data. Chapman and Hall, London.

Zhao, J., Wang, J., 2009. Robust testing procedures in heteroscedastic linear models. Communications in Statistics—Simulation and Computation 38, 244–256.

Zhao, S., Bakoyannis, G., Lourens, S., Tu, W., 2020. Comparison of nonlinear curves and surfaces. Computational Statistics & Data Analysis 150. https://doi.org/10.1016/j.csda.2020.106987.

Zhao, X., Lynch Jr., J.G., Chen, Q., 2010. Reconsidering Baron and Kenny: myths and truths about mediation analysis. Journal of Consumer Research 37, 197–206.

Zheng, Q., Gallagher, C., Kulasekera, K., 2016. Robust adaptive lasso for variable selection. Communications in Statistics—Theory and Methods 46, 4642–4659.

Zhou, W., 2008. Statistical inference for $P(X < Y)$. Statistics in Medicine 27, 257–279.

Zhou, W., 2009. Robust dimension reduction based on canonical correlation. Communications in Statistics—Simulation and Computation 38, 1292–1307.

Zimmerman, D.W., 2004. A note on preliminary tests of equality of variances. British Journal of Mathematical and Statistical Psychology 57, 173–182.

Zou, C., Liu, Y., Wang, Z., Zhang, R., 2010. Adaptive nonparametric comparison of regression curves. Communications in Statistics—Theory and Methods 39, 1299–1320.

Zou, G.Y., 2007. Toward using confidence intervals to compare correlations. Psychological Methods 12, 399–413.

Zou, G.Y., Huang, W., Zhang, X., 2009. A note on confidence interval estimation for a linear function of binomial proportions. Computational Statistics & Data Analysis 53, 1080–1085.

Zou, H., 2006. The adaptive lasso and its oracle properties. Journal of the American Statistical Association 91, 258–266.

Zou, H., Hastie, T., 2005. Regularization and variable selection via the elastic net. Journal of the Royal Statistical Society, B 67, 301–320.

Zu, J., Yuan, K.H., 2010. Local influence and robust procedures for mediation analysis. Multivariate Behavioral Research 45, 1–44.

Zuo, Y., 2003. Projection-based depth functions and associated medians. Annals of Statistics 31, 1460–1490.

Zuo, Y., 2010. Is the t confidence interval $\bar{X} \pm t_\alpha(n-1)s/\sqrt{n}$ optimal? American Statistician 64, 170–173.

Zuo, Y., 2013. Multidimensional medians and uniqueness. Computational Statistics & Data Analysis 66, 82–88.

Zuo, Y., 2019. A new approach for the computation of halfspace depth in high dimensions. Communications in Statistics—Simulation and Computation 48, 900–921.

Zuo, Y., He, X., 2006. On the limiting distributions of multivariate depth-based rank sum statistics and related tests. Annals of Statistics 34, 2879–2896.

Zuo, Y., Serfling, R., 2000a. General notions of statistical depth functions. Annals of Statistics 28, 461–482.

Zuo, Y., Serfling, R., 2000b. Structural properties and convergence results for contours of sample statistical depth functions. Annals of Statistics 28, 483–499.

Zuo, Y., Cui, H., He, X., 2004a. On the Stahel-Donoho estimator and depth-weighted means of multivariate data. Annals of Statistics 32, 167–188.

Zuo, Y., Cui, H., Young, D., 2004b. Influence function and maximum bias of projection depth based estimators. Annals of Statistics 32, 189–218.

Zuo, Y.J., Lai, S.Y., 2011. Exact computation of bivariate projection depth and the Stahel-Donoho estimator. Computational Statistics & Data Analysis 55, 1173–1179.

Zuur, A.F., Ieno, E.N., Meesters, E., 2009. A Beginner's Guide to R. Springer, New York.

Index